Springer Series in Design and Innovation

Volume 6

Editor-in-Chief

Francesca Tosi, University of Florence, Florence, Italy

Series Editors

Claudio Germak, Politecnico di Torino, Turin, Italy

Francesco Zurlo, Politecnico di Milano, Milan, Italy

Zhi Jinyi, Southwest Jiaotong University, Chengdu, China

Marilaine Pozzatti Amadori, Universidade Federal de Santa Maria, Santa Maria, Rio Grande do Sul, Brazil

Maurizio Caon, University of Applied Sciences and Arts, Fribourg, Switzerland

Springer Series in Design and Innovation (SSDI) publishes books on innovation and the latest developments in the fields of Product Design, Interior Design and Communication Design, with particular emphasis on technological and formal innovation, and on the application of digital technologies and new materials. The series explores all aspects of design, e.g. Human-Centered Design/User Experience, Service Design, and Design Thinking, which provide transversal and innovative approaches oriented on the involvement of people throughout the design development process. In addition, it covers emerging areas of research that may represent essential opportunities for economic and social development.

In fields ranging from the humanities to engineering and architecture, design is increasingly being recognized as a key means of bringing ideas to the market by transforming them into user-friendly and appealing products or services. Moreover, it provides a variety of methodologies, tools and techniques that can be used at different stages of the innovation process to enhance the value of new products and services.

The series' scope includes monographs, professional books, advanced textbooks, selected contributions from specialized conferences and workshops, and outstanding Ph.D. theses.

Keywords: Product and System Innovation; Product design; Interior design; Communication Design; Human-Centered Design/User Experience; Service Design; Design Thinking; Digital Innovation; Innovation of Materials.

How to submit proposals

Proposals must include: title, keywords, presentation (max 10,000 characters), table of contents, chapter abstracts, editors'/authors' CV.

In case of proceedings, chairmen/editors are requested to submit the link to conference website (incl. relevant information such as committee members, topics, key dates, keynote speakers, information about the reviewing process, etc.), and approx. number of papers.

Proposals must be sent to: series editor Prof. Francesca Tosi (francesca.tosi@unifi.it) and/or publishing editor Mr. Pierpaolo Riva (pierpaolo.riva@springer.com).

More information about this series at http://www.springer.com/series/16270

Luis Agustín-Hernández ·
Aurelio Vallespín Muniesa ·
Angélica Fernández-Morales
Editors

Graphical Heritage

Volume 2 - Representation, Analysis, Concept and Creation

 Springer

Editors
Luis Agustín-Hernández
Escuela de Ingeniería y Arquitectura
Zaragoza, Spain

Aurelio Vallespín Muniesa
Escuela de Ingeniería y Arquitectura
Zaragoza, Spain

Angélica Fernández-Morales
Escuela de Ingeniería y Arquitectura
Zaragoza, Spain

ISSN 2661-8184   ISSN 2661-8192 (electronic)
Springer Series in Design and Innovation
ISBN 978-3-030-47985-5   ISBN 978-3-030-47983-1   (eBook)
https://doi.org/10.1007/978-3-030-47983-1

This Springer imprint is published by the registered company Springer Nature Switzerland AG
The registered company address is: Gewerbestrasse 11, 6330 Cham, Switzerland

Preface

This book, titled 'Graphical Heritage,' contains the contributions to the XVIII International Congress of Architectural Graphic Expression, held at the School of Engineering and Architecture of the University of Zaragoza. The call for papers was answered by 239 contributions, which underwent a blind peer-review process. The initial analysis of the extended abstracts and, subsequently, of the full contributions discarded 59 papers and finally accepted 180. As a complement to the review performed by international experts, the contributions have been submitted to an anti-plagiarism software check and an examination of their English language academic style, in order to obtain a publication of the highest scientific level.

The geographical origin of the accepted contributions is very varied. With 126 from Spain and 54 from an international origin, it can be said that it is a clear exposition of the most contemporary knowledge in the Architectural Graphic Expression field of research.

As previous congresses, the main theme has been established. This theme has been heritage, understood as architectural heritage, as graphic heritage and as the graphics of heritage; its study, documentation, intervention, conservation, inventory and recovery ultimately point out its value. In support of this theme, invited conferences will be held, including speakers such as Rafael Aranda of RCR Arquitectes, Antonio Almagro, Professor Emeritus of the CSIC or Asunción Hernández, professor at the University of Zaragoza, as well as Luis Franco, architect, Ricardo Usón, PhD architect and Javier Ibargüen, architect, experts in restoration and intervention in architectural heritage.

This congress not only addresses the matter of heritage but every scientific research line related to the area of knowledge, geometry, architectural drawing, color, formal research, representation, diagrams, conceptualization, modeling, rendering, BIM, GIS, cartography, landscape, virtual reality, augmented reality, without forgetting teaching research.

To collect all these initiatives, the following topics have been considered:

1. Heritage and history.
2. Representation and analysis.
3. Concept and creation.
4. Mapping, cartography and landscape.
5. Teaching innovation.

From drawing, the creator's first decision in architecture, painting or sculpture, as the ability to transfer an idea onto a support, the ability to direct the hand in a specific line, as the first creative element, to virtual reality, the virtual building, far beyond reality. It all is reflected in the communications included in this book and in this congress.

The chapters included in this book, the second in a set of three volumes, are organized into two main sections: *Representation and analysis* and *Concept and Creation*.

Representation and Analysis

The importance of drawing as a means of studying and disseminating architecture has given rise to theories and principles that have set trends and played a fundamental role in their transmission. The approach to the drawing and the image of architecture from a theoretical perspective allows making a historical tour, contributing to recompose the history of the current forms of representation and analysis.

The reflection on the graphic representation arises from its ability to interpret architectural works and spaces through their reflection or codification using different languages. Together with the drawing in all its aspects—technical, sketches, etc.—the digital tools—renders, photomontages, etc.—which provide with essential concepts and a new imaginary in the current communication of architecture, are analyzed.

The study of graphic analysis is interpreted from the conceptual and formal decomposition of the work of architecture. This allows knowing the intention and inquisitiveness that originated it, the link with the artistic movements and lines of thought that influenced it, the relationship with the physical and historical context that surrounds it, the criteria that rule it and the configuration and articulation of the elements that constitute it.

Descriptive geometry and representation systems, which have accompanied the practice of architecture and conditioned its representation and structural systems, and therefore its shape, are also considered as evolving means that have allowed for the exploration of new possibilities for the graphic narrative.

Finally, together with the graphic means, the scale model is considered as a three-dimensional representation in order to work the materiality and spatiality of the architectural work or to synthesize the project idea. The models made with traditional systems currently coexist with virtual models (BIM) and digital manufacturing prototypes, which have contributed to new ways of working and communicating.

– Theories, principles and masters
– Graphic representation
– Graphic analysis
– Geometry and projectivity
– Mock-ups, models and prototypes

Concept and Creation

This line refers to the idea and its graphic materialization in heritage, in its broadest sense, the graphic heritage and the graphics of heritage. This broad topic is divided into five main sublines.

The first subline is 'built and unbuilt graphic concepts.' Ideas do not differentiate between what is built and what is not built, and the concept does not distinguish between what is built and what is drawn. As Vasari argued, 'drawing is nothing more than an apparent expression and materialization of the concept that is within the artist's mind and that which provokes in those who contemplate it.'

The second refers to 'graphic concepts of heritage.' It should not be forgotten that the biggest graphic evolution in the past years has occurred in the heritage field, due to the creation, development and use of new representation technologies.

Of course, concept and creation also hold a third subline related to 'artistic creation.' This is understood as the graphic materialization of the idea.

The fourth is 'phenomenology, perception and interaction.' Both, concept and creation are closely linked to perception and phenomenology in the architectural graphic representation.

Finally, the concept of 'shape grammar' will be approached. It includes the grammar of the form and the new parametric ways of design. It is important to work on the concept of the different rules that generate these new languages.

– Built and unbuilt graphic concepts
– Graphic concepts of heritage
– Artistic creation
– Phenomenology, perception and interaction

Luis Agustín-Hernández
Aurelio Vallespín Muniesa
Angélica Fernández-Morales

Organization

Honor Committee

Giuseppe Amoruso	Politecnico di Milano, Italy
Massimiliano Campi	Università degli Studi di Napoli Federico II, Italy
Enrique Solana	Universidad de Las Palmas, Spain
Carlo Bianchini	Sapienza Università di Roma, Italy
Pilar Chías	Universidad de Alcalá, Spain
Ángela García	Universitat Politècnica de València, Spain
Cesare Cundari	Sapienza Università di Roma, Italy
Mario Docci	Sapienza Università di Roma, Italy
Francesca Fatta	Università degli Studi di Reggio Calabria, Italy
José Antonio Franco	Universidade da Coruña, Spain
José María Gentil	Universidad de Sevilla, Spain
Emma Mandelli	Università degli Studi di Firenze, Italy
Carlos Montes	Universidad de Valladolid, Spain
Pablo Navarro	Universitat Politècnica de València, Spain
Rossella Salerno	Politecnico di Milano, Italy
Javier Seguí	Universidad Politécnica de Madrid, Spain

International Scientific Committee

Luis Agustín	Universidad de Zaragoza, Spain
Salvatore Barba	Università degli Studi di Salerno, Italy
Stefano Bertocci	Università degli Studi di Firenze, Italy
Ignacio Cabodevilla	Universidad de Zaragoza, Spain
José Calvo	Universidad Politécnica de Cartagena, Spain
Mara Capone	Università degli Studi di Napoli Federico II, Italy
Eduardo Carazo	Universidad de Valladolid, Spain
Noelia Cervero	Universidad de Zaragoza, Spain

Cesare Verdoscia	Politecnico di Bari, Italy
Emanuela Chiavoni	Sapienza Università di Roma, Italy
Javier Domingo	Universidad de Zaragoza, Spain
Ernesto Echeverría	Universidad de Alcalá, Spain
Angélica Fernández	Universidad de Zaragoza, Spain
José Javier Gallardo	Universidad de Zaragoza, Spain
Antonio García-Bueno	Universidad de Granada, Spain
Aitor Goitia	Universidad CEU San Pablo de Madrid, Spain
Elsa Gutiérrez	Universidad de Las Palmas, Spain
Luis Hermida	Universidade da Coruña, Spain
Iñigo León	Universidad del País Vasco, Spain
Mercedes Linares	Universidad de Sevilla, Spain
Jorge Llopis	Universitat Politècnica de València, Spain
Carlos L. Marcos	Universidad de Alicante, Spain
Barbara Messina	Università degli Studi di Salerno, Italy
Antonio Millán	Universitat Politècnica de Catalunya, Spain
Marta Quintilla	Universidad de Zaragoza, Spain
Javier Raposo	Universidad Politécnica de Madrid, Spain
Ernest Redondo	Universitat Politècnica de Catalunya, Spain
Simona Salvo	Sapienza Università di Roma, Italy
Roberta Spallone	Politecnico di Torino, Italy
Manuel Ródenas	Universidad Politécnica de Cartagena, Spain
Karen Sanabria	Universidad Tecnológica de La Habana José Antonio Echeverría, Cuba
Miguel Sancho	Universidad de Zaragoza, Spain
Aurelio Vallespín	Universidad de Zaragoza, Spain

Peer Reviewers Committee

Luis Agustín	Universidad de Zaragoza, Spain
Ángel Allepuz	Universidad de Alicante, Spain
Antonio Álvaro	Universidad de Valladolid, Spain
Antonio Amado	Universidade da Coruña, Spain
Antonio Luis Ampliato	Universidad de Sevilla, Spain
Federico Arévalo	Universidad de Sevilla, Spain
Salvatore Barba	Università degli Studi di Salerno, Italy
Stefano Bertocci	Università degli Studi di Firenze, Italy
Daniela Besana	Università di Pavia, Italy
Antonio Bixio	Università degli Studi della Basilicata, Italy
Alberto Bravo	Universidad de Las Palmas, Spain
Ignacio Cabodevilla	Universidad de Zaragoza, Spain
Marianna Calia	Università degli Studi della Basilicata, Italy

José Calvo	Universidad Politécnica de Cartagena, Spain
Massimiliano Campi	Università degli Studi di Napoli Federico II, Italy
Mara Capone	Università degli Studi di Napoli Federico II, Italy
Eduardo Carazo	Universidad de Valladolid, Spain
Noelia Cervero	Universidad de Zaragoza, Spain
Mauro Chiarella	Universidad Nacional del Litoral/CONICET, Argentina
Stefano Chiarenza	Università San Raffaele Roma, Italy
Emanuela Chiavoni	Sapienza Università di Roma, Italy
Giuseppe D'acunto	Università IUAV di Venezia, Italy
Pierpaolo D'agostino	Università degli Studi di Napoli Federico II, Italy
Giuseppe Damone	Università degli Studi della Basilicata, Italy
Francesco Di Paola	Università degli Studi di Palermo, Italy
Jorge Domingo	Universidad de Alicante, Spain
Ernesto Echeverría	Universidad de Alcalá, Spain
Angélica Fernandez	Universidad de Zaragoza, Spain
José Ángel Fernández	Universidade da Coruña, Spain
Juan José Fernández	Universidad de Valladolid, Spain
Fausta Fiorillo	Politecnico di Milano, Italy
Riccardo Florio	Università degli Studi di Napoli Federico II, Italy
Noelia Galván	Universidad de Valladolid, Spain
Antonio Gámiz	Universidad de Sevilla, Spain
Rodrigo Garcia	Universidad del Bío-Bío, Chile
Antonio García-Bueno	Universidad de Granada, Spain
Francisco García	Universidad de Alicante, Spain
Josefina García	Universidad Politécnica de Cartagena, Spain
Javier García-Gutiérrez	Universidad Politécnica de Madrid, Spain
Jorge García-Valldecabres	Universitat Politècnica de València, Spain
María Teresa Gil	Universitat Politècnica de València, Spain
Manuel Giménez	Universitat Politècnica de València, Spain
Andrea Giordano	Università degli Studi di Padova, Italy
Aitor Goitia	Universidad CEU San Pablo de Madrid, Spain
Roberto Goycoolea	Universidad de Alcalá, Spain
Francisco Granero-Martin	Universidad de Sevilla, Spain
Alessandro Greco	Università di Pavia, Italy
Alberto Grijalba	Universidad de Valladolid, Spain
Elsa M. Gutiérrez	Universidad de Las Palmas, Spain
Luis Hermida	Universidade da Coruña, Spain
Francisco Hidalgo	Universitat Politècnica de València, Spain
Ricardo Irles	Universidad de Alicante, Spain
Sonia Izquierdo	Universidad CEU San Pablo de Madrid, Spain
Pedro Miguel Jiménez	Universidad Politécnica de Cartagena, Spain
Pablo J. Juan	Universidad de Alicante, Spain
Emanuela Lanzara	Università degli Studi di Napoli Federico II, Italy

Iñigo León	Universidad de País Vasco, Spain
Massimo Leserri	Universidad Pontificia Bolivariana, Colombia
Marco Limongiello	Università degli Studi di Salerno, Italy
Mercedes Linares	Universidad de Sevilla, Spain
Placido Lizancos	Universidade da Coruña, Spain
Jorge Llopis	Universitat Politècnica de València, Spain
Massimiliano Lo Turco	Politecnico di Torino, Italy
Daniel Lòpez	Universidad de Valladolid, Spain
Concepción López	Universitat Politècnica de València, Spain
Margarita Lorenzo	Universidade da Coruña, Spain
Clara Maestre	Universidad CEU San Pablo de Madrid, Spain
Elena Teresa Clotilde	Politecnico di Torino, Italy
Carlos L. Marcos	Universidad de Alicante, Spain
Ángel Martínez	Universidad Politécnica de Madrid, Spain
María Luisa Martínez	Universidad de Las Palmas, Spain
Rodolfo Mejías	Universidad de Costa Rica
Ángel Melián	Universidad de Las Palmas, Spain
Barbara Messina	Università degli Studi di Salerno, Italy
Antonio Millán	Universitat Politècnica de Catalunya, Spain
Cosimo Monteleone	Università degli Studi di Padova, Italy
Carlos Montes	Universidad de Valladolid, Spain
María José Muñoz	Universidad Politécnica de Madrid, Spain
María Lucía Ojeda	Universidad de Las Palmas, Spain
Justo Oliva	Universidad de Alicante, Spain
Lia M. Papa	Università degli Studi di Napoli Federico II, Italy
Sandro Parrinello	Università degli Studi di Pavia, Italy
José Javier Pérez	Universidad del País Vasco, Spain
Inés Pernas	Universidade da Coruña, Spain
Francesca Picchio	Università degli Studi di Pavia, Italy
José Pinto	Universidade de Lisboa, Portugal
Francisco Pinto	Universidad de Sevilla, Spain
Juan Carlos Piquer	Universitat Politècnica de València, Spain
Enrique Rabasa	Universidad Politécnica de Madrid, Spain
Javier Francisco Raposo	Universidad Politécnica de Madrid, Spain
Manuel Alejandro Ródenas	Universidad Politécnica de Cartagena, Spain
Daniel Rodríguez	Universidad de Guadalajara, México
Eva Juana Rodríguez	Universidad CEU San Pablo de Madrid, Spain
Pablo Rodríguez-Navarro	Universitat Politècnica de València, Spain
Andrea Rolando	Politecnico di Milano, Italy
Gabriele Rossi	Politecnico di Bari, Italy
Adriana Rossi	Università degli Studi della Campania Luigi Vanvitelli, Italy

Michele Russo	Sapienza Università di Roma, Italy
Maialen Sagarna	Universidad del País Vasco, Spain
Karen Sanabria	Universidad Tecnológica de la Habana José Antonio Echeverría, Cuba
Miguel Sancho	Universidad de Zaragoza, Spain
Alberto Sdegno	Università degli Studi di Udine, Italy
Marina Sender	Universitat Politècnica de València, Spain
María Senderos	Universidad del País Vasco, Spain
Juan Serra	Universitat Politècnica de València, Spain
Ana Torres	Universitat Politècnica de València, Spain
Aurelio Vallespín	Universidad de Zaragoza, Spain

Contents

Concept and Creation

Representation and Analysis

Drawing as an Instrument for Geometric Control and Analysis in Architectural Heritage. The Case of La Piedad Chapel in Sevilla

Francisco Granero Martín[✉] [iD]

University of Seville, Seville, Spain
fgranerom@us.es

Abstract. Once the splendor and prosperity of Seventeenth century's Seville vanishes, at a relevant enclave of the river bank named "Monte Baratillo", the tiny chapel of La Piedad is completed in 1694. It stands for the immaterial heritage of the town in exchange issues with the Americas and lies in the vicinity of the most important buildings of the capital. The chapel bears traces of refined baroque architecture but the frontispiece is neoclassical. It was erected on the same spot in which an ancient pedestal holding the holy cross could be found, such cross now surmounts the dome of the chapel. It was conceived as a burial and oratory place in remembrance of the deceased as a consequence of 1649's terrible plague.

In the enlargement of the Chapel in 1724, the dome is projected in an octagonal fashion but a mistake appears at the drawing of the stakeout with an important rotation and lateral deviation from the vertical which, with the passing of time, provoked a defective situation threatening its stability.

In this paper we discuss the role of drawing as a tool for geometric control and analysis. It is deemed useful for the reconstruction project of a chapel in Seville commissioned to the author after a competition entry.

Keywords: Architectural heritage retrofit · Architectural drawing · Sevillian Heritage

1 Introduction. Immaterial Heritage and Location

Considerations on architectural heritage from reflection and analysis of the location in time.

The restoration methods to recuperate Heritage began to be organized since the nineteenth century through the search for definite principles (Capitel 1988, pp. 17–21).

Located in the city of Seville in the seventeenth century, "Monte Baratillo" can be understood as the intangible heritage of such city, or as a hinge between it and the reconstructed section of the Guadalquivir that precisely originated it and made it evolve over time: the port-town, maritime, commercial, place of trade (shipyards, olive oil cargoes to the Roman Empire, discharge of gold, etc.,) meeting point of urban topologies, creation of landmarks and references that, for centuries, have occurred in its vicinity, outlining

L. Agustín-Hernández et al. (Eds.): EGA 2020, SSDI 6, pp. 3–16, 2020.
https://doi.org/10.1007/978-3-030-47983-1_1

buildings as important as the Royal Yards called Atarazanas (centuries 9th to 12th), the hospice of la Caridad (17th), the Mint (16th), Customs (17th), the Loggia of the Merchants (16th), the royal Maestranza bullring (18th), and so forth. Thus the enclave of the "Baratillo" becomes an intangible heritage, echoing social and cultural vicissitudes from the remote Phoenician pre-existence, merchandise hub to the Roman Empire, the drama of Viking raids (9th century), the prime of the Muslim epoch (Atarazanas) and above all the American trade between centuries 16th and 18th. From the 16th century this place of the town of Sevilla, grew around a memorial cross that marked the spot as a landmark and oratory for arrivals and wishing-well for mariners that enrolled in the journey to the new world (Granero 1992, p. 39) and the great enterprise of circumnavigating the world initiated by Magellan and Elcano.

2   Historical Outline of the Urban Sector Around the Foundation Cross

As soon as the river arm that traversed the town from the place known as "la Barqueta" to the surroundings of the Torre del Oro (tower of the gold) was drained, the AlmohadIsbilya (Seville) of the 12th century acquired a new physiognomy with the reconstruction and enlargement of its walls that were to stand until the 19th century, making it the largest walled-city in Europe.

The reason for this protracted drainage was mainly defensive as episodes of siege had taken place in the Visigoth town of the 6th century and also it had been devastated by the Viking raids of the year 844 that reached Seville ascending by the river.

The urban sector known as "Arenal" (sandbank) was left outside and worked as a strip of land between the city and the riverbank in which soon enough appeared sundry quarters: "El Baratillo" (Fig. 1) was the real sand bank and mound of soil (afterwards a place for sailors, fishermen and minor trade and whence its name which in Spanish means something like "cheapo"); but also the quarter of Cesteria/Carreteria traced with string, that little by little acquired the fabric of typical neighborhoods, biased over all by the importance of the shipyards of Alfonso the Tenth (later enlarged by Pedro the First in the 13th century) and the associated naval industries. Therefore, till the end of the 18th century (Olavide's plan of 1771), even if the sector of the Carreteria was formed as a well traced and completed quarter (without significant changes from that time to our days), the Baratillo became the true sand quarry possessing the landmark of the oratory with the said cross on plinth and, from 1730, the rectangular bullfighting place after a royal concession by Phillip the Fifth to the Maestranza de Caballeria. Between the boat-bridge (Fig. 1) as communicational axis of Sevilla, with Trianaor the distant Aljarafe and the Torre del Oro, in the midst of Arenal, the most expansive activity of the town was held there in the 16th century. A consolidated activity since Sevilla undertook the monopoly of trade with the Americas after the foundation of the Casa de Contratacion of the Indies (in 1503, by the Catholic Kings). With the former came the regulation and control over all exchanges of commercial, fiscal or judiciary nature; even such monopoly was not affected by the edict of Charles the First (1529), by virtue of which, nine other ports where open for commerce with the Americas (Cádiz, Málaga, Cartagena, Bayona, La Coruña, Avilés, Laredo, Bilbao and San Sebastián). It was mandatory for every embarkation to return to the harbor of Seville and thus, records show that only in the year 1549 up to 101 ships departed from it and transported cargoes of more than 30.000 tons.

Fig. 1. of the view of Sevilla from Triana (in the center of the painting the Mount Baratillo is depicted). 1726. Oiloncanvas. Sevilla. SevilleTownhouse.

The topographic conditions offered by the mound of earth, soil and debris from all kind of port activities at el Baratillo, defined it as a place devoid of constructions, preserved beyond the presence of so much bustle and hassle of transactions. A plinth topped by a wrought iron cross, dominated the front of the river as place of prayer for sailors and fishermen who ventured in and out of the city for their daily labor. Being a typical feature of the urban landscape it is rare that we do possess any written source informing of its original presence and we must rely solely on the analysis of the historical iconography. A cross over a pedestal in such an open space as the Arenal would be an important element to be depicted in the etchings or paintings, however we have not found any trace of it neither in the drawings of Joris Hoefnagel (1565), nor the views from Triana by Anton von den Wyngaerde (1567), not even in the one representing the visit of Phillip the Second (Fig. 2). Therefore, we ought to situate it by the end of the 16th century, as the mound was settled with the first erection (the cross) and became a place of devotion and cult as it appears (though under a different aspect in the canvas attributed to Sánchez Coello (Fig. 3).

Fig. 2. The entourage of King Phillip the Second passes by the Royal Gate (Puerta Real). 1570. Woodcut by A. Escrivano. British Library London.

Fig. 3. View of Seville by the end of the 16th century. Oil on canvas attributed to Sánchez Coello. Madrid. Museum of the Americas.

To the Spanish political weakness and the decadence of the realm in the 17th century (especially with Phillip the Fourth and the War of the Thirty Years and later, with Charles the Second) we can add the aggressive strategy of trade with the Americas, performed by countries such as Holland and England. Even Spain was forced to settle a commercial Treaty that induced the demise of Seville's Monopoly from 1680 on. Another fundamental reason to this decline was the increasing necessity of boats of bigger draft that were impractical for the Guadalquivir river. Thus the harbor of Cadiz gained importance and finally the Casa de la Contratacion (House of Hiring) was transferred there in 1717.

3   Architectural Heritage as a Town-Generatorlandmark

The heritage attributes are an added value that society deems attached to the evolution of a place whose preservation seems desirable (Castillo et al. 2009, p. 141). Mount Baratillo, became one of the "charnel grounds" for burials in the city during the bubonic plague extending from Africa of 1649. The population of Seville was decimated by 46% (more than 60,000 people) and had to administer a quick burial place to the deceased. The cross on the pedestal, as the only existing element at that mound in an otherwise flat city and on the banks of the Guadalquivir, was erected in the center of the makeshift cemetery of that episode as a place of prayer and devotion giving rise to the Brotherhood of the Cruz del Baratillo (1693). As a consequence, the primitive Chapel (chief architect Bernardo de Bustamante) was erected as a towering architectural landmark of the territory of said urban sector. The works were completed in May 1696. It was, in principle, a very simple construction, consisting of a rectangular floor of 4.60 by 8.50 m^2 in the interior, with two load-bearing walls perpendicular to the facade which possibly formed the line of the future morphological trace of the head of the urban block. Its front wall was topped by a magnificent altarpiece (by Juan Fernández Iglesias) completed in 1707.

In summary we have to place the construction of the Chapel in the rich historical context of Sevillian Heritage with eminent figures such as Velázquez (1599–1660), Murillo (1617–1682), Zurbarán (1598–1664), Valdés Leal (1622–1690), Martínez Montañés (1568–1649), or Juan de Mesa (1583–1627). As introduction we would mention amongst the important buildings erected in the vicinity the House of Mint (Casa de la Moneda by Juan de Minjares 1584) that contributed to a great extent to complete the repertoire of relevant constructions in the Arenal neighborhood, the Loggia of Merchants –today Archive of the Indies– (Juan de Herrera 1572) and the interventions of Hernan Ruiz on the Cathedral and the topping of the Giralda Tower.

All the former constitutes the prime of 16th century's Seville and at the same time became a landmark for its history. At the time in Sevilla the church of the Sacristy had been built annexed to the Cathedral (1618–1663, by Zumárraga, Oviedo, Fernadez Iglesia, Alonso de Vandelvira and Cristóbal de Rojas); in 1644 Pedro Sánchez Falconete started the new chapel at the Hospital de la Caridad and later on, Leonardo de Figueroa completed its portal; He was building its ample and magnificent repertoire of baroque architecture in the town: Hospital de los Venerables (1687–1697), Church of la Magdalena (1691–1709), el Salvador (1696–1711), San Luis de los Franceses (1699–1730), the Sacristy at Santa Catalina (1721) and his contribution to the palace of San Telmo. Successive events took place in the city that adorned itself in a sort of disguise which was hiding its economic decadence, described by Quevedo and Calderon: the canonization of Saint Ferdinand in 1671, depicted in an etching by Matías de Arteaga (1672), the procession to transfer the mummy of the said Saint (14/5/1729) illustrated by Pedro Tortolero; the festival o f masks-joc-would be consecrated by scholastics of the College of Saint Thomas Aquinas, an event of 1742, and became the matter of illustration for the copper engravings of Agustín Moreno, and the festival of 1746 to commemorate the coronation of Ferdinand the Sixth, with the parade of chariots organized by the Royal Tobacco Factory and painted by Domingo Martinez (Granero 1992, p. 60 and following).

4   Geometric Analysis of the Piedad Chapel and Defects of the Original Stakeout

As a relevant issue we present, in summary, the thorough analysis performed (plan and section) from which we can deduct a consistent search by the builders of harmonic proportions and structured order in spite of the reduced volume of the church. It is remarkable the perfect control of measure in the relationship between the height of the church and the diameter of the dome: 2.5 times from the interior and 3 times at the exterior. Invisible lines of proportion between elements underlie in the whole design of the chapel, mainly in the height of the cornices and the skewback of scallops. The plan is deployed by means of a first rectangular body (1694) divided in pilasters and two transverse arches, which is covered with a cannon vault. A second body (expansion 1724) forms the transept by insertion of two pillars as support of the dome over four scallops (reform of 1755, after suffering Lisbon's Earthquake) which possess a diameter of 4.60 m. Behind the altarpiece wall, the sacristy opens in a recess of 3.50 m. The formal rhetoric offers great subtlety with wide cornices in superposed orders and molded balusters which offer a baroque harmonic work, unifying the original volume of the 17th century with the enlargement of the 18th (Figs. 4 and 5) by means of a neoclassical façade.

The roof of the dome is an irregular octagon with its angles reinforced by pilasters, and topped by pinnacles in order to buttress the horizontal thrust from the timber frame (Fig. 6). As a history relevant fact for the Arenal Quarter, we remind that the gabled roof of the primitive chapel of 1649 was crowned by the previously mentioned foundational cross made in wrought iron. In this way the towering cross is raised to the category of Faiths' Triumph. Each elevation forms a regular square showing a baroque idiom of recessed portholes, cornices and masonry eaves oriented to the four cardinal points. The extreme care put in every detail of the project and design of the chapel is not unprecedented since at that epoch (17th and 18th century), Seville oozed art and culture by all her pores. Economic decadence and poverty joined with epidemics led to a strong piety and religious feelings which are the distinctive feature of Sevillian baroque in architecture, painting sculpture and imagery. They are often linked to a somewhat lurid taste for contrast and paradox.

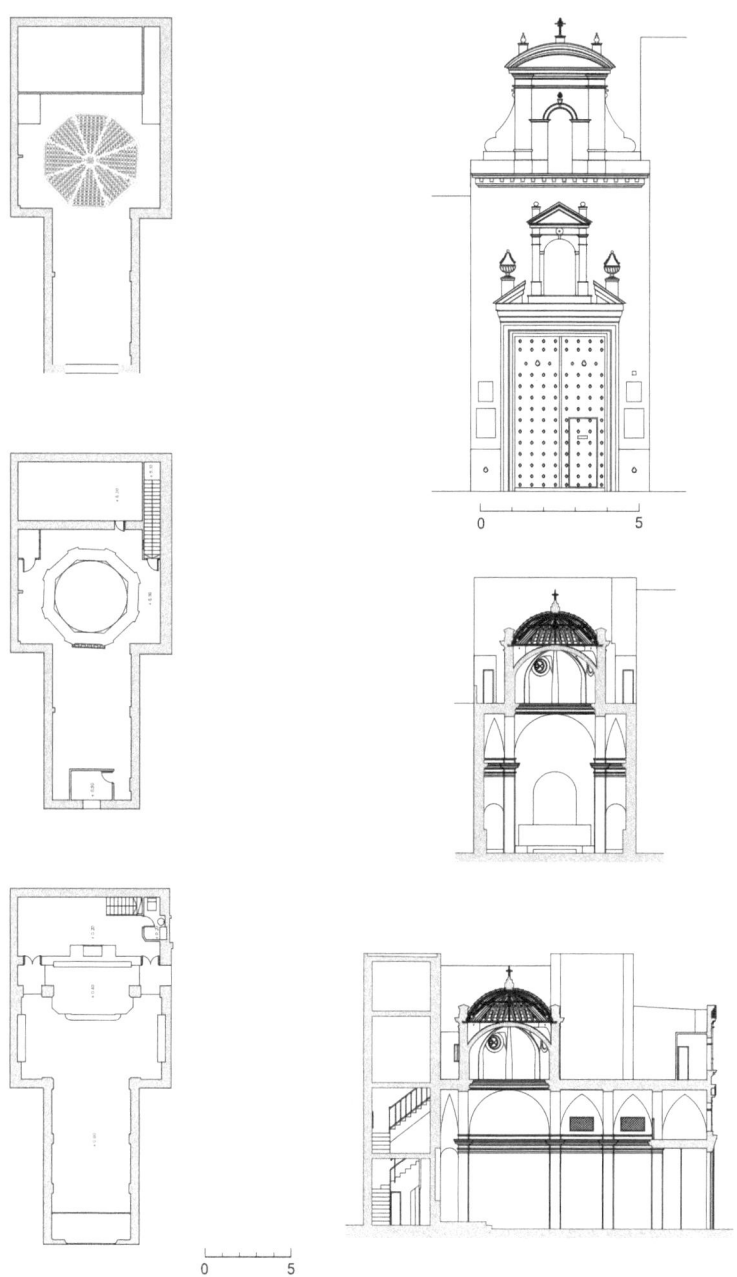

Fig. 4. Drawings of the Piedad Chapel. 2009. Source: F. Granero.

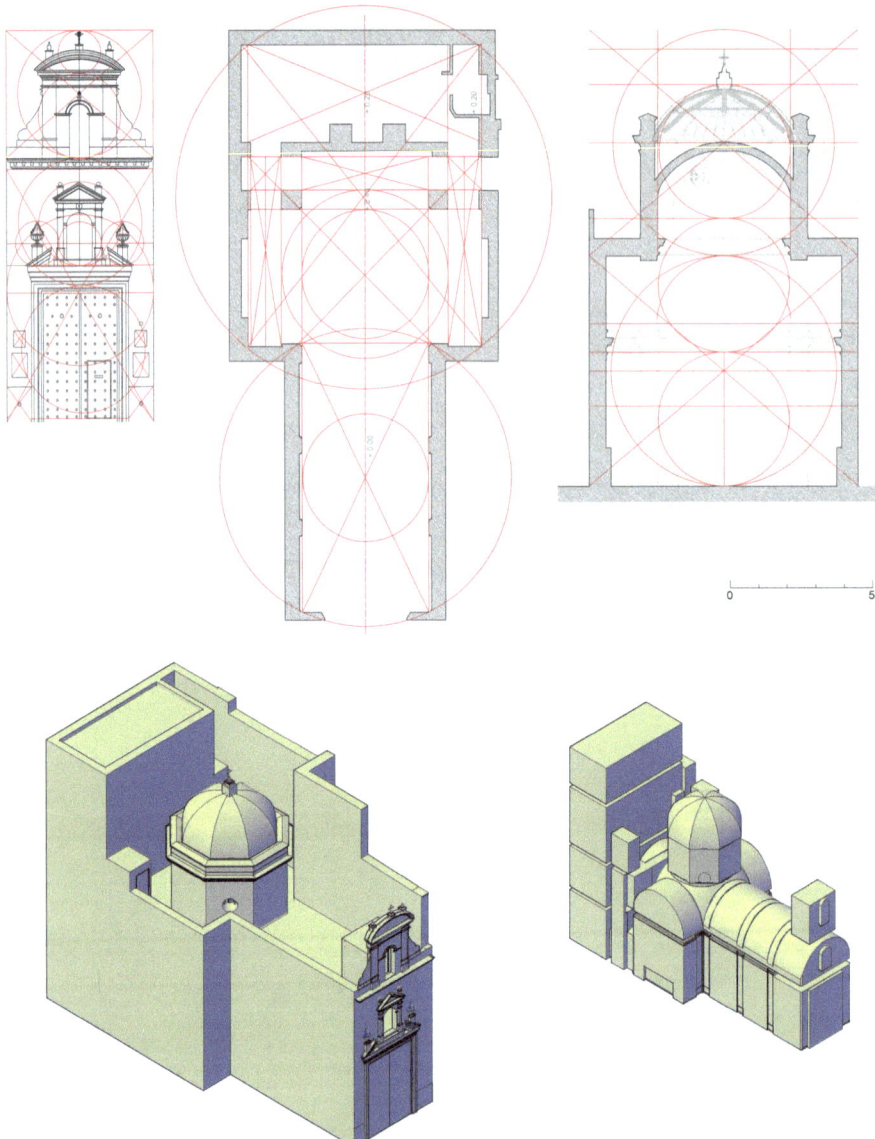

Fig. 5. Geometric analysis in plan, elevations and 3D. Exterior and Interior of the Piedad Chapel. 2009. Source: F. Granero.

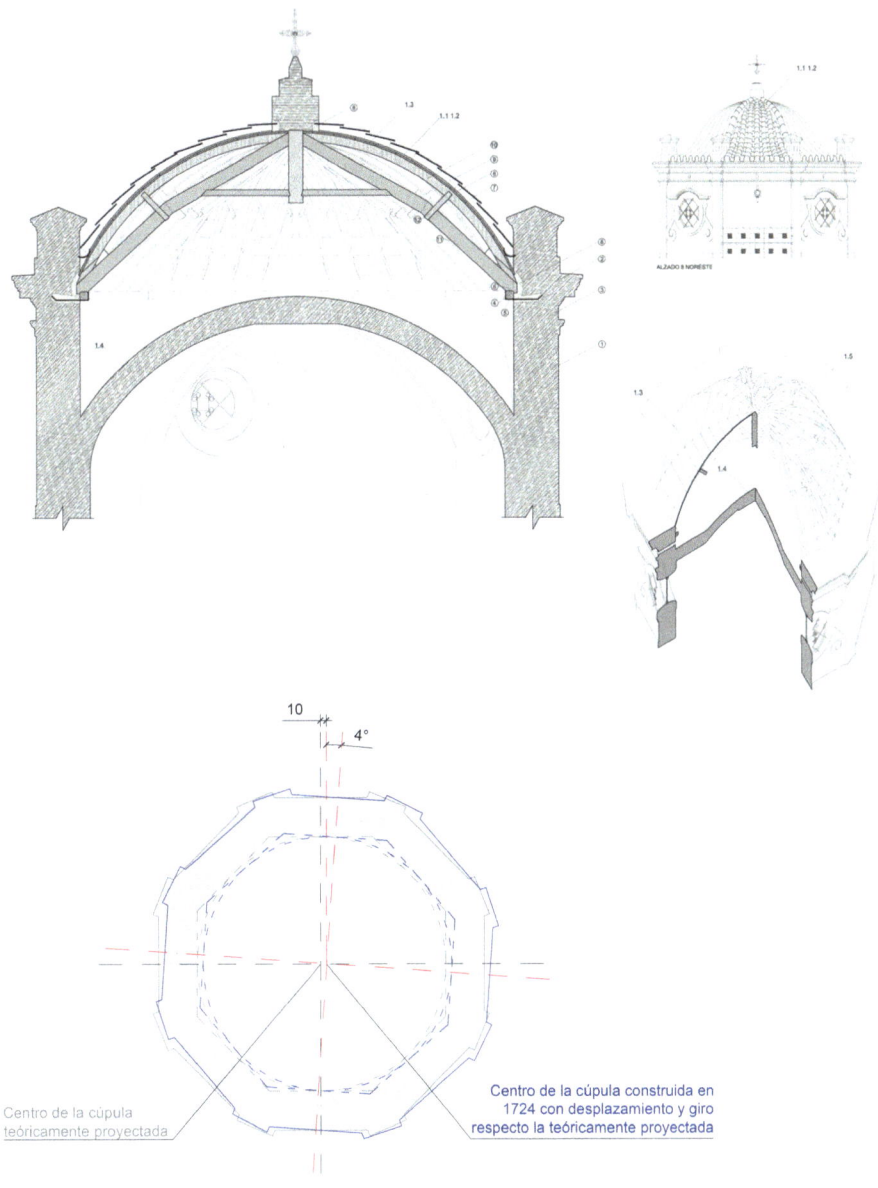

Fig. 6. Drawings of the dome and analysis detailing the rotation and lateral deviation from the axis as a possible mistake in the works of enlargement of the Piedad Chapel in 1724. Excerpts from the retrofitting project of 2009. Source: F. Granero.

In the geometric aspect, we find that the Northeast buttress had lost its verticality as a witness of the break of the timber frame and subsequent structural collapse. There happened a loss of section of the abutment, sliding of the roof and failure of the cover deck with the ensuing cracks in the pinnacles of the drum and the structural frames, all of which composed a veritable catalogue of defects.

As a methodology for the ensuing intervention on the Chapel, taking into account that it was not possible to open and dismount the elements, since this would endanger the mobiliary heritage of the building, we proceeded to draft detailed plans of the complex, depicting the current status of the situation and its graphic comparison with the original project. Such operation offered scientific knowledge on the deviation of the measurements of the project from the actual works of enlargement of the Chapel and the construction of its dome in 1724, which resulted decisive and pivotal to establish guidelines for the draft of the executive project (Fig. 6).

About the error on site of the stakeout drawing and the deviation in the service loads we can state the following. By means of the detailed drawings produced, the defects of the original disposition were determined for the dome on four scallops (the architect was Marcos Sancho in 1724). As a consequence, the causes that motivated the inclination and opening of cracks became clear. (Various Authors 2009, p. 29). Both the rotation on the vertical axis of the dome, estimated in 4 degrees, and the lateral sliding of 10 cm. Produced uneven overloads on inconsistent soil, mainly composed of disaggregate infills (Baratillo mound), which were aggravated by the devastating earthquake of Lisbon in 1755 that seriously damaged the dome.

The timber structure was well designed in origin but its construction was funded by economic allowances solely donated by the members of the Brotherhood and thus the budget was inconsistent and led to employing wooden remains of different origin and size (parts of chariots and other debris). This situation produced a swift process of decadence to reach the limits of support and the risk of structural collapse by ruin (Fig. 7). This is the reason why, in this Project, drawing has been a useful tool to control and analyze the causes that determined the damages and allowed for the retrofit of Heritage.

5   Retrofitting of Heritage and Preservation of the Soul of a Place

The required integration of retrofitting inside the protection process, makes it necessary that challenges, tendencies and hopes find their transcript in the domain of Heritage restoration, by means of criteria related to the interventions on real estate (Hernández 2016, p. 27).

During the years 2009 to 2010 the competition project (for the Andalusian Government) was finished and the works of consolidation and retrofitting were executed. The result is a remarkable baroque oratory of great social and architectural significance. Due to limitations in the publication, we have not described in detail the damaged status of this emergency operation, nor the restoration projects. We have just focused on the plans

Fig. 7. Restoration process of the wooden structure of the dome. Timber frame consisting mainly of debris and fragments of the carts from the ancient harbor. 2009. Source: F. Granero.

of the analysis-project and the pictures of the retrofitting process that is based on the geometric control with the main objective of providing a durable solution as close to the original heritage as possible.

Besides, after obtaining a second competition (from Seville's town Hall), a replacement for the foundational cross on pedestal over the old "Monte Baratillo" was projected and executed. Thus originating a new urban sector that accrued to the ample and rich historical memory of the place, and purveyed the ensuing drawing for the original holy cross from the 16th century that currently tops the dome of the Chapel. In this way, the square was reorganized from an urban point of view and the whole operation was completed by recovering the foundational cross that originally presided over the so-called "charnel ground" from the interments of the plague epidemics of 1649. This summoned to the restoration of the oratory consolidates the baroque repertoire within the old town of Sevilla (Figs. 8, 9 and 10).

Fig. 8. Restoration process of the dome and final stage of external Works. Piedad Chapel. 2009/10. Source: F. Granero.

Fig. 9. Restoration process of the works at the Piedad Chapel. 2009. Source: F. Granero.

Fig. 10. Final Stage of the restoration of the Piedad Chapel. 2010. Source: F. Granero.

6   Conclusions

As a consequence of the analysis performed, the place in which the Piedad Chapel is located, can be ascertained as immaterial heritage of an historical sector with relevant activities for the town of Seville from the 16th to the 18th centuries. Our research has focused on the importance of precise geometric 3D drawings to check potential errors of the original stakeout, present in the construction of the dome and its enlargement. It was executed in a "heart rules over mind" fashion, that it is to say, designed under the sound project of the architect Marcos Sancho but poorly built by employing residual materials from the harbor carts and other debris, which originated a decay that the devastating earthquake of Lisbon in 1755 was only to aggravate. The axis lateral deviation and rotation provoked a structural asymmetry sufficient to originate a number of increasing defects which in turned menaced the stability of the building.

With all the former we have tried to explain a double fold relationship between Drawing and Heritage: one is the importance of the stakeout in the Works of the dome (enlarged in 1724). The ensuing errors can be identified as the main flaw which produced and emergency situation that necessitated urgent consolidation and restoration. As discussed, the author conducted these works successfully after being selected in competition.

The second one shows us the potential of drawing as a tool of analysis, control and detection of the said errors in order to mend them and permit the retrofit and restoration procedure giving this oratory to the town as an important part of its original Baroque Heritage.

References

Various Authors: Seminario La Aplicación del Código Técnico de la Edificación a la Intervención en el Patrimonio Cultural. Col. e-Ph cuadernos IAPH. C. Cultura, Junta Andalucía, Sevilla (2009)

Capitel, A.: Metamorfosis de monumentos y teorías de la restauración. Alianza, Madrid (1988)

Castillo, J., Cejudo, E., Ortega, A.: Patrimonio histórico y desarrollo territorial, UNIA, Sevilla (2009)

Granero, F.: El Corral de los Olmos de Sevilla. Antiguos cabildos secular y eclesiástico de la ciudad. Sevilla. Sus orígenes, funciones, compilación de transformaciones y demolición, COAAO, Sevilla (1992)

Hernández, A. (ed.): Preserving the past projecting the future. In: Tendencies in 21st Century Monumental Restoration, Proceedings. I. Fernando El Católico, Diputación Zaragoza, Zaragoza (2016)

The Graphic Heritage of Fifteen Years of Covers of *Arquitectura* (1959–1973)

Amparo Bernal López-Sanvicente(✉) 📵

Universidad de Burgos, Burgos, Spain
amberlop@ubu.es

Abstract. The covers of the journal *Arquitectura*, published by the Institute of Architects of Madrid since 1918, are a hallmark of the journal and the most visible expression of its editorial line. This paper analyzes the graphic legacy of the editorial period between 1959 and 1973 under the direction of the architect Carlos de Miguel.

At the beginning of this new period of the journal, one of the goals of the editorial team was to incorporate other disciplines related to architecture so that the publication could turn into a multidisciplinary forum. This criterion, in addition to conditioning the structure of its contents, was reflected in the composition of the covers. During the first years of this period, the editorial team chose photography and, in the mid-sixties, graphic design began to be incorporated into the logo of the header and the composition of the cover.

The quantitative analysis of the images published on the front pages of the journal at this period will allow us to prepare a detailed inventory of all the published creations. Subsequently, the images will be classified according to the dominant technique in their composition to delve into the critical assessment of this graphic legacy.

Keywords: *Arquitectura* journal · Covers · Bibliometric analysis · Architectural graphic expression · Photography

1 Introduction

The journal *Arquitectura* is one of the main documentary sources for the study of modern Spanish architecture. The volume of its publications and the breadth of its scope of dissemination among architects have guaranteed its status as a reference throughout its one hundred editions.

In 2018, on the occasion of the centenary of its foundation, the Institute of Architects of Madrid (abbreviated as COAM in Spanish) organized a series of commemorative events to spread its legacy. Among them, a retrospective exhibition was organized at the institute headquarters, where original copies of the different editorial periods of *Arquitectura* where shown next to the architecture plans and models that were the sources that supported its graphic content. A commemorative issue of *Arquitectura* was also published, highlighting the hundred years of its editorial cycle. In this review, the most

L. Agustín-Hernández et al. (Eds.): EGA 2020, SSDI 6, pp. 17–27, 2020.
https://doi.org/10.1007/978-3-030-47983-1_2

important milestones that have turned the journal's documentary legacy into a reference for the study of the history of architecture in Spain were adapted to the current content structure of the journal grouped into the following sections: radar, highlights, labs, neighbourhoods, suburbia, techniques, invariant and editorial [1].

In addition, in order to facilitate the access to its contents in digital format, the Historical Service of COAM has carried out an exhaustive process of scanning the journal accessible through the corporate website [2]. In a first phase, the journals of *Arquitectura* published between 1918 and 1936, and between 1959 and 2018 were available. Afterwards, this digital archive was completed with the journals of the *Revista Nacional de Arquitectura* published between 1941 and 1958.

So, once this phase was completed, the archive of the digital journal of the one hundred years of *Arquitectura* was classified in sixteen periods according to the different editors-in-chief or editorial teams of the journal. The open access of this legacy enables the application of new research approaches for the study of the journal and its contents.

Arquitectura, as an architecture journal and publication of a professional association, is the expression of an editorial project that has been defined at each of its periods through the graphic appearance and the editorial line. This paper focuses on the analysis of the graphic legacy of the covers of *Arquitectura* when the architect Carlos de Miguel, was in charge as Editor-in Chief of the journal, between 1959 and 1973 (Fig. 1).

revista Arquitectura 1959-1973

Fig. 1. Composition of covers of the period 1959–1973. Source: own elaboration.

This research has been approached from the analysis of an exhaustive catalogue of the graphic legacy of the covers. A more critical review of this inventory will allow us to determine the uniqueness of its content and delve into its study. The aim of this research is to verify if the journal *Arquitectura*, between 1959 and 1973, was a coherent editorial project in its whole. Our aim is to get to know if the covers of the journal truly expressed the character of the architectural journal, if they anticipated the contents and if they helped to frame and define the editorial line of the publication regarding the editorial scope in the context of architectural journals.

2 Methodology: Quantitative and Qualitative Graphic Analysis

The graphic appearance of a journal is a set consisting of its header, cover, format, typography, stain, page composition, illustrations, relationship between text and images, paper quality and print quality [3]. Among all these factors that make up the image of a publication, the cover is, without any doubt, a hallmark. Its design synthesizes the graphic and compositional text criteria of the layout of the journal.

The methodology implemented in the quantitative and qualitative graphic analysis of the cover images employs further and more extensive bibliometric analysis as a reference, such as the one conducted on *Arquitectura* in the sixties [4] and the studies carried out by professor Fernando Linares on the journal *EGA: Expresión Gráfica Arquitectónica* [5] and [6].

The first data of the quantitative graphical analysis were obtained from the temporal delimitation of the research. In the period between January 1959 and June 1973, 175 numbers of *Arquitectura* were published in 171 journals, because four of these published journals were double issues.

In order to address the qualitative analysis of this legacy, five categories have been defined. The images have been classified with regard to the dominant technique in their composition. According to this criterion, the covers of the journal can be classified as photographs, graphic design compositions, architectural drawings, illustrations or prints, and artworks of painting or sculpture. From the data of this first classification, the general guidelines that were obtained allowed to carry out deep critical analysis (Table 1).

Table 1. Classification of cover images.

110 fotografías
27 diseños gráficos
17 dibujos de arquitectura
10 obras de arte, pinturas o esculturas
7 ilustraciones o grabados

171 portadas de Arquitectura 1959-1973

3 Photographic Image

Photographs were the predominant compositional resource on the covers of this period (64.33%). Therefore, this first classification is insufficient to qualitatively assess its presence in the journal and it was therefore necessary to carry out a detailed study of the photographers and the photographic genre to which the images belong, depending on their function and their reference.

As for the photographers, the photographs of Paco Gómez predominate since he was the photographer of the journal during all this period. Paco Gómez, a photographer of the *Escuela de Madrid* and a founding member of the group *La Palangana*, began collaborating with the journal in 1959 after submitting a portfolio with some of his creative photographs to Carlos de Miguel. From 1959, his work starred in the image of the journal and was published on the covers and on the inside pages. Likewise, his previous creative works were published and his photographic language started to specialize in architectural photography due to his collaboration with the journal [7]. Forty-eight of the one hundred and ten photographs published on the cover have been identified as of his own authorship (43.6%). Many of the thirty-seven photographs whose author has not been identified could also possibly belong to him (Fig. 2).

Fig. 2. Paco Gómez. 2A. *Arquitectura* 51, 1963, cover. 2B. *Arquitectura* 53, 1963, cover. 2C. *Arquitectura* 61, 1964, cover. 2D. *Arquitectura* 69, 1964, cover.

Some other photographs found belonged to renowned contemporary photographers such as Kindel, who had been a photographer of the previous editorial period such as *Revista Nacional de Arquitectura* [8], Catalá Roca, photographer of *Cuadernos de Arquitectura*, edited by the Catalonian Institute of Architects, and other renowned contemporary photographers such as Schommer, Pando, Masats, Lladó and Paco Ontañón, among others (Fig. 3).

The photographs on the covers of the journal were only the visible side of the photographic content of the interior architecture articles. In the sixties, architecture photography predominated in architecture documentation with regard to drawing and became the predominant language showing the panorama of contemporary Spanish architecture on the journals such as *Arquitectura, Hogar y Arquitectura, Nueva Forma* or *Cuadernos de Arquitectura*. Hence, the photographic content of *Arquitectura* has been one of the essential documentary sources in the documentation of modern architecture photography in Spain [9].

As for the photographic genre, all types of photographs were published on the covers of Arquitectura, although architectural photographs prevailed. Thus, the set of one hundred and ten photographs could be classified as: photographs of artistic creation, modern architecture, architectural heritage, photographs of models, portraits, and even other narrative and instrumental genres of photography as social documents and cartographic documentation (Fig. 4).

Because of its graphic interest and its correspondence with the expression of contemporary architecture, photographs of models prevailed. When analysing the role of

Fig. 3. 3A. Plasencia y Dachs. *Arquitectura* 3, 1959, cover. 3B. Schommer. *Arquitectura* 48, 1962, cover. 3C. Pando. *Arquitectura* 50, 1963, cover. 3D. Paco Ontañón. *Arquitectura* 62, 1964, cover.

Fig. 4. Paco Gómez. 4A. *Arquitectura* 95, 1966, cover. 4B. *Arquitectura* 119, 1968, cover. 4C. Pando. *Arquitectura* 133, 1970, cover.

the model in the methodology, representation and communication of the architecture project, our discourse was necessarily based on the photographs of the models, because the physical materiality of a model, its size, materials and the nature of its construction, condition its conservation.

In Spain, the sixties were especially prolific in the production of models as a reflection of the importance of the models in the architectural production of that period and also due to the proliferation of architectural competitions in this decade. The architecture of the competitions reached its maximum expression in its models and, sometimes the only possible one –unless the awarded project got to be built. Through the journal *Arquitectura* we have known the models of the annual editions of the National Architecture Prize, the competition for the Pavilion of the New York Fair (*Arquitectura* 52, 1963) and competitions such as the Opera Theatre and the Palace of Congresses and Exhibitions held in Madrid in 1964 (*Arquitectura* 71, 1964). All of them left the testimony of the architecture of the sixties in Spain built on models that have survived over time, thanks to the photographs.

The journal *Arquitectura* used model photographs on the cover up to fourteen times. These include: the monument to José Battle (*Arquitectura* 6, 1959), the Higueras Chapel (*Arquitectura* 9, 1959), the third prize of the Peugeot International Competition presented by Bravo, Fernández Plaza and Pintado (*Arquitectura* 40, 1962), the project of Playa Blanca in Lanzarote by Higueras and Miró (*Arquitectura* 70, 1964), as well as some other models of the winners of the National Architecture Award (*Arquitectura* 28, 1961), the Kursaal competition in San Sebastián in 1965 or the competition for the architecture scholarships in Rome in 1966 (*Arquitectura* 93, 1966) (Fig. 5).

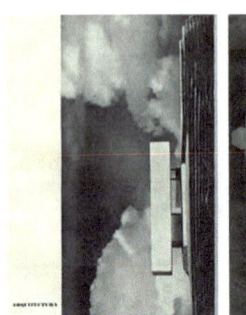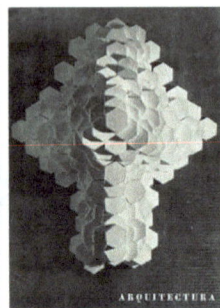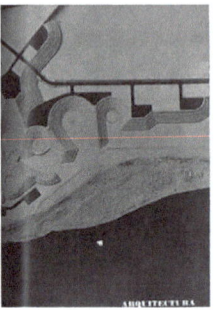

Fig. 5. Unknown photographer. 5A. *Arquitectura* 6, 1959, cover. 5B. *Arquitectura* 9, 1959, cover. 5C. *Arquitectura* 70, 1964, cover. 5D. *Arquitectura* 78, 1965, cover.

4 Graphic Design

A 15.79% of the journal's covers of this period are graphic designs where architectural plans or photographs were included as secondary elements of the design. Their publication is mainly limited to 1964 and 1966, a period coinciding with the incorporation of the graphic designer Juan José Morales to the graphic design management of the journal. Juan José Morales began working in *Arquitectura* after having collaborated with the architect Javier Feduchi in the renovation of the image of Galerías Preciados for the opening of the new building in Plaza Callao in the city centre of Madrid [10]. From November 1964, graphic compositions began to be published using the design of colour and geometry of the acronym AR adopted by *Arquitectura* as logo for its header (Fig. 6).

In addition to strictly geometrical compositions, there was also an important evolution in photographic design, understood as a creative manipulation of the photographic image [11]. Juan José Morales launched this technique on the covers of the journal, making photomontages and typographic compositions using the photographs as background. However, some of the most unique examples belong to the architect Rafael Fernández del Amo and some other unknown authors (Fig. 7).

Juan José Morales stopped collaborating with the journal in February 1968. From that moment, collaborations with graphic designers were sporadic. That same year, the cover o fthe journal dedicated to the architecture built in Mexico for the Olympic Games reproduced the design of the American graphic designer Lance Wyman and the architect

Fig. 6. Juan José Morales. 6A. *Arquitectura* 71, 1964, cover. 6B. *Arquitectura* 73, 1965, cover. 6C. *Arquitectura* 74, 1965, cover. 6C. *Arquitectura* 76, 1965, cover.

Fig. 7. 7A. Paco Gómez - Unknown graphic designer. *Arquitectura* 120, 1968, cover. 7B. Paco Gómez - Unknown grahic designer. *Arquitectura* 147, 1971, cover. 7C. Rafael del Amo. *Arquitectura* 168, 1972, cover.

Fig. 8. 8A. Lance Wyman and Eduardo Terrazas. *Arquitectura* 116, 1968, cover. 8B. Manuel López. *Arquitectura* 130, 1969, cover. 8C. José Mª Cruz Novillo. *Arquitectura* 151, 1971, cover.

Eduardo Terrazas (*Arquitectura* 116, 1968). The rest of the singular collaborations with graphic designers included works of Manuel López (*Arquitectura* 130, 1969; 131; 132) and José Mª Cruz Novillo, (*Arquitectura* 151, 1971) (Fig. 8).

5 Works of Art by Artists and Architects

Reproductions of works of art amount to 19.88% of the total. This group includes architectural drawings, sculptures, paintings, and other illustrations or engravings. The presence of works of art on the covers of the journal is an expression of the multidisciplinary editorial project carried out during this period including other disciplines related to architecture among the sections of the journal. The structure of the summary had been defined in 1960 according this editorial line by establishing three new sections dedicated to art, economics and philosophy [12]. The works of art published on the front pages of the journal came from the graphic content of the art section, called "Notas de Arte", which was usually related to the main topic covered in the articles.

5.1 Architecture Drawings

Unlike the previous editorial period as *Revista Nacional de Arquitectura* that stood out for its graphic content [13], most of the architectural drawings published on the covers since 1959 until 1973 were selected from the graphic content of the articles.

The architectural drawings and plans make 9.94% of the total of the journal's covers at this period. A percentage that matches the low presence of the architectural drawing on the journal, compared to the photographic reports of the architectural models or the photographs of built architecture.

Among the few examples published on the cover, the sketch made by Fernando Zóbel, painter and patron of Spanish abstract art, is worth noting, (*Arquitectura* 86, 1966). The drawing of his house in the city of Cuenca was part of a delicious set of sketches of the city, its people and landscapes made by the painter when he settled in the city in order to carry out his project to install a museum to exhibit his collection of Spanish abstract art in the Hanging Houses of Cuenca.

The architect Joaquín Vaquero Turcios did only make a drawing for the cover with the map of the Island of Paris turned into the letter "a" of the journal's header (*Arquitectura* 88, 1966). An unusual presence in comparison with the nine covers he had published in the *Revista Nacional de Arquitectura* where he usually collaborated as draughtsman when he was an architecture student, publishing many drawings for the journal's articles.

The architectural drawings of the brothers José Luis and Efrén García Fernández were on the cover of the journal twice. The first cover was the publication of the graphic study of the architectural, urban and landscape of Castropol condemning the need for protection and catalogue of its anonymous architecture (*Arquitectura* 98, 1967). The second cover was on the issue dedicated to the Partial Planning of the Old Town of Lugo made by this architects (*Arquitectura* 134, 1970). On this occasion, José Luis and Efrén García Fernández designed a new logo for the header of the journal that continued to be used until the end of this editorial period (Fig. 9).

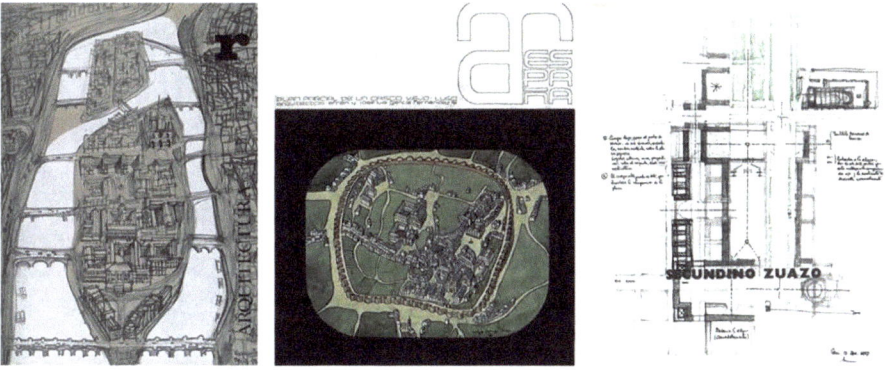

Fig. 9. 9A. Joaquín Vaquero Turcios. *Arquitectura* 88, 1966, cover. 9B. José Luis and Efrén García Fernández. *Arquitectura* 134, 1970, cover. 9C. Secundino Zuazo. *Arquitectura* 141, 1970, cover.

With regard to the rest of the architectural drawings, the most unique, according the artistic treatment on the cover was the sketch of the floor plan of Nuevos Ministerios by Secundino Zuazo (*Arquitectura* 141, 1970). Other less outstanding architectural drawings published were the facades of the Alfonso XII neighbourhood made by Juan José Morales (*Arquitectura* 100, 1970), the facades of the traditional architecture of Calatañazor made by the French architect Denis Pilven, during a scholarship of the Casa de Velázquez he enjoyed (*Arquitectura* 156, 1971), and the plans of a student's project of the School of Architecture of Madrid for Units of Emergency (*Arquitectura* 1570, 1972).

5.2 Paintings and Sculptures

Carlos de Miguel chose the image of a classical sculpture to start this period of the journal in January 1959. It was the bronze copy of the sculpture of Carlos V made in Carrara marble by Pompeo and Leone Leoni (circa 1553) located in the Prado Museum. The sculpture presided over an exhibition dedicated to the emperor in order to commemorate the IV centenary of his death in 1958, held at the Palace of Santa Cruz de Mendoza, in Toledo.

He once again chose the image of a classical sculpture to close this editorial cycle and remember the twenty-five years he had run the journal of the Institute of Madrid. On this occasion, it was the sculpture "Architecture", one of the twelve sculptures that decorate the main facade of the Prado Museum building, made in stone by Valeriano Salvatierra in 1830.

Apart from this voluntary parallelism that Carlos de Miguel wanted to establish between architecture and sculpture, the presence of sculptures on the covers of the journal was only repeated three more times. Chillida's works were published twice as a reflection of the close relationship between the sculptor and the architects, and the identification of architects with his spatial language (*Arquitectura* 30, 1961) and (*Arquitectura* 167, 1972) (Fig. 10).

With regard to the paintings, we can highlight those that were made *ex professo* for the composition of the cover. In some cases, they were total or partial reproductions

Fig. 10. 10A. Chillida. *Arquitectura* 30, 1961, cover. 10B. Paco Gómez-Chillida. *Arquitectura* 167, 1972, cover. 10C. Valeriano Salvatierra. *Arquitectura* 169–170, 1973, cover.

of engravings or paintings by avant-garde artists, such as Manolo Millares or Alberto Sánchez Pérez, known as Alberto (*Arquitectura* 162, 1972) (*Arquitectura* 139, 1970). The publication of these covers was always linked to the contents of "Notas de Arte" (Art Notes), the section run by the art critic Juan Ramírez de Lucas since July 1961. These notes were dedicated to publish contemporary Spanish art in the journal (*Arquitectura* 31, 1961).

The most unique of this set are the graphic artworks deliberately made for the composition of the cover of the journal. Among these, the works made by Benjamín Palencia (*Arquitectura* 143, 1970), Joan Miró (*Arquitectura* 125, 1969) and the cover made by Pep Bonet for the journal dedicated to the *Escuela de Barcelona* (*Arquitectura* 121, 1969) can be highlighted.

5.3 Illustrations and Engravings

The percentage of illustrations and engravings amounts to a 4.09% of the total number of covers. Except for the engraving of Juan de Herrera by Mariano Brandi for the series of portraits of the illustrious Spaniards (1791–1819) (*Arquitectura* 30, 1961) and an antique engraving related to the history of medicine as the cover of a journal dedicated to hospitals (*Arquitectura* 56, 1963). The rest of the covers classified under this heading are illustrations that do not correspond to architectural drawings.

6 Conclusion

After the renovation of the editorial project of *Arquitectura* in 1959, the journal decided to incorporate other artistic disciplines into its contents with the aim of spreading contemporary Spanish art and boost the integration of the arts into architecture. This criterion of turning the journal into a multidisciplinary forum has left an important graphic legacy on its covers.

The set of covers published between 1959 and 1973, stands out for its photographic character, where the photographs of Paco Gómez stand out as the leading ones. However, there are also other photographs of contemporary Spanish photographers.

In the mid-sixties, the incorporation of Juan José Morales to the editorial team of the journal as a graphic designer led to the renewal of the cover composition patterns and the experimentation in the design of the header logo.

The quality of the photography, the graphic design and other techniques of composition and manipulation of images, such as photographic design on the covers of the journal *Arquitectura*, framed its graphic legacy in the editorial and advertising field of the sixties in Spain as a specialized journal of architecture and an open space to all the artistic disciplines that can be integrated in architecture.

References

1. Revista *Arquitectura* (378) (2019)
2. Revista Arquitectura 100 años. *Arquitectura* Etapa 1959–1973, (1)–(175). COAM http://www.coam.org/es/fundacion/biblioteca/revista-arquitectura-100-anios. Accessed 25 Oct 2019
3. Alau Massa, J.: Arquitectura: un ropaje ecléctico. En: de San Antonio Gómez, C. (ed.) *Revista Arquitectura* (1918–1936), pp. 192–195. Ministerio de Fomento, Madrid (2001)
4. Bernal López-Sanvicente, A.: Las revistas Arquitectura y Cuadernos de Arquitectura: 1960–1970. Unpublished Ph.D. thesis. Escuela de Arquitectura de Valladolid (2011)
5. Linares García, F.: EGA: Revista de Expresión Gráfica Arquitectónica. A bibliometric analysis following twenty years in publication. *EGA Expresión Gráfica Arquitectónica* **20**(25), 36–47 (2015). https://doi.org/10.4995/ega.2015.3702
6. Linares García, F.: Twenty five years of EGA: latest indications. *EGA Expresión Gráfica Arquitectónica* **23**(34), 264–275 (2018). https://doi.org/10.4995/ega.2018.10848
7. Bernal López-Sanvicente, A.: Paco Gómez: fotógrafo de la revista *Arquitectura*. *RA revista de arquitectura* (14), 81–88 (2012)
8. Zarza Balluguera, R.: Las palabras nos ocultan la arquitectura. En *Kindel. Fotografía de Arquitectura*, pp. 7–14. Fundación COAM, Madrid (2007)
9. Bergera, I.: *Fotografía y arquitectura moderna en España, 1925–1965*. Fundación ICO, La Fábrica, Madrid (2014)
10. PRAG Arte Gráfico Español. Grafismo en una empresa española (Galerías Preciados, S.A.). *PRAG Arte Gráfico Español* (2), 31–35 (1965)
11. Costa, J.: Fotografismo, Fotografía y Diseño. *Convergências–Revista de Investigação e Ensino das Artes* I (1) (2008). http://convergencias.esart.ipcb.pt. Accessed 25 Oct 2019
12. Editorial. *Arquitectura* (13), 1 (1960)
13. Mendoza Rodríguez, I., Álvaro Tordesillas, A., Montes Serrano, C.: Drawing architecture in 1940s Spain: a study based on the *Revista Nacional de Arquitectura*. *EGA Expresión Gráfica Arquitectónica* **22**(29), 170–179 (2017). https://doi.org/10.4995/ega.2017.4168

The Okoshi-ezu (起絵図) of the Tea House: The Duplicity of Representation

Cristiana Bartolomei[ID] and Caterina Morganti[(✉)] [ID]

Department of Architecture, Alma Mater Studiorum, University of Bologna, Bologna, Italy
`{cristiana.bartolomei,caterina.morganti4}@unibo.it`

Abstract. The Japanese tea house, called *chashitsu*, is attributed to the tea master *Sen-no Rikyū*. It is the place dedicated to the traditional tea ceremony, beloved to the Japanese people. Japanese tea houses are relatively small pieces of architecture, 8.2 m^2, equivalent to about 4.5 tatami. They base the construction process on parameters defined in line with Zen Buddhist philosophy. Each element that makes up the architectural structure has a precise meaning and function and only when everything is arranged in a very particular manner, the traditional tea ceremony can begin. The position of the room for making tea (*mizuya*) and that for serving it, the layout of the garden, the entrances, the furnace and the alcove as well as the position of the windows and the choice of materials are never random. The internal architectural space therefore assumes a fundamental role, which must be immediately clear and legible already in the design phase. The internal architectural space is usually documented through representation methods such as orthogonal projections, using plans, elevations and sections or with axonometric or perspective models. These techniques, however, represent the interior space in an abstract way, whereas the three-dimensional representation instead allows to bypass the difficult reading of the drawings in orthogonal projection. In Japan, therefore, *okoshi-ezu* were born to represent the interior of the architectural space especially of the tea houses. They are folded drawings capable of becoming three-dimensional. These drawings offer a different way of representation: double, clear, synthetic and dense which allows for a better understanding of the space.

Keywords: Tea house · Japan · Models

1 Introduction

In ancient times in Japan, the constructions were entrusted to the master carpenters who planned and followed the construction of the work. The design was determined by practices and conventions, for example deriving from the stereotomy of the wood rather than from actual design intentions. These conventions were incorporated into *kiwari* systems with proportions and modules that guided traditional practice so much that an expert builder was able to build an entire building using only a schematic plan and measuring tools; detailed architectural drawings were therefore not necessary. Functional drawings, understood as graphic elaborations aimed at the realization of what they represented,

L. Agustín-Hernández et al. (Eds.): EGA 2020, SSDI 6, pp. 28–39, 2020.
https://doi.org/10.1007/978-3-030-47983-1_3

therefore had a secondary role (Evans 1995). For this reason there aren't many examples of drawings of this type in the archives: both for the nature of these drawings, little used, and for the fragile supports used (mainly paper), which certainly didn't help to preserve the drawings. In the past, the transmission of knowledge, especially for the construction of tea houses, took place through books and manuals that schematically reported the proportional systems that were the basis of the design of the buildings. The drawings were simple and essential, drawn in ink on cardboard, based on systems of known proportions and usual forms. They reported only the position and type of openings, pillars, verandas and gardens, including the minimum number of dimensions and written notes possible (Fig. 1).

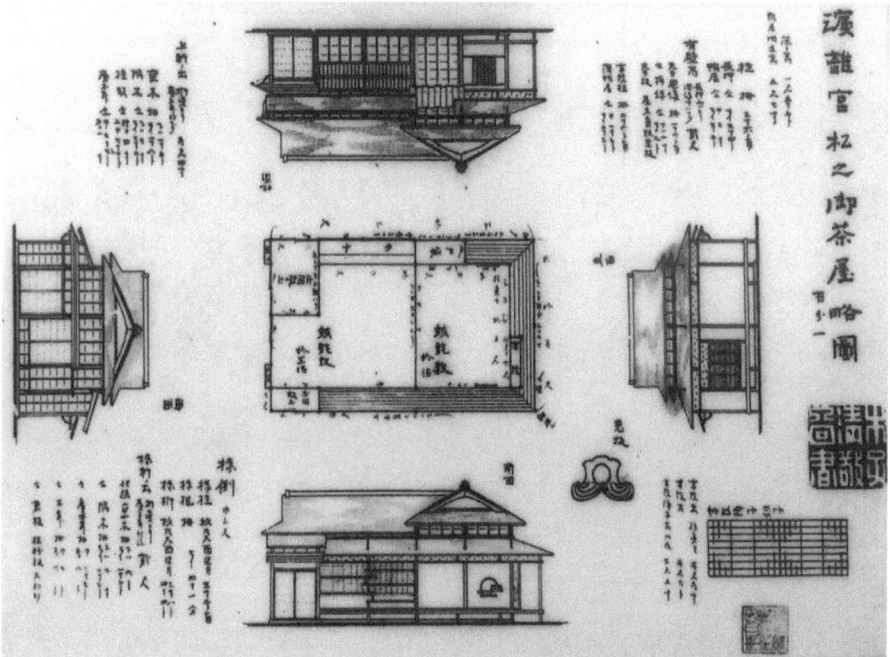

Fig. 1. Example of simple and essential drawings, drawn in ink on cardboard, based on systems of proportions and modules.

Wrongly, it's believed that the usual module used was the tatami (Engel 1985) but actually the correct measurement system of the traditional Japanese house is based on the ken module. The ken was developed by Japanese carpenters during the Middle Ages in Japan, and has since been adopted nationally as a universal measurement system for the construction of all traditional Japanese residences. Its size varies between 6 and 6.5 shaku. Shaku is equivalent to the English foot, which is approximately twelve inches.

The ken module is based on a measurement system that determines the interval between two columns in traditional Japanese residential architecture. Only a few buildings built in the sukiya style came out of this schematic design method, although the tatami is still used as a proportioning module: the tea houses, exactly. In fact, in the Edo period predetermined models are set aside and greater attention is paid to details thus giving rise to an architectural style referring, in particular, to a type of building or pavilion reserved for the tea ceremony: the sukiya style. The elaboration of the sukiya style was inspired by conceptions related to the spiritualism of the school of Zen Buddhism and also to Taoism, for its formal reference to basic types of rural architecture (Girardot 1983).

2   The Spatiality of the Tea House

The tea house (*Chashitsu*), as it is understood today, is attributed to the tea master *Sen-no Rikyū*, who codified the rigid rules of ceremonial, while also innovating with the spaces that were already present in Japanese architecture and that found their definition in the *Edo* period (Horiguchi and Chashitsu 1963–1967) (Fig. 2).

Fig. 2. Example of reconstruction of a tea house in Tokyo.

Since the Muromachi period (1336–1573), the samurai pastime was a tea tasting contest called a tōcha. The samurai had to guess the varieties of tea that were served while engaged in other recreational activities in the kaisho, a typical space destined for these meetings.

These spaces, although not exclusive, represented the first Japanese tea houses anyway. Following the development of traditional residential architecture, the *shoin-no* (tea lounge) developed and from the end of the 15th century, a new style of tea ceremony known as *wabi-cha* was born, based on an overall design and where the details were mainly related to the function.

A construction similar to a "straw hut" that created a "refuge in the city", offering a taste of the countryside in an urban setting.

It was precisely the tradition of *wabi-cha* that was perfected by *Sen-no Rikyū* in the second half of the 16th century. The tea house, strongly influenced by these intimate philosophies, required greater awareness in the design and construction because the arrangement of the peculiar elements of the tea house, such as the garden, the entrances, the floor, the furnace, the alcove, the room for the preparation of tea, the position of the windows and the choice of materials could not be entrusted to the simple carpenter. Instead, the figure of the designer was necessary, to fix his detailed and specific design intentions (Tanaka and Tanaka 1998).

Furthermore, the skills of a carpenter, a painter, a plasterer, a roofer, a tatami maker, a craftsman for doors and windows and a gardener were needed to have the guarantee of a good realization.

As clarified by now, the tea house is a space built specifically to celebrate the tea ceremony, but it isn't enough to build all the parts it consists of to create an optimal tea house; the internal space must guarantee a certain atmosphere, inherent in the sensitivity of the Japanese people (Fig. 3).

Fig. 3. Example of room used for the tea ceremony.

The sensitivity of these people is reflected more than ever in the tea house (Kuffel 2000). The tea house is a decidedly introverted space, it has only a minimum of openings, which are mostly covered by translucid paper screens (shoji), so that only soft light is let in. The world of tea is monochromatic. The building surfaces are free of artificial colors. Columns, walls, ceilings and floors preserve the original fabric of the materials with which they are built.

Chashitsu's architecture clearly rejects any effect achievable with color combinations. The structure of the Japanese tea house usually consists of two rooms: one is the Mizuya, tea and food preparation room and the other is the main room, where tea is served. In Japan, the construction of architecture is based on human behavior: the standard size of the Chashitsu is 8.2 m^2 (4.5 tatami).

The size of the tatami varies from region to region and ranges from 1.5 to 1.9 m^2. Chashitsu with lower than 4.5 tatami are called Koma while the larger ones are called Hiroma.

The space that at first glance seems bare, is instead designed around a specific focal point: the Tokonoma (床 の 間), a raised alcove in which Japanese works of art are displayed on a hanging parchment (Kakemono) or on a floral arrangement. This particular focal point comes to life from the void that surrounds it, which amplifies the space and is the result of an asymmetrical balance.

The space of the tea house is expressed in two dimensions. Walls, floor and ceiling are specific elements that can be read and understood separately, but forming a harmonious three-dimensional whole.

3 The Representation of the Okoshi-ezu

Usually, the architectural space is represented in the form of orthogonal projections, that is, plans, elevations and sections possibly supported by axonometric representations.

This form of representation was considered very limiting in the design of the tea house, as it didn't immediately show spatiality or interiority. The western perspective method could have done it, but in Japanese architectural representation this method was not used at all. To explain the design of a tea house, steeped in its atmosphere, which was to be built by different workers, then it was necessary to resort to a particular type of design, the okoshi-ezu (起 絵 図).

The okoshi-ezu method seems to have originated just before the beginning of the Edo period - exactly when and what's also not clear, however - and the oldest surviving examples date back to the second part of the Edo period, between 1800 and 1868 (Barrie 2010, pp. 62–71).

What is certain is that the models of tea houses were designed by great iemoti (masters) of the tea ceremony, such as Sen-no-Rikyu (1522–1591), Takeno Joo (1502–1555), Genpaku Sotan (1578–1658), Koshin Sosa (1613–1672), Genso Sosa (1678–1730) and Kawakami Fuhaku (1716–1807) (Hammad 1988, pp. 315–325).

The first models are tea houses designed by *Sen-no Rikyū* that seem to depict its tea rooms of 4.5 *jo*, 2.5 *jo*, 1.5 *jo*, where each *jo* represents the size of a tatami: 1,653 m^2 or 3.13 × 6.27 ft.

This is a unique representation technique, based on pop-up drawings that bend to create a completely three-dimensional model, in which there are descriptions, dimensions and graphic conventions as they appear in traditional technical drawing (Burton 2005), normally on a sheet slightly larger than an A3 (Pallasmaa 2009).

The model represents content and a function in a nutshell, through a material and formal configuration.

It's a type of design capable of being at the same time useful for the construction and for controlling the shape, providing information about the work and important for its disclosure.

It's an important model for acquiring cognitive experience on spatial transformations, search for shapes and as a means of investigation for design and structure.

It's a type of drawing whose purpose is not only to illustrate the project, but to communicate the requirements of the construction. In order to be carried out, it requires in-depth knowledge of the construction techniques necessary, good knowledge of representation, the ability to read plans and elevations, the ability to use scale ratios, the ability to simplify complex volumes in others that are simpler and the ability to develop the volumes requested through the design in such a way that they can be reconstructed in three dimensions.

Thus, the model created includes a double meaning in itself: it is a sort of manual with recognized authority and is at the same time a sort of guideline for the evolution in the project.

It defines the way of working that leads to the solution, avoiding any drastic reduction in the multiplicity of choices compared to the real demand. In fact, these designs are marked by a particular feature: the two-dimensional nature of the supporting material capable of ensuring usability and readability of the contents.

Only by folding the parts of the sheet does the paper acquire the model status with its three-dimensional character. They are not to be confused with origami, as origami remains folded, while *okoshi-ezu* have the characteristic of cyclicity: they can be made by folding the parts and remade indefinitely.

They are also real executive drawings that were used by the master builders on the field together with the meter and the plumb line. These drawings were the tools used by the site managers. So it was also important to design the model itself like an object. Therefore the format, the type of paper, the color of the ink in relation to the color of the paper, the folds, the choice of where to put the writing or the definition of the rightness text in relation to drawing were fundamental elements that one could not ignore.

The descriptive text on *okoshi-ezu* identifies the structural dimensions, the spatial measures, the original materials and the textures used (Barrie 2004).

The drawings gain their three-dimensionality by, starting from orthogonal projections in plan, section and elevations, precisely "building" through the cutting and folding of the sheets that allows for the perception of the internal space (Allen 2009) (Fig. 4).

It's up to the designer to decide where and how the cuts and creases should be positioned on the two-dimensional sheet of paper so that a three-dimensional model is generated when closing. While the architectural design in orthogonal projections fragments the building to produce knowledge of its parts, this model aims to validate the spatiality of the building.

The fold, the effect of folding or the point where something folds has always aroused interest in various areas, from an architectural and structural points of view.

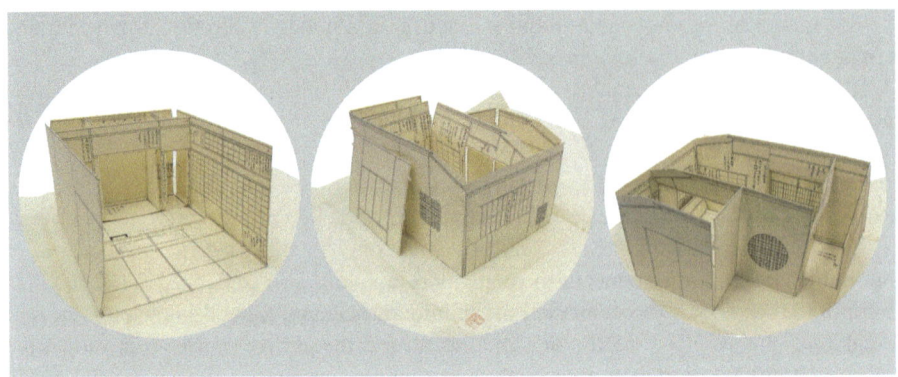

Fig. 4. Examples of *okoshi-ezu* (起絵図).

The fold can be understood as an architectural modeling device or tool. You can conquer the depth through the geometric reversal, a plane that turns upside down, allowing the passage from the plane to the space, from two to three-dimensions.

If we consider the sheet of paper to be a generic rectangular surface divided by n folds, parallel to each other, there are two main configurations: initial and final.

In the initial configuration, the surface is completely flat. The folds remain a series of parallel lines that lie in the same plane. The surface has the maximum number of dimensions in the horizontal plane and the minimum in the lateral one.

In the final configuration the surface is completely folded on itself. The overall area on the horizontal plane is reduced to a simple line, while the area projected on the lateral plane has its maximum extension.

Geometric relationships are then established between the various parts of the model. These relationships, in the *okoshi-ezu*, occur through the translational and rotational motion of the surfaces.

During the folding process, no surface changes its shape: the developed configuration is congruent with the completely folded one.

In fact, the internal space is produced through a foldable surface of a drawn surface.

The drawings of the walls of the building are included on both sides of the sheets of *washi* paper (Fig. 5), a handmade paper of good consistency, resistant and translucent (Klanten et al. 2009).

Cellulosic material becomes available as an essential tool to develop the project "thinking with your hands". It is an indispensable material for making the okoshi-ezu of scale study and simulations of assembly and operation of the tea house easily and quickly. Therefore, the physical commitment required to the designer, which takes place in the act of bending the parts to create the new physical form, shouldn't be underestimated: modeling directly engages the hands and the head.

We couldn't quote from memory those who said that *"The three-dimensional material model speaks to the hands and body as powerfully as to the eyes, and precisely the process of making a model simulates the construction process"*. But we take it on loan because this statement highlights the importance of working the space with your hands and proving the physical consistency of the elements that will build that space.

Fig. 5. Paper *washi*.

The generated space will be intimately linked to the experience of the architect who created the *okoshi-ezu* with his own hands.

While generating the model with your hands, you are firstly led to know the dimensional consistency, both of the whole and of its parts, and secondly you are forced to think deeply about the operations to be performed.

In fact, the *okoshi-ezu* is connected to a large plan that represents the floor, drawn with black ink in which the position of the tatami is highlighted, which again is a proportion code often used for proportioning of much of Japanese architecture (Snodgrass 2004, pp. 65–85).

The openings of doors and windows are made through holes and elements such as risers, floors, steps, shutters are fixed in position on the walls, while roofs and stairs are separate elements that can be positioned on the drawing.

The refinement of the *shoji* windows (Fig. 6) and the internal *shoji* walls, as well as the details of the *tokonoma* are appreciable. The entire tea house is flat when the entire structure is laid flat, to become three-dimensional when the tea house is erected by folding the walls and connecting them.

This particular type of depiction becomes fundamental in the representation of the tea house because it has the characteristic of giving priority to space and not to the structure (Linzey 2010, pp. 31–39). It is capable of encoding the complexity of the spatial articulation. The centrality of the space is indeed one of the characteristics of sukiya architecture. Once assembled, these models are similar to split models, able to quickly perceive the internal space (Frascari et al. 2007).

The architectural sign in these models is capable of showing and communicating its content synthetically. The model has an advantage over two-dimensional representations because of its three-dimensionality and its intrinsic legibility for those who are not designers. In fact, the reduced technical competence required to evaluate an architectural project represented with this system has practically remained the same in time, despite

Fig. 6. Windows *shoji*.

the alternation of architectural currents so different from each other that have produced Japanese architectural culture.

In the methods of canonical representations, the perception of space is produced by its representations and therefore interiority is presumed. In *okoshi-ezu* interiority is perceived and felt in another way: it's the representation of interiority. The *okoshi-ezu* possess this characteristic thanks to their objectivity and ability to incorporate volume and spatiality (Fig. 7).

This combines the documentation of the spatiality of an interior with the mode of representation that possesses its interiority. The link of these elements can be indicated as the duplicity of representation. In this way an alternative representation is offered to the construction of the interior. Interiority is like an "immersion space" in which architecture is enveloped and internalized. These drawings are easy to read and extremely understandable, and do not tend to lose readability as information increases.

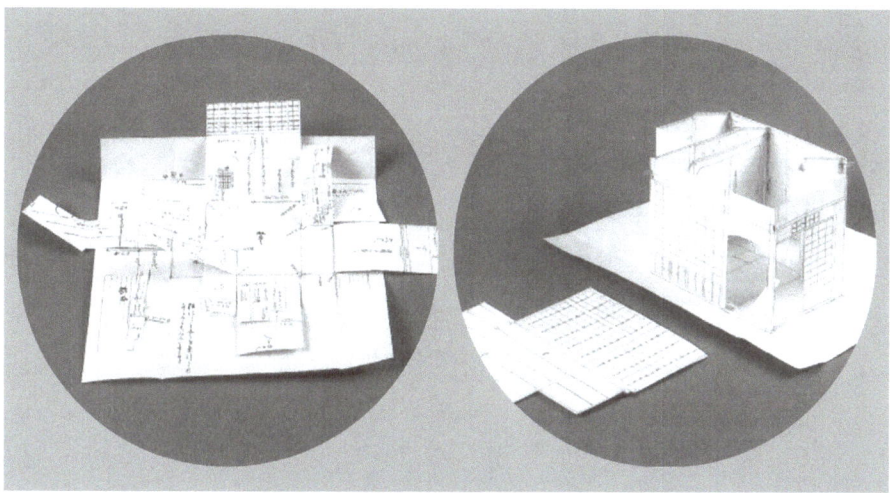

Fig. 7. Examples of *okoshi-ezu* (起絵図).

They are made in ink and its technique does not require retouching. Everything is decided at the moment when the inked brush touches the paper. Since it cannot be retouched, the ink drawing is an art of the moment.

4 Conclusion

The fact of having to "build" the building in order to read these drawings, makes the *okoshi-ezu* a singular representation technique in the field of constructions (Fig. 8). The information that the *okoshi-ezu* includes can be used as construction drawings for its realization (Schmidt and Stattmann 2009).

Fig. 8. Tea house design with *okoshi-ezu* technique.

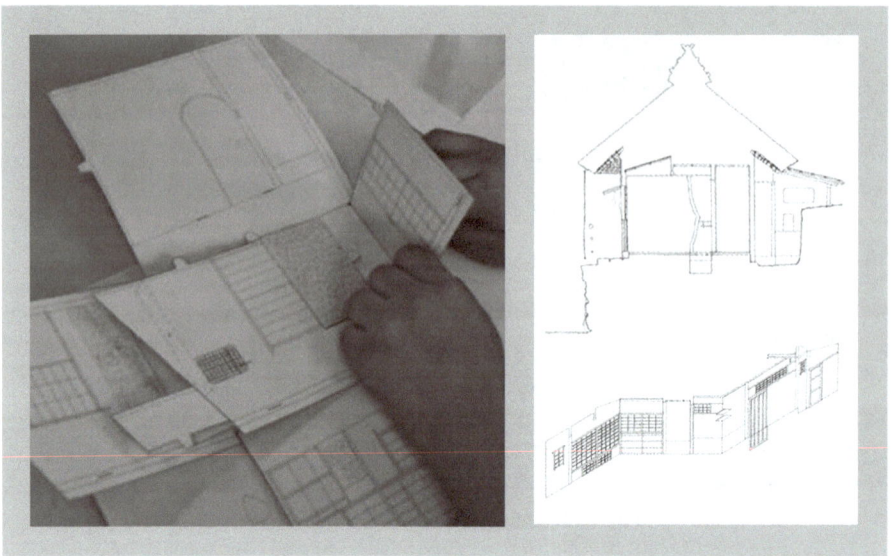

Fig. 9. Example of reconstruction of a tea house in Tokyo.

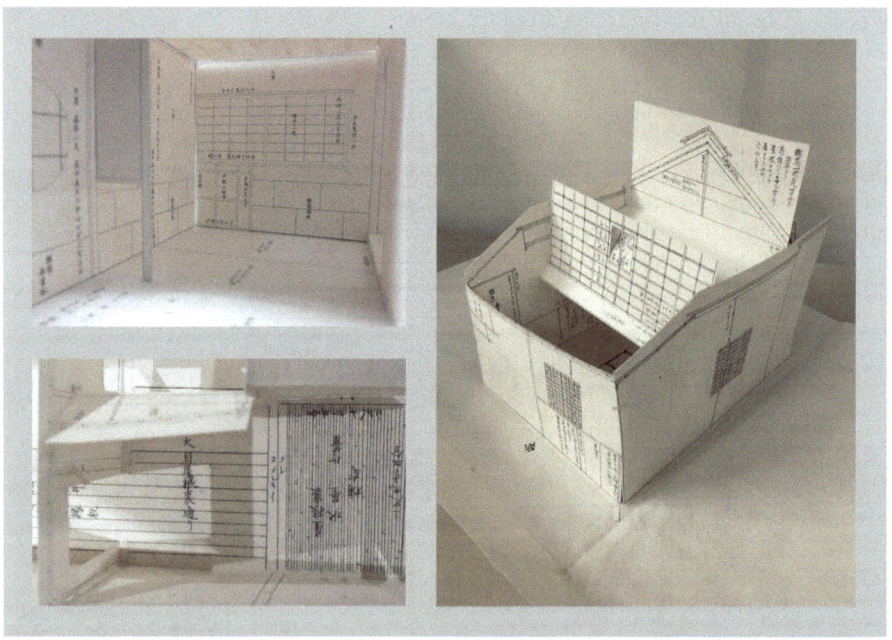

Fig. 10. Example of reconstruction of a tea house in Tokyo.

They were very similar to the architecture that they were meant to represent: thin walls wrapped around cubic spaces to create refined spaces with particular compositions of materials, and light (Sloman 2009). Reading these drawings allows the viewer to travel within the imagined space and to grasp the inspiring principles of the tea house space by acting as a medium for the transmission of knowledge (Fig. 9, Fig. 10).

The study of systems of representation of space has been one of the most important research topics faced in the field of representation due to the difficulty that exists in transferring the three-dimensional image of reality and the information content into a bidimensional space. The drawing of the *okoshi-ezu* is a modeling technique, at least in the sense understood by Tomas Maldonado when he writes: *"drawing, especially drawing to design, is a type of modeling which, as cognitive psychology teaches us today, poses a series of hard-to-answer questions. Because drawing to design manifests itself at the same time as drawing during designing and designing during drawing. It is this interacting co-presence between the means (drawing) and the end (designing) that allows us to move forward towards the solution sought and only found sometimes"* (Maldonado 2007).

The *okoshi-ezu* respond to this in an exemplary way: through the model therefore, you can shape the design thought and also technically inform about how to get to the project.

References

Allen, S.: Practice: Architecture, Technique and Representation. Routledge, London (2009)

Barrie, A.: Tracing Paper. Monument 64 (2004)

Barrie, A.: Okoshi-ezu: speculations on thinness, interstices. J. Archit. Relat. Arts Traction Drawing **11**, 62–71 (2010)

Burton, J.: Vitamin D: New Perspectives in Drawing. Phaidon, London (2005)

Evans, R.: The Projective Cast: Architecture and its Geometries. MIT Press, Cambridge (1995)

Frascari, M., Hale, J., Starkey, B.: From Models to Drawings. Routledge, London (2007)

Girardot, N.J.: Myth and Meaning in Early Taoism: The Theme of Chaos. University of California Press, Berkeley (1983)

Hammad, M.: Teaism aesthetics and architecture. Ekistics **55**, 315–325 (1988)

Engel, H.: Measure and Construction of the Japanese House North Clarenton, p. 34. Turtle Publishing Inc., Vermont (1985)

Horiguchi, S., Chashitsu, O.T.: Folding Drawings of the Famous Tea Rooms. Bokusui Shobo, Tokyo (1963–1967)

Klanten, R., Ehmann, S., Meyer, B.: Papercraft: Design and Art with Paper. Gestalten, Berlin (2009)

Kuffel, J.: Japanese Tea House: Plant Inventory Evaluation. Temple University (2000)

Linzey, M.: Architectural drawings do not represent. J. Archit. Relat. Arts Traction Drawing **11**, 31–39 (2010)

Maldonado, T.: Reale e Virtuale, p. 102. Giangiacomo Feltrinelli Editore, Milano (2007)

Pallasmaa, J.: The Thinking Hand: Existential and Embodied Wisdom in Architecture, p. 56. Wiley, Hoboken (2009)

Schmidt, P., Stattmann, N.: Unfolded: Paper in Design, Art, Architecture and Industry. Birkh, Basel (2009)

Sloman, P.: Tear, Fold, Rip, Crease, Cut. Black Dog Publishing, London (2009)

Snodgrass, A.: Thinking through the gap: the space of Japanese architecture. Archit. Theory Rev. **9**(2), 65–85 (2004)

Tanaka, S., Tanaka, S.: The Tea Ceremony. Kodansha International, Tokyo (1998)

José Luis Picardo's Artistic (and Drawn) "Action"

Juan Utiel González[(✉)] [iD]

Universidad San Pablo CEU, Madrid, Spain
jutiel@ceu.es

Abstract. José Luis Picardo was part of a large group of architects who, away from the spotlight, contributed to raising the bar of 20th-century Spanish architecture. An "artist" architect or, even better, an architect artist. He was capable of handling with ease completely different architectural languages with his talent and his extraordinary graphic capacity as his only weapons, which he developed not only in the field of his profession, but also as a cartoonist, illustrator, muralist and painter.

This article aims at highlighting his nature as a painter and draughtsman which was decisive in his way of interpreting and acting upon heritage, producing drawings through which he appropriated very different architectural elements, allowing him to have a wide formal repertoire with which he tackled projects with extraordinary freedom, in a sort of "artistic action" that was mainly supported, as stated above, by his excellent graphic output.

This is evident throughout his singular architectural trajectory, which ranges from the School of Equestrian Art in Jerez to the building of the new headquarters of the Juan March Foundation in Madrid and goes right through his very interesting projects for the Paradores Nacionales.

Keywords: Drawing · Artistic action · Paradores

1 Introduction

José Luis Picardo developed his work as an architect in the shadow of other great masters of his generation such as Sáenz de Oiza, García de Paredes or Corrales and Molezún. Without any doubt, he was an outstanding member of a middle class of the highest category that contributed decisively, with their talent, their professionalism and their silent work, to the level reached by Spanish architecture in the mid-20th century.

Born in Jerez de la Frontera, although he would settle in Madrid from a very young age, his approximation to Architecture came from the hand of Luis Moya who, upon seeing his abilities, encouraged him to abandon his law studies and enter the Madrid School of Architecture where he was a professor. It would be here, where he would finally obtain his degree in Architecture in 1951. After many years in the field, he was appointed as a member of the Academy of Fine Arts of San Fernando in 1997 and was awarded the Camuñas Prize in 2001.

L. Agustín-Hernández et al. (Eds.): EGA 2020, SSDI 6, pp. 40–50, 2020.
https://doi.org/10.1007/978-3-030-47983-1_4

When observing his works, the heterogeneous nature of his architectural production is striking, ranging from the neo-Renaissance style School of Equestrian Art in Jerez de la Frontera to the elegant Headquarters of the Juan March Foundation in Madrid, -his best work according to himself-, going through the medieval-inspired works carried out for the Paradores Nacionales. It is surprising the creative freedom and the lack of prejudices, above styles and schools with which Picardo unfolded, his naturalness and ease in the handling of different scales and the use of very different architectural languages, sometimes almost antagonistic.

2   José Luis Picardo: Draughtsman and Architect

2.1   Draughtsman, Painter and Illustrator

If we had to look for a common thread that could be used to unify his professional career, we would undoubtedly have to refer to his extraordinary ability to create high quality graphic documents. An outstanding student at the School of Architecture, a cartoonist for various publications, a successful mural painter, an illustrator and, above all, a talented and tireless draughtsman in the practice of his profession as an architect: sketches, perspectives, quick sketches, details and countless plans in which his drawing skills can be clearly appreciated.

While still a student, in 1950, the Revista Nacional de Arquitectura published a selection of drawings by students from the School of Architecture, among which a work by Picardo stood out almost as a prelude to what was to come and which perfectly reveals his technical mastery, his spatial vision and the varied nature of his formal repertoire. The drawing gathers, almost like a collage, a Greek temple, a port with several ships, statues, a Roman arch, a Renaissance-looking city in the background and, at the top, a steep hill crowned by an ancient city with a winged man flying over it. A fantasy

Fig. 1. José Luis Picardo. Ink drawing for an exhibition by Architecture students. Madrid. 1950. Source: *Revista Nacional de Arquitectura* No. 98

that could remind us of Piranesi's drawings and which is a display of imagination and virtuosity (Fig. 1).

He was very interested in popular and historical architecture, to the extent that he was a member of the group of architects, together with other figures such as Chueca Goitia, Miguel Fisac, Secundino Zuazo and Cabrero, who attended the sessions held in Granada in 1952 that resulted in the Alhambra Manifesto. He travelled around Spain accompanying Professor Leopoldo Torres Balbás, for whom he made a multitude of drawings, both sketches (Fig. 2) and details of singular elements of vernacular architecture that were included in his publications on popular Spanish architecture.

Fig. 2. Pencil sketch of a trip with his mentor, Torres Balbás. Albarracín (Teruel). 1947. Source: Pérez Arroyo, S., 2003. *Los años críticos: 10 arquitectos españoles*. Madrid. Antonio Camuñas Foundation.

His activity as a cartoonist was developed mainly in the *Revista Nacional de Arquitectura* (Fig. 3) and in the *Boletín de la Dirección General de Arquitectura*, directed by Carlos de Miguel, with whom he also collaborated at a professional level. His illustrations, with a very varied theme ranging from travel sketches to humorous drawings, could

Fig. 3. José Luis Picardo. Cartoons. Ink. Madrid. 1960. Source: Drawings by José Luis Picardo. Ed: *Dirección General de Arquitectura*.

be compared with those produced by international figures such as the English authors Osbert Lancaster and Gordon Cullen, or the Romanian-born American Saul Steinberg, who published their works in publications such as The Daily Express, The Architectural Review or The New Yorker.

Occasionally, in the same *Revista Nacional de Arquitectura*, he performed works of increasing sophistication as an illustrator, with drawings of greater complexity and extraordinary quality. Of particular significance are those that portray Madrid in the 1950s and 1960s, such as the one entitled *"Verbenas y ferias"* published in 1949, and Picardo's drawings for the Spanish Protectorate in Morocco, to which two monographic issues were dedicated in the 1940s, wherein Picardo perfectly recreated life in the Spanish Protectorate and which have been compiled and analysed by Isaac Mendoza and Marta Úbeda [1], those collected in the double issue dedicated to the Canary Islands in 1953 where he made a large number of sketches of various places on the islands, both modern buildings and details of vernacular architecture or different corners and squares and where photographs are even published of Picardo drawing in the streets surrounded by the public, as if he were a street artist (Fig. 4).

Fig. 4. José Luis Picardo. Ink sketch and photographs of the author drawing in the Plaza de la Catedral of Las Palmas. Source: *Revista Nacional de Arquitectura* No. 140–141 (1953)

He received first commissions as a professional thanks to his skills as a painter, working as a muralist, where he demonstrated his mastery in the use of colour and techniques such as watercolour and oil. His works at Luis Santamaría's stand for the National Exhibition of Decorative Arts or in establishments such as Los Cisnes de Jerez Hotel or the Plaza Hotel, the Fígaro Cinema, the Villamagna parking lot and the Mayte Commodore restaurant, all of them in Madrid, are particularly outstanding. For the latter, he made some remarkable decorative mural paintings on curved wall panels representing the cities of Madrid and Cadiz. In 1954 he participated in the competition of sketches of paintings for the *Hospedería de los Reyes Católicos* in Santiago de Compostela and, although the category to which he presented himself was deserted, he was awarded a prize highlighting the accuracy of the composition and appropriate scale of the figures.

2.2   Architecture and "artistic action"

During the 60s and 70s José Luis Picardo had the opportunity to work on the recovery of historic buildings and their rehabilitation into *Paradores Nacionales*, Spain's luxury hotels. It was in this field that he was able to give free rein to his craft and versatility with surprising ease. It gives the impression that the architect was really comfortable handling the complex commissions that were entrusted to him by the Ministry of Information and Tourism in its endeavour to provide the country with a network of truly unique hotels by restoring historic buildings that were badly deteriorated or in a state of abandonment, projects that have been extensively documented and studied by María José Rodríguez Pérez [2]. He was therefore an important figure in the expansion of the *Paradores* network, which recovered numerous castles, fortresses and religious buildings in ruins to serve a new function, something that was objectively positive in the first place, but which was carried out, on many occasions, with rather controversial intervention criterion.

The issue of intervention in heritage is undoubtedly controversial and has been since the nineteenth century. During the last century, great efforts were made to articulate a "restoration theory", since the Italian Restoration Charters (1831–1931) to the Venice Charter (1964), which was mainly conservationist in content, through the Athens Charter (1931), which already rejected the concept of "pastiche". It is not the purpose of this paper to judge Picardo's theories on heritage intervention or the policy followed by the State during those years when promoting the restoration of historic buildings. Evidently Picardo's approaches in this field are difficult to assume today and even more so if they are analysed in isolation from their historical context, since they sometimes seem to be conceptually closer to the principles of Viollet le Duc than to the positions prevailing today, which are much more respectful towards pre-existing elements. Beyond the theoretical approach, the focus of this paper is on the essentially creative attitude of the architect when dealing with the rehabilitation of this type of building. The aim is not to portray Picardo as an expert architect in restoration, but rather as a good professional who always tried to find the best solution to the challenges he faced. In such a way that, and this is what makes him a special case, without complexes or fears, and sometimes bordering on recklessness, Picardo, sheltering himself in the "artistic action", had no inconvenience in reconstructing, modifying or enlarging the historical remains that passed through his hands in the search for a satisfactory final result. As pointed out by Chueca Goitia, who defined Picardo as *"an architect who conceives architecture as art"*, his *Paradores* are *"essentially creative architecture, since they are old palaces or castles manipulated with a singular intelligence"* [3]. An idea of "artistic action" claimed by Picardo himself when speaking of his interventions in the *Paradores* in contrast to what he called *"scrupulous archaeological action"*, which gave him a shield against moral and ethical questions and allowed him to handle historical heritage in a way that was both free and unbiased: *"Art is eternal, and this is why it is always up-to-date. The reconstructions of the castles, although really false, if they are Art, are justified and if they are not, they are truly condemnable"* [4].

In an often somewhat scenographic conception of restoration, he combined work involving traditional materials and crafts with the unreserved use of new construction techniques, as long as they remained hidden and did not alter the image sought. The end, therefore, justified the means, in such a way that in his work we can find an impressive rib vault supported by a hidden metallic substructure, a coffered ceiling suspended from a concrete slab or a stone retaining wall with a reinforced concrete core. Faced with the dilemma of adopting a mimetic and conservative attitude or a more modern and disruptive approach, Picardo resorted once again to Art, claiming that *"in Art everything is possible; a good architect will know how to weigh up both solutions and his sensitivity shall dictate his choice"* [4].

Within this conceptual framework, in a territory where inspiration is taken from the old but formal freedom and compositional power prevail, Picardo will move with surprising ease, relying on his extraordinary graphic and design capabilities. There is a multitude of high-quality drawings in the *Paradores* projects: meticulous data acquisition and extensive collections of plans wherein spaces, rooms or singular elements such as coffered ceilings or strapwork, are outlined in great detail. In most of his interventions he made axonometric sketches of the set, showing the before and after of his work, freehand, ink drawings with bird's eye views, which tell us of his great capacity for synthesis, of his great spatial vision and of his extraordinary skill as a draughtsman.

In the Parador of the Castle of Santa Catalina in Jaén, after a thorough prior study, the decision was made to locate the new construction over the pre-existing structures of lesser value, leaving the so-called *Alcázar Nuevo* intact, in a sensible and respectful approach. In addition, Picardo addressed this work from an almost territorial scale, in such a way that his sensitivity allowed him not only to put a value on the remains that he found, but also the surrounding landscape and the picturesque itinerary of arrival. The concern for the image of the new building as perceived from the town and even when approaching it is evident. This is corroborated by the existing multitude of drawings, among which the elevations in line with shades (Fig. 5) in which the steep land and the perspectives are incorporated (Fig. 6) stand out.

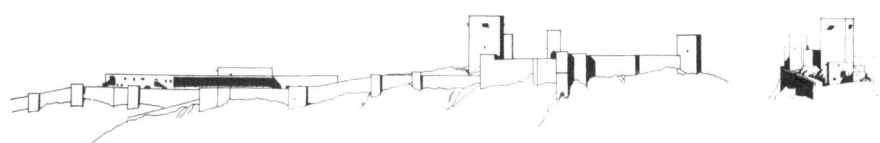

Fig. 5. José Luis Picardo. Elevations with shades of the Parador of the castle of Santa Catalina. Jaén. Source: *Revista Arquitectura* No. 108 (1967)

The reconstructed rib vault in one of its rooms is very interesting. It is false in its structural operation, but it achieves a spectacular spatial sensation. The central room built with brick arches is also quite remarkable. There is an extensive collection of carefully delineated plans of both the complex and the innumerable details (Fig. 7).

The work for the *Parador* of Guadalupe was intense, but less complex, as the very structure of the monastery offered greater compatibility with the new hotel use. From a graphic point of view, Picardo made an impressive aerial view of the set (Fig. 8) in which both the magnitude of the project and the extraordinary drawing skills of the architect can be appreciated.

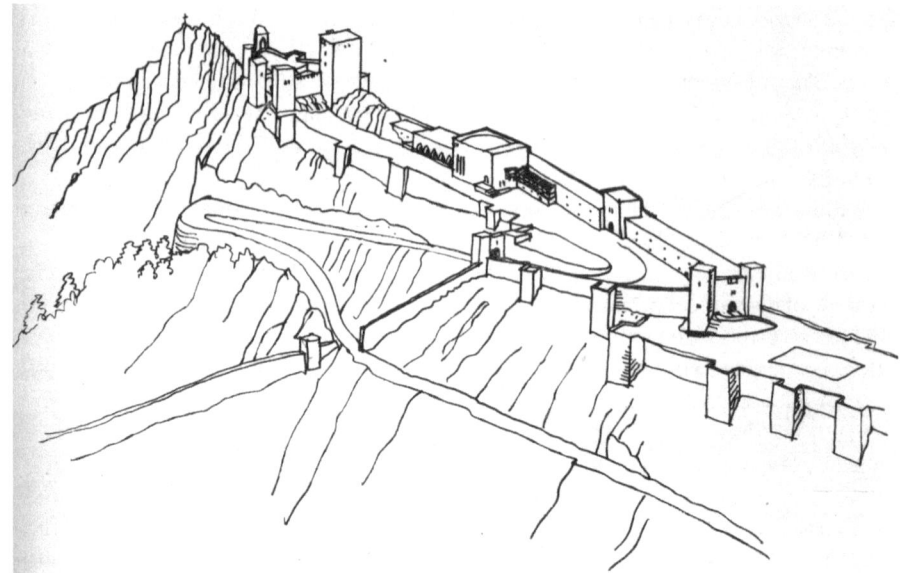

Fig. 6. José Luis Picardo. Aerial view of the Parador of the Castle of Santa Catalina. Jaén. Source: *Revista Castillos de España* No. 77 (June 1967)

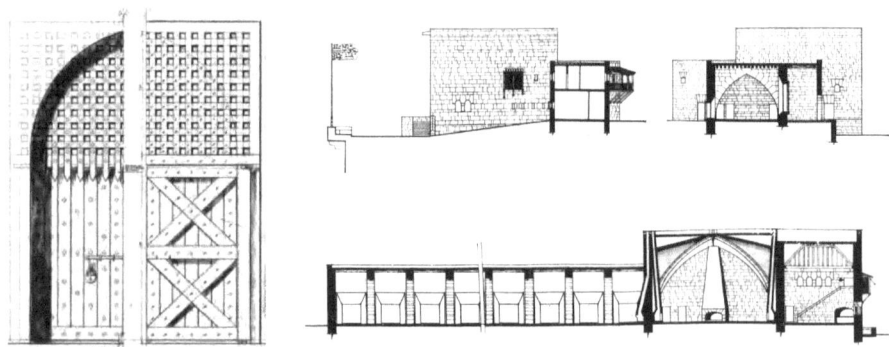

Fig. 7. José Luis Picardo. Detail of the Alcázar door and sections of the project for the Castle of Santa Catalina. Jaén. Source: Ministry of Information and Tourism and *Revista Arquitectura* No. 108 (1967)

In the *Parador* of Sigüenza, the architect's starting point is a fortification entirely destroyed in the Civil War and in a state of complete abandonment, an ideal setting for displaying his creativity through his wonderful drawings. The project was carried out with an eminently empirical and pragmatic approach, beyond theoretical considerations, somehow as if it were a new work, something that is not surprising given the author's enthusiasm and the state of the building. Thus, through different axonometric sketches (Fig. 9), Picardo made a series of outlines of the different solutions proposed where it is

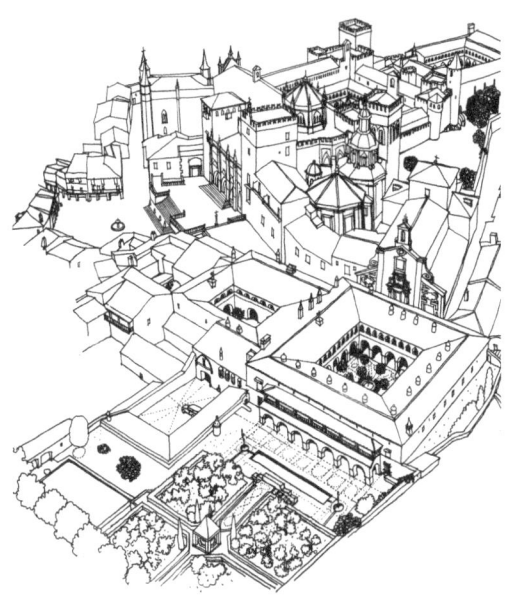

Fig. 8. José Luis Picardo. Aerial view of the Parador of Guadalupe. Cáceres. Source: *Revista Arquitectura* No. 108 (1967)

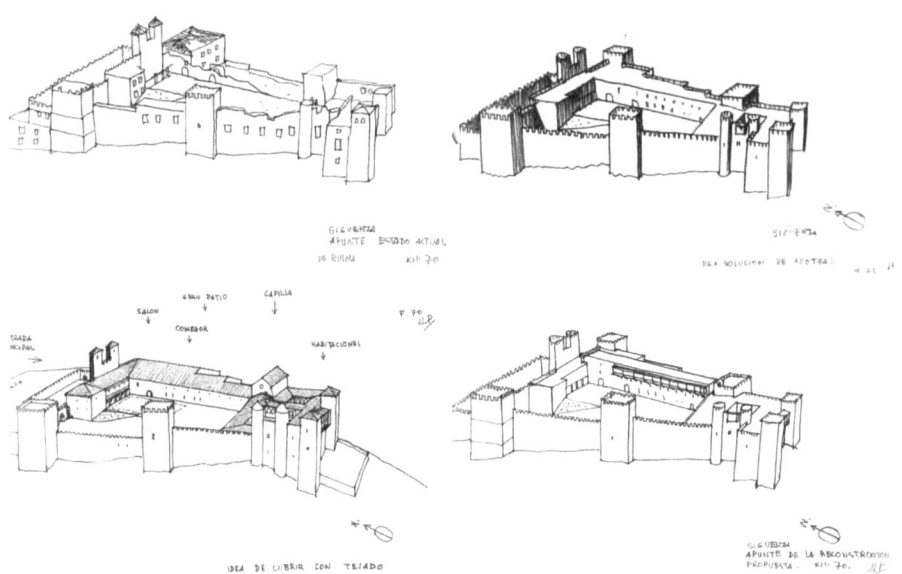

Fig. 9. José Luis Picardo. Axonometric sketches of the different options proposed in Sigüenza. Source: *Cuadernos de Arquitectura y Fortificación.* No. 1 (1970)

possible to once again appreciate, besides his skills with drawing, the degree of freedom with which he approached his interventions, without feeling intimidated by the weight of history or the great proportions involved. Initially two options were considered, one with sloping roofs of tile and another, "more medieval", with flat roofs. Of the latter option, the architect made two variants, which differed in the arrangement of the battlements within the set. Picardo even proposed, surely driven by his passion for design, the construction of a tower that had never existed and whose construction was not finally approved and, therefore, was not carried out.

It is, therefore, an imagined evocation of what the building could have been, seasoned by the author with new incorporations, in a risky and controversial exercise, but of undoubted architectural and also graphic interest, since there are innumerable plans, details and hand-drawn sketches of the project (Fig. 10). It is a restoration where historical rigour takes second place and where the final result, as Mª José Rodríguez points out, has something of *"a collage, of a sum of decisions inspired by the place itself, many of them of an anecdotal nature"* that do not achieve a completely harmonious and unitary image of the building, making it very clear, perhaps in this *Parador* more than in others, that the restoration carried out is the result of the author's "artistic action", generating a result that is configured as a *"theatrical scenography destined for tourists"* [5].

As pointed out by Antonio J. de Ulled, this is a rather outdated way of acting that, if it had become generalised, could have turned our country into a kind of large theme park, something like a great medieval setting, although, as the aforementioned author remarks, in the case we are dealing with here, we fortunately counted on *"the good work of this architect, who dreamt up in his imagination the historic moment of the construction of the Castle of Sigüenza"* [6].

The *Parador* of Carmona was probably the culmination of all Picardo's restoration approaches, which had been put into practice in his previous projects. Here he was decisive even in the choice of the site and, in it, his versatility, his eclecticism, his mastery of construction and his formal knowledge would reach their ultimate consequences. The

Fig. 10. José Luis Picardo Sketch of the Castle-Parador of Sigüenza after the restoration. Source: *Restauración y Rehabilitación*. No. 1 (1994)

result was a building with a resounding presence, with an imposing exterior image enhanced by the buttresses and a very careful interior with details of Hispanic-Muslim inspiration in lattices, floors and coffered ceilings.

Picardo combined his restoration activity for the *Paradores Nacionales* with commissions and competitions linked to modern architecture. In them he applied his "artistic action", his formal ideas and his graphic talent with equal enthusiasm and ease. It is curious and also paradigmatic of his way of designing that in the project for a small block of houses built in 1958 near Madrid's Plaza de Las Ventas, the architect, once again, pulled from his repertoire to, in his own words, *"arrange a veranda on the top floor in a solution inspired by Canarian architecture"* [7], that Canarian architecture that he knew so well, and drew, in his stay with the *Revista Nacional de Arquitectura* for the release of a double issue five years earlier.

In 1970 he won the competition for the headquarters of the Juan March Foundation in Madrid, to which he was apparently invited after Juan March's own stay at the Parador in Jaén, beating two great architects such as Mariano García Benito and Javier Carvajal. As Salvador Pérez Arroyo points out, *"controlling proportions and spaces with complete ease and achieving one of the best buildings in the recent history of Madrid"* [8]. The result is a very interesting building, undoubtedly peculiar in Madrid's architecture, with its imposing structure of white marble, its rounded corners and its ribbon windows. A dense and heavy building, full of details and high-quality solutions, in which dark woods, stone and brass predominate.

3 Conclusion

Analysing the whole of Picardo's graphic production and work, it is almost inevitable to conclude that his nature as a virtuous artist and painter determined his way of understanding architecture. It seems that he drew, he apprehended and incorporated everything into his formal repertoire, using it later with great freedom.

Somehow we may conclude that Picardo's unprejudiced way o f dealing with the commissions he received takes an experimental approach, in continuous challenge, an almost playful attitude, by which the architect accepts the rules of the game proposed to him and uses his weapons, acquired through drawing, to achieve the best result. In this sense, the words of the architect himself, pronounced in a meeting of architects that took place in the Canary Islands in 1953, stand out as a true declaration of intentions: *"after the exhaustion of the forms used, we have to look for new ones or use the new ones found by others, which are in line with our problems and which we feel them artistically as our own"* [9].

Indeed, his works include his trips with his mentor Torres Balbás, his drawings from his youth and his murals, his vignettes, his illustrations, but also his travels around the world and his attraction for the new architecture, something that probably represented a certain moral dilemma for him. In 1967, in an article in the *Revista Arquitectura* magazine about the Parador of Jaén [10], the author himself gives us some clues about his creative process, emphasizing that the architect had needed *"imagination, sensitivity and… containment"* in order to move on, almost as a form of self-justification, to asking himself *"what architect has not felt the temptation to enjoy medieval strength?"*. The article ends with a few lines in which Picardo says *"in this work, as in no other, the architect has experienced the presence of his mentors: Torres Balbás, Moya, Sota and Luis Santamaría. None of them has seen it. What will they say?"*

Although we will never know what his mentors would say, what seems clear is that it would be impossible to understand the trajectory and figure of José Luis Picardo only by way of his projects, isolating and separating them from his "artistic action", his very rich formal and architectural repertoire and, above all, his magnificent drawings.

References

1. Mendoza Rodríguez, I., Úbeda Barranco, M.: José Luis Picardo: arquitecto, muralista, dibujante e ilustrador. Revista EGA **24**(36), 163–173 (2019)
2. Rodríguez Pérez, M.J.: La rehabilitación de construcciones militares para uso hotelero: la red de Paradores de Turismo. Tesis Doctoral. Escuela Técnica Superior de Arquitectura de Madrid (2013)
3. Chueca Goitia, F.: José Luis Picardo IX Premio Antonio Camuñas. Restauración y Rehabilitación. (79), 16 (2003)
4. Picardo Castellón, J.L.: Sobre la teoría de la restauración. Restauración y Rehabilitación (1), 64–66 (1994)
5. Rodriguez Pérez, M.J.: Escenografía medieval para un alojamiento turístico. El Parador Nacional de Sigüenza (Guadalajara). Cuadernos de Arquitectura y Fortificación (1), 143–161 (2013/14)
6. Ulled Merino, A.J.: La recuperación de edificios históricos para usos turísticos. La experiencia española. Ed. Tecniberia. Madrid (1986)
7. Picardo Castellón, J.L.: Edificio de viviendas en Madrid. Revista Nacional de Arquitectura (202), 103 (1958)
8. Pérez Arroyo, S.: Los años críticos. 10 arquitectos españoles. Madrid. Fundación Antonio Camuñas (2003)
9. Picardo Castellón, J.L.: Sesión crítica de Arquitectura. Revista Nacional de Arquitectura Núm. **140–141**, 103 (1953)
10. Picardo Castellón, J.L.: El Parador de Santa Catalina en Jaén. Revista Arquitectura (108), 35–40 (1967)

The Social Poetry Drawing of Lina Bo Bardi

Alfonso Ippolito[1(✉)], Marcelo Mott Paulini[2], and Martina Attenni[1]

[1] Sapienza University of Rome, Rome, Italy
{alfonso.ippolito,martina.attenni}@uniroma1.it
[2] Faculdades Integradas Jaú - Estado de São Paulo, Jaú, Brazil
mpaulini@uol.com.br

Abstract. Within the complex and articulated architectural heritage of the Modern Movement, we are still amazed by the architecture of Lina Bo Bardi. Alive and modern, they never seem to age. The present study allows to better understand this vitality through her drawing. Spaces, geometries, functions, paths, views, perceptions, sensations, are grafted into an ever-changing urban space. Lina's architecture, at any scale, never seems to lose its freshness in a city in continuous transformation. This study reads them not only in the formal aspect, but also as the result of very strong elaborations and cultural evolutions of the designer, which concretize her "social poetics". Her consistent personality is enriched with all the peculiar aspects of the Brazilian culture, an immense country characterized by the fusion of Western, African, indigenous American culture. Lina takes hold of those cultural emotions that she was lucky enough to share with the Brazilian avant-garde of the time, and makes them her own. Here she makes them reappear, as if pulled out of a magician's hat, through her exciting architecture.

Keywords: Lina Bo Bardi · Brazil · Architecture · Social antropology · Drawing · Representation

1 Introduction

Achillina Bo (Rome 1914-São Paulo 1992), better known as Lina, graduated from the Faculty of Architecture at the Sapienza University of Rome in the 1930s. In full swing rationalist, Gustavo Giovannoni and Marcello Piacentini run her school. In 1940, she moved to Milan, a city in which the influences of modern European architecture circulated more freely. She worked with Gio Ponti, founder of the Domus magazine, mainly in the publishing sector. In the same period, she opened her own architectural studio, destroyed in a bombing in 1943. Some years later, she became a member of the National Liberation Committee, with the intention of opposing the fascist government and the Nazi occupation in Italy. At the end of the war, Lina, Bruno Zevi and Carlo Pagani founded the magazine "A – Cultura della Vita" (Culture of Life), with the first release on February 15, 1946. The magazine, politically on the left, became a place of experimentation and interaction with the public. Lina used architectural drawing as the way to express the relationship between architecture and everyday life, known through surveys, services and questionnaires. The use of photography, with a strong visual impact, gave

L. Agustín-Hernández et al. (Eds.): EGA 2020, SSDI 6, pp. 51–62, 2020.
https://doi.org/10.1007/978-3-030-47983-1_5

the magazine a markedly Neorealist imprint. Says Bruno Zevi, interviewed by Fabio Brunetti in 1983: "In the aftermath of the liberation of Milan, Carlo Pagani and Lina Bo had the idea of launching an architecture weekly, a popular body, widespread, focusing mainly on problems of the House. They talked about it with Gianni Mazzocchi of the Domus who, fearing the competition of other publishing initiatives that, in that moment, were swarming in Milan accepted…. Lina Bo proposed "Home"; after long discussions "A" (architecture, art, living, love) was chosen … "A- Cultura della Vita" (Culture of Life), a fragrant, varied, multidimensional organ capable of connecting urbanism and architecture to politics and costume …" (Copyright Adachiara and Luca Zevi 2018; Copyright Eredi Fabrizio Brunetti 2018).

In 1946, Lina married Pietro Maria Bardi, a critic affiliated with the fascist regime. They moved to Brazil in 1947, invited by the entrepreneur Francisco de Assis Chateaubriand for the inauguration of the Masp (Museo de Arte de Sao Paolo), the first museum modern non-profit country. This museum, with Pietro as director and Lina as developer of the architectural and exhibition project, was transferred to Avenida Paulista in 1968, in its current version. In 1951 Lina chose Brazilian citizenship. She never leaves her editorial, architectural, planning and didactic research, despite the negative influence of political events. In 1957, after two years of teaching at the Faculdade de Arquitetura and Urbanismo of São Paulo, she is denied the professorship.

2   Cultural Interaction

In the late 1950s, Brazil experienced a democratic and total growth period, under the leadership of President Juscelino Kubitschek, who demanded a development of "fifty years in five". Brasilia was built, Bossa Nova and Poesia Concreta were born, three icons of modernity. In this optimist atmosphere, the country lives its golden years in this period.

Far from the axis Rio de Janeiro - São Paulo, as the center of the political and economic power of the country, Salvador (the ancient Cidade do São Salvador da Bahia de Todos os Santos, first capital of Brazil), lived a period of intense cultural turmoil, forced by the doctor Edgard Santos, rector and founder of the Universidade da Bahia. He believed in the social function of the university, endowed with an open and dynamic vision. He invites to work in the city the main important scholars and people from different artistic fields, linked to avant-garde experiences and theories [1, 2].

Edgard invited Lina to hold several conferences at the Federale University of Bahia (UFBa) in 1958; the following year he moved to the Bahian capital, where she directed the Museu de Arte Moderna da Bahia (MAMB). During this period, she shared cultural confrontation with famous and cultural depth figures, as the set designer and theater director Eros Martim Gonçalves, the painters Mario Cravo Jr. and Carybé, the German-born professor and composer Hans-Joachim Koellreutter, the photographer and French ethnologist Pierre Verger, the rector Edgard Santos. Finally "Donna Lina", as she was called by her friends, will carry out an intense and creative cultural activity with the other artists invited by the rector, who catalyzed the avant-garde na Bahia, as defined by the writer Antonio Risério [3].

The Bahian avant-garde adventure should have begun to be castrated since the military coup in 1964. But this did not happen: the group's experiences were already advanced. This group is full of important people like Caetano Veloso and Gilberto Gil, with the Tropicália, and Glauber Rocha, with the Cinema Novo. The cultural engine of all these artists is linked to the ideas launched by the Brazilian modernist group of 1922, in particular by the writer Oswald de Andrade and his Manifesto Antropófago.

Lina's ideas are the result of her Italian cultural formation and her aesthetic, political and social sensitivity. She lives in Salvador a more lively contact with the poverty and the backwardness of a society historically built on exclusions and social differences [4].

Her conception of art, architecture and culture becomes defined [5], and she shows it through multiple activities, such as the articles in the Diário de Notícias de Salvador, theatrical sets of works by Brecht and Camus, directed by Martim Gonçalves, and the restoration of the Solar do Unhão to build the headquarters of the MAMB, inaugurated in 1963. In the same year, she designed the Nordeste exhibition, and she wrote: "Matéria prima: o lixo. Lâmpadas queimadas, recortes de tecidos, latas de lubrificantes, caixas velhas and jornais. Cada objeto risca or limite do "nada" from miséria. They limit and a contínua and martelada presença do "útil" and do "necessário" é que constituem or valor desta produção, its poética das coisas humanas não-gratuitas, não criadas pela mera fantasia. It is neste sentido de moderna realidade, que apresentamos criticamente esta exposição. Como exemplo de simplificação direta de formas cheias de eletricidade vital. Formas de desenho artesanal and industrial. Insistimos na identidade objeto artesanal-padrão industrial baseada na produção técnica ligada à realidade dos materiais e não à ab-stração formal folclórica-coreográfica".

The Bahian dream ends with the advent of the military dictatorship: the MAMB comes closed and the exhibition in Rome of Brazilian popular art, organized by Lina, was deleted.

Lina leaves Salvador and moves to São Paulo, where she will perform some of her most important architectural works. MASP, built in 1968, is certainly its best-known realization. In the SESC Pompeia unit (Serviço Social do Comércio) of 1982, she was able to develop her social and libertarian vision on a space of cultural and recreational conviviality, sports practices, manual arts laboratories, recovering an ancient factory. In 1976, she designed the Church of Espírito Santo do Cerrado, built with the participation of the local people (Ub-erlândia, a town within the state of Minas Gerais), on which Lina will says: "Pode ser que a grande obra seja a capelinha miserável de Uberlândia. Feita sem dinheiro, com os padres franciscanos and prostitutas. O Masp is important for this eagle".

On March 20, 1992, Lina died in San Paolo, in the Casa de Vidro, a residence designed by herself in 1951, where she lived most of her life.

3 The Poetry Drawing

Lina Bo Bardi's heterogeneous activity has always been characterized by a constant use of drawing, through which she expressed her cultural intentions and her way of conceiving architecture and the city. As an architect, she did not want to avoid considering the social, political and economic context in which she act, expressing an immense admiration for popular culture. Lina believes in a space built for the people, an unfinished space filled with everyday activities. Her architectural language remains impressed both in the urban space and in the identity of São Paulo, and on paper, which marks freeing the trait from any formalism or technical rigidity, and enriches with color and writing. Drawing is not only a tool for building, but also the means that allows it to represent the freedom in which it deeply believes, and which it has decided to share with Brazil.

The huge use of representation, however, is not a prerogative of her Brazilian period. Since the early years of academic training, Lina had the habit of drawing anything. Gabriella Bo Valentinetti, director of the Lina Bo and Pietro Maria Bardi Institute, says "[…] She did things that are not built […] She drew objects, worked with magazines, made drawings of illustrations, taught, wrote in magazines feminine […]". In her vast graphic production, there are designs of jewels, sets, theater costumes, watercolors, landscapes, architecture. Her project reflects this concept of design as an expression of freedom and emancipation, they are a means of being useful to society, far from the concepts of beauty or ugliness. In an interview on the MASP, Lina tells how her interest was not focused on the aesthetics of the form but on the search for a space that considered the political and social context of the country. Imagine a space created by free men, who generate free spaces. She then creates what he calls "a non-formal architecture", a suspended volume, with 75 m light, eliminating any obstruction in the square below, which thus defines a large space for the community.

The same principle guides the restoration of SESC Pompeia, defined as "Fábrica do Poe-ma" by Waly Salomão in a song set to music by Adriana Calcanhotto. She designs a structure that considers a living organism that gathers innumerable functions and for everyone, a space without separations characterized by an uninterrupted visual continuity.

The architecture created by Lina Bo Bardi expresses her conception of space, which changes radically when she encounters the soul of the Ibero-American continent. She considers time as "an invention of the West, […] a wonderful overlap so that, at any moment, it is possible to select points and invent solutions, without beginning or end" [6]. After having resigned as an Italian in her own way and chosen Brazilian nationality in 1953, Lina began working on that hybrid idea of "folk art", revealed through the profound sense of connection and discovery that shines through her numerous drawings. Her huge graphic production has more than 6000 drawings, most of which are currently collected in her personal archive or kept in the house she designed for her husband in Sao Paolo.

Although she wrote quite a lot during her life, Lina leaves just few notes regarding the subject of her drawings and sketches. However, they show that she consider the drawing much more than the sum of only graphic signs (Fig. 1). As the main language and form of communication, it constitutes the most powerful means to transform reality and to solve the problems of society, through projects of wide cultural and social importance. The drawing for Lina Bo Bardi is never just a graphic exercise, although her illustrations have always regarded heterogeneous subjects, from objects of common use to the interpretation of pictorial scenes, from architecture projects to plants (Fig. 2, Fig. 3). Whatever the purpose of the representation, it is never configured as an isolated episode; rather it is a process that expresses the multi-scale relationships established between territory, culture, city, society, architecture, intentions and objects. This is the reason why it is not possible to trace a linear path that regards her graphic elaborations,

Fig. 1. Museum do Instituto Butantù, 1964. Instituto Bardi/Casa de vidro.

Fig. 2. Casa Circular, 1962. Instituto Bardi/Casa de vidro.

although it is nevertheless possible to identify some features that relate the chronological order to the stylistic layout [7].

The first performances, those of the period of formation in Rome and the first steps of the Milanese career, show the emotional and artistic relationship between Lina and the drawing. She develops different themes and techniques. She used thin and delicate lines and strong colors to represent collective moments of everyday life, or to make illustrations with a fairy-tale character. She often represent scenes the central perspective or axonometric views, characterized by the presence of human figures, outlined with a few lines. These first drawings, which reflect the oscillation between her academic training and the values of modernity, combine the tradition of drawing with her desire for

Fig. 3. Biblioteca Encosta, 1962. Instituto Bardi/Casa de vidro.

innovation. The expression of the principles of realism does not leave that spontaneous, instinctive and free character of the illustrations of the following years.

When she moves to Brazil, she arrives at a vision of more mature architecture, in which she leaves some of the conventions and artistic artifices.

The drawing, as a synthesis of her thoughts, her academic training, and the adoption of rationalist ideas, becomes a true act of love, a way to understand the world and our way of living in it. Her desire for innovation is explicit in considering design as a value within the continuous changes that involve the political and social pattern. The representations with clear lines and luminous tones explain the vision of the world to which Lina wants to give color, going beyond the gray tones. In the Brazilian period, particularly in the years from 1958 to 1964, lived in Salvador de Baia, her attention was turned towards the design of antithetical worlds: well-being and poverty, formalism and spontaneity, representing lights and shadows of a population that distinguishes only "the aristocracy and the people". In recent years, Lina produced a huge series of illustrations for residential buildings. Some of these, practically unpublished (Fig. 4, Fig. 5), describe the way in which Bo Bardi understand the project, its conception of a multiscalar space, the link between the italian tradition and the transposition of new values aesthetic and cultural. The Baian period is a phase of strong experimentation opens up for Lina. Since 1958, she has been developing housing projects (Fig. 6). The Cirrell House (Fig. 7) constitutes an important change in the conception of project design: while maintaining the usual modes of expression, through two-dimensional models, the overall language of its operation changes.

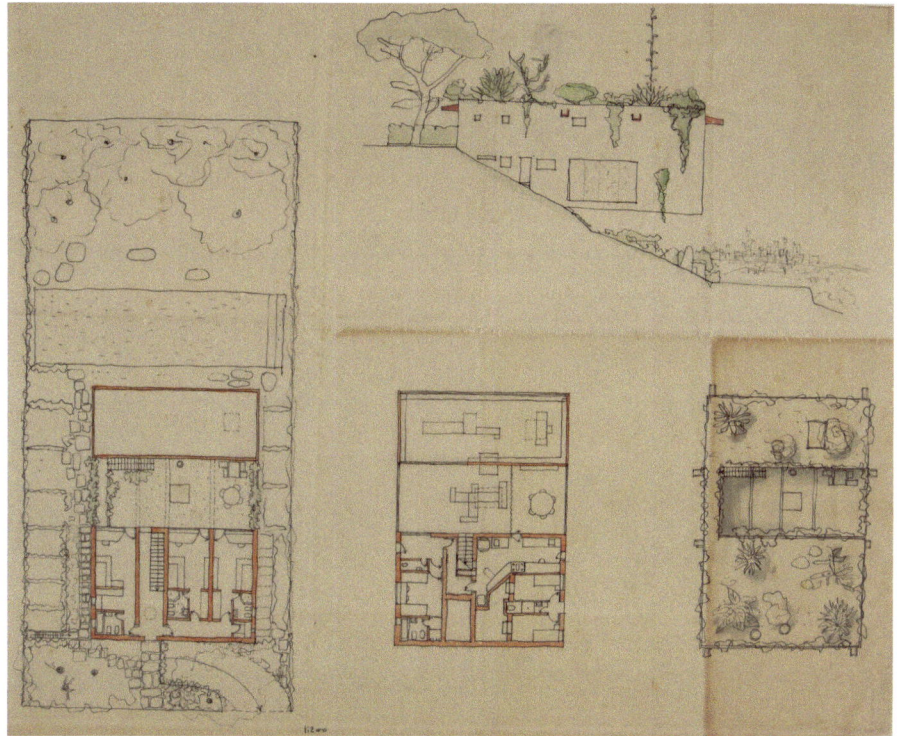

Fig. 4. Casa Ferraz, 1962. Instituto Bardi/Casa de vidro.

Regular geometric figures describes plans, elevations and sections, but no longer express the principles of rationalism. On the contrary, they evoke the organic architecture of Frank Lloyd Wright, with the fireplace it constitutes the fulcrum of the house.

In Casa Cirrell's project, Lina experiments with the insertion of natural elements; her drawings refer to daily activities, carried out within spaces in which to work for the achievement of concrete objectives, and at the same time to dream of individual and social improvement. The design therefore conveys not only the technical knowledge and the intuitions of the architect, but also that ethical value that for Lina begins with artistic modesty, and which attempts to demonstrate through daily professional practice and her love for design, in the search for an architectural and social space, accessible to all.

Fig. 5. Casa das Lanzeras, 1962. Instituto Bardi/Casa de vidro.

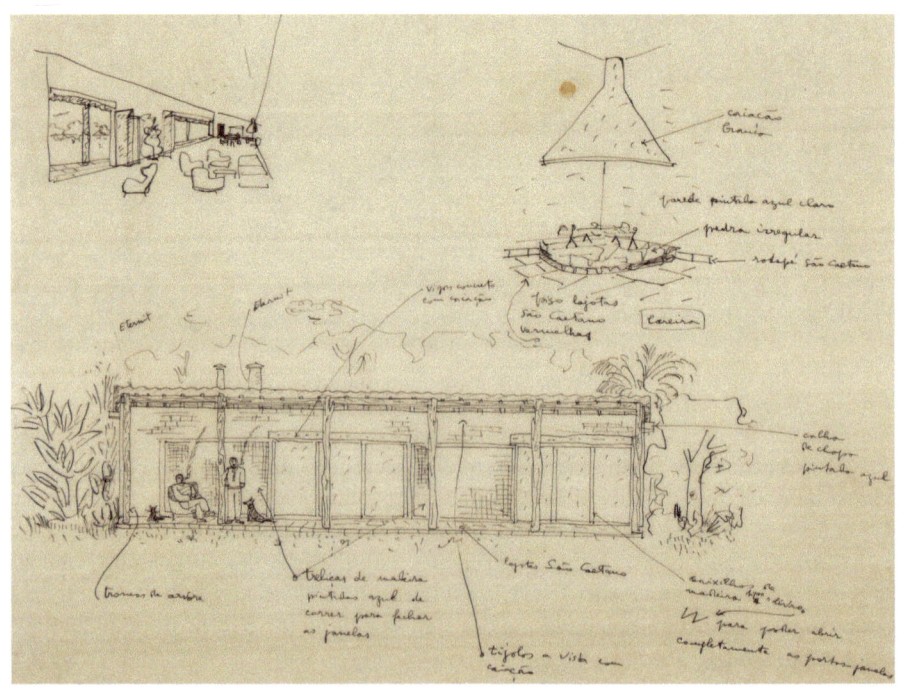

Fig. 6. Casa Shed, 1958. Instituto Bardi/Casa de vidro.

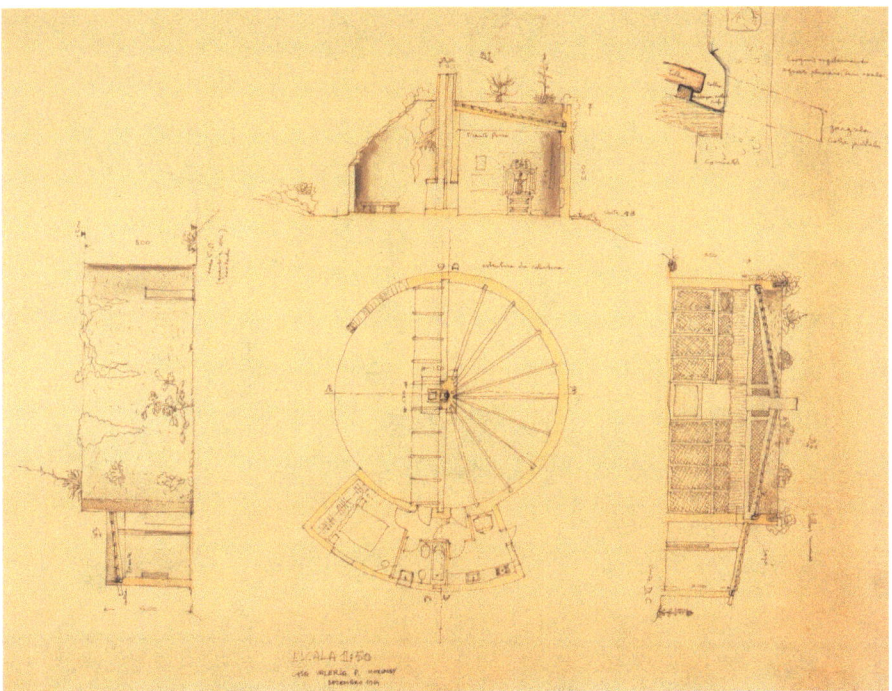

Fig. 7. Casa Cirrel, 1958. Instituto Bardi/Casa de vidro.

4   Conclusion

Through the mastery of drawing as a language, Lina creates a personal dialogue with the graphic expression methods that define her iconographic style. The analysis of Lina Bo Bardi's drawings constitutes an important element to better understand not only her figure, but also her work, which many have tried to bring back to dialectical opposites such as modernity/tradition, East/West.

Lina's language develops through different modes of expression, which pass from the representation system to the observation design, to the design drawing, and to the technical drawing. The first is a freer, looser drawing, in which the relationship between the architect and the place becomes more explicit. The second involves more the imagination of the recipients of the project, or of the clients: perspectives, photomontages, comics and digital simulations direct the attention towards his idea of place (Fig. 8, Fig. 9). The third, on the other hand, almost totally absent in the Bo Bardi representations, is accurate and consistent with the rules dictated by production and production requirements.

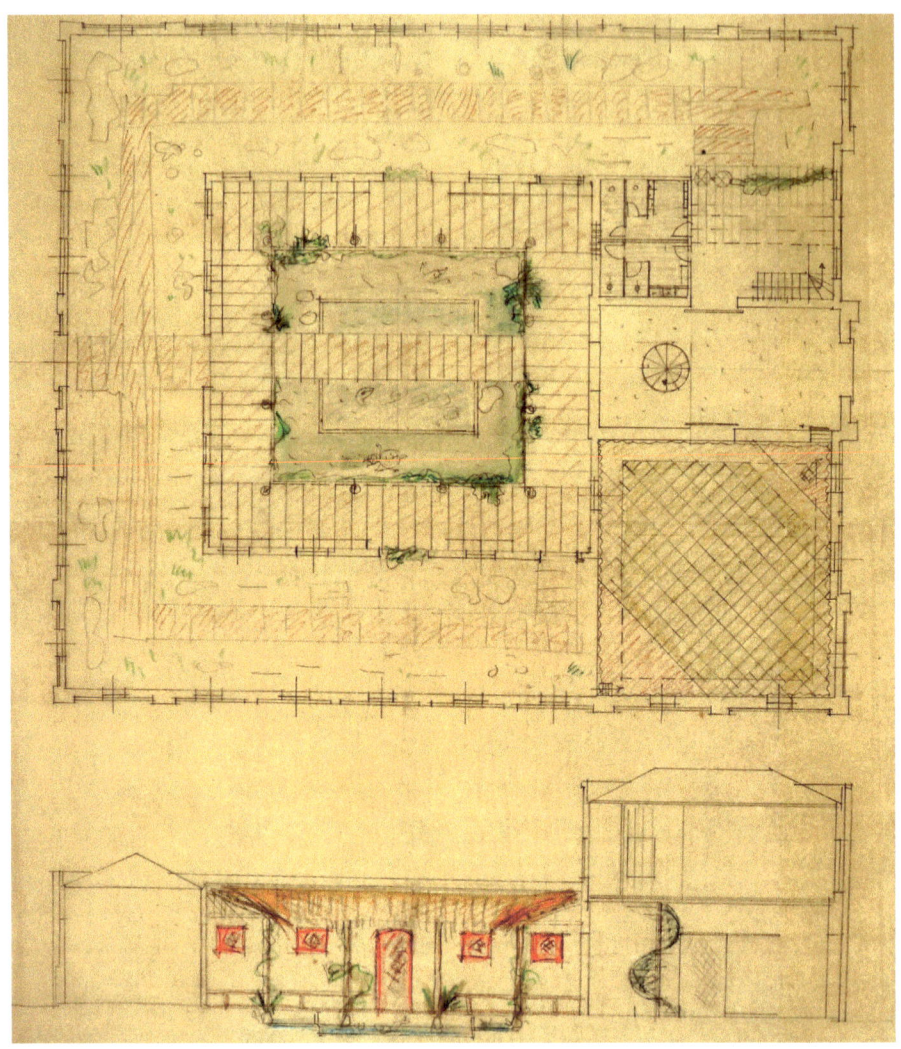

Fig. 8. Pana Elaser Salvador, 1964. Instituto Bardi/Casa de vidro.

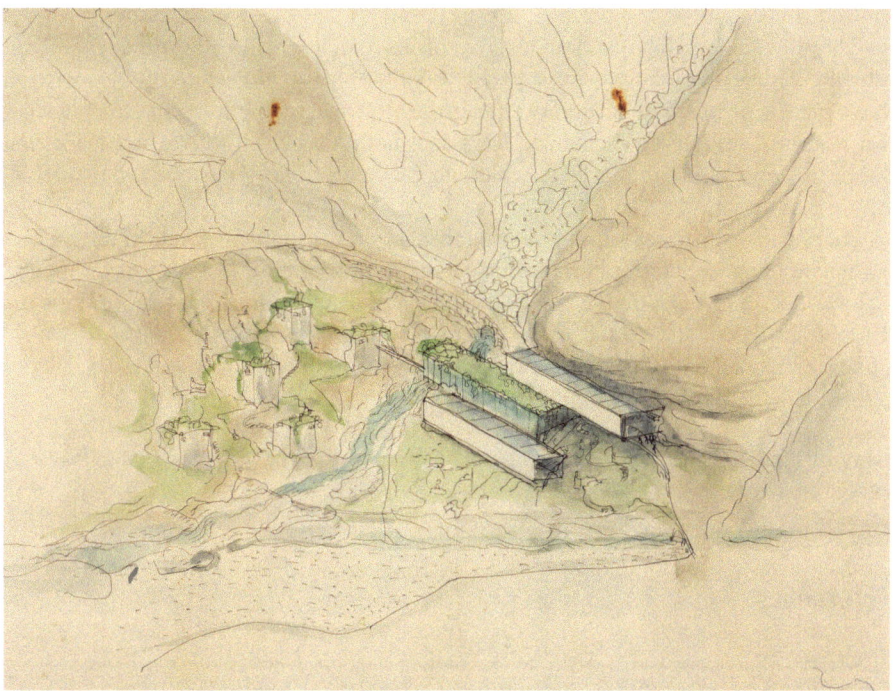

Fig. 9. Museo Do Marmore, 1963. Instituto Bardi/Casa de vidro.

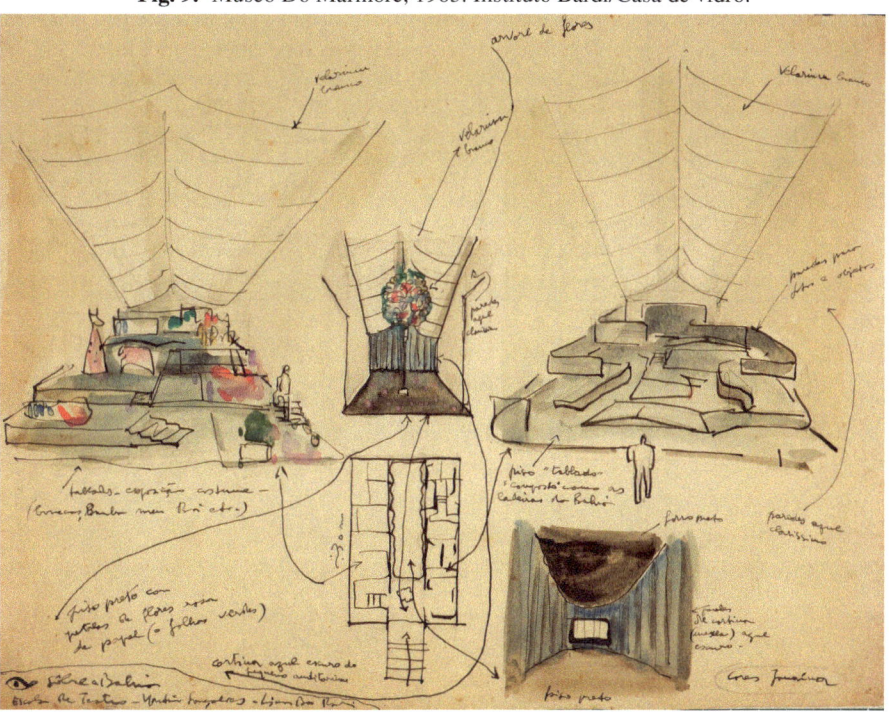

Fig. 10. Sao Paolo Bahia, 1959. Instituto Bardi/Casa de vidro.

The drawing therefore plays an heterogeneous role: it is at the same time process and outcome, object and relationship, creative process able to tell the needs of a society in which there is no hierarchical order between the historical tradition and the emergence of new wishes. Through representation, Lina does not limit herself to creating spaces, but defines relationships and places of which people appropriate themselves, thus making them never sterile (Fig. 10).

All of Lina's interventions, even those that are not currently in good condition, such as Ladeira da Misericórdia in Salvador de Bahia, still retain the charm of her poetry. Lina was there, involved in the spaces, in the material and immaterial relationships that allowed her to emphasize the enchantment of certain places in a country where her influence leaves the strongest signs. Her architecture continues to be, still today, a hymn to freedom.

Acknowledgments. Thanks to Instituto Bardi foundation - Casa de Vidro Sao Paulo/SP – Brazil for the reproduction of Lina Bo Bardi's drawings.

References

1. Subirats, E., Vanguarda, M.: Metrópoles. Studio Nobel, São Paulo (1993)
2. Luz, V.: Ordem e Origem em Lina Bo Bardi. Giostri Editora, São Paulo (2014)
3. Risério, A.: Avant-garde na Bahia. Instituto Lina Bo e P.M. Bardi, São Paulo (1995)
4. de Almeida Alves, A.A.: Um projeto para o Brasil: arquitetura e política na trajetória de Lina Bo Bardi no Brasil, 1946–1977. Risco - Revista de pesquisa em arquitetura e urbanismo, 20, Instituto de Arquitetura e Urbanismo (IAU-USP), February 2014. https://www.revistas.usp.br/risco/article/view/117439. Accessed 04 July 2019
5. Grinover, M.: Uma Ideia de Arquitetura. Escritos de Lina Bo Bardi. Annablume Editora, São Paulo (2018)
6. Ferraz, M.C. (ed.): Lina Bo Bardi. Instituto Bardi/Casa de Vidro/Romano Guerra Editora, quinta edizione, São Paulo (2018)
7. Criconia, A., Essaïan, E. (eds.): Lina Bo Bardi. Enseignements partagés/Insegnamenti condivisi. Archibooks + Sautereau Éditeur, Paris (2017)

The Architecture of Melvin Villarroel as a Source to Understanding the Landscape of the Costa del Sol in the Seventies

Jesús Estepa Rubio[1]([✉]) and Antonio Estepa Rubio[2]

[1] ER Arquitectos, Jaén, Spain
oficina@erarquitectos.com
[2] San Jorge University, Villanueva de Gállego, Zaragoza, Spain

Abstract. The description of the landscape that marks the Costa del Sol is conditioned by the constant oppression it has suffered by the architecture for leisure executed along the coastline over the last decades. Nonetheless, there are some sources of consultation that by its morphologic analysis through the graphic revision of the planning that it describes, help in the comprehension of the description of those elements that compose the primary qualities of this territorial enclave.

The effort to contain this uncontrolled transformation, devoid of any control or planning has led, in many occasions, to the use of technical instruments brought about and introduced by the public administrations that have not really generated true results, or at least, not the most efficient. It is for this reason that other forms of approaching this situation may bring forth appropriate methods that, in some way, will be used as tools for this purpose even though they consist of an academic approach of a theoretical nature.

The architecture of Melvin Villarroel is one of these descriptions that three decades later has become a source of contemporary composition and formal inspiration. The revision of the works of this architect, from the prism of abstract creativity still intact in his designs, becomes an exercise in analysis that become the mis a scene of the basic patrimonial fundamentals that are to be considered alongside the coastal segment o fthe province of Málaga.

Keywords: Villarroel · Costa del Sol · Marbella · Landscape · Design · Vegetation · Empty

1 Index of Contents

In the decades of Spanish development, and, as a result of the application o f economic and social development and expansion plans put in force by the technocratic Franco régime, especially after the assignment of Laureano López Rodo as Minister for Development Planning, a natural phenomenon emerged; repopulation, that centered most of its activity in redefining coastal landscapes.

Also, upon the arrival of Manuel Fraga Iribarne as Minister of Information and Tourism a more relaxed institutional approach took place which led to capital and investment, be it either public or private, enabling a profound transformation of the territorial

L. Agustín-Hernández et al. (Eds.): EGA 2020, SSDI 6, pp. 63–72, 2020.
https://doi.org/10.1007/978-3-030-47983-1_6

and landscaping structures and configurations of the Peninsula, to optimize the lucrative returns obtained from the tourist market exploitation (Martín López 2011).

In so doing, architecture would become a key tool that would serve to revise all that existed, good and bad, with the sole purpose of redirecting the focus towards the changing national economic plan and turn towards a more efficient model with quick results.

The architecture for leisure and free time were the protagonists of the action plans that motivated this change and with them the change in construction techniques and the import of model typologies from the international scene, all of which were included in the new generational renewal and the in-depth analysis of the syllabus in the Schools of Architecture. In this respect, macroeconomic investments in great projects became the purpose giving form to the decade of transformation; of course, always serving the productive interest and quick return on profit for the investors.

Many are the examples of constructions built to the service of this utilitarian economy but, despite the proliferation of anonymous and anodyne creations that did away with the autochthonous qualities of the landscape that characterized the Costa del Sol, others, timidly but with a strong will, appeared to enhance and protect the rich essence of such a unique territory without necessarily having to look away from the new conditions imposed, or the use of good practices and influence that came from technological and conceptual advances that other foreign architectures had managed to present, farther away from our geographical emplacement; even farther than the frontiers of old Europe.

Without any doubt one of these renovators was Melvin Villarroel. His architectural production in Marbella, even though born amidst the Spanish circumstances of his time, knew how to recognize and appreciate the values and opportunities that he was presented with. In his particular case, through a graphic intellectuality based on the versatile use of vegetation and the creation of intermediate spaces amongst the built and empty elements which resulted in an original and beautiful language capable of describing with great success the key elements of the local and peculiar workplace.

It is incredibly instructive as well as engaging to revise all the pictures and be able to comprehend how you can find the hidden intentions that firstly in a romantic way but also, with some subtle objections and protests claiming respect towards the original values of the autochthonous environment, to become, later on, descriptive elements of how the identity of the Mediterranean landscape has been conceived (Chueca Goitia 1981).

Project designs are the ideological expression for Villarroel, more than a simple tool to depict the formal or constructive task necessary in the design phase (Fig. 1). On purpose, the elevations and sections of the buildings seem to be blurry, undefined and are diluted and confused with the surrounding vegetation until the geometry is practically an excuse to create a green atmosphere; big sets of vegetation are planned from a complicated and changing morphology, that serve to reforest or repopulate the plain approach imposed by the logic of current urban town planning under the control of general interests foreign to Architecture.

The graphic expression that portrays this thought is depicted in the representation of white canvases on the facades and adjacent clumps of vegetation, threading a geometric patchwork or fusion that can be recognized in the simple differentiation of the

Fig. 1. Melvin Villarroel. Longitudinal section of the Hotel Puente Romano in Marbella. 1978. Source: Villarroel 1997, p. 97

chiaroscuro, the contrast of light and shadows of each division created by the sun. The light and shadow or visual perturbation which in each case is provoked by the architect, make Villarroels's projects not understandable without the luminary variable, which carries a phenomenal condition that reflects on the differentiating quality of his architectural work (Fig. 2).

Fig. 2. Melvin Villarroel. Transversal Section of Hotel Puente Romano in Marbella. Marbella. 1978. Source: Villarroel 1997, p. 100.

All through the seventies many studies were carried out on international impact that served to reflect, profoundly, on the development opportunities that were emerging in the Costa del Sol.

In relation to the above, Villarroel's contribution becomes more important when compared to the studies mentioned since it was analyzed on a smaller scale and brings light to the domestic territory and its typological identity, and hence, has become the most recognizable image of the architecture, patrimony and landscape of the Costa del Sol.

Thus, this analysis is to serve the purpose of making conclusions with which to use the graphic material of Villarroel as a comparative tool permitting to evaluate the values and quality of complex architectural projects that have transformed the landscape of the Costa del Sol.

2 Town Planning as an Intent to Control Tourism and Its Effect on the Territory and the Landscape

With the frenzied changes during the decades of the 50's and 60's along the Andalusian coastline as a result of the new transformation ideal of the image of the country brought about through foreign investment and capital through the tourism operation, the consequences observed was the irrefutable consideration that both society and the territorial and urban infrastructures were not sufficiently prepared nor adapted for the future that was about to come. This has not stopped today and despite the passing of the years the situation has not changed in a convincing way.

The discovery of this source of richness and opportunity, apparently inexhaustible made by the local administrations and its immediate transformation effects on the customs, traditions and way of life of the citizens of this territorial area, generated a permissive ecstasy that incentivized short term economic actions that unconsciously had consequences, not good, that were to flourish later on, heavily mortgaging future generations.

It took a short period of time to come to the realization that cities that had become tourist havens were neither prepared nor ready and that their infrastructure and morphological territory was not designed to grow in such a manner. A revision was necessary to analyze the situation from a strategical perspective to help channel the problem and therefore obtain effective diagnosis and amendments.

Thus, during the decade of the 50's and the early 60's a (Gavilanes Medraz de Medrano 2014, pp. 1–19) a tedious and complex set of urbanistic planning norms were elaborated to achieve a purpose that to date is still undefined. The first was published in 1955, *The Study for the Cost del Sol Zone Planning*, which indicated the need to elaborate a set of regulatory norms that would protect the landscape and natural spaces and even established a parallel stripe running along the coastline with a 2 5 m width limiting construction to certain private initiatives for public use and not for residential purposes.

This study already implied the need for public investment that would set the minimum conditions that would permit these private initiatives to focus on this area so that the result would fit into a different regulatory norm to the exiting one at the time.

Hence the appearance of the first Zoning Plan for the Costa del Sol in 1960 coordinated by the architect Luis Blanco Soler at the General Urbanistic Department, concentrating its activity in relevant improvements of general infrastructures of the territory at the same time dividing the planning protocol for the municipalities of Torremolinos, Benalmadena, Fuengirola, Marbella and Estepona, forgetting the areas in between that soon became the objective of many productive and speculative actions in the course of tourist development of the area.

Soon it became obvious that these regulations were clearly insufficient; 1956 brought approximately 70.000 tourists with an income of 4.000.000$ and by 1962 we received 850.000 tourists and 47.000.000$; it was clear that an in depth revision was in order.

Consequently the Presidency of the Spanish government requested the Greek architect Constantinos Doxiadis[1] the *Study for the tourist development of the Costa del Sol – Cabo de Gata 1963*, also known as the Doxiadis Study. This Study had a double objective; on the one hand to dictate a number of principles that would bring in the maximum number of tourists in an organized fashion and, on the other hand, to guarantee an orderly and controlled development that would protect in the short term the values and qualities of the territorial area under study.

The study proposed a careful and respectful conservation of the urbanistic structures of the existing municipalities to avoid uncontrolled growth that would disrupt the essence and, in parallel indicated that new developments were to grow starting from the existing ones but at a fair distance.

As explained previously this fragmented development made it necessary to connect traditional areas with the new ones coming into the picture through a classified rank of roads from the most important to the less. This was complimented with the regulation of building heights with respect to the landscape to avoid visual interference with sea views, either new build or preexisting ones. Also, the study supported the idea that not all development could come from state funding so collaborations or participations from private income would have to be incorporated gradually.

And so, after many a study with an analytic and informative purpose more than an effective one, we come to the *Plan Comarcal de Ordenación Urbanistica de la Costa del Sol Occidental of 1967* (Regional Urbanistic Plan of the Western Costa del Sol, 1967), coordinated from the General Directorate for Urbanistic and Town Planning by the architect Juan Gómez y García de la Buelga, that proposes, according to the trend of the time, to plan by zoning and to call attention to the motorized traffic at different speeds as well as modern rational functionalism, requested by his internal organization (Gómez y García de la Buelga 1964, pp. 41–50).

This meticulous plan establishes the basic criteria for the architects work that, later on, they would use to create, ex-novo, the areas of growth and repopulated urbanizations of the Costa del Sol, in compliance with certain parameters that in general terms may be seen repetitively in all municipalities of the Andalusian coastline.

At this point we are in a situation to now determine inversely and by graphic analysis of the work of certain architects of the area, the key values and significant qualities most relevant to the landscape of the Costa del Sol and the common denominators inherent to its architecture (Vazquez Avellaneda 2016, pp. 313–328).

---

[1] Constantinos Doxiadis (1913–1975) was a Greek urbanistic architect of great international prestige recognized by the International Union of Architects, that severely critized the small participation of architects in the elaboration of town planning codes and their margination from the development of the cities of the twentieth century. He led the movement *Ekística* (the science that conceives human establishments from all points of view, enabling the development of techniques that solve all of their inherent problems, voicing the idea that, in the future, urban areas and megapolis would merge into one sole world town, given the growth in development and population.

3 Comprehension and Transformation in the Landscape of the Costa del Sol Through the Architectural Language of Melvin Villarroel

Given that planning instruments analyzed as case studies have not been effective to satisfy the need to preserve the morphologic conditions and landscape of the Costa del Sol, it is possibly more interesting to elaborate a strategy based on the abstract revision of the architectural qualities of the projects that surged from these zoning plans (Pié y Nano, y Rosa Jiménez 2013, pp 303–326). Clearly, it is not feasible to take any project as a starting point for this analysis but, with the passing of time and a neutral approach it is clear that there are splendid examples of architecture in the Costa del Sol that, with the help of magnificent architects, will allow us to understand the limitations of the area of compulsory observance for their work that today have a specific and very relevant importance and therefore are of priceless value.

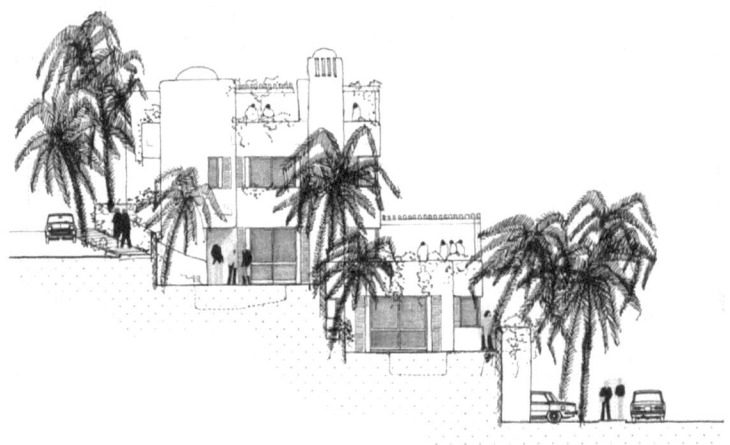

Fig. 3. Ramón Vázquez Molezún. Transversal Section of the urbanization Marbella Hills Club. Marbella. 1981. Source: Legado Vázquez Molezún in the COAM. Available at: http://legados. fcoam.eu/#arqA001717.

We find interesting examples such as Urbanización Las Lomas de Marbella Club by Fernando Higueras, Urbanization Marbella Hills Club by Ramón Vázquez Molezún (Fig. 3) or the proposal for the public tendering of the new urbanization at Elviria (Reyero Fadrique and Arias García 1961, pp. 12–13) with the participation of eminent architects of the time such as Jorn Utzon, Miguel Fisac, Antonio Vázquez de Castro, Jose Luis Picardo or Ricardo Bofill (Martínez y Sánchez Arjona 1961, pp. 14–44). These studies centered on issues related to territorial scale and under infrastructure limitations; the focus was on projects for population and occupation with formulas for global zoning or planning and not so much on specific formal issues (Lampreave 2005, pp. 28–39).

The work of Melvin Villarroel is to be found amongst this group of prestigious projects that with the quality and precision in the results may be considered as a landmark or reference to decipher common values that, inversely, may help to understand the idiosyncrasy of the landscape of the Costa del Sol. This support may lead the way to the elaboration of strategies that aim to preserve and permit the survival of the local landscape, especially against the threats arising from the vertiginous development of the tourist and leisure segment.

The architecture of Villarroel uses the basic concept of "empty"[2] throughout his work as a strategy to organize space, using this concept also for the formal configuration of his design and, therefore, the alteration of nature that is inherent to his execution (Fig. 4).

Fig. 4. Melvin Villarroel. Planning of Hotel Puente Romano in Marbella. Marbella. 1978. Source: Villarroel 1997, p. 98

---

[2] As described by the architect, through various texts collected in his official website at his office, the concept of "empty" is understood as essential to his creative essence, understanding this concept in its multiplying effect, offering the possibility of enlarging the number of elements that participate in the Project. Villarroel understands "empty" as a creative design strategy that originates in the concept of architecture itself, as a recipient that collects from the inside the forms of the outside, opening itself to the elements and views of the environment and projecting a continuous space.

Villarroel's primary organizational criteria is not spontaneous; each project emerges from the detailed analysis of the potential offered by the location, revealed by specific graphical studies where all inputs are considered and, on location, force the disposition and formal qualification of the architecture that must respond to the programmed demands made by the developer (Salgado de la Rosa, Raposo Grau y Butragueño Díaz-Guerra 2019, pp 204–205) (Fig. 5).

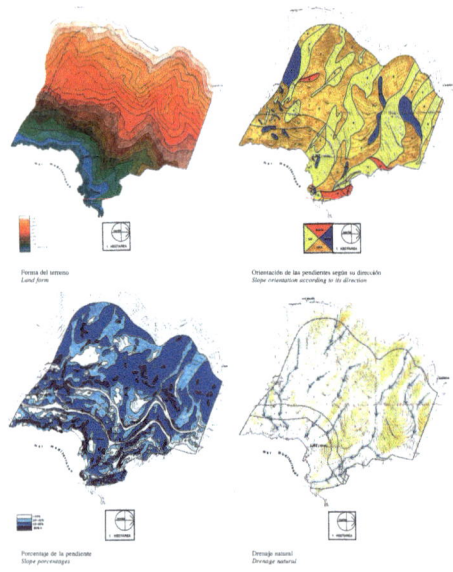

Fig. 5. Melvin Villarroel. Diagrams of territorial potential. Marbella 1978. Source: Villarroel 1997, p. 155.

The abstract graphic proposal (Seguí de la Riva 2018, p. 61) of the limitations and opportunities of the territory are a constant obligation for the Bolivian architect at the moment of initiating each project and today, for us, having contrasted the value and success of some of his projects, serve as references to take note and conjure ideas and concepts that transform into descriptions of the landscape of the Costa del Sol (Fig. 6).

The urbanistic limitations have always been only of relative importance to Villarroel even though he committed meticulously with every one of them and if they were the key features for the majority of architects in the design of their projects, the architecture of Villarroel goes farther, going in-depth into the concepts and only using them on the tangent to solve problems on the board.

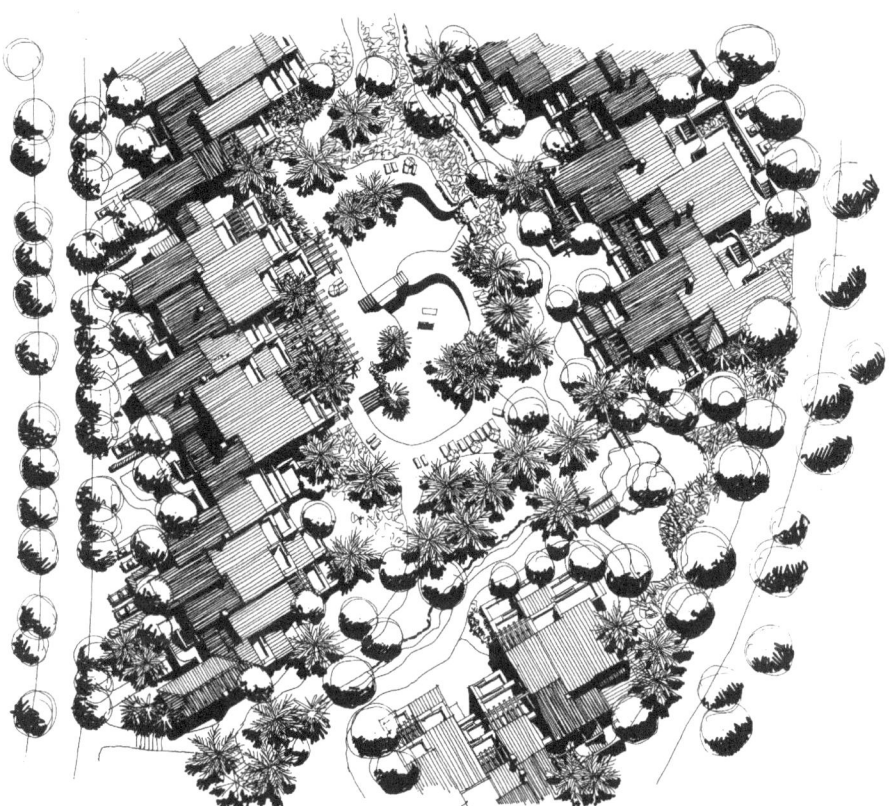

Fig. 6. Melvin Villarroel. Floor plan of empty space interior of Hotel Puente romano in Marbella. Marbella. 1978. Source: Villarroel 1997, p. 100.

4 Conclusion

After a close review of the key historic instruments used for urbanistic planning and architectural control in the Costa del Sol we have been able to observe that, in fact, they have been scarcely efficient. Even today it is not far-fetched to say that we are still lacking in some issues that still persist.

This brings us to argue that the architecture brought about by tourist interests, to date, has not been conveniently controlled; and in so doing has ignored and undermined the landscape richness and value of the areas provoking the loss of the characteristic distinction inherent to the andalusian coastline territory.

Evidently, town planning cannot attend to individual requirements at least not as a category but, on the other hand, given the positive results observed in some of the individual projects carried out, amongst others, those of Melvin Villarroel, we may consider revising the graphic tools used in his projects through a rational abstract logic, for areas of interest that magnify characteristic themes of the Costa del Sol.

So, once these themes are agreed, they may be used and incorporated in codes and norms that regulate new builds, including volumetric limitations, formal and tectonic conditionings and consequently the impact on the scenery and landscape.

The way Villarroel designs empty spaces and his treatment of scenery, landscaping and nature is of notable interest. In this respect his planimetry is a collection of plastic solutions that from the strict observance of graphic rigor, inform on how one may venture, in a future moment, into the landscape, at the same time connected to the exploitation of the tourist market in Costa del Sol.

As a corollary we may comment that certain aspects of Melvin Villarroel's work should, or must be catalogued and protected as patrimonial elements of interest; not only for the architectural richness or the freshness that still emanate from his project solutions, but also for his respect to the valid models of design that, apart from reversing the destructive mechanism of the andalusian landscape, help to protect and improve.

References

Chueca Goitia, F.: Invariantes castizos de la arquitectura española. Invariantes de la arquitectura hispanoamericana: manifiesto de la Alhambra. Editorial Dossat, Madrid (1981)

Martín López, D.: Controversias, turismo y estética: Africanidades explícitas en la arquitectura contemporánea canaria. AACADigital: Revista de la Asociación Aragonesa de Críticos de Arte (15) (2011)

Martínez y Sánchez-Arjona, J.M.: Concurso Elviria, Marbella-Málaga. Revista Arquitectura (27), 14–44 (1961)

Gavilanes Vélaz de Medrano, J.: Primeros Planes en la Costa del Sol (1955–1967): La escala inter-media frente al crecimiento concéntrico. In: XVIII Congreso AECIT. INVAT-TUR, Benidorm, pp. 1–19 (2014)

Gómez y García de la Buelga, J.: Plan General de Ordenación Urbanística de la Costa del Sol Occidental. Revista Arquitectura (65), 41–50 (1964)

Pié y Ninot, R., Rosa Jiménez, C.J.: La cuestión del paisaje en la reinvención de los destinos turísticos maduros. Málaga y la Costa del Sol. ACE: Arquitectura, Ciudad y Entorno 9(25), 303–326 (2013)

Reyero Fadrique, J.A., Arias García, P.: Elviria como problema. Revista Arquitectura (27), 12–13 (1961)

Salgado de la Rosa, M.A., Raposo Grau, J.F., Butragueño Díaz-Guerra, B.M.: El panorama grá-fico arquitectónico de los ochenta. Arquitecturas de escape y dibujos de Resistencia. Revista Expresión Gráfica Arquitectónica (36), 198–209 (2019)

Sánchez Lampreave, R.: Del jardín al paisaje: Elviria, ciudad nueva. Revista Arquitectura (339), 28–39 (2005)

Seguí de la Riva F.J.: Proyectar, proyecto; dibujar, dibujo. Revista Expresión Gráfica Arquitec-tónica (34), 56–73 (2018)

Vázquez Avellaneda, J.J.: Arquitectura y fetiche en la Costa del Sol. In: IDPA 2016: Investiga-ciones Departamento de Proyectos Arquitectónicos, pp. 313–328. Escuela Técnica Superior de Arquitectura. Universidad de Sevilla, Sevilla (2016)

Villarroel, M.: Arquitectura del Vacío. Ediciones GG., México (1997)

Sitio web de Melvin Villaroel. http://www.melvinvillarroel.com/pressroom.php. Accessed 23 Oct 2019

Design Drawings as Cultural Heritage. Intertwining Between Drawing and Architectural Language in the Work of Aldo Morbelli

Roberta Spallone(⊠) 🆔 and Giulia Bertola

Department of Architecture and Design, Politecnico di Torino, Turin, Italy
roberta.spallone@polito.it

Abstract. Increasingly, the archives of 20th century architecture are attracting the interest of scholars as a source for in-depth understanding of architectural works and the trajectories of artistic movements and individual designers. The autograph drawing, in the different phases of the elaboration of the project, uses methods and techniques that reveal the link with the architectural language. The rediscovery, analysis, highlighting and interpretation, with traditional and innovative tools and techniques typical of the discipline of representation, can contribute to the recognition of architectural archives as Cultural Heritage. The authors of this proposal are following this path about the drawings of Aldo Morbelli's archive, consulting them extensively, identifying characteristics and peculiarities and, finally, choosing a case study, through which to test the heuristic potential of representation for analysis and interpretation of the design. In this paper we propose graphic, manual and digital analyses related to the design of "Due case a Capri" (1942).

Keywords: Drawing · Design · Graphical analysis · Interpretation · Aldo Morbelli

1 Introduction (Roberta Spallone)

More and more 20th century archives of architecture arouse the interest of scholars as a source for a deep understanding of the architectural work and the trajectories of artistic movements and individual designers. The autograph drawing, in the various phases of design process, assumes physiognomies that reveal the interweaving with the architectural language; the design drawings made up in the cubist and neoplastic period are striking examples of this.

Rediscovering, analysing, highlighting, and interpreting, through traditional and innovative tools and techniques typical of the discipline of representation, can contribute to the acknowledgement of architectural archives as Cultural Heritage. The authors of this proposal are following this path on the drawings of Aldo Morbelli's archive, extensively consulting them, identifying characteristics and peculiarities, and, finally, choosing a case study, through which to test the heuristic powers of representation for the analysis

L. Agustín-Hernández et al. (Eds.): EGA 2020, SSDI 6, pp. 73–85, 2020.
https://doi.org/10.1007/978-3-030-47983-1_7

and interpretation of the project. In this work we propose graphic readings, manual and digital, relating to the project of the "Two Houses in Capri" (1942).

2 Drawing and Architectural Language: Aldo Morbelli in the Italian Context and in the Panorama of Turin (Roberta Spallone)

Aldo Morbelli (1903–1963), born in Orsara Bormida in Piedmont, is an Italian architect who graduated from the Faculty of Architecture in Rome and established his professional studio in Turin from the 1930s. In over thirty years of activity Morbelli produced about one hundred architectural projects, mainly concerning single-family houses in the countryside of Alessandria, holiday residences in seaside and mountain resorts, social housing for the INA-CASA program, entertainment buildings, company head-quarters, and post-war reconstruction works. The most important assignment was for the project for the reconstruction of the Teatro Regio, obtained in 1937 following the victory of the competition, which has been drafted in numerous variations with the colleague Robaldo Morozzo della Rocca, but never saw the realization and, upon his death, passed to Carlo Mollino.

The incomplete Morbelli's archive, kept in the Biblioteca di Architettura "Roberto Gabetti" of Politecnico di Torino, contains valuable proof of these projects, many of which have been built. It also contains youthful drawings, travel and body sketches, and illustrations. These last works are very interesting for the understanding of the progressive construction and formalization of the graphic language of the author. He line-draws with a greasy pencil, the ink pen of different tips, and the felt-pen, and paints using watercolor, felt-pen and tempera.

Morbelli's figure has been little studied in recent years, although some of his projects have been published in magazines such as Casabella (project of bathing establishment in Levanto, 1932), Atti e Rassegna Tecnica (restoration of Palazzo CEAT, 1946), L'architettura: cronache e storia (Palazzo RAI, 1962), and the coeval critics has devoted to his single-family houses a monographic number of L'architettura Italiana, 1942 [1].

He began working at the studio of Annibale Rigotti when he was seventeen and in 1923 he enrolled at the Scuola Superiore di Architettura in Rome, where the personalities of Marcello Piacentini and Gustavo Giovannoni, but also Arnaldo Foschini and Enrico Del Debbio stood out. In the Roman School Morbelli had as fellow students Adalberto Libera, Mario Ridolfi, Luigi Moretti, Luigi Piccinato. Very talented as a violinist, a passion that he cultivated in parallel with architecture, Morbelli was also a frequent member of the Turin intelligentsia: the philosopher Norberto Bobbio, who was among his clients, the sculptor Mario Giansone, and the painter Enrico Paulucci, who sometimes supported him in the realization of the artistic apparatus of the projects.

Emanuele Levi Montalcini, speaking about Morbelli's attitude, in the early stages of his career, to intellectual play and a free and abstract invention, emphasized his double relationship of attraction-repulsion for the new: on the one hand the definitive detachment from the architecture of the past, on the other hand to fall once again "into false shapes and false theories" [2].

The national and international context from the Thirties, passing through the period between the two Wars, to the second post-war period, is the background to the formulation of contemporary languages, as well as the construction of drawing standards updated on the growing complexity of the project. Some recent critical studies bear witness to this [3–6].

As Mezzetti observed, in such a period of time the path of Italian architecture and its representation often saw the drawing techniques coinciding with linguistic and compositional approaches. This led to the search for essentiality and cleanliness of drawing: abandoning all graphic decorativism, the architects concentrated on plastic volumes and masses through the use of charcoal and then tempera, tending to focus more and more attention on the building, isolating it, and making it autonomous from the context [4].

The fil rouge between design and drawing in that period was admirably outlined by an illustrious witness and protagonist of the time, Luigi Moretti who was a contemporary, as we have seen, of Morbelli. Moretti, in a 1936 paper, associates graphic techniques with the protagonists of the architecture of the previous and contemporary decade. In this way modern architecture passed from Limongelli's charcoal, to Aschieri's Wolf, to Libera and Ridolfi's tempera, to the pen of Le Corbusier's school [7].

In the same years, in Italy, while the rationalist current linked to the neoplastic poetics, was embodied by figures like Giuseppe Terragni and Alberto Sartoris, and the Pragmatist one oriented to an architectural reform that did not break the ties with tradition, was represented by Mario Ridolfi, the Turin scene appeared different, marked "by extremely interesting personalities, but sometimes a little isolated" [8]. In Turin, where Carlo Mollino is beginning to assert itself as a master, whose story intertwines several times with that of Morbelli, the latter passes unscathed between regime architecture and the 'post-resistance' tendencies of the 1950s [9] and his poetics seems far removed from rationalist and pragmatic tendencies. He pursues processes of formal simplification, control of decorative apparatus, mixing of syntactic elements taken from different contexts, and research into relationships and balances between empty and full spaces in the conviction that each design theme requires a different and appropriate treatment. In 1945, while in Rome the Associazione per l'Architettura Organica and in Milan the Movimento di Studi per l'Architettura were founded, in Turin 26 architects of heterogeneous leanings met in the "G. Pagano" group, dissolved in '48, in which the figures of Carlo Mollino, Gino Levi Montalcini, Ettore Sottsass stand out. Morbelli did not adhere to this movement.

Morbelli's drawings document the graphic talent of the architect, who moves with ductility from conventional technical drawing to expressive one, in which he uses a highly tactile and sinuous line, which could be defined Baroque, very original and recognizable, that seems to find in the interior perspectives, enriched by furniture, frills, figure, the freest place of expression. The exceptional skills in drawing and the mastery of methods and techniques for controlling space and architecture make use of disparate techniques, chosen case by case in the urgency of expression of the idea: from pencil, pen, felt-pen, sometimes enriched with watercolor and tempera.

Fig. 1. Morbelli, real-life drawings of Mediterranean architecture and objects. (Archivi della Biblioteca Centrale di Architettura "Roberto Gabetti", Politecnico di Torino, in the following BCA. Fondo Aldo Morbelli).

3   Two Houses in Capri: Analysis Drawings of Places and Design Synthesis (Giulia Bertola)

The two unrealized houses in Capri are a significant example of Morbelli's production, either for the synthesis between the characteristics of the architecture of the context and the plastic design research, and for the iconographic materials produced for the invention, definition and communication of the design process.

The theme of rural architecture in Campania, which was part of the national debate on the origins of Mediterranean and indigenous architecture since the late twenties, has been the subject of interest from the architecture historian Roberto Pane, who represented the subject with writings, drawings, and oil and acrylic paintings.

The Campania's rural architecture represents for Pane an important reference for modern architecture [10].

The latter must not imitate its forms, but can be inspired, while maintaining the autonomy of its own figurative language, by that abstract sense of the chiaroscuro values present in rural artefacts "which generates by the overall relationships of the whole and not by the single isolated parts" [11].

Fig. 2. Morbelli, fantasy drawings. (Archivi BCA. Fondo Aldo Morbelli).

Pane describes the urban morphology of Capri: "no longer the scattered houses with only ground floor but more complex houses, bizarre architecture with arches or half arches, strange climbing stairs that lead to terraces or balconies supported by corbels" [12]. Within this context, the repertoire of Morbelli's Capri drawings seems to fit in, one of which dates back to 1940, which we imagine to be preparatory for the design of the two houses.

The practice of life drawing appears to be an almost instinctive component of Morbelli's poiesis, which translates into the need to draw and express oneself: an inevitable process to understand where, during the design process, one starts from and where one wants to arrive. The drawing is in fact used by Morbelli to know, understand, and design an architecture determined to free itself from stylistic repertoires and able to express the nature of places and the perception of the environment through the observation and recovery of antique shapes of construction mirroring ancient anthropological values.

The visual notes are made up of sketches and drawings made on spolvero pad or white paper sheets and depict both rural homes and views of the Capri landscape up to linger on some details of furniture (Fig. 1). Other references come, instead, from the personal iconographic repertoire and are the result of intense activity in the areas of design, interior design and illustration (Fig. 2).

The two houses in Capri, designed between 1941 and 1942 (Figs. 3 and 4), embody the concept of the Mediterranean house that Morbelli had already experimented with in his first houses in Alessandria countryside and whose interest is also demonstrated by the collection of copies of drawings of glimpses of Capri (1926–28) and by the autograph drawings of Capri, as mentioned above. The buildings, defined by Morbelli as Large house (located lower down) and Little House (higher up), are thought to be situated on land among olive groves at the foot of Mount Tuoro, in "La Cercola" region. The steep slope to the west orients the layout of the main rooms towards the valley. The design theme, commissioned by Dr Peisino, is solved with a combination of compact volumes that fit into the steep terrain through the connection with sinuous walls that follow the contour lines and tie the buildings to the surrounding environment, while steps and pergolas are used as an extension to the outside of the two houses.

Fig. 3. Morbelli, study sketches for the two houses in Capri. (Archivi BCA. Fondo Aldo Morbelli).

Morbelli sometimes breaks up the compactness of the volumes through the insertion of classic elements, such as the opening of the entrance with corinthian columns, while the reference to local architecture is resolved in the choice of elements such as the low arches, colors and materials for walls and coverings such as the white plaster, Vietri ceramics, and pieces of furniture such as oil jars.

Even in the interior spaces, the intention to give the rooms a plastic sense is evident: the walls and the distribution elements, such as the "S" staircase of the Large house, adapt to the layout of the plan.

The Large house is on two levels, the lower level is intended for the living room, with the living and dining area at double height that occupies the entire front with sea view and that of the services developed up the mountain, while the higher level is intended for the sleeping area, with four bedrooms.

The Small house is spread over three floors, with a basement with tavern and bathroom, a ground floor dedicated to the entrance and living area and the first floor with three bedrooms.

The composition by volumetric masses, interrupted by a few holes carefully distributed on the façade, is expressed through linear drawings and a sketch gone lost

Fig. 4. Morbelli, studies for the two houses in Capri. (Archivi BCA. Fondo Aldo Morbelli).

but documented photographically, which highlights the relationships between full and empty.

The collection of design drawings, ideally connected to the series of travel and study sketches made from life on the island of Capri and the Amalfi coast mentioned above, consists of a total of twelve graphic works.

Five of these are sketches and study drawings, two are plans with the allotment, the remaining five are technical plates (Figs. 5 and 6), four of which are about the two houses, drawn up in October 1941 in a scale of 1:100 and accompanied by some perspectives and studies of variants, one, dated March 1942, is a specific study of the relations between the two houses in elevation and section along the slope. The project is flanked by a lost plastic model; of which we have memory through four photographs of the time. The project is published in six pages in a monographic issue of L'architettura italiana [1] dedicated to Aldo Morbelli, which enriches the image of the model with more photos than those available today. Among the techniques of representation, the sketch is established as a constant tool of design. During the conception phase, Morbelli works on the interiors, creating elevation views of the most significant rooms, while the views of the exteriors are mainly inserted in the technical tables. Another important tool is the plastic model, which is a very useful device for the design prefiguration [13]. In the archival photographs of the model for the two houses at Cercola it is possible to read the plastic characteristics of the two buildings in relation to the morphology of the site thanks to the rich play of chiaroscuro that clarifies the relationship between volume, matter and space.

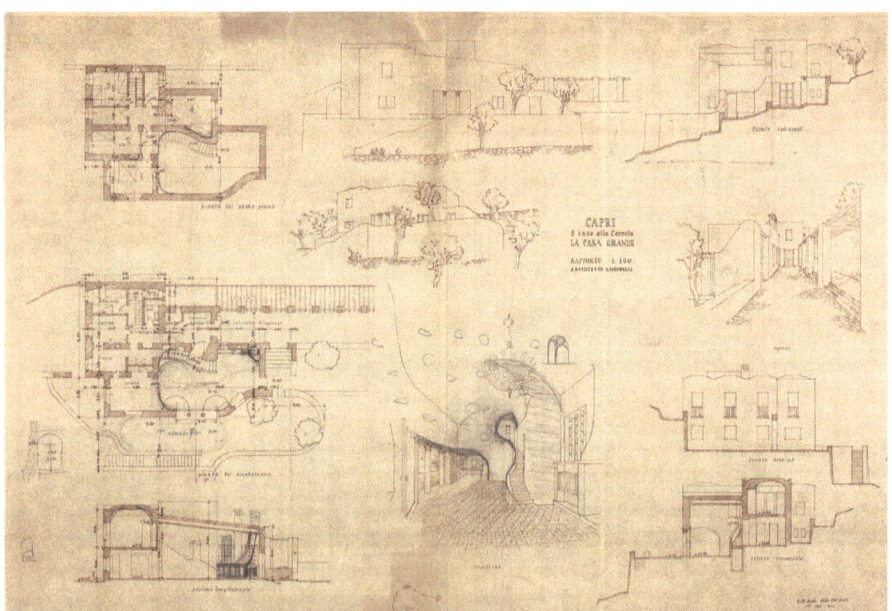

Fig. 5. Morbelli, the Large house in Capri, technical drawing. (Archivi BCA. Fondo Aldo Morbelli).

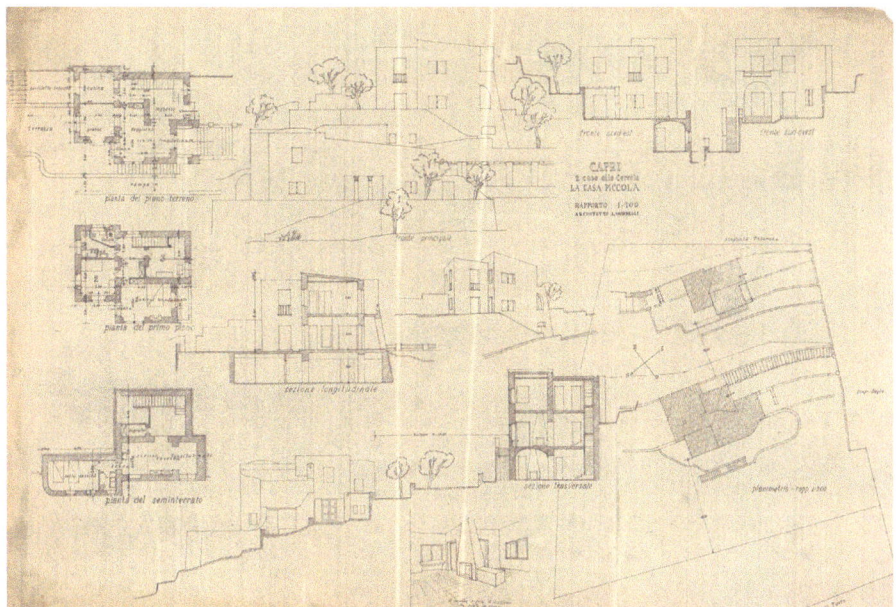

Fig. 6. Morbelli, the Little house in Capri, technical drawing. (Archivi BCA. Fondo Aldo Morbelli).

4 Two Houses in Capri: Redesign and Modeling for Interpretation (Giulia Bertola)

"There is an immense difference between seeing something without a pencil in your hand and seeing it drawn. Or rather, they are very different things that you see. Even the most familiar object in our eyes becomes quite different if we start to draw it… we realize that we didn't know it, that we had never really seen it" [14].

The characteristics of the complex of buildings have led us to integrate, in the interpretative representation, the freehand re-drawing and the reconstructive digital modeling. In this case study, in fact, the re-drawing, with the exploration of the master's graphic techniques, allows the de-composition of the project aimed at the analysis of its peculiarities (Figs. 7 and 8), while the re-composition through digital modeling allows to control the consistency between the elements from the reconstruction of unrealized architectures characterized by a strong volumetric character.

Morbelli made his first sketches on tracing paper with a low hardness graphite pencil. Both the choice of the support and the choice of the material allowed him to investigate the architectural structure through wide and fluid gestures, in which the sharp line, which does not exist in reality, is set aside to leave space for the nuance.

This effect is lost on the glossy support, allowing the articulation of the areas of light and shadow on the surfaces.

The re-drawing proposes the same techniques. This procedure allows the understanding of the articulation of volumes in space and their proportional relationships.

In Morbelli's drawings there is a strong gesture and a particular stubbornness of the graphic line on some architectural elements. These are straight elements which,

Fig. 7. Interpretative drawings of the two houses in Capri, exterior views (Drawings: Giulia Bertola).

Fig. 8. Interpretative drawings of the two houses in Capri, interior views (Drawings: Giulia Bertola).

through the insistence of the sign, become unreal shapes and cannot be traced back to regular geometric elements. Especially in the interiors, it seems that he "excavates" inside the rock, spontaneously raising the architecture from the earth, thus promoting the relationship between man and nature and underlining the link with the surrounding context.

The sign, intended as a plastic expression of form and color, is the main means of expression of its inner motion.

The same chiaroscuro used by Morbelli in the freehand drawing, and re-proposed in the interpretative re-drawing, is transposed in the architectural volume of the sketch and in the reconstructive digital one (Fig. 9). From the latter, the compositional complexity of the volumes emerges from the first phases of modelling. The building develops freely in space, showing its volume and the rich articulation of the surfaces, shaped almost like sculptures. The plasticity is given by the alternation of full and empty spaces, by the play of light and shadow and by the arrangement of the architectural elements.

All this helps to avoid giving the building a compact and flat effect and allows the architecture to be shaped through interaction with light. The harmony of the composition derives from a careful balance between the different parts. Morbelli exploits the slope

Fig. 9. Morbelli, two houses in Capri, external views. (From top: re-drawing: Giulia Bertola, photograph of the original model, digital model: Giulia Bertola).

Fig. 10. Comparison between the original model and reconstructive digital model (Digital model: Giulia Bertola).

of the ground using a succession of parallel plans/scenery flats, which advance in space both horizontally and vertically and on which walls and floors are arranged.

The space is thus divided into internal spaces arranged on several levels, external courtyards and stairs hidden by masonry curtain (Fig. 10). A personal sensitivity emerges, expressed through the conscious adoption of plastic and fluid forms, alternating with rectilinear and regular forms, characteristics that can be found in the surrounding Mediterranean architecture and that already emerge from some of his travel sketches dating back to the '20s.

5   Conclusion (Roberta Spallone)

The use of an architectural language, measured and rigorous, alien to fashion, as critics pointed out, is accompanied in Morbelli by a graphic exuberance just contained, yet controlled, which allows him to manage the spatial problem with confidence and, as the graphic reconstructions have shown, to solve it, perhaps, really thanks to the drawing.

Graphical analysis, re-drawing and reconstructive digital modeling, even more so in the cases of unrealized projects such as the one in the case study, affirm their complementary nature and offer new possibilities of investigation, critical re-reading, interpretation.

Retracing the conceptual phases manifested in the initial sketches, using the same techniques, tools and graphic materials, trying to reproduce the flow of the line, the chiaroscuro, the chromatic rendering, has allowed to deepen and hypothesize design paths and rethinking, in the slow and progressive definition of the shape.

This, not being achieved in the building, finds in the digital modeling a rigorous instrument of shape consistence control, somehow anticipated in the original plastic models not preserved but documented photographically, which hypostatizes the process of knowledge made, opening it to possible new developments.

Indeed, digital modeling lends itself to possible different solutions, capable of opening up to new reflections on the part of scholars and to enhance an archival heritage that is sometimes little known and/or at risk of obsolescence. It also offers the potential to become a vehicle for the dissemination of knowledge, the sharing of the results and their updating.

References

1. Melis, A.: Architetti italiani. Aldo Morbelli. L'architettura Italiana **3**, 49–72 (1942)
2. Morbelli, G.: Aldo Morbelli architetto a Torino, 1903–1963. Atti e Rassegna Tecnica della Società degli Ingegneri e degli Architetti a Torino **38**(6), 208 (1984)
3. Magnago Lampugnani, V.: La realtà dell'immagine. Disegni di architettura nel ventesimo secolo. Comunità, Milano (1982)
4. Mezzetti, C. (ed.): Il Disegno dell'architettura italiana nel XX secolo, p. 24. Edizioni Kappa, Roma (2003)
5. Millán Gómez, A.: Equivalencias: arquitectura, diagramas y el rechazo de lo icónico. In: Congreso Internacional de Expresión Gráfica Arquitectónica EGA, pp. 299–303 (2010)
6. Montes Serrano, C.: Un posible canon de los dibujos de arquitectura de la modernidad. EGA. Revista de expresión gráfica arquitectónica **16**, 44–51 (2010)

7. Diemoz, L.: Propositi di artisti: Luigi Moretti architetto. Quadrivio **IV**(3), 5 (1936)
8. Sacchi, L.: Il secondo dopoguerra: dal disegno "utile" al disegno "inutile". In: Mezzetti, C. (ed.) Il Disegno dell'architettura italiana nel XX secolo, p. 202. Edizioni Kappa, Roma (2003)
9. Morbelli, G.: Aldo Morbelli architetto a Torino, 1903–1963. Atti e Rassegna Tecnica della Società degli Ingegneri e degli Architetti a Torino **38**(6), 193 (1984)
10. Pane, R.: Tipi di architettura rustica in Napoli e nei campi Flegrei. Architettura e arti decorative **7**, 529–543 (1928)
11. Picone, R.: Capri, Mura e volte. Il valore corale degli ambienti antichi nella riflessione di Roberto Pane. In: Casiello, S., Pane, A., Russo, V. (eds.) Roberto Pane tra storia e restauro. Architettura, città e paesaggio, pp. 312–319. Marsilio, Napoli (2008)
12. Di Liello, S.: Isole come utopie: architettura mediterranea e modernismo nel golfo di Napoli durante il Novecento. In: Spesso, M., Porcile, G.L., Servente, D. (eds.) Italia 45/00 Storie/Progetto, Discipline in dialogo. Franco Angeli, Genova, pp. 112–118 (2018)
13. Scolari, M.: Il disegno obliquo. Una storia dell'antiprospettiva. Marsilio, Venezia (2005)
14. Valèry, P.: Scritti sull'arte, p. 27. Guanda, Milano (1984)

Drawing Moveable Architecture: The Case of Piñero and Escrig's Deployable Structures

Pedro M. Jimenez-Vicario$^{(\boxtimes)}$ ⓘ, Manuel A. Ródenas-López ⓘ,
Martino Peña Fernández-Serrano ⓘ, Adolfo Pérez Egea ⓘ,
and Pedro García Martínez ⓘ

Universidad Politécnica de Cartagena, Cartagena, Spain
pedro.jimenez@upct.es

Abstract. The development of deployable structures began its journey in the early 1960s. Spain's input in this area was pivotal on a global level, with the architects Emilio Pérez Piñero and Félix Escrig leading the charge. Unfortunately, the graphical legacy underpinning their work is somewhat scant and in the main relates to the specifications of the patents for which they applied. That said, the drawings submitted along with their patents reflect a complex architectural school that forms part of Spain's architectural heritage from the second half of the 20th Century. Our goal in this paper is to analyse the drawings and sketches contained in their patents that help define architectural projects in which geometric development is key to their dynamic functioning. From this material, a graphical depiction using parametrical design tools will be devised, representing their kinematic behaviour. The findings address aspects related to the graphical representation of an architectural legacy specifically linked to the representation of its moving parts.

Keywords: Parametrical design · Deployable structure · Moveable architecture

1 Introduction and Background

During the 1960s, architects such as R. B. Fuller or Frei Otto were working with prototypes that were important contributions to developing a new architectural typology that reached its zenith in that decade and the next (Fenci and Currie 2017). The first writings about deployable structures emanated from Zuk and R. H. Clark (1970) in their publication entitled Kinetic Architecture. In it they describe a new concept – architecture that is capable of morphing in order to adapt itself to a new context or environment:

"Surely our present task is to unfreeze architecture – to make it fluid, vibrating, changeable backdrop for the varied and constantly changing modes of life. An expanding, contracting, pulsating, changing architecture would reflect life as it is today and therefore be a part of it".

As the above quotation states, a new architectural paradigm was being sought, but there were several inconveniences to deal with when it came to graphical depiction, in both the design and narrative processes. geometry, kinematics (the unfolding process) and structural behaviour are significantly inter-related fields (Alegría Mira 2014).

© The Editor(s) (if applicable) and The Author(s), under exclusive license
to Springer Nature Switzerland AG 2020
L. Agustín-Hernández et al. (Eds.): EGA 2020, SSDI 6, pp. 86–96, 2020.
https://doi.org/10.1007/978-3-030-47983-1_8

Spain played a central role in this process from the outset. Emilio Pérez Piñero (with his 1961, 1965 and 1976 patents) is considered one of the pioneers in the field of deployable systems. On the other hand, we have Félix Escrig, who patented his prototypes (in 1984) based on the generalised insights acquired from Piñero's findings.

The drawings and sketches that appear in these graphical documents define deployable systems, their foldability being a feature born of their geometry (see Fig. 1 and Fig. 2).

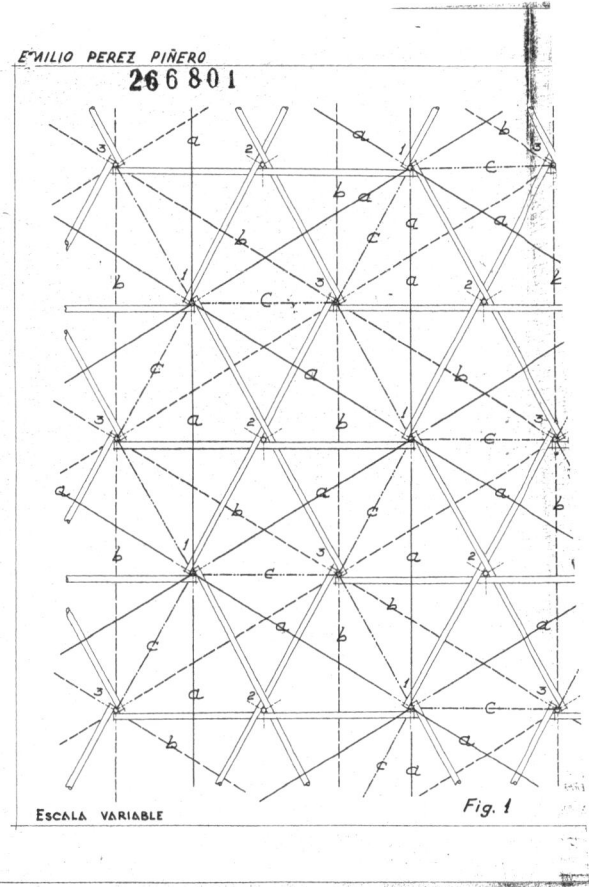

Fig. 1. Emilio Pérez Piñero, *Estructura Reticular Estérea Plegable* ('Folding Reticular Stereo Structure') patent. Spain, 1961. Source: Spanish Ministry of Industry.

Approaching this issue from a contemporary perspective has several advantages. While, in the aforementioned decades, these architects always had to build their prototypes as real models in order to demonstrate the kinematic attributes of their structures, nowadays this sort of modelling may be tested leveraging digital and parametric tools in what is known as an infographic lab. Algorithm-aided design has been recently applied to the study and development of these types of structures (Pérez et al. 2019). Parametric modelling allows the designer to understand the different variables during the creative

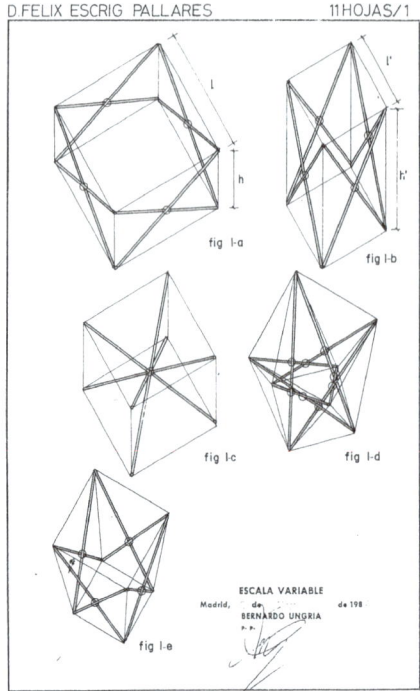

Fig. 2. Félix Escrig Pallarés's patent, Sistema modular para la construcción de estructuras espaciales desplegables de barras ('Modular system for the construction of spatial deployable structures using bars'). Spain, 1984. Source: Spanish Ministry of Industry.

process and helps them to improve the dynamic efficiency. Consequently, being able to view the graphical possibilities in real time also deepens the designer's understanding of what may be achieved (see Fig. 3).

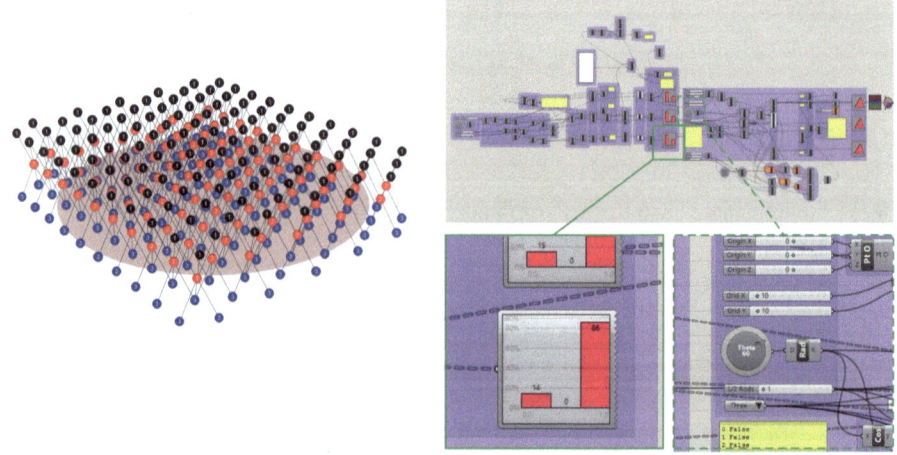

Fig. 3. Modelling of one of Emilio Pérez Piñero's patents, using Grasshopper software. Source: author's image.

This paper therefore seeks to analyse, from a graphical point of view, Emilio Pérez Piñero and Félix Escrig's original patents, in order to develop them using contemporary parametrical design tools. Comparing the results that we obtain with the architects' original sketches will allow us to establish several conclusions with regard to optimal graphical representation, when it comes to dealing with architectural projects in which the drawing of moveable elements presents the biggest challenge.

2 Methodology

In the last few years, parametrical/procedural design and its applications (Grasshopper, Dynamo, etc.), have enabled the generation and geometric formation of complex shapes. Previously, these had to be drawn with CAD tools, i.e. they start out digitally, and afterwards they become physical objects, thanks to 3D-printing. In our study, models drawn on paper correlating to the designs and descriptions appearing in Piñero and Escrig's patents will be taken as reference points. Modelling these digitally will then enable us to ascertain their operability, by creating different operations and combinations of the models.

Broadly speaking, the process is as follows:

1] Generating a three-dimensional model from every defined module, consisting of a number of bars and joints, as well as the folding system. We will look at two modules from Piñero's patents (Tb1 and Tb3) and two from Escrig's ones (Ss1 and Ss3). As part of this, we will need to analyse their original patents and understand each module´s geometries and its dynamic features. Thus, we can determine that the geometry of Piñero's basic module is similar to that of a beam of bars – this is known as a 'Tube-bundle' (Tb), because it consists of three or more rods articulated in a central connecting joint. In the case of Piñero's first module (Tb1), the total volume can be registered in a straight antiprism whose bases are equilateral triangles. The basic module of the second system (Tb3) is similar to the previous one but consists of four rods articulated in a central joint, which are contained in a straight prism with a square (or rectangular) base.

For its part, the basic module of Félix Escrig's systems is obtained by combining pairs of articulated bars in a central joint, forming a pair of scissors or crosses (the 'scissors' system) (Pérez et al. 2019). In this case, the basic module of the system known as Ss1 is contained in a straight prism whose base is an equilateral triangle, while the Ss3 system is defined by a prism with a square base (see Fig. 4).

Three-dimensional modelling is carried out using Rhinoceros v5 software (Robert McNeel & Associates), with the support of the Grasshopper plug-in (Daniel Piker) for parametric design. The introduction of bars and nodes is defined by parameters that will allow us to modify the structures and introduce their dynamic appearance. Among these parameters we find the bar length, opening angle and node insertion points.

2] Repetitive addition of the module to obtain a complex foldable geometry (see Fig. 5). This allows us to cover different surfaces with shapes in plan view, varying the boundary conditions. The geometric efficiency of the different structures represented can thus be evaluated (Ródenas et al. 2019).

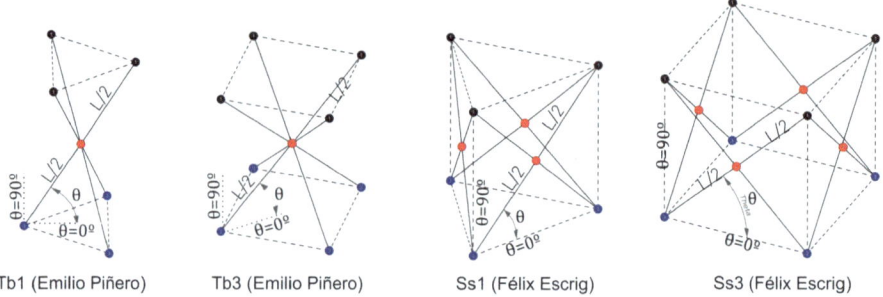

| Tb1 (Emilio Piñero) | Tb3 (Emilio Piñero) | Ss1 (Félix Escrig) | Ss3 (Félix Escrig) |

Fig. 4. Emilio P. Piñero and Félix Escrig's modules, and definition of the variables. Source: author's image.

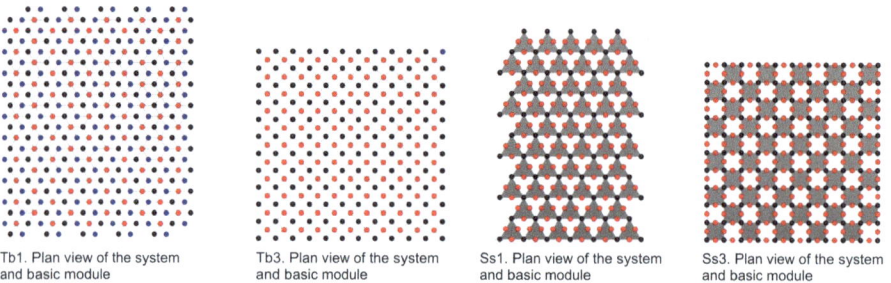

| Tb1. Plan view of the system and basic module | Tb3. Plan view of the system and basic module | Ss1. Plan view of the system and basic module | Ss3. Plan view of the system and basic module |

Fig. 5. Plan view of complex geometries generated from the repetition of each module in the horizontal plane. Source: author's image.

3] Analysis and dynamic study of the behaviour of the structures created. The dynamic behaviour of the structure will be determined by the number of modules and the parameters that define them. Specifically, the Grasshopper interface allows the generation of visual diagrams, based on nodes that control, edit or generate the geometries to which we are referring. In our case, we have generated four diagrams, which reproduce the geometries of the structural systems in our study (Tb1, Tb3, Ss1 and Ss3) and through which it is possible to observe how such systems fold and unfold.

4] Generation of graphical results. The production of orthogonal and axonometric views relating to the parameters considered (opening angle, bar length, position of the knots) allows us to prepare the necessary material for hard-copy representation. The production of videos provides a relevant graphic resource for understanding this using digital media.

5] Producing physical prototypes via 3D-printing. Devising the printing criteria for the 3D models, after they have been verified using Z-Suite, the proprietary software from the manufacturer of the 3D printer Zortrax M200 used. The prototypes' construction is then carried out using ABS as the plastic printing material for the elements that form the articulated elements of the folding structures (see Fig. 6).

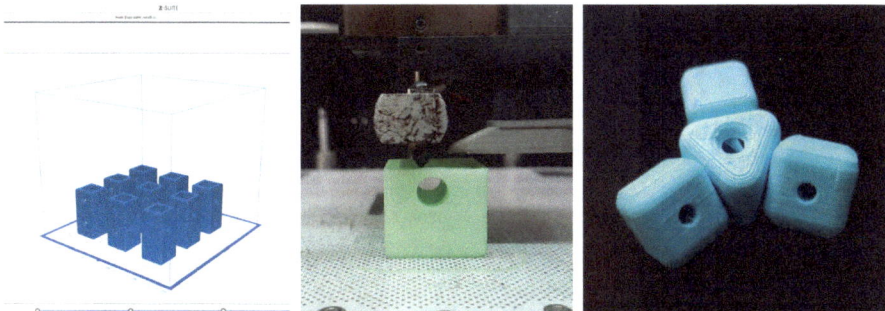

Fig. 6. Node manufacturing process. From left to right: (1) setting the printing criteria of the 3D models using the software pertaining to the printer, (2) printing with ABS plastic, (3) the resulting node. Source: author's image.

3   Evaluation of Findings

The difficulty in representing the movement of a deployable structure is an aspect to consider at its conception stage, as well as once it has been developed, when it is necessary to produce graphical material to explain the thinking and processes behind it. Generating videos (from the digital model), building physical models through the same manual means used by Piñero and Escrig in their day, and leveraging the 3D-printing technology available to us today, are suitable methods through which to understand the behaviour of these structures. The challenge arises when the graphical support is paper-mediated. Our choice of representative system and our selection of views (i.e. whether plan, elevation or sectional views) continue to prove decisive in the graphical definition of these types of structures, as well as their combined use. The representation looks even more complex when graphic criteria to define the geometry not only of the bars but also of the nodes need to be established (XXX 2019). By way of example, in the attached sheets we use ideal geometries for bars in perspective diagrams and simplified nodes as spheres in axonometries (see Figs. 7 and 8).

The descriptive scheme that we have shared involves the combination of several graphic representation systems, accompanied by texts that explain not just the dynamic behaviour of the module-structure, but also its history, origin, applications or built examples (see Fig. 9).

The traditional representation systems used are the dihedral and axonometric, which allow us to observe 'actual moments', as if they were frames (d). For this, the opening and closing limit states, in addition to an intermediate state, are chosen, and this sequence follows a chronological order. This choice involves the distinction of the remaining infinite possibilities between the two limit states in question: deployed structure and folded structure. This criterion allows us to understand the movement for both an isolated module, as well as for a set of modules (b). The use of the computer tools described above facilitates our ability to obtain views and perspectives for their representation on paper, in addition to the generation of videos that simulate the dynamic behaviour of the structures without the need to build them.

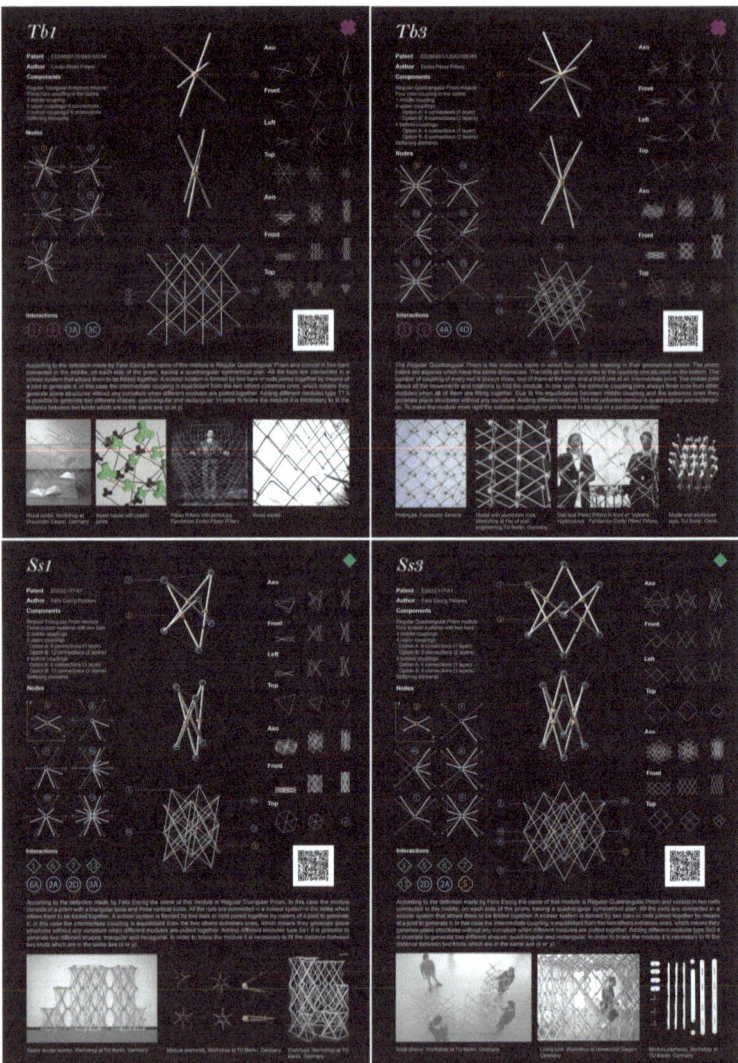

Fig. 7. Graphical representation of the four modules (Tb1, Tb3, Ss1, Ss3) and the structures they generate. Source: author's image.

These videos, in which the moving structures are visible, are accessible via QR codes (b) that link from the digital representations (see Fig. 9).

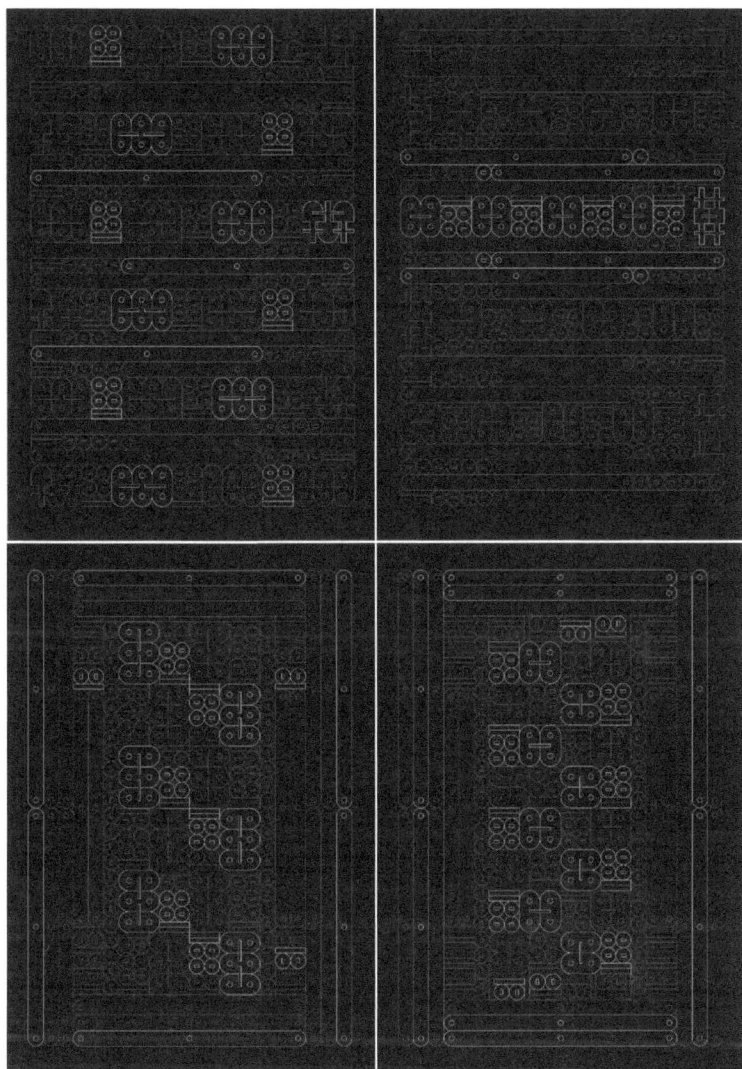

Fig. 8. Exploded view of bars and nodes of the four modules (Tb1, Tb3, Ss1, Ss3) for construction on paper. From left to right and from top to bottom: Tb1, Tb3, Ss1, Ss3. Source: author's image.

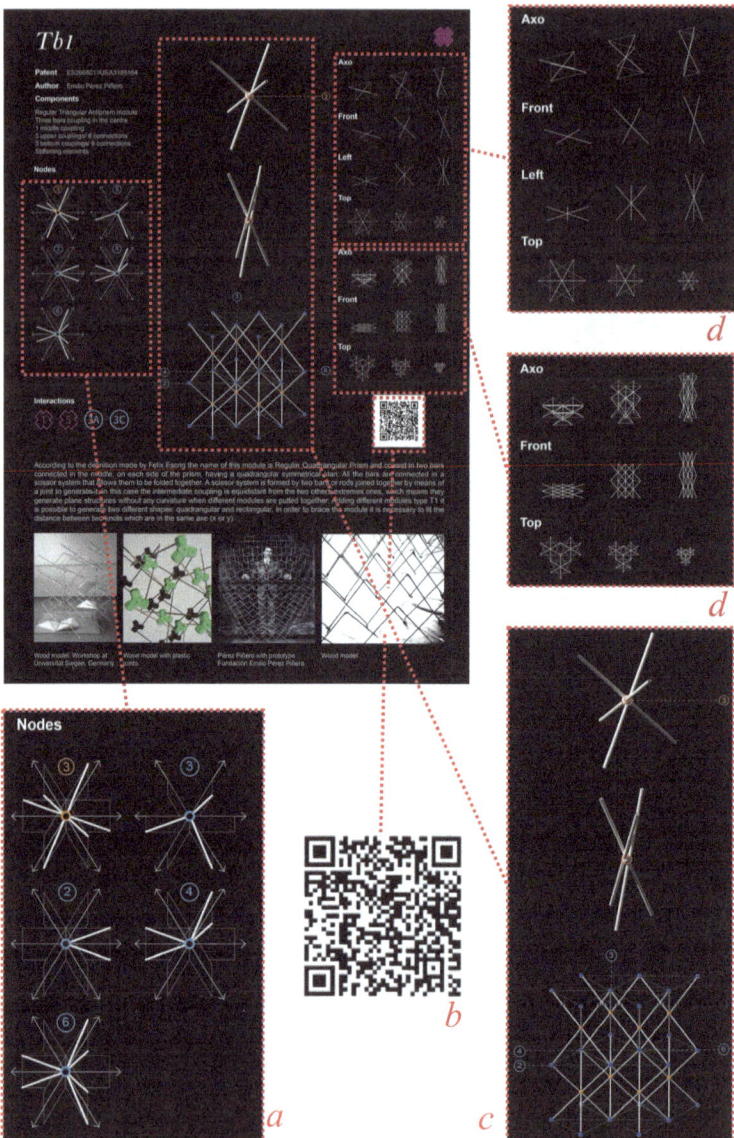

Fig. 9. Graphical representation of Emilio P. Piñero's Tb1 module. Combination of views to aid in its definition. Source: author's image.

4 Conclusion

The use of tools based on parametric design opens up a spectrum of possibilities in the field of deployable structures. They allow us to model complex structures in which a multitude of variables come into play in real-time. These variables are responsible for the structures' dynamic behaviour: bar length, thickness, node design, the relative position of nodes and bars, etc.

The architects behind the patents in our study created their prototypes with the graphic media that were available at that time. They developed an architectural-structural system too ambitious for the graphic tools of the age.

Traditional representation systems (perspectives, dihedral views) helped them not so much in the ideation-conception phase, but during a later stage of representing the model. In this sense, we can deem traditional graphical representation systems to be unsuitable for these early phases, given the variables involved in the modelling of deployable structures.

Our findings lead us to conclude that the physical mock-up route best suits the designer's purposes. Nor does the traditional drawing method lend itself to a more straightforward understanding of the model per se, as it represents specific individual states from within the infinite possibilities that might be available to us. Finally, in relation to the parametric design applications used, we are able to confirm their suitability for use during the ideation and creation phase. However, the challenge remains of representing a dynamic architectural object of the type that we are analysing here via a static medium such as paper, albeit to a lesser extent, since the advantages proffered by these programs in generating results (drawings) are evident.

Using contemporary tools to graphically represent a legacy as specific as the one being considered here has allowed us to evaluate the abilities of the architects behind it and the lengths they went to in order to make their ideas understandable.

Acknowledgments. This publication forms part of the Support Plan for Scientific Research, Development and Technological Innovation, dated Universidad Politécnica de Cartagena, with reference 2017_2453 – "Application of parametric design and generative design for analysis and optimisation of deployable spatial structures".

References

Alegria Mira, L., Thrall, A.P., De Temmerman, N.: Deployable scissor arch for transitional shelters. Autom. Constr. **43**, 123–131 (2014)

Escrig-Pallarés, F.: Patente n. 532.117. Ministerio de Industria y Registro de la Propiedad Intelectual (1984)

Fenci, G.E., Currie, N.G.: Deployable structures classification: a review. Int. J. Space Struct. **2**(32), 112–130 (2017)

Pérez Piñero, E.: Patente n. 266801. Ministerio de Industria y Registro de la Propiedad Intelectual (1961)

Pérez Piñero, E.: Patente n. 3185164. U. S. Patent (1965)

Pérez Piñero, E., Belda Aroca, C.: Patente n. 3975872. U. S. Patent (1976)

Pérez Egea, A., Jiménez Vicario, P.M., García Martínez, P., Ródenas López, M.A., Peña Férnandez-Serrano, M.: Geometry and efficiency in the joints of deployable structures. In: Polyhedra and Beyond. Congreso Internacional Geometrias 2019. Departamento de Matemáticas de la Facultad de Ciencias, Oporto (2019)

Ródenas López, M.A., Peña Fernández-Serrano, M., Jiménez Vicario, P.M., García Martínez, P., Pérez Egea, A.: Geometric Evaluation of deployable structures using parametric modelling. Nexus Netw. J. **22**, 240–272 (2019)

Zuk, W., Clark, R.H.: Kinetic Architecture. Van Nostrand-Reinhold, Nueva York; Londres (1970)

Epigraphy, Typography and Architecture: Control Over Graphic Design Beyond Drawing

María Senderos Laka[(✉)], Iñigo Leon Cascante, and José Javier Pérez Martínez

Architecture Department, UPV/EHU, San Sebastian, Spain
maria.senderos@ehu.eus

Abstract. Architectural drawings and architecture have a complex but undoubtedly very interesting relationship. Writing used in both has different functions, and it can even be sometimes included in the drawing category.

Typography has evolved slowly over the centuries, with slight modifications resulting from improvements in writing tools and supports. The changes that took place in early 20th century were the revulsive that led to the appearance of a "new architecture", and of different attempts to achieve a "new typography". Buildings from the thirties and forties are intriguing due to their rationalist style, but, undoubtedly, the signs, with their different styles of writing, moulded and secured to factories and theatres, workshops and garages, - writing converted into drawing and architecture -, makes them unique.

The different types of relationship that occur between architectural drawing, typography and architecture, - relationships of coherence and incoherence -, are addressed in this article through the analysis of 20th century buildings, plans and inscriptions. The work of architects, such as Frank Lloyd Wright, Richard Neutra or Enric Miralles is studied, as well Archigram advert-graphic design, advert-buildings like Googie's, the Goya petrol stations, the Carrion building or the Coca-Cola factory.

Writing used in plans or drawings offers us clues that help to identify the era, the author or their purpose. These meticulous and "modern" letters form part of architectural drawing, and occasionally of architecture, reflecting architects' interest in extending their own graphic design beyond drawing.

Keywords: Architectural graphic expression · Modern typography · Modern epigraphy

1 Introduction

Architecture has been historically linked to other types of visual arts, and the kinds of typography included in old buildings, hieroglyphics, the "archetypal inscriptions of a Roman architrave", or the "ubiquitous tattoos over a Giotto chapel" contain messages that go beyond their "ornamental contribution to architectural space" (Venturi et al. 1972, p. 24).

Early writing systems, such as pictograph systems, consisted in drawing concepts or ideas. Buildings were used as canvases to draw upon, and walls were used in the same

© The Editor(s) (if applicable) and The Author(s), under exclusive license
to Springer Nature Switzerland AG 2020
L. Agustín-Hernández et al. (Eds.): EGA 2020, SSDI 6, pp. 97–107, 2020.
https://doi.org/10.1007/978-3-030-47983-1_9

manner as they are today with graffiti. In Roman times, the writing carved on buildings, - monumental writing -, possessed a precise manufacturing process as described by Mallon (López 1993, pp. 28–29): a first text on papyrus or parchment in common writing, the preparation of the work with the execution of auxiliary lines, called *ordinatio*, and finally the definite incision with the chisel in capital writing. The process was carried out by *lapicidas,* stone-cutters, and the slight variations in letters was mainly the result of the physical characteristics of the surface on which the incision was made, as well as the type of tool used. Monumental epigraphic writing, which was normalised and canonised in times of Emperor Augustus, can be seen on the façade of the Pantheon, but it lives on today (Romero et al. 1995, p. 60). This type of writing is typical of the time and place, but has no relationship with the person who plans the building, and it is repeated in different Roman constructions with the typical variations found in artisan work. The importance of these epigraphs is much more than just the text inscribed on the buildings.

Important changes occurred in the Middle Ages, such as the appearance of universities, the creation of new monastic orders or the invention of paper. In this context, Gothic calligraphy emerged, and Carolingian writing techniques and forms were adopted (Romero et al. 1995, pp. 65–66). This new writing was gradually consolidated in manuscripts, as well as in inscriptions, but in the latter case more due to a change in taste rather than for technical reasons. Therefore, the stone-cutter carried out a meticulous job in each composition, with decorative motifs and colour, attracting the public's attention. These exuberant letters, natural when written with a pen, would be much more complicated to write on stone, due to the existing contrast between fine and thick lines. For this reason, workshops had specialists for each part of the work (García 2014, p. 5). Gutenberg would use the Gothic letter to create the movable-type printing press. The copies of the Bible, - his most famous work -, were printed and decorated later by hand. This mass production of books would permit the dissemination of culture, and in turn, would favour the arrival of the Renaissance.

Roman writing better represented the values of this new movement, and, in 1525, Albrecht Dürer wrote a treaty on its geometric regularisation. As a Renaissance man, he was heartened to work in different facets of arts and sciences, and consequently, he created an identifiable signature, which has remained for posterity as the first logotype carried out, an anagram with his name. The inherent individualism of the time led him to seek coherence, to the need to master all the artistic facets without leaving anything to chance.

Latin continued to be used for centuries to produce inscriptions, seeking to emulate the values of the classical world. Even though the people could not understand the written messages, arcs of triumph, public buildings or mausoleums were marked with "Roman style lapidary inscriptions" due to their great symbolic value (Satué 2007, p. 14). Thus, during the Classicism, Roman printing would be modernised, based on the aesthetic and philosophic patterns of classical Greece.

During the final stages of the 19th century, the Arts and Crafts movement was another style that attempted to set individuality against the homogenisation imposed by industrialisation, "channelling the desire to see designers raised to similar categories as painters or architects" (Satué 2007, p. 24). For this reason, there would be a return to Gothic typography, representing what is done by hand, craftwork. Shortly afterwards, the Art

Nouveau movement would develop. This movement would attempt to create a young art both in painting and architecture, and in graphic arts, furniture or jewellery. The Eckmann typography would be the most characteristic and in it "…the organic forms of Jugendstil were combined with the German Gothic letter tradition", and the memory of the mediaeval quill could be sensed, which represented a tremendous metal letter production effort (Blackwell 1998, p. 22).

The Viennese Secession's exhibition pavilion (Fig. 1), designed by Joseph Maria Olbrich in 1898, became a decalogue-building of the movement. Here, painting, sculpture and architecture were combined in a totally coherent manner, following the group guidelines. Each message on the façade is studied within the composition, the type of writing is modernist, and the text is not a plot that decorates the façade, but rather the messages contain the basic principles of the movement.

Fig. 1. Joseph Maria Olbrich. Poster for the II exhibition of the Secession. Vienna 1898. Source: Mr & Mrs. Leonard A. Lauder Collection.

2 New Typography

The architectural renovation of the first half of the 20th century was coupled with a renovation in the illustration and writing used on plans. The revolution of visual arts permitted a new way of understanding architectural graphic expression, where experimentation in drawing and typography went hand-in-hand with experimentation in planning. In this sense, Tschichold (2003, p. 14) set out that "anyone who has recognised the deep underlying similarity between typography and architecture, and has understand the true nature of the new architecture, can no longer doubt that the future will belong to the new typography and not the old."

Although the more orthodox modern architects tried to eliminate ornaments from their buildings, abandoning "an iconological tradition where painting, sculpture and graphic design were combined with architecture" (Venturi et al. 1972, p. 24), some of them did not relinquish the use of a "new typography". If, in residential buildings, the letter was generally restricted to the graphic representation of the buildings, in the case of public and industrial buildings, that new writing took a leap forward to reality, with large signs, which, even today, are an example of the aesthetic of an era. As Behrens would say, "typeface is one of the most eloquent means of expression of any era or style, and similarly to architecture, it typically portrays a period and the most severe testimony of the intellectual level of a country" (Satué 2007, pp. 28–33).

The changes that occurred in this period generated, in architects, the need to intervene in all the facets of their work, designing furniture for their buildings, luminaires, logotypes and even typography. They tried to express their architecture through modern compositions, through photography, collages and drawings. Typography forms part of this personal drawing style, and although some architects were happy with the use of templates, Letraset sheets, or the classical Leroy sign systems, those architects who insisted on being coherent, both in terms of their architecture and the graphic design used to represent it, took risks, designing or modifying the existing types of letters.

3 Modern Architecture vs Modern Typography. Coherence in Graphic Design

Hand drawing facilitated the creation of personal typographies, and it even permitted playing with them (Fig. 2).

Architects, such as Frank Lloyd Wright, created an example of own drawing, with a personal stamp and very easily identifiable typography. His drawings could be recognised by his graphic style, by the colours used in some of his compositions, by the type of architecture, and also by the writing. The latter takes on considerable importance in his compositions, with letters that frame the perspectives, rounding off drawings of an exceptional quality. These emblematic features can be seen on the façade of one of his most representative buildings, the Guggenheim building in New York (Fig. 3). Typography not only appears here on the building plan, and on the composition of the plan, but also in the actual museum as corporeal letters. The architect controls everything, from the design of the building through to the end.

Fig. 2. Frank Lloyd Wright, Heinrich de Fries, Stamo Papadaki, Pauline Schubart, Yen Liang, Dorothy Johnson Field, Edgar Tafel. Taliesin, Volume 1, No 1. 1934. Source: The Wright Library.

Fig. 3. Frank Lloyd Wright. Guggenheim Museum. New York. 1937. Source: pixabay. com.

Richard Neutra is another architect whose font style has remained to posterity, thanks to the recent creation of the Neutraface typography, based on this style. In the case of Neutra, the letters are closely related to his architecture, forming a whole project concept. Consequently, he decides to design his own typography, which is used on drawings and in the architecture. This typography forms part of the composition of his drawings, and it has a clear graphic vocation; but it is also used on the signs of buildings, forming part of the architectural composition. In general, he uses the same typography on drawings and on buildings. This is the case of San Bernardino Medical Group or San Pedro Hacienda, two public buildings with corporeal letters on the façade. In the case of the Norwalk petrol station, the typography differs from the architect's own, but Neutra manages to adapt it to integrate it into the project. The different typographies that appear here are clearly legible, the architect's own and that of the trademark (Figs. 4 and 5), although, in an orderly manner, typical of the planner, managing to endorse the design imposed by the client.

Enric Miralles is, perhaps, one of the most extreme examples of coherence between the graphic structure used, the typography and the architecture. His personal planning style was accompanied by unique plans, recognisable for their singularity, with a "drawn letter" that evolved throughout his career. The first writing used, Viaplana, was drawn by hand, and he learnt it at the Viaplana-Piñon studio. During a second phase, he carried out the abstraction and geometrisation of this typography with Marcia Codinachs, and finally, a typography designed solely by Miralles, which adapted entirely to the plan (Vidal 2010, pp. 34–45). This singularity in the Miralles graphic style, his unmistakable personality, converted his graphic design into a trademark that was so easily recognisable that his participation in anonymous architecture tenders was relatively controversial.

The importance granted by the architect to typography, understand more as a plot or form than as a letter with meaning thus becomes clear. Used on his plans as part of

Fig. 4. Richard Neutra. Norwalk Service Station. Bakersfield, California. 1947. Source: Hines 2005, p. 239.

Fig. 5. Richard Neutra. Norwalk Service Station. Bakersfield, California. 1947. Source: Hines 2005, p. 239.

graphic design, the layout of the letters rotates with the composition of the different views, in an overlapping of axes, auxiliary lines and projected lines that complicate the drawing so superbly that is difficult to replicate (Fig. 6). The typography of the great architect can be found on many buildings, such as the Hamburg School of Music, or the Vasco da Gama square.

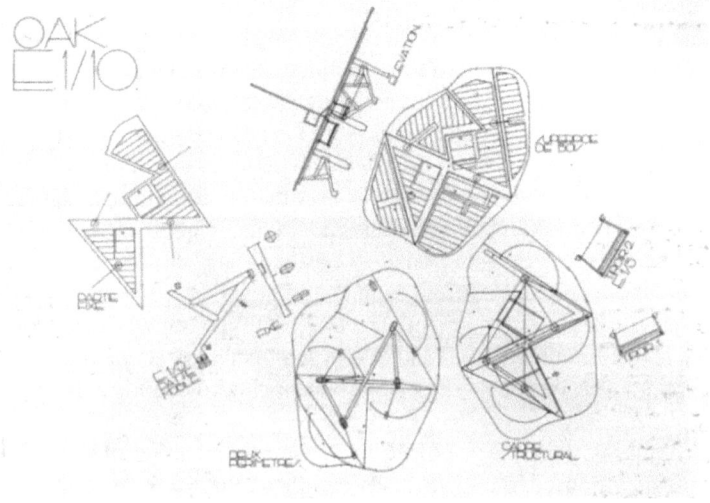

Fig. 6. Enric Miralles, Benedetta Tagliabue. Ines-table Table. Barcelona. 1993. Source: El croquis 100/101, p. 46.

4   Incoherence in Graphic Design. Advert-Drawings

The Archigram group had an impact on the architecture of its time due to its subversive message, but also due to its novel and exaggerated graphic design, based on comics and with a profuse use of letters. The richness of these letters, in different sizes, with flexible placements and impacting typographies, fostered the forcefulness of the drawn messages, and masterfully accompanied the new ideas of the group. Collages and comics are present in almost all the designs, both in black and white, and in colour; typed letters and handwritten letters; photographs, axonometries and drawings; sketches and diagrams. All of this information was composed in an attractive and irreverent manner, to show the change in architecture - graphic design at the service of renewed architectural ideas.

Its idea of futuristic and pro-consumerist architecture, drawn with a pop-art aesthetic (Fig. 7) and using collages with profuse letters, playing with different types of typography, has no direct relationship with the buildings constructed, thus emphasising the utopic nature of its proposals. Graphic design, here, is a means to critically analyse the "existing contradiction between technical rationality and restriction of individual autonomy" (Agudo-Martínez 2013, p. 2). Novel graphic techniques are used to claim, "the ephemeral, the dynamic, and the continuous and necessary flexibility in the functions of the urban environment", but there is no relationship with the constructed architecture (Valfiel 2013, p. 136).

Fig. 7. Ron Herron, Dennis Crompton, Peter Cook, Warren Chalk. Control and choice dwelling. Exhibition for the Paris Youth Biennial. 1967. Source: Chalk et al. 2018, p. 200.

It is, therefore, an advert-graphic design, with a highly virtuous and original drawing method, which enriches the graphic inheritance of architecture but that is not reflected in the building. The dissociation between the typography of plans, and typography in architecture, or lack thereof, is absolute. In this case, the existing decomposition is voluntary, as it is not a question of lack of control over the drawn and constructed work, but rather, that the designer optionally decides not to use graphic design/typography in his constructed work, maintaining drawing as an isolated part of the architectural construction.

5   Incoherence in Graphic Design. Advert-Buildings

In the opposite corner, we can find buildings such as Googie's (Fig. 8), the cafeteria designed by John Lautner in 1949, which would give its name to a futuristic and somewhat gaudy style, which used architecture as a lure. These advert-buildings, with their

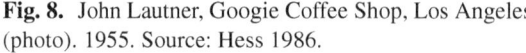

Fig. 8. John Lautner, Googie Coffee Shop, Los Angeles (photo). 1955. Source: Hess 1986.

Fig. 9. José Luis López de Uralde and Francisco Alonso Martos. Goya service station. Vitoria 1935. Source: Docomomo ibérico.

large signs and strange shapes, resorted to the typical advertising letters, typical of the brands they promoted. Therefore, the disconnection between the project writing and the writing designed for the signs is absolute. The architect cannot intervene in the typography or the signage in this type of building, and the final result will depend exclusively on the quality of the publicised advertisement. Architecture passes to a second plane and "it is more a communication than a spatial architecture; communication dominates the space…" (Venturi et al. 1972, p. 25).

The mass production of the automobile in early 20th century brought about changes in the roads and in the speed of vehicles. Buildings, which until then, were located on the roadside were moved further away. Likewise, the signage that was used in the new buildings had to be revised to cope with the speed and the new distance between them and the spectators, consequently taking on considerable relevance. In this sense, the petrol stations and service stations are another type of building where writing adopts different functions through the use of large letters with a representative and aesthetic function. When driving along the Cadena Eleta road next to Goya Petrol Station (Fig. 9), one can only admire the rationalist workmanship of the construction. The name of the building, materialised in corporeal letters, advances as one travels along the road, so that "The enormous letters create a scale and a unit that are appropriate for a public place, and contrast with the unavoidable individual scale of multiple tables and booths…" (Venturi 1995, p. 184).

Another example of a roadside building is the Coca-Cola factory of San Sebastian, which was constructed in 1958 by the architect, Alvaro Libano (Fig. 10). Originally, the Coca-Cola logotype was situated on the marquee over the main entrance, in the form of corporeal letters, in an orderly and well-thought out way. However, years later, the eastern façade of the building was converted into an advertising front, with an impacting result. The building, situated on national road 1, has remained forever in the memory of travellers as the milestone that marked the entrance to the city. The questions asked in *Learning from Las Vegas* would be relevant in this case, "Is the sign the building or the building the sign? These relationships, and combinations between signs and buildings, between architecture and symbolism, between form and meaning, between driver and the roadside, are deeply relevant to architecture today…" (Venturi et al. 1972, p. 98).

Fig. 10. Álvaro Líbano. Coca-Cola building. San Sebastián. 1958. Source: Zaldua 2012, p. 223 (original photograph of Santi Yániz treated).

Fig. 11. Luis Martínez-Feduchi; Vicente Eced. Carrión building. Madrid. 1933. Source: wikimedia.org.

A different model of dissociated typography can be found in the Capitol building of Madrid (Fig. 11), as part of the façade has been used as a hoarding whose advertisements have changed over time as well as the types of typography used. Although in this case, the graphic sign does not dominate the architecture, it is an example of totally differentiated typography and architecture, due to the fact that the publicity makes use of the building.

6 Conclusions

Typography has been a rather misunderstood discipline in the study of architectural graphic expression. However, the graphic design of some of the most important architects is accompanied by a meticulous, personal and identifiable typography, that gives indisputable personality to their compositions. This typography is recognisable in their plans but also in the signs of the constructed buildings, giving them a signature that identifies them as their own, with the aspiration of preserving the coherence of the architectural design through to the end.

The more orthodox architecture of the modern movement may perhaps reject the ornament, and generate a new architecture resulting from an "iconography based on the American industry". But it is also true that simple, modern buildings, "volumes under the light", were the perfect element on which to place some signs, which, due to the absence of ornament, participated of the composition and essence of the building (Fig. 12). Some of the signs would be deliberately architectural, others deliberately commercial, others simply personal, as a prolongation of the architect's graphic style, but all of them had a considerable impact on the final result, on the collective imaginary, and on the creation of a new iconography that is, per se, an essential value of architectural heritage.

Fig. 12. Walter Gropius. Bauhaus building. Dessau 1925/26. Source: Droste 2006, p. 257.

References

Agudo-Martínez, M.J.: La casa como cápsula: planteamientos conceptuales del Grupo Archigram (1961-1974). In: Jornadas Internacionales de Investigación en Construcción. In: International Conference on Construction Research, (1-9). Eduardo Torroja Institute of Construction Sciences, Madrid (2013)

Banham, R.: La Atlántida de Hormigón. Nerea, S.A. (1986)

Blackwell, L.: Tipografía del siglo XX. 3rd edn. Gustavo Gili, S.A., Barcelona (1998/2004)

Bravo de Laguna, A.: Algunas arquitecturas de la A a la Z. EGA Expresión Gráfica Arquitectónica **18**(21), 150–161 (2013)

Chalk, W., Cook, P., Crompton, D., Greene, D., Herron, R., Webb, M.: Archigram, the Book. Circa Press, Londres (2018)

Droste, M.: bauhaus. TASCHEN (2006)

El croquis: Enric Miralles/Bendetta Tagliabue. no. 100/101 (2000)

Escoda, C.: Lugar, dibujo y arquitectura en Wright. EGA Expresión Gráfica Arquitectónica **15**(16), 132–139 (2010)

García, A.: La escritura gótica publicitaria del S. XIII en la provincia de Burgos. PROGRESSUS, Rivista di storia, no. 2, pp. 1–28 (2014)

Hess, A.: Googie: Fifties Coffee Shop Architecture. Chronicle Books, San Francisco (1986)

Hines, T.: Richard Neutra and the Search for Modern Architecture. Rizzoli, New York (2005)

Lamprecht, B.: Neutra. Taschen, Germany (2004)

López, P.: Epigrafía Latina. Tórculo edicións, Santiago (1993)

Musper, H.T.: Albretch Dürer. Les Grands Peintres, Editions Cercle D´art, New York (1989)

Pelta, R.: El pensamiento tipográfico moderno. http://www.monografica.org/04/Art%C3%ADculo/5824. Accessed 06 Oct 2017

Romero, M., Rodríguez, L., Sánchez, A.: Arte de leer Escrituras Antiguas. Paleografía de lectura. University of Huelva, Huelva (1995)

Satué, E.: Arte en la tipografía y tipografía en el arte. National Library of Spain, Germán Sánchez Ruipérez Foundation, Siruela editions, Madrid (2007)

Tschichold, J.: JAN TSCHICHOLD. LA NUEVA TIPOGRAFÍA. Campgràfic editors, S.L., Valencia (2003)

Valfiel, M.: La Arquitectura Pop. De la razón al significado, pasando por la existencia. EGA Expresión Gráfica Arquitectónica **18**(21), 128–139 (2013)

Venturi, R.: Complejidad y contradicción en la arquitectura, 8th edn. Gustavo Gili, Barcelona (1995)

Venturi, R.: Iconography and Electronics Upon a Generic Architecture: A View from the Drafting Room. MIT Press, Cambridge (1996)

Venturi, R., Scott Brown, D., Izenour, S.: Aprendiendo de las Vegas, El simbolismo olvidado de la forma arquitectónica, 3rd Spanish edn. Editorial Gustavo Gili, Barcelona (1972)

Vidal, R.: La letra dibujada en prosa de Enric Miralles. Tesina de Máster en Tipografía Avanzada. Eina, Escola de Disseny i Art, Autonomous University of Barcelona (2010)

Zaldua, J.: Norbega S.A./Coca-Cola. Patrimonio Industrial en el País Vasco, vol. 1, pp. 223–227. Central publications service of the Basque Government, Vitoria (2012)

Zaragoza, I., Esquinas, J.: Enric Miralles: desdibujando límites entre re-presentación y proyecto. EGA Expresión Gráfica Arquitectónica **20**(26), 160–169 (2015)

Iberian Docomomo website. http://www.docomomoiberico.com. Accessed 20 Oct 2019

The Wright library website. http://www.steinerag.com/flw/Books/chicagosch.htm. Accessed 06 June 2019

The Archigram archival projects website. http://archigram.westminster.ac.uk/projects.php. Accessed 09 Oct 2019

Building the Walden 7. The Model as Patrimony of the Design Process

Raquel Álvarez Arce$^{(\boxtimes)}$ ⓘ, Noelia Galvan-Desvaux ⓘ,
and José Manuel Martínez Rodríguez

Valladolid University, Valladolid, Spain
`raquel.alvarez.arce@uva.es`

Abstract. The Walden 7 building is presented as a *rara avis* on the spanish housing scene of the 1970s. The Taller de Arquitectura team uses the model as a fundamental graphic tool throughout the entire project process, in order to develop the proposal of this unique building.

We will check how the chameleonic character of the model will define each phase of the project, with a different model capable of narrating in a different way each phase - work, structure or dioramas models, building the Walden 7 on a small scale. Therefore, the models are an intangible graphic heritage that shows us the project process, allowing us to reread the finished building from a new point of view.

Keywords: Taller de Arquitectura · Walden 7 · Model · Architecture · Project

1 Introduction

The Walden 7 house building by Taller de Arquitectura is presented as a *rara avis* within the Spanish architectural landscape of the 70s. The team headed by the figure of Ricardo Bofill, was formed by a multidisciplinary group, and was influenced by the thinking of the french *gauche divine*, the free love and May 68 ideas, and also by the architecture's international scene. Among the different architectural currents of the late 60s we can highlight the influence in the Taller of the structuralists, the thought of Louis Kahn or the ideas of the Team X, and in some level the architectural utopias of Archigram.

The Walden 7 building is located in the town of Sant Just Desvern, in the surroundings of Barcelona. At the beginning, the project in the outskirts of the town, since it is built on the old land of a cement factory called Samson. With this building, Taller de Arquitectura aims to put into practice the ideas that they had been developing since their publication *Towards a formalization of the city in space* in (1968). Taller de Arquitectura could not test their ideas in the previous project for the City in the space of Moratalaz, in the surroundings of Madrid, so Walden 7 was their first trial (Fig. 1).

This work aims to analyze the different models that appear parallel to the development of the Walden 7 project, showing that darwinian capacity of the model, which allows it to adapt chameleonically to each new circumstance (Carazo Lefort 2018 p. 168). This

L. Agustín-Hernández et al. (Eds.): EGA 2020, SSDI 6, pp. 108–117, 2020.
https://doi.org/10.1007/978-3-030-47983-1_10

Fig. 1. Photograph of the access façade of the Walden 7. Source: Autor's photograph

set of models defines a graphic heritage of the different phases of the Walden 7 project process since different purposes have always been producing different types of models.

We will also discuss another feature of the model, in terms of its ephemeral and provisional nature. There is no record of whether the different models that were used during the Walden 7 project process are physically preserved, since it is understood that architects cannot store multiple work models after their professional journey, due to the obvious need for space in the workshop or study, in addition to the inexorable passage of time. We will therefore use the photographs of the different models that have been found in the archives. The process of overflying over these images will take us into a parallel story of the creation of Walden 7, and will even bring us closer, not only to the pass history, but to that one that could have been (Bergara 2016 p. 34); this is because the models show us the possible forms that the building could have taken, the different structural systems and even the following phases of the project that were never built.

2   Taller de Arquitectura and the Modular Housing

The Taller de Arquitectura group is formed in Barcelona in the mid-60s around the figure of Ricardo Bofill, an architect who receives most of his academic training outside Spain, specifically in Switzerland. Alongside Bofill were Xavier Bagué, Ricardo's cousin who also receives his architectural training in Paris (although he will leave the team before the work of Walden 7), Peter Hodkinson, an English architect trained in the AA, where he coincided with some members of Archigram, Manuel Nuñez Yanowski, a stage designer born in Uzbekistan, writer Juan Agustín Goytisolo, Ricardo's little sister Anna Bofill, and the italian actress Serena Vergano. To this diverse group more members will be added along the course of the workshop, such as Salvador Clotas, writer and well-known Catalan politician or Ramón Collado.

The Barcelona of the late 60s lived more intensely the openness that was being experienced throughout the country of Spain, which made it easier for this young group to be influenced by the ideas of May 68. These ideas are reflected in the first sketches for the Walden 7 housing complex: the project sought to break with the traditional idea of family, proposing a modular housing whose size would respond to the number of individuals living there and to the new ways of life. Faced with the attitude of the political regime of the moment, that defended the traditional family structure, Taller de Arquitectura opts to propose homes in which the individual is the center of everything. In this way, the housing-study of an individual was the size of a module, which corresponds to an area of about 30 m^2. In the event that two individuals wanted to buy a home, the best option was to buy a two-module home, with an area of 60 m^2. This way, although there were intermediate solutions, in order to respond to the number of individuals that form the family unit, the number of modules that formed the house should be increased, with a maximum area of 120 m^2, formed by the 4 basic modules that were able to define a single study by themselves. Thanks to this modular housing system, the Taller presented their idea, defending that the basic unit of coexistence is not the family, but the individual.

Walden 7 is not the first experience of the Taller with modular housing. The team had already experimented with housing cells in their projects for Castell Kafka in Sitges, the Xanadu and the Red Wall in Calpe, and their unbuilt proposal of the *City in the space* of Moratalaz. However, Walden is the first experience in which a module was able to generate a home by itself. The combination of the different modules resulted in different types of housing for the different family units, and these in turn had to be grouped together to form the final set of Walden, and that is where the first models of the project appear.

3 *Ars combinatoria* or the Work Models

While the final model sends a more or less figurative message of what the architecture aspires to be, the work models are an intermediate state of the latent condition of their architecture and put the value in the representation of the methodology in architecture (Bergara 2016 p. 37 and 38) as well as showing us how the architect is working at that moment.

From the combinatorial of the different housing cells Taller de Arquitectura would establish a basic module consisting of 4 cells, attached two to two in the lower and upper part. The result is cube with which Taller de Arquitectura could "play".

As if they were the famous children's game of Froebel pieces that Wright always referred to, the Taller will look for different compositions, stacking the modules on each other, moving them on one of the axes at each level... in definitive, giving faith of that Huizinga's *homo ludens* to which the model is directly attached (Fig. 2).

In this part of the process the model becomes a working mechanism, its first function is not to communicate an idea, but to think, as Campo Baeza says, with the hands, although in this case it is not with a pencil but in a spatial way. Another architect who uses this way of spatially thinking is Herman Hertzberger, who, even he insists on the need to draw everything to understand it, also shows a particular interest in models, to complement that learning process from hand to mind (Rodriguez-Prada 2016 p. 101). For example, in the case of its project for the Beheer Central, Hertzberger takes as a structural unit the office units that, through its repetition in the three dimensions, produce the global image of the building (Rodriguez-Prada 2016 p. 109) (Fig. 3).

Fig. 2. Photograph of one of the cube models with which Taller de Arquitectura start to make different combinatorics. Source: Curtesy of Serena Vergano

Fig. 3. Photograph of the model that Herman Hertzberger made for the Forum magazine in 1959 in which the Dutch architect used matchboxes to represent the housing unit. Source: Nederlands Architectuurintituut http://schatkamer.nai.nl/en/projects/herhaalde-wooneenheden

In the case of the Taller, the structural unit is the basic housing module, from which the architects wanted to establish the image of their city in space. We can imagine that during this process Taller de Arquitectura would generate multiple models for different configurations in their search of the final Walden's image.

These models would show us the different configurations that Walden might have had. There is a photo of one of these models in which the set presents a shape that resembles a ziggurat, a configuration that, apparently, had always interested Ricardo Bofill (García Hernandez 2011 p. 225). The pieces form a series of columns that are approaching each other to form towers. We assume that this provision would be one of the first discards, since the final form does not resemble this first layout at all.

Gradually the different prototypes would show the rules that would mark the layout of the modules inside the building and their final configuration. Pieces that move on one of the axes, generating a kind of vertical sine curve that ends up defining the characteristic façade of the Walden 7.

These models of pieces would introduce other types of models, whose objective is not to distribute the modules of the houses, but to understand the way on how the loads are distributed (Fig. 4).

Fig. 4. Photograph of one of the cube models in which the arrangement resembles a Zigurat. In the model we can see the first displacements of the modules on one of the axes. Source: García Hernandez (2011) *La agregación modular como mecanismo proyectual residencial en España: El taller de Arquitectura* Tesis Doctoral. Universidad Ramón Llull p. 225

4   Strings and Nots

Another advantage of the model is that it can also serve as an exploration process for the construction and structural design (Carazo 2014, p. 64). This kind of models, like the ones with the cube combinatorics, are still models of the process since it helps the architect visualize how the building is sustained, in short, it helps the architect to think about the "bones" of the building (Fig. 5).

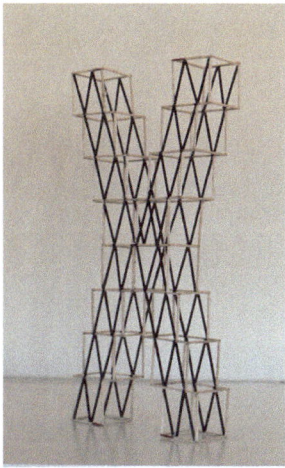

Fig. 5. Photograph of one of the structure models in which Taller de Arquitectura analyzes the efforts of the layout of the main elevation of Walden 7. Source: Curtesy of Serena Vergano

The photos of this part of the process's models put us in direct contact with the famous string models of someone who Ricardo Bofill has repeatedly shown his admiration, the catalan architect Gaudí. The images show us how the massive volume of the building is reduced to lines that, in this moment, are telling us about the support of Walden 7.

However, Gaudí's models were prior to the layout of the building, since the architectural form is subject to the structure, which arises, in this case, as a result of an internal and autonomous process only dependent on the mechanical laws to which the resulting geometry is contingent (Ubeda Blanco 2002 p. 141). In the case of Walden 7, the structure models respond to the search of the building support system, after the formal configuration is decided in the previous phase of the project.

The models seem to propose large metal trusses, in which the elements of the structure are classified by colors: wood for the vertical and horizontal elements, black for the diagonals that work under compression, blue for the diagonals that work under tension, and finally red for flights that flex.

There are other photos of the models in which the efforts of the different modules arrangements are analyzed. Photos of the cross arrangement that defines the main elevation of the building and also of the interior elevations that define the central courtyards. String and knots help the architect understand the efforts of the elements and they act as another work tool.

This analysis is carried out by Taller de Arquitectura and Joan Margarit, who was a doctor architect of structures in Barcelona's School of Architecture. Taller de Arquitectura tried to define a steel structure that would be able to support the building. But despite this laborious process of work, a reinforced concrete system with some steel reinforcement would be chosen (García Hernández 2011 p. 151) with which Walden 7 has finally been built (Fig. 6).

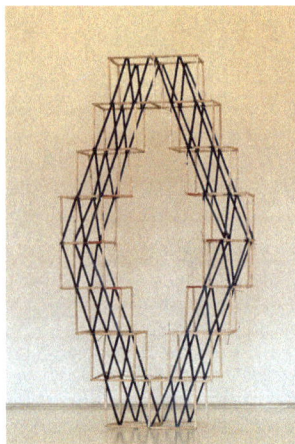

Fig. 6. Photograph of one of the structure models in which it is observed how Taller de Arqui-tectura classifies by color the different elements of the structure, according to the efforts to which they are subjected. Source: Curtesy of Serena Vergano

5   Testing and Dioramas

We could say that the last models of the process are those to checkup what has been design. As a way of making sure, the Taller seems to build these models to see if what supported the paper was also supported by reality.

One of these models shows the configuration of the Walden 7, with its streets in the air, balconies, and stairs. This model does not seem to be the one for the divulgation of the project, since it is unfinished. Only half of the elevation presents all the elements of the facade, in which we can see the balconies and the windows of the houses, the walkways of the common spaces... The opposite elevation has only modeled some walkways and stairs of the lower half of the building. Despite this, the model allows us to imagine the final appearance of the building. We could define it as an trial, as a preview model, where photography becomes an analytical instrument of the project process, the image checks asserts or rejects the architect's conceptual expectations (Bergara 2016 p. 38) before making a final decision (Fig. 7).

The one that looks like the last model of the project is the only one that shows us an utopian reality. It is the representation of the whole of the Walden Island. The building we know today as Walden 7 was the first phase of a complex in which, as we have said before, Taller de Arquitectura wanted to test their theories about the city in space. After the construction of the first phase, two other towers of similar shapes were to be built, connected by two linear blocks of smaller height that closed the contour of the plat, overlooking the main road. The interior spaces of the plat were to be occupied with a garden. This space was also occupied by the remains of the old factory, which the Taller intended to reuse turning them into common spaces for the neighbors, occupying them with music halls, libraries, and even swimming pools. One of these old constructions is the one that has finally been destined for Ricardo Bofill's house-studio, and which the architect has named *La Fábrica*.

Fig. 7. Model of the Walden 7 in which only half of the elevation is represented completely. Source: Curtesy of Serena Vergano

This final model shows the buildings, the garden areas and the surrounding streets with tiny cars. The model could be defined as a diorama, understood as a realistic representation, almost kitch (Carazo 2018 p. 169) of the imagined by the Taller. This model's aim is to seduce the future buyer of a Walden's home, showing him a city in space in the municipality of Sant Just Desvern, where he can enjoy an innovative home surrounded by green areas (Fig. 8).

Fig. 8. Full model of the Walden Island. The model shows cars, vegetation… a diorama to convince a future buyer of a Walden home. Source: Curtesy of Serena Vergano

6   Multiple Realities

The models of the Walden complex form an important graphic heritage about this work of modern Spanish architecture. These models allow us to know more about the project, giving us information about its conception, and about the possible realities that could have been carried out. An intangible heritage, the project process, which is supported by the graphic heritage formed by the photographs of the project's models.

Fig. 9. Photograph in which we can see Serena Vergano taking photos of the model at the construction site. In the background, it can be appreciated the final phase of the construction of Walden 7 and Ricardo Bofill's house-studio, which was already under construction. Source: Ricardo Bofill Archives, Curtesy of Pedro Garcia Hernández

But the value of these models, and their photos, increases, since they not only give us information about the project process, but also about realities that can no longer exist. Although we can find the plans of the complete proposal in the historical archive of the municipality of Sant Just Desvern, the model allows us to see how this city in space could have occupied the old land of the cement factory. The photographs of this model show the existence of an unbuilt modern architectural heritage, whose only and most reliable expression is the photographic report, which constitutes a bank of images of the side B of the history of our modernity (Bergera 2016, 38) (Fig. 9).

In one of the photos of what looks like the end of the construction of the Walden 7 we can see Serena Vergano photographing the model of the Walden Island set in the worksite. Of this report are the "photomontages" in which the proposal of the set is superimposed on the built Walden 7. A graphic heritage of an impossible reality that invites us to ask if these photos were the last attempt of the Taller to show their proposal, of that dream imagined during the 60s, of their city in space.

Acknowledgments. The authors would like to thank Serena Vergano and Pedro García Hernandez for handing over the photographs that have made this research possible.

References

Bofill, R., Goytisolo, J.A., Ponç, J.: Hacia una formalización de la ciudad en el espacio Editorial Blume; Barcelona (1968)

Bofill, R., James, W.: Taller de Arquitectura. Ricardo Bofill, Taller de Arquitectura: edificios y proyectos, 1960–1985 Gustavo Gili, Barcelona (1988)

Carazo Lefort, E.: La maqueta como realidad y como representación. Breve recorrido por la maqueta de arquitectura en los 25 años de EGa Expresión Gráfica Arquitectónica EGA 25 años número 34, pp. 158–171 (2018)

Carazo Lefort, E., Galván Desvaux, N.: Aprendiendo con maquetas. Pequeñas maquetas para el análisis de arquitectura. Expresión Gráfica Arquitectónica EGA número **24**, 62–71 (2014)

García Hernandez, P.: La agregación modular como mecanismo proyectual residencial en España: El taller de Arquitectura Tesis Doctoral. Universidad Ramón Llull (2011)

Norberg-Schulz, C., Futagawa, Y.: Ricardo Bofill: Taller de Arquitectura Tokyo ADA Edita Tokyo (1985)

Rodriguez-Prada, V.: La generación del estructuralismo holandés a través de sus maquetas. El caso de Herman Hertzberger, 1958–1968 PPA Maquetas, no. 15, pp. 101–111 (2016)

Solé i Ubeda J.L., Amigó J.: Walden 7 i mig Ajuntament de Sant Just Desvern; Barcelona (circa 1998)

Ubeda Blanco, M.: La maqueta como experiencia del espacio arquitectónico Valladolid Secretariado de publicaciones e intercambio editorial de la Universidad de Valladolid (2002)

Bergera Serrano, I.: Retratando sueños. Fotografías de maquetas de arquitectura moderna en España PPA Maquetas, no. 15, pp. 30–41 (2016)

Carlo Scarpa. Negozio Gavina
Time Represented in Two Acts

Alberto Grijalba Bengoetxea(✉) and Julio Grijalba Bengoetxea

Valladolid University, Valladolid, Spain
agrijalb@arq.uva.es

Abstract. Moneo and Cortes select the drawings of the floorplans and the sketches for the Negozio Gavina, 1961–63 by Carlo Scarpa, a project in which the idea of a fragment sets up a whole, to include them in the book *Comments on 20 drawings by current architects.*

Scarpa transmits in these sketches the idea that the fragments possess a unity of their own while at the same time they are capable of keeping an intact memory of the irredeemably lost order of the model, that we regard as completely impossible to be reconstructed.

In the sketches we discover the traces of a creation process that remain etched in the building. It represents the acknowledging of time and personal memory as the architect's work material and their graphic representation.

The nature, the position, the materiality and display of the five columns that make up the space are a very good example.

Keywords: Time · Sketch · Process · Fragment · Urban landscape · Heritage

1 Introduction

Moneo and Cortes choose the sketches for Negozio Gavina, 1961–63, by Carlo Scarpa to be included in their book *Comments on 20 drawings by current architects.* From a close reading of the book this may seem as a surprising inclusion since, for the authors, in their seeking for "what is essential to Architecture Representation", the sketches are considered too subjective, too intimate; something that belongs to the creative process. Moneo and Cortés aim was to "show the way Representation in Architecture holds within, though in some ways restricted, its future reality". This way they try to exclude something as subjective as creative processes from the disciplinal in Architecture.

However, they find it very difficult not to notice the depth irradiating from the drawings by Scarpa. They are the only sketches included in the book (Moneo and Cortés 1975).

1.1 The Five Floor Plans

In their book, Moneo and Cortés only include five floorplans numbered 1 to 5. No perspective or other detail is given, the authors do not seem interested in elevation and

© The Editor(s) (if applicable) and The Author(s), under exclusive license
to Springer Nature Switzerland AG 2020
L. Agustín-Hernández et al. (Eds.): EGA 2020, SSDI 6, pp. 118–127, 2020.
https://doi.org/10.1007/978-3-030-47983-1_11

section drawings that shape the image as well as the space. They rely on the floorplans as the essential tool of architecture to solve what they refer to as "the problem", which was no other than the structural change from a linear system to a one consisting of pillars; seeking for a figurative fluid space which was what the Gavina business needed to display their produce. The problem was solved by overlapping the two realities, the current one and the one we are aiming for at the same time. This overlapping is always regardful of a specific reality, from the constructive to the material (Fig. 1).

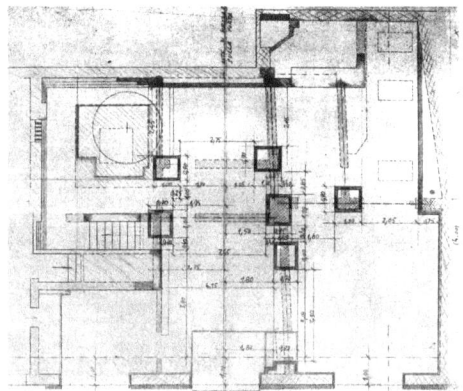

Fig. 1. Sketch Carlo Scarpa *Negozio Gavina* 1961–63. Bologna, Italia. Sketch 1

1.2   In Between Technicality and Ideation

Moneo and Cortés do not intend to make a narrative discourse of the project process. They do not try to make an hypothesis on the probable time line either. They number the sketches according to the publication date that is all. The first two ones are not proper sketches, since they simultaneously show the definition of the future reality technical drawing and the hand drawing. The technical drawing provides the dimensional reality, the constructive reality and the spatial reality, whereas the hand drawing, in soft pencil and in color convey the necessary imprint of the creative drawing, of the seeking drawing; of the act of thought, of an unhurried reflection at the drawing board (Fig. 2).

The other 3 floor plans do show the features of an ideation sketch despite their draft planimetry base. The floor appear as past remains in some of them, another one is a study of the program, but in all of them the solution to the problem of structural change and its relation to space seems to slowly be getting into shape, and eventually the pillars get to that figurative state Scarpa was seeking for (Fig. 3).

1.3   Space and Ceiling

Finally, in the sketch n 2 a surprise awaits us; any constructive element, any reference to existing walls, any hint of the treatment of the floors, have disappeared. The analysis

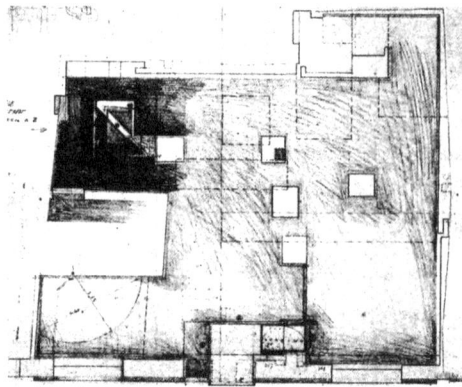

Fig. 2. Carlo Scarpa *Negozio Gavina* 1961–63. Bologna, Italia. Sketch 2

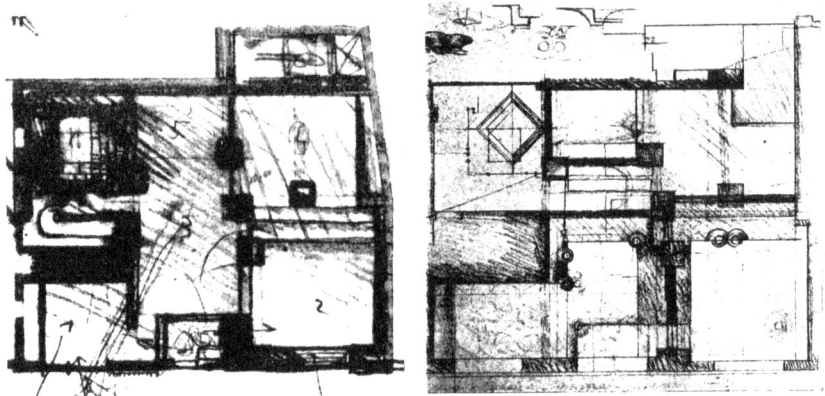

Fig. 3. Carlo Scarpa *Negozio Gavina* 1961–63. Bologna, Italia. Sketches 4 y 5

accurately refers to it as a continuous space drawing. But we believe there is something more to this floor plan. Under the freehand shading that fills the space there is a second structure drawn in dots. It is nothing but the remains of the beams and the horizontal support systems. Two different times overlap.

2   First Act. The Fragment

It is in the 50s that Scarpa introduces the concept of fragment in his work. With the inclusion of these fragment elements he develops his determined personal poetry. New signs and new formal suggestions turn his works into a peculiar field of expression. The space structure seems to disappear giving way to multiple, entwined and open paths.

In 1950 Scarpa restructures two small businesses in Venice: the boutique "A la piavola de Franza", and the very interesting antique shop "Ongania". We could say that it is in this two works when he starts using the multi-referential fragment as a creative method.

In the first one he achieves to create a new space with very few elements, in the second one he anticipates the way in which geometry and new non structure interweaving result into open arrangements.

At first sight the floor plant reminds us of the one for Negozio Gavina. Located in the archway of a street adjacent to Piazza San Marcos, it extends along two bays. Scarpa restores the original spatial structure by overlying an oriental inspired multi-barred frame that seizes all the space. We can appreciate the remains of a previously existing building at the same time that the new fragments take and enhance the space offering us a completely new sight (Tafuri 1984) (Fig. 4).

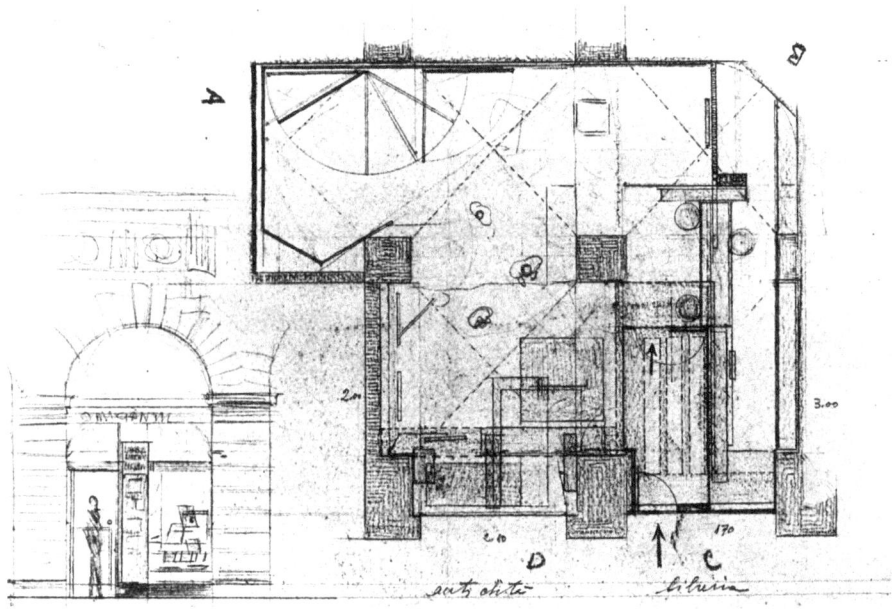

Fig. 4. Carlo Scarpa *Negozio Ongania* 1950. Venezia, Italia. Floor plan

2.1 Nostalgia, Lost Order and the Non-finite

In what Scarpa proposes, there is an abundance of fragments and of indetermination of the open paths which evokes the particular, the unique. So much so that Tafuri suggests understanding Scarpa's works with a wistful tone as a modern allegorical translation of a baroque memory (Tafuri 1984).

Through his work we can perceive the yearning for an ideal lost order, a bereavement act. The fragment suggests something has been irredeemably lost, it is an invitation to nostalgia at the same time that it conveys its uniqueness within the entirety. What is finished and the "non-finite", both, question the pretended stability of architecture and let the vulnerability of the world exposed. This leads us to regard history not according to permanence and unity, but as a succession of irreversible stages. It is in this context that the fragment appears replete with a legion of highly enriching values, which certifies its dramatic and unforeseeable quality.

Scarpa tries to convey, particularly in Negozio Gavina, the idea that the fragments are a unity in their own at the same time that they are capable to keep an intact memory of a reference to an irredeemably lost order, assuming that it would be impossible to be rebuilt.

The view of the whole would be created by means of fragmented and multiple partial glimpses. It is the recognition of time and history as factors in the creative process. The observer assumes a new roll, articulated around a subjective and partial approach, in which the new views and frames of reference acquire nuances hitherto unnoticed (Linazasoro 2013).

2.2   On Ground Level and Vertical Elements

In the book by Moneo and Cortés, there is no chronological order or date we can use to identify the floor plans, it is impossible to tell which is the one first conceived. The creation process order has never been established in any of the publications of the images. Not even the floor plans published in 1962 in *Zodiac and Domus* magazine are the same ones include in the book *Comments on 20 drawings by current architects.*

Nevertheless, we can assume that the first to be created, the seed to the project is the one that appears under number 3. The reason behind this assumption is that in this sketch we can identify all the elements present in the ultimate floor plan: the two simultaneous times, the structural change and the seeking for a geometrical order in the plan (Fig. 5).

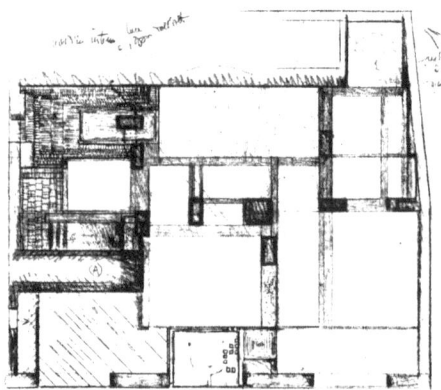

Fig. 5. Carlo Scarpa *Negozio Gavina* 1961–63. Bologna, Italia. Sketch 3

In Sketch 3 we can appreciate how the lineal elements are replaced by pillars, but they still pose a memory of the walls they substitute for. They look like fragments of the existing walls, elongated pillars that go along with the beams structure. There are nine of them, but only one of them is square. These fragments, together with the orthogonal geometry connecting them, have not got free of the spatial memory of the preexisting layout yet. Those fragments are dependent on the past and even follow the outline of the walls of Negozio Gavina.

The almost final sketch, the number 1, has gone beyond this dependence stage. It is a drawing of the actual layout with all the walls and partitions, we can appreciate just six pillars, one of which would be later removed; these fragments are no longer elongated in memory of the elements they substitute for. They have become independent fragments, since only the connection to each other is marked down. The outline of the segments of the existing walls that have been kept subtlety differ from the one of the ones suppressed. The fragment reaches the boundary walls. Some of them are coated, some are standardized, and some others remain in place. The two dimensional concrete sheet of the elevation appears as an independent element for first the time (Fig. 6).

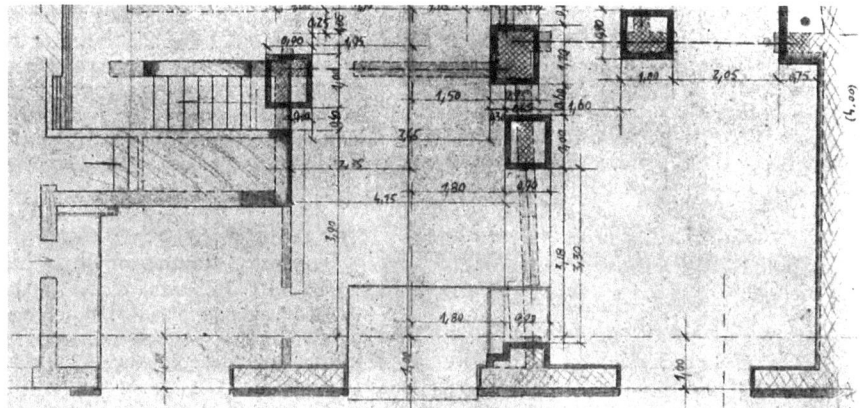

Fig. 6. Carlo Scarpa *Negozio Gavina* 1961–63. Bologna, Italia. Sketch 1. Detail

2.3 A New Fragmented-Poem-like Order

This extreme element simplification reflects on the powerful presence of the new five pillars. They represent the new order both cryptic as well as soothing that seems to shroud a mystery; a mystery pending to be deciphered.

It is in the same sense that we can say a fragmentary poem is not the same as an unfinished poem, but that it opens to another way of knowledge and that it cannot be related to oneness (Blanchot 1969) In the project for Negozio Gavina what seems fragmentary constitutes a narrative, that goes beyond its fragmented lay out, and that is what defines Scarpás way of doing.

3   Second Act. The Process

3.1   Trace and Path

Just as Eisenman reread Terragni's projects considering time, the process and their representation as part of the analytical discourse (Eisenman 1998) our aim is to disclose the traces of a similar process in Scarpa which remain etched in his architecture.

This is why we introduce the terms "trace" and "path" as remains of the dynamics the process and generation followed when designing the projects. These words refer to the idea of change, thus, to the perception of time. A past time restrained in the remains and their traces, but not less important or less explicit when we try to get comprehensive understanding of his body of work.

It is a palimpsest where we can discern the gone load bearing walls, the series of new spaces and the presence of the five pillars.

3.2 Trace. Kintsugi and Time

The process in which Scarpa selected the fragments, the multiple entwined open paths and the overlapping of times, is deeply linked to the generation of the floor plan. The Sketches 2 and 3 are the evidence.

The sketch 3, which we identify as the initial one, focuses to a large extent on the vertical structure elements as we have already analyzed. The orthogonal geometry intertwining the wall fragments is very evident. This presence may be regarded as a geometric help, a search for certainty in the ideation process- But there is something else. If we look into the upper left of the drawing we notice that the space is more subtly treated, it appears more definite. The entrance, still with an odd shape, gets organized and defined. We can see this as a clear sign that Scarpa, at the same time that he is reflecting on the structure vertical fragments, is also projecting the space and the ground level. Just as this inherited geometry helps him transit from an existing reality to a future one, it turns into a path, a transformation, a memory of the past.

The walls have disappeared, but their remains are visible. It is a kintsugi, that oriental technique so appreciated by Scarpa, in which breakage and repair are part of the project he himself drew (Albertini and Banoli 1988). They are part of the same memory of space. As an oriental craftsman, he wants to repair what has happened, he does not want to strip away the breakage, and he wants to show it, to incorporate it to his project making the transformation more evident. This is his way to show beauty of the traces of the process linked to the passing of time.

He soon abandoned this idea, none of the other known sketches of this project reflects it. The floor, of brown PVC slabs, only shows the subtle memory of this stage in the recreation of the oversized pillars with an exquisite stone frame. But time and memory are still present in the space, in the delicate ceiling of the sketch 2 of Negozio Gavina.

3.3 Trace. Path and Section

In the sketch 2, apart from showing a continuous space, with an open path marked in relation to the pillars, we can appreciate, overlapped, the horizontal ceiling structure. Scarpa is still reflecting on the two realities. Once abandoned the exquisite Kintsugi proposal, he still feels the necessity to make the trace evident. His implementing proposal consists on reflecting the memory of the conformation of the ceiling structure. There is a subtle memory of the original walls. The ceiling does not only mirror the space, it is a surface that shows its scars in their transformation and memory. It is a ceiling plan.

If overlapping of vertical systems becomes evident in the sketch 1, the overlapping of horizontal systems can be also appreciated. Only the pillars and the walls leave their traces. Despite its graphic design and at the same time that the architect studies the possibility of opposing the flat surface of the floor to a changing spatial reality, on the ceiling we can see is a sketch where the two realities also overlap.

Scarpa anticipates the spatial solution: the ceiling plan. This is a proof that he trusts the ability to create the figurative space he is seeking, not only the pillars, as fragments. Scarpa, with the overlapping of times and realities through the fragmentation of the ceiling finally accomplishes his task to create a figurative landscape. He does all this with his peculiar way of representing all in a single plan. (Los 1994).

The unusual narrative of the creative process and its representation by Scarpa is, thus, told as a landscape of happenings in past and present times that emerge simultaneously. As Paul Virilio said: "It is now a matter of urgency that we reform the whole dimension of general history so as to make way for the fractal" history of the limited but precisely located event" (Virilio 2000) (Fig. 7).

Fig. 7. Carlo Scarpa Negozio Gavina 1961–63. Bologna, Italia. details

4   The New Reality

Scarpa proposes an architecture where there is no room for chrono-philia, nor for chrono-fobia. Different times share the same space. The way he projects, trace and path are evident, and fragments as something unfinished, recover a soothing, wistful order, just like his admired L. Kahn did (Marciano 1985).

In Gavina, he turned the vertical elements into the oversized main focus of his work. As he himself stated, space is defined by physical phenomena, substance and gravity (Scarpa 1984a). This is why he opposed the whitewashed walls to different pillars in different colours and materials: blue cobalt stucco, mortar, silver leaf and black plastic. The black plastic one has a bilobed perforation, like the one in the facade, this is a game to help appreciate the heaviness of the structure and how oversized the pillar is (Fig. 8).

Fig. 8. Carlo Scarpa *Negozio Gavina* 1961–63. Bologna, Italia. Office and perspective of the perforated pillars

As we have analyzed, spatial relationships dynamically occur between the open spaces, the polychrome pillars, the intertwined paths and the traces on the ceiling. Gavina is a clear example of architecture of fragment and of Scarpa's need for a vertical architecture. He once declared himself a bit Byzantine deep inside, at a conference when praising Josef Hoffmann (Scarpa 1984b).

5 Conclusion

Scarpa proposes an open formal articulation in Negozio Gavina, with its interrupted phases of elements But when we confront the work as a whole, those elements, the trace, and the path change.

The multiplying of elements, the dissemination of his image discourse, the materiality stimuli, the overlapping of colours and the intertwined paths make u p a landscape. Landscape in two acts.

The emerging of the walls turned into pillars, their labyrinthine layout, the dimensional relation game, and the subtle visual tensions they all create, change the way we see the fragments that we regard as figures. A landscape of figures with endless constellations, as Manfredo Tafuri defines Scarpa's body of work.

References

Albertini, B., Bagnoli, S.: Scarpa: Architecture in Details. Architecture Design & Technology, London (1988)

Blanchot, M.: L'Entretien infini Gallimard. Madrid (1969)

Tafuri, M.: El Fragmento la Figura, el juego. En Dal Co, F., Mazzanol, G., Polano, S.: Carlo Scarpa 1906–1978. MOPU-Electa. Milán (1984)

Eisenman, P.: Giuseppe Terragni: Transformations, Decompositions, Critiques. Rizzoli, New York (1998)

Linazasoro, J.I.: La memoria del orden. Parábolas del sentido de la arquitectura moderna. Abada, Madrid (2013)

Los, S.: Scarpa, Taschen, Köln (1994)

Moneo, R., Cortés, J.A.: Comentarios sobre dibujos de 20 arquitectos actuales. ETSAB, Barcelona (1975)

Marcianó, A.F.: Carlo Scarpa. GG, Barcelona (1985)

Scarpa, C.: Leccion inaugural, 1964–65. IUA de Venezia. En Dal Co, F., Mazzanol, G., Polano, S.: Carlo Scarpa 1906–1978. MOPU-Electa. Milán (1984a)

Scarpa, C.: Conferencia en Madrid, 1978. En Dal Co, F., Mazzanol, G., Polano, S.: Carlo Scarpa 1906–1978. MOPU-Electa. Milán (1984b)

Virilio, P.: A Landscape of Events. MIT, Cambridge (2000)

The Colors of Porto: Watercolor as a Technique of Representation of Porto's Architecture

Marta Úbeda Blanco[1(✉)], Daniel Villalobos Alonso[2], and Sara Pérez Barreiro[2]

[1] Architectural Graphic Expression, Urban Planning and Territorial Planning, University of Valladolid, Valladolid, Spain
martaubeda@gmail.com

[2] Theory of Architecture and Architectural Projects, University of Valladolid, Valladolid, Spain

Abstract. Colour is one of the elements that most strongly marks the character of a city. And if something characterizes the city of Porto, it is its colours. From the explosion of colour on the façades of its houses, to the blue reflection of its tiles, through the red of its roofs, the green of the moss that is accumulated on its stones, to the black and white of the cobblestones that extend along its pedestrian pavements.

The colours of its streets, the graffiti of its artists who paint everything from walls, to electricity boxes and facilities. Everything is colour. Colours that change with the light of day and night, the light of the seasons, the reflection of the city in Douro's waters and the influence of the fog and the humidity that comes from the Atlantic. Everything in Porto refers to water and light broken down into colours. Therefore, we represent Porto's architecture with watercolour technique, using these same elements.

Keywords: Porto · Colour · Light · Water · Watercolour

1 The Colors of Porto: Watercolor as a Technique of Representation of Porto's Architecture

Although there are multiple elements that define the city of Porto, if we had to choose the ones that characterize it, we would say without a doubt that they are its colours. In the architecture of Porto, and paraphrasing Le Corbusier, "there are no more elements than light and its façades that reflect it" (Le Corbusier: Towards an architecture).

The colour of Porto seems to have been taken directly from Newton's experiment, since his light seems to split up into thousands of colours that splash throughout the city. Thanks to the light, a close relationship is established between the colour and the architecture of the city. This connection provides chromatic visual sensitivity, and causes visual pleasure, because "each pure colour has a register and a timbre, like any instrument of an orchestra" (Polano 1989, p. 101).

© The Editor(s) (if applicable) and The Author(s), under exclusive license to Springer Nature Switzerland AG 2020
L. Agustín-Hernández et al. (Eds.): EGA 2020, SSDI 6, pp. 128–141, 2020.
https://doi.org/10.1007/978-3-030-47983-1_12

Porto is very close to the water. It is next to both the Atlantic and its Duero river; a river we share and that when it flows into Portugal, its name changes to Douro. Thanks to the amount of rainfall that bathes Porto, it appears with a special whitish light that decreases the intensity of its shadows during most of the days. You have to wait for the clouds to vanish to be able to see the contrast of shadows on its facades. And as Tanizaki would say: "beauty loses its existence if the effects of the shadow are suppressed" (Tanizaki 2009, p. 69), a "shadow that provokes inexplicable resonances" (Tanizaki 2009, p. 36). The sunlight with its natural movement produces shadows and effects that make the facades shine, the colours break down into all their tones and create reliefs and contours that offer a volumetric view of the city.

The cause may be in the abundance of water in this river. It acts as a dissociative prism of colours that remain trapped in the Portuguese architecture, doubling the city in its reflection and painting it on its surface.

This is how we propose to capture the image of the city of Porto, with colours and water that are the main qualities of the watercolour technique.

1.1 Analysis of Porto's Colours

As we began to prepare the color palette to paint the beauty of this city, we saw that there were certain colours that were constantly repeated. It was then that we faced the need to proceed with its analysis and classification.

Two different parameters can be considered: natural and artificial. We find natural colours, in those that come naturally from the own materials and do not respond to any human action. On the other hand, colours that are artificially applied or superimposed on the architectural elements.

1.2 The Color that Is Painted: Applied Colours

We call applied colours to all that are not on the material they colour, but are painted afterwards. We look at that colour palette that we see on the façades of the houses of one of the oldest neighborhoods in the city. La Ribeira, a medieval quarter which with its reflection dyes the Duero with colour, Fig. 1.

Fig. 1. Watercolour of Marta Úbeda. Houses of La Ribeira, Porto 2019.

Despite having been a remodeled area, it still retains its medieval structure of old houses that accommodate the steep and narrow streets. A maze of stairways and alleys that ascend up the hillside while the remains of the Fernandina wall embrace them. With its scale and proportions, its overlapping roofs, the colour of its façades and balconies, the rhythm of its colourful doors and windows, they appear as a great backdrop in an open-air theater, where its inhabitants and city visitors are the actors on stage. The walkers are integrated into this great masterpiece that causes a mirror image in its reflection on the water.

Since colour depends on the light, the reflections on the Douro depend on it as well. The reflection also creates perceptions of shape, scale, colour, materials, chromatic and volumetric contrasts on the appearance of the city.

The colors of its doors and windows, its balconies and its clothes stretched out like flags moved by the wind; rightfuly names the place where they are found, "La Pena Ventosa".

Thie beauty is found in the simple harmony of these vivid colours that seem to rise from the sun when crashing against the façades. They are absorbed, in order to catch in a diffuse clarity, the deep concordance between their different shades: "the beauty is not a substance itself, but only a drawing of shadows, a game of chiaroscuros produced by the juxtaposition of different substances" (Tanizaki 2009, p. 69).

1.3 Blue and White

Blue is another characteristic colour of Portugal and therefore of Porto, and along with the white colour we also consider it as an applied colour since its metallic pigments are added as a paint on a tile. Although it is a ceramic material, in turn, it is placed on the smooth surface of the facades.

Historically, the use of tiles in Portugal dates back to the fifteenth century. It is a tradition that was initially functional, safeguarding its facades because it is a country that is very exposed to moisture due to its proximity to the ocean and its abundant rains. Covering the facades with tiles made them more waterproof. This tradition over the years ended up becoming an art.

"Azulejo", a word that comes from the Arabic *az-zulaiy*, which means brick, and which in turn comes from *zallaja*, that in other words means to leave smooth and slippery, is considered an Arab invention. But, it was already used by the Assyrians and the Persians. From the Spanish-Arab palaces, and especially through Dutch potters who worked for the Portuguese aristocracy, the white walls of the manor houses, churches and palaces of Portugal began to be decorated.

In the beginning of the usage of this chromatic system, the work was more colourful. It was influenced by the Chinese porcelain of the Ming dynasty (1368–1644), which brought a touch of distinction and elegance to the buildings.

However, this technique was economically very expensive, only feasible for the noble class and wealthy aristocrats. Hence, the typical Dutch Delft tile, characterized by having only two colors: white and blue, began to be used, thus, reducing the cost significantly.

Thanks to this tradition, strolling through Porto is like walking through tiles that tell "pixelated" stories and traditions in white and blue. Large "puzzles" such as those that appear on buildings that are very representative of the city such as the Cathedral, whose cloister is decorated with tiles of scenes from the Metamorphosis of Ovid. The Sao Bento Station, has 20,000 tiles in its lobby it is a work of Jorge Colaço. It includes scenes from the history of Porto and murals with country themes, allegories about the railroad and geometric and stylized elements, typical of Art Nouveau. The chapel of Las Almas decorated with 16,000 tiles by Eduardo Leite, showing religious scenes, or the facade of the church of Los Carmenes, a work of Silvestre Silvestri, where the foundation of the Carmelite order is narrated. These are some of the many examples that can be found in the city, Fig. 2.

Fig. 2. Watercolour of Marta Úbeda. Mosaics of the Cloister. Cathedral of Porto. 2019.

1.4   The Colour of the Material

The rest of the colours of Porto express the sincerity of the materials, that is, they are natural colors and therefore are not applied, but found in the materials used for the construction of the buildings.

Red. The red of Porto is found in the most usual materials of its daily architecture, in the bricks and especially in the tiles. The geographical shape of the city allows its contemplation, because it sits on a rocky hill, and as you ascend its slope it is impossible not to admire the sea of tiles that extends to the river. Thus, when looking at the city from the highest areas we can see a large red carpet made up of hundreds of small roofs that stain the city red, Fig. 3.

Fig. 3. Watercolour of Marta Úbeda. Rooftops of Porto. 2019.

Cream. This delicate colour appears on the stone façades of the main buildings of Porto. The vast majority are dependent on the Gothic style, and reminiscent of English style. Others represent magnificent examples of Art Nouveau, such as the famous Café Majestic, founded in 1921. With its marble façade, its columns and stylized decoration, it represents the bourgeois atmosphere of the time (Fig. 4).

It is also found in the remains of the Fernandina wall built in the fourteenth century. It was completed during the kingdom of D. Fernando (1730). It is nine meters high and crowned with triangular battlements. The wall surrounded the city and extended towards the river. It was guarded by several square-plan watchtowers. It began to be demolished in the eighteenth century due to the new expansion of the city limits. There are still

Fig. 4. Watercolour of Marta Úbeda. Majestic Café. Porto. 2018.

several parts remaining, such as Trecho dos Guindais, a tower, Torre do Barredo and a door, La Puerta de Carvào, Fig. 5.

Murelle Teuandina
Monasterio da Serra do Pilar
Porto . marzo 2019 —

Fig. 5. Watercolour of Marta Úbeda. Remains of the Fernandina Wall, Porto 2019.

1.5 The Colour of the Metal

We can see the differences between two types of color that are associated with the metals of this city: Gray and Green.

Gray. Porto is known as the city of bridges, and its metallic material brings a new color: gray, represented in the most characteristic image of the city, with the large iron bridge at its feet.

The first bridge that was built in Porto was the Ponte das Barcas, dating back to 1806. It was made up of twenty wooden boats tied together with steel cables. Despite its simplicity, it could be opened to make way for river traffic. It disappeared in 1809, when Napoleon's troops invaded the city, and the weight of the population that tried to flee chaotically by crossing the bridge caused its destruction. This bridge was replaced by the Pensil bridge or suspension bridge of Doña María II. You can still see the feet of the two pillars on which two obelisks were hung and the guard's house on the bank of the Ribeira.

The iron bridge, the first one in Porto, was built in 1886 on the remains of the Doña María II bridge. Dedicated to Don Luis I, it was designed by Teófilo Seyrins, a disciple of Eiffel. At the time, it was the longest bridge in the world with two overlapping boards and an arch of 172 m, Fig. 6.

Fig. 6. Watercolour of Marta Úbeda. Bridge of Don Luis I. Porto. 2018.

Green. We mainly find this colour in the rust of the statues, such as the equestrian of Count Vimara Peres, the reconqueror of the city. He was considered the Portuguese Cid. In 868 he expelled the Moors, controlling the northern region between Miño and Duero. This region was the first county in Portugal and is considered the foundation of Portugal as a country. Vimara Peres was appointed governor of the county as a reward for his deed. He gave the name to the city of Vimarais, the current Guimaraès. The copper statue of this character of historical significance presides over the space that precedes the cathedral, Fig. 7.

Fig. 7. Watercolour of Marta Úbeda. Vimara Peres. Porto 2018.

But there is another green that draws attention to our watchful eye. The colour of the moss that adheres to the stone in walls, façades, walls and roofs of Porto. This tiny vegetation that very slowly, little by little and without rest, colonizes the stone in an attempt to return the great windy Pena Ventosa to its original appearance, the one it had before the appearance of man upon it. The moss is what gives the city one of its most characteristic colours: green, Fig. 8.

Fig. 8. Watercolour of Marta Úbeda. Fountain next to the cathedral. Porto 2018.

Adding the green of the lawns that cover a large part of the Pena Ventosa, or the one that is rooted upon the roofs of some buildings such as the San Sebastiao market. The concrete platform structure on its roof, makes it appear like a stepped garden in front of the Sé Catedral, Fig. 9. It is safe to say that green is the main color of Porto.

Fig. 9. Watercolour of Marta Úbeda. San Sebastiao Market. Porto 2018.

1.6 Black and White

The last colours we will refer to are black and white. We can see them when we look down, since much of the ground of the city of Porto is upholstered with stone pavers of these two colours. The pavement of the city seems to follow the recommendations of the neoclassical designer Peacock, who suggested using darker tones for the pavement: "… for the ground a deeper tone, imitating the carpet of nature" (Cruickshank 1989, p. 92).

This type of exterior flooring called "Calçada Portuguesa" is a type of paving stones composed of limestone and basalt that adorns pedestrian areas as a mosaic. It is typical in Portuguese-speaking countries and forms decorative patterns using the colour contrast of diaclases or cobblestones.

This technique emerged in Lisbon in the 19th century. It was a work done by the inmates of the Castillo de San Jorge, later called "mestres calçeteiros". The paving was done not only to promote its elegance, but also for functional reasons. It is a type of pavement that is applied very well in the deformities of the land with little technical difficulty and adapts perfectly to the irregularities so characteristic of the rock on which the city is settled. That is the way Porto is structured, featuring steep alleys with crossroads, stairways, etc.

But its streets were also paved with drawings to tell stories, traditions, customs and parts of the country's history. Like the Greek *tirreme* from the pavement of the street Sampaio Bruno that leads iconographically directly to the foundation of the country. It represents the ship of the Argonaut Cale, who reached the mouth of the Duero and founded a commercial enclave there. However, the settlement known by the Greeks had very poor conditions for navigation. It was the Romans who moved it inland, where a port was built, the so-called *Portus Cale*. There the etymological origin of the city is revealed: Porto. Therefore, Portus Cale became *Portucale*, the current Ribeira, giving the country its name: *Portugal*, Fig. 10.

Fig. 10. Watercolour of Marta Úbeda. Pavement in the street Sampaio Bruno. Porto. 2019.

In conclusion, we would like to highlight the importance of colour in the architecture of Porto, since "colour is part of the process of representation of reality by allowing a closer approach to nature and is directed to the senses" (Gallego 1990, p. 151), providing volume, depth, realism and above all life. This great artistic work that is Porto reminds us of the works of Luis Barragán (Villalobos 2002) where we are allowed to enjoy the pleasure of visual perception of colour and to experience aesthetic emotions at the confluence of light and its reflections on the water. Not only can we contemplate it, like in a painting from outside of the composition, but it also allows us to be part of this urban experience as a masterpiece. Being inside and contributing to that constant transformation. Following the thoughts of the artist and architect Theo van Doesburg, "placing man within the work itself" (Theo van Doesburg 1929).

On the other hand, and taking into account Goethe's colour theory (Goethe 1987), where he stated that colours are in matter and make us feel emotions, we check that the colours of Porto are not a simple play of light. They are there "simple, pure, serene, silent, expressing the presence of the invisible essence" (Soares and Simões 2009).

On the other hand, in the work of graphic representation of the colours of Porto, we state that the main principle of the technique of watercolour, is the guiding thread of the illustrations. These coloured notes are made with the same characteristic elements of the city: water, light and colours. A silent manifestation of light and colour through water that gives the city its character and makes Porto so charismatic.

References

Barriada, E.M.: Empedrados artísticos de Lisboa: A arte da calçada-mosaico. Lisboa (1986)

Bellido, S.: El río Duero: influencia del entorno natural en la conformación del paisaje humanizado: un análisis gráfico arquitectónico. Universidad de Valladolid, Valladolid (2005)

Le Corbusier. Hacia una arquitectura

Cruickshank, D.: Soane y el significado del color. Arquitectura **277**, 86–100 (1989)

Cullen, G.: The Concise Townscape. Architectural Press, London (1971)

Gallego, P.: La experiencia del color: armonías e interacciones. Dibujo y Realidad. El problema del parecido en las artes figurativas 151–169 (1990)

Goethe, J.W.: Esbozo de una teoría de los colores. En Obras completas, Madrid (1987)

Laranja, J.: Porto Forma Urbis, Redesenho Cartográfico do Porto. Tesis, Doctoral inédita Universidad de Oporto (2007)

Moireau, F.: París. Cuaderno de viaje. Anaya, Madrid (2011a)

Moireau, F.: Londres. Cuaderno de viaje. Anaya, Madrid (2011b)

Polano, S.: La nueva cromo plástica en la arquitectura. El color del estilo holandés. Arquitectura **277**, 100–115 (1989)

Soares, H., Simões, Z.: The construction of a scenic imaginary in the Porto city through ambiences of light and colour. Sidney (2009)

Tanizaki, J.: El elogio de la sombra. Siruela, Madrid (2009)

Van Doesburg, T.: The Stijl en "Hacia la pintura blanca". París (1929)

Villalobos, D.: El color de Luis Barragán. Morés, Oviedo (2002a)

Villalobos, D.: En la Ruta de Oriente. Cuaderno de dibujos de viaje. Universidad de Valladolid, Valladolid (2002b)

Villalobos, D.: Hasta los pies del Himalaya. Cuaderno de dibujos de viaje. Universidad de Valladolid, Valladolid (2004)

Luigi Moretti. Solidified Space

Jaime J. Ferrer Forés[✉]

Universitat Politècnica de Catalunya, ETSAB, Barcelona, Spain
jaime.jose.ferrer@upc.edu

Abstract. Italian architect Luigi Moretti (1907–1973) analyzes the void in architecture as an inversion of the solid by means of a series of analytical scale models in which he solidifies the space. In his theoretical writings published in *Spazio. Rassegna delle arti e dell'architettura* journal, Moretti resorts to the construction of a series of drawings and scale models as an analytical tool of the void in architecture. The solidified space of the scale model fosters the understanding of architectural space as a specific element of architecture. The void becomes solid in these scale models illustrating the spatial organization of the Project as an extension of the inner continuum, reflecting the relation of the different spaces through the negative of the space sequences made in clay. The scale model becomes an instrument of analysis of the built heritage as well as a conceptual model of the project that vindicates the primacy of the void in architecture. The scale model defines a way of seeing and Moretti's work puts the creative emphasis on the void in architecture, which deeply permeates his entire career.

Keywords: Luigi Moretti · Space · Scale model · Analysis · Heritage

1 History in the Training of the Architect

The trajectory of the Italian architect Luigi Moretti (1907–1973) flows from the academic classicism with which he builds his first works with academic accuracy to the abstract purity of the strict geometries of the *Casa della Gioventú del Littorio*, which holds on to the monumental severity of the regime. He transits as well from the functional pragmatism and plasticity of the Milanese collective housing complex in Corso Italia or the *palazzina "il Girasole"* in Rome, to the progressive abstract and expressive language that characterises his architectural maturity.

Graduate of the *Regia Scuola Superiore di Architettura* in Rome in 1930, Luigi Moretti completes his training as an assistant to the History Chair with Vincenzo Fasolo and Gustavo Giovannoni. His early interest in construction and history is illustrated in his works, where he analyses how Roman or Renaissance and Baroque architecture expresses itself using the ornament, as a display of its very constructive materiality (Moretti 2006, pp. 70–79). To Moretti, ornament is an essential element of architecture as an instrument for the representation and accentuation of architectural logic (Rostagni 2006, pp. 80–85).

Moretti's project design exploration is developed simultaneously to the study of heritage. To Luigi Moretti, the understanding of the interpretive keys of the built heritage

© The Editor(s) (if applicable) and The Author(s), under exclusive license
to Springer Nature Switzerland AG 2020
L. Agustín-Hernández et al. (Eds.): EGA 2020, SSDI 6, pp. 142–152, 2020.
https://doi.org/10.1007/978-3-030-47983-1_13

provides with an endless repertoire for the study that guides his professional career. His particular interest on construction and art is illustrated in *Spazio. Rassegna delle arti e dell'architettura* journal (1950–1953), which he founded in Rome. His constructive and structural reflection is shown in the articles published in the *Spazio* journal, displaying his persevering exploration of the inner logic of the architectural project. In his articles, he looks into the spatial, constructive and plastic nature of architecture, the theoretical fundamentals and its formal and perceptual principles.

Moretti's first analytical scale models focus on the relationship between form and construction. The model of Michelangelo's Laurentian Library vestibule in Florence looks into the correspondence between form and structure, the nature of construction and its representation as a mechanism for intensifying its inner logic (Fig. 1). In this scale model, Moretti analyses the structural fundamentals and its constitutive identity and reveals how the facade accentuates its autonomy regarding the supporting system. Thus, his first analytical models explore the representation of the construction through the analysis of the line elements of the structure.

Fig. 1. Luigi Moretti. Analytical scale model of Michelangelo's Laurentian Library vestibule structure. Source: Bucci and Mulazzani 2000, 9.

2   The Representation of the Construction

In the *Casa della Gioventú del Littorio* that he builds in Rome's Trastevere (1933), Moreti conceives a three-dimensional concrete grid with outer walls. By way of linear elements, the supporting structure becomes a formal element that establishes the order of the isotropic space. In an article titled *"Struttura come forma"* published in the *Spazio* journal, Moretti argues, "architecture is essentially structure" (Moretti 1951b, p. 30). To Moretti, the image of the structure's spatial grid, as a product of the technical, functional and expressive order, is "reality and representation", thus identifying architecture as a representation of construction. The constructive reality of the reinforced concrete framework becomes the decisive compositive aspect of the project.

Using the structural expression as governing system, the scale model of the structural framework reveals the modernity of reinforced concrete structures and the isotropy of the architectural composition. Moretti seeks an architectural expression in accordance to the innovative possibilities of technology and the timeless values of structural order. As a product of the technical and functional order, the cross-section scale model of the project shows the emphatic articulation of the supporting system (Fig. 2).

Fig. 2. Luigi Moretti. Scale model of *Casa della Gioventù del Littorio*, Rome (1933). Source: *Archivio Centrale di Stato*. Archive Moretti.

3   Clay Scale Models

To Moretti, the scale model is an instrument with which to experiment and determine the architectural project. The model allows establishing the ideas throughout the development of the project and verifying the progress in the design process. Made with different materials, initially cardboard and wood, and of several scales, the model is an essential form of expression for the development of the project and its ultimate presentation. Among them, the clay models built early in the competition for the *Palazzo del Littorio* in Rome (1937) stand out. Moretti resorts to the construction with clay of the model, using an easily pliable material, with which to model and verify the project.

The cross-section scale model of the *Casa della Gioventù del Littorio* in Trastevere (1933) is the effective instrument to represent the coherence between the architectural space and the supporting structure. The model looks into the constructive and material dimension of architecture. In the clay model of the *Palazzo del Littorio* (1937), the building arises from the modeling of the clay, a material according to his plastic interests (Fig. 3). The photographs show the work to shape the primary forms in clay arranged on a plinth that foreshadow the final architectural configuration and the scalar gradation of the whole characterised by symmetry, mass and monumentality. Based on the subtraction and accumulation of clay and the characterisation of the surfaces of the scale model, Moretti formalises an architectural expression that takes full advantage of its expressive strength.

Fig. 3. Luigi Moretti. Scale model for the *Palazzo del Litorio* competition, 1937. Source: Bucci and Mulazzani 2000, 15.

4   The Understanding of Space

Moretti's stubborn reflexive and theoretical work is introduced by a number of papers published in the seven issues of *Spazio. Rassegna delle arti e dell'architettura* journal in which he focus on the understanding of structure, form and void in architecture (Pierini 2016, pp. 62–71). Moretti develops his theoretical reflection on architecture in the papers he publishes in *Spazio* journal, among which stand out: *Eclettismo e unità di linguaggio* (1950), *Genesi di formme della figura umana* (1950), *Forme astratte nella scultura barocca* (1950), *Colore di Venezia* (1950), *Trasfigurazione di strutture murarie* (1951), *Discontinuità dello Spazio in Caravaggio* (1951), *Valori della Modanatura* (1951) and *Strutture e sequenze di spazi* (1953).

This exploration of the discipline based on the inner logic of the project illustrates the permanent transfer of issues originated in the analysis of heritage. The models of the Roman architect Luigi Moretti that illustrate some of the papers are intended to represent the specificity of architecture, space (Bucci and Mulazzani 2000; Rostagni 2008). According to Cornelis van de Ven, "space is the true essence of architecture" (1981, p. 75). Its reason for being is to introduce and make evident the most immediate. Visualize what is hidden, what goes unnoticed. To Bruno Zevi, "taking possession of space, knowing how to see it, constitutes the key to the understanding of buildings" (Zevi 1976, p. 20). In the book *Architecture as Space. How to look at Architecture*, Bruno Zevi reflects on the representation of space (Zevi 1976). In the floor plan of Saint Peter in Rome, the layout of the walls lends them the character of figure. In the inversion of the drawing of the floor plan, this character is attributed to space by means of that spatially defined area (Fig. 4). As Stan Allen points out, "form matters but not so much the form of things as the form between things", space (Allen 2009, p. 218).

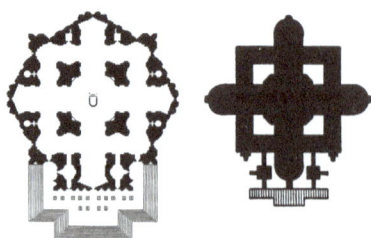

Fig. 4. Bruno Zevi. Floor plan of St Peter's Basilica in Rome and the representation of the void. Source: Zevi 1976, p. 20.

Through the abstraction inherent to the construction of a scale model of the void, Moretti intends to insist on the understanding of space in architecture by conferring it the role of figure, a leading element, and not that of background (Fig. 5). Thus, the works shown in the paper entitled "*Strutture e sequenze di spazi*" published in the journal *Spazio* (Moretti 1953) display a common resolve. In order to unify their representation, he carries out a series of models showing the void in the works analyzed.

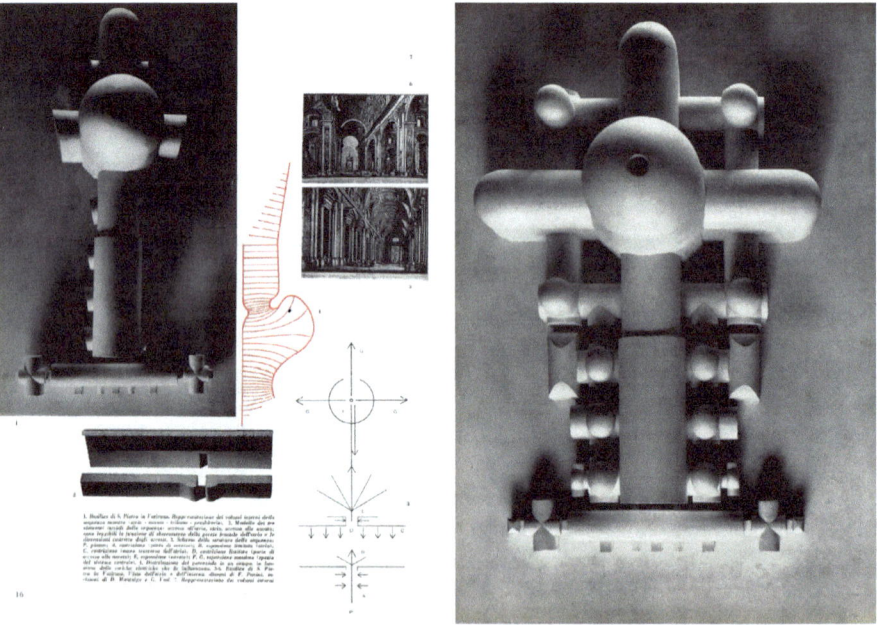

Fig. 5. Luigi Moretti. Scale model of the void in St Peter's Basilica in Rome. Source: Moretti 1953, p. 16, 17.

The figurative void becomes a solid in the scale model that represents the inner spatiality. By permeating space with an apparent solid feature, he displays the inner void. Salvatore Vitale points out in *L'estetica dell'architettura* that "space, while maintaining its essential character of pure extension, that is, of emptiness, somehow manages to conquer a corporeal appearance and solidify" (Zevi 1976, p. 167). The scale model is an element that solidifies the space. The model of the void in St Peter's Basilica in Rome shows this corporeal appearance. The photographs of the scale models illustrate the critical and theoretical reasoning of the paper in which he looks into the spatial form (composition), the corporeal form (definition of limits) and the visible form (definition of light).

5 Spatial Solids

With the spatial solids built by means of these analytical models of historical and modern buildings, Moretti illustrates the idea of space as a solid volume. The void contained within their physical boundaries is the material with which Moretti analyses architecture. He ignores both structure and exterior enclosure, and the scale models, as Peter Eisenman states, "seemed to deny a relationship to the exterior" (Eisenman 2011, p. 31). However, in Moretti's analysis, the dialectic between interior and exterior is emphasised. The models illustrate the conceptualisation of space when transforming the void into a solid and the importance of the profile that delimits the space is confirmed and developed in a paper published in *Spazio* entitled *"Valori della modanatura"* (Moretti 1951a, pp. 5–12). The character of the outer surfaces determines the atmosphere of the interiors. In the paper entitled *"Forma come struttura"*, Moretti states: "The internal and empty spaces of an architectural work are ruled by a structure, a very significant one in this case. In fact, the structure of the empty spaces opposes, as a mirrored value, symmetrical and negative, as a negative matrix, to the ensemble of fundamental structures linked to the matter put into play in the architectural works. It opposes, above all, to the structure of constructivity that, through isomorphism, grants some specific links" (Moretti, 1957).

The volumetric models to represent the projects extend throughout the whole trajectory of Moretti with the scale representations of a particular architecture. However, with the construction of abstract spatial solids, through a conceptual representation, the scale models become analytical instruments and constitute an important lesson in the work of Moretti (Fig. 6). An analysis of the architecture largely defined from the object to space. To Peter Eisenman, "Moretti's models inverted this convention by taking space, rather than its enclosing surface, as a starting point for analysis" (Eisenman 2011, p. 31).

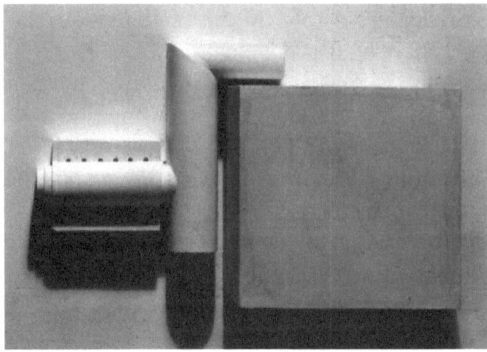

Fig. 6. Luigi Moretti. Scale model of the void in *Palazzo Farnese*. Source: Rostagni 2008, 84.

6   Structure and Sequence of Spaces

By means of the scale models of the void Moretti looks into the rhythmic basis of architecture and the organic coherence of the interior. The immersion of Moretti in the spatial field in the search for the rhythmic and binding substrate is shown in the paper entitled *"Strutture e sequenze di spazi"* published in the journal *Spazio* (Moretti 1953).

Moretti analyses in the article the spatial continuity and the unitary space of Guarino Guarini's project for S. Filippo Neri's church in Casale Monferrato or the organic expression of McCord House by Frank Lloyd Wright (Fig. 7); the spatial sequence of the Villa Adriana complex in Tivoli; the articulated sequence of the cloistral organisation of the rooms of the Palazzo Ducale of Urbino by Luciano Laurana and Francesco di Giorgio; the sequence of different geometric shapes and volumes of the Palazzo Thiene in Vicenza by Andrea Palladio; the sequence of spaces of Villa La Rotonda by Andrea Palladio (Fig. 8); the study of the rhythmic and spatial nature and the representation of the sequence of the internal volumes of St. Peter's Basilica in Rome; the sequence lobby, porch, courtyard of Palazzo Farnese in Rome by Antonio da Sangallo and Michelangelo; the spatial organization of the House of the Tragic Poet in Pompeii; the volumetric representation of the interior spaces of Santa Maria della Divina Provvidenza's church project in Lisbon by Guarino Guarini (Fig. 7); the interpretative analysis of Michelangelo's project for San Giovanni dei Fiorentini's church in Rome; Moretti's own work in the comprehensive and appraising model of the Fencing Academy at Foro Italico (1934–1936) or the analytical drawings of Tugendhat House by Mies van der Rohe.

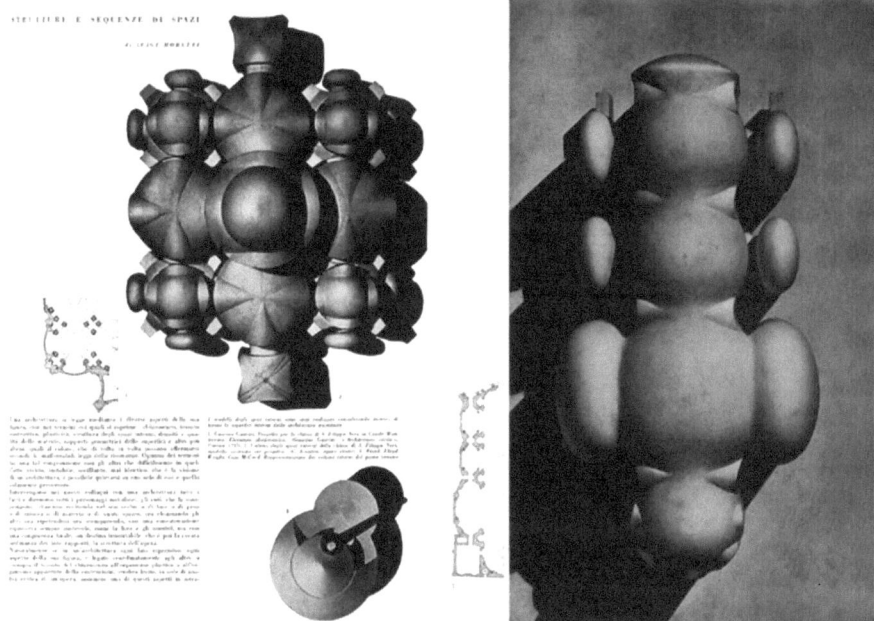

Fig. 7. Luigi Moretti. Scale model of the void in Guarino Guarini's project for S. Filipppo Neri's church in Casale Monferrate, McCord house by Frank Lloyd Wright and S. Maria della Divina Provvidenza in Lisbon. *"Strutture e Sequenze di Spazi"*. Source: Moretti 1953, 9, 19.

This arrangement of the volumes in continuity, contrast or aggregation defines the variety of the spaces involved and the sequence as an additive phenomenon, underlining the shape of the spaces and that of the intervals separating them. The morphological distinction of the spaces is distinguished in the model, being arranged accordingly to an underlying structural grid. To Moretti, "The internal space of San Pedro continues to be a composition of uniquely individual elementary volumes, linked to each other by passages or lesser spaces" (Moretti 1953, pp. 16–17). The scale model represents the means that channels, governs and supports architecture. Their assembling leads to a sequence that articulates the spatial interaction of the different areas.

This research guides Moretti's continuous search that is reflected in the scale model of the Fencing Academy project at Foro Italico (1934–1936) which he defines as "one of the first attempts at a totally unitary spatial modulation that still makes use of the complete range of light, dimensions and shapes" (Moretti 1953, p. 20). To Moretti, architecture evolves from merely organising functions to arranging spaces.

The analyses of historical architecture show Moretti the value of construction in architecture. The solidity of Roman architecture is always a fixed value in his work. The abstract and contained volume of the Fencing Academy at Foro Italico (1934–1936) combines the abstract vision of classical values with the modernity of reinforced concrete porticoed structures and spatial dynamism. Evocative of the classic precedents, the smooth Carrara marble cladding symbolizes the emulation of Ancient Rome by the Italian fascist regime.

Fig. 8. Luigi Moretti. Scale model of the void in the project of Luciano Laurana and Francesco di Giorgio for Palazzo Ducale in Urbino and Andrea Palladio's Palazzo Thiene and La Rotonda. *"Strutture e Sequenze di Spazi"*. Source: Moretti 1953, 12, 14.

The cross-section of the main hall extends linearly into the space as a straightforward extrusion. The cross-section is composed of the curvilinear profile of two corbels of uneven heights that delimit a large window hidden from the inside. The interior space is characterised by the effect of the continuous surface, the richness of the perceptual effects and its plastic strength. To Moretti, the space in architecture is a plastic phenomenon. Moretti builds the spatial solid in which the void solidifies. The scale model aims to make visible the interior of architecture and represent its depth (Fig. 9). According to Naum Gabo and Antoine Pevsner, "depth is the form of space" (Micheli 1999, 325–328).

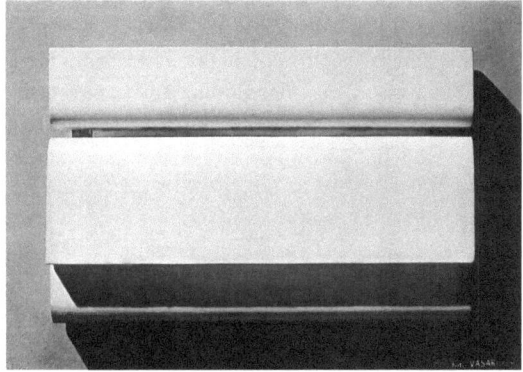

Fig. 9. Luigi Moretti. Scale model of the void in the Fencing Academy at Foro Italico, 1934–1936. Source: Bucci and Mulazzani 2000, 14.

7   Limits

In the paper entitled "The Functional Tradition and Expression" published in *Perspecta* journal, James Stirling describes Moretti's scale models as materialising a solidified space: "By treating the external surface and the inner constructions of a building as a three-dimensional negative or mould, he was able to obtain solidified space" (Stirling 1960, p. 91). The model solidifies the intense internal stillness of the work.

As Enrico Tedeschi points out, "The architectural space, because it is limited, cannot get rid of its limits nor ignore them, and, because it can be traversed, it cannot be separated from the presence of those who go across" (Tedeschi 1969, p. 245). Moretti introduces the character of the facades of the building on the surface of the model. The scale model, as an expressive device, reflects everything around it. Moretti thus recognizes the importance of the profile and the potential of creating with light, by controlling its effects and characterising the architectural space.

The form of the space limits is recognized in the scale model. The interior is activated by the light falling upon and space understood as the physical atmospheric volume delimited by the constructive elements in which the light reveals perceptually its properties to the observer that moves about.

The characterisation of the limit defines the adjectivisation of space and the external envelope determine the features of the interior spaces. The planes that circumscribe the interior are projections of spatial attributes. Physical limits enrich spatial pulsation and multisensory perception. Moretti captures and analyses the form understood as a sensitive manifestation of the internal order (Santuccio,1990). Constrained by the inner envelopes, the scale model collects the imprint of the envelope and the contour, the *modanatura*. In the paper entitled *"Valori della modanatura"*, published in *Spazio*, Luigi Moretti reflects on the notion of *modanatura* as an accentuation of the constructive and spatial system (Moretti 1951a, pp. 5–12). As Heinrich Wölfflin points out "Classic taste works with tangible limits (profiles, contours), definite lines; every surface has its precise edge; each volume reveals as a fully tangible form; nothing inapprehensible in its corporeity exists" (Wölfflin 2013, p. 101). Moretti's spatial solids come closer to the notion of the corporeal volume with which Wölfflin defines architecture. He evokes the works of Bruce Nauman, such as "A cast of the space under my chair" (1965–1968), where he solidifies the space under his work chair, or the most recent works of Rachel Whiteread that continue to make visible what surrounds us, what encompasses us, space.

8 Conclusion

The space exists within a delimitation. In Moretti's models, the plastic form characterised as void represents space. The captured space becomes a corporeal solid in the scale models. By building that plaster void, space becomes a reality. Luigi Moretti's analytical models do not resort to express their usual meaning, a scale representation of the built reality, but they make visible the empty space within, as a guiding and organising element of the architectural work. The analysed works share a common determination and through these scale models an underlying physical world is revealed that manifests itself in the models of solidified space. Moretti's models transfigure the ordinary into something extraordinary. The scale models assimilate the delimited physical volume, the inner latent space.

References

Allen, S.: Practice: Architecture, Technique + Representation. Routledge, Londres (2009)
Bucci, F., Mulazzani, M.: Luigi Moretti. Opere e scritti. Electa, Milano (2000)
Eisenman, P.: Diez edificios canónicos 1950–2000. Gustavo Gili, Barcelona (2011)
Micheli, M.: Las vanguardias artísticas del siglo XX. Alianza, Madrid (1999)
Moretti, L.: Vallori della modanatura. Spazio **6**, 5–12 (1951a)
Moretti, L.: Struttura come forma. Spazio **6**, 21–30 (1951b)
Moretti, L.: Strutture e sequenze di spazi. Spazio **7**, 9–20 (1953)
Moretti, L.: Forma come struttura. Spazio (1957)
Moretti, L.: Canovaccio per un saggio sul l'architettura di Michelangelo e del Borromino e su quella barocca in genere; e intorno alla natura dell'architettura e alle possibilità di una nuova critica architettonica, Roma, 1927. Casabella **745**, 70–79 (2006)
Pierini, O.S.: Continuity and discontinuity in Casabella and Spazio. The 1950's architecture magazines directed by Luigi Moretti and Ernesto Nathan Rogers. Cuadernos de Proyectos Arquitectónicos **6**, 62–71 (2016)
Rostagni, C.: Moretti, Michelangelo e il barroco. Casabella **745**, 80–85 (2006)
Rostagni, C.: Luigi Moretti. 1907–1973. Electa, Milano (2008)
Santuccio, S.: Luigi Moretti. Zanichelli, Bologna (1990)
Stirling, J.: The functional tradition and expression. Perspecta **6**, 88–97 (1960)
Tedeschi, E.: Teoría de la arquitectura. Nueva Visión, Buenos Aires (1969)
van de Ven, C.: El espacio en arquitectura. Cátedra, Madrid (1981)
Wölfflin, H.: Conceptos fundamentales de la Historia del Arte. Austral, Madrid (2013)
Zevi, B.: Saber ver la arquitectura. Poseidón, Barcelona (1976)

Re-vision of the Drawn Heritage of Alison and Peter Smithson: Make and Remake = Innovate…

Isabel Zaragoza[✉] and Jesús Esquinas-Dessy

Universitat Politècnica de Catalunya UPC, Barcelona, Spain
isabel.zaragoza@upc.edu

Abstract. This paper seeks to show an interpretation of the drawings done by Alison and Peter Smithson to organize and present the exhibition "Painting and Sculpture of a Decade, 1954–1964", in order to evaluate how the graphic resources used in the project are a reflection of a concrete ideology in facing his work and how it is based on understanding the architectural space from its adaptation to human activity, and in this case, from the more feminine approach of Alison Smithson.

The diversity of the Smithsons architectural work is widely documented and disseminated by numerous writings and audiovisuals of the authors themselves in addition; it has been studied through an extensive bibliography. However, from our point of view there are still facets of their way of representating architecture that arouse a certain curiosity such as the relationship between the visual resources used and the architecture they conceived.

The research focuses on offering a personal perspective of the archived graphic heritage of the project, focused on its illustrated plan drawings and synthesis diagrams. This exploration manages to identify a series of graphic techniques that show the innovative intention that guided this Project.

Keywords: Alison and Peter Smithson · Drawing · Innovate · Graphic heritage

1 Introduction

"Our intention has always been –conscientiously since the Doorn Manifiesto in 1954- to turn architecture towards particularity… of place, person, activity: the form to arise from these.

That is why there has been so much effort…projects, histories, novels, films, essays, furniture, exhibitions." [1].

The term heritage, which comes from Latin means "what is received by the father's line", located in the discipline of architectural representation, the first reading suggests inheritance of drawings or documents done by recognized authors that imply a record of deep theoretical reflections. The outstanding work of Alison and Peter Smithson in the field of architecture and urbanism has left a rich graphic legacy, as a result of

L. Agustín-Hernández et al. (Eds.): EGA 2020, SSDI 6, pp. 153–165, 2020.
https://doi.org/10.1007/978-3-030-47983-1_14

their intense professional and teaching activity in the world of avant-garde culture from the mid-twentieth century. The continuous self-critical and reflective nature of their work, deliberately disseminated, sometimes reappeared intentionally reverting to the new particularity of the project.

A significant particularity of the Smithsons' work is their link to the art world. Its participation in Independent Group is remarkable, where together with a group of young artists and critics, in the 1950s; they organized actions to make art more inclusive, as a challenge to face the dominant elitist culture of that time. All types of human activities were the object of their attention and they assumed an anthropological definition of culture [2]. In this sense, it seems highly appealing to go into their way of doing through a project related to the art world (Fig. 1).

 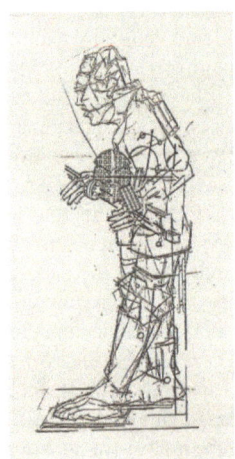

Fig. 1. Works of some Independent Group members. Left: Henderson, N. *Head of Man*. London. 1956. Source: Tate Gallery. Right: Paolozzi, E. *Sir Eduardo Paolozzi ('The Artist as Hephaestus')*. London. 1987. Source: National Portrait Gallery. London

With resources from the "Special Collections" of the Frances Loeb Library Harvard Graduate School of Design, we have based our work of Smithsons original documents [3]. When examining the documents, we were struck by the uniqueness of the drawings done by hand in Chinese ink from the project for the exhibition "Painting and Sculpture of a Decade, 1954–1964", made by Alison Smithson.

After verifying that it had not been investigated it, we thought it might be of interest to reflect on this heritage, so carefully inventoried, and that allow gathering various documents and drawings, looking at them carefully, as well as finding connections with which to advance in the knowledge of their work on graphic representation. Their particular way of "making", "remaking" will lead to confirm the innovative intention that guided the project.

In 1962, the Calouste Gulbenkian Foundation proposed to organize an exhibition whose objective was to show the art of the last ten years, in a sample unprecedented at the time, both for quality and extension, and for the international character of its

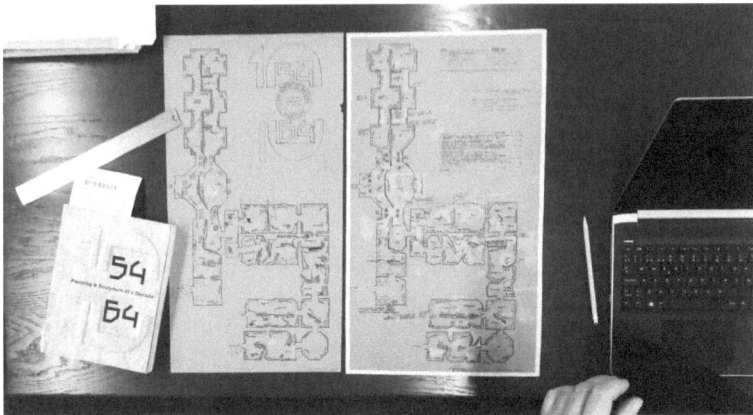

Fig. 2. Image of the working table in the Special Collections of the Frances Loeb Library, Harvard Graduate School of Design, with original project materials by Alison and Peter Smithson and the exhibition catalog [4]. Own source

selection. A totality of 354 works by some 161 different artists including paintings and sculptures were shown; coming from both workshops of the artists themselves, such as private collections or different museums, not only from Europe but also from the United States. The exhibition brought together, among others, the work of "old masters" such as Picasso, Braque, Léger or Lipchitz, or the young Tàpies, Pollock, Hockney, Oldenburg, Chillida, or Tinguely. The innovative nature of all of them reflected in some way the establishment of certain permissive and reflective attitudes towards these new forms, already recognized by a growing number of people, and from different countries [4].

The project occupied several spaces of the neoclassical building known as the National Gallery of British Art, where the Tate exhibited both its modern and classic collections. Alison and Peter Smithson intention was focused on the study of the spaces transformation that on the one hand enhanced the dialogue between the works, and that on the other hand prevented their intervention from losing prominence against the imposing architecture of the hall, with the presence of the orders of the repetitive pairs of columns or the spectacular vaults of the galleries. In this line, the monumental main entrance to the building, located on the axis of the façade with its staircases, entablature and Ionic colonnade was disabled, and it was preferred to organize the access to the exhibition at one of the edges.

> "Our intention was t o present the works in such a way as to clarify the relationships that governed their selections. The paintings and sculptures were only meant to speak. No presentation tricks were used, and none of the Tate detailed architecture was allowed to obstruct." [5].

Therefore, they devised the exhibition transforming the Tate spaces with a montage that the authors called "milky way". The artworks were presented on spectacularly illuminated white screens to thereby achieve great contrast effects with the darkness of the existing building roofs, in that way they visually blurred the historical architecture.

2 Materials from the Archive

2.1 Drawings

The project is inventoried in the folder "DES, 2003, 0001, 009455761, J000, Volume: J135". When we open it, we find eight hand-drawn drawings made in Chinese ink on tracing paper at different scales, which include distribution plans, diagrams and a green-hatched transparency [3].

We will examine the set of plans drawn up at 1/16 scale that illustrate different aspects of the architectural process of the project, all of them drawings in plan. The interest of this series of drawings lies in the artisan character that reveals and suggests a detailed involvement of the architects in the peculiarity of the work. It is likely that they practiced freehand drawing of the plan as an integral action that would help them become more familiar with the artworks to be included in their project, also made with their hands.

Focusing on the drawing "Detailed layout dwg n° T6400" (Fig. 3), at a first glance, the absence of the existing architecture is surprising. It is possible to think that the Smithsons deliberately omitted it to underline the particularity of their project: that suit tailored to the works to be presented, in which the conformation of each of their folds would be given by the needs of space to view them and which would later be adjusted to the Tate geometry.

During the process, they investigated in detail the space needs of the individual paintings and sculptures, as well as the possibilities of groupings, paying special attention to the visitor movements and visuals. Peter Smithson explained in a lecture that [6]: "the articulation of the plan is based on the works" also stressed, "it follows the shape of the spaces, which is guided by the nature of the artwork." The formal result is a kind of labyrinth with dilations of diverse geometry that organize groupings of the works in a total of 24 subsections, and which in allows you to explore the entire exhibition through various alternative routes.

In the study process, which took them a year and a half, they worked from photographic miniatures of each of the works. With all this, they created a pattern of spaces consistent with their main intention: to serve those particularities that concerned them related to the experiences and emotions of the visitors.

Another aspect to note is the graphic strategy used to indicate the groupings of the works. On the one hand, handwritten lettering carefully designed with a "flexible" character that adapts to the new geometry and follows the entire exposure perimeter. It is accompanied by "flexible" arrows, graphic resource that facilitates the understanding of the different groups and that helps raise the hypothetical visual relationships with visitors. If you look closer at the drawing, you can see some "stars" that link to the idea of the "milky way" discussed above. This observation leads us to deduce that the Smithsons made a previous assessment of the works/artists and awarded three categories: those of a star, those of two stars, and those that had not been awarded any stars. In the first category we find, among others, Picasso, Braque, Lèger, Albers, Matisse, Morandi, Avery, Moore or Miró. In the second we find Bacon, Dubuffet, Jorn, Vasarelly, Koonig or Rothko. Younger or then emerging artists such as Tàpies, Paolozzi, or Pollock are not accompanied by the aforementioned "bright" badges.

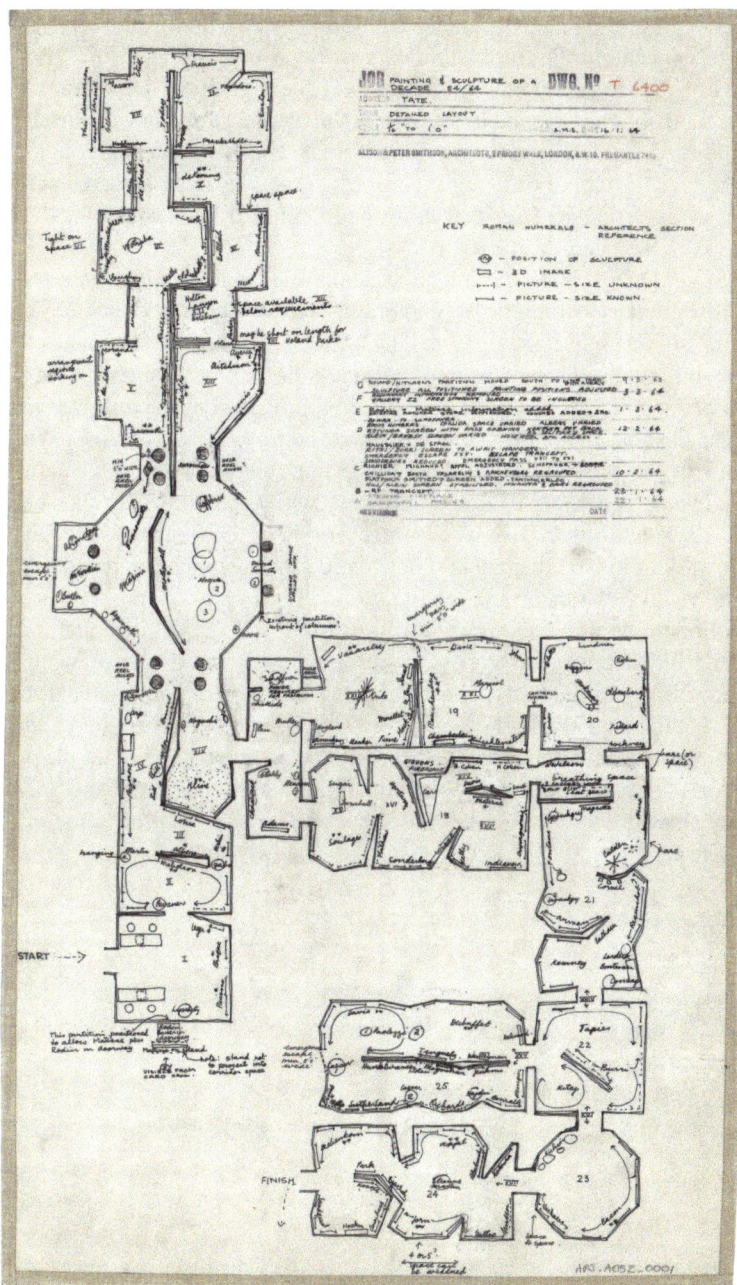

Fig. 3. Alison and Peter Smithson. "Painting and Sculpture of a Decade, 1954–1964" "Detailed layout" dwg n° T6400. Chinese ink drawing on tracing paper. Dimensions 35 × 60 cm. London. 1964. Source: [3]

The strategy of the "galaxy" elaboration proposed by Smithson seems to have the intention to concentrate the most luminous work, which is that of the "old masters" in the area of the main entrance as the basis of the exhibition. However, the artwork with different distinctions is interspersed in a timely and strategic way through the rest of the exhibition spaces, having just formed that "milky way" as a conglomerate of constellations with stars of different intensities, some already recognized and others to recognize, as an articulated set that is formed fleeing from chronological logics and thought to keep the visitor's attention.

Peter Salter [7], explains - as a regular contributor to the Smithsons -, the care and detail of the drawings, calling them "exploratory" in which it also "includes production information". In this sense, the rigor and intention of the drawing that invites to stop and discover their thoughts through the multiple annotations, precise situations of the works, intentions of grouping or guessing of the visitors' movements, should be emphasized. You can also observe all kinds of comments in response to the possibilities of expansion, or incidents of the existing space: thus, for example, it includes emblematic symbols specifying in a personal way by drawing "an eye with a directional arrow" the visibility intentions of Matisse's work from the entrance, or the visual connection of Giacometti's sculptures. In addition to other annotations of order closer to the detail and to the human scale of the visitors, the incorporation of question marks point to possible accesses.

Special attention deserves the care in recording the written information that they incorporated into the project process, as an integrated part of the drawing itself. To facilitate this aspect, different types of stamps were prepared to stamp the drawings and thus to be able to manually marked each item of the document. Three types can be seen: the stamped heading box to complete manually the basic information of the project, the revision box in which all the changes and additions were dated on hand drawn, or the circular stamp with the brand of the office stamped with red ink in some of the plans (Fig. 4).

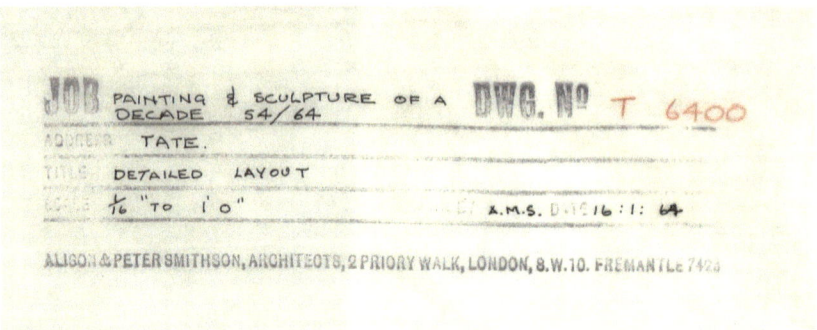

Fig. 4. Alison and Peter Smithson. Fragment of Fig. 3 illustrating in detail the heading box. The numbering of each drawing is highlighted manually with red graphite. The acronym "A.M.S" corresponds to Alison Margaret Smithson. Source: [3]

Salter [7] also adds "the drawing became an active of thought processes". In this sense, we can see that a detailed record of the various changes and implementations is included with brief descriptions, all of them with the specific indication of the day, month and year. It i s followed by a need for registration derived from the intention to evolve the drawing on constant occasions. In addition, the elaborated revision box is conveniently ordered and numbered with correlative letters. It can be seen that the period in which the different actions are elaborated runs since March 1963 to January 1964, therefore it can be understood that the perimeter of the whole plane is adequately protected with a thick ribbon. It is relevant being able to appreciate the doubts and modifications carried out during the entire project process. The ease of having the original plans physically has allowed us to appreciate these changes in more detail. In this sense, it has been possible to arrange the plane of tracing paper on a dark background. In this way, the change in texture produced by erasing the ink using the classic "scratching" technique, stands out on the darkened tracing paper. It is the proof that drawings are constantly modified and evolved in search of the intended innovation (Fig. 5).

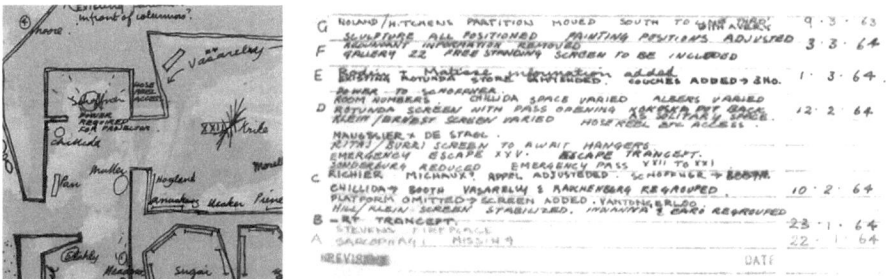

Fig. 5. Alison and Peter Smithson. "Painting and Sculpture of a Decade, 1954–1964". Left: fragment of Fig. 3 with details of the modification recorded on date 10.2.64: "C Chillida Booth Vasarely & Rochenberg reagruped". Right: revision box in which all changes and additions were noted. Own source

In front of the plan drawing with all the process details (Fig. 3), we find another general plan (Fig. 7), also hand drawn on the same scale but, in this case, without registration data. Presumably it would have been a redrawing with the final set of alterations made after construction, in which the variations that had occurred in the execution process were implemented, as an "as built" drawing. It is also freehand drawing, with all the information of the artworks and probably "it was taken very seriously, requiring days to be spent remeasuring parts" [7] of the grouping of exhibit spaces. This plan incorporates as well a drawing in its upper right, with the anagram of the exhibition. This design will be presented on the cover of the exhibition catalogue (Fig. 2) [4]. In addition, the map is proudly stamped in red with the circular seal from the Smithson's studio, which is a leading part in the composition of the drawing.

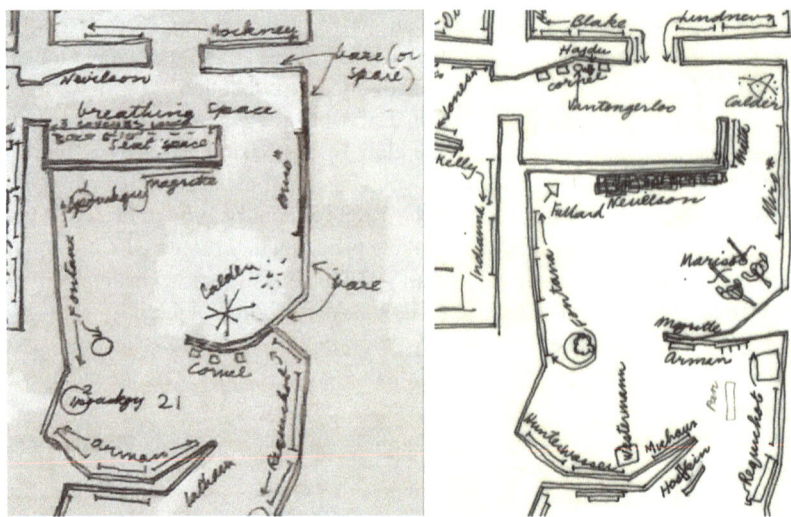

Fig. 6. Alison and Peter Smithson. "Painting and Sculpture of a Decade, 1954–1964". Left: fragment of the "Detailed layout dwg n° T6400" plan (Fig. 3) of subzone 21. Right: fragment of the general plan (Fig. 7). It is observed that Miró's work is marked with a star, and maintains the same position, instead, Calder's work has shifted. (fragments reproduced in real scale). Own source

It is interesting to compare "twin" fragments of one plan (Fig. 3) and the other (Fig. 7), as the evolution process can be observed by close views. Peter Smithson explained in the Climate Register conference [6] that there is a high percentage of work carried out in situ in exhibition assemblies, an issue that can be verified by carefully visualizing both drawings (Fig. 6).

Another noteworthy aspect is the way to graph the accesses to the exhibition with emblematic symbols: the drawing of as multiple concentrated arrows at the entrance and exit probably register the multitude of visitors that presumably would be at the beginning, and would stop at the exit to comment the exhibition (Fig. 7).

Peter Smithson offered several clues about the importance of exhibitions in the exercise of good practices, considering it as an opportunity to experience 1:1 scale aspects of interest that could be exportable to projects. To support this concept, he provided some examples of the three generations of architects of the modern movement, which he considered remarkable because, as he claimed, his emerging ideas proved in a real space and with ephemeral materials [8].

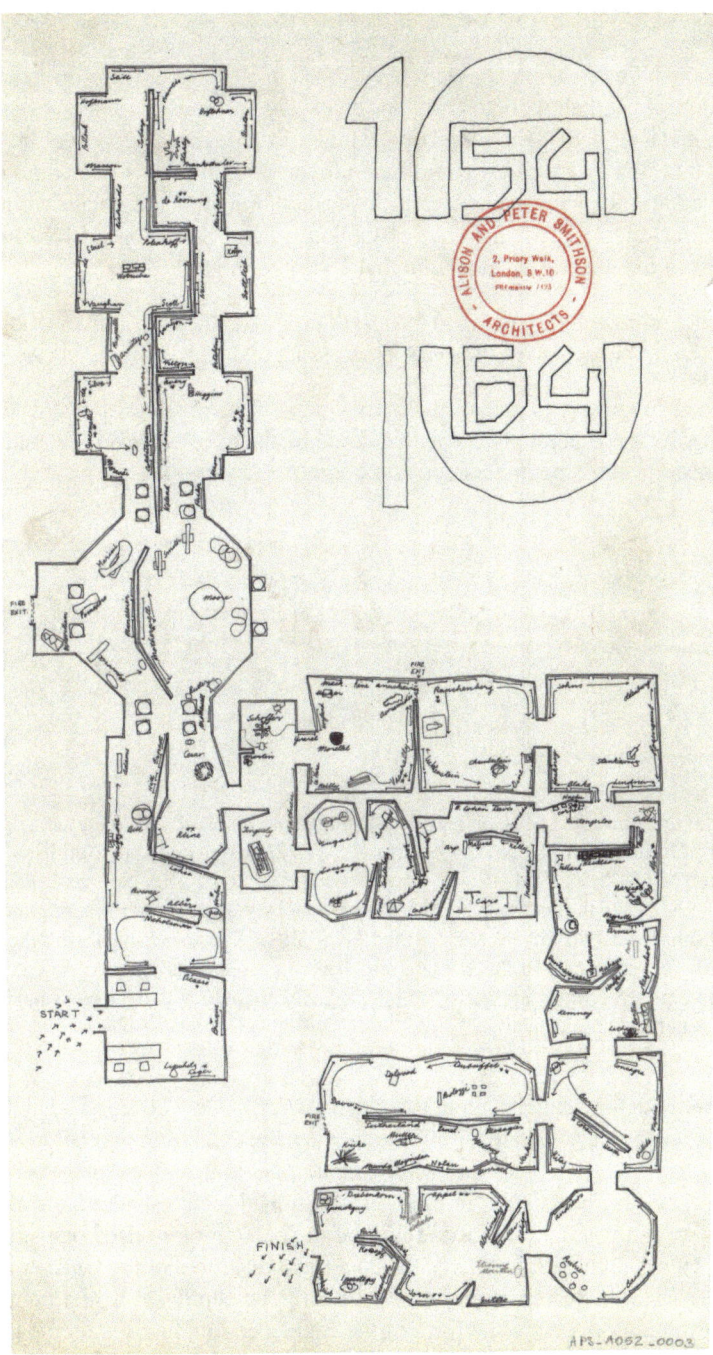

Fig. 7. Alison and Peter Smithson. "Painting and Sculpture of a Decade, 1954–1964", General Plant. Chinese ink drawing on tracing paper. Dimensions 35 × 60 cm. London. 1964. Source: [3]

2.2 Diagrams

The curiosity of finding some diagram plans dated in 1995, after the project, leads us to research more about their work to find some clue. Unglaub and Spellman [9] state that the Smithsons had a particular way of incorporating the diagrams in support of the projects, and added that "they were a new contribution to the discipline of architectural representation." Diagrams showing interconnections and combinations of connected networks, which demonstrated a particular way of thinking. Peter Smithson [10] visualized the foundations of his work in an ideogram that appeared in *Team X First*:

"(The) net of human relations (is) a constellation with different values of diferent parts in an immensely complicated web crossing and recrossing. Brubeck!"

It seemed suggestive to gather all the diagrams of the project, as well as to overlap all of them with the general plan. The first could be the different layers or configuration lines; the overlay could be understood as a complete ideogram of the project (Fig. 8).

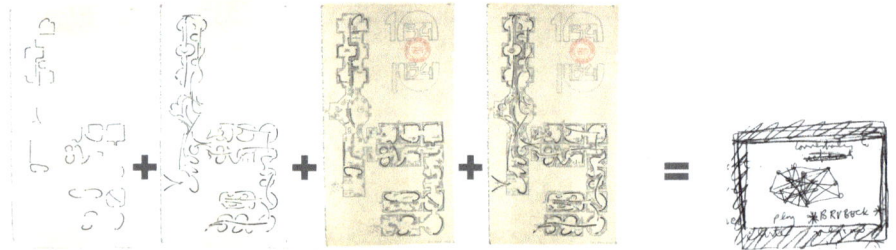

Fig. 8. From left to right: 1) Alison and Peter Smithson. "Painting and Sculpture of a Decade, 1954–1964". Groups of works. NGD 1995. Source: [3]. 2) Alison and Peter Smithson. "Painting and Sculpture of a Decade, 1954–1964". Visitor movements. NGD 2005. Source: [3]. 3, 4) Superposition of 1 and 2 with the general plan (assembly traced by the authors of the communication). 4) Peter Smithson, 1962. Play Brubeck. Published in the Team X Primer manual with the following note from Alison Smithson: "Brubeck: Ideogram of a human relations network. A constellation of parts of different value, in an immensely complicated network of lines that intersect and intertwine. Brubeck! From here a plot can arise."

Together with the diagrams there is a transparency made with a green hatch inserted into an acetate base, also dated in 1995, years after the exhibition was held, as well as a paper copy reproduced in black and white of that time. In the adventure of investigating the subject, we discovered that in the aforementioned lecture [6], a slide with green holograms was shown, and was illustrative of the perception of used paths or ways in the access spaces to the classrooms of the Architecture School of the University of Bath that had been built 6 years before. In the speech Peter Smithson would point out that the way of thinking to obtain the form of the Bath corridors focused on the practice to design the form based on the groupings in the exhibition of the Tate Gallery. Consciously or unconsciously the Smithsons had practiced in the creative process of the exhibition, a way of projecting that they would practice years later moved into a different state, in other words, innovating. In his comments on Bath's hologram, he focuses not only on the expected behaviors of future users, but also on their current use, the forecast of the activities to be developed. In addition, it coincides that it also intervened in an existing building, such as the Tate project (Fig. 9). Looking for their writings, we find more thoughts on it in "Conglomerate ordering. Restating the Possible" [8], as a way to approach to the project by establishing complex relationships at different levels, always with the idea of focusing on the main purpose, which were the experiences in the projected architectural space. The innovative concept of the conglomerate order, according to Millán-Gómez, was introduced by Peter Smithson in 1986 in the ILAUD workshop in Urbino [11], an occasional continuation of the theoretical debates of the influential CIAM or Team X.

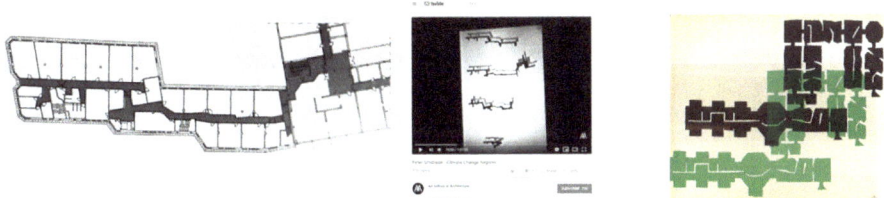

Fig. 9. Left: Alison and Peter Smithson, 1984–1988 Building 6 east, University of Bath. Level 2 floor. Center: Holograms Building 6 east, University of Bath, presented by P.S. at the conference [6]. Source: Architectural Association. Right: "Painting and Sculpture of a Decade, 1954–1964". Transparency with green trimmed adhesive pattern and 4 anchor points drawn in Chinese ink and paper copy. Overlays re-assembled by the authors of the communication. Own source.

3   Conclusion

To conclude, an aspect to highlight is that the Smithsons took especially care of the authorship of each of the project documents. In the publication of their complete works [12], they specify that each drawing, diagram or photograph was properly dated and attributed to the person who had produced it, but not to its evolution, which was always the responsibility of Alison or Peter Smithson.

Although one could already intuit for the delicacy and sensitivity of each one of them, curiously all the drawings of the project that concerns us, have been drawn by female hands: on the one hand, the illustrated floor drawings of the project bear the acronym of AMS, 1963 (Alison Margaret Smithson), and on the other hand, the synthesis diagrams of the architectural spaces, drawn after three decades of ending the exhibition, in this case by NGD, 1995 (Nuria García de Dueñas). It was not difficult to intuit the feminine approach in the elaboration of these careful documents.

To close this interpretation of the graphic work of the exhibition, it seems appropriate to quote a final note by Peter Smithson in the publication "The charged void", which confirms that his "make-draw-think", along with a continuous effort "remake-redraw-rethink", has resulted in the constant discovery of innovative concepts. This is an interesting material as an example of significant graphic heritage for an innovation teaching in architecture [12] (Fig. 10):

Fig. 10. Alison Smithson and Enric Miralles at the UPC School of Architecture, in the framework of the "Upper Lawn" workshop (1986). Source: ETSAB UPC Graphic archive

"The follow-up drawings, to explain better or further evolve an unbuilt project, can sometimes be dated forty years after the project's initiation"

Acknowledgements. The drawings from Alison and Peter Smithson Archive are courtesy of Frances Loeb Library Harvard University Graduate School of Design. The authors would like to thank to Inés Zalduendo and Ann Whiteside for their collaboration.

This investigation was supported by the research group ADR & M-Architecture, Design: Representation and Modeling of the Polytechnic University of Catalonia UPC BarcelonaTech and the organization Consortium for Advanced Studies Abroad (CASA). Isabel Zaragoza, one of the authors, is a Serra Húnter Fellow.

References

1. Smithson, A., Smithson, P.: The Charged Void: Architecture, vol. 1, p. 316. Monacelli Press, New York (2001)
2. Webser, H.: Modernism Without Rhetoric: Essays on the Work of Alison and Peter Smithson, p. 20. Academy Editions, London (1997)
3. Frances Loeb Library, Harvard University, Graduate School of Design. The Alison and Peter Smithson Archive: A descriptive inventory of the archival holdings in the Frances Loeb Library, Graduate School of Design. Frances Loeb Library, Harvard University Graduate School of Design, Cambridge, Mass (2005)
4. The Calouste Gulbenkian Foundation: Painting & Sculpture of a Decade 54-64, 2nd edn. Tate Gallery, London (1964)
5. Smithson, A., Smithson, P.: The Charged Void: Architecture, p. 316. Monacelli Press, New York (2001)
6. Smithson, P.: Climate Change Register. Conferencia en AA School of Architecture, London (1994). https://www.youtube.com/watch?v=WXGNH3x89bg. Accessed 21 July 2019
7. Salter, P.: Strategy and Detail. En: Architecture is Not Made with the Brain: The Labour of Alison and Peter Smithson, pp. 44–49, Architectural Association, London (2005)
8. Smithson, P., Webster, H.: Modernism Without Rhetoric: Essays on the Work of Alison and Peter Smithson, pp. 183–185. Academy Editions, London (1997)

9. Smithson, P., Spellman, C., Unglaub, K.: Peter Smithson: Conversations with Students: A Space for Our Generation, p. 39. Princeton Architectural Press, New York (2005)
10. Krucker, B.: Complex Ordinariness. En: Architecture is Not Made with the Brain: The Labour of Alison and Peter Smithson, pp. 84–90. Architectural Association, London (2005)
11. Millán-Gómez, A.: Notas sobre el orden conglomerado. EGA Expresión Gráfica Arquitectónica n. 11, 91–93 (2006). https://polipapers.upv.es/index.php/EGA/article/view/10318/10055. Accesed 21 July 2019
12. Smithson, A., Smithson, P.: The Charged Void: Architecture, vol. 1, p. 597. Monacelli Press, New York

Louis Meunier: Engraving and Heritage in the Decline of the Spanish Golden Century

Ángel Martínez Díaz$^{(\boxtimes)}$ (iD) and María Teresa García Sánchez (iD)

Escuela Técnica Superior de Arquitectura, Universidad Politécnica de Madrid, Madrid, Spain
angel.martinez@upm.es

Abstract. The research, which result is briefly showed here, approaches the French draftsman and engraver of the XVII century Louis Meunier. Its main purposes are as follows: to collect his Spanish theme work; to analyze it attending its content and formalization; to context it; and to explore its later influence in both the engraving production throughout the following centuries, and the Spanish architectural and urban heritage historiography. His production has been classified according to its principal features, distinguishing two wide groups. The first one is formed by the engravings mainly gathered in his work called Diversas vistas de Las casas y Jardines de plazer del Rei despana… and in other minor collections, which are devoted to a particular building, place or environment. The second group of works, for its part, is composed by the large panoramic views of Spanish cities, whose attribution should take into account the editorial contribution of Israël Silvestre. The text brings to light Meunier most remarkable features regarding themes and aspects strictly graphic. However, it takes part of a wider project related to the architecture and city drawing history, which is being developed in several complementary directions.

Keywords: Meunier · Engraving · Drawing

1 Introduction

Little is known about the Louis Meunier life, sometimes called Meusnier, Musnier o Le Meunier. Dumesnil placed on record this fact in his monumental catalogue (1841, volume V, pp. 245–299), deducing no more than scarce biographical notes by the engravings observation. In some of them, the dates of 1665 (El Escorial) y 1668 (Sevilla, (Fig. 4, left)) appear, thus Meunier could be born between the second and third decade of the century. It is not known the date of his death either. He worked in Paris and other areas of France and, according to his legacy, we might conclude he traveled to Italy, Spain and Portugal. His work bears similarities and coincidences with those by Perelle-Gabriel, the father (1604–1677), Nicolás (1631–1695) and Adam (1638–1695) - e Israël Silvestre (1621–1691) (Fig. 1). From the latter, Meunier could be a disciple and collaborator (Facheux 1857, p. 29). Indeed, they both share, as we will see, the attribution of an essential part of his work.

L. Agustín-Hernández et al. (Eds.): EGA 2020, SSDI 6, pp. 166–177, 2020.
https://doi.org/10.1007/978-3-030-47983-1_15

Fig. 1. Israël Silvestre. "Veüe du Chasteau de Chambor du costé de l'entrée". Ouvre de Silvestre: collection of engravings representing the Versalles royal celebrations, the royal palaces, views of France and Italy…/engraved by Israël Silvestre. Source: BNE ER/2974

2   The Work of Louis Meunier

According to Dumesnil, Meunier's work is composed by 86 engravings (81 excluding the covers, titles and headings). They are etched and almost always finished by dry point or burin. Most of them (all but 9) feature views of diverse places of Spain and Portugal.

2.1   The Views of Buildings and Places

The most significant group of engravings (55) appears on a volume titled *Diversas vistas de Las casas y Jardines de plazer del Rei despana dedicado a La Reina por Louis meunier* (Figs. 2, 3, 5, 6 and 10 left). With an awkward Spanish orthography, it is also written in French. Most of them were reedited in subsequent printings with minor changes of labeling. The volume is divided on four sections with their correspondent heading covers. The first one is devoted to Madrid and its surroundings. Specifically, they are the Alcázar (3), the palace of El Pardo (1) and the Zarzuela (1), including also several views of emblematic places of the capital city (5), of the Aranjuez palace (1) and of its gardens (5). The second section aims also Madrid, in the Buen Retiro palace (5) and Casa de Campo (1), while the third one attends to El Escorial (7). The last part gathers views of Granada (10), Toledo (3), Segovia (2), Sevilla (5), Cádiz (1) and Lisbon (1). Sometimes, it is accompanied by other 8 engravings of Lisbon theme that are not made by Meunier but by the Flemish painter and engraver Thierri Stoop (better known as Dirk Stoop). Specifically, they are related to the wedding of the infant Catalina Enriqueta of Braganza to the king Carlos II of England in 1662.

Fig. 2. Louis Meunier. "La grande fouente en arangoise". c. 1660–70. Diversas vistas de Las casas y Jardines de placer del Rei despana dedicado a La Reina por Louis meunier. Source: BNE, ER/5844.

Furthermore this central corpus of views, other minor collections of engravings exist. Despite some of them bring certain novelties, the majority are copies or versions from the previous ones. One group is devoted once again to Aranjuez, with 9 views and one cover; the other was promoted by the company of the Flemish editor Van Merle, who was active in Amberes in the middle of the century, and addresses scenes of several places. Among them, maybe the most interesting one is a novel view of Cádiz from the sea. The others are two copies of images of the Alhambra, with one or other license, appeared in *Diversas vistas...*, one of Sevilla, and two more of Lisbon, one of which is a copy of the Stoop's original.

In general terms, this part of Meunier's work refers particular themes and places intentionally selected. Regarding *Diversas vistas...*, the title itself clarifies that it is focusing on the royal palaces and their gardens. Even though other singular buildings are represented and not all of the royal residences are there, we might say that the majority of engravings follow that purpose. They offer a way of looking neither completely homogeneous nor equivalent for all the royal places, varying their attention from the buildings to the gardens depending on the specific site and its most outstanding highlights. Regarding the palaces, Meunier offers almost always their general external appearance, understanding this procedure as an essential element of the monumental valuation expected. It is made closely or with the sufficient distance to integrate the palace into an environment that is usually emphasized. This is the case of the fortresses of Madrid, Toledo or Segovia (Fig. 3), the Zarzuela, Pardo, Aranjuez, El Escorial, the Alhambra, Cádiz or Lisbon. In other cases, however, that external appearance is more difficult to be transmitted due to the own features of the building, as in Buen Retiro or in the Alcázar of Sevilla. In few cases, Meunier shows the palaces' interior and, when he does it, he approaches yards or similar spaces, as in the case of Madrid, El Escorial and the Alhambra (Fig. 6). In addition, his fine intuitive narration gives essential prominence to the garden in Aranjuez (Fig. 2) and even in Retiro. Moreover, whereas it is used as a context in the Zarzuela, and as a sole object of interest in Casa de Campo, it is forgotten in all other cases.

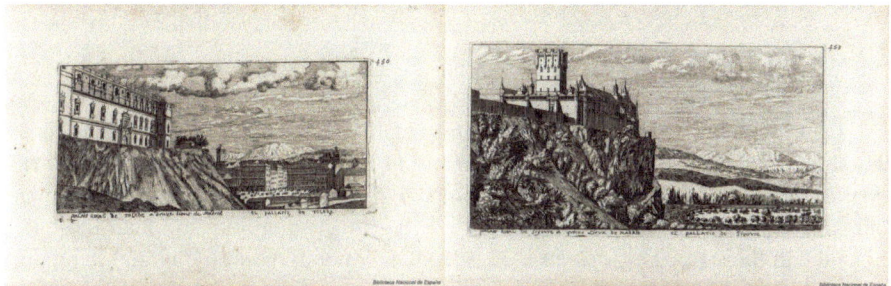

Fig. 3. Louis Meunier. Left: "eL PALLATIO de Sigovie" and right: "EL PALLATIO de TOLEDE". c. 1660–70. *Diversas vistas de Las casas y Jardines de placer del Rei despana dedicado a La Reina por Louis meunier.* Source: BNE, ER/5844.

Despite this selective prominence of palaces or gardens, Meunier's engravings have a clear will to go further when it is considered necessary. The city and the landscape appear in a lot of images, especially when the urban or natural references are unavoidable features for the monument understanding. Mountains, rivers, scarps, vegetation…but also streets, squares or paths are recurrent elements to underline the character of fortresses and palaces. Beyond this subsidiary ambient value, sometimes the site seems to play the main role of the view, while the object becomes one more piece of a complex space engaged in dialogue. This fact affects the place of the building within the print composition, which changes its usual central position by a supporting one that blocks at times the whole vision of the building itself (fortresses of Madrid (Fig. 10 left) or Toledo (Fig. 3 left)). The very fact happens when Meunier draws the fountains of Madrid. Here, his main interest is not the fountains themselves but the space that surrounds them in the squares of Sol, Santo Domingo or Cebada. Furthermore, he does not need almost pretext when he looks from afar the Alhambra in Granada, or when he draws the San Francisco Square in Seville or the *grand place* of Cádiz. The same occurs in the subsequent series to *Diversas vistas*…Without any referral need for the royal houses, he returns one more time to the Guadalquivir facing the city from Triana, or orients himself toward the bay sighting the oriental facade of Cádiz (Fig. 4).

Fig. 4. Louis Meunier. Left: "Veüe et perspective dela grand Eglise de Seville en Espagne et dune partie d ela ville", 1668, and right: "Veüe dune partie de la ville de Cadis en Espagne, du costé du port", c. 1660–70, *Vistas de España, Portugal, Francia e Italia.* Source: Biblioteca Universitaria UPM, ETS Arquitectura, depósito Cebrián T-91.

The engravings size of *Diversas vistas...* and the subsequent series is not always the same, but is around 240 mm width and 130 mm height. The images appear alone in separate pages, except at four of them included in *Diversas vistas...* devoted to Aranjuez fountains, which are gathered in pairs for being much more narrow than the others (around half the size). The formats are horizontal, which forces the framings and points of view many times and, in some extreme cases, compels Meunier to lie blatantly as in the facade of the Toledo cathedral, where the tower appears really flatten to fit in the available space.

All these engravings are conic perspectives. There is no plans, elevations or sections; neither axonometries (in view of the date, it would be better to say parallel perspectives). The inherent perceptual vocation of the conic projection has several registers, from those aspects directly engaged with the strict appearance of the building to others concerned by the place and the environment. The decisions on the perspective construction involve these issues, varying the height and position of the point of view, the horizon line and the angle of vision relatively freely. However, it is possible to find some characteristic invariants. Without doubt, Meunier prefers the central perspective and uses it frequently, forcing the subject if it is necessary, even when the reality is obstinately "oblique". The use of this type of perspective results quite natural to represent building fronts with a vocation of elevation and also spaces with central plans or an orthogonal geometry (for example, the meridional facade of the Alcázar or the Plaza Mayor of Madrid). However, this procedure becomes less natural when it is focused on a building foreshortening as a priority, forcing one side to keep itself parallel to the plane of projection independently of its extension. As regards this issue, which is almost a constant in Meunier's way of working, there are many examples such as the Alcázar of Madrid, Segovia or Escorial, El Pardo or the cathedral of Sevilla (Fig. 5). Depending on the case, this use makes more rigid the resultant image to certain extremes difficult to be credible, specially when it is translated to urban space. In that case, this recurrent "orthogonalization" - no matter its fidelity to reality - does not contribute to generate a realistic perspective image. In addition, the horizon line is often raised with respect to the proper height of an immersive vision. Meanwhile, there exists only one strictly aerial perspective. It is devoted to El Escorial monastery and imitates the engraving by Perret. In this bird's eye view, we can distinguish the whole building frontally from the western. However, this procedure is not so obvious when the vision concerns wide surrounding extensions (in case of El Escorial or Granada for example).

Fig. 5. Louis Meunier. Left: "VISTA DE LA CASA REAL DEL PARDE CASA DE PASER DEL REI DESPANA A DEUX LEUXAS DE MADRID" and right: "La granda iglesia de Seville". c. 1660–70. *Diversas vistas de Las casas y Jardines de placer del Rei despana dedicado a La Reina por Louis meunier.* Source: BNE, ER/5844.

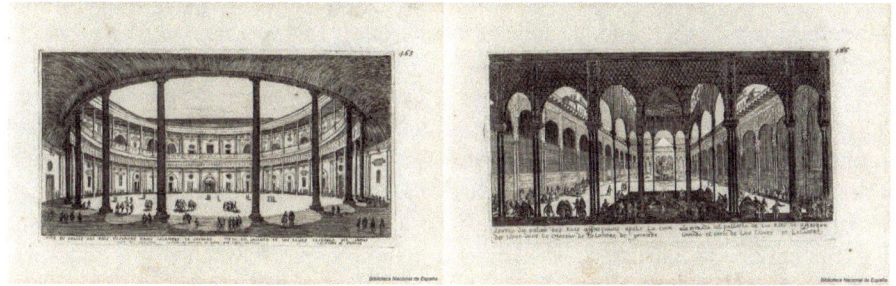

Fig. 6. Louis Meunier. Left: "VISTA DEL PALLATIO DE LOS REYES DESPANA PERDINTRE LALAMBRE DE GRENADE" and right: "LA ENTRADA DEL PALLATIO DE LOS REES DE AFFRIQUA LIAMADO EL PATIO DE LOS LIONES EN ALAMBRE". c. 1660–70. *Diversas vistas de Las casas y Jardines de placer del Rei despana dedicado a La Reina por Louis meunier.* Source: BNE, ER/5844.

There are other resources, in Meunier's engravings, that tend to counteract in a certain way the rigidity derived from the perspective construction. One of the most efficient, in order to involve the spectator in the views, is the introduction of a first plane clearly defined as depth indicator. It can be a tree, figure, group, architectural element or unevenness. Almost always he shows them backlighting, going out of framing in many cases. Regarding this procedure, his views of interior yards are quite significant, when he is placed in the interior of a gallery coming into view the open space behind the corresponding colonnade (courtyards of the Lions, Arrayanes, and of the Palace of Carlos V in the Alhambra (Fig. 6)). The use of light also contributes to introduce the spectator in the views, by means of quite pronounced contrasts and a very rigorous treatment of own and projected shades. Moreover, a certain intention engaged with the line gradation value according to the depth helps to make more realistic the representations, specifically those with wide horizons. Among them, it is able to read frequently ambiental factors related to the weather and its effects on nature. Clouds, water, wind, aridity, leafiness… are fluently evoked by means of the line and with a skilled use of the grain and texture.

Meunier populates the majority of its views with figures that enliven and illustrate the life of the palaces, gardens and cities. In addition, they contribute to give scale to the buildings and spaces showed, almost always lying the reader by means of growing their dimensions, which "monumentalizes" the perception.

2.2   The Views of Cities

There exists another very important and homogeneous group of engravings with a quite different character to the aforementioned ones. This is composed by cities' panoramic views of large format that try to reflect both the exterior image of the architecture and its relation with the surrounding territory. The authorship of them has been controversial due to the engravings - except one - lack the sign or determining physical evidence. Robert Dumesnil (1841, p. 292) considers they *sont indubitablement de lui* and list them including Madrid (712 × 282 mm), Toledo (564 × 272 mm), Segovia (548 × 276 mm), Badajoz (544 × 275 mm), Zafra (556 × 280 mm), Sevilla (693 × 274 mm) and Granada (546 × 278 mm), apart from Lisbon (1.398 × 360 mm), which has both another format with respect to the others and the authorship reference: *designe au naturel eT graue par Louis Meunier*. By his part, Faucheux (1857, p. 76) adds to the list two more cities: Écija (558 × 140 mm) and Turín (558 × 140 mm), stating that these and the previous ones are *Pièces gravées et dessinées par Louis Meusnier sous la direction de Silvestre*. Indeed, all of them appear in one compilation of Silvestre's work, except Lisbon. The pages of the copy preserved in the Biblioteca Nacional of Spain are formed by curious pairs: Granada and Toul, Toledo and Zafra (Fig. 7), Badajoz and Segovia, Sevilla and Madrid (Fig. 8); and one astonishing trio: Écija, Turín and Florencia. Meunier's attribution of the no signed engravings seems plausible if we attend to the graphic resources used (broadly similar to those mentioned above) and the views' geographical distribution (the authoring engravings approach all the cities except Badajoz, Zafra and Écija, which are placed in the path between Madrid, Lisbon and Seville).

Assuming that this attribution is true, the group of views can be considered as one Meunier's contribution as relevant as his other engravings with a minor territorial ambition. His intention here is clearly perceptual, as it occurs in the other cases. Thus, he tries to show quite faithfully the cities' appearance, despite that he abuses the very "frontality" tendency detected in other cases, he exaggerates the natural features that define each city enlarging the proportion or impression of scarps, accidents, mountains or rivers, and he amplifies the size of the more representative buildings in urban skyline. He does not pretend neither to give an idea about the general form of the city plan, nor cover all its perimeter manipulating the view point as it was made before by other draftsmen dealing with the same problem. The horizon line is placed slightly high, as we have seen previously, but not to the extent of raising a bird's eye view. The main purpose here is to make a strong urban portrait with realistic attributes, as it could be reached by one visitor approaching it from the exterior. His delicate use of graphic variables, with a

PROFIL DE LA VILLE DE SAFRA, EN ESPAGNE

PROFIL DE LA VILLE DE TOLEDE CAPITALLE DV ROYAVME DE LA VIEILLE CASTILLE .

Fig. 7. Louis Meunier. PROFIL DE LA VILLE DE SAFRA/PROFIL DE LA VILLE DE TOLEDE CAPITALE DU ROYAUME DE LA VIEILLE CASTILLE. In Israël Silvestre: collection of engravings representing the Versalles royal celebrations, the royal palaces, views of France and Italy…/engraved by Israël Silvestre. Source: BNE ER/2974.

more or less convincing expression of human figure, of the lighting, materiality or depth by means of treatments of line, scratches or textures, are applied in an unequal successful aerial perspective. As in many other engravings, here Meunier always uses a very strong foreground combined with a heaven that appears cloudy at the top and empty of attributes at its encounter with the horizon and the city or mountains silhouette, causing a clear reading of the drawing depth.

The very horizontal format contributes to the purposes of the author, who chooses the more characteristic city side. According to this, it is worth emphasizing the balanced way in which Meunier places himself in front of cities of quite different character, and his attention to orography. Predictably, Madrid shows its cornice with a very grown Manzanares. Likewise, the Guadalquivir, Guadiana or Genil decide the position of Sevilla (with Cartuja interposed), Badajoz or Écija (with the corresponding bridges

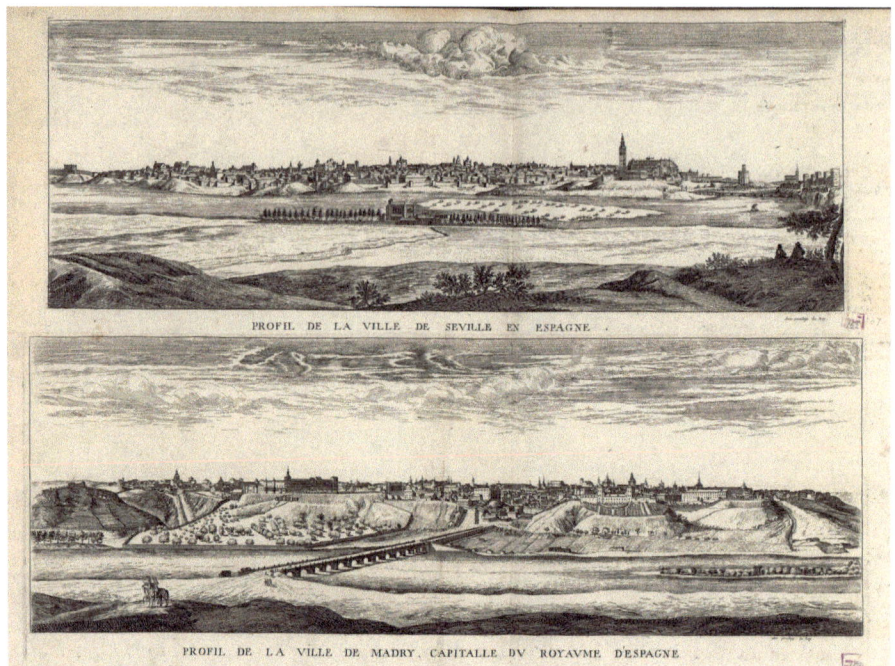

Fig. 8. Louis Meunier. PROFIL DE LA VILLE DE SEVILLE EN ESPAGNE/PROFIL DE LA VILLE DE MADRY CAPITALLE DU ROYAUME D'ESPAGNE. In Israël Silvestre: collection of engravings representing the Versalles royal celebrations, the royal palaces, views of France and Italy…/engraved by Israël Silvestre r. Fuente: BNE ER/2974.

very present). In Segovia, the Eresma River gives up its prominence for the scarps, crossing longitudinally the framing but placed in a second plane behind its own vegetation. In Toledo, and despite the strong landscape influence of the Tajo, the northern facade is the most dignified, with the wall and door of Bisagra, It conditions the view orientation, as in the Greco's painting, pushing the river into a symbolic lateral position. In the view of Granada, the Darro disappears due to the difficulty to include it in a general vision able to reach Sierra Nevada in its full extent. In Zafra, with no river, the dialogue between the Dukes of Feria palace and the Candelaria church, together with the entrance to the urban enclosure, decides the orientation of the view.

3   Before and After Meunier

We cannot support the originality of Meunier's work neither on his obvious knowledge of the drawing and engraving craft, nor on the use of the graphical and narrative techniques involved in the perspective views similarly applied by his French contemporaries, nor even on the unequal quality of his production. His essential role in the Spanish panorama derives from his temporal contextualization: he was a precursor, because of both his view intention and the systematic character of his work. There were several registers of the image of the Spanish architectural heritage, but not so many. Indeed, the number of cases

Fig. 9. Pieter Van der AA. Diverse views of Granada copied of Meunier's originals: *La galerie agreable du monde, Tome II, du Roïaume d'Espagne*. Source: BNE ER/2055.

decreases when we search for the engravings only. This is not true at all considering the image of the city as a whole. As regards, the monumental *Civitates Orbis Terrarum*, its 6 volumes (1572, 1575, 1581, 1588, 1598 y 1617) and its sequelae (Meisner (1626); Janssonius (1657) among others), well provided by a large number of views of Spanish cities, most of which were admirably drawn by Hoefnagel and edited and re-edited many times, should be incorporated as main precedents. Passing by the shy - and rough - contribution of Pedro de Medina (1548), we find Meunier together with other two similar systematic attempts, only drawn: the work of Wingaerde, one century before, and the contemporary work of Baldi (Martínez Díaz and Muñoz de Pablo 2014). It would be worth to mention the work of Texeira for Felipe IV describing the coastal cities of his kingdom.

However, the contribution of Meunier becomes fundamental in the closest scale. His work related to the royal houses and gardens, and the urban spaces engaged with them, is the first representation of a relevant heritage site and with a clear systematic vocation to place value on the object represented itself. Before Meunier, the engravings with great buildings can either be occasional, or have a supporting role accompanying texts, maps, plans o rgeneral views, or appear as reference, pretext or scenographic background of an exaltation, narration or specific event. The engravings of Perret-Herrera of El Escorial, a Spanish rarity in the strict sense of the term, are the sole exception due to the other significant group of engravings we could mention: the cathedral of Sevilla, included in the work of Fernando de la Torre Farfán (1671), was edited on occasion of the Fernando III canonization and is probably later to Meunier's work (Cea Brea 2013, pp. 42). By tracking each city we find how the engravings produced before Meunier had other nature and sense: Madrid with the view of the Alcázar by Vermeyen (1534), the Guadalajara door in the work of Diego de Colmenares (c. 1629), Toledo with the approaches to the Alcázar and cathedral included by Hoefnagel in the second view of the city in *Civitates...,* or the images of the plan of Portocarrero (1681) slightly after Meunier, or Sevilla and the details of the Giralda in *Civitates...*

Meunier's engravings were copied with more or less fidelity throughout more than a century. They became a relevant piece of the commercial mechanism of consuming printed images. Van der Berge (c. 1700 and c. 1717), María Coronelli (1706) or van der Aa (1707) (Fig. 9) with his most known edition under the authorship of Álvarez de Colmenar (1715), maybe are the most relevant samples of the utilization of Meunier in general works (Gámiz 2008, pp. 127–146). However, the view displayed by an anonymous Italian engraving preserved in the Museo de Historia of Madrid (IN. 22137) could be its most poetic transformation. Here, the inexistent sea of the capital city laps the scarped risks where the Alcázar is settled, converted into cliffs over a harbor occupied by great sailing boats.

Fig. 10. Left: Louis Meunier. "el PALLATIO de MAdRId de La paRTe de Los canpos". c. 1660–70. *Diversas vistas de Las casas y Jardines de placer del Rei despana dedicado a La Reina por Louis meunier.* Source: BNE, ER/5844. Right: Anonymous. "Vedutta del Palazzo reale di MADRID del lato della Campagna". siglo XVIII. Source: Museo de Historia of Madrid IN. 22137.

4   Conclusions

The relevance of Meunier in the Spanish historiographic panorama mainly derives from the testimonial value of his work. It has allowed developing and justifying hypothesis involved in the physical life of buildings and architectural sites of primary importance, specifically those that are now disappeared or highly transformed. However, beyond this incontestable instrumental historiographic value, it would be also worth to precise the criteria for analyzing his production, thus nuancing a certain tendency towards the severe judgement of its quality regarding a perspective not only strictly graphic but also related to its precedents, immediate vicinity and subsequent repercussion.

References

van der Aa, P.: Beschyving van Spanjen en Portugal… por Pierre Vander AA. Leyden Versión francesa: La galerie agreable du monde, Tome II, du Roïaume d'Espagne. Leide (1707)

Álvarez de Colmenar, J.: Les Delices de L'Espagne & du Portugal… por Pierre Vander Aa, vol. 6, Leide (1715)

Van der Berge, P.: Heatrum Hispaniae… Amsterdam (1700)

Braun, G., Hogenberg, F. (eds.), Hoefnagel, J. (dib): Civitates Orbis Terrarum. Tomo I Colonia y Amberes. 1575 Tomo II. 1581 Tomo III. 1588 Tomo IV. 1598 Tomo V. 1617 Tomo VI. Colonia (1572)

Cea Brea, M.: La memoria visual de la arquitectura española en los grabados de la Edad moderna. Anales de Historia del Arte, vol. 23, pp. 37–50. Núm Esecial (2013)

Coronelli, V.M.: Teatro della guerra. Gran Bretagna, Spagna, Portogallo, vol. II (1706)

Robert Dumesnil, A.-P.-F.: Le peintre-graveur français, ou Catalogue raisonné des estampes gravées par les peintres et les dessinateurs de l'école française: ouv- rage faisant suite au Peintre-graveur de M. Bartsch. Tomo 5. Imprimerie de Bouchard-Huzard, París (1841)

Faucheux, L.-É.: Catalogue raisonné de toutes les estampes qui forment l'œuvre d'Israel Silvestre: précédé d'une notice sur sa vie. Bonaventure et Ducessois, París (1857)

Gámiz Gordo, A.: Alhambra. Imágenes de ciudad y paisaje (hasta 1800). Fundación El Legado Andalusí, Granada (2008)

Janssonius, J.: Theatrum In quo visuntur Illustriores Hispaniae Urbes, Aliaque Ad Orientem & Austrum Civitates Celebriores…. Francofurti (c1657)

Martínez Díaz, Á., Muñoz de Pablo, M.J.: Wyngaerde y Baldi, ¿dibujante o arquitecto? Dos miradas viajeras a ciudades españolas. 15 Congreso Internacional de Expresión Gráfica Arquitectónica. Las Palmas de Gran Canaria, pp. 541–548 (2014)

Medina, P.: Libro de las Grandezas y cosas memorables de España. Sevilla (1548)

Meisner, D.: Thesaurus sapientiae civilis, sive vitae humanae ac virtutum et vitiorum Theatrum: symbolis aeri incises (1626)

Berge, P.V.D.: Theatrum Hispaniae... Amsterdam (1717)

de Colmenares, D.: Historia de la insigne ciudad de Segouia y conpendio delas historias de Castilla. Engraving of the Door of Guadalajara (Madrid), Diego de Astor fecit 1629, p. 175. Diego Díez, Segovia (1637)

Leonardo, Juan Francisco, Toledo (Arzobispado), map known as Plano de Portocarrero (1681)

de la Torre Farfán, F.: Fiesta de la Santa Iglesia Mctropolitana y Patriarcal de Sevilla al nuevo culto del señor rey Fernando. Sevilla (1671)

Vermeyen, J.C.: Le chasteu de Madrid et l'aqueduct de Segovie (1534)

The Heritage of the Unbuilt: The Value of the Drawing

Noelia Galvan-Desvaux⬛, Antonio Álvaro-Tordesillas[✉]⬛,
and Marta Alonso-Rodriguez[✉]⬛

Valladolid University, Valladolid, Spain
noelia.galvan@uva.es

Abstract. The graphic heritage is part of an architectural legacy that we must recover, especially in the case of those works that, for various reasons, were not built but were preserved through drawings.

Recovering them means reviving them in some way, for which we must imbibe their rules and understand their architecture from their design process. With this desire, we will address the restitution of a series of houses of the modernity, which we will study from their drawings to the subsequent three-dimensional modeling.

The new technologies are a fundamental tool that allow us to achieve, not only the establishment of the different version of the houses, but to create a tangible element through its recreation in three dimensions. With this we will be able to understand the project process of those works that were born with the desire to be built.

The subsequent analysis will bring up the architecture and, with it, all those ideas that, in addition to being understood through this study, will make us see the trajectory of some of the best-known architects in a different way.

Conceiving the drawing of the unbuilt as a heritage to be recovered, allows us to configure the present and, to a large extent, innovate towards the future in our teaching.

Keywords: Unbuilt architecture · Drawing · Idea · Analysis

1 About the Failed or Unfinished

Addressing the issue of heritage from the study of the unbuilt may, a priori, be a contradiction. It is true that what we understand by heritage refers to what belongs to us or what we inherit with a characteristic value. From this premise, to reason that the drawings of an architect are a fundamental part of his legacy and, therefore, that the drawing can constitute by itself an inheritance of great value, bases the foundations of what this study intends to be.

The architect, as an artist, creates. And that work becomes the heritage for the one who perceives and studies it. So, our task is to take care of those unfinished works and, in a way, try to revive them as an inheritance received from the previous architects.

L. Agustín-Hernández et al. (Eds.): EGA 2020, SSDI 6, pp. 178–188, 2020.
https://doi.org/10.1007/978-3-030-47983-1_16

We should then work with the history and heritage understood, not only as a collection of facts and constructions, but also as those failed or postponed moments that endow the architectonic timeline with meaning and understanding. This is what Tafuri (1997, 384–385) understands when he affirms that "today we are forced to recognize in history not a large deposit of codified values but a huge collection of utopias, failures and betrayals (…) it is rather the architecture, in its constitution, in its change, in its attempt to create again the new objectives and the values themselves, which gives history meanings in perpetual metamorphosis". The interpretation we make of these works will be the one that allows us to value their contribution to the history of architecture and will also allow us to create new perceptions in that future that underlies the critique of architecture (Lizondo 2016, 29).

Now, the key to this question is how to understand and transcribe those works that only exist in the drawing. Gallego Jorreto (2004, 27) says that "an architecture in plans or drawings, when not yet built, is not architecture. When we are interested, it is by its power of suggestion (…) We imagine the architecture that evokes with us inside it, feeling it and experiencing it. How would we be surprised if it was built? We will never know. How would you build it? We can only imagine it." (Fig. 1).

Fig. 1. Model of the house for Josephine Baker, Adolf Loos, 1928. Source: Albertina Gallery © Armin Linke, 2012.

And so, we can find the famous definition of William Morris in 1881 "the architecture is the set of modifications and alterations introduced in the earth's surface in order to meet human needs, except only the pure desert" (Morris 1999); To August Perret's categorical statement that "an architect is a poet who thinks and speaks in the language of construction", we always find implicit the reciprocity between architecture and construction, as if the one without the other did not exist.

But there are other approaches that address this question from a broader vision, which allows us to focus on the issue that we are dealing with. Louis Kahn says "(…) that architecture does not exist. What does exist is each concrete work of architecture. And a work is an offering to architecture, with the hope that this work could become part of the architectural treasure. Not all buildings are architecture" (Kahn 2003, 228). An architect, therefore, is a creator of worlds that somehow improves the lives of those who inhabit them. Kahn proposes, based on a somewhat metaphysical view, that a work to be considered architecture should meet other conditions outside its physical realization. In particular, it refers to the need for it to serve the institution for which it was created, to respond to it so that the truth of architecture can arise.

Therefore we cannot assess a project based on its physical materiality, knowing that sometimes there are unfinished works that have greater values than others that do exist (Mariano Bayón in Vellés 2004, 11). This is due to the poetic capacity of the drawing that could suggest and recreate from our own imagination.

2 Architectures *ausentes*

In the etymology of the spanish *ausentes* term we find the root of the verb "to be" that, through its compound, produces the Greek word abesse, whose meaning is "to be far". It is important to denote that in this term, with which we define the architectural group to be addressed, two ideas appear - that of being and that of staying- that refer us to the fundamentals of the problem to be treated (Fig. 2).

Fig. 2. Restoration of the brick wall house of Mies Van der Rohe, 1924–25, of which only images of three original drawings are preserved. End of degree work. Source: Alberto Galán Muñóz.

According to Mariano Bayón, we can say that these absent projects are: "an elucidation of several coincident architectures, all of them in the fact that they are nonexistent today in a strict physical sense. (..) All of them however, belonging to the twentieth century, continue still today to be an authentic unfinished light of suggestion and reason for the architecture of this twenty-first century" (Vellés 2004, 11).

Then the need arises to verify what happened with all those projects that were stored, discarded or reused, or even in some cases built and destroyed (although this would be another investigation).

Investigating these absent architectures reminds us of well-known works such as Koolhaas proposal for the Villete Park, Mies's Glass Skyscraper, the Venice Hospital of Le Corbusier, or other visionary proposals such as those of Archigram or Ledoux. There are revolutionary works in them, experimental and visionary projects, and none of them are vain architectures.

These absent architectures[1] are far apart, as the root of the term explains itself, because although they are in the drawing, they have lost their being. An architect as prone to unfinished projects as Louis Kahn indicated that "everything that is not built is not really lost. Once its value has been established, the demand for presence is undeniable. He is just waiting for the right circumstances" (Lobell 1979, 84) (Fig. 3).

Fig. 3. Infographics of the house behind courtyards by Mies van der Rohe, 1934. 3D modeling and post-production. Source: Diego José Hernández Julián.

So recovering the assets of these projects means taking advantage of the circumstances and reliving them from the suggestion and evocation, as Gallego points out "through its re-view, to re-view not to review" (2004, 27).

The first danger appears here, and when we interpret a project, we inevitably generate an inference. As if it were Schrodinger's famous paradox, that absent architecture is in a superposition of possible states until we intervene as observers.

Thus, the interpretative speculations that we make about the work will crystallize in a definitive state in which, without remedy, a change will take place even if it is almost

---

[1] When we talk about absent architectures, we cannot fail to refer to the exhibition held at the Archery of New Ministries in Madrid in 2004, which was entitled "Architectures of the twentieth century" that shows the unbuilt works of well-known architects from the history of architecture, such as the Silkeborg Museum by Jorn Utzon or the Congress Palace of Venice by Louis Kahn.

imperceptible. But this reaction is not something negative if what we are looking for when investigating a work, in addition to understanding, is an answer.

It is then about the intentionality of the drawing, the main element when dealing with this type of project. This intention that the author himself has already set, and that conditions us initially, and that proposed by the analyst, which does not have to be the same. In this way we will be - now yes - building through our intentional drawing, an architecture that does not exist physically. And it is through the observation of the architect's original drawings that we will obtain many of the own characteristics that reflect the intentionality of the work (Ochotorena 1996) because, as Cortés (1992) states "The architect, when is drawing, is already building its architecture".

3   Methodological Strategies

The problem lies in addressing the graphic heritage. We will be focusing on housing as the main topic of research, to rescue these unfinished aspirations from their ideas, as pieces that explain and complete the trajectory of the most famous architects.

The premise that we establish is to adopt the house as a work theme, understood as the origin of the architecture -Laugier- and as an eternal project -Kahn. The modern house appears as well as the materialization of the need to live and the field of experimentation of the architects of the twentieth century. In short, and now with a teaching nature, we choose the house because it restricts more complex attributes of other typologies, and modernity because it is the tradition to which we belong and in which our work takes place.

The next step would be to select the projects on which to work from the heterogeneous group of unbuilt houses of modernity. From well-known homes, such as patio houses or the brick wall house of Mies van der Rohe, to more experimental ones, such as Kahn's Adler house; or even projects in which the documentation appears fragmented or difficult to access, such as the Loos house for Josephine Baker. They all share a premise: we have enough documentation so that, through the drawings, we are able to recreate them.

Once the work object has been defined and its graphic vestiges collected, it is about understanding the history of the project. Cataloging and ordering the drawings allows to integrate the work within a timeline and show, not only the final moment in which it was interrupted, but the multiple crystallized states of its course, in order to understand the development of each house as a seriation of versions.

At this point, the exploration drawing is the perfect tool for personal transcription of the originals, which in some cases are no more than some sketches or partial plans. The graphic language becomes the ideal instrument of expression, but also the tool to generate knowledge capable of providing new meanings.

Redrawing the unbuilt architecture is a principle of uncertainty that, as we have said, comes from a dilemma about how we recreate the original object through our preconceived ideas. When we propose a new truth for that unfinished work there is no doubt that we do it from a specific interest and through an intentional language. Deepen into the architect's drawings, in his schemes and sketches, in the delineated plans and the models, even in the letters and interviews, does nothing more than allow us to access the genesis of the project (Fig. 4).

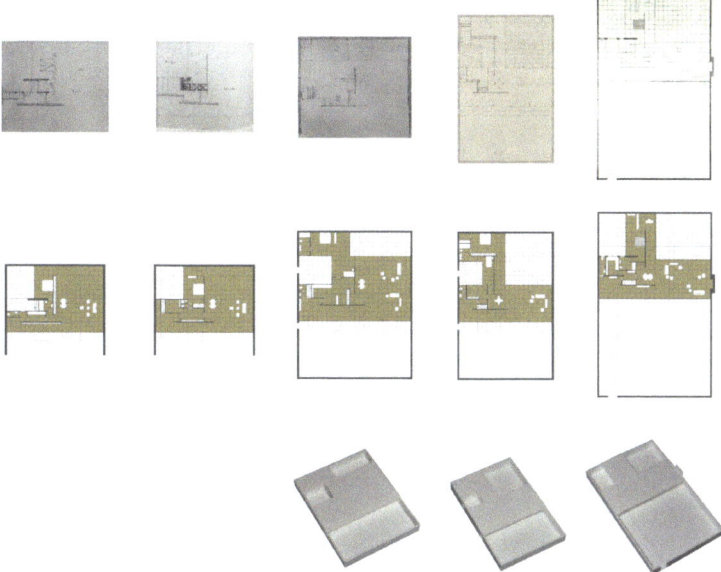

Fig. 4. Different versions of Mies Van der Rohe's three patios house 1934. Source: Diego José Hernández Julián.

The conclusions we obtain from this, passed through the filter of our drawing, should bring new visions and modify the initial vision that we had of that work, or the architect's trajectory. The truth about the work, therefore, is likely to be found, but also to be created, when through our drawing we build these unfinished architectures.

The need arises then to verify the new materiality of that reborn entity through different graphic tools, and to establish clear working rules so that the uncertainty of the interpretation does not take over the recreation of the original. In the words of the philosopher Rorty (1991) almost anything can be considered "good or bad, important or insignificant, useful or useless, (simply) rewriting it."

4   Extrapolations Looking for Architecture

As of this moment, we are already able to configure simultaneous visions that, thanks to the homogeneous code used in the graphic reinterpretation process, allow us to draw conclusions. Helio Piñón considers that in this type of research "the reversal of the usual process or, in other words, given a building to look for the architecture" (Piñón 2006) should be carried out.

The purpose is to reconstruct, based on the documentation consulted and, on the architect's, constructed work, the course of events to establish some bases from which to obtain conclusions for our investigation.

The digital materialization of the works has been experienced through various systems linked to the new means of architecture representation. A recreation by 3D printing was planned based on basic digital models, generating volumetric models and a subsequent rendering, from REVIT or 3DStudio with Photoshop postproduction, even some trials with augmented reality (Fig. 5).

Fig. 5. Infographic of Le Corbusier's Villa Meyer 1925–26. 3D modeling and subsequent postproduction. Source: Raúl Villafáñez Marcos.

The three-dimensional modeling poses, on occasion, problems that we must solve from an interpretation of the sources and a previous study of other projects carried out, where we can clarify details that in the work, as it hasn't being carried out, has not yet been defined.

It is our job, as architects and at the same time as teachers, to emphasize the creative character of the tools, which can provide visions that enhance the experimental and emotional when it comes to showing these unbuilt projects. In short, it is about digitally resuscitating non-built works so that they can be part of the global heritage, whether scientific or non-specialized, and from this point what has to do with experimentation is fundamental.

There are many experiences in this regard, especially as far as infographics are concerned (Larson 2000). In this sense, the results, in general, are too artificial or realistic, when, from our point of view, the image to be offered should be close to the model or at least maintain a certain dose of abstraction and a link with the intentions of the architect that allows to recreate an architectural atmosphere. We must flee from those media that present the images as dead natures, and in that sense the video and the movement would allow us to introduce the aspect of personal experience (Fig. 6).

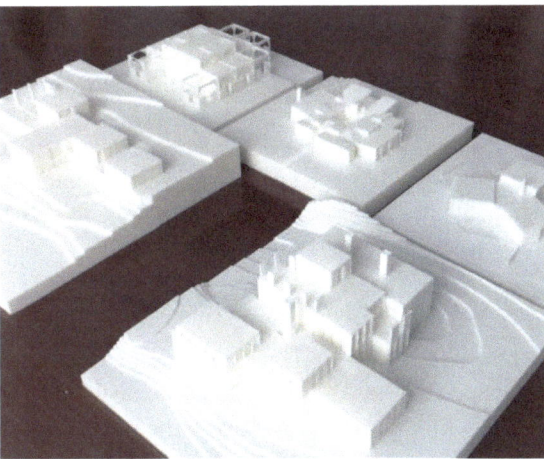

Fig. 6. Models made with 3D printing of several unbuilt homes of Louis Kahn between the fifties and sixties. Source: Noelia Galván Desvaux and Luis Matas Royo

The navigability of digital models and environments has been generally treated through two means: 3D printing and augmented reality. As for the physical recreation of the work at scale, 3D printing allows us to obtain small models - initially volumetric - of the different versions of the house and reproduce them under the same code, achieving the infinite cloning of the piece and the exploration in the project process (Carazo 2014, 64). Thus, we will be able to use the printed model quickly and economically, in the phase of graphic verification and research, but also as a representation of architecture, on a small scale. We speak of manageable models that, due to their size, allow us an almost axonometric vision, approaching the real experimentation of the architectural object under the light (Carazo 2014, 66).

Regarding the use of augmented reality, we have worked with free software and with an existing topography on which we project various three-dimensional models that provide not only physical verification but also analysis. It is then about the superposition of information and the possibility of perceiving different readings of the same work (Fig. 7).

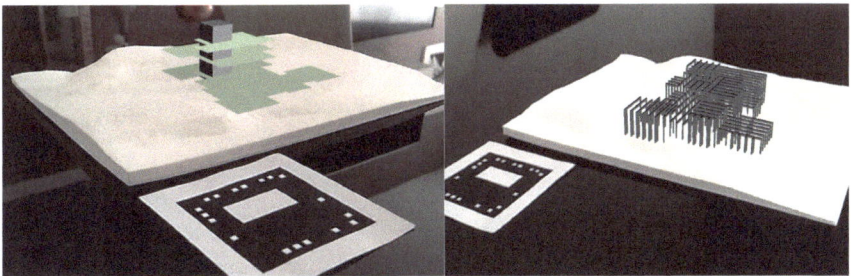

Fig. 7. Topographic model printed in 3D and augmented reality for the analysis of the Morris house by Louis Kahn, 1958. Source: Álvaro Martín González.

So the fundamental contribution of this work is not the compilation or recreation of houses, but the search for ideas from which to found analytical categories suitable to make us understand the architectural process and experiment with the graphic to achieve it.

The idea is to have a total vision of the architect's work, and through the analytical drawing and the oral discourse, to deepen what it suggests by introducing these unborn projects, unknown in many cases, that explain in a new way those that were built.

5 Analytical Categories

Up to this point we have dedicated ourselves to rebuild, to redraw, to rethink the architectures under study with the means in our reach. The analytical phase, which arises later, consists in looking with new eyes, in interpreting and articulating readings that allow us to establish some categories. These arise thanks to this graphical-analytical instrumentation and will be able to contribute ideas and theories to the architectural critique.

The analytical categories approach the project process in that clear binomial that arises between idea and analysis. The architecture does not appear only as a mere transcription drawing, but arises to develop a narrative discourse where "color distinguishes and separates, analytical schemes follow each other, models acquire an objectual character and references to history or context are explicitly reflect" (Grijalba 2004).

The concepts are also linked to projects already built and are singled out and settle here, through these circumscribed works. The drawings used are limited to the scope of the diagram, proposing comparative analysis of the different versions or criteria that generate information by themselves since they synthetically house the intentions of who projected it (Carazo 2017).

When defining the categories we have to establish links between the concepts that, after the graphic rereading, seem relevant to us. These categories are going to become instruments that will guide our analysis from a theoretical formulation (Fig. 8).

The topics discussed inquire about issues that have to do with place, form, space organizations or circulation among others, and are specified in different categories among which we can not miss the *raumplan* (if we refer to Adolf Loos), the transparency (if we deal with Mies van der Rohe) or order (if we do it with Louis Kahn). These, which come from a historical analysis close to the one formulated, are grouped with other specific ones of the work to be treated. And so, the categories are based on drawings that turn these unbuilt architectures into diagrams, compare each other and expose theories that were obtained with the analysis.

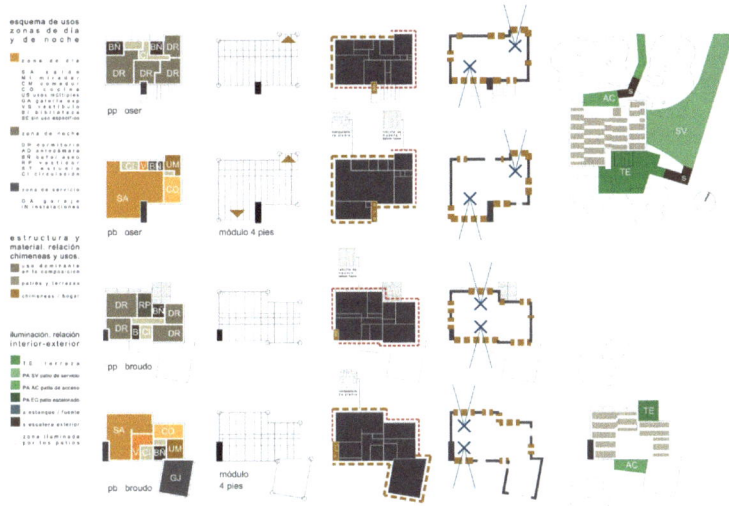

Fig. 8. Diagrams of the project versions of the Broudo house by Louis Kahn, 1941. Source: Noelia Galván Desvaux.

6   Conclusion

The topic discussed here may raise different points of view. It is true that we can agree or not with the system used, but the undeniable thing is that the study and recreation of unbuilt projects produce an intellectual value, as scientific research is based on parameters that have to do with rigor, innovation and contribution.

We work from a continuous line of research. Each new project restored - we have already added more than twenty projects to our work - complements the previous ones and redefines the vision of the next project. Thus, these works that are missing links will be transformed for us into new blocks that fill the wall of knowledge, connections and influences, which allow us to know the past but also to relate it to the current architecture.

The use of new digital resources when carrying out this work raises doubts and problems, also triumphs, but above all it contributes to the enrichment of the graphic heritage. The link between drawing, project and historical research relates our scientific practice and the teaching we exercise, being a promising example of synthesis.

The drawing is thus the guardian of the memory of our architectural heritage. Those strokes of a past world revive when used to build the present and innovate in the future. The opportunity to vivify those unbuilt projects with the new digital tools makes them recover the will to exist witch with they were born in their drawings.

References

Carazo, E., Galván, N.: Diagramas: del isotype al gif notas para una didáctica del análisis gráfico de la arquitectura. EGA- Revista de Expresión Gráfica Arquitectónica **30**, 30–41 (2017)

Carazo, E., Galván, N.: Aprendiendo con maquetas. Pequeñas maquetas para el análisis de arquitectura. EGA Expresión Gráfica Arquitectónica **24**, 62–71 (2014)

Cortés, J.A, Moneo, J.R.: Comentarios sobre dibujos de 20 arquitectos actuales. ETSA, Barcelona (1992)

Gallego Jorreto, M.: Alejandro de la Sota: viviendas en Alcudia, Mallorca, 1984. Rueda, Madrid (2004)

Grijalba, A.: Dibujar lo que no vemos. En: X Congreso Internacional de Expresión Gráfica Arquitectónica. Granada (2004)

Larson, K., Scully, V., Mitchell, W.J.: Louis I. Kahn: Unbuilt Masterworks. The Monacelli Press, Nueva York (2000)

Lizondo, L., Santatecla, J., Salvador, N.: Mies en Bruselas 1934. Síntesis de una arquitectura expositiva no construida. VLC arquitectura 3(1), 29–53 (2016)

Lobell, J.: Between Silence and Light: Spirit in the Architecture of Louis Kahn. Shambhala, Boston (1979)

Morris, W.: On Art and Socialism. Dover Publications, Dover (1999)

Ochotorena, J.M.: Sobre dibujo y diseño. T6 ediciones, Pamplona (1996)

Piñón, H.: Teoría del proyecto. Ediciones UPC, Barcelona (2006)

Rispa Márquez, R.: Arquitecturas ausentes del siglo XX. Tanais, Madrid (2005)

Rorty, R.: Contingencia, ironía y solidaridad. Paidós, Barcelona (1991)

Tafuri, M.: Teorías e Historia de la arquitectura. Celeste Ediciones, Madrid (1997)

Vellés, J., Casariego, M.: Louis I. Kahn: Palazzo dei Congressi, Venezia, 1968–1974. Rueda, Madrid (2004)

Inhabited Drawings: A Century of Manners of Representing the Architecture Experience

María Teresa García Sánchez[(⊠)] 🆔 and Ángel Martínez Díaz 🆔

School of Architecture, Technical University of Madrid, Madrid, Spain
`mariateresa.garcia@upm.es`

Abstract. This research aims to elucidate how the architectural drawing expresses the experiencing architecture. To this end, it approaches around one hundred of drawings of the last century as cases of study in which the human figure appears. By means of their close observation, it establishes a conceptual matrix composed of spatial and temporal elements. The size, position and direction of the bodies within space, the number of actors, the graphic techniques, the actions of the figures and even the moment of their insertion in the drawing, structure our systematic study of the drawings and allow to raise compared analyses among them. Beyond this point, the research addresses those drawings inhabited by other kinds of occupiers such as buildings, sculptures, pieces of furniture or machines that likewise evoke a particular manner of experiencing architecture.

Keywords: Architecture · Drawing · Human figure

1 Introduction

This research approaches the architectural drawing as operating analogy of reality, suggesting its study as if it was a sort of stage; a space with its own spatial and temporal coordinates disposed by the architect, a s asort of theatre director, to be experienced by all kind of occupiers (Rasmussen 2004, 17).

To start with, we might say that each actor inhabits the drawing dually. On the one hand, it occupies a stable and univocal place, which is the result of the relationships of coexistence among the physical elements that define it. On the other hand, it recreates a space as mobility crossing, which appears and disappears depending on how the place is put into practice by the occupier. Taking into account the vectors of direction, speed quantity and even the time factor, the flows, actions, tensions or encounters that result of the actors' experience make the drawing as a frame of a certain geometrical eloquence (Delgado 2007, 71). The figure addresses, appeals and confronts corporeally that well-disposed stage, establishing different levels of proximity depending on the affected sense (Pallasmaa 2012, 57). During this continuous interaction process, the actor integrates the place represented into a complex net of internal references. Doing so, he strengthens his identity as being in the drawing world (Bloomer and Moore 1982, 62).

But, to what extent the architectural drawing has tried to evoke the sensible experience of reality? with which intentions? by means of what graphic devices? To sum up, what

L. Agustín-Hernández et al. (Eds.): EGA 2020, SSDI 6, pp. 189–196, 2020.
https://doi.org/10.1007/978-3-030-47983-1_17

sensible values adds incorporating the human figure to the graphic expression of the architectural form? To approach these questions, the research deals with more than one hundred of drawings, made throughout the last century. Their selection and study has tried to be as much systematic as possible regarding authors, themes, historical periods, systems of representation, techniques, etc. From their comparative analysis, this paper presents orderly the main results of research. The very structure of the following text would like to be read as one more result of the study; maybe it could be its most relevant achievement.

2   Inhabiting the Drawing

The following study reads the drawing architecture as scenographic device. Using the figures inside each example as guideline for discussion, it considers spatial and temporal aspects only pertinent during the action moment of its inhabitants (Joseph 1999, 92). The results obtained are ordered in two parts. The first one is devoted to the size of the characters, their number and hierarchy into the drawing, their position and direction, and the technique applied for their execution. The second one addresses their potential movement during the development of a particular action, the moment in which the figures are inserted in the drawing, and even the speed of their trajectories. By means of these issues, the research approaches the graphical mechanisms that are used to express the sensible experience of the drawing form, from the perspective of both the character in action and the spectator itself.

2.1   The Size of the Characters

This point sets out two substantially different situations. In the first ones, the size of the figures is scrupulously proportional to the dimension of the form that holds them. On the contrary, there are situations without this kind of concordance (Fig. 1).

Any alteration of the figures measurement provokes scalar interferences in perception. Size reductions add a certain monumental character to the drawing form. In these cases, the space seems to expand proportionally to the time required by the actor to experience it corporally. Getting bigger, the form seems to mute; it becomes less apprehensible sensorially by the figure that inhabits there.

2.2   The Number and Hierarchy of the Figures

Three ways of inhabiting the drawing structure this point: situations with only one character, as a soloist; dialogues between two actors; or choirs formed by groups of figures (Fig. 2). Whereas the incorporation of a sole character directs our look towards a restricted area, the entrance in scene of two or more actors urges our eye to travel.

The attributes of each actor within the stage guide our look towards one side or another. In continuous sway, it evaluates elements and formal properties linking together the figures as soon as they share similar features. Forms, sizes, weights, colours and textures, together with the position and distancing from the rest of figures, define the length, direction and sense of the chain. Each path chosen by the eye among all the unlimited

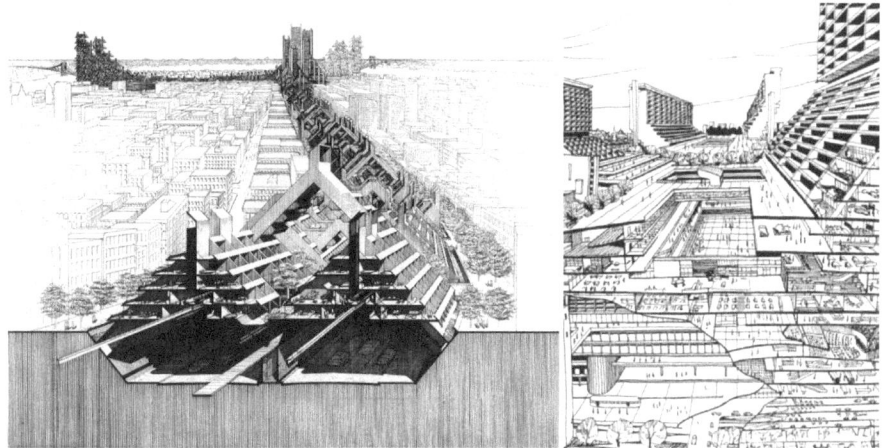

Fig. 1. Left: Paul Rudolph. Byrd's-eye perspective section of the Lower Manhattan Expressway. New York. ca. 1970. Source: Paul Rudolph Foundation. Right: Rem Koolhass. *The Asian City of Tomorrow*. 1995. Source: OMA.

ones, adds new quantities of space and time to the drawing. From the perspective of the spectator, these times and spaces overlap with those above mentioned from the perspective of the actor.

2.3   The Position and Direction of the Body

Regarding the figure insertion point into the graphic field, three basic procedures can be discerned. In some cases, the figures fulfill the empty areas, as if they yearn for being able to sound out the capable maximum of space (Fig. 2 right). In other cases, they seem to spread sporadically with no evidence of any kind of insertion pattern (Fig. 1 right) or trying to balance the visual weights of the graphic field (Fig. 2 centre). A last procedure aims to sign, delimit and establish a certain hierarchy for a specific space, offering the figure an immediate acting field separated from the rest (Fig. 2 left).

In addition, the position and direction of the actors' bodies orient our look. Sometimes the characters present head-on. At other times, however, they appear sideways establishing relationships of tactile coexistence with the rest of occupiers. By looking, grazing, touching, or crossing each other, our characters reinforce the sensible dimension of the place that holds them as well as justifies them. We might say that it is a place of *exposure*, a scenario of exhibition and risk, where the actor works out the partial conflicts that fade him being unaware of the director's hidden purpose (Fig. 3).

Moreover, the very position of the drawing author is considered as susceptible of attention.

When the author places himself in the graphic field, two different situations arise. In some cases he shares a common horizon with the actor. This procedure fixes and apprehends the figure within the drawing form. However, in other cases there are not these kinds of correspondences. There, the figure is freed from any type of action pattern that could be established beforehand. Whereas the former orient the emotive quality of

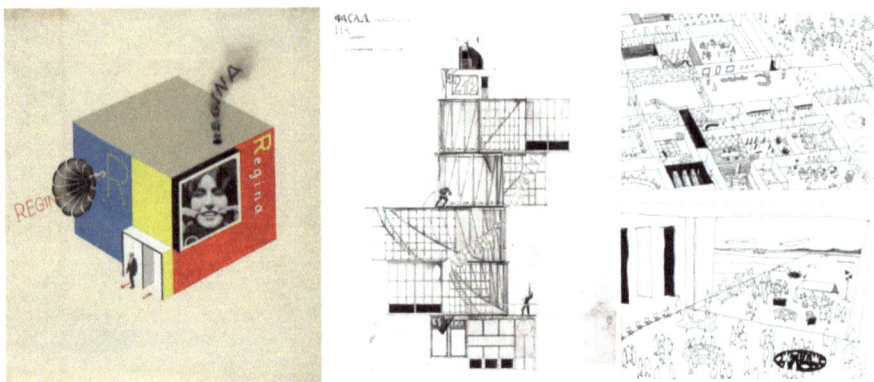

Fig. 2. Left: Herbert Bayer. Kiosk. 1924. Source: Harvard Art Museums/Busch-Reisinger Museum. Centre: Konstantin Melnikov. *Leningrado Pravda*. 1924. Left: Ryue Nishizawa. *De Kunstlinie*. Almere. 2005. Source: SANAA.

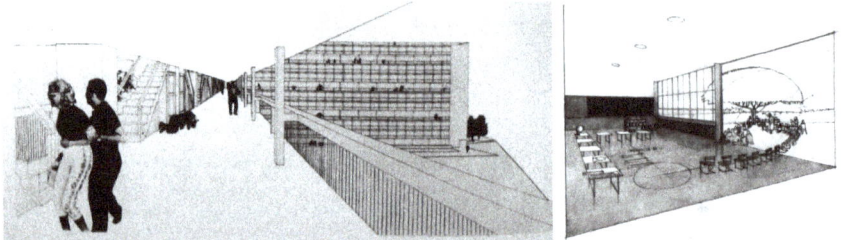

Fig. 3. Left: Peter Smithson. *Golden Lane*. London. 1952. Source: Smithson Archive. Right: Richard Neutra. Emerson Primary School. California. 1937.

form towards the intimate and close proximity, making space as stable refuge of univocal positions, the latter leave the actor immersed in the fortuitous risk indirectly derived from the lack of director.

Beyond this, the author can appear within the graphic field expanding the drawing form spatially and inciting a split look. Without any interruption, we go in and out the drawing crossing overlapped layers: the represented space of the figure, the graphic field of the author, and the very place from which we see. However, the author's hand turns its interest in some of these auto quotes. It seems to get lost in a detailed author characterization, becoming unconcerned with the space itself. Indeed, sometimes a formal recognition is unachievable (Falcón 2015, 37), making difficult a fluid linking between those spaces referred before.

2.4 The Graphic Technique

Equally, the graphic technique of the figures evokes particular manners of inhabiting. We can obtain camouflages of powerful effect by working character and form with the same graphic register. The same type of line, of spot, colour or texture fade the figure into the air of the paper (Fig. 4 left). In other cases, however, the actor takes a materiality in contrast with the appearance of the set design that lodges him, as if he inhabited the space between the drawing and the paper (Fig. 4 right). Doing so, the character seems to confess in a certain way his no natural belonging to that specific place that calls him for a moment (Fig. 6 left).

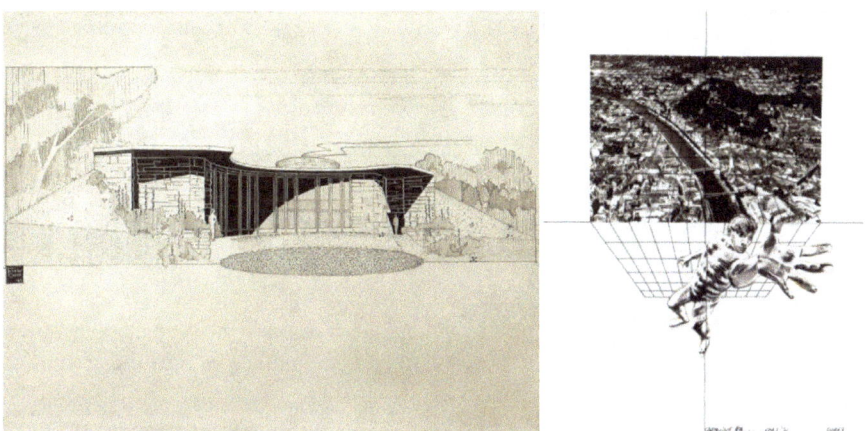

Fig. 4. Left: Frank Lloyd Wright. Herbert A. Jacobs House. Middleton, Wisconsin. 1944. Source: The Art Institute of Chicago. Right: Superstudio. Graz. 1971. Source: Superstudio.

2.5 The Action

Our actors observe, walk, go in and out, up and down, touch, stand or rest on surfaces and walls, *feeling* the form in movement (Lynch 1984, 131). Mostly, these actions have a programmatic intention: they try to explain the specific use of each space (Fig. 2 right). Although, they allow us to set out other readings (maybe less pragmatic) depending on the way the figure comes into contact with space by means of the specific action developed. Whereas those static situations seem to distance the figure from space (Fig. 4 left), these dynamic ones make this relation closer and more inclusive (Fig. 5).

2.6 The Moment of Insertion

A last aspect of pure temporal nature closes the analysis. In some cases, the figures are incorporated by the author synchronically, or at the same time in which the drawing is created. Other examples, on the contrary, refer a diachronic time. Here, the figures are included or even removed by a very different hand, thus enabling the transfiguration of the represented space. The sensible quality of the drawing architecture turns in one direction or another by means of all sort of tricks of scale (size and number of figures), position, graphic technique or action. Spaces formalized by lines, inert and silent, enter in resonance (Fig. 6 left); likewise, those others delimited by planes, in constant vibration, become mute once they are occupied (Fig. 6 right). Changing its quality, the space reveals

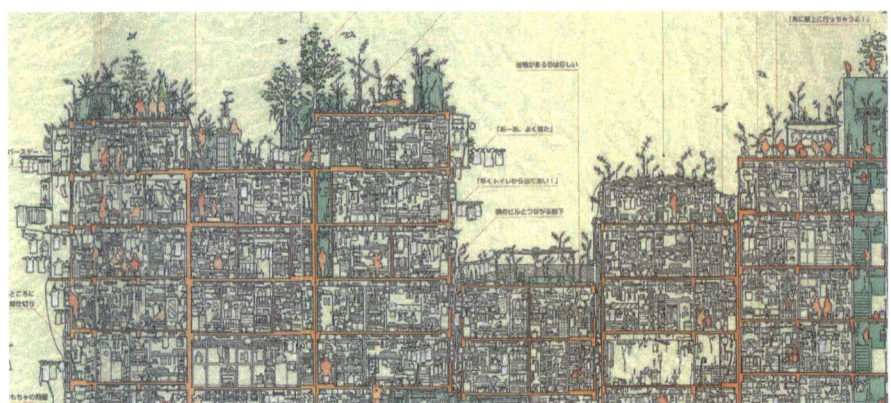

Fig. 5. Teresawa Kazumi. Detail of Kowloon. Hong Kong. 1997. Source: Kani, H. *et al*. 2003. *Kowloon larged illustrated (1997)*. Hong Kong: Iwanami Shoten.

its hidden self dressed of overlapped meanings, as if it was a new or maybe different costume. Theses mutations are in debt to the fine understanding of how the architecture ultimately results from the inhabiting act itself.

Fig. 6. Left: Fred Scott. *Interior*. House for Artisans by Le Corbusier (1923) occupied by the characters of the painting *The Egg Dance* by Jan Steen (1674). 1975. Source: Betts Project. Right: Ugo La Pietra. *Inmersioni – Uomouovosfera*. 1969–70.

2.7   Other Manners with Other Occupiers

Beyond human figure, occupiers with other kind of natures allow us to set out new reflections that, at this point, we would like to sign out briefly.

Perennial inhabitants with the skin of architectures (Fig. 7 left) or stealthy sculptures, suggest apprehension times extremely slow or quasi frozen. In contrast, inhabitants shaped like mobile objects, as furniture elements or machines (Fig. 7 right), play inhabiting manners more vivacious and lighter.

Fig. 7. Left: Haus-Rücker. *Cover*. Krefeld. 1971. Right: Mies van der Rohe. *Deutsche Bauausstellung*. Berlin. 1931.

Regarding the latter, the types of machines (planes, balloons, helicopters, cars, bicycles, etc.), and even the speed insinuated by the way they are represented, indicate the level of sensorial participation of their incognito occupiers with the space around them (Hall 1972, 217) (Fig. 8).

Fig. 8. Left: Aldo Rossi. Perspective. 1961. Source: Casabella 258. Right: Francisco Mujica. *The City of the Future*. 1929. Source: The Getty Research Institute.

Furthermore, those drawings without explicit presence of an actor equally suggest manners of existence. While the furniture visibly located evokes those characters out of stage (Colomina 2010, 166), deliberate mismatches of position give away the type and character of the action and the moment of its advent. Forced locations, unusual displacements or unexpected changes of direction of this or that designed element enable to reconstruct the past or future of the instant trapped in the drawing. However, in other cases the action concerns the drawing *present*. Calculated graphic marks, as *trails* made by sinuous lines or arrows, become tangible traces of the stay of anonymous characters, which could be so fast for our eyes (Fig. 9 left) or maybe remain hidden behind the paper (Fig. 4 right and Fig. 9 right).

Fig. 9. Left: Louis Kahn. Traffic Study Project, Philadelphia. 1953. Right: Alvar Aalto. Muurat-salo Experimental House. 1953. Source: Alvar Aalto Museum.

3   Conclusion

From a chronological perspective, we do not find a clear tendency towards a greater or lesser attention to the sensible qualities of the drawing architecture over the years. The trend of thought ascribed by the author (architect or draftsman) frame these kinds of issues. However, his hand finally guides the drawing in one direction or another transferring to the paper a unique and personal way of understanding and expressing the manner in which the architecture is apprehended by its occupiers.

Despite this, throughout the study we have seen the fecund depth of the architectural drawing as operating analogy of a reality amplified in space and time. Each drawing studied offers an interlaced dialogue among different spaces: the director-author space, the actor-figure space, the stage-paper space, and even the very space of the spectator. In parallel, they give tuned counterpoint to all sorts of overlap times: the action duration, the insertion moment of the figure, its potential movement, or the specific time described by the drawing. On the whole, we hope all these issues could be read as a modest contribution to the way in which we study, project and teach the *experiencing* architecture.

References

Bloomer, K., Moore, C.: Cuerpo, memoria y arquitectura: Introducción al diseño arquitectónico. H. Blume, Madrid (1982)

Certeau, M.: La invención de lo cotidiano I. Artes de hacer. Universidad Iberoamericana, México (1996)

Colomina, B.: Privacidad y publicidad. La arquitectura moderna como medio de comunicación de masas. Colegio de Arquitectos, Murcia (2010)

Delgado, M.: Sociedades movedizas. Pasos hacia una antropología de las calles. Anagrama, Barcelona (2007)

Falcón, J.M.: El espacio gráfico habitado: la figura humana en la comunicación visual arquitectónica. Arquitectura Revista **11**(1), 31–45 (2015)

Hall, E.: La dimensión oculta. Siglo XXI, México (1972)

Joseph, I.: Retomar la ciudad. El espacio público como lugar de la acción. Universidad Nacional de Colombia, Medellín (1999)

Lynch, K.: La imagen de la ciudad. Gustavo Gili, Barcelona (1984)

Pallasmaa, J.: Encounters 2: Architectural Essays. Rakennustieto, Helsinki (2012)

Rasmussen, S.: La experiencia de la arquitectura. Reverté, Barcelona (2004)

Drawing with Louis Kahn in Venice

Hugo Barros Costa⃝, Pedro Molina Siles⃝, and Irene López Moya(✉)

Universidad Politécnica de Valencia, Valencia, Spain
{hubarda,pmolina}@ega.upv.es, irlomo1@arq.upv.es

Abstract. Throughout history, scholars and artists have traveled the world to draw and learn about architecture. One of the most travelled destinations is Italy, a visit that became popular from the 17th century onwards. This trip known as Grand Tour, has also been made by great architects of the twentieth century: Le Corbusier, Alvar Aalto, Arne Jacobsen and Louis Kahn among others.

The aim of this article is to continue the research we have been carrying out in recent years on the non-projectural pictorial work of Louis Isadore Kahn.

The architect's first two trips to Italy were a fundamental source of inspiration for his future projects.

Based on the work published by J. Hochstim in 1991: The paintings and sketches of Louis I. Kahn, we have tried to identify and characterize each drawing made by the American architect, through the elaboration of unpublished plans and worksheets. In addition to getting to know the spaces crossed by the great architect and establishing an eventual relationship with his work, we have also made our own drawings in the same spaces where Kahn drew.

In this article, we will focus on drawings, plans and worksheets related to the city that Kahn represented the most: Venice.

Keywords: Kahn · Travel · Drawing · Italy · Venice

1 Article

Travel in Europe for the training and inspiration of artists and wealthy young people became widespread in the 17th and 18th centuries. Travelers were mainly looking for destinations in the southern countries of the old continent such as France and Italy. Philosophers, literati and architects made this itinerary known as Grand Tour, which allowed those who traveled to become familiar with Greek and Roman culture and art.

Innumerable architects have found travelling as a way of training and as a source of inspiration to create their own architecture, which is why the Grand Tour was, and still is, although in a very different way, a common practice among architects. Before the generalization of photography, those who wanted to capture their experiences, landscapes and architecture on their travels had no other form of visual representation available than drawing or painting.

Many architects made the Grand Tour in the twentieth century, expressing various interests and concerns when it came to drawing. Narrowing down our study field and focusing on the comparison of great Nordic architects who travelled to southern Europe:

L. Agustín-Hernández et al. (Eds.): EGA 2020, SSDI 6, pp. 197–207, 2020.
https://doi.org/10.1007/978-3-030-47983-1_18

Alvar Aalto, Arne Jacobsen, Gunnar Asplund and Louis Kahn, we can easily distinguish the graphic work of each of them, both on a technical level and the objectives they seem to pursue.

The Finnish Alvar Aalto made his trip to the South at a late age, so he can say he drew t o emember, capturing sensations and focusing on graphic composition and balance, leaving aside the usefulness for his architectural projects. For Arne Jacobsen, the scenario inspired plastic creation, a graphic composition that describes a feeling arising from the model observation. On the contrary, G. Asplund was faithful to the models, detailing them rigorously since his drawings were made to know through analysis and to use that knowledge in his later architecture. While for L. Kahn drawings were a tool for reflection, which interpreted reality, without capturing it literally. His drawings were a source of inspiration to form and create the basis of his own architecture (Bosch Espelta 2003).

The American was critical to the travel sketches of other contemporary architects due to their fidelity to the model and lack of creativity: "Trying to imitate accurately is worthless; if that is our purpose, photography will be most useful"… "I have learned not to consider moving mountains and trees, or changing domes and towers to suit my tastes, as a physical impossibility" (Latour 2003).

Valuing expression as opposed to realistic representation would be one of the reasons why it has been so difficult for us to identify some of the drawings he made. We believe that, eventually, he used buildings from different locations, to compose some drawings made in the Amalfi Coast, in his first trip to Europe.

Kahn explored part of European continent mainly in two different stages, separated by about two decades. It is curious and difficult to explain why he barely drew in all the visited countries before arriving in Italy. "He landed in England, and crossed successively the Netherlands, Germany, Denmark, Sweden, Finland, Estonia - where he stayed for almost a month visiting his relatives - Latvia, Lithuania, Poland, in August again Germany - where he visited the social housing of the Siedlungen - Czech Republic, Austria and Switzerland" (Montes 2005). However, once he had crossed the Italian border he spent a considerable part of his stay drawing compulsively.

Nevertheless, the extensive pictorial work of the great architect is hardly known today, although certain drawings such as those made in Siena, Egypt or Greece are widely disseminated from scientific publications to generalist pages on the web. The publication of Jan Hoschstim: The paintings and sketches of Louis I. Kahn (1991), is the only exhaustive work (except for the catalogues of some exhibitions) dedicated to the non-projectural drawings of the architect. However, Hochstim's magnificent publication contains some ambiguities about the location of certain drawings made in Italy by the American architect. Some of them, located on the Amalfi coast, have been clarified by Professor Carlos Montes Serrano (Montes 2005).

In Louis Kahn's (1928–1929) first trip to Italy, we can see, in his notes, obvious influences of certain styles contemporary to those dates (Fig. 1). Common characteristics in his first travel drawings are strong contrasts between light and shadows, flat and angular shapes, graphic abstraction and simplification of forms, among others.

Fig. 1. Louis Kahn. San Giorgio Maggiore. 1928–29. Source: The paintings and sketches of Louis I. Kahn.

During his second stay in Italy (1950–1951), Louis Kahn would achieve an even more personal graphic style (Fig. 2), combining his vision, his thoughts, his feelings and his imagination in the representations he was creating. We can appreciate an evolution in this second journey when it comes to choosing the subject to his drawings. During the first journey, he seemed to limit himself to drawing landscapes and buildings, always from his point of view and manipulating forms to create above all a certain composition, showing no concern for being faithful to reality. While in the second journey, his attention was focused on capturing atmospheres and urban landscapes that reflected sensations; these drawings seem more sensitive to the representation of less physical and tangible elements than the ones of his first trip.

Fig. 2. Louis Kahn. St. Mark's Square. 1950. Fountain: The paintings and sketches of Louis I. Kahn.

We adopted the research of Jan Hoschtim and Carlos Montes as a reference for our investigation, and then went through the spaces drawn by Louis Kahn on his two major trips to Italy, carried out in the first half of the last century. After the location identification, in situ (not always easy) of each perspective perpetrated by the Estonian-born master, we proposed drawing them from the same point of view; as well as trying to understand the reason for the selection of each of the points of view chosen by the American architect, taking as a reference his written, graphic and built work.

Fig. 3. Hugo Barros Costa. Basilica of Sant'Ambrogio. Milan. 2016.

Thus, we decided to "walk", in several stages, from South to North Italy, in the opposite direction to the original Kahn journeys, the villages, cities, ruins, gardens, landscapes or beaches where the architect had been drawing (Fig. 3). Not without a certain emotion and nostalgia, since the drawings relating to his first journey turned 90 years old this very winter (2019).

Finally, all the drawings from "The paintings and sketches of Louis I. Kahn" (1991) done in Venice have been classified, as well the approximate point where each sketch was drawn.

2 The Trip

A three-month research stay at the University of Salerno in 2016 made it possible to develop a large part of the drawings that were intended to be formalized, finally giving impetus to this research (Fig. 4).

Fig. 4. Hugo Barros Costa. Saint Mark's Square, Venice. 2016.

Thus, after the numerous trips we made to Italy in different phases, from 2012 to 2018, we created an extensive amount of personal graphic information that was organized and collected. This research, together with the orientation of the aforementioned reference work, as well as certain scientific articles, among which we highlight the one of Vincent Scully in Lotus Magazine, allowed us to create a database of the travel drawings made by Louis Kahn.

In 1991 Scully wrote: "Soon, with a book by Jan Hochstin in the process of publication, we will be able to see all of Kahn's travel drawings gathered together and define more precisely the totality of their meaning" (Scully 1991).

Thus, following Scully's prophecy, with Hochstin's magnificent book in our hands, we have rectified and reorganized the data collected in this book. So, in an unpublished way, we created worksheets for each city, where we compiled dates, references to the sources we used, drawing techniques, subject or location, among others, hoping with this information to support future research on the topic.

Before, we questioned why Kahn practically did not draw during his long journey in Europe, from England to Switzerland, contrary to his time in Italy, where he made a large number of sketches (87 drawings published in Hochstim's book).

In addition, after cataloguing and analyzing all the drawings for this research, we wondered why, in years as decisive for modern architecture as the end of the 1920s, Kahn did not sketch any example of that architectural style; an opinion also pointed out by Professor Carlos Montes in one of his magnificent articles.

"The strangest thing about Kahn's study trip is his lack of interest in avant-garde architecture, which indicates that Kahn's attention and training during the previous years must have been somewhat deficient, practically linked to the academic tradition of the Philadelphia school and his dedication to drawing." (Montes 2005).

This leads us to think that the American architect would pretend to draw mainly Italy's classical architecture. However, after analyzing the data we compiled for this article, we realized that only about 10 percent of his drawings made in the Italian peninsula were seeking for classical art. The architect seems to have been interested above all in Italian landscapes and vernacular architecture, especially in the South, which we deduce has caused him a great and unforgettable impact.

"There are some facts that make us think so: on the occasion of his intervention at the CIAM Congress held in Otterlo (Holland) in 1959, Kahn referred to a village in Europe that he visited in his youth, thanks to which he discovered what architecture was. Although he did not refer to any specific town, I like to think, when reading this passage, that Kahn is remembering, thirty years later, his visit to Ravello" (Montes 2005).

Eight drawings of the small town of Ravello and twenty-five drawings in total of the Amalfi Coast are known to have been made. They represents about thirty percent of all the sketches done in Italy, which shows the enormous empathic reaction that the vernacular architecture of "La Costiera" and in particular Ravello, will have produced in Louis Kahn.

The architect drew the humble Saracen towers in the adjacent villages of Amalfi and Atrani, as well as the views from the beach at Amalfi, where the expressive lines of the graphite clearly distinguish the group of modest houses that adapt to the powerful topography. As we toured these villages, we could not help but question Kahn's eventual lack of interest in monumental architecture, such as the impressive Amalfi Cathedral or the omnipresent Collegiata di Santa Maria Maddalena Penitente in Atrani.

Very close to Amalfi we can find the important Greco-Roman cities of Paestum and Pompeii, two of the most important references in the classical world, visited for example by Piranesi and also by Kahn. However, we have only identified one drawing of each of these impressive sites, compared to the eight made in Ravello.

As for the graphic techniques employed by Kahn, the predominant ones are graphite and watercolor. Both of them allow the author to quickly and imprecisely represent contrasting sensations and planes, avoiding the pictorial precision required by other techniques.

In our research we also tried to find out why Kahn chose certain spaces in particular and also how he characterized them (colors, lights, shadows, details, tonal values,…).

Cataloguing that information, besides compiling the charts and plans that characterize each drawing, allied with the experience of drawing the same landscapes or buildings that Kahn apprehended, seems to us to be the most effective way of obtaining information to be able to link his drawings with his architectural work.

It is undeniable that the main two trips Louis Kahn made to the European continent left an imprint on his way of conceptualizing architecture, and will have an influence on his later works, hence their importance.

3   Kahn and Venice

In 1959, the mayor and director of the Venice Biennale, Giovanni Favaretto Fisca invited Kahn to participate in the 34th edition of the Biennale looking for a great architectural signature, emulating another competition promoted four hundred years earlier by Duque Andrea Gritti. A few months later, the commission for the construction of a congress hall in the Castello district is completed. (Navarro-de-Pablos and Pérez-Cano 2019).

The architect, having travelled around and drawn the city on his first two trips to Italy, recognizes the character of the connecting thread between East and West in the capital of Veneto. This is demonstrated by the first sketches of the project, showing a suspended building in the form of a suspension bridge, such as Palladio's proposal for the Rialto Bridge 400 years earlier (Fig. 5) (Fig. 6).

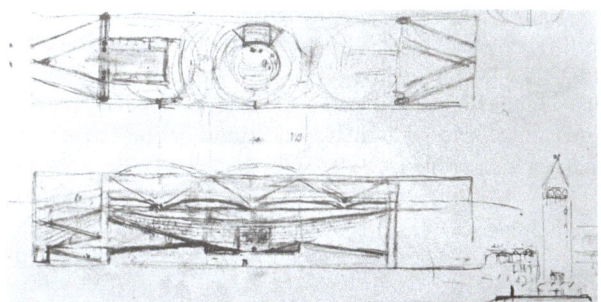

Fig. 5. Louis Kahn. Project for the Venice Congress Centre. 1972. Source: Unbuilt Masterworks.

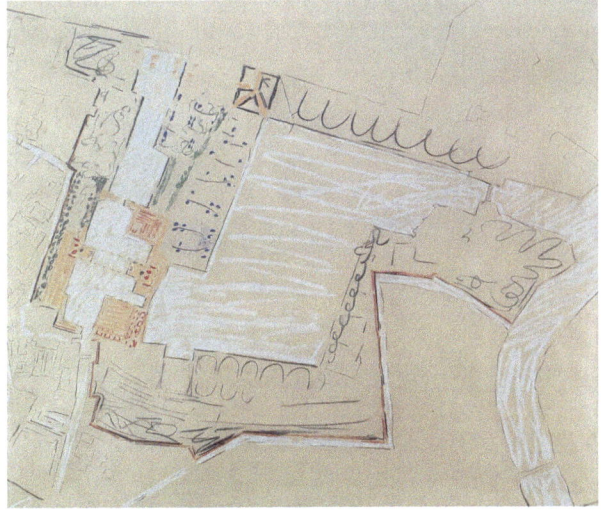

Fig. 6. Louis Kahn. Project for the Venice Congress Centre. 1972. Source: Unbuilt Masterworks: Louis I. Kahn: complete work 1935–1974.

On December 26, 1968, Kahn and his collaborator Vallhonrat traveled to Venice to study and prepare the public presentation of the project. This had been a rather active, frustrating and complicated year for the American architect who was facing setbacks in several projects scattered between the United States and Asia (Philips Exeter Academy, Kimbell Art Museum, the Dhaka Parliament, the Institute in Ahmedabad, Dominican Mother House, The Memorial to Six Million Jewish Martyrs, Hurva Synagogue in Jerusalem…).

Kahn produced a large number of sketches and drawings of the Venetian project, presented together with two models, in the Doge's Palace, however, there are no known sketches made, at that time, in this city. Nevertheless, at one of the plates exhibited of the project of the Palace of Congresses, one can see some Venetian buildings drawn by the master, accompanying a section (Fig. 5).

In fact, the drawings made in 1959, when he travelled briefly to Europe to attend the C.I.A.M., are the last travel drawings we know of. However, he continued to use his graphic skills in his projects, both in the conceptual phase as well when communicating with his clients, as could be seen in the exhibition at the Doge's Palace in the last century, and more recently in the exhibition Louis Kahn and Venice in Mendrisio (2018).

Finally, Kahn's suggestive "bridge project" could not be carried out due to the lack of liquidity of the Venetian coffers. In 1972, the year in which the Kimbell Art Museum was completed (1967–1972), Kahn visited Venice for the last time to give a series of lectures at the 26th University. Once again, drawings from this trip are unknown.

4   The Drawings from the Venice Trip

Among all the locations visited by Louis Kahn in Italy, this article has focused on systematizing the drawings he made in Venice (Fig. 7). After analyzing the charts we made, we have been able to conclude that this was the city most drawn by the master.

	año	viaje 1	viaje 2	nº dibujo	pag. libro	ubicación	tema	color/blanco y negro	ténica gráfica
1	1929-28	x		97	106	Canal de San Marcos			
2	1929-28	x		99	107	Canal de San Marcos			
3	1929-28	x		98	106	Canal de San Marcos			
4	1929-28	x		101	109	Plaza de San Marcos			
5	1929-28	x		102	109	Cúpulas de San Marcos			
6	1950		x	349	252	Canal de San Marcos			
7	1950		x	350	253	Plaza de San Marcos			
8	1950		x	352	255	Plaza de San Marcos			
9	1950		x	351	254	Basílica de San Marcos			
10	1951		x	353	256	Basílica de San Marcos			
11	1951		x	354	257	Ca' d'Oro			
12		x		100	108				
13		x		103	110				

Fig. 7. Classification of the drawings made b y L Kahn in Venice 2019. Irene López.

Likewise, he visited the capital of Veneto on his two major trips to Italy, developing a wide pictorial production where the graphic evolution of the architect can be appreciated.

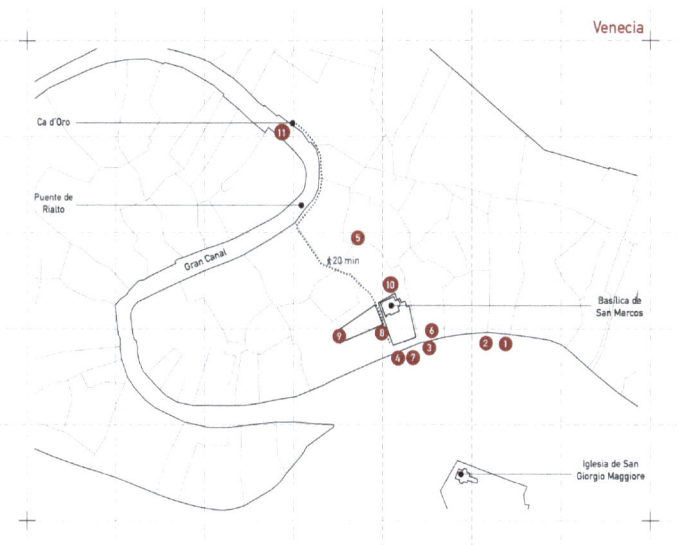

Fig. 8. Location of the drawings made by L. Kahn in Venice. 2019. Irene López.

Thus, since it would not be possible to exhibit all the drawings, plans and tables that we produced of each city visited by Kahn, we incorporate at the present article only the documents related to Venice (Fig. 7) (Fig. 8).

5   Conclusion

This article aims to summarize part of the research we started doing on our trip to Italy, following the drawings made by Kahn on his travels to Italy. Given the large number of drawings and information, we have focused on the city of Venice, where we have produced an unpublished plan, indicating where the American architect drew, as well as a chart that characterizes each of these drawings, according to their date, technique, location and theme (Fig. 7) (Fig. 8).

The presented data adds new information to the study of Louis Kahn's travel sketches. Their organization and synthesis throughout the charts and symbols, has helped to obtain graphs and percentages to obtain conclusions about his pictorial work in his trips to Italy and more specifically in Venice, the city most drawn by Louis Kahn in his "Grand Tour".

The data we present about Venice, for the XVIII Architectural Graphic Expression Congress in Zaragoza, allows, through the unpublished plan, the rapid identification of each drawing made by Kahn. In this way, we add a small grain of sand in the research of the graphic work of one of XX century masters, which can serve as a database for further research into the graphic work of this architect; or simply for anyone who wants to know where the drawings were made, or represent them in situ, like us.

Fig. 9. Hugo Barros Costa. San Giorgio Maggiore, Venice. 2016.

According to Álvaro Siza, travelling is the best way to learn: "the best learning process for an architect is travelling, seeing things live. You can't create things out of nothing" (Gámiz 2003).

Without a doubt, Kahn's trips to Italy changed the conception of his later projects.

In a letter sent from Rome to his collaborators he wrote categorically: "I have realized, firmly, that architecture in Italy will remain a source of inspiration for the works of the future" (Johnson and Lewis 1996, p. 72).[1]

As we wandered through the spaces that Kahn walked through, we became aware of the monumentality of Venice and the importance of the contrasting light that defined forms between the buildings (Fig. 9).

Michael Graves, after a conversation with Kahn, wrote: "it was not until he went to Rome and saw the Domus Aurea and other Roman villas in the Forum that he realized the strength of the wall, the power of the chiaroscuro, the interest in light" (Johnson and Lewis 1996, p. 73).[1]

When drawing the same spaces drawn by the Master, we could not help but find formal and conceptual relationships with his projects.

"Khan found (in Italy) an architecture that was designed to house rites and human activity" (Larson and Kahn 2000, p. 31).

While we were drawing, we were moved, thinking that there, in that same space, stood once Kahn with his graphite or pastels.

Might San Marcos have been a reference for the Salk institute in La Jolla? Might Kahn have considered the functional space and planimetric organization of the Basilica of Santa Maria della Salute, when he was making the first sketches of the Trenton Bath House, together with Anne Tyng?

It was very complicated to do some of the drawings, given the enormous amount of tourists, which would be less in 1950. Thus, we were surprised that Kahn searched in Venice mainly touristy spaces and monumental architecture, when for example in the Amalfi Coast he drew particularly landscapes and popular architecture.

We would have wished to find some drawings made from the roofs of Venice, that Kahn loved so much to visit.

We will keep on researching….

References

Barros Costa, H., Hidalgo Delgado, F.: Conversando con… Frank DK Ching. EGA. Revista de expresión gráfica arquitectónica **20**(25), 20–31 (2015)

Barizza, E., Falsetti, M.: Rome and the Legacy of Louis I. Kahn. Routledge, Abingdon (2018)

Bosch Espelta, J.: El arquitecto viajero. EGA. Revista de Expresión Gráfica Arquitectónica **8**, 66–69 (2003)

Goethe, J.W.V.: Viaje a Italia (1891)

Granero Martín, F.: Viaje a través de la mente: La idea. Fundamentos del dibujo. EGA: revista de expresión gráfica arquitectónica **22**, 60–67 (2013)

Hochstim, J.: The Paintings and Sketches of Louis I. Kahn. Rizzoli, New York (1991)

Johnson, E.J., Lewis, M.J.: Drawn from the Source: The Travel Sketches of Louis I. Kahn. The MIT Press, Cambridge (1996)

Larson, K., Kahn, L.I.: Unbuilt Masterworks. The Monacelli Press, New York (2000)

Latour, A., Sainz, J.: Louis I. Kahn: escritos, conferencias y entrevistas (2003)

Louis Kahn and Venezia: Silvana editoriale (2018)

Montes-Serrano, C.: Louis Kahn en la costa de Amalfi (1929). escritos, conferencias y entrevistas (No. 72 Kahn). El Croquis Editorial (2005)

Montes Serrano, C., Galván Desvaux, N.: Las litografías de Louis Lozowick y su influencia en Louis Kahn. EGA. Revista de expresión gráfica arquitectónica **21**(28), 92–99 (2016)

Moreno Mansilla, L.: Apuntes de viaje al interior del tiempo. Fundación Caja de Arquitectos (2002)

Navarro-de-Pablos, J., Pérez-Cano, M.T.: El alfabeto de la memoria. Puentes, tiempos y tipos: John Hejduk, Andrea Palladio y Louis Kahn en Venecia. rita_revista indexada de textos académicos (11), 28–35 (2019)

Navarro, P.C.: La representación de la ciudad, del territorio y del paisaje en la Revista EGA: mapas, planos y dibujos. EGA Expresión Gráfica Arquitectónica **23**(34), 106–121 (2018)

Ronner, H., Jhaveri, S. Louis I. Kahn: complete work 1935–1974. Birkhäuser (1987)

Scully, V.: Marvelous Fountainheads + Kahn, Louis, I. Travel Drawings. Lotus International (1991)

An Unpublished Legacy of Antonio Rubio Marín (1884–1980), An Architect Between Aragon and Madrid

Enrique Castaño Perea[1]([✉]), Gonzalo García-Rosales González-Fierro[1], and Miriam Martín Diaz[2]

[1] University of Alcalá, Alcalá de Henares, Spain
enrique.castano@uah.es
[2] University of Castilla-La Mancha, Ciudad Real, Spain

Abstract. An unpublished legacy of the architect Antonio Rubio Marín has been found in the library of the University of Alcalá. He was an important an architect of the first half of the twentieth century in Madrid and Aragon, being responsible in Zaragoza for the Post Office building and the Grand Hotel.

Still in a state of certain abandonment, the legacy has great interest for its documentary quality, being a fairly complete set of the activity of an architect of the early twentieth century. Antonio Rubio combined other notable merits, such as his training as a mathematician and cartographer, or his work as dean of the College of Architects of Madrid between 1948 and 1954.

The inventory and cataloging process of the material found by researchers from the School of Architecture of the University of Alcala is currently being carried out, with the collaboration of his family and those responsible for the archiving o fthe same University.

The legacy contains an important variety of drawings of different types. Most of them are freehand drawings in ink, in different formats and supports, highlighting large format ones using watercolors. These drawings are of great interest as they are a project commissioned by the city of Zaragoza in which the proposal included the square and the Basílica del Pilar. All these documents show the work of a not very well known architect, but of undoubted value through his interesting drawings.

Keywords: Unpublished legacy · Antonio Rubio · COAM

1 Introduction

In the 21st century, dominated by the world of digitalization, where information is stored and preserved in a virtual cloud, it is unusual to find forgotten legacies of totally unpublished architects. In the library of the University of Alcala we located some cardboard boxes with numerous architectural drawings corresponding to a mid-20th century unknown architecture office. Nobody knew where they came from, nor their authorship. After a thorough investigation, it was discovered that they belonged to Antonio Rubio

© The Editor(s) (if applicable) and The Author(s), under exclusive license
to Springer Nature Switzerland AG 2020
L. Agustín-Hernández et al. (Eds.): EGA 2020, SSDI 6, pp. 208–215, 2020.
https://doi.org/10.1007/978-3-030-47983-1_19

Marín, a Madrid adoption architect, little known but of notable importance in Madrid and Aragon in the mid-twentieth century, since, among other merits, he became dean of the College of Architects of Madrid in the 50s (Fig. 1).

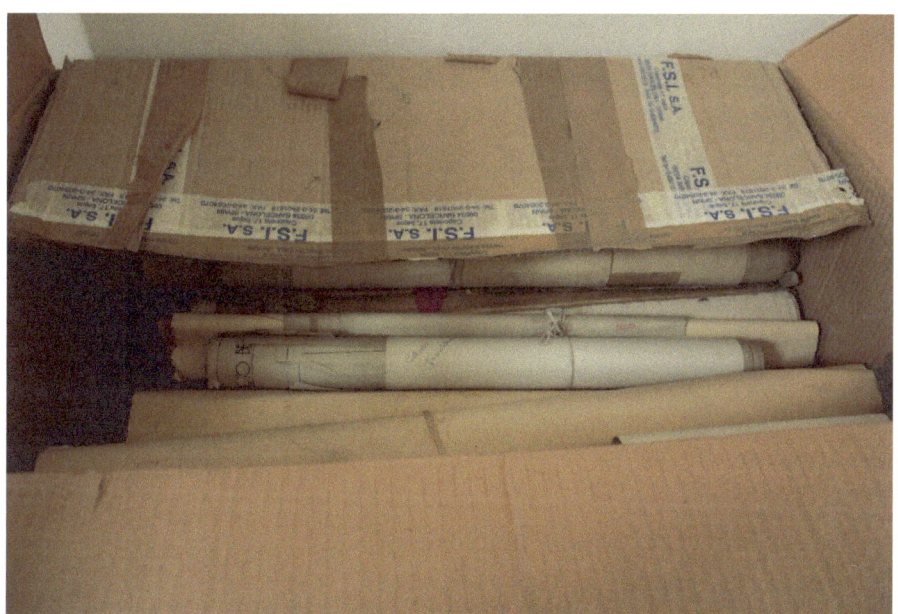

Fig. 1. Boxes containing the legacy of the architect Antonio Rubio Marín. Alcala University, Madrid.

From the important localized graphic material it was necessary to inquire about the character and the origin of the material. Different inquiries were made among several contemporary architects, such as Antonio Fernández Alba, Gaspar Blein, Vicente Sanchez de León or Manuel Sainz de Vicuña. This work was completed with an intense internet search. This process allowed the location of two of the architect's grandchildren, who were willing to collaborate in the contextualization of their grandfather's legacy. At that time the rights and responsibilities of the parties were successfully established. From then on we were able to face the study of the architect legacy in its entirety, but by then some questions needed urgently to get solved:

From a legal point of view, what about the copyright of an abandoned legacy for a few dozen years and now recovered? Who owns the legacy?, Who owns the reproduction rights?, Who pays inventory and classification expenses?

From a conservation point of view, how should we act with those sheets of vegetable paper rolled in a not good state of conservation? What to keep?, What to expunge or throw?, What to restore and at which level?

We will try to solve all these questions both in this communication and in different works developed by the research team in other forums devoted to architecture archives [1] (Fig. 2).

Fig. 2. Antonio Rubio Marín. Drawing of the casino project in Teruel. Source: Antonio Rubio Marin Legacy. University of Alcala.

2   The Architect Antonio Rubio Marín

Antonio Rubio Marín was born in Granada in 1884, and before entering the School of Architecture he graduated in Mathematics. He later graduated as a Cartographic Engineer, a profession he developed for years and that was an important technical-graphic background that would later result in his work as an architect and in his knowledge of the territory. In fact, part of the legacy includes plans of the territory of the Aragon area [5].

His work as an architect was mainly developed between Madrid and Aragon. In Zaragoza, the Post Office building built in 1926 on the site of the former Pignatelli Theater, and the old Grand Hotel of Zaragoza built in 1929 were two of his most important works [4]. In Teruel, he built the Casino headquarters and the Marín Theater opened between 1918 and 1920 [2]. Finally, his most outstanding work in Madrid was the "Musical Union" headquarters in the Carrera de San Jerónimo on the corner of Echegaray Street, a building representative of the eclectic style in the center of Madrid. He also designed some schools and numerous residential buildings, highlighting those of Castelló and General Oraá streets. Another of his important interventions was the "chalets" colony he built for Doctor García Tapia in the town of Riaza in the northern mountains of Segovia. From his office at Alcalá street in Madrid he took part in more than 200 projects, many of them collected in the inventory that we are going to carry out (Fig. 3).

In 1949 he was elected as dean of the College of Architects of Madrid, COAM, and held the position until 1953. His election as dean was the first elected position of the reborned College after the Civil War, since the designation had so far been nominal by the political authorities. During the period of his mandate, and under his initiative, the library was developed, incorporating a significant amount of bibliographic funds. He

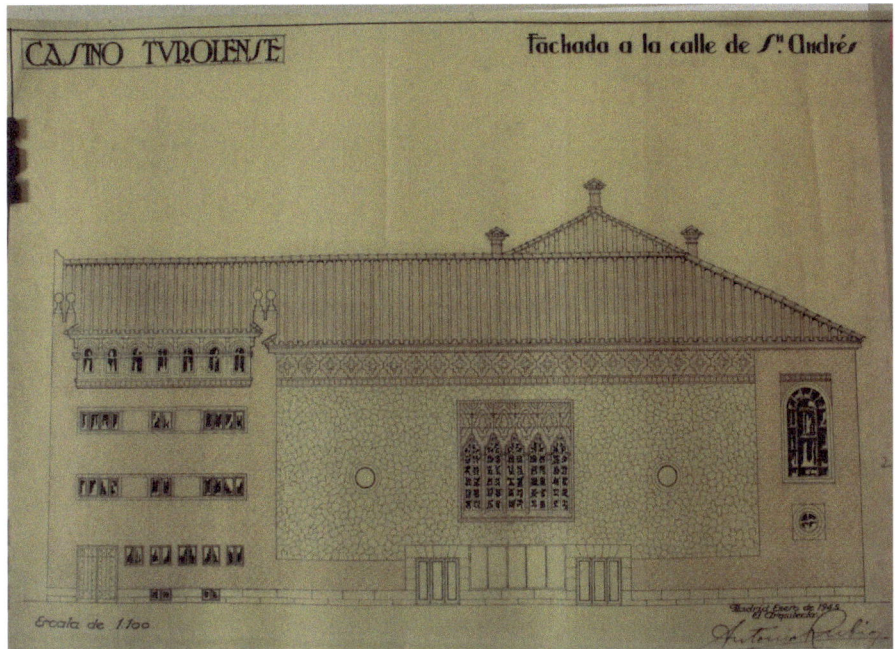

Fig. 3. Antonio Rubio Marín. Drawing of the casino project in Teruel, 1945. Source: Antonio Rubio Marin Legacy. University of Alcala.

also established the figure of a professional librarian as director of the library, since until then the library had been managed exclusively by a member of the board, so that with this impulse the institution has perpetuated to the present, standing out for its development of library and archival architecture services.

3   The Legacy and Its Administrative Process

The processing of such a legacy has important administrative issues to resolve. First, the ownership of the material and who can use it must be defined. As a first step is to get in touch with the author is necessary; in this case with his heirs. This process started with the internet search from the few data we had. The first contact was with the architect Antonio Fernández Alba since part of his legacy donated to UAH coincided in date with the found boxes of Antonio Rubio. Fernández Alba confirmed that the shipment had been made from his office, but could not remember where they had come from or who had provided them. Then, through forums on the Internet we contacted one of his granddaughters, with whom we agreed to organize a meeting. In this session, which was attended by two grandchildren and a great-granddaughter, we were able to establish the guidelines of what will be the future record of donation of goods. From the will of the family as donors, it was necessary to organize the complex administrative procedures of the University as a recipient, where the Legal Services, the General Secretary, and finally the person in charge of the Archive should participate. This process has already been resolved in recent months, now it is up to technical decision making (Fig. 4).

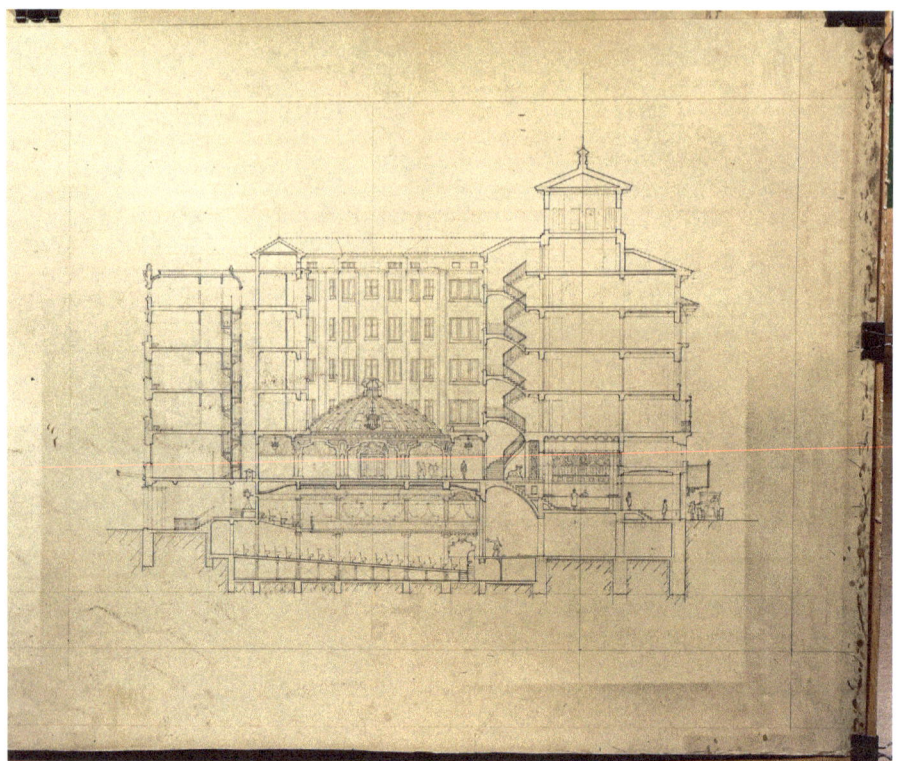

Fig. 4. Antonio Rubio Marín. Drawing with pencil. 1927. Source: Antonio Rubio Marin Legacy. University of Alcala.

4   The Order

The legacy is mainly composed of paper rolls that group plans corresponding to the same project. There are approximately one hundred rolls of different thicknesses and sizes. Along with the plans, there was also a record book containing numerous chronologically ordered projects. It will be necessary to check if the list corresponds to the projects found and to what extent the information is complemented, and if all are included in the list, or if others are missing or there are some incomplete ones. In a first exploration work, we have already found some disruptions of plans, therefore, it must proceed in order, generating records of each of the recognizable architecture projects, both at the preliminary project level o r executed project [3], as well as its complete documentary study.

5   The Processing of the Legacy and Its Treatment

The material is in an acceptable state of preservation, although many of the plans are rolled and without protection tubes, so it will be necessary to first stretch on large tables to assess their state of conservation. We have the collaboration of a restorer who will establish guidelines for document restoration based on dirt, the presence of metallic, plastic or fatty elements and the existence of tears [3].

The legacy is made up of drawings on different supports: cartons, tracing papers, cloth paper, sketch paper, so each of them will need a different treatment and protection. Many of them are ink drawings in an acceptable state of conservation, others are graphite, degrading with the handling process, so it will be necessary to protect them with appropriate papers. Some of them will only need to be flattened, others will require restoration work. Finally, all of them must be filed in tubes or in filing cabinets in such a way that their degradation is prevented.

6 Antonio Rubio's Drawings

The set of drawings found in this legacy could be classified as corresponding to architectural drawings of an architectural office in the beginning of the twentieth century.

The graphic documents include sketches, preliminary projects, basic projects, execution projects. These last ones are very far from what would currently correspond to an execution project if we consider the current requirements of the technical code and the new technologies devoted to an execution project.

The materials are diverse: cardboard paper, tracing paper, sketch paper, tissue paper, ammonia copies, these are in different colors, blue or brown, and in positive and negative formats. As for the techniques, the most frequent are: graphite pencil, black ink, red ink and watercolors, apart from copies on paper or tracing paper. The sizes of the sheets do not follow a recognizable norm by the current systems UNE, but do maintain approximate standard dimensions surely conditioned by the drawing scales and the boards where they were drawn.

Within the set, some large-format drawings that correspond to an unbuilt building of the municipality of Zaragoza stand out, on a plot next to the Pilar Basilica in Zaragoza. These are drawings made in watercolor with the tempera technique of flat and opaque colors, almost without glazes. These are three large-format drawings up to two meters long, where the residues found in their trasdos indicate they had been stuck on a wall. This project must have been important in the architect's production given the size of the drawings, the care with which they were made and their extreme quality.

There are also some very interesting and large-format drawings (close to the current din-A0) that correspond to a project of the Archbishop's Palace of Madrid next to the Almudena Cathedral, a project that was not carried out (Fig. 5).

7 Conservation and Digitization

Once the inventory and the cataloging of the material have been completed, the digitization work will proceed. This process will progressively be carried out depending on the resources and the needs of using these images. The use of non-invasive and non-contact systems is proposed. For this reason the use of a scanner should be rejected since deformation of the support paper would be an added difficulty, so a scientific photography system will be used. For this work an appropriate photographic studio has been designed for scanning all the material. Large format supports have been made where the plans can

Fig. 5. Antonio Rubio Marín. City Hall project in Zaragoza, 1921. Source: Antonio Rubio Marin Legacy. University of Alcala.

be extended. Next, adequate lighting has been arranged with several strategically located spotlights so as not to cause shadows and with a temperature as neutral as possible for the best reproduction of the colors of the drawings. Finally, a high quality digital camera with a good lens will be used, with angular objectives so as not to distort the lines and the quality of the images, all acquired with the advice of experts in digitalization.

The system used to hold the plans and tension them in the best possible way, will be carried out using a metal support where the plans will be held with clamps at the ends and magnets in the central areas to avoid bulges. Finally, clamps and magnets will be eliminated in a post-production process through digital image processing tools, such as Photoshop.

All images will be made leaving a free space around the plan of at least a centimeter wide so as to visualize the entire document and the support in case of irregularities or tears. On this empty space three scales will be placed, a graphic one, a color one and a ten tones of gray one, allowing on the same document all the necessary data to reach, once manipulated or modified, the maximum fidelity in future reproductions.

8 Conclusion

The knowledge, from the primary documentary sources, of a legacy like the one we are lucky to know first-hand, will let us have an idea of how an architecture office worked in those complicated years of the mid-twentieth century. We have samples of Antonio Rubio's architecture before and after the Civil War, which will also help us to know the evolution of society in those years (Fig. 6).

Fig. 6. Antonio Rubio Marín. Drawing of the proposal for a ville, 1945. Source: Antonio Rubio Marin Legacy. University of Alcala.

References

1. Martín, M., Castaño, E.: The forgotten legacy of Antonio Rubio Marín. International Congress on Architecture Archives, Braga (2019)
2. Biel, M.P., Hernández, A.: La arquitectura neomudéjar en Aragón. Institución Fdo. el Católico, Zaragoza (2005)
3. Cagigas, Y., et al.: El Archivo General de la Universidad de Navarra, Príncipe de Viana (PV), 266, Pamplona, pp. 1193–1233 (2016)
4. VVAA: Gran Hotel Zaragoza. Arquitectura SCDA, no. 4-1931, pp. 120–125 (1931)
5. VVAA: Zaragoza. Arquitectura. Siglo XX. Antonio Rubio Marín **2**(5), 99–110 (2016). https://zaragozaarquitecturasigloxx.com/category/rubio-marin-antonio/. [2019, 17 de junio] e la revista

Digital Reconfiguration of the Project of Brasini for the Soviet Palace in Moscow

Caterina Palestini[1][(✉)] and Laura Carnevali[2]

[1] Architecture Department, G. d'Annunzio University, Pescara, Italy
palestini@unich.it
[2] DSDRA Department, Sapienza University, Rome, Italy

Abstract. The digital language makes it possible to reproduce non-materially constructed places, making them realistically perceptible and usable in the virtual dimension. The possibility of reconfiguring non-existing works in three dimensions is a graphic opportunity to investigate the contents expressed by the cultural heritage contained in projects not completed.

The contribution in this sense focuses on the study of the graphic heritage related to unrealized projects, referring to the analysis of documents left on paper that constitute an important cultural background to understand the historical and architectural events that underlie the failed achievements.

In particular, the survey methodology is applied to an emblematic international comparison, that relating to the competition for the construction of the Soviet Palace in Moscow which constitutes a salient episode for the history of twentieth-century architecture and can refer to new aspects through drawing. It is one of the first competitions that the Soviet Union opened to the West with the participation of many well-known architects and among them Le Corbusier, Erich Mendelson, Walter Gropius, Hans Poelzig with avant-garde proposals to which the monumental classicism of the utopian project of the Russian Boris Iofan, expression of political power.

The research conducted from the point of view of representation traces, through graphical elaborations, the requests formulated by the notice for the great building, that was never completed, of which some of the significant project proposals have been examined.

In particular, the survey focuses on the designs drawn up by Armando Brosino, the only Italian participant and teacher of Boris Iofan, who strongly influenced the style of the winning project.

The main objective of the contribution is to communicate the values of the graphic heritage expressed in the design solutions, analyzed, reconfigured and explored three-dimensional in the digital dimension.

Keywords: Design · Digital models · Moscow

1 Introduction

The conception of a "Labor Building" conceived as a multi-purpose center and place of activity of the Soviet should have been the symbol of Soviet power on a world scale, for

L. Agustín-Hernández et al. (Eds.): EGA 2020, SSDI 6, pp. 216–226, 2020.
https://doi.org/10.1007/978-3-030-47983-1_20

this reason the elaboration of the great project, after a phase of internal confrontation started in 1992 from the Association of architecture of Moscow in which the principal objectives were delineated, was opened to international level (Samonà 1976, pp. 84–85).

In June 1931 the international competition for the Building of the Soviet was run by a special council set up by the Stalin government, in the idea of the party, it had to be a multi-purpose building to be built near the Kremlin, so it was decided to sacrifice the Church of Christ the Savior.

The site in which the Christian temple stood corresponded to more requirements, it would have allowed the enlargement of the square in front of it, favoring an urban planning solution suitable for the transit of crowds, of the twenty-five thousand people that the building was supposed to host, and change the function of popular reception, transforming the sacred area into a place of secular worship.

The building was to be configured as an administrative and congress center, modern and multipurpose with libraries, concert spaces, large rooms to set up mass events capable of hosting performances and collective parades.

One hundred and sixty designers joined the competition, including a minority of foreign technicians, present with twenty-four proposals relating to the Building in the significant urban context. Famous architects of the international scene such as Le Corbusier, Walter Gropius, Auguste Perret, Erich Mendelsohn, Hans Poelzig and among them the Italian Armando Brasini participated (Pisani 1996, pp. 40–48).

Brasini is invited to participate by his student Boris Michailovic Iofan, and after resulted to be the winner of the competition.

The style of the selected project raises many controversies, especially compared to the great proposals of the other participants.

The proposal of Boris Iofan, of classicist derivation like that of his Roman teacher who receives a mention of merit for the occasion, will never be made for the onset of the Second World War. Only the foundations of the building were built on the site of the Cathedral of Christ the Redeemer, demolished in 1937, from 1958 to 1960 the Building was then converted into the largest open-air swimming pool in the world, to then give space at the end of the nineties to the reconstruction of the suppressed church (Vigano 2002, www.larici.it).

The contribution examines the values of the graphic heritage produced in this symbolic contest in which the documentary apparatus has assumed a preponderant role, due to the paradigmatic demands and the failed implementation.

The drawings make it possible to trace the compositional intentions of the illustrious authors in relation to the related ideological and political influences.

Through a series of comparisons and graphical analyzes, some of the significant projects presented that reveal the architectural qualities and the no less significant cultural and social implications, have been examined.

In particular, the competition results highlight an analogy between Rome and Moscow which, although motivated by different political convictions, manifest similar celebrative architectural choices adopted to express the greatness of the respective regimes.

In this sense, among the many possible, it was decided to re-configure Armando Brasini's project in three dimensions, explored through the spatiality of digital study models.

2   The Contest Requests

The brief description of the facts and the insolvency requests allows us to understand the proposed design solutions and the investigations carried out through the analysis.

The competition that foreshadowed linear and contemporary forms in its intent, required the creation of a unitary structure articulated in a series of rooms, two of which were large, joined by galleries and accessory rooms with restaurants, exhibition spaces and library, all to be arranged in a large square able to accommodate popular assemblies.

A pharaonic project at the limits of reality aimed at transposing the ideals of communist politics into architectural forms (Hoisington 2003, pp. 41–68).

The Council, designated for the construction of the representative building, specified the destinations of use of the environments and their capacities.

The main room for communicating collective work methodologies and Soviet reforms was to have at least fifteen thousand seats, arranged sectorally around a stage.

The smaller room with six thousand five hundred seats, for congresses and assemblies, should instead have included a gallery reserved for two thousand guests with an independent entrance. The special accesses allowed the possibility of approaching the speakers in hierarchical order, to the personalities who alternated in the mass events. Special posts were also reserved for the press, which had the task of divulging political activities.

Of these mass indications, the designers developed their selected proposals with awards, the first assigned was assigned to Boris Iofan, the second to George Hamilton and the third to Ivan Vladislavovic Zholtovskij.

In reality, it was not unequivocally to the choice of a definitive project, but to a succession of improvements, carried out on the first classified proposal, derived from the solutions of the projects considered best.

3   The Design Proposal of Some Illustrious Participants

Among the significant proposals, Le Corbusier occupies an important place, despite not having received any awards, he has been universally recognized by the commissioners and critics as a "masterpiece of functionalism" that is able to fit into the context with an antithetical expression of urban renewal.

Modern linear forms, state-of-the-art technologies and style counterbalance monumentality (Samonà 1976, pp. 40–44).

The project is based on the arrangement of the two trapezoidal-shaped rooms that mirror each other on the large square of the popular assemblies; the latter structured on several levels and crossed by an elevated element that acts as a link introducing the speaker platform. The smaller room towards the Kremlin perceptually directs the vision of the larger one highlighted by a system of septa that frame the large external stage (Fig. 1).

Fig. 1. Projects by Le Corbusier and Gropius, images of the models of the competition

On the contrary, Gropius's proposal is an isolated architectural element enclosed in the circular form that defines the organizational structure of the space.

The two rooms are in this case circumscribed within a cylindrical sector which occupies opposite parts of the generating circumference and find the contact point in the center of the composition. Externally it appears as a compact volume that does not open in the city but i s dstinguished from it, the contact occurs with the infrastructures that cross it on several levels, with the subway that cuts the base circle and the roads that enter along the rays.

Another exponent of the German architectural renovation Hans Poelzig, with ideas opposed to those of the Regime, intervenes in the competition to affirm the spatial and technological possibilities offered by the new materials. The designed solution distributes the building volumes on the large square, turning the smaller room towards the Kremlin with a horseshoe shape that extends with the lateral bodies up to the hall of the main block which opens like a fan. The compositional unity is resolved through the modular scanning of the light septa that cover the entire complex.

An innovative project is also that of Moisei Ginzburg, an exponent of Soviet constructivism, who proposes a planned relationship with the area of intervention.

The solution is characterized by the presence of a main element that stands out from the context with a large and technological dome. The structure, intended to house the main hall of the building, was served by a spiral ramp and included an auditorium with a large stage on the inside. The courtyard and terraces complete the interesting proposal incorporating an articulated system of connections that connect the different spaces. Everything is enclosed within a square plate where the trapezoidal minor room is located and the assembly square is arranged (Fig. 2).

Fig. 2. Projects by Ginzburb and Poelzig, images of the plastics of the competition

The purity of the forms of Ginzburg, the rational linearity of Poelzig, the proportionate spatial organization of the volumes designed by Le Corbusier can only contrast with the pletorical solution of Jofan that with its 500 m in height, 80 of which occupied by the towering statue by Lenin, represent the ostentation of power, also readable through the graphic evolution of the drawings of the selected project.

The initial proposals of the architect, elected as the designer of the most prestigious work in Moscow, undergo requests for modifications that completely transform the compositional aspect. The changes requested transfigure the style, the statues magnify, the colonnades widen the already monumental composition, ironically defines as "wedding cake" (Fig. 3).

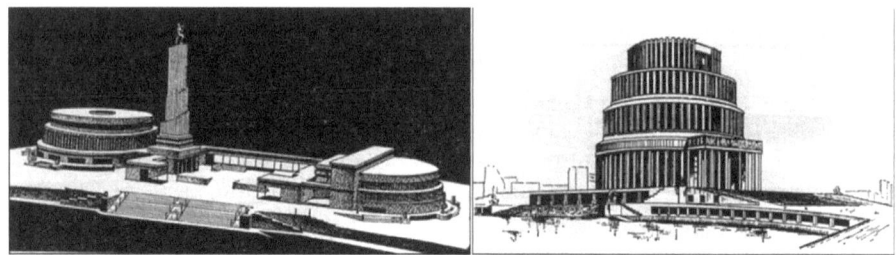

Fig. 3. Boris Jofan, images of the initial design solutions (1931–32)

The overlapping of volumes does not help the tormented project that has no definition, becoming for Jofan a political imposition from which he will not be able to free himself of.

4   The Project and Armando Brasini's Role

It is appropriate to consider the role of Armando Brasini, an Italian participant and Boris Jofan's teacher, to understand the educational influence from which the outcome of the competition arose.

The Roman architect at the service of the Duce who expressed his artistic and visionary skills mainly through the design from which, in reality, he received his best results.

His graphic talent led him to draw from the monumental architectures of the past different figurative repertoires, amalgamated with the mastery of the sign.

The theatricality of the designed proposals translate on paper the ideas indicated by Mussolini for state architecture that push him to enhance monumentality. The latter suggested by the dictatorship is further emphasized by Brasini who through a mixture of styles, with a preponderance of Baroque and neo-Gothic, manipulates his eclectic compositional lexicon. This compositional trend influences the training of the young Jofan who studies and works in Rome collaborating for a decade in the study of Armando Brasini.

The style of the winning project reflects the teaching received that is well suited to the ideological demands of the theme, the classic language is assumed for the symbolic representation of power. The conceptual and architectural circumstances that arise between the two capitals thus create an unequivocal orientation in the design proposals of the two architects (Sedova 2006).

Fig. 4. Boris Jofan, final project 1936 and proposal by Armando Brasini

Brasini's solution shows clear references to the Roman Empire at the Pantheon, the Forums, the Colosseum reinterpreted in the drawings of the Palace of the Soviets which, in perspective views, appears as a compact monolithic block with soaring pyramidal truncated towers, from which beams radiate of light converging towards Lenin's soaring statue. From the two projects emerge similarities and contact points that are not very investigated from a graphic point of view (Fig. 4).

5   Digital Reconfiguration of the Armando Brasini Project

The first phase of the digital reconfiguration work involved the analysis of the original designs produced by Brasini for the competition which, through a sequence of graphic designs, describe the imposing layout and the compositional choices adopted.

The scenographic impact of the three external perspective views, amplified by the chiaroscuro effects made of graphite and charcoal illuminated by white lead on large format cardboard, show the appearance of the majestic palace presented like a mighty fortified structure. A sequence of buttresses dominated by volumes, similar to sentry boxes, surrounds the building that appears as a turreted citadel. The first level covered in rustic rustication is delimited by protruding spurs that mark the entrances to which great importance is assigned: the lateral ones were reserved for delegates who had separate entrances based on the political role, the main one welcomed the crowds sorted by means of a circular pronaos that follows the shape of the planimetric system introducing in the central courtyard and in the halls for congresses.

The perspectives that frame the rear, shown in a front and a side view, highlight the two curved ramps that allow the access of the cars inside the structure, capable of receiving even the aircraft welcomed in the large central arch, pivotal of the dynamic flows that permeate the building (Fig. 5).

The presence of the vehicular transit is confirmed in the perspective views of the interiors, which show the "carriageable portico" leading to the group B hall with the spectacular theatre and "Large concert hall for 15,000 seats" on which an airplane appears.

Fig. 5. Project of A. Brasini, perspective views of the rear and lateral front

The ostentation of grandeur is reaffirmed in the main front with the monument of Lenin that stands on the majestic central tower, pyramidal in shape, radiated by beams of light from the two lateral towers.

The perspective representations aim to surprise the observer by enhancing the monumentality accentuated also by the setting, by the groups of people and vehicles that like ants go towards the entrances. An out of scale loved by Brasini several times used in the visionary representations for "Urbe Maxima" for the visionary drawings conceived for Rome as the capital of fascism (Orano and Brasini 1917, p. 58).

In the Palace of the Soviets, the Italian architect shows off his entire figurative and symbolic repertoire which, through his extraordinary graphic skills, is well suited to indulging the megalomania of Soviet power.

The citations of the ancient are revealed above all in the grandeur of the interior spaces which compared to the more perched volumes modeled for the exterior, in which medieval echoes are recognized, clearly follow the architecture of the Roman Empire.

The meticulously designed plants seem to re-propose the sequence of the thermal environments which in this case house the numerous rooms that make up the multifunctional complex (Fig. 6).

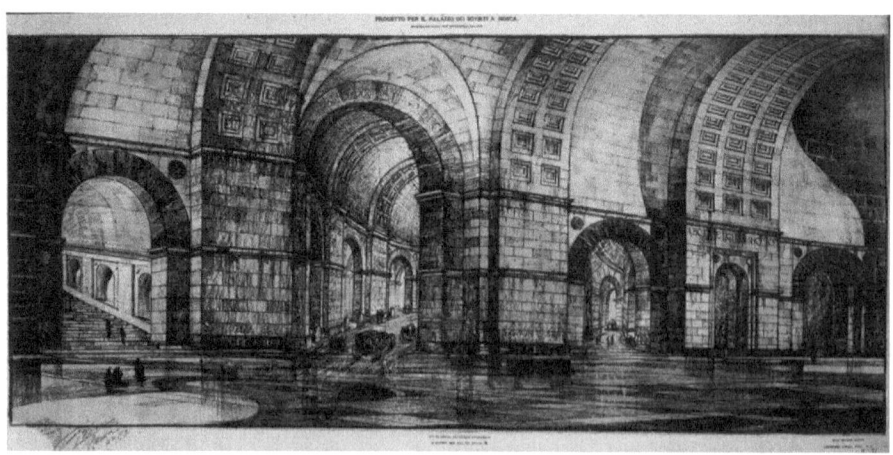

Fig. 6. Project of A. Brasini internal view o fthe carriageable porch

The basement was used for the numerous services necessary for the operation of the complicated structure: power plant, heating, kitchens, parking for 400 cars which was a record for the time, rooms for drivers, mechanical workshops. The ground floor was supposed to have a distribution plan capable of managing the network of important flows, relating to the aforementioned entrances hierarchically organized so as not to generate disturbances in large demonstrations, meetings and political parades.

Characterizing elements of the system were the two main rooms: the circular "A" placed on the main front with 6,500 seats, equipped with podiums, access ramps to the stalls and the stage with annexed exhibition rooms and library; the semicircular "B" on the opposite side was to contain 15,000 spectators and to have access from the large arcades, which can be travelled by cars. The two fulcrums were connected by a large open courtyard which lightened the compact structure allowing direct lighting to the rooms, functionally distributed in five levels. The upper floors follow the scheme of the underlying floors with smaller rooms, galleries, libraries organized in the spaces left free by the bulky volumes of the two main rooms extended over the entire height of the building.

Returning to the citations, the dome of the Pantheon with a similar design of the coffered ceiling is proposed again in the circular room of group A, where it doubles in height, rising in the upper part up to the connection with the truncated pyramidal tower from which particular light effects originate. The latter, filtered by the openings arranged around the main tower, was ingeniously addressed to obtain suggestive irradiation both in the conference hall and in the upper part where the statue of Lenin is placed, which was to appear suspended between the light beams, as drawn in the sections.

The stenographic light effects are represented in the perspective views that allow you to perceive the solemnity of the rooms, resembling Piranesian spaces reproduced with Baroque emphasis (Fig. 7).

Fig. 7. Project of A. Brasini internal views of Congress Halls B and A

A déjà vu (once seen) which summarily unites and amalgamates a concentrate of exemplary monuments. The results is a hyper-monument that responds well to the megalomaniac requests of the competition and aims to generate a hegemony of spaces, undoubtedly quantitative, but arguably qualitative, especially when compared with the modern linearity of the other design proposals and with the history of the architecture.

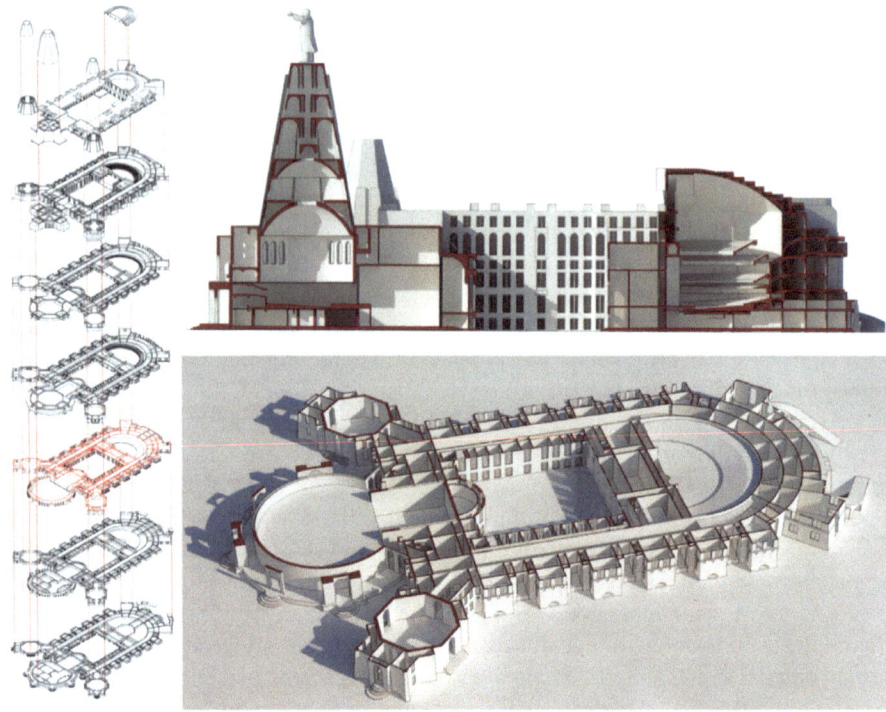

Fig. 8. Digital reconfiguration. Three-dimensional study models

The sensation of the already seen also accompanies the pharaonic space for meetings, theater and large concert hall of the group B of semicircular shape, in which they find themselves in addition to the replicated dome of the Pantheon, in this case halved to be adapted to the planimetric system, the bleachers that echo the shapes of the Theatre of Marcello, mixed with the arches of the Colosseum. An amalgam of quality ingredients readjusted if necessary through the mastery of the design.

The suggestion provided by the chimerical images characterize the work of Brasini and justify the low consideration of the critics who assessed the design contents, in evident contrast to the modern proposals of the other European participants, but which from a graphic point of view constitute the excellence of the character, a value for "graphic heritage" or for "heritage graphics".

To conclude using the words of Paolo Portoghesi who defines him "…one of the great intruders of twentieth-century architecture" but considers him "…a great designer, capable of creating architectural tables that attest to him as one of the greatest artists and visionary designers of the twentieth century in Italy".

Fig. 9. Digital reconfiguration. Three-dimensional study models

6 Conclusion

In conclusion, the contribution through graphic analyses carried out on the project drawings, through comparisons and analogies with the other designers involved in the singular competition of the Palace of the Soviets, proposes a review, an investigation conducted by means of the representation that leads from traditional documents to digital explorations, highlighting further results on international competition expressed through modern visualizations (Fig. 8).

The construction of three-dimensional study models allow us to restore the spatiality of the forms conceived for the Palace, to understand its design, the volumes, to enter the rooms visually perceiving the architectural language of the heterogeneous compositions (Fig. 9).

Failure to carry out the work extends the intangible values of the graphic heritage left on paper, the many solutions that, with the help of today's increasingly realistic and immersive technologies are able to communicate important contents and addresses, often dictated by political-cultural choices.

References

Hoisington, S.: "Ever higher": the evolution of the project for the palace of Soviets. Slavic Rev. **62**(1), 41–68 (2003)

Pisani, M.: Architetture di Armando Brasini. Officina, Roma (1996)

Samonà, A.: Il Palazzo dei Soviet 1931–1933. Officina, Roma (1976)

Sedova, I.: Il palazzo italiano dei Soviet. Architettura Italiana e sovietica tra le due guerre. Catalogo Mostra Museo dell'Architettura "Schusev" (MuAr) (2006)

Palestini, C.: Il disegno e la storia dell'architettura. Letture attraverso un'opera di Armando Brasini. Opus. Quaderno di Storia Architettura e Restauro, no. 12, pp. 431–442. Carsa edizioni, Pescara (2013)

Viganò, D.L.: L'invisibile visione del Palazzo dei Soviet: Iconografia e architettura negli anni Trenta sovietici (2002). http://www.larici.it/culturadellest/arte-architettura/vigano/index.html

Orano, P., Brasini, A.: L'Urbe Massima: l'architettura e la decorazione di Armando Brasini. Formiggini, Roma (1917)

Brasini, L.: L'opera architettonica e urbanistica di Armando Brasini: dall'Urbe Massima allo stretto di Messina, Roma (1979)

"L'inganno degli occhi". Borromini's Perspectival Niche for the Casa dei Filippini in Rome

Marco Carpiceci and Fabio Colonnese(⊠)

Sapienza University, Rome, Italy
{marco.carpiceci,fabio.colonnese}@uniroma1.it

Abstract. The RCIN 905602 sheet of the Royal Collection, Windsor contains an orthogonal projection presentation drawing made by Francesco Borromini in 1627 and showing one of Gianlorenzo Bernini's proposals for the Tomb of Urban VIII. The apsidal frame rendered in perspective can be related to the perspective niche that Borromini himself placed above the fictitious entrance in the center of the facade of the Casa dei Filippini in Rome a few years later. The sheet is studied here in the context of the influence that the diffusion of representation methods and practices has had on the forms of architecture as well as their visual perception. After discussing the reason for the construction of such a perspective device, its representation in the design documents and its reception in the images of the square after its construction, the authors present the results of a geometric analysis and reconstruction after photo-modeling of the niche itself, aimed at determining the position of the ideal observer in the square considered in designing the device, and they discuss the relationship between this spatial stratagem and the graphic trick adopted in the tomb design.

Keywords: Francesco Borromini · Solid perspective · Photo-modeling

1 Introduction

The heterogeneous, contingent and personalized visual code the architects had inherited from the Middle Ages, still significant in the illustrations of the 15th century treatises by Filarete and Francesco di Giorgio, was gradually replaced in the early 16th century by the projective canon emerging in the Roman context, derived from Vitruvian readings, Bramante's experiences and, indirectly, Leonardo da Vinci's studies. This canon is expressed in Raphael's prescriptions for Roman monuments' survey (Di Teodoro 1994) through the combination of plan, elevation and section, the "inside view" which ideally replaces the Vitruvian "scenography", to be left to painters.

Starting from Rome and St. Peter's construction site, the epicenter of applied research in architecture, this canon quickly established itself first in Italy and then, thanks to treatises and travelers, in Europe. However, important derogations persist, both in the monuments' survey and the projects' design, fields in which the semantic efficacy and

L. Agustín-Hernández et al. (Eds.): EGA 2020, SSDI 6, pp. 227–237, 2020.
https://doi.org/10.1007/978-3-030-47983-1_21

the economy of the marks often suggest arbitrary contamination between different views. For example, orthogonal projection drawings show parts in central projection and parts arbitrarily developed in order to maximize the synoptic quality. This is the case of some of Pirro Ligorio's drawings after ancient monuments, but contaminations are visible also in the designs for St Peter's cathedral. In sheet 70r at the Uffizi, Antonio da Sangallo the Younger presents his ideas on the future basilica in an elevation divided by the axis of symmetry between section and elevation (Carpiceci and Colonnese 2018). On the right half, niches are represented both in orthogonal projection and in perspective, with the curved frame and shades to show the visual effect of chiaroscuro (Fig. 1).

Fig. 1. Antonio da Sangallo il Giovane, Design for St Peter's church, detail. Firenze, Gabinetto Disegni e Stampe degli Uffizi, f. 70r.

This kind of drawings reveals the need to design the internal and external parts of the building in a coordinated way and, at the same time, the intent to anticipate the visual effect, resorting to occasional projective derogations and graphic treatments in this sense. Over the decades, this practice contributed to shape the relationship between design and construction, and encouraged a mutual inspiration between the two areas: on the one hand, the anthropomorphic and projective conception of architecture, committed to apply the architectural order in increasingly "difficult" conditions, has oriented models and graphic formats of representation; on the other, graphic elaborations and experiments, especially in the field of illustration for printing (Carpo 2001), have constituted an assortment of visual types that influenced the work of the designers themselves.

Press promoted the dissociation of the images from the context of ideation and production, offering them to a wider fruition and interpretation. Think of the decorative theme applied to vertical rectangular fields formed by a circle in the middle and semicircles at the top and bottom. Long before decorating doors and furnishings, this motif had been applied to the pilasters and the reason is quite simple: it clearly derives from the representation of the architectural order as it was codified in the 16th century, with the overturning of the key horizontal sections at the base, in correspondence of the entasis, and at the collar; properly geometrized, this graphic code was translated into a decorative motif applied first to the supports - in particular, the pilasters that have a representative and non-structural function - and then to any rectangular element. This can already be seen at the end of the 15th century in Mauro Codussi and Jacopo Sansovino's facade of the Scuola Grande of S. Marco in Venice (1490) and in the corner pilaster of Palazzo Turchi di Bagno in Ferrara by Biagio Rossetti (1492) (Fig. 2).

This dynamic context of communicating vessels between design practice, systems of representation and images of architecture is here assumed as the context to investigate Francesco Borromini's drawing for the funeral monument of Urban VIII designed by Gian Lorenzo Bernini. In particular, Borromini's choice to draw explicitly a curved cornice, as if represented in perspective, sounds odd in a period and place in which architectural projective code is shared and consolidated. It raises some questions, especially considering the perspective niche Borromini created later in the upper part of the facade of the Casa dei Filippini in Rome, where the cornice is curved indeed. Is the frame in the tomb design just a graphic expedient or does it allude to a real niche in solid perspective to optimize the visual perception of the statue?

This perspective stratagem is investigated in the representations and in the facade for the Filippini, which has been surveyed and reconstructed through photo-modeling and analyzed to evaluate the relationship between the actual shape of the frame and the shape perceived by the observer in the square designed by Borromini. The results of this analysis are then used to judge whether the perspective niche of the Tomb is only a graphic device or even a spatial one.

Fig. 2. M. Codussi e J. Sansovino, Scuola Grande of S. Marco, Venezia, 1490; B. Rossetti, Palazzo Turchi di Bagno, Ferrara, 1492.

2   The Perspective Niche: Graphic and Perceptive Stratagem

In 1627, Gian Lorenzo Bernini, who was already involved in the construction of the Baldacchino Vaticano, was commissioned to design the tomb of Pope Urban VIII (Fagiolo Dell'Arco 2007). The tomb, which was designed in 1627 but was not completed until 1647, is located inside a niche of one of the four main pillars of St Peter's basilica.

The RCIN 905602 sheet of the Windsor Royal Collection, which represents one of the projects, shows the design of the niche flanked by Corinthian columns, in which the statue of Urban VIII stands upon a base flanked by the figures of Charity at left and of Justice at right (Fig. 3). Borromini is supposed to have completed this accurate ink-made presentation drawing in which Bernini himself added the human figures and the winged skeleton (Blunt and Cooke 1960). Although most of the design respects a rigid orthogonal projection, the cornice of the niche appears curved as if it were represented in perspective. This unusual detail seems out of place, even when comparing this design with others produced over the years by Bernini's crowded atelier.

Fig. 3. Gian Lorenzo Bernini and Francesco Borromini, Design of a Tomb for Urban VIII, 1627. Windsor, Royal Collection, RCIN 905602.

2.1   The Perspective Niche for the Casa dei Filippini

In the late 1630s, Borromini transformed the curved cornice from a simple graphic stratagem to an architectural solution in the niche located above the central doorway of the Casa dei Filippini in Rome (Fig. 4). This perspective device is rather difficult to decipher from below: even a representation expert would be persuaded that the cornice is horizontal. It was to be seen from the square designed by Borromini himself, which has been reshaped by the opening of Corso Vittorio Emanuele in the early 20th century. Borromini applied it to accentuate the apparent depth of the niche while remaining within the wall thickness and contributing to the virtual aestheticism of the innovative concave facade. This device is connected to studies on complex geometries, on solid perspective (Sinisgalli 1998; Colonnese 2016; Colonnese 2018) and on the so-called "oblique architecture" celebrated by Juan Caramuel Lobkowicz (Iurilli 2017). However, it seems to be closely related to the practice of architectural drawing and to the sense of a theatrical representation of architecture itself (Camerota 2006). As Connors commented (quoted in Thelen 1959, p. 53), "The central niche and the illusion of depth it entails can be traced back to Borromini's extraordinary ability to manipulate the conventions of architectural drawing".

Fig. 4. Roma, Filippini's facade with the perspective niche, 2019 (foto di M. Carpiceci).

Two stories about the Filippini's façade intertwine: the story of the project and construction, revealed by Connors (1989) and Downes (2014), and the story of the reception and reproduction of its image (Chiavoni 1996). According to Connors (1989), "The facade of the oratory had developed [...] with changes made during the construction [...] as a living organism" and years after the end of building of his assignment, Borromini was still trying to control the way his work was perceived: he did it both physically, by reshaping the square (Bonadonna Russo 1965) to ensure a better view of the building (Fig. 5), and mediatically, through idealized pictures conceived for his *Opus Architectonicum*, which were to influence the work of artists and engravers.

In 1650, Fioravante Martinelli's (1650) *Roma ricercata* offers a description of the Casa dei Filippini with three pictures designed and engraved by Dominique Barrière. In one of these, the facade appears isolated from both the church of S. Maria in Vallicella on the right and the left corner. According to Connors (1989, p. 373), "the degree of idealization present in the print supports the hypothesis that this, like other engravings in the series, was based on a drawing provided by Borromini himself: the print in fact preludes to the project revisions, which go in the direction of greater drama, shown in a drawing of 1660 [...] and in a number of prints that derive from it". Connors refers here to the semi-perspective (RC 5595) at the Royal Collection of Windsor which can be placed not so much in the design process as in the years following the construction site, between 1550 and 1560.

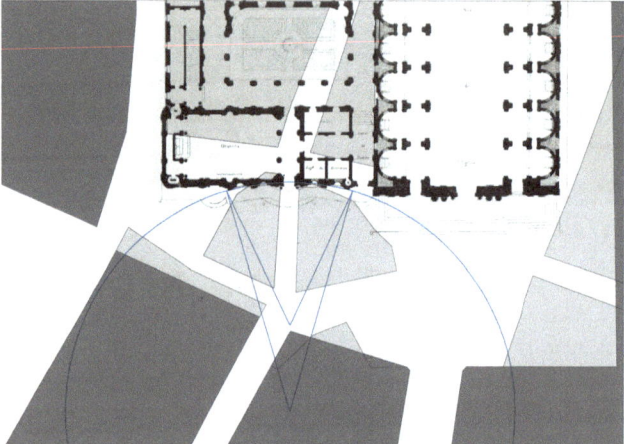

Fig. 5. Plan of the square designed by Borromini. In light gray, the arrangement of the insula at the time of the project; in blue, the 200 palms radius of curvature of the facade of and the actual point of view aligned with the fictitious entrance (drawing by F. Colonnese)

In the same collection there is also an original elevation (RC 5594), dating back to 1638, which was also used as a model by various artists who reproduced the facade, and which shows the curved frame. However, this curvature is generally ignored by those who have dealt with illustrating the square. For example, Giovan Battista Falda's view of 1665, which frames the facade in the square, shows a straight cornice. Quite the same can be said of Giovanni Giacomo de' Rossi's (1684) *Insignum Romae Templorum Prospectus*. In 1684 he created an idealized facade of the oratory based on the pictures owned by Bernardo Borromini, the artist's grandson, made by Dominique Barrière on behalf of Borromini. In addition to the general stiffening of the decorative elements reported by Connors (1989, p. 399), De Rossi's elevation, although enriched with chiaroscuro treatments aimed at revealing the curved surfaces, prefers to ignore the stratagem of the curved cornice (Fig. 6). For decades, the artifice was reserved to a direct experience of the building. Only in the 18th century did the stratagem begin to be revealed. Sebastiano Giannini, who published Borromini's *Opus Architectonicum* in 1725 (Borromini and

Giannini 1725), creates four views of the facade, also inserting a perspective of how Borromini intended it, "with more ornaments not carried out" (Martinez Mindeguia 2004, p. 21). Its orthogonal projection facade, always idealized, is based on the engraving by Borromini and Barrière, too, but the curvature of the cornice is explicit, as in Borromini's original drawing of 1638. Almost a century had passed since the project and, over time, the *inganno* had been accepted, understood, explained and reproduced.

Fig. 6. Giovanni Giacomo de' Rossi, *Insignum Romae Templorum Prospectus*, 1684, pl. 31.

3 Rilievo e Analisi Della Nicchia Prospettica

The upper niche of the facade of the Filippini, with its perspective gapped basin, has been subjected to a photographic survey, from both below and the windows of a building nearby. A first result, invisible from below but evident from the pictures taken at the balcony height, is the evidence of the deformations affecting the upper part of the portal (Fig. 7). The complex design of the portal top appears to be further manipulated according to an inverted arch curvature, at least up to the door lintel, whose key height is 4 cm lower than the shutters'. This deformation, which in some parts is almost unperceivable but still present, documents the care the artist took in his deception game. In the end, the portal top is the main reference an observer, more or less unwittingly, adopts to evaluate the shape of the upper niche: if this had been straight, it would have compromised the artifice even from below. By comparing this data with the graphic documents, this deformation turns out to be visible in the detail made by Borromini (Connors 1989, pp. 304–305) but is instead censored in the Table XX of the *Opus* edited by Giannini.

To evaluate the degree of deformation applied to the elements of the niche, the authors applied a photo-modeling procedure to get a questionable model, the key sections (Fig. 8) and other general considerations. The model shows, for example, that the overall curvature of the facade at the height of the second order refers to a center about 45 m away, indirectly confirming the radius of 200 palms (1 Roman palm = 22.34 cm) Borromini also refers to in his drawings and which roughly corresponds to the size of the facade itself.

Fig. 7. Rome, Oratorio dei FIlippini, niche detail, 2019 (photo by M. Carpiceci).

The curvature of the facade, with its 45 m-long radius, does not refer to a real point of view in the square, from which the effect of the curvature would have been somewhat impalpable, but only to a geometric center. Along the axis of the niche, an observer walking in the square designed by Borromini could place himself or herself about 25 m-away from the façade, surely perceiving the general curvature, which was a peculiar intent of Borromini. Martinez Mindeguia (2004) demonstrated how the client did not accept the curvature proposed by Borromini who had to reduce it and straighten the frames of the first order, only to propose it again in all its drama in the semi-perspective of 1660, reducing the radius from 200 to 150 palms (Connors 1989, p. 376).

Fig. 8. Horizontal and vertical section on the perspective niche (model by M. Carpiceci)

3.1   Looking for the Point of View

The perspective niche is almost four-meters wide and about 50 cm-depth by the axis. Its vertical profile reveals all the expertise necessary for the construction of the lacunaries in a solid, *stiacciata* perspective, aimed at arranging an illusory image. Its horizontal profile can be assimilated to a semi-oval construction with five centers, resulting almost flat in the central part.

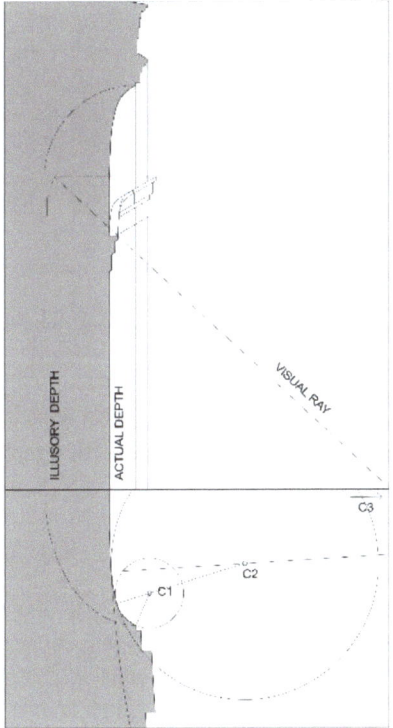

Fig. 9. Section and plan of the niche. The section highlights the relationship between the actual shape and the perceived one; the plan shows the centers of the oval (drawing by F. Colonnese).

With respect to the horizontal plane, the capitals supporting the arch are inclined by about 30° and the cornice itself drops about 70 cm by the bottom of the niche. The cornice of the niche is located about 25 m above the ground and could therefore be viewed from the front from about 25 m away.

Unlike other coeval solid perspective devices, which have the task of virtually expanding a room by taking up and deforming in the foreshortening elements perceivable in their true form - think of Bernini's perspective sacellum designed to house the statue of Filippo IV of Spain (Colonnese 2018) or the famous perspective gallery of Palazzo Spada by father Bitonto and Borromini - here it is difficult to state how deep the niche is expected to appear. The authors conjecture that the niche basin, a portion of an ovoid surface, has an axial section close to a circular arc. Consequently, the point of view necessary to carry out this prospective artifact has been carried out. After building a simplified section starting from the data collected in the digital model, the position of the point of view on the ground has been verified. It is useful to remind that the optical deception is activated when an observer standing considers, more or less consciously, that the edges of the cornice on the side capitals are as high as those on the bottom are at the same height, both of them belonging to a horizontal plane. Given this premise, the line that connects the upper drip of the frame actually on the bottom of the niche with its virtual position on the horizontal plane passing through the same drip on the capitals, identifies the inclination of the visual ray useful for stating the position of the point of view (Fig. 9). The identified visual radius, whose inclination is about 45°, identifies, along the axis of the building entrance, a point that is located on the opposite limit of the square designed by Borromini. Based on this hypothesis, the perspective niche, which looks more than 120 cm deep, would have been conceived for an observer who walks in the square from the alley and sees in all its extension the facade right on the axis of the fake entrance and of the upper niche - a view prevented today by the large plane tree that partially shades the facade itself.

4 Conclusion

The perspective niche Borromini elaborates for the Filippini is based on perceptive expectations of an audience educated in the canon of the Renaissance architecture. It finds its reason in saving the space behind the wall and works particularly well for a number of reasons. First of all, it is very high, and the size of the square did not allow people to go too far and to take advantage of a more "orthographic" vision to discover the *inganno*. Moreover, the curved cornice is in continuity with that above the neighboring windows and the deformations of the portal reinforce its illusory depth. Conversely, the conditions under St Peter's dome, where the tomb of Urban VIII is located, are radically different. Here the observers would have been too close to the curved cornice and, unlike the square, their visual field would not have been limited. Eventually, there seems to be no need to optically expand the depth of the concavity intended to house the statue. In that case, Borromini's design must therefore be still considered as a mere visual stratagem, capable however of inspiring, over time, an innovative architectural solution.

References

Blunt, A., Cooke, H.L.: The Roman Drawings of the Seventeenth and Eighteenth Centuries in the Collection of Her Majesty the Queen at Windsor Castle. Phaidon, London (1960)

Bonadonna Russo, M.T.: Il contributo della Congregazione dell'Oratorio alla topografia romana: Piazza della Chiesa Nuova. Studi romani **13**, 21–43 (1965)

Borromini, F., Giannini, S.: Opus architectonicum, Roma (1725)

Camerota, F.: La prospettiva del Rinascimento. Arte, architettura, scienza. Electa, Milano (2006)

Carpiceci, M., Colonnese, F.: Between antinomy and symmetry. Architectural drawings of presentation and comparison in the XVI century. In: Marcos, C. (ed.) Graphic Imprints. The Influence of Representation and Ideation Tools in Architecture, pp. 66–78. Springer, Cham (2018)

Carpo, M.: Architecture in the Age of Printing: Orality, Writing, Typography, and Printed Images in the History of Architectural Theory. The MIT Press, Cambridge (2001)

Chiavoni, E.: Il disegno nell'analisi degli organismi architettonici: l'oratorio dei Filippini in Roma. Disegnare **12**, 33–42 (1996)

Colonnese, F.: Perspective, illusion and devotion. The chapel of S. Agnese in Sant'Agnese in Agone. In: Zirpolo, L.H. (ed.) The Most Noble of the Senses: Anamorphosis, Trompe-L'Oeil, and Other Optical Illusions in Early Modern Art, pp. 87–110. Zephyrus Scholarly Publications LLC, Ramsey (2016)

Colonnese, F.: Reconstructing the illusion: virtual heritage on Bernini's solid-perspective for Felipe IV. In: Ippoliti, A., Inglese, C. (eds.) Analysis, Conservation, and Restoration of Tangible and Intangible Cultural Heritage, pp. 286–320. IGI Global, Hershey (2018)

Connors, J.: Borromini e l'Oratorio romano. Stile e società. Einaudi, Torino (1989)

De' Rossi, G.G.: Insignum Romae Templorum Prospectus, Roma (1684)

Di Teodoro, F.P.: Raffaello, Baldassar Castiglione e la Lettera a Leone X. Nuova Alfa Editoriale, Bologna (1994)

Downes, K.: Averlo formato perfettamente: Borromini's first two years at the Roman Oratory. Agric. Hist. **57**, 109–139 (2014)

Fagiolo Dell'Arco, M.: Il gran teatro del barocco. De Luca Editori d'Arte, Roma (2007)

Iurilli, S.: Trasformazioni geometriche e figure dell'architettura. L'«Architettura Obliqua» di Juan Caramuel de Lobkowitz. Firenze University Press, Firenze (2017)

Martinelli, F.: Roma ricercata nel suo sito, Tani, Roma (1650)

Martinez Mindeguia, F.: La visione frontale dell'Oratorio di San Filippo Neri. Disegnare **29**, 10–24 (2004)

Sinisgalli, R.: Una storia della scena prospettica dal Rinascimento al Barocco. Borromini a quattro dimensioni. Cadmo, Firenze (1998)

Thelen, H. (ed.): 70 disegni di Francesco Borromini dalle collezioni dell'Albertina di Vienna. Gabinetto nazionale delle stampe Farnesina, Roma (1959)

Drawing as Heritage

Francisco Xabier Goñi Castañón(⊠) ⓘ and Inmaculada Jiménez Caballero ⓘ

Department o f Theory, Projects and Urbanism, University of Navarra, Pamplona, Spain
`{fgonic,ijimenez}@unav.es`

Abstract. The use of drawing as a form of language has always been linked to the history of humankind; at the same time it has been one of its basic development tools, being the very trigger of creativity. Relegated many times to a tool plane at the service of the arts, in the case of architecture, it is not only the indispensable path to make art real – it is also in charge of safeguarding it, thanks to the efforts applied and progress made in the area of architectural heritage. Architecture is the creation of humans, bringing greater knowledge about the history of civilisations; its constructed works constitute a very large segment of enclaves declared to be world heritage sites. Given the value of the architectural drawing as a previous 'iteration' of the built works (and subsequent recreation of many missing works), with this article, we propose to rethink this quality of drawing as an essential component of the architectural works. Through analysis of its techniques, its characteristics and its qualitative and stylistic content, we aim to establish whether this medium meets conditions to qualify, in certain cases, as part of humanity's cultural heritage.

Keywords: Architecture drawing · History of representation · Architectural drafting · Cultural heritage · Graphic styles

1 Introduction

Given that our common graphic heritage is the proposed topic of this edition, it seems appropriate to begin by reviewing the meanings of both terms: *heritage* and *graphic*.

Heritage: 'Set of assets and rights acquired by any title.' This interpretation suggests diversity in the considerable assets implied by such terms as 'heritage' – clearly a reference that extends well beyond the more restrictive category of material goods alone. Another meaning that broadens the meaning is: 'Goods that someone has inherited from their ancestors' in relation to its etymology *patrimonium* (father's inheritance). This interpretation can have an appropriate application to architecture and its drawing – if we use a symbolic figure, understanding the built architecture as a formalisation inherited from its ancestor(s) (i.e. the drawing that engenders it). According to this consideration, the effect of paternity of the work accomplished is attributed to the architectural drawing, an adequate attribution, since – at the end – the graphic documents of a project contain all information necessary to 'illuminate' a new reality; they are the genetic code of the constructed work. It is a unique and nontransferable code that belongs only to the architecture itself, distinguishing, identifying and allowing it to intervene in it if necessary, due to suffered pathologies or preventive actions.

L. Agustín-Hernández et al. (Eds.): EGA 2020, SSDI 6, pp. 238–248, 2020.
https://doi.org/10.1007/978-3-030-47983-1_22

We can find an interpretation also applicable to architectural drawing in its correspondence to the work it originates, in two more definitions of the *RAE*. The first: 'set of goods belonging to a natural or legal person, or affected by a purpose, susceptible to economic estimation' and, if historical heritage is involved: 'a set of goods of a nation accumulated over centuries, which for their artistic, archaeological significance, etc., are subject to special protection by legislation'.

In both cases, the definition conforms to the architectural drawing since, in the first case, it is a set of goods that are subject to economic purposes, this estimate being interpreted in different ways and that, over time, acquires a diverse meaning always deserving of protection. For the term 'graphic', we read: 'said of description, an operation or a demonstration: which is represented by figures or signs'; also, 'manner of speaking: exposes things as clearly as if they were drawn'. Both interpretations agree because the architectural drawing is the representation by means of signs, typical of a graphic language, of an operation that is to build a building with all its complexity; it is a graphic language that also serves as a 'way of speaking' of unparalleled efficiency. From this definition, we highlight the implicit evidence that drawing is the most effective means of transmitting any message, idea or interpretation with great clarity – especially the more complex that reality is. This consideration is very suitable for architecture, which was already conceived in Greek antiquity as a compendium of *Syngrafhai*, *Anagrafhai* and *Paradeigma* (Ruiz De La Rosa 1987), clarifying that *Syngrafhai* alluded to the written word – including descriptions and contracts, *Anagrafhai* to the drawing and *Paradeigma* to real scale models.

It is that second idea *Anagrafhai,* that today we recognise as the drawing that conceives and describes architecture that we want to make the subject of this research. The work of architecture, as stands before existing materially and being able to become a patrimonial subject, exists in the drawings that originate, describe, and/or recreate it after its disappearing.

It seems appropriate to point out that there are architectural structures that exist only in the created drawings – either because they only existed in his/her mind (Fig. 1), were never built (Fig. 2), or ultimately disappeared (Fig. 3). These drawings preserve the ideas with which they were conceived, the way in which they existed or could have existed, and the cultural and technical features of the historical era to which they belonged. The drawings, in a sense, contain the genetic code of the originator's existence and acquire a cultural value that fits well in the interpretations of heritage that we have analysed.

Fig. 1. Giovanni Battista Piranesi. Engraving of the Carceri d'Invenzione series. Italy. 1745. Source: Berggrum Museum, Berlin.

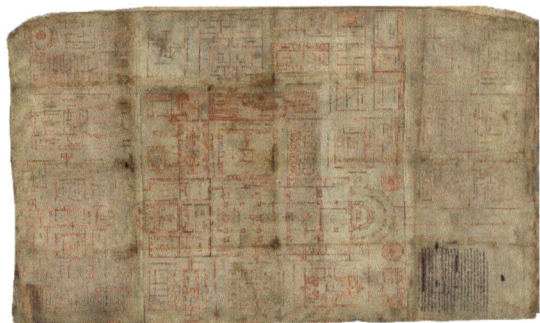

Fig. 2. Probable authorship of Bishop-Abbot Heito. Plan of the monastery of Saint Gall. St. Gallen, Switzerland 825. Source: San Galo Abbey Library.

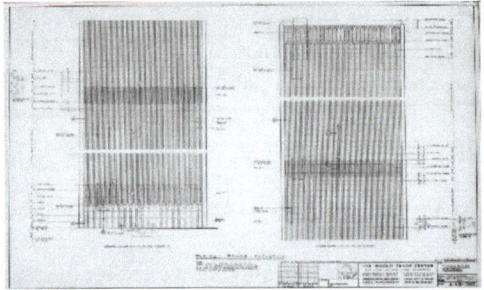

Fig. 3. Minoru Misaki Yamasaki & Associates. World Trade Center facade elevation (North Tower). NY. 1967. Source: Wright, Auctions of Art and Design.

2 Objectives

There are many authors who have pointed out the need to prepare a systematised study of the history of the representation of architecture in Spain. *EGA* magazine was born in 1993 with the purpose of collecting all the scientific productions in the area of graphic expression; indeed, this publication served as the primary reference journal for any study on the graphic representation of architecture.

We agree with the editors of the magazine (García Codoñer 2016) and some other authors (e.g. Mendoza Rodríguez et al. 2017) that the history of architectural representation needs yet to be written – a lack perhaps motivated by the difficulty of establishing exactly what we mean when we talk about graphic representation of architecture (i.e., what is the status and the nature of architectural drawings?). The representation of architecture requires various systems of graphic representation, which, based on analytical and ideation drawings, advances through successive graphic levels, with different interpretative keys and representation systems, without excluding those corresponding to missing works and expressive fantasies (Sainz Ávila 2005).

The genetic character that the drawing acquires in the creative process gives it a certain operational autonomy by which the author takes it to the finished form (Ibáñez Langlois 1964).

With the emergence of new and more sophisticated methods of architectural documentation, there are representational tools that even completely dispense with the image. This is a consequence of the technological development of digital procedures. This circumstance is an added factor of interest in preserving the existing documents of this specific language of architecture: drawing.

The drawings provide an existence more or less parallel to that of the buildings, that we could enunciate as cultural or mental life and still another, that is, their existence or graphic destiny (Ortega Vidal 2010).

This paper, through analysis of the specificities of the architectural drawing, proposes that it be considered as a historical heritage object of protection (which could rise to the category of cultural heritage in other, related categories, similar to other monuments thus recognised).

The 1972 UNESCO General Conference on the protection of world, cultural and natural heritage arose out of the need to identify a part of the priceless and irreplaceable works of nations. The loss of any of these creations would represent an invaluable loss for humanity as a whole:

Article 1

For the purposes of this Convention, 'cultural heritage' shall be considered as:

- The monuments: architectural works, (...) that have a universal exceptional value from the point of view of history, art or science.
- The sets: groups of constructions, isolated or assembled, whose architecture, unity and integration in the landscape.

As of 2003, the category of 'intangible cultural heritage' was also established, which refers to 'all that heritage that must be safeguarded and consists of the recognition of the uses, representations, expressions, knowledge and techniques transmitted from generation to generation that infuse to communities and groups a sense of identity and continuity, thus contributing to promoting respect for cultural diversity and human creativity.'

As architectural drawings serve as a graphic heritage linked to the real or imaginary architecture to which they refer, it seems natural that this class of material would receive a degree of priority, as provided by UNESCO's declaration, as a cultural heritage. On the other hand, the drawing of the architect's trade explains his/her cultural presence. It is the result of a logical and coherent evolution formulated with scientific criteria recognising uses, representations, expressions, knowledge and techniques transmitted from generation to generation with a feeling of identity and continuity – which also allows it to be identified as intangible heritage.

3   Background

UNESCO has established a series of categories for items that can be protected and listed as goods of humanity; these are summarised below (Fig. 4):

Architectural drawings can also be part of this selection of elements within the category of *tangible heritage* or, alternatively, within that of *intangible*, oral or immaterial

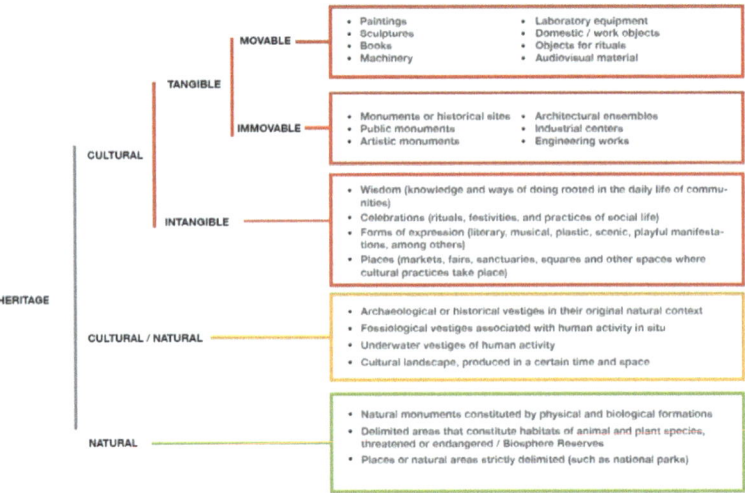

Fig. 4. Albert Macaya Scheme of types of patrimony (translation). Source: Dr. Albert Macaya University of Girona.

in terms of justification (referring to creations based on the tradition of a community and among others, the technical knowledge related to handicrafts, included on the list of heritage items in 1989). The architect's trade, as a set of technical knowledge-based skills related to 'graphic arts' could well be considered appropriate for being considered also as cultural heritage within this category.

Although, based on the description of intangible cultural heritage, architectural drawings certainly fit, given their material reality as physical objects, the natural thing is to refer them to tangible personal property – yet this nevertheless poses a problem.

Movable goods are those objects of cultural heritage testimonies of human creation or the evolution of nature and that have a certain historical, archaeological, artistic, scientific or technical value that can be moved and thus transferred without losing their value.

When the World, Cultural and Natural Heritage Convention (established at the 17th meeting held by UNESCO in 1972) established the category of movable cultural heritage, it did so only as a recommendation for its protection in Annex I. Individual states were left responsible for the care of such items (conceiving as an aid to the prevention and coverage of risks for the museums whose high insurance costs exceed what many public and private museums can afford). Subsequently, what is understood as movable cultural property was defined, with up to 11 different categories established. Architectural drawings, arguably, fall both in the *ix* and the *vi* designations, which are interesting to elaborate here:

vi) Goods of artistic interest, such as paintings and drawings made entirely by hand using any medium and on all kinds of materials.

But paradoxically, this category was established simply to distinguish goods that may be elevated to World Heritage from movable goods that may not be.

After this, movable property is only cited again as a reason for the direct exclusion of a possible candidacy (UNESCO 2005, p. 48). The possibility that a 'good' may be susceptible to becoming a 'movable good' represents the automatic dismissal of the good assumed to be considered World Heritage.

However, in 1992, the Memory of the World Programme (MoW) was created. This initiative was aimed at preserving the world's documentary heritage (housed in libraries, archives and museums), as a symbol of the collective memory of humanity. Since its creation, there has been only one registered case (UNESCO 2012a) in which architectural drawings have been catalogued as such. This involves the incorporation of 425 drawings of Gothic architecture (Fig. 5) into the Department of Drawings and Engravings of the Academy of Fine Arts in Vienna in 2005 (Böker 2005).

Fig. 5. Nordseite des Nordturms (The north side of the North Tower). Vienna. 1465. Source: Vienna Academy of Fine Arts

4 Method and Development

With this background, we will try to document the value of architectural drawings that make them worthy of the world heritage category and, based on that justification, or at least, try to establish with them, a new specific section within the classification of Memory of the World. We would like to focus on drawings of modern architecture, so as not to be affected by the respect that the passage of time gives to objects, sometimes making it difficult to maintain an objective eye with respect to them. We will take as reference the classifications and criteria already established (UNESCO 2005) preparing a list of characteristics of architectural drawings that are well covered by those and adding two more elements – themselves a consequence of the current historical moment.

The first additional category relates to the disappearance of many buildings as a result of climate change, war, armed conflict, terrorism, uncontrolled urban developments or property rights over the buildings that predate their collective heritage value (Bevan and Guinart Palomares 2019).

The second relates to the interest of conserving 'a bank' of architectural creations of history; In the same way that the risks of unsustainable development have led to the creation of banks, where to preserve valuable copies of contemporary civilisation, it would be interesting to create an archive of architectural drawings that enable preservation of those imagined/intended buildings, such that they would be retained for posterity in their drawings.

Let us take the drawings of Louis Kahn's projects for the Phillips Exeter Academy library as an essay (Fig. 6) and apply the valuation criteria that can make it worthy of belonging to a tangible movable cultural heritage. To do so, we will review the requisite characteristics and attributes.

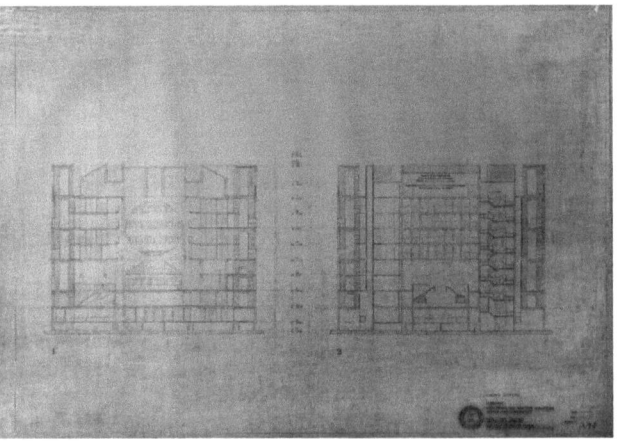

Fig. 6. Louis Kahn blueprint of the library section plan for Philips Exeter Academy, Exeter. 1968. Source: Archive of the Cooper Union of New York.

This document represents the work of a master of human creativity, insofar as it departs from the formal, constructive and spatial compositions typical of the era in which it was created. It is a testament to a set of values that gave rise to an architectonic, urban landscape, rich with technological design elements. Indeed, the project in Kahn's drawing is giving us information about a stage in history, the second half of the twentieth century, in Western culture, on the American continent, and what is the way to develop cities and their endowments, as referents to – and reflections of the way of life of the society to which it belongs.

At another level of analysis, we know the uses and constructive traditions, and even more, the cultural meaning of additional forms that refer to the influence of previous cultures that remain part of the current one.

But if, when analysing this architectural drawing, our interest is directed towards the handicraft of this graphic representation, we could also think of validating it as an intrinsic part of an oral and intangible heritage.

It also reflects constructive techniques and knowledge that have been received from previous generations and instil a sense of identity that deserves respect.

Only in the time elapsed since this project by Louis Kahn, we verify that the construction techniques, residential modes, and graphic craft of the architect have been completely transformed, thus reinforcing the idea of identifying it as a heritage worthy of being conserved in all its values.

The same analysis of the drawings of the Mies van der Rohe project for the Promontory Apartment Building provides us with the same information as the differences attributable to a residential typology – devised this time by a European. On the other hand, it allows us to verify similarity in the graphic-representation profession with almost equal methods that only distinguish the personal and calligraphic style that language wants (Fig. 7).

Fig. 7. Mies van der Rohe. North and west elevation of the Promontory Apartment Building. Chicago, 1947. Source: Chicago History Museum Research Center.

We propose one more argument that links the classification of the work of architecture to its drawing and vice versa, especially as applied to modern architecture. Recently, in July 2019, UNESCO declared eight Frank Lloyd Wright projects World Heritage Sites, after the previous elaboration of a nominative list in February 2015. It seems altogether reasonable to associate the protection this classification affords with the drawings that shaped them, and for that matter, also provide this classification for them (Fig. 8). The opposite holds as well: in the event some drawings enjoyed a priori values worthy of preservation as cultural heritage, the works they gave rise to also had it.

Fig. 8. Frank Lloyd Wright. Elevation of the Solomon R Guggenheim Museum, NY. 1956. Source: Avery Drawings & Archives, Columbia University.

In summary, we propose to introduce certain architectural drawings, through the normative procedure, in any list of the three mentioned categories that UNESCO applies (Fig. 9).

Fig. 9. UNESCO. Logos of the World Heritage and Intangible Cultural Heritage of Humanity and Memory of the World of UNESCO programme.

The preparation of such a list would be applied to the graphic material selected in archives that conserve works of architectural heritage and that – according to ICOMOS – must acquire, over time, value greater than the original (clarifying that this value can be cultural or emotional, physical or intangible, historical or technical). We have already mentioned how, on certain occasions, architectural drawings can acquire greater value than the work itself – for a range of reasons (Fig. 10).

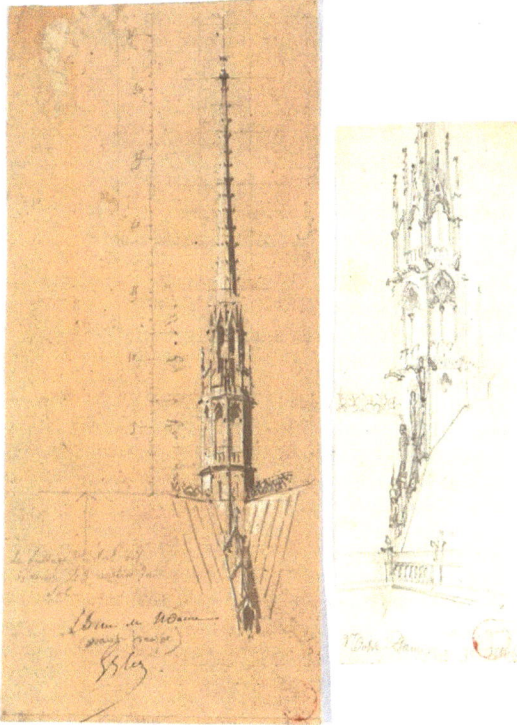

Fig. 10. Eugène Viollet-le-Duc. Drawings of Notre Dame's spire of Paris (now lost). 1860. Source: Guilles Mermet.

5 Conclusion

Through a series of representative examples of architectural drawings taken from reference archives in different countries, a reasoned scientific argument has been presented that allows considering architectural drawing as a historical heritage – with the capacity to be classified as universal historical and cultural heritage in its category of tangible movable, or intangible or oral and immaterial (which is adjusted by the values that serve to define them but not in the denomination). In this case, a new heading should be established that could include architectural drawings and other similar items; these could also be included in the Memory of the World classification, creating for them an exclusive section within this category.

With this consideration of architectural drawing, the built heritage finds a new method of protection; this graphic documentation – in addition to being a tool or archival cataloguing material, of interest to researchers versed in the area – would also be an exhibition and museum object with an interest of its own, similar to the work built, of great historical or cultural value for the history of Europe, its institutions and citizens as well as with a fundamental role in its compression.

The professional drawing would gain more attention and help us to recover the consideration of artists who, using an artistic trade destined to create pieces of art, managing to make architectural works real.

References

Bevan, R., Guinart Palomares, D.: La destrucción de la memoria. La Caja Books, Algemesí (2019)

Böker, H.: Die Architektur der Gotik. A. Pustet, Salzburg (2005)

García Codoñer, Á.: Aproximación histórica al dibujo de arquitectura en España en el siglo XX. EGA Expresión Gráfica Arquitectónica, no. 21, pp. 30–31 (2016)

Ibáñez Langlois, J.M.: La creación poética. Rialp (Biblioteca del pensamiento actual: 126) (1964)

Mendoza Rodríguez, I., Álvaro Tordesillas, A., Montes Serrano, C.: El dibujo de arquitectura de los años cuarenta en España: un estudio a partir de la Revista Nacional de Arquitectura. EGA Expresión Gráfica Arquitectónica, no. 22, pp. 170–179 (2017)

Ortega Vidal, J.: El dibujo del patrimonio y su vida gráfica. In: Documentación gráfica del patrimonio, pp. 46–63. José Manuel Lodeiro, Madrid (2010)

Ruiz De La Rosa, J.A.: Traza y Simetría de la Arquitectura. En la Antigüedad y Medievo, pp. 117–124. Publicaciones de la Universidad de Sevilla, Seville (1987)

Sainz Ávila, J.: El Dibujo De Arquitectura. Nerea, Madrid (2005)

UNESCO: Directrices Prácticas para la aplicación de la Convención del Patrimonio Mundial, p. 47. Centro del Patrimonio Mundial, Paris (2005)

UNESCO: Memory of the World: The Treasures That Record Our History from 1700 BC to the Present Day, pp. 134–135. NOOK Book (2012a)

Tamés and the Drawing of Heritage Buildings

Monica Gómez Zepeda[1]([⊠]) and Juan Carlos Ortiz Tabarez[2]

[1] University Center of Art, Architecture and Design of the University of Guadalajara, Guadalajara, Jalisco, Mexico
moni_monigz@hotmail.com
[2] Autonomous University of Guadalajara, Zapopan, Jalisco, Mexico

Abstract. This article intends to reflect on the importance of the record of heritage buildings through6 observational drawing; this by knowing, describing and disseminating how drawings by Mexican architect Jorge Tamés y Batta, which refer to the architectural heritage of the 16th century in the state of Morelos, are a means of study and dissemination of heritage architecture.

In order to address the topic of architectural representation through observational drawing, we will make a brief review of this type of drawing in the history of architecture as an antecedent that can give us the framework to approach Tamés as a contemporary author of this way of representing architecture.

We believe that observational drawing by Tamés has a double value: the value it possesses as graphic heritage, and also the value as a record of the heritage it is representing.

We support that, thanks to this type of graphic actions, it has been possible to hear about some heritage buildings that no longer exist or that have been modified; therefore, we consider important to disseminate them so we can know them and preserve them as well.

Keywords: Jorge Tamés y Batta · Observational drawing · Architectural heritage

1 Introduction

1.1 About the Observational Drawing of Heritage Buildings

"A line that takes shape emerges, the shape takes shape and becomes architecture."
Tchoban

To explain the topic of architectural representation through observational drawing, we will address some authors in the history of architecture as background to later show how Mexican architect Jorge Tamés y Batta is a contemporary author of this way of representing architecture.

Jorge Tamés y Batta is a graduate and former director of the School of Architecture of the National Autonomous University of Mexico (UNAM), who has a master's degree in Architecture: Research and Teaching. He has experience in architecture projects, urban

L. Agustín-Hernández et al. (Eds.): EGA 2020, SSDI 6, pp. 249–261, 2020.
https://doi.org/10.1007/978-3-030-47983-1_23

development and projects of different genres, specializing in the design of hospitals, hotels and offices, and having more than 50 works of his own. In addition, he has received several national and international awards (UNAM Foundation 2015). For decades, he has consolidated the discipline of drawing every day, tackling diverse topics: imagination drawings, travel drawings, observational drawings or natural drawings, and lately he has been drawing with the Urban Sketchers group.

In the 16th century, Italian painter, architect and writer, Federico-Zuccari, considered that drawing was the expression of a divine idea (concetto) that God himself lent to the artist; therefore, drawing is an autonomous genre (Tchoban 2015). Through history, drawing took its own relevance, not only as an auxiliary sketch for paintings, but due to its own expression and content (Fig. 1).

Fig. 1. Jorge Tamés y Batta. Side Gate from Church of the 3rd Order of St. Francis. Cuernavaca, Morelos. 2018. Source: Courtesy of the author.

For his part, during the 18th century Piranesi was one of the pioneers of *observational drawing*, as Nicholas Penny comments: "Based on his knowledge of ancient civilizations, many of which derived from his own archaeological excavations, he presented an intensely personal interpretation of the ancient ruins of Rome as they appeared in the 18th century." Nicholas Penny documents each stage of Piranesi's restless career with over a hundred illustrations, many of his best-known and representative drawings and engraving series. He measures the impact of the pictorial traditions of the native Venice of Piranesi and the paintings of masters as ancient as Magnasco, Canaletto and Panini on his varied work in papal Rome (Penny 1988). Piranesi has had a significant influence on generations of architects and set designers, such as the creators of the Harry Potter films (Tchoban 2015).

Just as Piranesi took advantage of his archaeological knowledge from the heritage buildings to which he had access, Tamés has also graphically compiled the heritage he has been able to access; with an urban archeological sense in his drawings, he explores and freezes heritage buildings with his graphic imprint. Our intention is then to make a brief compilation of Tamés' observational drawings concerning the 16th century heritage buildings, in Cuernavaca, Morelos, Mexico.

On the other hand, Sergí Tchoban mentions in his article *"The Art of Architectural Drawing"* that drawings of British neoclassics such as Joseph Gandy (1771–1843), Robert Adam (1728–1792), and Sir William Chambers (1723–1796) are well-known all around the world, same as the main members of the Rossica group: those foreign artists who worked in Russia in the 18th and 19th centuries, such as Giacomo Quarengi (1744–1844), Thomas de Thomon (1760–1813) and Pietro di Gottardo Gonzaga (1751–1831), whose works were also based on the study of ancient shapes. The twentieth century and new movements such as art deco, art nouveau, expressionism, futurism, Russian avant-garde, constructivism and the Bauhaus have not only had a significant influence on the *architectural style*, but also, of course, on the *architectural drawing*. They have also mainly characterized modern architecture. Architectural drawing increased in importance after the 19th century: art in drawing has become a point of reference for artistic and architectural drawing that has been established as an integral part of architectural study (Tchoban 2015). Following Tchoban, we can then say that *architectural drawing* has evolved along with history, thanks to the fact that in each era drawing has been representing different styles, and this has also allowed us to understand the spirit of each era: with its instrumentation, its expression, and its way of approaching it (Fig. 2).

As contemporary on the subject of observational drawing, Francis D. K. Ching, a professor at the University of Washington, in his book *"Drawing and Project"* he explains: "Despite the subjective nature of perception, vision continues to be the most important sense to gather information about our world. Thanks to the vision process we are able to extend through space and locate the boundaries of objects, examine surfaces, feel textures and explore space. The tactile, kinesthetic nature of the drawing, which is a direct response to sensory phenomena, sharpens the awareness of the present, dilates the visual memories of the past and stimulates the imagination when designing the future." (Ching 2004). In this sense, Tamés' drawings cover aspects of perception of heritage buildings; he collects information, he draws and delays or minimizes the possibility that said heritage buildings are forgotten.

In Lorraine Farrelly's saying: "this type of perceptual sketch is a detailed observation that allows us to first absorb and then understand what we are seeing and represent it in a drawing. It is a skill that is learned. The perceptual sketch is a perspective view,

Fig. 2. Jorge Tamés y Batta. Temple and former Franciscan Convent of Apostle Santiago. Cuernavaca, Morelos. 2018. Source: Courtesy of the author.

Fig. 3. Jorge Tamés y Batta. *Church of the 3rd Order of St. Francis.* Cuernavaca, Morelos. 2018. Source: Courtesy of the author.

but it also describes the scale and mass of the buildings, the use of colors suggests the idea of the materials and the context of the view.", the author carries on, "they are used to describe aspects of the buildings, to explore in detail materials or space." (Farrelly 2008). As we have already said, it is possible through these observational drawings of heritage buildings to have a record of a certain temporality and space (Fig. 3).

Since 2007, the Urban Sketchers group develops the practice of observational drawing, and in words of its forefather Gabriel Campanario, he mentions on its website: "We are an international non-profit organization dedicated to fostering a global community of artists who practice *drawing on-site*. Our mission is to *raise the artistic, narrative and educational value of drawing on-site*, promoting its practice and connecting people from all over the world who draw in the place where they live and travel." (Urban Sketchers 2019). Tamés as well as this group, shares the philosophy of drawing on-site, fulfilling the function of telling the story of his surroundings, frozen at a specific time: in which they were drawn. This gives the possibility of disseminating them and promoting them among people around the world, so that both the characteristics of the place and the building in question are known simultaneously.

2 The Importance of the Graphic Record of Heritage Buildings

As an objective, we can say that this publication intends to reflect on the importance of the record of heritage buildings through observational drawing; which helps us to know, describe and disseminate how drawings of architect Jorge Tamés y Batta, referring to the

Fig. 4. Jorge Tamés y Batta. Parish of Our Lady of Guadalupe. Cuernavaca, Morelos. 2018. Source: Courtesy of the author.

architectural heritage of the 16th century in the state of Morelos, are a means of study and dissemination of heritage architecture.

It is intended to promote this type of drawing to encourage students to practice and use observational drawing, and that this practice of drawing leads them to know and appreciate heritage buildings; and thus promote the conservation of both graphic heritage and heritage buildings (Fig. 4).

3   Tamés' Drawings as a Record of Heritage Buildings

As a hypothesis, we can say that observational drawings by architect Jorge Tamés, which he has made about the temples of the 16th century in the vicinity of Cuernavaca, are a means of studying heritage architecture that promotes its historical, artistic and narrative record. In their time, some teachers did it with their drawings throughout the history of architecture, and thanks to these drawings it has been possible to have graphic news of such architecture that in some cases no longer exists or that has been modified.

The importance of drawing heritage buildings lies in being a means of study and diffusion of architecture, which can represent a tool that can be taken to other areas of knowledge; thus, socializing it in other areas such as history, sociology, landscaping, urban planning, etc. The drawings of heritage buildings by Tamés are already by themselves a graphic heritage, since their value lies in their graphic quality, as well as in their way of proceeding, and what they are representing (Fig. 5).

Fig. 5. Jorge Tamés y Batta. *Parish of St. John the Apostle.* Cuernavaca, Morelos. 2018. Source: Courtesy of the author.

During the direct interview conducted in December 2018 to architect Tamés, he told us about his link with drawing, which emerged from his childhood: "No one is born with a talent, discovering it is difficult, but if you do not practice, then you will never discover it, everything is given by means of practice.", and this practice is reflected in his work, which he demonstrates with excellent execution and expertise. Architect Tamés continues commenting when he was asked about what drives him to draw: "Discipline drives me when I am most bored." (Gómez 2018). Therefore, in relation to practice, we can see that the discipline of drawing is part of his routine, because this activity has been present throughout his life. Thanks to his discipline, he has achieved a vast graphic production, which is to be recognized for drawing has not always been his only activity, as he has also combined it with teaching and professional practice.

About his current creative process when drawing, architect Tamés tells us: "Morning is the best time to draw, but sometimes it changes, it can also be when the day is already ending. When drawing, one becomes very observant, on a street in Cuernavaca there was a horrible building that has already been demolished, and then I discovered there a house. Inspiration is on the go. Any environment is favorable for drawing, which does not bother me. The trick is to observe. The thing that can make me stop is not finding a place to draw, which is why I usually carry a chair to sit on." (Gómez 2018) (Fig. 6).

Fig. 6. Jorge Tamés y Batta. Calvary and Shrine of St. Joseph. Cuernavaca, 19th and 20th centuries. 2018. Source: Courtesy of the author

Regarding the supports and instruments he uses to capture his work, architect Tamés tells us: "I do not have a technique. I never studied watercolor. I studied drawing, but there is nothing like practicing it, as long as a technique works for you. I have an arsenal of things and I do not know what I am going to use in the drawing. I have half letter notebooks that I take with me on a trip and I make a travel log. For example, I was in Tuscany to draw. The travel log is a life log too. What I really use is anything. I'm prepared with certain instruments that I carry in a little bag." (Gómez 2018). Here we can see that he is open to using different techniques. He does not focus on one only, this has allowed him to experiment and explore different mixed techniques, where he has shown a wide variety of possibilities. For example, the use of ink in the line drawing is recurring in his work, and provides him with the expression with different shades by applying what he has at hand: whether it is watercolor, coffee or red wine.

4 The Relevance of Observational Drawing

"… the reality that, when drawn, is analyzed.", Tamés

The current discussion about the contribution of Tamés' drawings, as a record of heritage buildings, is relevant in three areas: in the field of *architectural heritage*; in the field of *graphic representation*, and in the *academic field*.

In the Field of Architectural Heritage. We find that photographs can illustrate the current state, efficiently, but a photographic record lacks from the mental exercise of understanding the architectural elements such as: proportion, layout, order, rhythm, sequence; which the practice of manual drawing provides through its elaboration process.

In this regard, architect Tamés tells us that: "Unlike taking a picture or filming a building, an internal space or an urban square, drawing implies more time. It is about recognizing the place with the five senses. It goes beyond a cycle of walking and living the space behind a lens. It involves the orders that the brain sends to the hand that draws and, without a doubt, the experience of living it fully." (Tamés 2019) (Fig. 7).

The author continues: "It is worth taking a photograph and comparing it to the drawing in a critical way, and not giving an opinion about its quality as a work of art, but as a mere exercise of observing what is transmitted from hand to paper." (Tamés 2019). Then we can reaffirm that photography can serve as an auxiliary of these graphic expressions of observing heritage buildings to verify what is drawn on-site or contrast it.

As for the Field of Graphic Representation. One of the great values of observational drawing is to obtain the graphic record of the place where they were elaborated; thus, narrating their temporality, freezing such state in the drawing, as a historical record. As for the process, we can say that different categories intervene: the instrumentation, the techniques, the supports, the time socially used during the elaboration, as well as each author's graphic imprint. Regarding these categories, architect Tamés refers to us: "In this regard, it is important to insist that, during these exercises in outdoor areas, travelling or vacations, having or not having time, various techniques are to be experienced. Eventually, a technique will become better for us than another. This will help us in the process, and allow us to enjoy the activity without anxiety, just like a moment to let the mind and hand flow. It can also be interesting and fun as another technique to draw and stain with coffee or red wine. What emerges from there may surprise us." (Tamés 2019). Therefore, the author recommends eliminating anxiety by experimenting with various techniques, without the pressure of trying to carry out a perfect execution of a technique (Fig. 8).

Fig. 7. Jorge Tamés y Batta. *Dolores Chapel.* Cuernavaca, 16th century. 19th and 20th centuries. 2018. Source: Courtesy of the author

Fig. 8. Jorge Tamés y Batta. *Saint Mary Chapel.* Cuernavaca, 19th century. 2018. Source: Courtesy of the author

In the Academic Field. We can suggest that by knowing and practicing more of this way of proceeding, different skills such as perception, observation, understanding, analysis and synthesis of heritage buildings are acquired, and that also methodologically, manual graphic skills are acquired to address any architectural object. In this regard, Tamés shares us: "The practice of drawing undoubtedly exercises our sensibility to detect in things and spaces what is not obvious, which only becomes visible through a careful, leisurely look, accompanied by the lines and traces related to them. In addition, drawing provides us with a place and time for a reflexive introspection. This moment allows us to realize the proportions, the scale and the perspective; to glance through the construction, to stop at the decorative elements, to feel the texture, lighting and shadows and even to detect a scent like moisture, perhaps through smell. This is, therefore, an exercise that combines the passive process of observing and reflecting, with the active process of thinking and drawing." (Tamés 2019).

5   Tamés and his Experience with Observational Drawing

Recently architect Tamés has been drawing the architectural heritage with the Urban Sketchers group, and by his own. For this, we celebrate the existence of initiatives such as Urban Sketchers, which produce graphic heritage from heritage buildings. To address these topics, we have the book by architect Tamés *"The Drawings. Different Experiences"*, recently published in 2019; in addition to the direct interview, conducted in 2018; as well as the graphic work that architect Tamés has provided us for this publication.

In his book, Tamés refers to his concept about drawing by explaining: "For me, drawing is an extension of man's thinking, eyes and hand. It is an ideal companion of our intellect, because it helps us to articulate ideas, exercise thought and, of course,

Fig. 9. Jorge Tamés y Batta. *St. Jerome Chapel.* Cuernavaca, Morelos. XIX century. 2018. Source: Courtesy of the author

produce dreams. Its objective is, therefore, to know and develop the architecture." (Tamés 2019). The author refers to the fact that during the process of drawing the architecture that is observed, we can also speculate and learn about how it is resolved (Fig. 9).

During the interview with architect Tamés, he shared that: "With the Urban Sketchers of Cuernavaca, we have been drawing for about two years on a Sunday-in and a Sunday-out basis. The places we go to draw are different." (Gómez 2018). He reaffirms, for what he says, his discipline in the exercise of drawing, which he has personally, and as a team, sustained for years.

Architect Tamés carries on: "I started drawing buildings about 40 years ago, when I was an undergraduate, by means of the exercises that teachers asked us to perform in watercolor or pencil. We use to draw the President's office building of the UNAM or we would go out to a graveyard and draw graves or some other important buildings, and in some cases where I couldn't be there at that time, at the Notre Dame Cathedral for example, before visiting Paris, I drew it from a photograph. And sometimes I do so, when I'm unable to go, I make drawings from photographs, but most of my drawings are on-site drawings, or to make some composition with some little fish." (Gómez 2018) (Fig. 10).

Fig. 10. Jorge Tamés y Batta. *Cuernavaca Cathedral.* Morelos. 2018. Source: Courtesy of the author.

About the experience of drawing on-site, he tells us that: "We have to take our time to contemplate, and then draw. We have to appreciate the value and meaning of each corner we step on, while we contemplate all of its architectural features, of which the books talk about, especially those that we only understand and apprehend by being in that place." Here he talks about each one of us giving ourselves the opportunity to live this experience, which can only be appraised through the lived experience.

And with respect to those lived experiences, he says: "Practicing urban and architectural drawing allows us to collect images, ideas, sights, sensations, colors, aromas, textures, shapes, paths, walls, etc. The ability to find these echoes, and to conceptualize connections and shape them relies on the student or architect. The tours that can be taken through the architecture of Mexico and through big or small cities from around the world are amazing." (Tamés 2019). Then, as architects, we cannot see the world as it was before, but we rather start perceiving these spatial characteristics through the drawing of heritage buildings.

6   The Perpetuation of Observational Drawing

As a result, we can point out that Tamés' drawings are a graphic heritage for future generations, for they can be used as *dissemination, historical record and/or teaching resources.*

As Dissemination. We consider them as heritage, because what his drawings are representing refers to the interest of many people, which is heritage buildings. We can also say that observational drawing can be an attractive resource for the public in general, since they can captivate the viewer, also as a work of art.

As Historical Record. Because when drawn, they freeze their current state, showing how they looked at that time.

As Teaching Resources. Because they assists during the teaching-learning process of architecture, allowing the understanding of constructive and stylistic details.

It should be noted that the value of Tamés' work lies in its graphic imprint, where he shows us the potential of contemporary manufacturing, without suspicion for digital technologies, since its numerous graphic production flows between mixed manual techniques.

7   The Significance of Tamés' Drawings

In conclusion, we can say that Tamés' drawings impact on three aspects of the architectural field: that of heritage; in the graphic representation field, and the academic field as well. Since through the drawings of heritage buildings produced by Tamés it is possible to *disseminate, learn-teach and maintain.*

Just as the Urban Sketchers intend to *increase the artistic, narrative and educational value of drawing* on-site; intrinsically, architect Tamés has achieved all these values. His graphic imprint, his discipline and his passion are a living manifesto that ***observational drawing*** is still prevailing, as a historical record of heritage buildings, as an architectural graphic representation and as a teaching resource in the teaching-learning process of architecture.

Acknowledgments.   Research Project: *Drawing Books as an Instrument of Architectural Ideation.* LABTER Representation Techniques Laboratory, Representation Department, Technology and

Processes Division, University Center of Art, Architecture and Design (CUAAD). University of Guadalajara (UdeG). Research assistants: David Méndez Hernández and Cesar Iván Pérez Hernández.

Dean of Design, Science and Technology of the Autonomous University of Guadalajara (UAG).

Translated by Francisco Prieto.

References

Ching, F.D.K., Juroszec, S.P.: Dibujo y proyecto (Drawing and Project). Editorial Gustavo Gili. Mexico City (2004)

Farrelly, L.: Técnicas de Representación (Representation Techniques). Editorial Promopress. Barcelona (2008)

Gómez, M., Ortiz, J.C.: Entrevista a Jorge Tamés y Batta. Inédita (Interview with Jorge Tamés y Batta), Guadalajara, Mexico (2019, unpublished)

Penny, N.: Piranesi. Editorial Bloomsbury Books. Printed in Yugoslavia (1988)

Tamés, J.: Los dibujos. Diferentes experiencias (The Drawings. Different Experiences). Editorial UNAM. Mexico City (2019)

Tchoban, S.: El arte del dibujo arquitectónico. Capítulo de libro publicado en Meuser, Natascha. (The Art of Architectural Drawing. Chapter from book published in Meuser, Natascha). Construction and Design. Drawing Manual for Architects. DOM publishers, Berlin (2015)

Urban Sketchers. Nuestra misión (Our mission). http://www.urbansketchers.org/p/our-mission.html. Accessed 31 Aug 2019

Fundación UNAM (UNAM Foundation) (2015). http://www.fundacionunam.org.mx/rostros/jorge-tames-un-arquitecto-muy-puma/. Accessed 15 Dec 2018

Geometric and Spatial Systems in Rafael Leoz

Noelia Cervero Sánchez[(✉)] [iD]

University of Zaragoza, Zaragoza, Spain
ncervero@unizar.es

Abstract. Rafael Leoz (1921–1976) devoted his life to the investigation of housing space and, therefore, architecture, which should lead to its harmonious systematisation, and on a later stage, to its industrialisation. He proposed studying combinatory spatial topology, which involves working with mathematical logic, and geometric and spatial systems, which come from the study of pure geometry. His space organisation systems are connected with the materialisation systems of rhythmic compositions, as he reflected in his book *Redes y ritmos espaciales [Spatial networks and rhythms]* (1969), and in his numerous articles and conferences in Europe and Latin America. In a second phase, cut short by his early death, he proposed to develop a study on the dimensions, materials and building techniques, which would lead to a new prefabrication industry. His incomplete career left us many drawings and scale models of great plastic interest that were key to develop his theory, and nowadays constitute his most valuable legacy, as they ensure its transmission and comprehension.

Keywords: Leoz · Spatial system · Geometry · Networks · Rhythms

1 Working System

Rafael Leoz (1921–1976) was, according to Moya Blanco (1978, p. 25), "a scientific researcher who could not forget that he was an architect", a scientist who rigorously sought to order abstract space, and an architect who attempted to set the space in which humans must live. He produced drawings and scale models of great plastic interest to establish his theory, as Juan Daniel Fullaondo pointed out in the journal *Nueva Forma* (1968, p. 41). This way they became an essential part of his legacy, which cannot be explained without them.

His work being internationally acknowledged, in Latin America from the Sao Paulo Biennial Expo in 1961 and praised by Le Corbusier and Jean Prouvé, who promoted his introduction in Paris in 1962 as part of the *Cercle d'Études Architecturales*, was not reflected in Spain (López Díaz 2012b). He was not able to materialise his theoretical research while alive and, after the homage exhibition held in 1978 at the Palacio Velázquez in the Retiro Park of Madrid, save some exceptions, his work was left to oblivion as abstract and limited to speculation theories (Leoz 1981).

His main objective, which was guided by his marked social awareness (Moya 1978, p. 7), was to raise the standard of science and technique for the solution of one of the

L. Agustín-Hernández et al. (Eds.): EGA 2020, SSDI 6, pp. 262–272, 2020.
https://doi.org/10.1007/978-3-030-47983-1_24

biggest problems of his time, housing. This purpose remained constant and fundamental since the start of his career at the end of the 1950s, when he took part in the policy of new towns in Madrid leading a project known as Poblado Dirigido de Orcasitas along with Joaquín Ruiz Hervás.

This first shared experience became an abstract exercise of architecture linked to the Modern Movement ideology, in its most formalist form, which, according to Justo F. Isasi in *La Quimera Moderna* (Fernández Galiano 1989, 120–121), thought like Van Doesburg that social evolution could be guided and addressed and motivated from plastic art.

At this time, Leoz became aware of the wide gap that existed between the building technique and the scientific and technological advances of engineering (Leoz 1960, pp. 705–708). He was convinced that architecture needed more efficient solutions that could be achieved from intuition and scientific systematisation. He caught a glimpse of the new theoretical principles that ranged from abandoning craftsmanship and overcoming conventional architectural practices to trusting industry as an "urgent and unavoidable" decision (Leoz 1969, p. 21).

In 1960, his consideration of the problem led him to change his professional office to create a "laboratory" investigation on "social architecture", which should achieve industrial production favouring efficiency and economy in building, starting with a theoretical study of space based on geometry and mathematics. His main aim was to fulfil general abstract space organisation principles to derive rules for ordering inhabitable architectural spaces.

Mathematics, which guided the methodologies of Le Corbusier and Mondrian in the 1920s, were once again taken up 40 years later by architects as Christopher Alexander who, in *Notes on the Synthesis of Form* (1964, p. 14), considered mathematics to be "an extremely useful tool to explore conceptual order". When the functionalism crisis occurred, Leoz proposed establishing a series of related components to formulate generating patterns of the global form, as the typical material of an extended language that adapted to the requirements of the time (Drew 1972, pp. 27–31).

The search for new patterns as a prior condition to re-establish an equilibrium took place in the 1960s by reinterpreting and acknowledging the limitations of modern architecture. In *Theory and Design in the First Machine Age* (1960), Reyner Banham went back to the sources in order to interpret mathematics as a means that the creators of the International Style used to devise a language with symbolic forms. Overcoming this artificial order by sacrificing the complexity and ambiguity of life to give way to the frankness of forms resulted in a new model, mainly in the housing and city domain, that moved further away from the CIAM's main principles. Such decadence of modernity, promoted by Team 10 and some members of the Third Generation of modern architects, was backed in Spain by a major sector of the profession.

Nevertheless, Rafael Leoz was theoretically linked to the endangered modern architecture based on the research of Gropius, Le Corbusier or Klein on functionalism in housing, as well as on the progress made by Prouvé to introduce industrial elements into social housing and architecture. Along these lines, Leoz went on to determine the fewest types of industrial components, whose use would provide the widest variety of architectural options. To develop such an open system, he believed it was necessary to consider a projective geometrical systematisation methodology for architectural spaces

(Leoz 1981, pp. 23–28). With this methodology, he sought intelligible beauty as a means to fulfil people's spatial and spiritual requirements.

2 Spatial Organisation System

Leoz's theoretical considerations started from the original form, the raw material of architecture, that is space and the way to organise it towards its industrial materialisation through complete and, at the same time, elemental "molecules". He proposed studying the structure, the only form that can be assigned to the architectural space, from the spatial combinatory analysis of mathematical logic and the architect's own artistic sensitivity. In his book *Spatial networks and rhythms* (1969), he explained this intention to overcome the exclusive arrangement of the space exclusive to each project in order to work with a constant modular system and to seek general rules, or in the words of Christopher Alexander, "to create systems that create systems".

Since ancient times, architecture has developed many processes to modulate space with regulating outlines, quite often with anthropometric intentions; in other words, by taking human proportions as measurement units. This was the case of the Greek's abstraction, which lasted up to Vitruvius and was taken up once again by Renaissance treatise writers, with independent systems based on modules that connected all parts with one another and the whole. Lecorbuserian Modulor also emerged from humankind's figure, that is from Nature, and took a "step forward" with its modular organisation on the classic regulating outlines of space by reconciling the human-proportion variables with transforming action or materialisation (Corbusier 1980; Leoz 1973, p. 11).

Although Leoz's system started from Geometry, it matched that of Le Corbusier as regards to the utmost importance of Mathematics via the proportion and numerical series that compose the Golden Ratio and the Fibonacci sequence. By taking the Modulor's blue and red series and the 0.12-m modulus as a key dimension in construction, he found similar relations to those that controlled his patterns. This enabled him to rely on a system of coordinated measurements, which he believed would provide enormous possibilities by embarking on a dialogue with a world of harmonious proportions (López Díaz 2012b). Leoz was convinced that a system with such characteristics, based on ideal theoretical conceptions and constructive conditionings, would become significant if implemented as compulsory for the whole building industry.

In his search for order backed by pure mathematics, Leoz resorted to the geometrical and faceted systematisation of mathematics as a spatial combinatory topology to organise the Cartesian three-dimensional space polyhedrically. Based on studies by Russian crystallographer E.S. Federoff, he considered four polyhedrons with central symmetry – a cube or regular hexahedron, a straight prism with a regular hexagonal base, a rhombic dodecahedron and a heptaparallelohedron or Lord Kelvin's polyhedron-, for their capacity to fill space. This allowed him to fill space without leaving gaps, to form spatial networks from their edges (Leoz 1966, pp. 1–26; Leoz 1969, pp. 61–66). These four spatial networks could endure geometrical deformations on one or several directions to create new spatial reticles.

To apply them to architecture, these spatial networks, which organised continuous space rhythmically, were decomposed by sectioning the four polyhedrons into flat networks, which Leoz divided into three kinds depending on their base polygon. A grid, of

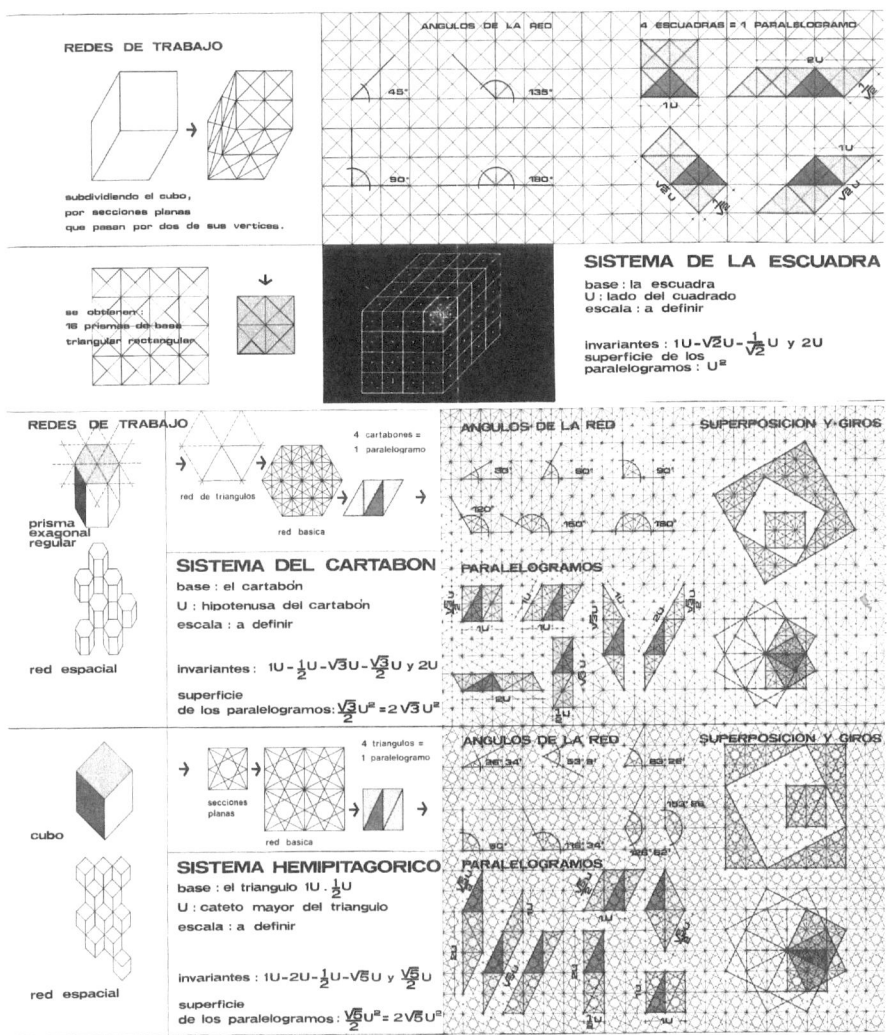

Fig. 1. Rafael Leoz. Work patterns. 1969. Source: Moya Blanco 1978.

45° set squares or right-angled triangles of equal catheti. A hexagonal network of 60° set squares or right-angled triangles with catheti forming angles of 30° and 60° with a hypotenuse and a squared double network. Hemi-pythagorean triangles or rectangles with catheti measuring one half the other, where the first and third could be superimposed on a single reticle (Fig. 1). These flat networks could stem from Lord Kelvin's polyhedron, a figure formed by six squares and eight equal hexagons in parallel to two-to-two and to seven planes or different directions. He considered all this to be extremely interesting by cross-cutting it through singular points lying in parallel to the squared faces (the reticle of the 45° set square) and to the hexagonal faces (the reticle of the 60° set square) (Leoz 1969, p. 150) (Fig. 2).

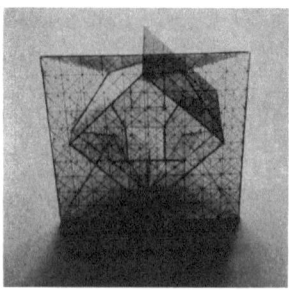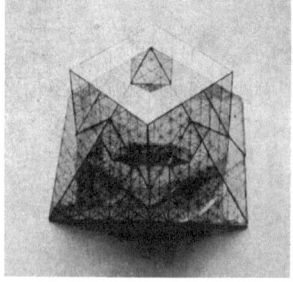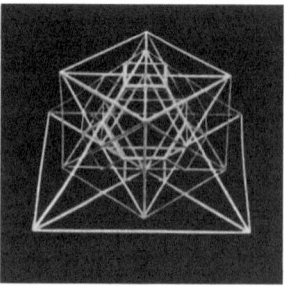

Fig. 2. Rafael Leoz. Topological complexes using Lord Kelvin's polyhedron. 1969. Source: Leoz 1969.

This methodical approach took Leoz to an architecture based on flat and spatial networks, taken as guidelines to obtain three-dimensional compositions of equivolumetric elements, to which it was necessary to add harmony and architectural sense.

3 Rhythmic Composition Materialisation System

All the infinite scales and forms used to order space that spatial networks enable comprised infinite rhythms with a numerical and combinatory capacity to overcome monotonous repetitions (Leoz 1969, pp. 94–99).

Fig. 3. Rafael Leoz. The HELE Module. 1960. Source: Leoz 1960.

The HELE Module (Fig. 3), an acrostic formed by the surnames Hervás and Leoz, was the first and simplest rhythmic composition that acted as materialised intuition in a typology of the Poblado Dirigido de Orcasitas project, which triggered all the subsequent theoretical research process (Leoz 1981, p. 43; López Díaz 2012c, p. 42). It involved a multiple asymmetric polyhedron composed of a minimum number of equal polyhedrons -three cubes in line and another at a right angle in an L-shape-whose proportions were considered beautiful as they established a connection with such a significant sequence within architecture's laws of proportion as the Fibonacci sequence (1-1-2-3). Leoz's interest lay in the maximum of combinatory possibilities that this allowed in symmetric or asymmetric compositions, in towers or extensive arrangements, by playing or doing away with repetition, etc. (Fig. 4). Leoz also trusted in the ease of such prefabrication, being based on a single module (Leoz and Ruiz Hervás 1960, p. 41): "In any case, we believe that having such a simple basic common element with so many possibilities can be a good step towards the purpose that we all wish to achieve: obtain good architecture economically".

Fig. 4. Rafael Leoz. Compositions with HELE Modules. 1960. Source: Leoz 1960.

Due to its great diffusion at the beginning of the 1960s in specialised publications, and even in the press, Leoz occupied a place at the forefront of Spanish architecture's modernity. After questioning the evolution of architecture in relation to other knowledge areas in the journal *Temas de Arquitectura* (1960), he presented the HELE Module in the journal *Arquitectura* (1960), with a lengthy article whose main features were drawings and photographs of scale models showing its versatility. Despite his professional colleagues not understanding his work, which clearly came over during the conference he gave in 1962 at the Colegio Oficial de Arquitectos de Madrid [Madrid Architects' Association] (ABC, 27 April 1962, pp. 62–63; Leoz 1962, pp. 15–21). The faith Leoz had in the need for this research led him to consider the module to be a means for practicing architecture from both the conceptual and practical viewpoints (Leoz and Ruiz Hervás 1960, p. 30). "HELE, being a scale model, will be an extremely useful tool for architects in their offices... by inflexibly setting intervals between parallel lines, which is what confers it a surprisingly beautiful rhythm that we come across in nearly all the compositions made with this new module" (Fig. 5).

The polyhedral containers of human life, whose understanding and manipulation were possible thanks to scale models and drawings, render necessary for their construction an outer covering, which turned them into hyperpolyhedrons. Leoz proposed that inhabitable spaces were surrounded by other service spaces, which would reproduce or provide with a new architectural form (Fernández Ordoñez 1973, p. 205). "I have recently discovered an interesting concept: housing is a theoretically closed element with a skin; it has a skeleton, which is the structure; it has a system of arteries and veins;

Fig. 5. Rafael Leoz. Combinatory possibilities with the HELE Module. 1969. Source: Leoz 1969.

it has an data centre and a transmission centre. So I found that the bodies, the polyhedrons, which I had handled until that time, considering them fundamental, were not what was fundamental because what was actually important were the hyperpolyhedrons formed by a polyhedron inside another polyhedron and a wrapping cover".

This figure, which Leoz obtained in his last research years, which allowed polyhedrons to be produced and related, helped to enrich his combinatory possibilities. To apply the rhythmic composition of hyperpolyhedrons to architecture, Leoz reflected on their suitability for the housing programme and was convinced that any conditioning factor could be solved by the wide range of forms he had developed, stating that his problem did not lie in creation, but in selection (Fernández Ordoñez 1973, p. 206). "What I have come across is most intriguing and starts with an almost infinite repertoire of forms. That is, if we have practically determined the form, due to the idea, for economic reasons, for social reasons, and even for aesthetic reasons, I know that it is extremely difficult to find a programme that cannot be accommodated in one of these infinite forms we have here".

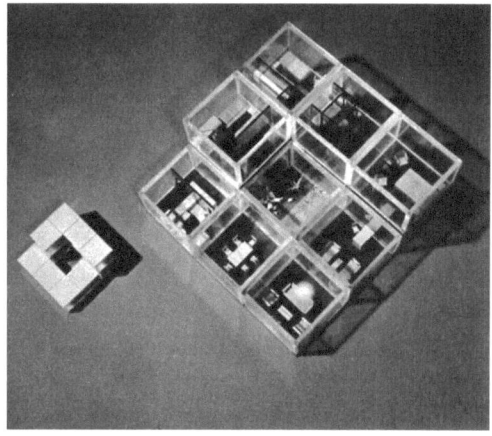

Fig. 6. Rafael Leoz. Functional units. 1969. Source: Moya 1978.

Introducing a programme in the hyperpolyhedrons allowed their material sizing to be added to the combination of modules and spatial/flat networks; in other words, the metric determination of the basic module for which he started from experimenting with the HELE Module. As mentioned in his book *Redes y ritmos espaciales [Spatial networks and rhythms]*, in 1968 Leoz started designing some housing prototypes by making a series of simplifications to the modular unit formed by grouping four straight square-based prisms, which were the equivalent to an average dwelling. He conducted sociological-type studies about domestic behaviours to establish which the most suitable functional combinations were; he defined four unit types -living, working, resting and services- and never forgot their project-based freedom and growth possibilities (Fig. 6). These groups of square-based hyperpolyhedrons varied in terms of their surface and volume until a large catalogue of floor plans was obtained.

These experiments with square-based networks or a system of 45° set-squares were also transferred to hexagonal-based networks or a system of 60° set-squares, which also provided applicable solutions and allowed analogies to be sought (López Díaz 2012a, pp. 65–66). Both research lines led Rafael Leoz to his two built works: the experimental housing at Torrejón de Ardoz (Madrid, 1973–1977), with a square-based module (Fig. 7) and the Spanish Embassy in Brazil (Brasilia, 1793–1976), with a hexagonal-based module (Fig. 8). In both cases, he applied architectural spatial systematisation by means of volumetric and highly experimented rhythms and took the first step towards modulating his structural systems.

Nonetheless, as Prouvé stated, Leoz was unable to find any industrial and financial backing to be able to put into practice his systematic industrialisation of architecture and, with it, the material construction of his theories. This last step, which would have confirmed not only the usefulness and possibilities of his geometrical developments, but would have also demonstrated their feasible application, remained a pending matter

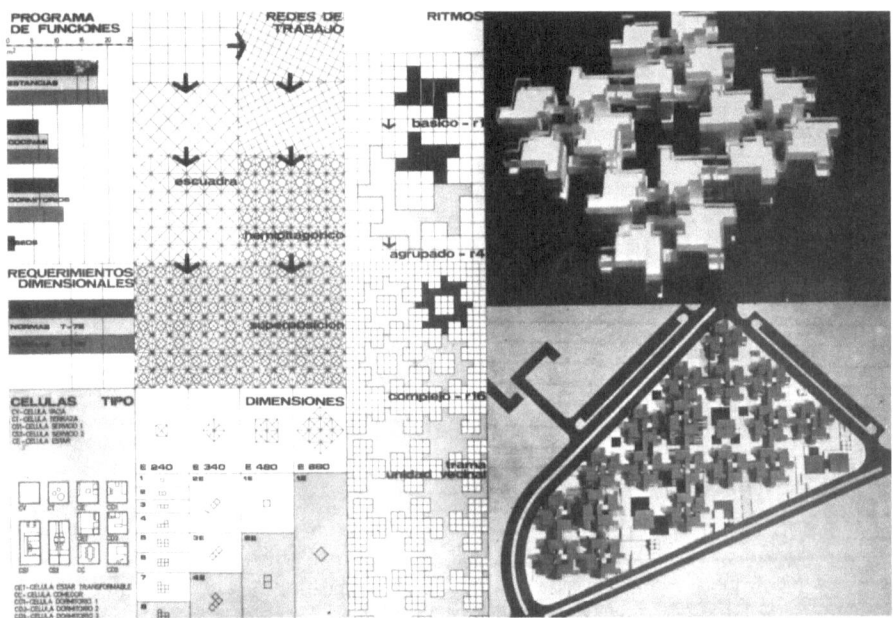

Fig. 7. Fundación Leoz. Project with a square-based module. 1969. Source: VV.AA 1978.

after his death in 1976 to be pursued years later by the Foundation named after him. His contribution to Architecture was, therefore, his theory of the harmonious systematisation of an architectural space using mathematical invariants and laws of harmony, which could not be understood without the graphical and volumetric pieces that constitute his most valuable legacy.

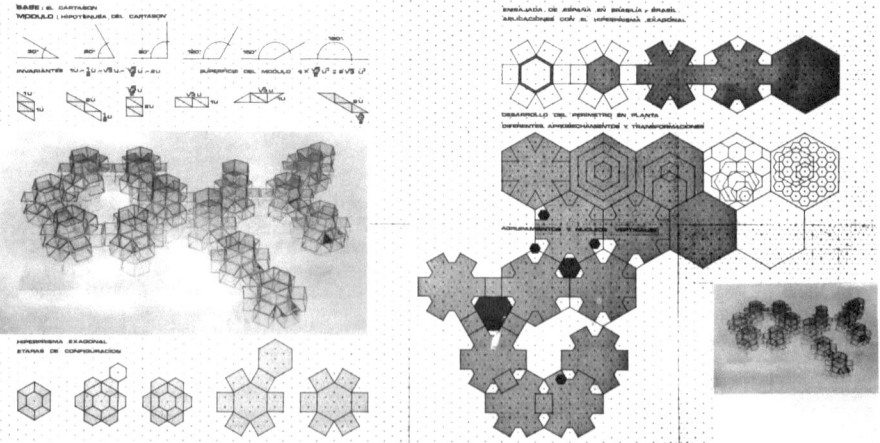

Fig. 8. Fundación Leoz. Project with a hexagonal-based module. 1969. Source: VV.AA 1978.

4   Conclusion

One constant feature of Rafael Leoz's experimental research was design and social housing production by the use of universal laws to organise architectural space. Drawing and making scale models allowed him to develop, and make visible, the systems for dividing space into volumetric rhythms capable of being industrially produced, whose combination in different ways would prevent monotony and dehumanisation in prefabricated housing.

Far from his principles falling in line with the new currents of thought, he took up again the theoretical system of Le Corbusier and the rationalist period before World War II, when what we now call Modern Movement came about, with Utopian abstract considerations and absolute trust in prefabrication systems. His theoretical consideration, far from this final production stage, originated from what is original; i.e., the raw material of Architecture: the essence of architectural space.

He benefitted from volumetric experimentation and projects on paper, which were in a very advanced stage when they were published in the book *Spatial networks and rhythms* in 1969, which based architecture on geometry. For practical and aesthetic reasons, he found in symmetry and proportion systems the conditions to outline an inhabitable space, relate it with mankind's scale, facilitate its construction and acquire harmonious beauty that was, assumedly, a reflection of ideal geometry.

Leoz was convinced that his "discoveries" about the systematisation of architectural space would be universal and revolutionary, and would start a "Renaissance of architecture" and its universal laws. The material construction of this theory proposed providing a "catalogue" with wide-ranging formal games and a fast execution by industry. As this last stage, which depended on the industry sector, is lacking, his most valuable contribution is the graphic and volumetric representation of his theories and prototypes, which have been shown in this communication.

References

Alexander, C.: Notes on the Synthesis of Form. Harvard University, Cambridge (1964)

Banham, R.: Theory and Design in the First Machine Age. The Architectural Press, London (1960)

Drew, P.: Third Generation. The Changing Meaning of Architecture. Pall Mall Press, London (1972)

Fernández Galiano, L., de Isasi, J.F., Lopera Arazola, A.: La quimera moderna. Hermann Blume, Madrid (1989)

Fernández Ordóñez, J.A.: Conversación con Rafael Leoz de la Fuente. In: Arquitectura y represión. Seminario de prefabricación, pp. 203–214. Cuadernos para el Diálogo, Madrid (1973)

Fullaondo, J.D.: Agonía, Utopía y Renacimiento. Nueva Forma **28**, 41 (1968)

Leoz, F.R.: Arquitectura e industrialización de la construcción. Fundación Rafael Leoz para la Investigación y Promoción de la Arquitectura Social, Madrid (1981)

Corbusier, L.: El Modulor. Poseidon, Barcelona (1980)

Leoz, R., Ruiz Hervás, J.: Un nuevo Módulo volumétrico. Arquitectura **15**, 20–41 (1960)

Leoz, R.: ¿Vamos por buen camino? TA: Temas de Arquitectura **18**, 705–708 (1960)

Leoz, R.: División y ordenación del espacio arquitectónico. TA: Temas de Arquitectura **39**, 15–21 (1962)

Leoz, R.: División y organización del espacio arquitectónico. Arquitectura **89**, 1–26 (1966)

Leoz, R.: Redes y ritmos espaciales. Editorial Blume, Madrid-Barcelona (1969)

Leoz, R.: Humanismo, investigación y arquitectura. Arquitectura **173**, 11–35 (1973)

López Díaz, J.: Rafael Leoz: el arquitecto y su legado. In: VV.AA., Rafael Leoz, arquitecto de la Embajada de España en Brasil, pp. 26–67. Briquet de Lemos, Brasilia (2012a)

López Díaz, J.: Tras los pasos de Le Corbusier: la modulación geométrica y la vivienda social en las teorías del arquitecto Rafael Leoz (1921–1976). In: Proceedings XVIII Congress CEHA, Santiago de Compostela, 2010, pp. 1850–1859 (2012b)

López Díaz, J.: El Módulo HELE de Rafael Leoz, una historia de contradicciones: del éxito internacional a la difícil relación con la arquitectura española, RA: Revista de Arquitectura **14**, 37–50 (2012c)

Moya Blanco, L., Leoz, R.: Ministerio de Educación y Ciencia, Madrid (1978)

Boden (coord.): Monográfico sobre la Fundación Rafael Leoz, no. 17. Boden (1978)

Lively Models

Scale and Three-Dimensional Representation in the Pavilions and Prototypes of Modern Architecture

Eduardo A. Carazo Lefort[(✉)] and Álvaro Moral García

Department of Urban Planning and Representation of the Architecture,
University of Valladolid, Valladolid, Spain
carazo@arq.uva.es

Abstract. Models have always been a useful instrument for prefiguring the construction, initially overcoming economic and technical problems, since the model is manufactured in general before the building, achieving a speed of execution that the technical complexity of the construction does not allow.

The aim of this work is to establish a parallel between the model, whose theoretical premieres have yet to be included in the history of architecture and representation, and other architectural experiments of the 20th century, such as the prototypes of modern housing and the pavilions of the exhibitions in which the architecture of the Modern Movement sought to make itself known to the general public.

Comparing these three so important elements for the three-dimensional definition of the architectural project, and so complementary to its own graphic definition, we intend to make a new reading for the study of these parallel mechanisms of ideation and diffusion of architecture.

Keywords: Model · Prototype · Pavilions · Exhibitions · Architecture

1 Introduction

In a recent writing we affirmed the persistent survival in time of the architectural model since its origins, which are lost in the remote past, up to the present day. And we wanted to interpret this survival from the multiple functions that the model or scale model had been able to assume over time in order to respond to a multiplicity of demands. This explains in first instance its success and also its demonstrated capacity to compete even with the current virtual models of the digital universe.

Thus, the model, generally based on a sought-after relationship of scale with respect to the building it represents, allows the client or end user to preview the form of the project much better than what would be achieved from a drawing: Given its manageability, its mobility, or its volumetric changes under natural or artificial light (Campo Baeza 2013). In other words, we could already point out, in principle, two characteristics of the model: relationship of scale and preeminence of form.

© The Editor(s) (if applicable) and The Author(s), under exclusive license
to Springer Nature Switzerland AG 2020
L. Agustín-Hernández et al. (Eds.): EGA 2020, SSDI 6, pp. 273–281, 2020.
https://doi.org/10.1007/978-3-030-47983-1_25

However, the changing mutability of the model to adapt to the different purposes we have spoken about, has also allowed it to overcome even those two basic premises of its own objectual nature. This implies that we can also refer to models that have wanted to prefigure not only the exterior form or appearance of the object to be represented, but also the interior form, more linked to the concept of spatial experience, and therefore of architectural space.

In this regard, we can think of the great models of the Renaissance in Italy (Millon and Lampugnani 1994) whose size allowed us to show their interior, even though the experience of space was not complete for the spectator, as these models lacked absolute habitability. We can also recall some expensive 1/1 scale models -which would be the limit at which the model could be confused with the building- such as that made by Michelangelo for the cornice of the Palazzo Farnese in Rome, that of Albert Speer for the New Chancellery in Berlin, or the ephemeral life-size model made by Mies van del Rohe for the Kröller-Müller House-Museum in Wassenaar (Holland), executed in 1912 with planks and fabrics, as a proof or test of the impact caused by the volumetry of the project in its immediate surroundings, and which unfortunately for Mies, meant his not being chosen to carry out the project (Windhorst and Schulze 2016, pp. 68–72), (Figs. 1 and 2).

Fig. 1. Model for the Kröller-Müller house, Mies van der Rohe, 1912–1918. Wood. Figure taken from AA.VV. Mies in Berlin. New York: Museum of Modern Art, MOMA, 2001, pp. 166–169.

Fig. 2. Scale model 1:1, placed on the actual site of the Kröller-Müller house, Mies van der Rohe, 1912–1918. Textile and wood. Figure taken from AA.VV. *Mies in Berlin*. New York: Museum of Modern Art, MOMA, 2001, pp. 166–169.

2 Discussion

If we continue to introduce ourselves at the beginning of the twentieth century, w e fnd new concepts that we can well relate to models, and their chameleonic capacity for survival and adaptation to new uses and new commands: We refer to the prototypes of modern housing presented in various exhibitions in the first half of the century, which sought to prefigure the ideas of the avant-garde, and which found in these mechanisms the fastest and most effective procedure for expanding the ideas of new architecture, long before architecture itself. In fact, many of the utopias that those prototypes announced never came true. Let's think, in that sense, of the whole series called under the heading "house of the future", developed with diverse examples in the first decades of the 20th century, trying to become models or habitat for the following decades, decades that have been overcome, without those premonitory proposals ever having been fulfilled.

Starting from the first decades of the 20th century, and in relation to the ideas that the heroic avant-garde wanted to introduce into the society of the time, we could mention, among other representative examples, the house of Mies van der Rohe for the 1931 Berlin exhibition, included in the set of 1/1 scale models of various architects for the exhibition Die Wohnung Unser Zeit curated by Mies Van der Rohe and belonging to the 1931 Exhibition of German Construction in Berlin (Lizondo et al. 2013); or also the L'Esprit Nouveau Pavilion at the 1925 Paris Exhibition of Decorative Arts - built as a living cell model of the Inmueble Villa; Marcel Breuer's temporary pavilion 'Exhibition House' or 'Gane Pavillion' for the Royal Agricultural Society of England Exhibition held in Bristol (United Kingdom) in 1936 (Fig. 3), or, already referring to the utopias of living for the future, the Future House by Alison and Peter Smithson (Fernández Villalobos, 2012) set up for the exhibition held in London in 1956 by the Daily Mail newspaper (Fig. 4).

Fig. 3. Gane Pavillion, Ashton Court, Bristol, United Kingdom. Marcel Breuer 1936. Stone, glass and wood.

In this sense, we could ask ourselves to what extent these representations, not only of form, but also of architectural space, can be considered and analyzed with the same

Fig. 4. Future House, Ashton Court, Bristol, United Kingdom. Alison y Peter Smithson 1956. Plywood and plaster stucco. Plastic punctual elements.

parameters as models, and included in a general theory of the model as a system of representation of architecture, which, by the way, has yet to be developed as such. In fact, the model has been treated as a marginal or curious object in the evolution of architecture, when the architectural critique of the last decades proposed to highlight its relevance within the theoretical, conceptual and graphic framework of the architectural discipline itself.

In order to continue with the examples we are dealing with, we should first analyse the differences and analogies between three very different elements: model, pavilion and prototype.

The model can be defined in a generic way, as the representation on a reduced scale of an architectural object, generally not yet built. The model also forms part of a world of craftsmanship, which uses its own materials, closer to sculpture or decorative arts than to construction; in such a way that it adapts or provides its small size to the possible use of those materials, which are generally not the same as the architecture it represents. Furthermore, although the model is generally fragile and difficult to preserve, it does not have an ephemeral vocation, it is not expressly made to be destroyed; and if no more examples have been preserved throughout history, it has been precisely because of the physical fragility that characterises it.

The model, given its small size, or its character of prefiguration of an architectural idea or project, is not habitable. And that is an inherent quality, regardless of its scale and size, that is, it does not admit its permanent use, as a building does. We can "get inside" some models, but we can't stay long, we can't live in them, inhabit them.

Exhibition halls and prototypes. The pavilions, understood as large containers for use and temporary location, derive from the nineteenth-century exhibition tradition, where they already represented a great technical revolution, as would be the case of the glass pavilions, the Crystal Palace of Paxton in London in 1851, or the Grand Palais of Paris in 1900. In the context of this work, Mies Van der Rohe's German pavilion would be paradigmatic for the 1929 Barcelona exhibition, where the most relevant thing was the capacity of the building - dismantled at the end of the exhibition - t o show the world the spatial capacities of the new modern architecture. In other words, the Pavilion itself was in reality the object to be exhibited, regardless of its content.

The prototypes, on the other hand, are 1:1 scale reproductions of architectural proposals ideologically advanced to their time. They acquired great importance in the first decades of the twentieth century, above all in the essential relationship that allowed establish with the new architecture -also called International Style (Hitchcock 1984 [1932]), or architecture of the Modern Movement- and the attempt to demonstrate that this new architecture was capable of responding to the housing needs of a new society.

Fig. 5. 1:1 scale models of houses by various architects, interior of the 1931 Exhibition of German Construction in Berlin. Figure taken from Der Baumeister Vol. 29 No 7. Verlag Georg D.W. Callwey, München, july of 1931. Propiety of The Museum of Modern Art, The Mies van der Rohe Archive, New York.

Within both categories, we could, in turn, establish some differences and qualities between the prototype and the pavilion.

On the one hand, we would have the so-called prototypes, as is the case of the 1/1 scale model of the 1931 Exhibition of German Construction in Berlin (Fig. 5), which involved a model to imitate and repeat, a test or experiment that could be tested and improved. In reality, a whole manifesto of modern architecture and a new way of life that these visionaries of the heroic epoch foresaw, and who laughed to make the society of their epoch see through of this great exhibition model.

Another significant example would be Marcel Breuer's 1949 MoMA garden house (Fig. 6), made as an archetype of the new American post-war house, ready to be built and repeated on any plot of urban outskirts of a U.S. city. The house was erected with light and non enduring materials, as a large model that wanted to show itself, which sought the illusion of a dream reality, of a new "modern" life, hygienic, rural and peaceful, for a society that came out of the great tribulation of war with the promise of a new and bright future for its citizens. What better than a representation of that ideal society through a visionary house, materialized in an infinite world of opportunities for a new way of living? Wouldn't a great model the object that could best reflect that promised optimism?

Finally, mention should be made about the Future House of the Smithsons, from the Daily Mail exhibition in London in 1956, which was conceived as the future suburban habitat for 1980. The proposal is framed within the architectural tradition of the first decades of the 20th century, aimed at envisioning the so-called house of the future (Figs. 7 and 8), probably rooted in nineteenth-century literatures, pioneers of futuristic prediction, and highly successful in the positive society of the new era of the machine.

This futuristic project, useless in the long term as any futurism seen from precisely that future - Ridley Scott's film Blade Runner announced in 1982 an apocalyptic future for 2019 that, at the end of the year, we have not yet seen… - wanted to propose not only a plastic house, but a new way of life, nothing less than for 1980! ….

To this end, the sponsors of the exhibition "Sixty Years Back/Sixty Years Ahead" proposed to the Smithsons the creation of a space of the future. The plastic, and the sinuous forms that this material produced, seemed the most suitable to dazzle the visitors, producing a completely antithetical effect with a domestic space to use. The effect was stretched to the maximum in the sample, if we consider that the house was not only uninhabitable -the prototypes never are- but that it was not even possible to go in. For

Fig. 6. Construction of the "House in the garden of the MoMA". Marcel Breuer 1949. Prototype of housing for the exhibition The House in the Museum Garden organized for the MoMa. The Museum of Modern Art, New York.

this, some actors were hired, who got dressed up like authentic foolish and gave the impression of living in the future.

But everything was really representation and play (Huizinga 2000 [1938]). The prototype of the house on display could not be made in plastic, since the moulding was singular and not serialized, and there was no time to make such singular and diverse moulds: it had to be made in two weeks, and on the basis of plywood boards covered with plaster, quite a contradiction if we understand it as "house", but very coherent if we see it as "model". This is also how we could understand those two couples of extravagant inhabitants, incarnated by actors dressed in futuristic fashion: they were not really, but the plastic dolls -now made of flesh and bone, and on a real scale, of course- that we still place in our models.

This merely demonstrative and propagandistic intention meant that the prototypes of modernity were not built with durable materials. In general, as they were made inside an exhibition hall, they avoided the inclemency of the weather and with it, they stole real constructive solutions, as usually happens with models. On the other hand, the prototype, as a test piece, a tentative element, almost a full-scale sketch, although momentarily habitable, is destined for a futile, brief life, until the test phase for which it is made ends, and then, it is dismantled; and either destroyed, or moved to another place to be shown, but never to be inhabited.

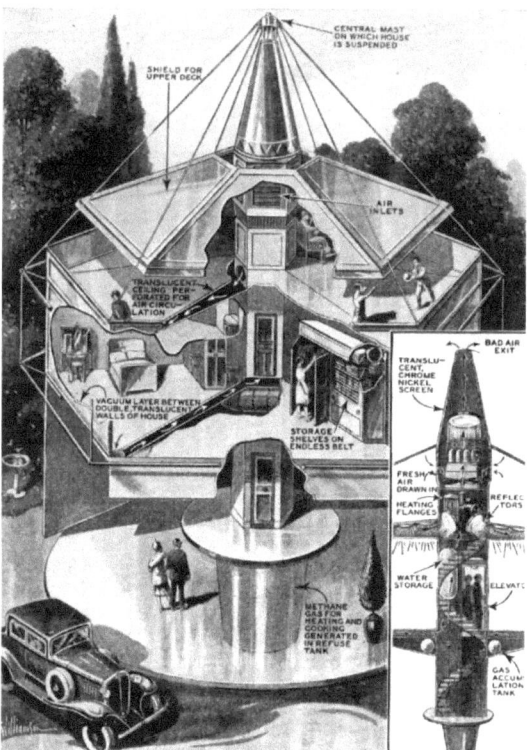

Fig. 7. House Dymaxion, Buckminster Fuller, 1929.

Though the pavilions - in the case of Marcel Breuer's Gane Show House or Bristol Pavillion in 1936, Le Corbusier's L'Esprit Nouveau Pavilion in 1925, or Barcelona's Mies Pavilion in 1929, among many others - are really built manifestos. And although they share the experimental and ephemeral character of the prototypes, they do not have that vocation of later repetition, and they constitute themselves a paradigm of their moment, of an element of propaganda, of transgression, of a visual and figurative icon. Nor do they attempt to resolve constructive problems, since the idea of *firmitas*, inherent to architecture -even to modern, prefabricated and industrial architecture- comes from the need for durability, which is the antithesis of an exhibition pavilion. In fact, although in the Bristol Pavilion, Breuer gives a lesson in the modern use of traditional materials with its large stone walls, all the pavilions mentioned were dismantled - not destroyed - at the end of each of its events. Although their iconicity, symbolism and a certain nostalgia for the heroic ideology that sustained them led to the restitution of Mies' in Barcelona and Le Corbusier's in Bologna in the last third of the twentieth century, they have only been so as a mimicry, an object of worship, stripped of their experimental and temporary character, perhaps converted into architecture, a complete contradiction with their initial nature (González Cubero 2011).

3 Conclusions

We have pointed out, to synthesize, some common characteristics, shared, between model, pavilion and prototype.

In all cases, scale is fundamental. On the scale model, it seems that it must be precisely smaller than the object it represents. In the other two elements, precisely the only possibility of scale is one by one, since they want to represent a habitat, not so much a form.

In fact, it is to inhabit another of the common characteristics. More precisely not inhabiting. In some models, the idea of space in pavilions and prototypes is intuited and can even be seen alive. But none of the three is inhabited. No one remains in them, only temporarily, fleetingly, even without entering, one can perceive the spatial cavity that they want to represent.

The experimental, innovative, provocative or future character is another of the common features between modern models, pavilions and prototypes. We should not forget that the first projects of Le Corbusier's houses were solid plaster models for the exhibitions of the Paris Autumn Salon of the 1920s (De la Cova Morillo-Velarde 2016).

Material is another matter to consider. Here differences and similarities open up. Model and prototype, they share simulation, representation and deception; nothing is what it seems, the materials are assemblages, systems adapted to the situation, to the scale in the model, and to the lightness and speed in the prototype. In the pavilion, things can, however, be built as in true architecture, even though its dismountable vocation also leaves its mark on construction systems.

This temporary vocation, not lasting, is also an essential characteristic of pavilions and prototypes, which clearly differentiate them in this case from the model, which, despite a difficult life and worse conservation, does not have the express will to make itself to be destroyed.

We can conclude, finally, that these architectural experiments could be included as a variant of the architectural model, since they share with it some fundamental characteristics, and be studied within this three-dimensional world of architectural representation. The problem for its inclusion in this field, however, would not be so much the transgression of its scale, taken to the limit of the real -which undoes in that limit the concept of architectural representation-, or its realization with futile materials, but rather their temporary habitability or their inhabitability in certain cases - which bring them closer to true architecture, to the house as the essential cell of living (Heidegger 1994 [1954]) - since that habitability is so questionable - for ephemeral, for impractical, for delocalised - that it does not allow them to be included in the genuine world of architecture and of the model.

References

Campo, A.: An idea in the palm of a hand. DOMUS **972**, 10–11 (2013)

De la Cova, M.A.: Maquetas de Le Corbusier. Técnicas, objetos y sujetos. Universidad de Sevilla. Sevilla (2016)

Fernández, N.: Utopías domésticas. La casa del futuro de Alison y Peter Smithson.: Fundación Caja de Arquitectos. Barcelona (2012)

Fullaondo, M.: Casa en el Jardín del Moma. Arquia-Fundación Caja de Arquitectos, Barcelona (2010)

González, J.: Doble, escenografía y clon del movimiento moderno: La imposibilidad de la copia física en arquitectura. En: Trevisan, A., Virtudes, A.L. (ed.) Apropriações Do Movimento Moderno. Centro de Estudos Arnaldo Araújo da CESAP/ESAP, pp. 168–184. Oporto (2011)

Heidegger, M.: Construir, habitar, pensar. Martin Heidegger. Conferencias y artículos. Ediciones del Serbal, Barcelona, pp. 127–142 (1994) [1954]

Hitchcock, H.-R.: El estilo internacional: arquitectura desde 1922.: Consejería de Cultura y Educación. Murcia (1984) [1932]

Huizinga, J.: Homo Ludens. Edición Española ed. Alianza Editorial. Madrid (2000) [1938]

Lizondo, L., Santatecia, J., Salvador, N., Bosch, I.: La idea Materializada en la muestra Die Wohnung Unserer Zeit de Mies van der Rohe. Proyecto, Progreso y Arquitectura (8), 28–41 (2013)

Millon, H., Lampugnani, V.M.: Rinascimento. Da Brunelleschi a Michelangelo. La rapresentazione dell'architettura. Bompiani. Milán (1994)

Windhorst, E., Schulze, F.: Ludwing Mies Van Der Rohe: una biografía crítica. Reverté, Barcelona (2016)

Frediano Frediani and the Santa Lucia Skyscraper Drawing and Re-drawing an Urban Utopia

Alessandra Cirafici[✉]

Università degli Studi della Campania Luigi Vanvitelli, Aversa, Italy
alessandra.cirafici@unicampania.it

Abstract. In the panorama of 20th century architecture and the great urban transformations that marked the city of Naples between the two world wars, Frediano Frediani is one of its "minor figures". Nevertheless, there are several works of significant architectural quality in his prolific production as a designer and consultant for large transport and infrastructure companies in Campania. It is worth merely mentioning the two stations in Piazzale Tecchio and Via Leopardi (1939/40) built for the Cumana railway in the context of his long collaboration with the EAV (Ente Autonoma Volturno) as well as the collaboration with Luigi Cosenza with the construction of the Fish Market and the Sannazzaro quarter (1929/30). However, his most utopian project remains without doubt the International Work Centre 'Santa Lucia' (1945/46), a magniloquent building imagined on an artificial peninsula to be built between Molo San Vincenzo and Castel dell'Ovo along the east coastline of the city. The work was never realized, but the project represents a significant and little-known episode of 20th century architectural production in Naples. The contribution proposes an analysis of the project and the vibrant debate that it generated, through an operation of re-drawing that, using the strategies of representation as a valuable critical tool, proposes an interpretation of the instances of the project and spatial thinking of a work that is presented with the force of a real 'provocation' in the context of post-war architecture in Italy.

Keywords: 20th century architecture · Architectural design · Redesign · Skyscraper · Frediano Frediani

1 An Introduction

The idea of the International Centre of Culture and Business 'Santa Lucia', designed by Frediano Frediani between 1945 and 1946, is part of the scenario of the great 'utopia' characterizing the urban transformations of the city of Naples in the immediate post-war period, with particular regard to the interventions imagined for the front along the coastline to the east of the city. In the aftermath of the 1943 bombardment, the city was devastated and the structures of the port area, a privileged target of the Anglo-American forces first, and the fury of the retreating German troops later, were completely destroyed. The damage was evaluated by an Italian commission the day after the end

© The Editor(s) (if applicable) and The Author(s), under exclusive license
to Springer Nature Switzerland AG 2020
L. Agustín-Hernández et al. (Eds.): EGA 2020, SSDI 6, pp. 282–296, 2020.
https://doi.org/10.1007/978-3-030-47983-1_26

of the hostilities and reported that over 80% of the docks, buildings and installations had been destroyed. An immediate reconstruction programme was necessary. It was the American Colonel Harold H. Towsend, officer of the Public Works and utilities of the Allied Military Command in Naples and President of the American Italian Development Enterprises, who inspired, among the various interventions, a solution which used the 'skyscraper' typology, still not very widespread in Italy.

The project proposed by Frediano Frediani is certainly its most complex, ambitious and visionary: it was to create the sumptuous venue for an international business centre, congresses and meetings, but at the same time a place of fun and recreation for businessmen, travelers and Neapolitan citizens, framing it in the more general plan aimed at starting the recovery of the country. The impact on the image of the coastline was considerable and, in spite of the approval of the competent authorities, the reactions to its possible realization triggered an intense debate among the public opinion and among the intellectuals of the city, so much so that there was talk of a veritable 'skyscraper scandal'. The ambitious project of the skyscraper just off Borgo Santa Lucia included the construction of a peninsula stretching out into the waters of the gulf for over 200 m long and 50 m wide, protected by a system of dams more than 600 m from the shore, in the stretch of sea between Molo San Vincenzo and Castel dell'Ovo (Fig. 3). Here Frediani had already designed the new location for Rari Nantes (1938), but the size and visual impact of its octagonal tower standing out in the middle of the sea was really a provocation to the imagination of the entire city and its famous 'panorama' (Fig. 1). So Frediani's project therefore remained a splendid utopia, and as often happens with unrealized projects little known and investigated.

Fig. 1. F. Frediani (1945), *Veduta dalle pendici del Vomero*. The photomontage shows the exact location of the *Centro Internazionale della Cultura e degli Affari 'Santa Lucia'* and its imposing octagonal tower in the center of the Gulf of Naples. (Frediano Frediani Private Archive).

The aim of this contribution is to critically read Frediani's project, through the analysis of the beautiful autograph drawings kept in the Frediani Archive, along with a cultured operation of redesign, is the aim of this contribution, proposing its point of view in the intense debate on the relationship between 'design of architecture' and 'architecture realized', supporting of an interpretation that sees in the graphic analysis and in the foreshadowing capacity of the Drawing the possibility of fully restoring to the sphere of architecture "the endless territory of the unrealized proposals, of the drawings that have remained so, understood as the ideal landscape of architecture and the architect" (Purini 1993), a privileged place of pure intentionality and of its expressive force. Many influential critical sources have maintained that "architectural expression is accomplished only when the work is realized" (Zevi 1972) only when it is possible to walk through its spaces. According to this interpretation, from a historical point of view, the project is not architecture, and the project design remains confined to the instrumental role of a stage in the creative process that has no value in itself if not associated with the work realized. On the contrary, the itinerary proposed here trusts in the aptitude of the design to configure a path of critical analysis as well as in its ability to prefigure space (Migliari 2004; Ugo 2004), in a process of reading, often inedited, in which it is possible to 'make visible' the design thought in its development, and to allow that cultural awareness necessary to interpret in a broad sense, the conditions that made the conception of such a singular project possible. A powerful and visionary project that should be fully included in the post-war season of Italian architecture and that, if realized, would have significantly modified not only the image of the city, but probably also its propensity to include the 'modern' in its urban transformation process (Belfiore 1994, Meridione Nord e Sud 2011).

2   Skyscrapers. Utopia and Realism in the Urban Transformation Processes of the City in the 20th Century

"The tower, one of the many towers of which Italy is rich, will emerge as a grandiose lighthouse to replace the old "lantern of the pier" demolished and will rise where it is now filled with all the rubble of Naples: this also has a symbolic value and is an austere for the rebirth of Naples and Italy, of which this tower will be a beacon of civilization." (Towsend 1945).

This is how American Colonel Harold H. Towsend described the S. Lucia in «La Voce» on November 24, 1945. A grandiose building with a strong symbolic value and a sure positive impact on the city's economy. There is no doubt that Frediano's project was influenced, both in the typological choices and in the functional destination, by that sort of 'cultural colonization' and fascination that the whole of Italy, in the aftermath of the war, suffered from that of the North American model. More. It is possible to summarize in the history of this project the traces of that itinerary in which «the 'Italian way' to the skyscraper and the desire to be 'American' tells the birth of a recurring feeling between the defense of one's own cultural and artistic identity, fear of colonization by other models and the need to find a necessarily unstable balance between local and global» (Molinari 2011, p. 38).

Frediani's proposal is therefore in the wake of that substantial reversal of the trend in the flow of symbols and models that for almost two centuries had travelled in the opposite direction: that which led from the old continent to North America, suggesting a way of looking at the classical world as a sort of great 'emporium' of materiality and symbols of classicism to be recombined in infinite solutions. It is precisely «the skyscraper that is the first technical and architectural artifact to reverse the motion in American-Italian relations, that secret virus that will explode after the Second World War, but that is incubated throughout the entire Twenty Years through the reading of magazines and the first trips overseas» (Molinari 2011, p. 36). However, the relationship with the image and the imaginary of the skyscraper lives, in Italy, an ambiguous path, of frightened aspirations and visions that worry above all our intellectuals, frightened by a machinist and industrial idea of our future (Faroldi et al. 2008).

Between 1920 and 1924, Piero Portaluppi, one of the prominent figures of Italian architecture in the first decades of the last century, produced a series of watercoloured Indian ink drawings depicting a skyscraper, sketched with a certain virtuosity. These were just studies, but somehow the way is traced, and the skyscraper as an urban and economic model capable of gaining space in the sky by saving it on the ground slowly takes hold, even under the pressure of new technical and technological innovations. In this imagery Frediani's idea was born to propose the typology of the skyscraper for the seat of the *Centro Internazionale della Cultura e degli Affari 'Santa Lucia'* (Fig. 2).

Fig. 2. F. Frediani (1945), *Palazzo del Santa Lucia guardato da Nord-Est. Ingresso su Via Nazario Sauro*. Perspective sketch. The monumentality of the building is narrated with attention to detail and attention to decorative solutions. The silhouette of Vesuvius in the background defines the relationship between the building and the landscape system. (Frediano Frediani Private Archive)

The project, presented at a press conference on 22 November 45, obtained the authorization of Mayor Fermariello and the Civil Engineers. However civil society, as well as its intellectual class, was not ready for such a challenge. The truth is that it was not just a matter of accepting the idea, in itself already rather disconcerting, of a skyscraper in a historical urban context, but of accepting a whole model of economic development as Vitale's words clearly demonstrate: «Those who had the good fortune to admire Naples from the top of the hill of Sant'Elmo or from Posillipo can judge, for example, in the simple light of memory, what a horrible outburst it would have constituted in that enchanting panorama, in the presence of that sky and that sea that seem to weave a continuous luminous dialogue, the brand-new skyscraper that an unconventional mimicry and a crude commercial mentality had planned to erect in Santa Lucia, almost ridiculous challenge of a people of Lilliputians to the dominating mass of the imminent Vesuvius» (Vitale 1947, p. 160). It was not worth the heartfelt defence of the project made by Frediani himself, nor the reassurance that the entire work would be financed with US funds to eliminate the prejudice towards that building so clearly inspired by the American model, nor to overcome that mistrust of the possibility that the modern would dialogue with the ancient in the historic centre of the city. A distrust that will very often re-emerge and accompany its urban development for many decades.

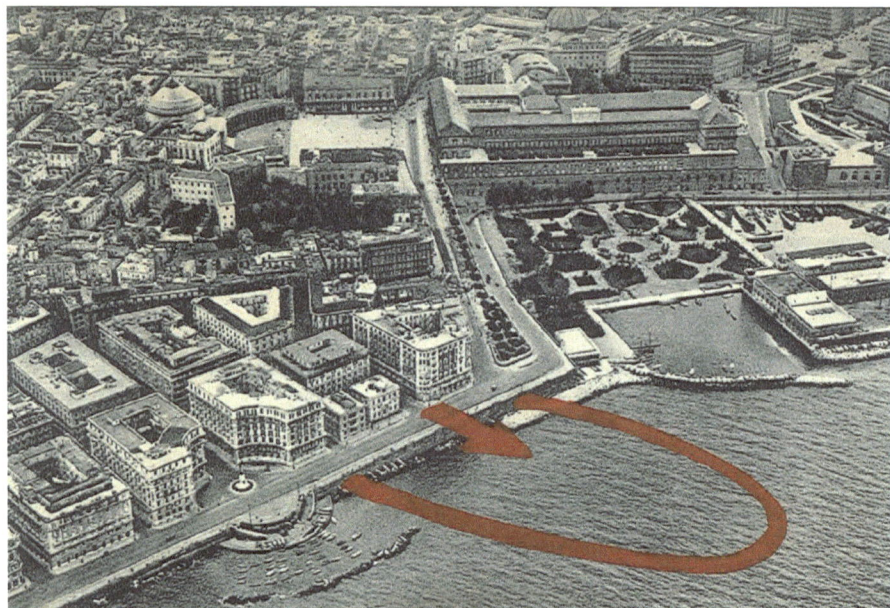

Fig. 3. F. Frediani, *Progetto di massima. La zona del Santa Lucia.* The image shows the exact location of the floating raft extending into the waters of the gulf for over 200 m, in axis with Via Petronio. The relationship between the installation and the surrounding urban fabric emerges: the hemicycle of Piazza del Plebiscito, the imposing mass of the Royal Palace and the regular 19th century settlement of Santa Lucia. The octagonal tower would have risen at the intersection of the two axes of Via Cesario Console and Via Petronio (Frediano Frediani Private Archive).

3   Project, Drawing, Architectural Document

Remaining therefore in the sphere of pure intentionality, Frediani's project offers itself today to our attention for a path of graphic investigation that retraces its vocations and potential with all the expressive force of 'project drawing'.

The first objective that we set ourselves was to frame the episode of 'Santa Lucia' in Frediani's prolific production and to define the meaning that the proposal of such a typology takes on in the more general history of the skyscraper and its use in Italy and in the world from the beginning of the 20th century until the 1970s. The series of beautiful perspectives of the 'Santa Lucia' skyscraper skilfully created by Frediano Frediani allows us to have a fairly clear idea of the design development while also appreciating its quality. But it also allows us to propose some considerations on the use of drawing as an expressive vehicle of architectural thought as it was declining in the context of Italian architectural culture at the turn of the Second World War. The wind of the avant-garde, which had marked much of the modernist architectural production in Europe, was late to take root in Italy, where Futurism and the unmistakeable graphic sign of Sant'Elia had to be awaited in order to identify a real element of novelty in the way the representation of architecture was understood. The Roman school that would soon express itself in the representations of Terragni, Libera, Ridolfi, Sartoris, just to mention the most famous, is certainly to be understood as a field of experimentation in the field of graphic innovation, where to work in search of linguistic coherence and a new way of understanding the drawing of architecture. A drawing almost always created for narrative purposes, i.e. as a description of a work aimed at the outside world, much more than an instrument of work aimed at communication during its realization. A narration that does not renounce the praise of the volume and that identifies in charcoal or graphite fat a technique perfectly consistent with its objectives, not without a certain virtuosity and without ever abandoning the perspective representation. The initiatory path of modernism that will lead to favour the absoluteness of axonometric representation (Reichlin 1979), to exalt the almost abstract quality of the pure geometric form without any ornamentation and, above all, will lead to the drastic elimination of the 'context' in which the representation of architecture is inserted, does not seem to interest the experience of Frediani, whose drawings, instead, fit well into the narrative vocation to which reference was made, with some elements of particularity.

The Santa Lucia project develops, as we have said, in aftermath of the war at that moment when critics place a season of architectural drawing that they define as 'useful drawing'; a season of drawing that, though in continuity with the previous period, is connoted by a particular pragmatic, instrumental, functional approach to a physical, but also cultural reconstruction of the country and its architectural consciousness. Seen from this point of view, Frediani's drawing for Santa Lucia shows an attitude towards the past and therefore towards the recent pre-war period, and the choice to use perspective massively, in addition to orthogonal projections, has an undoubted significance: to verify the volumetric dimension and above all the scale of his intervention, succeeding in communicating to the client, as well as to public opinion, the monumental and symbolic quality of his courageous design idea and the attempt to build a dialogue with the pre-existing context (Fig. 4).

Fig. 4. F. Frediani (1945). The two *Vedute prospettiche dalla via Cesareo Console* enhance the monumentality and expressive power of the octagonal tower. The Perspective Sketch from the south clearly recounts the articulation of the overall layout by aggregation of overlying volumes culminating in the slender volume of the tower. (Frediano Frediani Private Archive).

Fig. 5. F. Frediani, (1945), *Schizzo schematico delle masse strutturale (veduta da palazzo reale)*. Realized with the technique of photomontage, the drawing documents the visual impact of the building seen from the Royal Palace. It is clear Frediani's intention to verify the building's relationship of continuity with the urban structure of the city and with the landscape context of the gulf.

The numerous views present in the Frediani archive, in the booklet dedicated to the project of the Santa Lucia, are all one-point perspective; with effects sometimes more realistic at other times with an extraordinary power of 'chiaroscuro' in the definition of volumes and light cuts, they define the material quality of the project, exalting together with the clarity of the volumes, the symbolic and monumental dimension with a precise choice of expressive poetics. A classicist character emerges in them that also recurs in Frediani's other works, a synthesis between the neo-romanity of the Italian rationalists formed during the Fascist period and the solemnity of American taste. It is difficult not to highlight the almost explicit reference to the image of the grooved column suggested by the tower shaft, an inevitable reference to Adolf Loos' project for the Chicago Tribune: look at the projection lines of the edges of the octagonal tower, together with the windows and frames that would have created a design of very thin vertical membranes similar to the grooves. Equally evident is the reference to the American model as can be seen in Colonel Towsend's report: «The building complex is architecturally sober, with glass dominating the masonry. Even the corners of the octagonal tower will be covered with luminous glass vertebrae throughout its height» (Towsend 1945). In all of the perspectives, the choice of the vertical picture highlights the monumental dimension of the tower much more than the dynamic volumetric articulation of the building by successive aggregations whose description is entrusted to the only aerial perspective that tells the dimensional development of the artificial peninsula destined to host the building complex. The greatest challenge is the insertion of the project into the landscape and the elegant work of the perspective views aims, above all, to insert the project into the context of the pre-existing landscape, whether it be the languid backdrop of Vesuvius or the dense urban fabric. Remarkable is the "perspective sketch from the slopes of the Vomero" is remarkable with its fast stroke seems to really represent a thought out loud, almost a verification for itself of the scale of the intervention that appears here in full evidence. Also of particular interest is the use of the photomontage technique, is also of particular interest, to be considered really experimental for the time. A technique, the latter, capable of combining modern architecture and urban planning, the photo of the urban context and the drawing at the stroke with a fascinating outcome (Fig. 5). Frediani's intention was obviously on one hand, to define the object designed in full plastic and compositional autonomy. While on the other, to describe its impact on the city, which clearly represented the most audacious element of the intervention, which is why in all probability the project remained an unexpressed potential of the city's destiny.

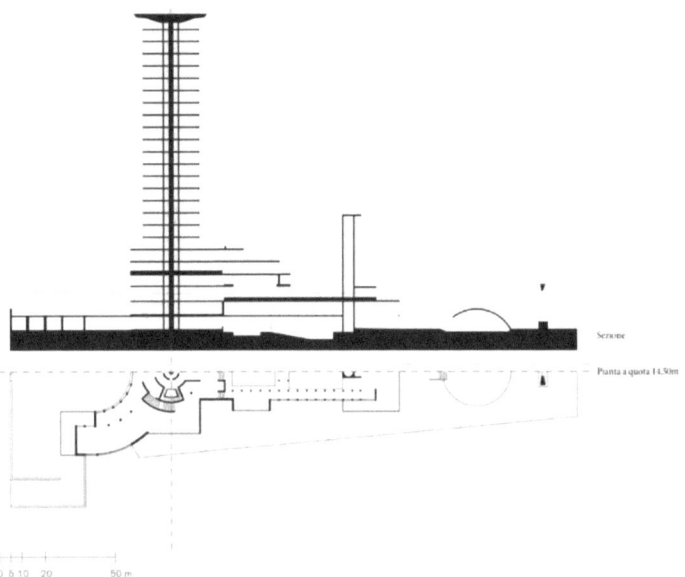

Fig. 6. The St. Lucia's building. Graphic analysis and critical reading. Longitudinal section and floor plan (drawings by arch.Teresa Esposito. Degree thesis in Architecture - Academic year 2018/19 supervisor prof. A. Cirafici)

The critical reading of the project can, however, also involve orthogonal projections, especially of the splendid plants elaborated by Frediani and richly accompanied by quotas and legends, which make it possible to understand the functional intentions and destinations of use provided by the project: the area intended to house the building complex, would stretch like a peninsula protected by a system of dams emerging from the waters of the gulf, between the San Vincenzo pier and the Castel dell'Ovo with a plant that would gradually become narrower, over two hundred metres from the shore, with sports facilities, swimming pools and administrative functions. The building complex, conceived as a set of aggregated volumes according to an ascending composition, would end with the slender figure of the octagonal skyscraper that would accommodate inside, arranged in a radial pattern along the perimeter, from the sixth to the ninth floor, apartments for illustrious guests and, from the tenth to the twenty-second floor, single lodgings and ateliers with access from a circular landing open around the nucleus of the lifts.

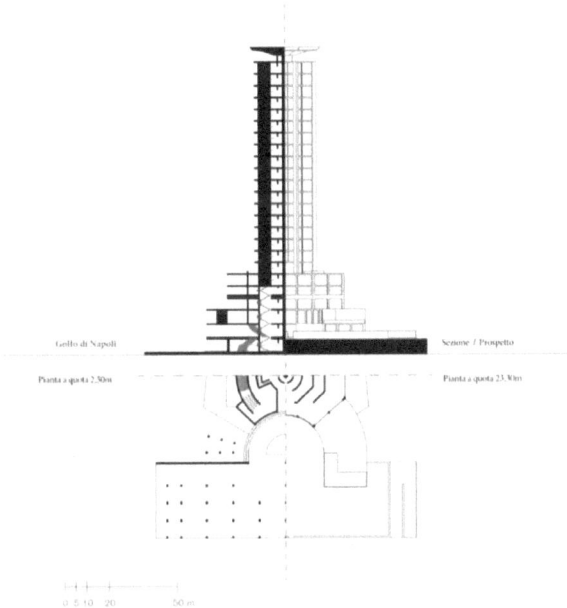

Fig. 7. The St. Lucia's building. Graphic analysis and critical reading. Croos section and portion of the plan (drawings by arch.Teresa Esposito. Degree thesis in Architecture - Academic year 2018/19 supervisor prof. A. Cirafici)

On the twenty-third floor there would have been the panoramic restaurant in which taste typical dishes of the Neapolitan tradition, while upstairs there was the belvedere terrace, from which to enjoy the 360-degree view of the gulf. Finally, on the top floor, there was a helicopter landing platform to crown the skyscraper. Precise lines and meticulous details make the plans, as Frediani's caption states, extremely effective in terms of representation. All of them on a scale of 1:200, often describing the building with a narrative approach, as in the case of the 'Planimetry of the typical lodgings and the panoramic restaurant' where the definition of the panoramic views in a radial pattern is particularly effective.

3.1   The Graphic Study: Re-drawing and Critical Reading

The re-drawing operation carried out on the "Santa Lucia" is part of a wider work of analysis and graphic representation of the works designed by Frediano Frediani in Campania, between Naples and Benevento, with particular attention given to those that have not had the luck to be realized or have, over time, been demolished. The graphic re-proposal of Frediani's projects is part of the process that allows for the critical reading o f awork 'in absentia' through a transcription that is not only a conventional re-proposal of the project drawings, but on the contrary it is the result of a selection of signs and elements of a metalinguistic system that in itself configures the features of a critical transcription. In fact, investigating architecture through the graphic transcription of the Drawing means identifying the design process, starting from its condition as 'drawn architecture', as if to demonstrate, to quote Gregotti (Gregotti 2014), how thin the line separating design and drawing is and how the two terms exchange and overlap, reflecting one in the other. In this sense, the re-drawing of Frediani's work has had the sense of searching not only for its formal values, but also for the traces that make it possible to identify the complex link that the designer has tried to weave between architectural text and urban-landscape context. The archive material relating to the 'Progetto di massima del Santa Lucia' is extensive and includes, as mentioned above, a series of preparatory perspective sketches and photographic images, a series of splendid perspectives of the exterior volume and perspective views of the interior, as well as a series of photographic insertions of the building into the profile of the city's gulf. The unfinished condition of the work finds in the strategies of representation and in particular of the digital one, its own expressive figure and an interesting way of further figural verification. Far from pursuing a hyper-realistic approach, digital representation has relied on the strength of line and its ability to synthesize form. It has privileged an action of progressive reduction of the signs up to the extreme synthesis, succeeding in reducing in a very short but effective way the elements characterizing the project: the exact axiality in its relationship with the urban fabric; the rigor of the geometry underlying each compositional element (Figs. 6 and 7); the concentric force in the articulation of the volumes of the tower; the synthesis in the synoptic reading of the orthogonal projections up to the abstraction of some images in which, the horizontal picture perspectives (Figs. 8 and 9). I propose again to the imaginary the ascending dimension and the spectacular view that could have been enjoyed from the 'Belvedere Italia' reconstructing the perspective views that Frediani had wisely designed to allow from the octagonal tower the vision of wide panoramic horizons open on the gulf. The re-drawing has allowed to add new elements of reflection to the 'already said' of the project, exploring from the inside the relationship of the designed space with the city and the landscape. The succession of planes and volumes is synthetically narrated with an expressive force through the axonometric view in which the entire articulation by successive overlapping is evident (Fig. 9). The construction of the digital model has, in fact, made possible a further verification of the inclusion of the monumental volume in the profile of the Gulf of Naples, verifying its perceptive dimension both in the view from the sea as well as from the land, confirming the reading that Frediani himself had well intuited in his studies in a relationship that has not substantially changed since then. The poetic dimension of Frediani's design has found resonance in the formal elegance of the drawings and the project has been narrated in all its formal and symbolic power.

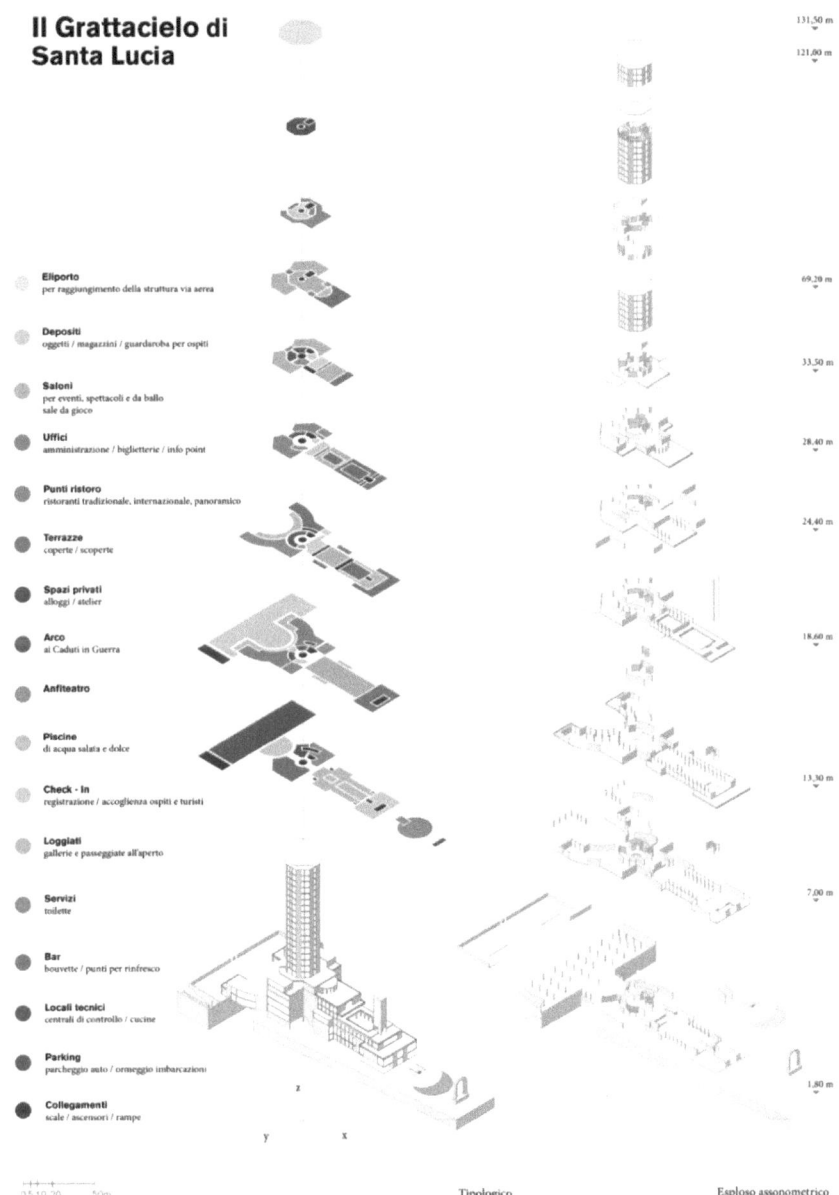

Fig. 8. The St. Lucia's building. Graphic analysis and critical reading. Exploded view. (Drawings by arch. Teresa Esposito. Degree thesis in Architecture - Academic year 2018/19 supervisor prof. A. Cirafici)

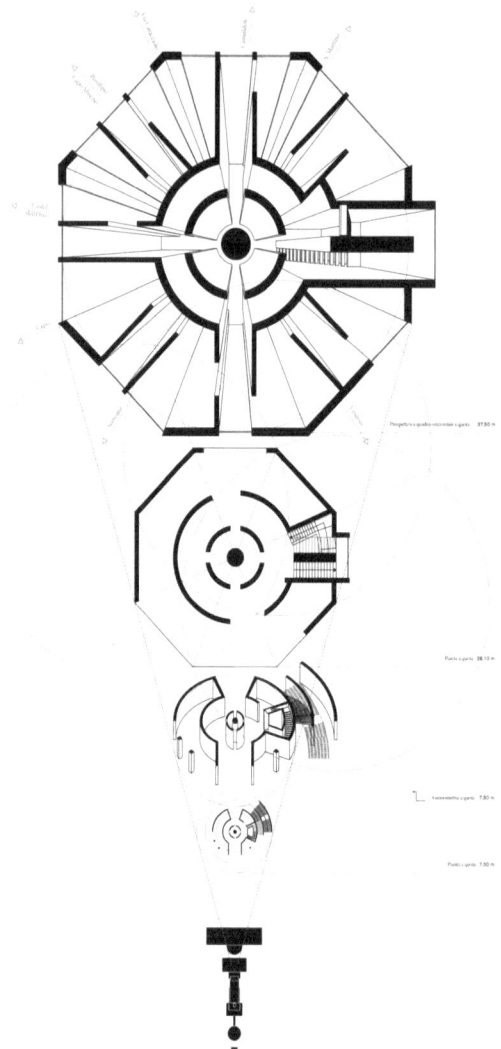

Fig. 9. The St. Lucia's building. Graphic analysis and critical reading the graphic composition enhances the geometrical matrix of the tower and emphasizes the concentric and ascending dimension of the plant, up to the narration of the panoramic view of the last level. (Drawings by arch. Teresa Esposito. Degree thesis in Architecture - Academic year 2018/19 supervisor prof. A. Cirafici)

Once again the Drawing, therefore, has taken on the character of a real text that is added to another text: the one of the architectural "body" investigated. Drawing is, the instrument that the architect needs «to 'meditate' and 'meditate again', an eventual operation projected into the practice of design whose aim is that of architectural realization» (Mezzetti 2003). 'Lineamenta' e 'structura': it is no chance that Alberti summarized the elements that define architecture as «the building is a body and like all other bodies, it consists of design and matter: the first element is, in this case, the work of genius, the other is the product of nature». The creative idea that Drawing preserves and represents can therefore be investigated precisely through the graphic sign, through a process of re-drawing that saves from the limbo of inutility the thousands of drawings that, just because they are not realized, risk being judged irrelevant. This is certainly not the story of the project for the skyscraper of Santa Lucia, which represents one of the most courageous and visionary episodes of thinking around the city that Naples has been able to produce since the war and in which the sensitivity and poetics of a refined designer like Frediano Frediani has found full expression (Fig. 10).

Fig. 10. St. Lucia's skyscraper. Sea view. Hypothesis of insertion of the digital model in the current profile of the coast. (Drawings by arch. Teresa Esposito. Degree thesis in Architecture - Academic year 2018/19 supervisor prof. A. Cirafici)

Acknowledgments. This study was initially part of the degree thesis in Architecture entitled: *Il Grattacielo di Santa Lucia di Frediano Frediani. Progetto/Disegno/Documento di Architettura*, by Teresa Esposito at the Department of Architecture and Industrial Design of the University of Campania Luigi Vanvitelli (a.y.2018/19 prof. Alessandra Cirafici).

Thanks to professor Gianluca Frediani for his kind collaboration in consulting and authorizing the use of the images taken from the Frediani Private Archive.

References

Alberti, L.B.: De Re edificatoria, Libro I, p. 18

Belfiore, P.: Dal dopoguerra ad oggi. In: Gravagnuolo, B. (ed.) Napoli, Architettura e Urbanistica del Novecento, Laterza, Roma-Bari, Belfiore (1994)

Faroldi, E., Gramigna, L.C., Trapani, M., Vettori, M.P.: Verticalità. I grattacieli: linguaggi, strategie, tecnologie dell'immagine urbana contemporanea, Maggioli Editori, Sant'Arcangelo di Romagna (2008)

Gregotti, V.: I l Disegno come strumento del progetto. Marinotti edizioni, Milano (2014)

Meridione Nord e Sud. Anno XI, n°4, ottobre –dicembre 2011, La Napoli degli Americani dalla liberazione alle basi NATO. ESI, Napoli (2011)

Mezzetti, C.: Il Disegno dell'architettura italiana nel XX secolo. Edizioni Kappa, Roma (2003)

Migliari, R.: Disegno come Modello. Edizioni Kappa, Roma (2004)

Molinari, L.: Italia e America, in La Napoli degli americani, dalla liberazione alle basi NATO, ESI, Napoli, p. 35 (2011)

Purini, F.: Il disegno di progetto dell'architettura. La rappresentazione delle idee, intervista di M. Unali, Tesi di Dottorato di Ricerca in Disegno, V ciclo, Università degli Studi di Roma La Sapienza, vol. III, p. 347 (1993)

Reichlin, B.: L'assonometria come progetto. In: «LOTUS», n. 22 (1979)

Towsend, H.H.: Una grandiosa costruzione a S. Lucia. In: «La Voce», vol. II, p. 277, 24 novembre 1945

Ugo, V.: Mìmesis. Sulla critica della rappresentazione dell'architettura, Maggiolo editore, Milano (2004)

Vitale, S.: Attualità dell'architettura, p. 160. Ricostruzione urbanistica e composizione spaziale, Laterza, Bari (1947)

Zevi, B.: Architettura in nuce, Sansoni ed. Firenze, p. 129 (1972)

Design and Graphical Analysis of the Church of Santa Maria della Misericordia in the Verano Cemetery in Rome

Laura Carnevali and Fabio Lanfranchi[✉]

Department o f Hstory, Drawing and Restoration of Architecture, Sapienza Università di Roma, Rome, Italy
{laura.carnevali,fabio.lanfranchi}@uniroma1.it

Abstract. This contribution describes the graphical analysis of the Church of Santa Maria della Misericordia, a building situated at the back of the scenic area of the first layout of the Verano Cemetery in Rome, which was designed by Virginio Vespignani during the pontificate of Pio IX in the first half of the 1800s. This religious building is therefore an integral part of a single 'urban complex' that also includes the four-sided portico [Quadriportico] situated in front, which acts as a porticoed square, and the building to access the cemetery area, a symbolic door situated at the border with the city of the living. The study was based on graphical results deriving from survey campaigns made in the 2012–2015 period, the purpose of which was to build knowledge and documentation regarding the current state of the complex. Considering the complete absence of original graphical material related to the elevations and the reduced reliability of the plan found in the archives - rather summary in that it is contextualized within a broader area - geometric/proportional studies had still not been made. The analysis also has value given the particular historical/cultural period. The designed space is regulated by a rigorous geometrical logic that, for as much as it lacks unique stylistic references overall, is posed as a solution typical of the Enlightenment formulated in the cultural wake of figures such as Poletti and Camporese, but interpreted and proposed anew by Vespignani 'on the human scale'.

Keywords: Church Santa Maria della Misericordia · Verano Cemetery · Graphical analysis

1 Introduction

1.1 Brief Historical Considerations

The current scenic layout of the first centre of the Verano Cemetery in Rome, i.e. the area including the entrance, the four-sided portico [Quadriportico], and the church, is due to the work of Virginio Vespignani.

Appointed architect of the factory in 1850, he proposed 'redeveloping the traditional model of the cemetery with the portico' (Cardilli and Cardano 2006, p. 11) and coordinated the work until 1871, 'the year of his resignation as artistic director of the complex' (Barucci 2006, p. 108).

© The Editor(s) (if applicable) and The Author(s), under exclusive license to Springer Nature Switzerland AG 2020
L. Agustín-Hernández et al. (Eds.): EGA 2020, SSDI 6, pp. 297–309, 2020.
https://doi.org/10.1007/978-3-030-47983-1_27

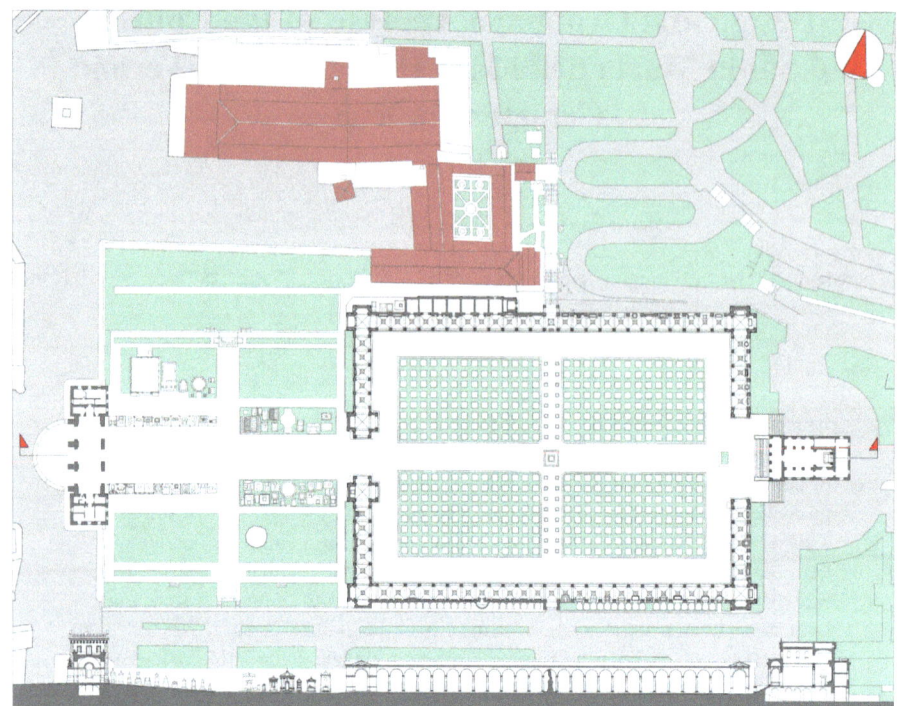

Fig. 1. Plan of the four-sided portico, with the longitudinal section below. To the west is the entrance building, the four-sided portico is in the centre, and the Church of Santa Maria della Misericordia to the east. To the north is the Basilica di San Lorenzo Fuori le Mura.

'And in effect, we maintain that the system composed of the four-sided portico, the Church of Santa Maria della Misericordia, and also the entrance building can be considered in the guise of an extended theatre system composed of a proscenium: the entrance building, an ideal filter between the space of the living and the city of the dead; a band free of significant architectural elements, that is, the 'paradoi'; the two series of arms of the four-sided portico in guise of parascenia and the church as the back-drop/altar' (Barlozzini et al. 2018, p. 1213) (Fig. 1).

As previously stated, the current urban layout of the Verano Cemetery is due to a project by Virginio Vespignani, who, however, based his formal results on previous studies made by three generations of architects. The first nucleus was based on the design hypotheses formulated by Giuseppe Valadier in 1811 and 1813, followed by those developed by numerous other architects, including Giuseppe Camporese, Raffaele Stern, Gaspare Salvi, and Paolo Belloni.

1.2   Background and Goals of the Study

We focus attention on the Church of Santa Maria della Misericordia (Fig. 2), and particularly on aspects relating to the analysis of the possible geometric/proportional design matrices.

Fig. 2. Photo showing the Church of Santa Maria della Misericordia (background), the four extremities of the four-sided portico (to the sides), and Ascension of Jesus by Leopoldo Ansiglioni (centre).

The study acquires particular value also in consideration of Virginio Vespignani himself, whose signature style is framed by the rigour of the Enlightenment already promoted by the generation of Giovan Battista Piranesi, Canova, and Valadier. In pre-Unification Rome, this climate found an important laboratory capable of both competing with neoclassical theories pertaining precisely to contemporary European culture and also measuring itself against the nineteenth-century culture of celebrating the terrestrial virtue of the dead, driving out death through the use of symbols and signs. 'Never as on this occasion does Vespignani so express himself with evidence of reference models that are not only uniform, but especially free of his usually stylistic mix, in defining a space regulated by geometric reason that admitted no alterations of any sort' (Spagnesi 1976, p. 34).

For the church, however, as for the four-sided portico and the entrance building, original graphical material was substantial absent from the archives (National Archives in Rome and the Library of Archaeology and Art History (Biblioteca di Archeologia e Storia dell'Arte, BiASA). The little material that exists relates especially to the elevations, beyond the scarcity and poor reliability of the plan - rather summary in that it is contextualized within the broader area of the first layout of the cemetery - made it practically impossible possible to carry out any geometric/proportional studies.

After 2012, following contact made with the manager of the Centro di Documentazione dei Cimiteri Storici di Roma, the basis of a collaboration useful for instituting a university research unit was suggested. Instituted by the City of Rome in 2003, this office deals with archiving the funerary monument heritage of the Verano Cemetery. In 2013 the Research Unit for Outdoor Museums was instituted at the Department of Architectural History, Design, and Restoration at the Sapienza University of Rome. Its purpose is to build knowledge for a project to conserve artistic and historical/architectural goods inherent to the Verano Monumental Cemetery in Rome.

Survey campaigns were made over a three-year period, which were programmed according to different phases and levels of investigation and made through the integration of different acquisition methods and techniques. Following this, an initial body of graphics was compiled, which were useful both for documenting the consistency and current state of the complex and for proceeding with a detailed, original graphical analysis of the entire complex, and the church in particular.

2   Treatise

2.1   Document Analysis and Related Conditions

As mentioned above, there was a nearly complete absence of graphical documentation about the Church of Santa Maria della Misericordia, especially relating to views such as elevations and sections. There are, instead, a series of plans of Vespignani's various design proposals that, while drafted on a scale appropriate for representing the wider context of the first cemetery nucleus, in later modifications show the planimetric layout of the church substantially in agreement with the final building.

In addition to those dated and signed by Vespignani, the drawings include others that, without a date or author, cannot be attributed with certainty to the designer. One example is the general plan of the Verano Cemetery shown in Fig. 3. This drawing is nearly entirely in agreement with the consistency of the individual buildings as hypothesized by Vespignani. Note, for example, the original design configuration of the four-sided portico, that is, without the double thickness of the building body of the northwest arm, which was most likely made by Agostino Mercandetti after 1871 as successor to Vespignani in the artistic direction of the building.

Despite this, the drawing displays two design proposals that were never realized: a long portico connecting the cemetery and Porta Tiburtina, and the plano-altimetric structure adjacent to the church that is completely different from what exists today.

We return to the formal analysis of the church. Beyond the considerations already expressed regarding critical metric aspects characteristic of the large scales of representation and the consequent difficulty of interpreting them exactly, it should be added that the current building 'incorporates' a series of transformations that followed over time and were made necessary for technological/static reasons.

Especially from the structural point of view, these building interventions have delivered an organism that in some measure, as will be investigated more in depth below, should be viewed as most likely different from what was originally conceived by Vespignani.

At the same time, the period of changing from the system of pontifical measurement still adopted by Vespignani at the time of the design to the advent of the metric system should not be underestimated. As we will see below, this consideration may have contributed to variations, in this case dimensional, between design and realization.

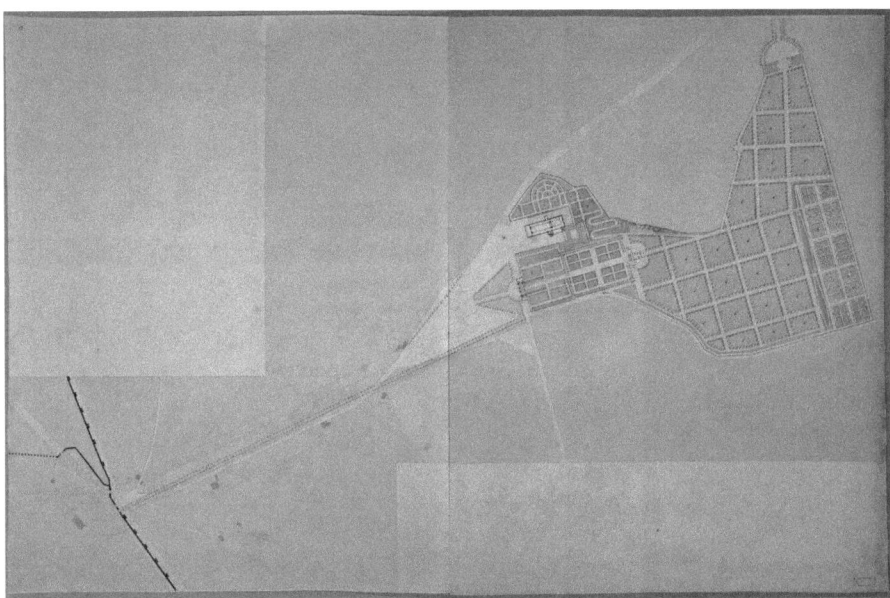

Fig. 3. General plan of the Verano Cemetery 1860–1874 (background). Anonymous. Source: Lanciani Collection – Rome Library of Archaeology and Art History (BiASA).

Construction of the church began at the end of 1855 and was 'finished in 1859 and consecrated on 29 October 1860' (Del Bufalo 1992, p. 34).

The survey campaign and rendering of the data obtained highlighted some formal differences that we hypothesize may be dependent, perhaps also in this case, on problems of statics or possible discoveries of ancient footprints. However, we disregard the aspects tied to static criticalities that were already revealed a few years after consecration, and which will not be considered here since they are not directly related to the specific topic.

In this respect, it is important to note that part of the current Verano Cemetery rests on an ancient burial area. 'The Ciriaca Cemetery, or Coemeterium Sancti Laurentii, on Via Tiburtina has a troubled history of conservation, which culminates in the destruction of many galleries following the construction of the Verano Monumental Cemetery starting in 1836' (Ferri 2016, p. 2225).

The document shown in Fig. 4 must certainly contribute to validating the hypotheses of possible modifications necessitated during the building's construction. The comment written to the side of the drawings states: 'This sepulchre should have been discovered by Vespignani in the excavations of the Cemetery of San Lorenzo' (Questo sepolcro deve essere stato scoperto da Vespignani negli scavi del Camposanto di S. Lorenzo). The figure shows the projection in plan and elevation of the sepulchre (above) and the transverse section of the repository plane of the Church of Santa Maria della Misericordia (aligned below). Here, it is possible to theorize about the event of the sepulchre's discovery precisely in the area of construction, with the consequent editing of the section of the design for appropriate planning evaluations.

2.2   Rendering the Survey Data

As mentioned above, the drawings used as the basis for the research and graphical analysis derive from survey campaigns made in the 2012–2015 period. The measurement operations regarded the survey and analysis of the first layout of the cemetery complex, the infographic representation of all its parts, experimentation, and the widespread use of the most advanced measurement and digital modelling techniques. The different acquisition methods were also integrated to compose both the necessary documentary repertoire and the best communicational effectiveness. The graphics produced in relation to the church include the plans of all levels, elevations of the façades and a series of transverse and longitudinal sections. For reasons of space, the present contribution reports only the drawings deemed most pertinent to exemplifying the studies made, both regarding the proportions and the possible geometric design matrices constituting the formal order of the building.

2.3   Graphical Analysis

From an initial analysis made only based on archival documents and excluding the crepidoma for access at the front, the planimetric layout of the church seems to be contained within the outline of a rectangle composed of the union of two squares measuring 14 m on a side (Fig. 5). More extensive analysis made on the survey drawings, while not metrically confirming this initial hypothesis, nevertheless leaves open the possibility that the system can be formulated through an aggregate system of two squares set next to each other along the long axis of the church. The survey drawings also reveal how the layout of the individual parts of the building are reconciled, at least along the essential lines, with a grid composed of a base module of 140 cm, that is, a tenth of the length of the side of the square.

   However, we have hinted at possible variations between the project and realization due to the particular period that saw a change from the system of measurement used by the Papal States and the metric system, which was then in the process of being adopted. The graphic in Fig. 4 (left) shows the graphical scale in architectural rods and palms, that is, adhering to the metric system then in use by the Papal States. The excerpt of the transverse section of the church on this scale shows a building body formulated on a length of 5 rods and 8 palms, with the length of an architectural rod equal to 2.23 m (10 palms) and a palm equal to 22.34 cm; the size in metric units is 13.49 m. 'Five arch. rods make an arch. or engineer's chain' (Commissione de' Pesi e Misure 1811, p. 101). The dimensional verification made on the thickness of the walls present in section view shows thicknesses proportionally less than what was surveyed. In detail, the intermediate division as designed indicates a thickness of 96 cm compared to the 101 cm surveyed; the external wall as designed is 68 cm compared to the 73 cm surveyed. A uniform comparison of the various values shows design measurements that are consistently larger than the built object, for which the average is about 1.05%. Applying the ratio of reduction to the 140-cm module identified gives 133.33 cm, that is, a measurement close to 6 palms (134 cm). These data seem to substantially confirm the hypothesis of a building made using a metric system different from the system used in the design phase. Obviously only theories may be made regarding the possible motivations. From the metric point

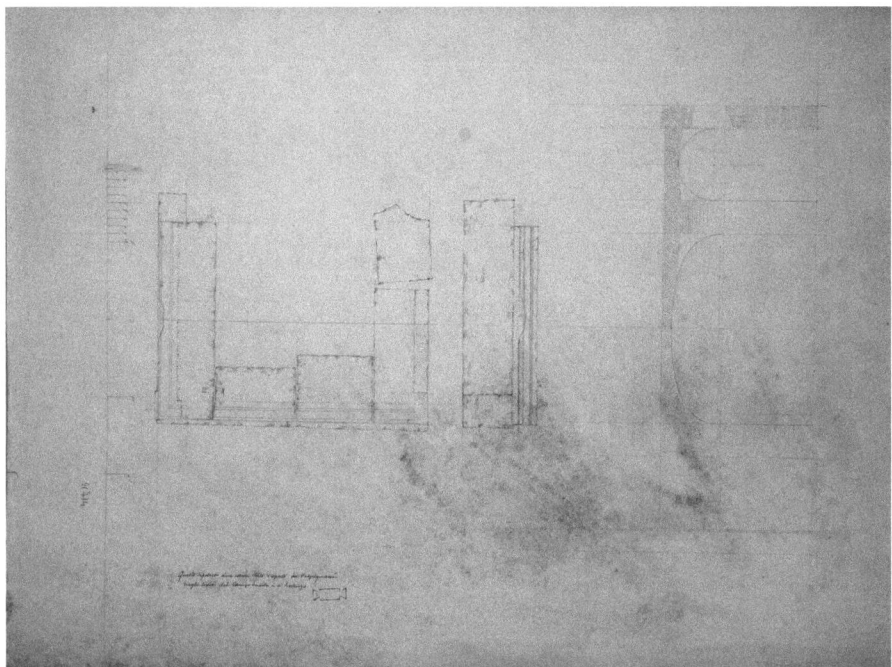

Fig. 4. Left: the surveyed plan and elevation of the sepulchre found in the Verano Cemetery. Right: transverse section of the repository plane of the Church of Santa Maria della Misericordia. Drawings by Virginio Vespignani. Source: Lanciani Collection – Rome Library of Archaeology and Art History (BIASA).

of view, we exclude the simplistic assumption that the measurements were converted incorrectly, assuming instead a necessary rectification that may have been required by the supply of blocks and lesenes for the narthex and interior of the church. As reported in documents stored at the Capitoline Historical Archives under Title 61, Sheet 225 from 1856, the work of stonecutters was contracted to reduce the height of the columns in eastern granite and pilasters in Greek marble for the chapel being built in the new cemetery. The materials, which were donated by the Papacy, came from the ancient portico of the Basilica of Saint Paul Outside the Walls, which was then being rebuilt. The dimensions of the materials recovered - the column blocks, for example, measure 65 cm in diameter - may have led Vespignani to increase the dimensions of the church, adapting the unit of measurement to the metric system at the same time. We return to the dimensional data acquired via the surveys to confirm what has previously been expressed both regarding the dimensions and shape of the outline of the church identified from the base module. That is, the real length of the church is not 28 m, that is, double the thickness of its building body as initially hypothesized by analysing the drawings in the archives, but rather 26.60 m, that is, 140 cm less. The lower part of Fig. 6 shows the grid formulated with the dimensional values hypothesized earlier overlaid on the survey map. It shows how the placement of the axes of the windows of the north and south elevations match only slightly with the hypothesized grid. In the figure, the lack of alignment is

Fig. 5. Excerpt of the general plan already presented in Fig. 3, north-east part of the four-sided portico. Note the rectangle composed of two 14-m squares that contain the plan of the Church of Santa Maria della Misericordia.

characterized with traced lines of reference. The offset, on the order of a few tens of centimetres, is highlighted especially near the central part of the elevation, giving rise to related, obvious repercussions also in the interior planimetric layout of the church.

In addition, as can be seen in the survey elevation and plan shown in Fig. 7, the lack of reconciliation between the layout of the constructed building and the hypothesized compositional framework is reflected even more in the area of the dividing wall between the narthex and interior of the church. The loss of a geometric link between the façade and grid also continues even more evidently on the side façade of narthex. This is characterized by evident dissonance in terms of architectural 'weight' with respect to the symmetrical block, congruent with the hypothesized grid, located on the right in the elevation. While being fully aware that we are moving in the circumscribed area of possibility and stimulated by the lack of correspondence between the results deriving from the survey drawings and the hypothesized 'ideal frame', we are directed towards the development of a new model whose results are shown in Fig. 7. The reworked plan and side elevation were formulated starting with the formal adjustment to the geometric grid hypothesized with the intention of remedying the critical points mentioned above. Once this phase was completed, attention focused on the search for possible geometric references related to the means of composing the entire structure. We have hinted about the possibility of a design formulation originally created by Vespignani based on setting

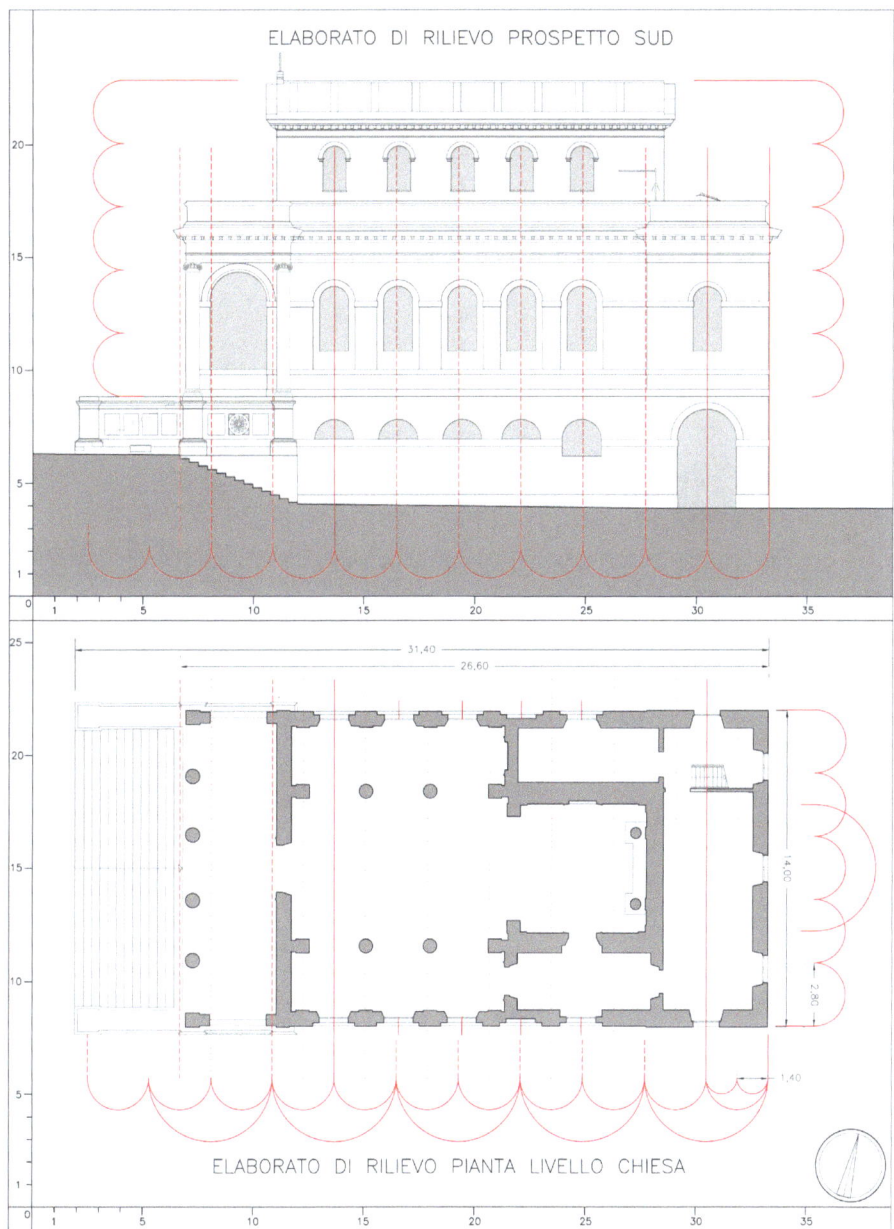

Fig. 6. Surveyed plan and southern elevation. The traces represent the alignments that diverge from the basic hypothesized matrix.

two squares side by side longitudinally, so the result of the reworked plan cannot be excluded. Rather, in our opinion, this idea is also validated considering the dimensional rebalancing of the side façade of the narthex.

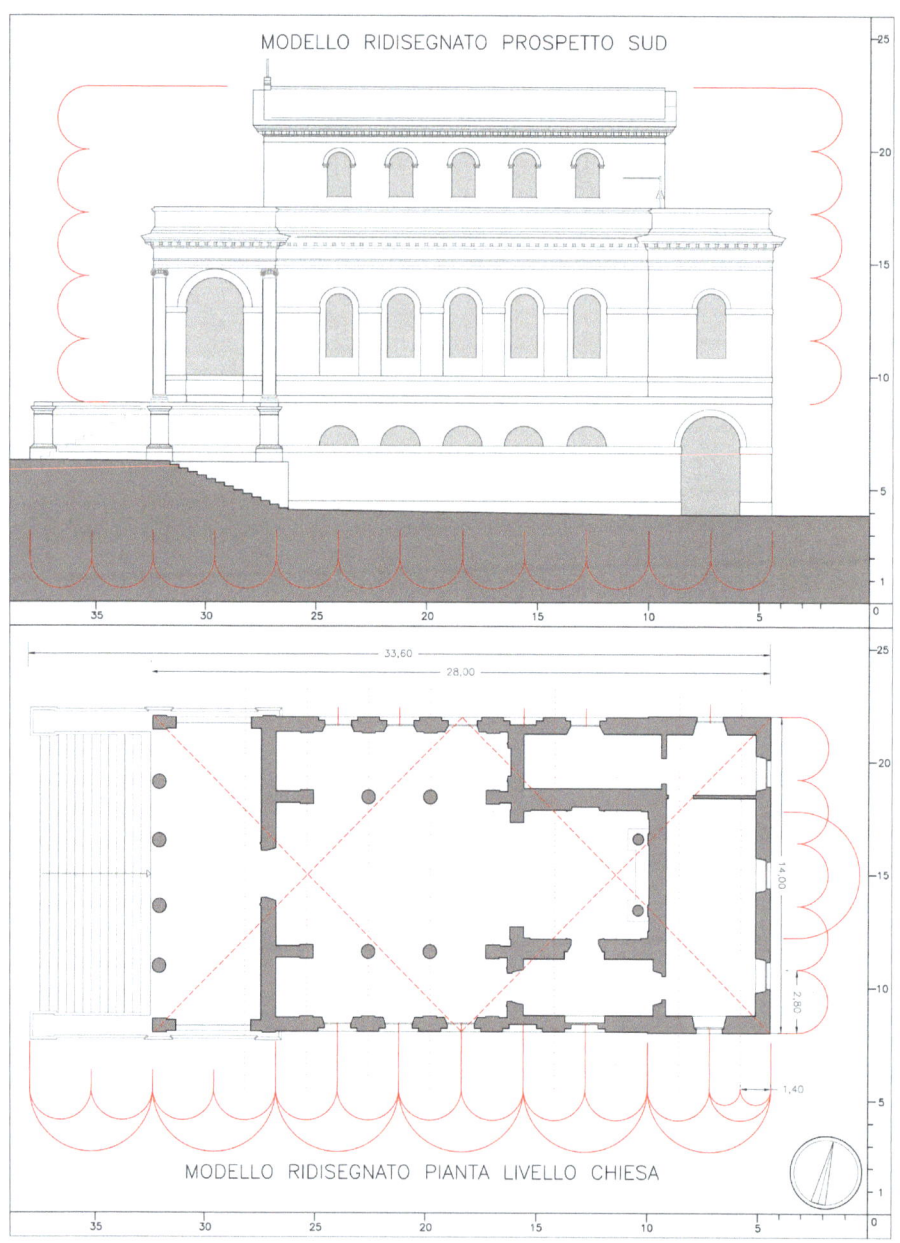

Fig. 7. Reworked plan and southern elevation.

Among the causes that we deem may have led to the realization of such a different building with respect to what we suppose may have been originally designed, we recall what was previously mentioned regarding the discovery of a sepulchre during the building's construction. 'The church and four-sided portico can nevertheless be considered

Fig. 8. Western façade of the Church of Santa Maria della Misericordia with the extremities of the four-sided portico to the sides (survey). As highlighted by the elevation, it is inscribed within a 14-m square situated on the level of the floor.

a single work, both for the neo-fifteenth-century purity that inspires them both, and, as shown by the design drawings, for pertaining to a unique moment in compositional synthesis' [Spagnesi 1976, p. 178].

In this perspective, the square represents a geometric reference that frequently recurs in Vespignani's architecture. In fact, in many of his projects, he resorts to this matrix characterized by its strong symbolic value: the square represents stability, terrestrial perfection, the geometric shape that more than any other adheres to proportion and symmetry in its Vitruvian acceptance. With regard to the elevations of the church, the graphical analysis reveals its coherence with the geometric results described for the plan of the building. Once the grid is superimposed on the main elevation of the building, it is clear how, starting from the level of the floor at the top of the crepidoma, the façade can be perfectly inscribed in a square that is 14 m on a side (Fig. 8).

Finally, another reflection merits what emerges from the general plan of the Verano Cemetery previously presented in detail in Fig. 5. The considerations reported below regarding the layout of the building in relation to its immediate surroundings may also in this case be described in detail under possible variations probably made during construction with respect to the original project.

In effect, the area of the Verano Cemetery was the object—even before the building interventions—of enormous movement of ground to level and terrace the terrain, which was originally rather steep. Vespignani himself already mentions this in a report about the state of the work dated on 2 September 1849 and stored at the Capitoline Historical Archives under Title 61, b.1, Sheet 35.

We return to the image presented in Fig. 5, which we recall is an excerpt of the general plan shown in Fig. 3, a rather faithful image of the definitive solution but lacking the architect's signature. In it, we note how in contrast to the current layout the ramps connecting the plane of the four-sided portico to the plane of the lowered country more to the east, are positioned to the sides of two terraces detached from the church.

Among the images of the area just realized, no documents were found that were useful for understanding the effective orographic situation except for the engraving by P. Petri from 1865 (Fig. 9). The view, made before the construction of the four-sided portico, offers a face-on view of the church from a vantage point situated to the left. Through careful observation, it is possible to distinguish the orographic layout at the sides of the church and the existing walls, which is very different from its current state. In fact, the level of the four-sided portico envelops the entire side elevation and not only the narthex. Nevertheless, this configuration differs from what is present in the plan shown in Fig. 5.

Fig. 9. Church in the Verano Cemetery. 18 cm × 25 cm. Engraving by P. Petri *The Church of Santa Maria della Misericordia Upon Completion*. Note the orographic layout represented without jumps in height at the sides of the church. Source: *Le scienze e le arti sotto il Pontificato di Pio IX*, Rome 1865, vol. I, Rome, BIASA, Dep. Banc.20 B 21.

However, we consider the general system of the building, imprinted on the double formal distinction of the front and rear, which is reiterated by a different use of finishing materials: marble columns for the narthex and plaster for the rest of the body of the building. Interpreted in light of what was hypothesized with the graphical re-working presented above, and at the same time considering the graphic related to the general plan mentioned, we cannot exclude the possibility that the area immediately surrounding the building may also have undergone construction modifications with respect to the design guidelines.

We therefore consider it reasonable to include the hypothesis that the drawing presented in Fig. 3, although lacking Vespignani's signature may effectively form part of the group of his design modifications.

3   Conclusion

Architectural drawings can be understood as a means and tool useful for conceiving and transmitting the design intent. At the same time, however, by graphically studying an existing building, the drawing becomes a tool for decoding. It serves as a strategic means useful to identifying regulating traces, proportions, and modularities that sometimes—following the most disparate causes—are only partially activated and thus explicit in the built architecture. They remain limited to a model that is solely and exclusively ideal only by the state of the design intent.

Redesigning, or rather, 'the drawing used as metalanguage' (Docci 1989, p. 149), can obviously not provide certainties but only hypotheses, that is, possible interpretational keys that, when lacking original documents, may become the only means to understand the creator's thought.

In conclusion, the formal results expressed by graphically studying the church do not and obviously should not claim to cross the borders of pure hypothesis, but are simply contextualized in the realm of possibility. It is, nevertheless, limited in 'light' of an architectural area characterized, in this case, by a stylistic signature that is strongly limited by rigorous compositional writing styles.

References

Barucci, C.: Virginio Vespignani, architetto tra Stato Pontificio e Regno d'Italia. Argos, Rome (2006)

Barlozzini, P., Carnevali, L., Lanfranchi, F., Menconero, S.: Il cimitero monumentale del Verano a Roma, da campo santo suburbano a città dei defunti. In: La Città Altra/The Other City, Cirice 2018, pp. 1211–1219. FedOA - Federico II University Press, Naples (2018)

Cardilli, L., Cardano, N.: Percorsi della memoria, il Quadriportico del Verano. Fratelli Palombi Editore, Rome (2006)

Commissione de' Pesi e Misure: Prospetto delle operazioni fatte in Roma per lo stabilimento del nuovo sistema metrico negli Stati Romani dalla Commissione de' Pesi, e Misure. Mariano de Romanis e Figli, Rome (1811)

Del Bufalo, A.: Il Verano. Un Museo nel verde per Roma. Edizioni Kappa, Rome (1992)

Docci, M.: Dalla parte del rilevatore. In: Finelli, L. (ed.) Luigi Moretti la promessa e il debito, architetture 1926–1973. Officina Edizioni, Rome (1989)

Ferri, G.: Alcune riflessioni sull'apparato decorativo del cimitero di Ciriaca. A proposito di due arcosoli superstiti. In: Costantino e i costantinidi: l'innovazione costantiniana, le sue radici e i suoi sviluppi, pp. 2225–2240. Acta XVI Congressvs Internazionalis Archaeologiae Christianae (Rome, 22–28 September 2013), Papal Institute for Christian Archaeology, Vatican City (2016)

Spagnesi, G.: L'architettura a Roma al tempo di Pio IX (1830–1870). Cassa di Risparmio di Roma, Rome (1976)

The Debut of the Section-Inner Elevation. Restitution Hypothesis of an Image of the XV Century BC

Adriana Rossi[1] and Daniel V. Martín-Fuentes[2]([✉])

[1] Dipartimento di Igegneria, Università della Campania Luigi Vanvitelli, Aversa, Italy
adriana.rossi@unicampania.it
[2] Dep. de Expresión Gráfica Arquitectónica, Universitat Politècnica de València, Valencia, Spain
dmartin@ega.upv.es

Abstract. On one hand, it seems hypothesized that the conventions of technical drawing are characteristics of the disciplinary development associated with the great Hellenistic treaties and writers. On the other hand, we know that the complexity and maturity of constructions, even in ancient Egypt, required, as witnessed by some papyruses, the ability to communicate technical information. Therefore, seems appropriate to analyse the translation of rules and conventions in the drawing that represents them.

We focus on the fresco of an Egyptian palace from the 2nd millennium BC found inside the tomb of Djenhutinefer, in Thebes. This drawing is of particular importance as it has a great level of detail and precision, given that sections, in contrast to the very detailed elevations and constructive plans in the history of drawing, have generally played a secondary role and almost exclusively for the indication o f heights.

The three-dimensional reconstruction and virtualization of the model will be based on iconographic and documentary sources, which are useful for methodologically motivating the critical interpretation. The resulting cultural importance will make the virtual use of the model a useful experience for the dissemination and knowledge of the inherited heritage.

Keywords: Initial forms of technical drawings · Origin of the section · Calculation of crenelated drawing · Tumba de Djehutinefer (Tebas. Dinastía XVIII)

1 Introduction

Until Gaspar Monge published his studies (1794–95) [1], the theoretical problem of architectural drawing was mainly limited to the possibility of establishing on the 'artifacts of the artifact' or, to say it with Vitruvius words '*adumbrationes*' [2], the quantities of intuitively recognizable forms [3]. Going back to the time of Octavian Augustus, Vitruvius also explicitly mentions the procedure already used to control the *dispositio*, or the correct positioning of parts. The *iconographia*, we read in his De Architettura

L. Agustín-Hernández et al. (Eds.): EGA 2020, SSDI 6, pp. 310–321, 2020.
https://doi.org/10.1007/978-3-030-47983-1_28

[4], is the "drawing of the footprints", as the word literally translates: the "plan" of the building. In turn the *orthographia*, the straight drawing as the author of the treaty says, represents the lines that are raised from the plant. Finally, the third way of drawing mentioned is the controversial *scaenographia*, that is the building represented according to what we could assimilate to the perspective, since all lines converge on the fixed rod of the compass (Dürer 1525 in Bessoni 1578).

Referring to all knowledge in the area during the Hellenistic period, Vitruvius vouches the existence of the need of a close link between the operations carried out on the drawing plane and the construction site. The only way to organise the arrangement within a well-structured schema so that it is not possible to add or remove any element without altering the stability, functionality and harmony of the whole, it is, for the ancient engineer of Caesar, a result conceived with the mind and controlled through the graphic models [5]: the *modŭlus* (from module precisely) serves as a measure for each element to be commensurate with the others and altogether with the whole (*symmetria*). To understand adequately how much Vitruvius certifies as current normality at the end of the Republican age, it is necessary to go back to the previous millennia, as this procedure seems to testify two famous papyruses now preserved in the Egyptian Museum of Turin. The oldest is related to the profile view of the sanctuary of Gurob, datable between 1550–1350 BC; the other more recent, a plan and an elevation at 1:28 scale of the tomb of Ramesses IV dating between 1152 and 1145 BC, both made in a technically flawless way. Prior to guaranteeing the geometric framework of the model, albeit embryonic or schematic, are the proportional relationships of the fortress depicted on the so-called Narmer Palette, 31st century AD (Egyptian Museum of Cairo - Identification: Cairo J.E. 14716, C.G. 32169) decoded by the same authors [6].

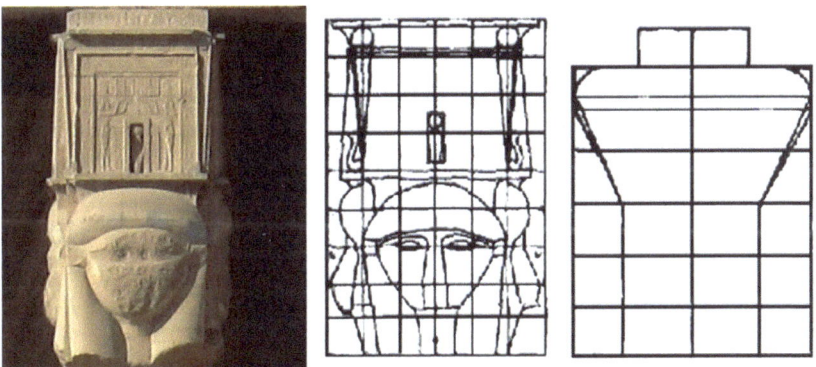

Fig. 1. These drawings were found by the French in the quarries of Gebel Abu Feida in 1789. They are chapiters of pillars, designed for the temple of Denderah which was built by Cleopatra in 70 and 69 BC and they were traced with red ochre in the rocky surface half of its real size. A quite old practice [13].

The most elementary system devised in the past to proportionally reduce the selected elements can be traced back to the use of orthogonal grids [7], the only ones traceable in prehistory due to the crudeness of the conceptual and operational instruments starting from the right angle whose manifestation is it is evident by observing a stone that falls and normally impacts the surface of a pool of water (Fig. 1).

Having recognized the concept of orthogonality as a consequence of the force of gravity, the question shifts to the need to establish when this concept was applied to the design of rectangular spaces. The squared geometries were in fact a real achievement for humanity rather than the roundish huts, difficult to be graphically represented without having adequate goniometric instruments or trilateration procedures. Much preferred started to be orthogonal quadrilateral shapes, whose contours translated into space acquired the characteristics of solids, similar to the cubic crystals of pyrite or sodium chloride, forms not unknown in primitive civilizations [8]. The question concerning the use of technical modes (considered that way because they provided instructions on the construction) concerns, therefore, archaeology: the scientific study of ancient civilizations, through the traces derived from the numerous excavations, provides the parameters of the respective cultures [9]. The answers constitute the lower end of our research, identifying the date, *post quem*, in which the transcription of the drawing into a graphic scale made its debut, analysing on conventional examples (elevations and sections) that have come arrived to us from the past. However, the findings extracted from the ancient world and assumed as a paradigm of the ways of drawing, reveal a chronological divergence.

Anticipating the conclusions of our reflections we can say that not only *ichnographia* (the plan) and *orthographia* (the elevation) anticipate by far the "drawing of the scene", but it appears undeniable that the "right drawing", intended as the main elevation, anticipates the "design of the footprints" by millennia, while the so-called vertical sections, seem to require an abstract capacity developed many centuries later than the drawing of the contours.

The reason for this ascertained temporal distance is certainly one of the interests that this essay proposes: with the instruments proper to the discipline (traditional and innovative), in fact, the authors question the debut of the section, or inner elevation, through a case study: the fresco found in the tomb of Djehutinefer (about 1,410 BC) one of the Tombs of the Nobles located in the so-called Necropolis Tebana, on the western shore of the Nile in front of the city of Luxor in Egypt, catalogued with the repertoire number (abbreviation) TT104 (Theban Tomb 104) [10] (Figs. 2 and 3).

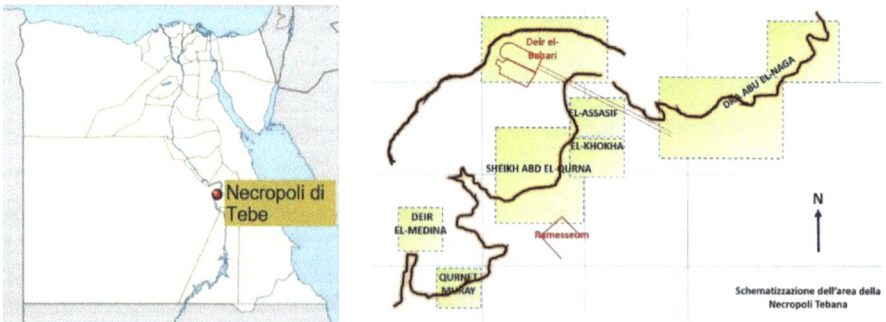

Fig. 2. Location of the necropolis of Thebes in Egypt and schema of the reciprocal positions of the necropolis of the Theban area, in Egypt (on a grid)

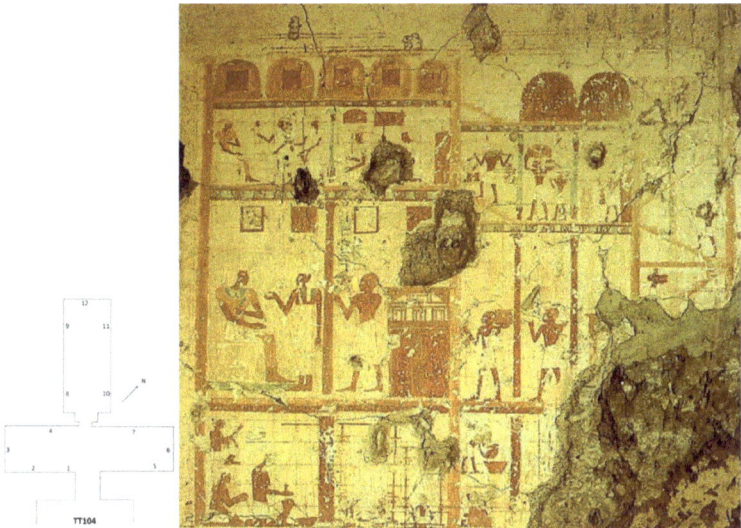

Fig. 3. Schematic planimetry (The numbering of the rooms and walls follows that of Porter and Moss 1927, pp. 217–218) and the fresco of the tomb TT104

2 Hypothesis

The most immediate and direct consequence of humans moving towards the object of interest, thereby progressively cancelling the depth, generates visual images that make possible, by virtue of tactile and perceptive awareness, conceptualizations similar to frontal elevations. In clear relation to the physiological perception and its objectification of perspective, the "right drawing" thus began over 20,000 years ago, perhaps in the darkness of the caves re-lightened by rudimentary torches, perhaps on rocks burned by the sun, where Palaeolithic humans traced the simple outline, initially coinciding with the real visible shapes against the light. This is a fundamental step in a technical acquisition, since it reproduces not only a two-dimensional representation, or an orthogonal projection of the shape, but preserves the correct proportions, thus providing the necessary measures for its reconstruction, even exclusively with a frontal view. Obviously those very remote artists could not even remotely suppose that their representations were 'in scale', since they kept the same factor as the corresponding dimensions of the real object.

The intuitive verisimilitude of the image is therefore a consequence with no relation to the exact reproduction of the relationships between the parts. Responding to a logic that we could define as spontaneous that arose from the imitative capacity, the criterion of proportion at the base of the technical drawing appears far before than the geometry itself. The assumptions of its invariance for over 5 millennia are evidenced by the aforementioned Narmer tablet or Narmer Palette.

The question for the "drawing of the footprints" arises quite differently. The plan or more properly the planimetry is, from a strictly technical point of view, a horizontal section of an artefact. An image that is certainly less frequent in everyday experience, being perceptible only during the first phase of a building, or when it ends up "at ground level", thus making it visible just above the foundation.

A greater capacity for abstraction is required, instead, for the elaboration of a vertical section: only the collapses that have occurred due to natural catastrophes or human action make them perceptible. Therefore, the vertical cutaways, object of our present discourse, constructible by pure conventions, were certainly among the last technical ways to be used. Although often containing more information than the side elevation, the designs of vertical sections are rare. Among the reasons, the operational difficulties do not follow the communicative ones: refined is the mental ability that they require both to bring back the exact proportions on the drawing compared to the real ones, and to recognize them through icons.

3   The Case Study Prodromal Example of Section-Inner Elevation

The drawing in section found in the sepulchre of Djehutinefer (about 1,410 BC), is sophisticated and particularly interesting for our purposes. Its singularity relies in the abundance of iconographic details that allow us to correlate information capable of restoring the measurements and a configurative hypothesis of the plan of the house of a royal scribe, the treasurer of Amenofi, pharaoh of the XVIII dynasty [11], buried in the Theban Necropolis[1].

The building is on three levels above ground. In the foreground, the thicknesses of the sectioned walls divide the overlapping images into bands, as required by the Egyptian conventions. The series that distinguishes and frames the image from the cultural point of view describes the domestic activities evoking the uses of the environments. In the semi-underground, the drawing of the warehouses suggests a restricted space for the servants' quarters who perhaps rested in beds placed in the commercial premises on the ground floor: from right to left there is a succession of spinners and weavers but also maize grinders and sieves of flour. It is assumed that the lost part should contain the entrance and access to the staircases that were mounted on the terrace, on which five granaries and two ovens can be distinguished: right there the doers prepared the lunch that served on the lower floors.

Greater in height than the others, the floor intended for the reception of guests, shows architectural details that reveal more sumptuous environments. The mobile symbol identifies the main room: the scribe sits on an armchair raised on a platform while the servants offer him food and flowers. In this room, just below the ceiling, four small light compartments ventilate the room rather than illuminate it. A double door leads into a lighted passage, then to the stairs to the third floor, to be considered a sort of office, according to the icon that represents the seated scribe always seated on his ventilated and revered table by his servants.

At the last level, the terrace-kitchen seems shaded towards the interior thanks to the volumes mentioned above (the characteristic granaries and ovens), whose round

---

[1] The first numeration of the tombs was from number 1 to 253. They were numbered in order of discovery and not geographically. The tombs from 253 on are also numbered in chronological order.

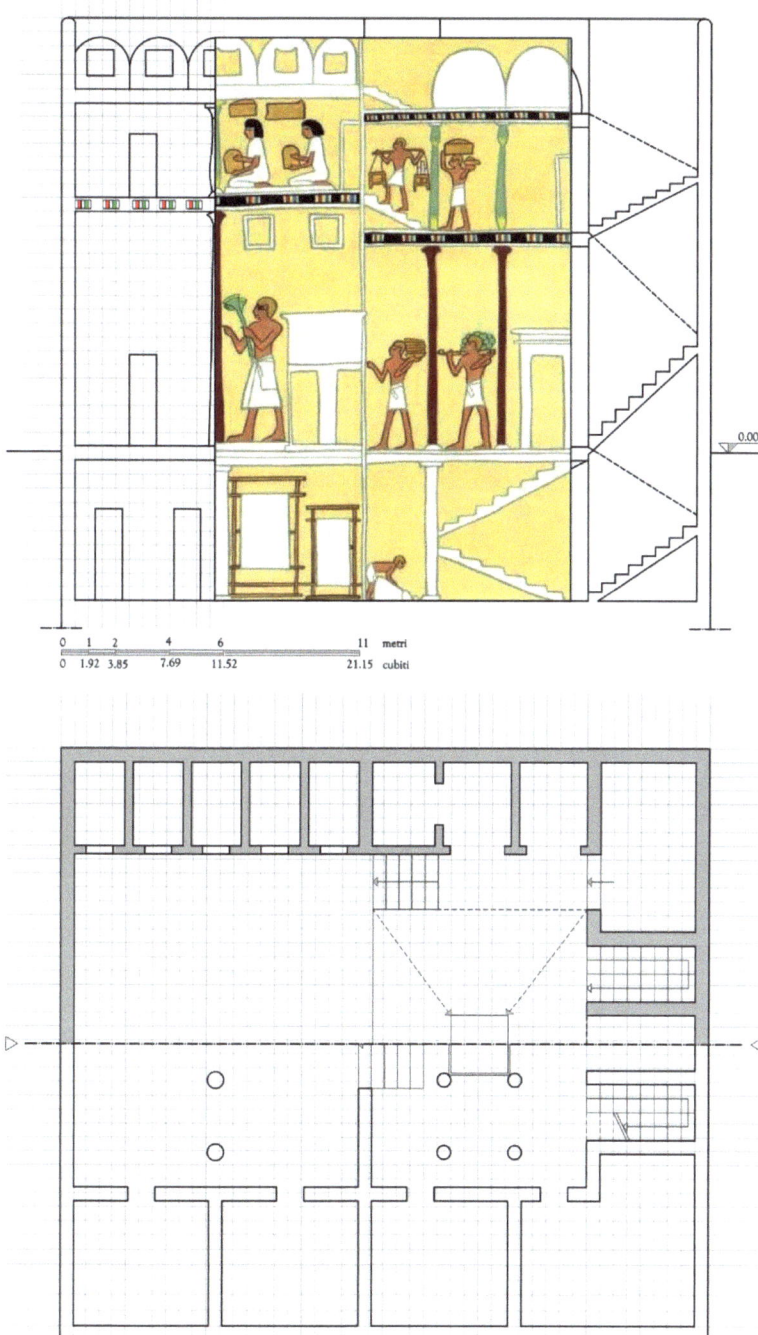

Fig. 4. The module measure and the modulation of the section-inner elevation in correspondence with the hypothesis of the plan.

covers collected the precious rainwater in capable tanks, as there is still nothing like the aqueducts.

In the absence of the plan, the studies of the archaeologist Jean-Claude Margueron (1993) which describe the conspicuous building renovation program that invested from the Old to the New Kingdom, even the private residences, lead the reconstruction of a configurable hypothesis of interior space. From them derive the characteristics that will identify the multi-storey buildings, houses whose owners were of high social rank and of considerable wealth (Oppenheim 1980).

4   Virtualization of the Graphic Model

The thickness of the sectioned walls is evident, and it serves as a datum used to organize the display of the plan in a biunivocal correspondence. The necessary proportional ratios have been derived from the size of the granary deposits located on the terrace: there are still similar ones in North Africa today, for example the grains of Medenine near Tunis (Fig. 5).

Fig. 5. Medenine near Tunis

The measurements in meters guided the choice of the module which, in cubits, makes the organization of the spaces proportional (Figs. 4 and 6); multiples and anthropometric submultiples make it easy to dimension surfaces and volumes.

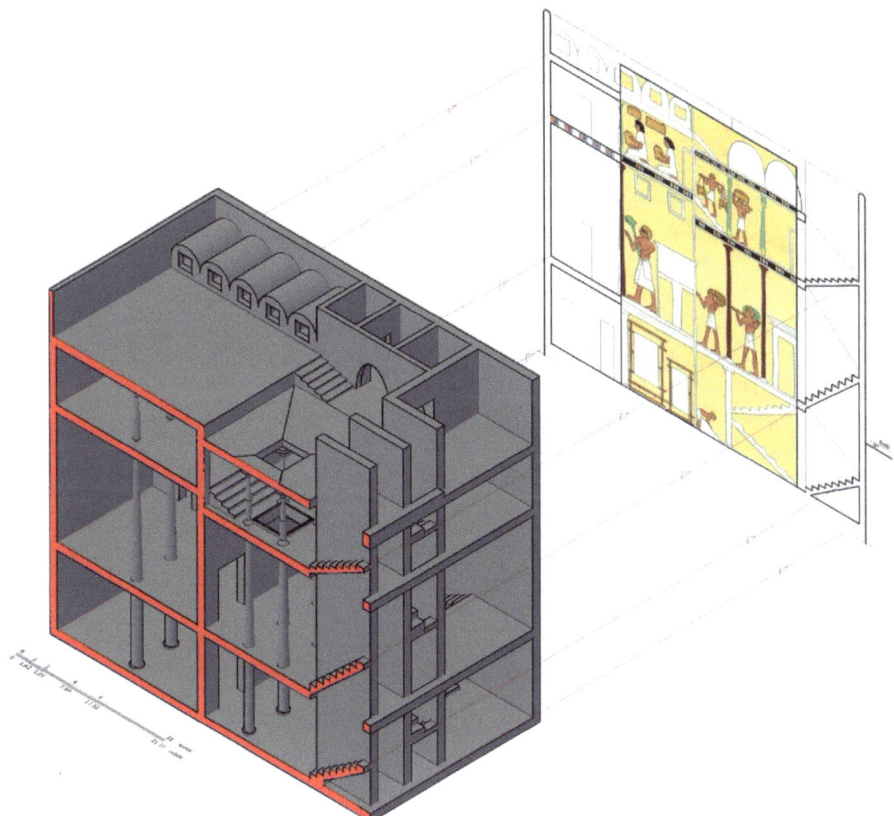

Fig. 6. Axonometric view in correspondence with the modulation and measures deduced by the fresco (drawing by the authors).

Since it is a load-bearing masonry construction, it must be imagined that the three levels follow the articulation of the ground floor possibly lightened upwards with cadenced openings so as to have more practicable spaces. The floors of all the rooms, as clearly visible in the painting, are supported by pillars, with three different types of chapiters: unadorned in commercial rooms and in servants' quarters, more elaborate on the noble floor, simplified on the upper floor. In this regard it should first be noted that the maximum light that could be covered by beams taken from local tree species could not exceed the distance of 3–4 m thus justifying the recursion of columns and chapiters.

The spaces, therefore, could only be narrow and long, a usual characteristic of archaic Mesopotamian architecture being, as aforesaid, a choice imposed by the weakness of the wood used which, among other things, also required the adoption of props diagonals, which further reduced the light Fig. 6.

With deductive logic, maximum estimations are obtained and confirm what is already present in the literature: excluding the stairs and all the additional rooms not represented, the surface area per level must have been about 90–100 m^2. The house of Djehutinefer was therefore not enormous when compared to the sumptuous houses of Hotep Senuseret which covered about 2000 m^2 but offered a comfortable and rational articulation of the spaces, functional for the activities of domestic life and to the representativeness of its social power (Fig. 7).

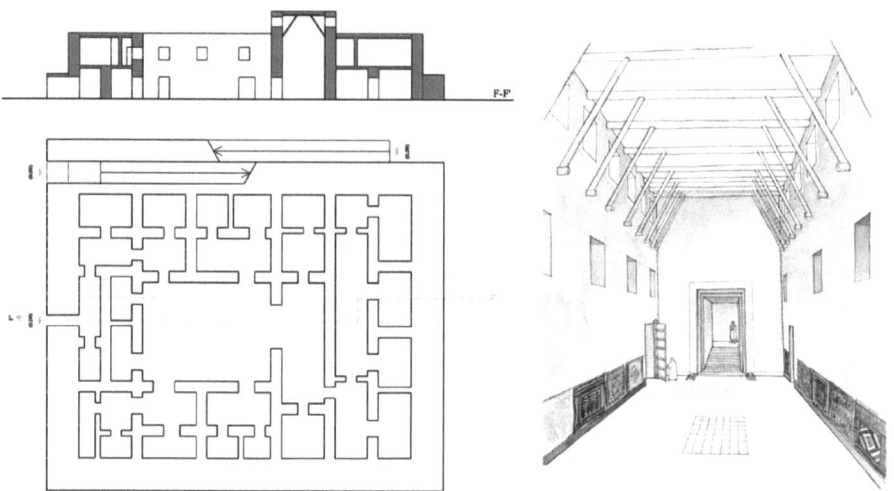

Fig. 7. Left, plan and section of Nur Adad Palace (drawing by the authors). Right, inner perspective (Margueron J.C. 1993).

The literature insists on the symbolic functions of the throne room and the typological metamorphoses derived from the variation of rituals and myths: from the memory of the "house of the god" or the temple accessible from the street, the room in which the Pharaoh, having become the god on earth, is transported to the upper floor by modifying the typological model of the royal palace that inspired the homes of his closest collaborators.

Focusing on the typological analysis of the model traditionally identified by the succession of the internal hall-courtyard-service space and throne room (Margueron 1993), the main environment in which the treasurer welcomed the guests on the upper floor, occupies a large portion of the surface of the noble floor. Repertoire images return environments controlled by specialists able therefore to give a cultural depth to graphic virtualization. The analytical reflections and the ascertained dimensions, if correlated to the constructive techniques of the epoch and to the materials in use then, justify in the methodological plan the reconstruction that the advanced techniques of representation have made virtually verifiable for communication and documentary purposes (Fig. 8).

Fig. 8. Reconstruction model of Nur Adad Palace. View of the Throne Room and the previous courtyard.

5 Conclusion

The technical drawing has the purpose of transferring building/construction information, so it is undeniable that it cannot be considered a target itself, as painting is, but a means to translate the idea into hypothesis of constructive feasibility or, vice versa, to abstract from the already built and extract ideas of new relations or conformative proposals. The present analysis has therefore looked at the exquisitely human ability to free the material from the sensitive connotations to describe and compare the categories of the building discipline. Based on a rigorous analysis of measurements and shapes, the investigation of the Theban fresco has focused on identifying the invariants that, in their articulation, have regulated the historical development of architectural forms as well as the communicative and documentary modes. The food for thought about the recognition paths of ideas provides some clarification to the questions set above and to some generalizations.

Belonging to the category of technical drawings finds satisfaction or in the sepulchral representation given the adherence to the scalar relationships that link the portrayed and the real structure. Furthermore, the way of evoking the verisimilitude of the life that unfolded in the rooms of the Djehutinefer palace induces, motivated, to consider the drawing a sort of forerunner section-perspective. In fact, cancelling the depth of the spaces according to a scaled reduction ratio, a procedure intuitively adopted, betrays in that age a discrete diffusion of the way of communicating. Therefore, shifting the attention from the object of imitation to the purpose of representation, it is well evident, in the case analysed, that the source element relies precisely in the divulgation. Which brings the wall fresco closer to a kind of survey drawing (Fig. 9).

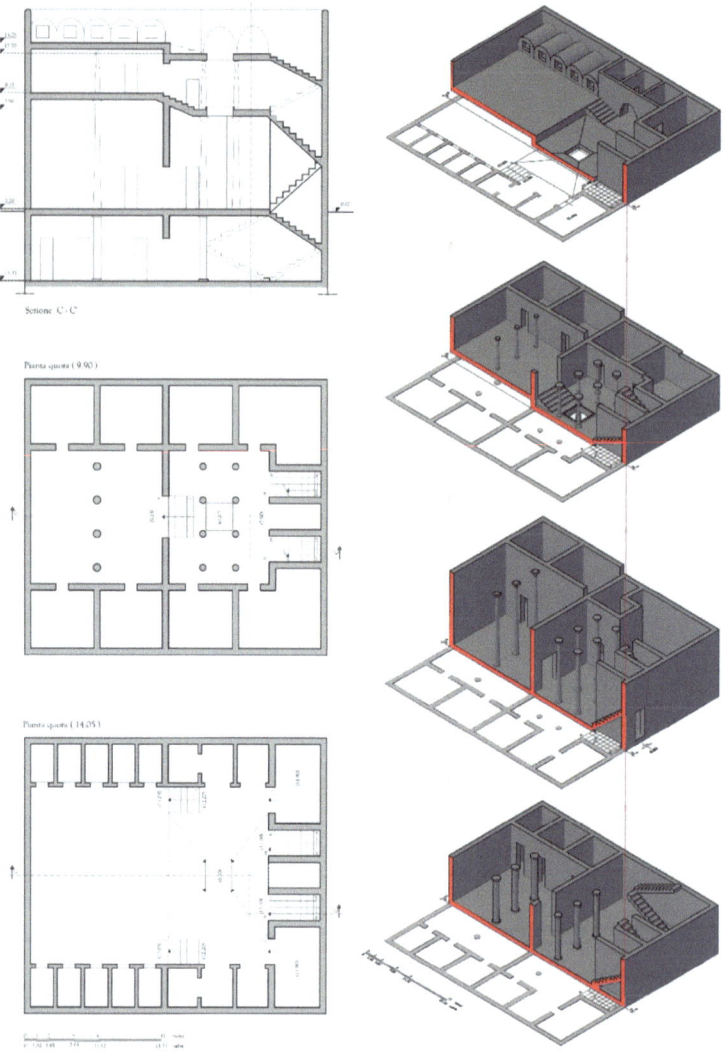

Fig. 9. Reconstruction in plan, sections and axonometric view thanks to the modulation stablished in the study of the fresco painting.

From this point of view the link found between proportional relationships and the way to reveal the architectural reference elements, confirms the reiterated conclusion, addressing the identification of universal ideas now transcribed in institutionalized conventions that are renegotiated over time, since even the technical language changes evolving with the progress of the sciences. Therefore, even the rediscovery of drawings similar to a sort of longitudinal and transverse cut of the artefact makes us to backdate not only the origin of the *iconographia* and *orthographia* but also of the vertical section. By reducing the distance between past and present, graphic virtualization makes what is supposed and demonstrated through the study of available sources.

References

1. Monge, G.: Géometrie Descriptive - Leçons données aux écoles normales l'an 3 de la Republique. J.M. Hachette, Paris (1796)
2. Docci, M., Migliari, R.: Scienza della rappresentazione, Fondamenti e applicazioni della geometria descrittiva. La Nuova Italia Scientifica, Roma (1992)
3. de Rubertis, R.: Il disegno dell'architettura, La Nuova Italia Scientifica, Roma (1994)
4. Vitruvio Pollio, M.: De Architectura, I, 2, 1: Species dispositionis, quae grecae dicuntur ideae, sunt hae: ichnographia, orthographia, scaenographia (sec. I)
5. Vitruvio Pollio, M.: De Architectura (sec. I)
6. Rossi, A.: The origin of technical drawing in the narmer palette. Nexus Netw. J. **19**(1), 27–43 (2017)
7. Argan, G.C.: Progetto e destino. Il Saggiatore, Milano (1965)
8. Vitruvio Pollio, M.: De Architectura, book: II l'origine dell'architettura (sec. I)
9. Høyrup, J.: Lengths, Widths, Surfaces: A Portrait of Old Babylonian Algebra and Its Kin, Studies and Sources in the History of Mathematics and Physical sciences, Berlin & London (2002)
10. Friberg, J.: A Remarkable Collection of Babylonian Mathematical Texts, Sources and Studies in the History of Mathematics and Physical Sciences, Berlin (2007)
11. Gardiner, A., Weigall, A.: Topographical Catalogue of the Private Tombs of Thebes. Bernard Quaritch, London (1913)
12. Franco, C.: Dizionario delle dinastie faraoniche. Bompiani, Milano (2003)
13. Gardiner, A.: La civiltà egizia. Oxford University Press, Torino (1997)

The Representation Through the Geometric Configuration of the Liparoti Village in the Royal Palace of Caserta

Enrico Mirra[✉]

Università degli Studi della Campania "Luigi Vanvitelli", Aversa, Italy
enrico.mirra@unicampania.it

Abstract. The contribution investigates the iconographic apparatus of the Liparoti Village, a former residential facility of the majestic monumental complex of the Royal Palace of Caserta and focuses on the theoretical and applicative connections of the process of understanding the data acquired for the understanding of geometry and its heuristic function. Through the drawing a new graphic reading is returned, with a geometric-configurative analysis of the system which allows the observer a new perception of spatial conformation. Furthermore, the considerable amount of data acquired through the survey allows us to formulate and verify different interpretative hypotheses on the genesis of the form according to a conceptual abstraction process that sees geometry as the tool to decipher the complexity of reality, rediscovering the design intentions and thus reconstructing the ideal model corresponding to the conceived shape.

Keywords: Royal Palace · Liparoti · Geometric analysis · Drawing

1 Introduction

The research work offers an important contribution to the knowledge of the majestic monumental complex designed and created by Luigi Vanvitelli in Caserta, constituting a precious tool for comparing architectural design and geometric analysis. For the representation work it was necessary to study the system of paths outside the Royal Palace, which is of fundamental importance for the functioning of the entire architectural complex. For the description of the vegetation system it is useful to understand the geometries present in the entrance of the Palace, when the green as far as the eye can see is framed between the vaults of the central gallery and the courtyards: an axial symmetry strongly desired by Vanvitelli, conceived for a - development of several kilometers. This naturalistic plant is a typical example of Italian garden, built by vast meadows, square-shaped flowerbeds and, above all, by a triumph of water games that flow from the numerous fountains. The vegetational architectural system was conceived by Luigi Vanvitelli and directed in the execution of the works by his son, Carlo with a large group of sculptors and stonemason's intent on adorning the statues and the fountains, setting them in the green geometry of the meadows and ponds. water, rows of trees and oak groves. The

L. Agustín-Hernández et al. (Eds.): EGA 2020, SSDI 6, pp. 322–332, 2020.
https://doi.org/10.1007/978-3-030-47983-1_29

present research studies the portion of Park to the west of the central axis, the oldest area called Bosco Vecchio, within which the extraordinary naturalistic and architectural context named Villaggio Liparoti arises (Fig. 1).

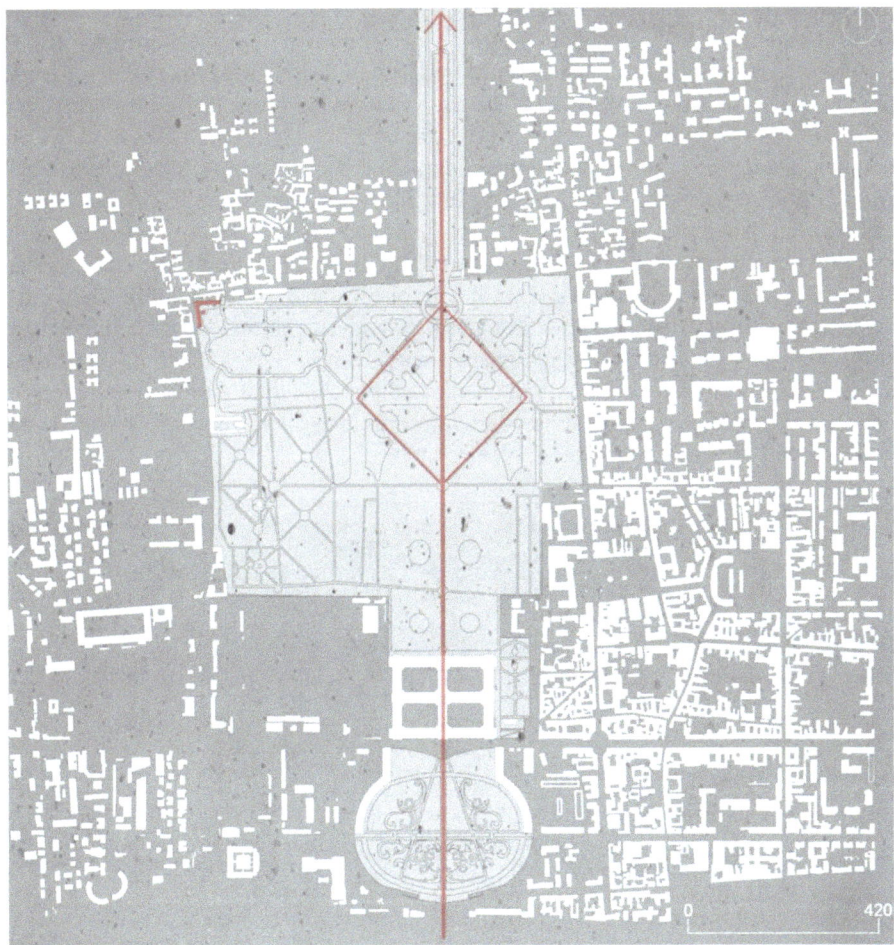

Fig. 1. Geometric configuration system with the planimetric identification of the Liparoti Village (in red) in the Royal site of the Royal Palace of Caserta.

Vanvitelli himself thought of inserting a vegetational context, within his project, maintaining unchanged alignments and asymmetries, in the new rigorously orthogonal plant.

2   The Geometric Representation of the Liparoti Plant

During the absence of Vanvitelli in Caserta, the figure of an architect stands out, who was able to give new expressive unity, in neoclassical style, to the legacy of the Master, the architect Francesco Collecini. He designed lodgings for the Sicilian sailors coming from the island of Lipari, in the archipelago of the Aeolian islands, called to act as extras for the war shows requested by King Ferdinand IV inside the Peschiera Grande. The Liparoti Village was built specifically to house a large number of sailors, coming from the island of Lipari, called to satisfy the King's needs through simulations of naval battles within the adjacent stretch of water. The design of the Liparoti is characterized by a strictly geometric compositional order and finds reason in the cultural formation of the Col-lecini through the Grand Master Vanvitelli. The main facades are characterized by a rhythmic scan of windows in which the ordering principle consists of an orthogonal mesh, whose module is a circumference inscribed in a square. In the front space, which is determined by the L-shaped geometry of the building, an additional square-shaped artifact is placed, serving as an island for the first two rectangular elements. All the buildings have only one level above ground and inclined pitched roofs. In detail, the slats are characterized by a double flap, while the angular body and the one isolated from flaps of various kinds; the construction type is in load-bearing masonry, with the internal walls having a mere partition function, they did not perform any static function. The rhythmicity represents, that possibility of arranging within the parts of the architecture one or more recurrent elements, with more or less coherent dimensions among them, concur to articulate the structural and linguistic times of an architecture of the last century (Figs. 2 and 3).

Fig. 2. The Royal Palace of Caserta, the Liparoti Village, view of the current state.

The rhythmicity of the structures connects to the lines by means of the interaxes of the figures or of those elements that beat the time of an architectural composition. As is known, in classical antiquity the rhythm of architecture was concretized in the joke

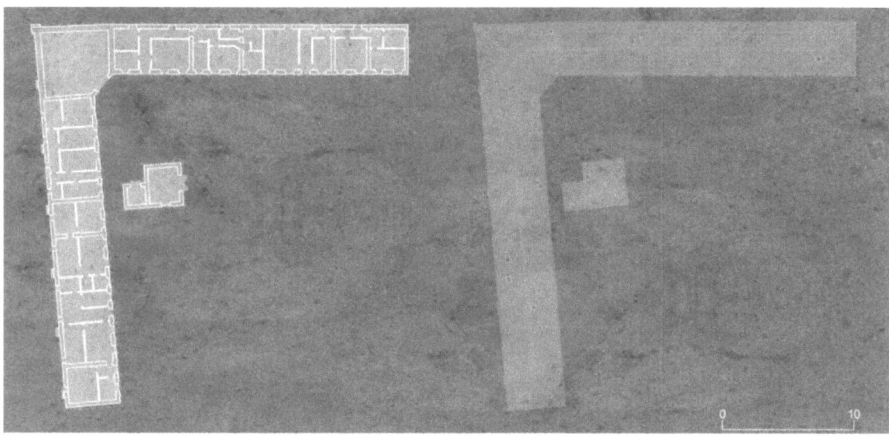

Fig. 3. The Royal Palace of Caserta, the Liparoti Village, ground floor plan (left) and roof plan (right).

of the colonnade of the Greek temple: the interaxis of the columns, for example, was provided by a measurement module represented by the diameter of the same columns that structured the four sides of the classical architecture archetype. The rhythmicity should not be read or detected only in the presence of full elements such as they are, but also through the presence of the voids and openings of the architecture, be they entrance halls or windows, porticoes or arch. Furthermore, this rhythm is not only expressed from right to left and vice versa but also from the bottom upwards of an architecture. This last rhythmic succession, in most cases of architecture, is realized through the use of bases, stringcourses, cornices or cornices and crowns: elements that underline with their presence the existence of another rhythmic order not of linear character but superficial. The vertical rhythmic scans are dominated by rectangles which, as we know from Euclidean geometry, can be golden or dynamic depending on whether the dimension of the longer side is provided by the measurement of the diagonal or by the length of the diagonal of the upper semi-portion of a figure square assumed as a basic generator module. All of these options, which refer to the architecture or to the way the elements and figures are articulated within the volumetric mass of the building. The basic morphology for architectural language, in classical antiquity, was proposed as a universal discipline through which to design and decode architecture (Fig. 4).

The rhythm of architectural elements, intended as a category capable of contributing to the attribution of beauty, is present not only in the architecture of the past, but also in many architectural examples of contemporary design culture. A culture that assimilates or changes the teachings inherited from the past through in-depth studies or distancing from the different modes of development and sequence of rhythmic accents identifiable both in typological systems (the design of plants) and in linguistic aspects (the drawing of elevations) of an architecture. The rhythmicity of classical architecture has always been based on the repetition, in varying shapes and sizes, of certain elements which, taken together, are combined according to regular and symmetrical patterns: the layout of the load-bearing structures in plan or 'alternation of full and empty spaces on the

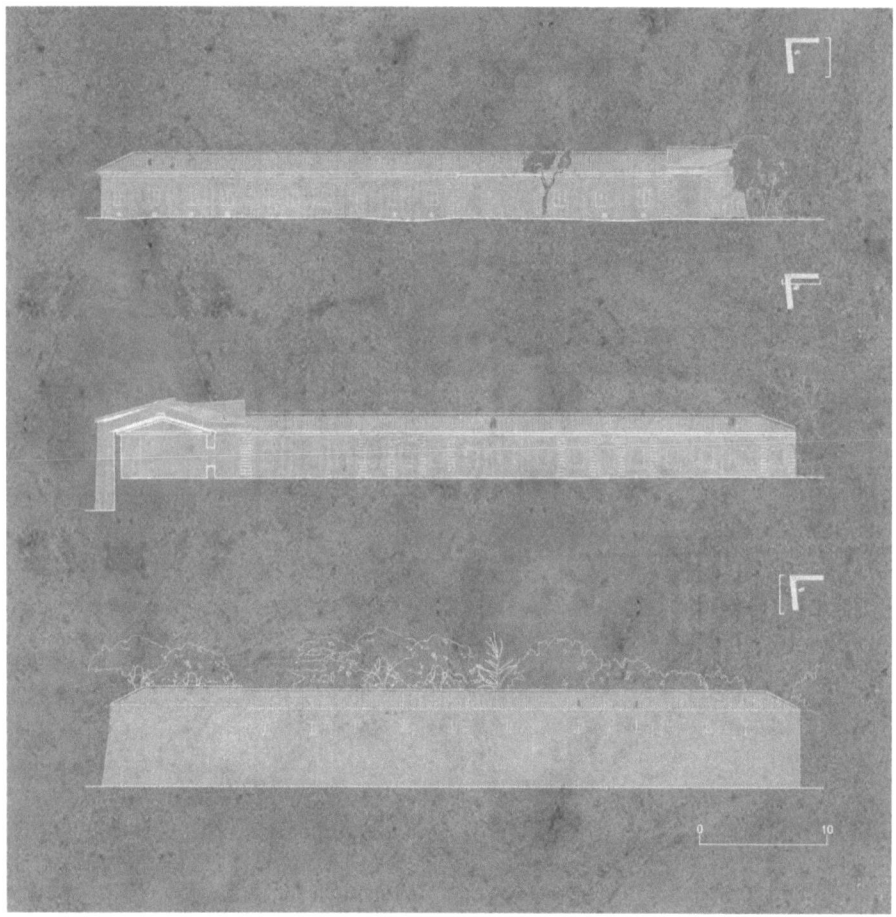

Fig. 4. The Royal Palace of Caserta, the Liparoti Village, graphic restitution of the elevations of the building along the longitudinal and transverse directions.

elevation; the planimetric concatenation of the interior spaces in their differentiation between served spaces and serving spaces or the superimposition of the structural elements in relief with the background decorations in the design of the facades represent the primary canvases on which the rhythmic texture of the architecture is interwoven through elements such as pillars, columns, masonry walls, windows, pilasters, pilasters, cornices and string courses. Contemporary rhythmicity proposes, on the other hand, differentiated type-morphological and linguistic solutions in which the concepts of cadability and temporality, strictly connected to measurability, are clearly legible in an architectural order attributable to those regular codes based on the repetitiveness of the classical or on the other hand, choices from which transpire, according to a spectacularizing emphasis, the complete dissonance with respect to those principles consolidated through the staging of irregular and asymmetrical scores: on the one hand, once again, the rational architecture connected to the legacy of the past and, on the other hand, architecture understood as a work of art, even transgressive, willing to accept in its form, in its typology and in its language the deconstructed, heterogeneous and fragmentary characters expressed by the globalized society. Not only. If the rhythm consists of a succession of accents and the accent represents the greatest importance that some elements assume with respect to others in the general composition of the architectural work then it is possible to identify not only more or less accented architectures but also buildings non-accented: this is the case, for example, of those architectures of silence, so-called minimalist, which, eliminating the secondary elements from the clear white surfaces of their facades, cancel any reference to rhythmicity. A real presence of absence capable of evoking an abstract symbolism, not connected to the material rhythm of architectural elements, defined by the essential configurative nature of a pure architectural form, most often provided with an interior universal rhythm (Fig. 5).

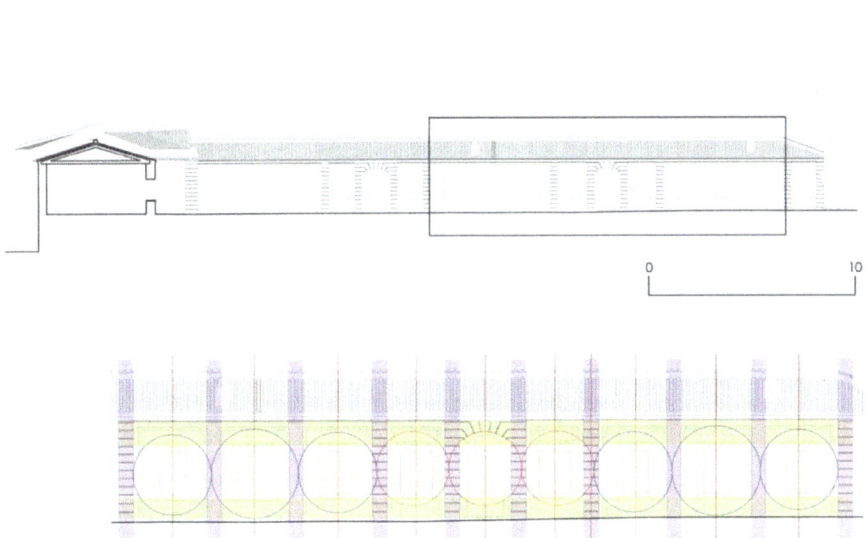

Fig. 5. The Royal Palace of Caserta, the Liparoti Village, geometric configurative analysis related to a particular of the prospectus.

3   The Survey of Villaggio Liparoti

Of considerable interest for research was the surveying activity that allowed to highlight a forgotten place that was never studied within that extraordinary Park which is the Old Wood. The overall shape of the planimetric plant is presented as two rectangular building slats placed perpendicularly, L-shaped, at the sides of which are the entrance doors, connected together, by a building hinge: a large trapezoidal-shaped room. In the front space that is de-terminated from the geometry of the building, another square artifact is placed, isolated from the first two. All the buildings have only one level above ground and inclined pitched roofs (Fig. 6).

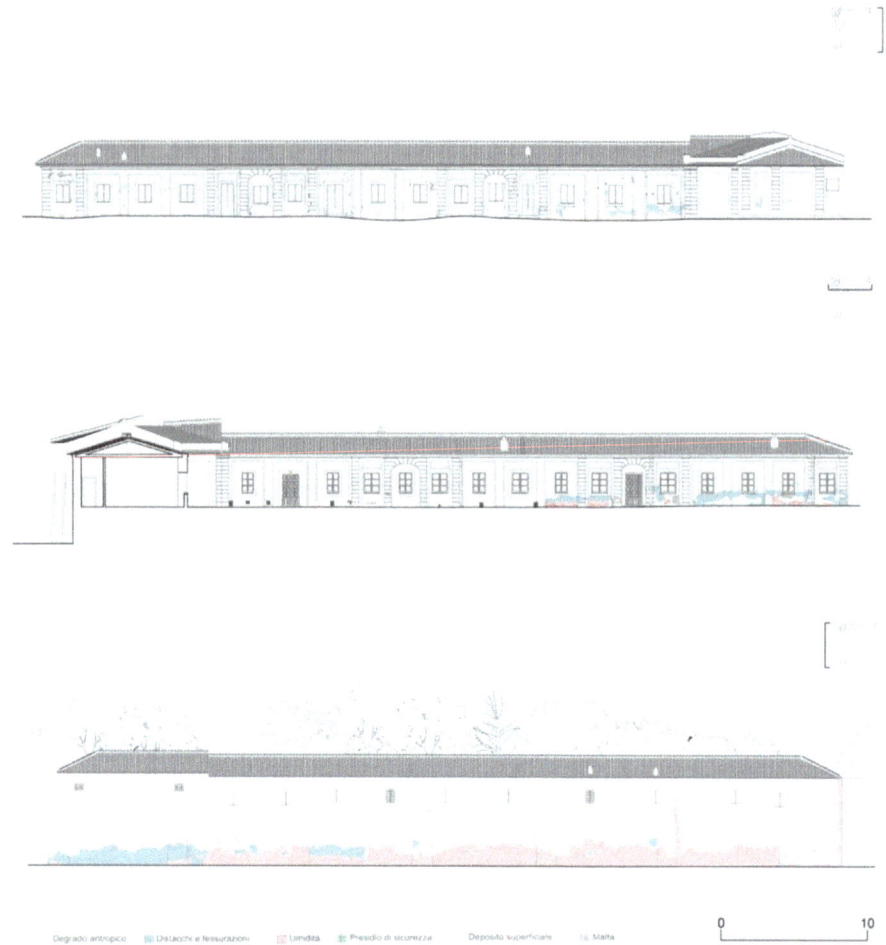

Fig. 6. The Royal Palace of Caserta, the Liparoti Village, graphic restitution of the picture of the degradation relative to the facades that make up the building.

In detail, the ribs are characterized by a double flap, while the angular body and the one isolated by several layers; the constructive typology is in load-bearing masonry with the internal walls with mere partitioning function, not fulfilling any status function. The center of the complex is instead intended for parking. In the survey campaigns, numerous traditional and innovative techniques have been implemented which are clearly recognizable in the design disciplines in order to illustrate the complexities of architectural and environmental problems that affect the protection of urban and extra-urban territories. An action, the latter, deriving from a process of knowledge, of the material and immaterial characteristics of the places in question. The knowledge of the natural and built environment can be explained only by a survey of reality capable of bringing out its evident qualities, but also the limits of a natural or man-made landscape. The graphs obtained are, from the survey of architecture and vegetation, the basis for the documentation, study and conservation operations of the building. This graphic instrumentation, using the conventional representative techniques, makes explicit the materialization of the idea, its understanding, and communication by reducing, through the relief, the three-dimensional space of the architecture in two-dimensional form (Fig. 7).

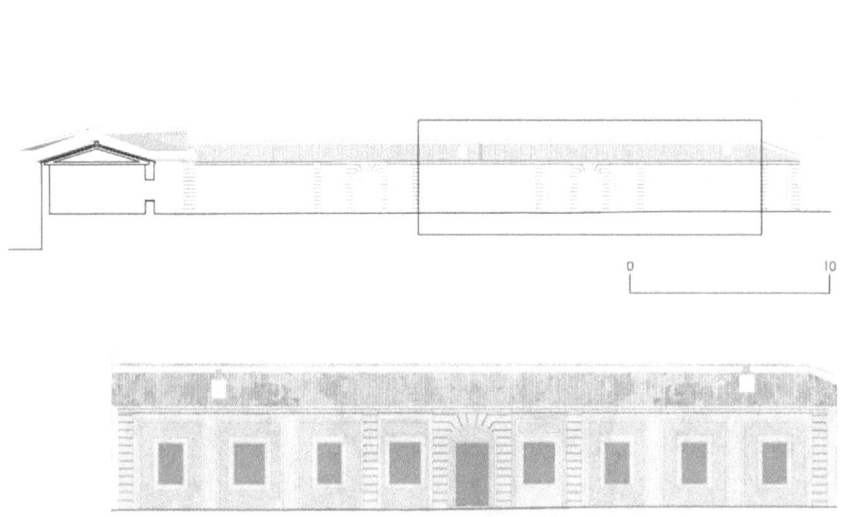

Fig. 7. The Royal Palace of Caserta, the Liparoti Village, chromatic analysis related to a detail of the front.

Considering reality means exploring, within those complex contemporary urban agglomerations, those places forgotten and abandoned by civic, social and collective use: urban places characterized both by the presence of heterogeneous, fragmented and disjointed building fabric and extraordinary architectural realities that have fallen into disuse from a functional point of view but not from a cultural one. Architectural realities rendered unusable by economic mechanisms that, by not attributing patrimonial value to the cultural aspect of the critical requalification of the testimonies of the past, correspond to the cancellation of substantial cultural, environmental landscape and industrial assets. Exploring the Mediterranean cities in search of architecture, urban areas, landscape contexts basically mean knowing, analyzing and passing on to others the result of this exploratory survey. A survey that, to be as broad as possible in terms of space and time, uses representation to document not only the reality that presents itself to the specific contemporary reading of the artifact or the context object of the investigation - through the survey - but also of its past history - through drawing. To extrapolate graphically from the urban mosaics of the Mediterranean cities those tiles or those abandoned, forgotten and degraded building inlays means to contribute to their potential revaluation and definitive valorization. The representation, in this perspective, is the cultural instrument most suited to making different and distant times interact, those of history with those of human destiny.

4   Methodological Notes of Applied Research

The disciplines of surveying, over the centuries, have refined measurement techniques by adapting them to situations and technological innovation: from the simplest methodologies, from direct surveying with elementary instruments, to those derived from descriptive geometry and finally to those today allowed by the electronic and computerized technologies of modern geomatics. In recent decades, the sudden development of technologies dedicated to the field of architectural heritage has brought new theories and new expectations that lead to a review of the methodological-logical processes. The second part of the research lays the foundation on the formulation of a proposed operational intervention protocol, which aims to analyze, deepen and investigate the morphological, metric, material, construction and structural aspects of the architectural complex, with particular reference to the intercurrent spatial relations with the neighboring urban context. The walls, despite being presented as geographically assonant architectural portions and still perfectly recognizable, have been affected over time by continuous transformations, consolidations and demolitions that have partially hidden the ancient structure undermining its historical and urban identity (Fig. 8).

Fig. 8. In Reggia di Caserta, the Village of the Liparoti, photopiano of the southern front of the building.

The critical reading of the fronts required a preliminary survey and an integration of methods to overcome the considerable logistical difficulties linked to the complex articulation of the spaces (presence of unsafe structures leaning against the walls, lack of depth of field, inaccessibility of the spaces in front. For the realization of graphic works finalized to the description of the wall texture, the photogrammetry technique was used which provided a documentation of raster metric images for rectification of sockets, integrated with topographic inter-section forward methods and irradiations for the survey of support points useful for the realization of the photopiano (Fig. 9).

Fig. 9. The Royal Palace of Caserta, the Village of the Liparoti, volumetric model of the building.

5   Conclusion

The activity of surveying and graphic representation of the Liparoti Village, as well as of the entire surrounding architectural and vegetation complex, allowed to bring to light a marginal place in the Vanvitellian complex of Caserta. This reflection, based on the disciplinary foundations of the Design aims at the specific protection and enhancement of the single heritage, but also that of the relative vegetational contexts of belonging. Three-dimensional modeling allowed us to gain a better awareness of the spatial configuration of the village and its elements. As a result, the subsequent visualization phase allowed to build more virtual views of the Liparoti and to place the same within the city, giving it a visual impact on the urban scale. The excursus, defined, defines the guidelines of a broader research path, proposing the cognitive analysis of the places. The use of semi-automatic and automatic tools defines not only the essential geometries but also the forms of the particular obscure odds in the manual survey campaigns. Placing the Royal site at the center of a reflection having as its objective not only the specific protection and enhancement as well as the knowledge of the relative contexts of belonging, it means transposing the relative architectural value from the inside of the urban perimeters outside their enclosures in order to reverberate it even in areas of belonging for the use of the users, whether they are residents or tourists.

References

1. Bertocci, S., Bini, M.: Manuale di rilievo architettonico e urbano. Città Studi Edizioni, Milano (2012)
2. Cundari, C.: Il rilievo per la conservazione, in "Il Complesso di Monteoliveto a Napoli. Analisi, Rilievi, Documenti, Informatizzazione degli archivi", Gangemi Editore (1999)
3. Cundari, C.: Fotogrammetria architettonica. Kappa, Roma (1983)
4. Cundari, C.: Teoria della rappresentazione dello spazio architettonico: applicazioni di geometria descrittiva. Edizioni Kappa, Roma (1983)
5. Cundari, C.: Il Palazzo Reale di Caserta: immagini e rilievi. Kappa, Roma (2004)
6. Docci, M., Gaiani, M., Maestri, D.: Scienza del Disegno. Città Studi Edizioni, Novara (2011)
7. Giordano, P.: Napoli, Guida Architettonica Moderna. Officina Edizioni, Roma (1994)
8. Giordano, P.: L'Albergo dei Poveri. Il Cimitero delle 366 Fosse, i Granili. Edizioni del Grifo, Lecce (1997)
9. Giordano, P.: Il Disegno dell'Architettura Funebre. Napoli_Poggio Reale, il Cimitero delle 366 fosse, il Sepolcreto dei Colerici. Alinea Editrice, Firenze (2006)
10. Giordano, P.: L'Albergo dei Poveri a Napoli. La scuola di Pitagora editrice, Napoli (2014)

11. Zerlenga, O.: La forma ovata in architettura. Rappresentazione geométrica. CUEN, Napoli (1997)
12. Author, F.: Article title. Journal **2**(5), 99–110 (2016)
13. Author, F., Author, S.: Title of a proceedings paper. In: Editor, F., Editor, S. (eds.) CONFERENCE 2016. LNCS, vol. 9999, pp. 1–13. Springer, Heidelberg (2016)
14. Author, F., Author, S., Author, T.: Book Title, 2nd edn. Publisher, Location (1999)
15. Author, F.: Contribution title. In: 9th International Proceedings on Proceedings, pp. 1–2. Publisher, Location (2010)
16. LNCS. http://www.springer.com/lncs. Accessed 21 Nov 2016

The Survey of the Architectural and Vegetational Heritage in the Royal Park of Tirana in Albania

Luigi Corniello[1](✉) and Lorenzo Giordano[2]

[1] Department of Architecture and Industrial Design, University of Campania "Luigi Vanvitelli",
Aversa, CE, Italy
luigi.corniello@unicampania.it
[2] Department of Architecture, University of Naples "Federico II", Naples, Italy

Abstract. The Royal Park, located on the hill of Mulleti, covers an area of 74
hectares and includes six architectures such as the Odeon, the Royal Palace, the
Chapel, the greenhouse complex, the generator house and the gardener's house, as
well as four green areas such as the avenue of cypresses, the avenue of oleanders,
the garden with the belvedere and the lake. The first hypothesis of the Royal
Garden, presumably commissioned by King Zog, was conceived by the architect
Florestano di Fausto (1890–1965), later modified by the engineer Giulio Berté
and the architect Gherardo Bosio (1903–1941) and finally realized by the architect
Ferdinando Poggi (1902–1986). The research included a conspicuous analysis of
the archival documentation - such as drawings, photographs, sketches and notes
- related to the examination of the project and executive drawings of engineer
Giulio Berté carried out in the Arkivi Qëndror Teknik i Ndërtimit (Technical
Archive of Construction) in Tirana. The study also covered the project drawings
of Architect Bosio including a representation of the Villa Reale in the centre of
a perspective with an amphitheatre at the base, demonstrating the emphasis on
the representative role proposed for the residence by the designer himself. There
are also numerous interventions by landscape architects and artists who have left
ample documentary evidence of the interventions carried out and proposed such
as Pietro Porcinai (1910–1986) and Antonio Maraini (1886–1963).

Keywords: Survey · Landscape · Albania

1 Introduction

The Royal Park is represented in three preserved plans respectively, the first dating
back to 1939 and the second from 1941 at the Arkivi Qëndror Teknik i Ndër-timit in
Tirana, the third from 1942 at the Gherardo Bosio Heirs Archive. In the first document is
represented the Park according to the hypothesis of 1939 where it is possible to find the
Royal Villa with its access avenue, the belvedere, some service rooms and, in the lower
part of the plan, the racecourse. The structure of the sports facility occupies most of the
planimetry, with particular attention paid to horse racing, i.e. to recreational activities.
Of great interest is the arrangement of the spaces in front of the Villa where an elliptical

© The Editor(s) (if applicable) and The Author(s), under exclusive license
to Springer Nature Switzerland AG 2020
L. Agustín-Hernández et al. (Eds.): EGA 2020, SSDI 6, pp. 333–342, 2020.
https://doi.org/10.1007/978-3-030-47983-1_30

parterre delimits the scenic and representative green portion of the one-way driveway. The spaces currently occupied by the Chapel are, instead, designated to serve the Villa itself with a private garden and a labyrinth, in the style of eighteenth-century European royal gardens. The second document defines the entire vegetation system in its almost definitive configuration: the thick foliage of the trees is represented with the definition of a first hypothesis of an Italian and English garden as well as the identification of the Villa alone. The service buildings are not designed but flat spaces are identified for their subsequent location. The altitude jumps and the pedestrian and driveways suitable to enjoy the entire Park have been de-ended: the avenue of cypresses pours into a pedestrian path that leads to the Italian garden, and the next flower garden (Fig. 1).

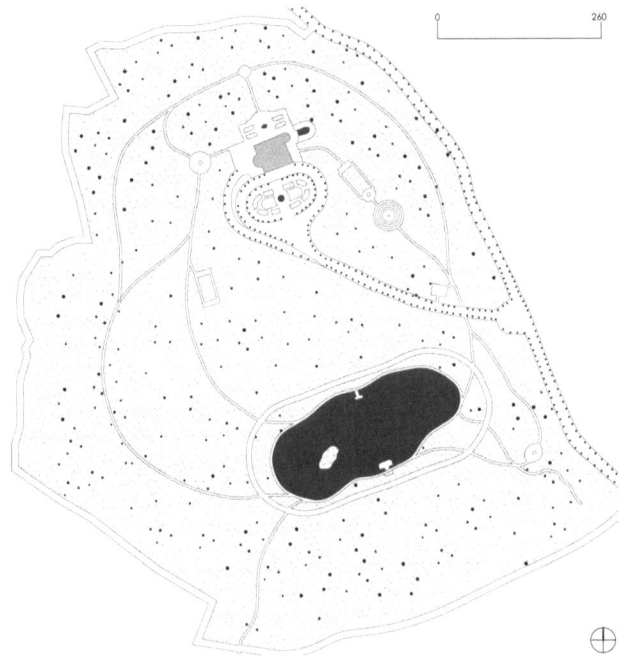

Fig. 1. The Royal Park of Tirana. Graphic reconstruction of the 1939 project.

The sports structures, having eliminated the hypothesis of the construction of the racecourse, are relegated to sports fields located in the lower part of the Italian garden. In the third planimetry, on the other hand, a project drawing in the section of 1942, the Park with the garden system is defined with greater graphic accuracy, as well as the pedestrian and vehicular paths such as the cypress avenue, the Villa with an initial proposal for the parterre in front, the functional buildings such as the generator house and the gardener's house and the military buildings along the perimeter. The Villa Reale is extended in the northern part with the winter garden, an architectural system on different levels shielded to the north by white concrete walls to protect the precious tree species. The scenic backdrop of the Odeon defines the curvature of the avenue of cypress trees, expressing the architectural strength of the megalithic architecture of the mid-twentieth century. The gardens are defined, both the Italian garden and the flower garden, and the belvedere, an elevated structure reachable by two flights of curved steps, able to guarantee the view towards the gardens and the forest below, as well as the entire city of Tirana. In the lower part of the plan you can see the barracks composed of two linear blocks: the first one is located near the main entrance, the second one is behind. The gardener's house, in the western part of the Royal Park, can be reached through three paths leading respectively to the Villa, the Parku i Madh and the sports facility. Through the graphic re-proposition, therefore, it is intended to propose the perceptive and codified ·interpretation of the existing reality, transmitting the heterogeneous and scientific content of the graphic image (Fig. 2).

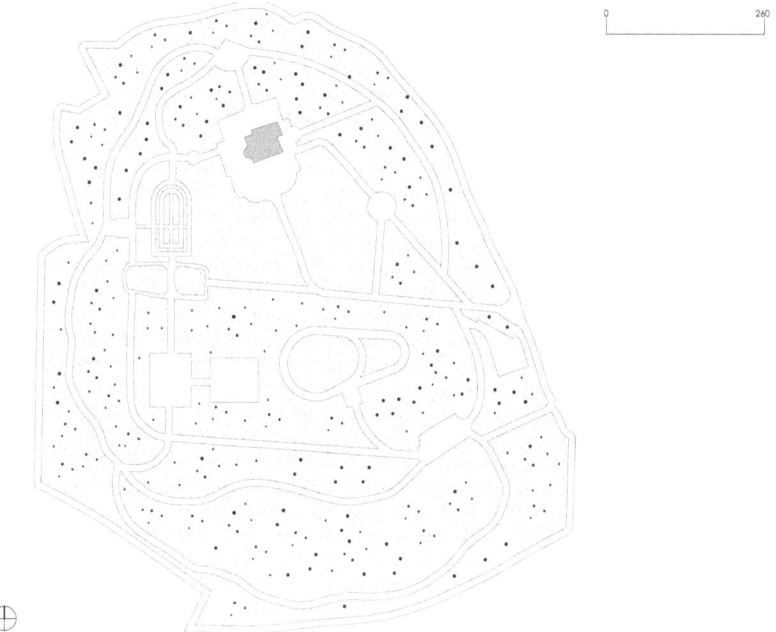

Fig. 2. The Royal Park of Tirana. Graphic reconstruction of the 1941 project.

2   The Royal Park: Notes on Architecture

The Royal Park, located on the hill of Mulleti, covers an area of 74 hectares and includes six architectures such as the Odeon, the Royal Palace, the Chapel, the greenhouse complex, the generator house and the gardener's house, as well as four green areas such as the avenue of cypresses, the avenue of oleanders, the garden with the belvedere and the lake. The original project, dated 1935, was signed by the architect Florestano di Fausto and included the Royal Palace, a Villa for the King and a building for the Princesses. After that date, the design work passed to the engineer Giulio Bertè who, in 1937, made use of the landscape architect Pietro Porcinai for the design of the Park, who worked out various solutions to define the design of the floors, the monumental staircase and some decorations. The engineer, very successful in Tirana, had created numerous residences for local aristocrats following the rationalist style and far from the classic forms of the European Royal Villas. The works, following the indications of Berté himself, began in 1938 but were interrupted in 1939 with the Italian invasion in Albania. In 1939 the intervention of the architect Gherardo Bosio was decisive. He reinterpreted the project by redesigning the geometries of the Villa and producing the executive drawings of the internal furnishings, the external and internal finishes and the final design of the garden with clear references to the European Residences (Fig. 3).

Fig. 3. The Royal Park of Tirana. General plan.

In 1942 the architect Ferdinando Poggi, after Bosio's death, revised again the project of the Villa, the gardener's house, the Cappela and the barracks and completed the works in a short time. The Villa Reale, was revised in the interior decoration (previously designed by Bosio both in the designs of the building materials and in the choice of the workers, materials and artists of the time) and in the architectural composition with asymmetrical geometries in plan and elevation as well as with absidat volumes facing west. The current morphological layout of the Villa is on four levels, one of which is underground, and a flat roof. The load-bearing structure is in reinforced concrete with brick paving, the facades plastered and perforations bordered by travertine jambs and closed by wooden frames. The main facade, exposed to the south, is characterized by an entrance with 11 steps leading to the travertine portal and wooden door (Fig 4).

Fig. 4. The Royal Park of Tirana. The Royal Villa, main elevation.

The Chapel inside the Royal Park, called Palatina, stands on a promontory and was built in 1939 at the behest of Vittorio Emanuele III (Giusti 2012). The current plane-metric layout is a single hall with a semicircular apse and stone paving. The main facade is preceded by a portico with four columns plastered and painted gray: the elevations are painted white and have four holes in the west, five windows to the east, one of which is on the bell tower (Fig. 5).

In the part below the flower garden to the west of the Villa Reale, instead, is built the generator house, a basement structure with flat roof from which, through a scenic ramp, you can access to a first terrace, covering a technical volume of the same architectural organism and, through a subsequent flight of stairs, to the square below with central

Fig. 5. The Royal Park of Tirana. The generator house, main elevation.

water tank. The western facade, which encloses the diesel generator for the production of electricity in the entire complex, is characterized by a central portal with canopy and two side entrances closed by wooden doors and windows that intersect laterally with the portal, square-shaped, are characterized by retractable shutters.

3 The Royal Park: Notes on the Vegetation Planting

In the southern part of the cypress avenue, sheltered from the north winds, there is the complex of greenhouses, built between 1941 and 1942 at the same time as the gardener's house. The design idea came from the landscape architect Pietro Porcinai who, as can be found in the archive documentation, outlines not only the geometrical shape but also the layout of the interior vegetation, the seasonal planting of vegetables and flowers, as well as the strategic layout of the wood behind (Fig. 6).

The greenhouses follow the path below the avenue of oleanders separated from it by a vegetation system with protective function from north winds and by a reinforced concrete containing wall on which the iron and glass structure rests. The complex is composed of four distinct identifiable volumes: the first to the west with a masonry base and glass roof, the second, raised 50 cm above the floor, in iron and glass, the third in masonry, with an opaque roof and large openings exposed to the south and, finally, the fourth, a service building with the function of storage for equipment, vases and for the storage of seeds (Fig. 7).

Fig. 6. The Royal Park of Tirana. The greenhouses, main elevation.

Fig. 7. The Royal Park of Tirana. The avenue of cypresses and the odeon.

In 1943, the architect Ferdinando Poggi defined the current configuration of the Park according to the new needs of the client and the guidelines coming from the styles of the European Gardner. The avenue of cypress trees was created, a driveway leading from the entrance on Via Elbasan to the Villa Reale, characterized by a regular geometric mesh of two trees interrupted by seats, the central fountain in front of the Villa with parterre, water chain and sculptural group.

From the same period as well as the work of the same architect are the avenue of oleanders, a pedestrian path of 46 meters, and the sequence of Italian gardens, flowers and the belvedere (Giusti 2012). The Italian garden is characterized by two polygons of squared vegetation placed in axis with the flower garden, the oasis of the entire park, where tree species with seasonal flowering are planted. The path ends with the raised system of the belvedere which is accessed through two symmetrical spiral staircases that follow the mound containing masonry from which you can admire the entire planting of the gardens and glimpse, through the thick vegetation, the underlying city of Tirana (Fig. 8).

0   10

Fig. 8. The Royal Park of Tirana. The generator house, elevation of the elevations.

Through a pedestrian path sloping westwards and far from the representative scenic system, stands the gardener's house, a building on two levels with adjoining external services such as ovens and storage rooms. The structure, now in a state of total abandonment, is invaded by weeds and presents the total collapse of the roof: it preserves the flight of stairs leading to the second floor and the supporting walls of the same level. The Poggi, redefines, moreover, the paths in the Park and the bordering of some parts of the lake. The original basin had three small lakes in succession, two of which have disappeared and can now be identified in the plan only through the identification of the marsh vegetation. Today's stretch of water has a reinforced concrete border in the south while to the north it is bounded by a boulder wall (Figs. 9 and 10).

Fig. 9. The Royal Park of Tirana. The generator house, main elevation.

Fig. 10. The Royal Park of Tirana. The Italian garden, cross sections.

4 Conclusion

The research work represents with attention and dedication, for the first time, the complete state of the architectural and environmental places of the Albanian Royal Residence. The activity carried out, with regard to the purposes and contents, have increased the thematic issues related to the knowledge, through the redesign of the archival documentation and the manual and instrumental survey of the architecture in the Royal Park, and the enhancement of the vegetation spaces. The Royal Park of Tirana is analyzed through the historical-chronological reconstruction of the events that, in the last century, shaped the Albanian territory and the subsequent architectural and environmental, manual and instrumental survey, as well as the verification of the state of conservation and digital modeling of the architecture.

References

Bulleri, A.: Tirana: contemporaneità sospesa. Quodlibet, Macerata (2012)

Corniello, L.: Il disegno del Parco Reale di Tirana. La scuola di Pitagora editrice, Napoli (2019)

Giacomelli, M., Vokshi, A.: Architetti e ingegneri italiani, Edizioni Edifir, Albania, Firenze (2012)

Giordano, P.: Il giardino inglese della Reggia di Caserta, il Petit Trianon di Versailles ed il parco di Worlitz: rilievi e disegni inediti. In: Salerno, R. (ed.) Rappresentazione Materiale/Immateriale. Drawing as (in)tangible representation. Atti del Convegno UID Milano 2018. Gangemi Editore, Roma (2018)

Giordano, P.: Il disegno della firmitas, la Scuola di Pitagora, Napoli (2015)

Giusti, M.A.: "Villa Reale" di Tirana: architetture, giardini, arredi, opere d'arte, dai progetti del ventennio al progetto di restauro, Edizioni Edifir, Firenze (2012)

Nepravishta, F.: Il progetto di Giulio Bertè per la Villa Nepravishta, Edizioni Edifir, Firenze (2012)

Posca, L., Barucci, C.: Architetti italiani in Albania (1914–1943), Clear, Roma (2013)

Rossi, M.: Strade d'Acqua. edizioni Mattioli, Fidenza (2004)

Vokshi, A.: Tracce dell'Architettura Italiana in Albania 1925 – 1943, DNA Editrice, Firenze (2014)

Piranesi at the Nymphaeum of Egeria: Perspective Expedients

Sofia Menconero[✉]

Department of History, Representation and Restoration of Architecture,
Sapienza University of Rome, Rome, Italy
`sofia.menconero@uniroma1.it`

Abstract. In the year of the birth anniversary of the important Venetian artist, the paper aims to contribute to the great amount of studies conducted on the figure and work of Giovanni Battista Piranesi.

After a brief presentation of the previous researches concerning aspects of the interpretation of Piranesian spaces, the contribution proposes a geometrical analysis of the *Veduta della fonte e delle spelonche d'Egeria fuor della Porta Capena or di S.Seb.no*, collected in the *Vedute di Roma* (1748–1778). The subject represented is the so-called Nymphaeun of Egeria, an archaeological structure still existing today, located in the Appia Antica Regional Park.

Through the analysis, based on the principles of the perspective method and compared with the Nymphaeum survey, it was possible to identify the fundamental elements (horizon line, ground line and projection centre) of the perspective approach adopted by Piranesi. Despite the approximations due to the nature of the data used, the author's attempt to manipulate perspective is evident. He, in fact, introduces the expedient of moving the projection center away from the picture plane, gradually considering parts of the structure closer to the observer.

The conclusions outline the possibility that Piranesi has adopted these stylistic choices in order to make the overall view and details of the archaeological site more comprehensible at the same time, implementing an operation that makes the side walls less foreshortened.

Keywords: Piranesi · Nymphaeum of Egeria · Geometric analysis · Perspective · Etching

1   Introduction

In the year that marks the birth of the great artist and engraver or, as he called himself, Venetian architect, we want to contribute with a small piece to the great amount of studies conducted on the figure and work of Giovanni Battista Piranesi. He was born in Mogliano di Mestre, or in Venice as agreed by the first two biographers (Bianconi 1779; Legrand 1799), on October 4, 1720. So in the year 2020 we celebrate his 300 years.

The publications on Piranesi are numerous and touch on different aspect of his life and art. The intention here is not to provide a complete overview of his bibliography,

L. Agustín-Hernández et al. (Eds.): EGA 2020, SSDI 6, pp. 343–356, 2020.
https://doi.org/10.1007/978-3-030-47983-1_31

but only to demonstrate, through a few examples, the vastness of the themes touched by the various researches that have taken place over time. Just think of Piranesi's immense artistic production, a collection of engravings that counts almost a thousand pieces according to the catalogue raisonné of the works drawn up at the beginning of the last century (Focillon 1918), to understand the equally immense list of books, essays and paper that derive from it. There are those who, as an architectural historian, talk about it in an aesthetic key (Tafuri 1980), those who focus on the historical-artistic analysis of some works (Gavuzzo-Stewart 1999), those who read the legacy of his visual imagination on contemporary architectural and artistic production (Purini 2008), those who trace the genius of his *cerveau noir* and its importance not only in the figurative arts (Yourcenar 2016), those who study the engraving technique (Pane 1938) and those who study the matrices (Mariani 2010, 2014, 2017), those who investigate aspects related to the production of furniture (Wilton-Ely 2007), and those who try to give a spatial interpretation of its settings (Marcos 2014).

This contribution, in line with the latter group, aims to propose an analysis of the space represented by Piranesi in one of his *Vedute di Roma* through the tools provided by the central projection method.

2 State of the Art on Geometric Studies of Piranesian Spaces

The first studies concerning the understanding of Piranesian spaces, avoiding the usual literary interpretations and approaching the issue from an analytical and geometrical point of view, date back to the late Fifties (Vogt-Göknil 1958). The main object of investigation in this case is addressed to the series of his *Carceri*, which stimulates, certainly more than the other collections, the desire for knowledge of the space, generator of such imaginative visions. In these tables, the observation points multiply, demonstrating how Euclidean geometry is not the only architectural solution for Piranesi, creating a sense of dizziness and discomfort in the spectator.

In a more recent study (Rapp 2008), which applies perspective restitution to some of Piranesi's engravings with views of bridges, it can be seen how the author multiplies the observation points, creatively using the geometric rule of central projection, this time not to destabilize the viewer, but to increase the readability of the architecture represented.

In both cases we see how the Venetian artist manipulates perspective according to the type of communication he wants to provide.

3 The Nymphaeum of Egeria from the Origin of the Name to the 19th Century

The work that is investigated in this study is the *Veduta della fonte e delle spelonche d'Egeria fuor della Porta Capena or di S.Seb.no.*

The subject represented, the so-called Nymphaeum of Egeria, is an archaeological site, still existing today, located in the Caffarella valley, within the boundaries of the Appia Antica Regional Park (Fig. 1).

The structure was mistakenly identified with the spring and the cave of the nymph Egeria since the early 16th century, when the antiquarian practice of finding matches

Fig. 1. Photograph of the current state of the Nymphaeum of Egeria at the Appia Antica Regional Park.

between literary sources and archaeological remains still visible gave rise to superficial and sometimes inaccurate attributions.

Legend has it that Egeria, an ancient Latin deity linked to the events of Numa Pompilius, after the death of the second king of Rome, melted in tears and gave life to a spring (Ovidio 1848):

> *Non tamen Ægeriæ luctus aliena lavare*
> *Damna valent, montisque jacens radicibus imis*
> *Liquitur in lacrymas, donec pietate dolentis*
> *Mota soror Phœbi gelidum de corpore fontem*
> *Fecit, et æternas artus tenuavit in undas.*

("But disaster other people suffered could not relieve Egeria's discontent. At the foot of a mountain she lay down, dissolving in her tears, until Diana, moved by Egeria's loyal sorrow, transformed her body to an icy spring and made her limbs an ever-flowing stream").

Excluding the identification with the *fons Egeriæ*, localized by recent studies near the ancient Porta Capena, the structure is more likely a *specus aestivus*, that is a place where banquets, moments of leisure, rest and cultural activities were hosted, obtained in an artificial cave in the outbuildings of the villa of Herod Atticus and his wife Annia Regilla, in the 2nd century A.D. The excavations, carried out at the same time as the jubilee restoration in 1999, demonstrated the artificial nature of the cavity, which was entirely built above ground near a hill, later called Sant'Urbano, and only later buried to simulate the appearance of a cavern. The architectural layout of this first phase remains: a T-shaped plan, oriented NE-SW, which identifies what were originally intended to be two rooms: a rectangular entrance room, oriented transversely with respect to the axis, whose roof still remains to be clarified today; and a rear room, also rectangular but oriented longitudinally, which still shows a monumental barrel vault in cement conglomerate (Fig. 2). The rear room, which has not undergone any structural modifications with respect to the original conformation, is characterized on the short side by a rectangular apse, on which there is a semicircular niche; while on the long sides there are three symmetrical niches: the two lateral ones with a rectangular plan and the central one, slightly higher than the others, with a semicircular plan. The entrance area, which is in a worse state of conservation and has undergone transformations and additions that make it difficult to stratigraphically read the walls, has two wings attested to the short sides. Very few decorative fragments remain of this first Hadrianic-Antonine phase (De Cristofaro 2014).

Fig. 2. Current plan of the Nymphaeum obtained through the point cloud of the photogrammetric survey.

In the 4th century A.D. the villa and its outbuildings were purchased by Emperor Maxentius, who made it a luxurious residential and political complex. The Massenzian restructuring of the Nymphaeum seems not to transform the original function. The largest morphological modification commissioned by the Emperor concerns the introduction of the central basin of the entrance hall, which is still present today. It is a rectangular structure with a small semicircular apse on the northern side, which was fed by a water conduit, largely preserved, consisting of terracotta *tubuli* that ran inside the western masonry of the vaulted room, at the base of the three niches (De Cristofaro 2014).

The next documented phase relates to the Renaissance period. In the early 16th century the estate was bought by the Caffarelli family, of whom it still bears the name. The first known graphic representation of the Nymphaeum is of these years: a survey sketch by Antonio da Sangallo the Younger commissioned by Giovan Pietro Caffarelli. The drawing, preserved in the *Gabinetto dei Disegni e delle Stampe degli Uffizi* (GDSU) (Fig. 3a), appears essential and shows the general ichnography of the structure. Worthy of note are the tracking of the aqueduct inserted in the cavity between the structure and the hill behind, which supplied water at the innermost niche in the eastern wall; and the attempt to hypothesize a cross vault on the entrance area. The two drawings attributed to Sallustio Peruzzi and preserved at the GDSU (Figs. 3b–3c) are also 16th-century, and perhaps referable to a renovation phase. In one of the two, for the first time, appears the fountain with corbels in the apse, still existing today, from which a system of floor ducts ran along the perimeter of the inner walls, and which was originally to be completed below by a basin consisting of a sarcophagus or an ancient basin. The statue depicting Almone, a river god, is ancient and probably recovered in the surrounding area. In 1536, with the Roman triumph of Charles V, the Nymphaeum hosted a banquet offered by the Caffarelli to the Emperor. For a long time a marble element of reuse used as a table and ancient capitals used as stools has been preserved (De Cristofaro 2013). The iconography between the 17th and the 19th centuries helps to understand the layout of the Nymphaeum in that period. One can see the permanence of the reuse furniture for

Fig. 3. Various iconographies of the Nymphaeum of Egeria: (a) plan by Antonio da Sangallo the Younger (first twenty years of the 16th century); (b, c) plans attributed to Sallustio Peruzzi (mid-16th century); (d) engraving by Herman van Swanevelt (1650–1655); (e) engraving by Bartholomeus Breenbergh (1640); (f) painting by Charles-Louis Clerisseau (c. 1760).

the libations of Charles V's triumph, though increasingly compromised with the passage of time (Fig. 3d). The construction of a small tavern in correspondence of the east wing of the entrance hall dates to the 17th century (Fig. 3f). In this period the iconography no longer shows the Massenzian basin and instead a dry ground hosts the popular outings. We find again the flooded floor, and a basin appears in correspondence of the niche where Sangallo had signaled the water capture from the 18th century (Fig. 3g).

With the rediscovery of the Renaissance antiquarian taste, the Nymphaeum is back in vogue and becomes a sort of reference for artists and architects thanks to its good state of preservation (Castellini 2018), as well as a popular place for parties and outings at least until the 19th century.

4   The Piranesian Representation of the Nymphaeum of Egeria

The previous chapter allowed to outline the main points of the history of the Nymphaeum of Egeria, in order to better understand the contexts of Piranesi's engraving, geometrically studied in the following chapter.

Of the many *facies* that the Nymphaeum has had, Piranesi records the one in the mid-18th century.

The *Veduta della fonte e delle spelonche d'Egeria fuor della Porta Capena or di S.Seb.no* is collected in the first volume of the *Vedute di Roma*, engraved between 1748 and 1778 (catalogued at n. 782 in Focillon 1918) (Fig. 4). This is an etching of considerable sizes, equal to 680 × 395 mm, signed at the bottom left with *Cavalier Piranesi F[ecit]*. The caption, integrated at the bottom right of the composition, contains, in addition to the title cited, references to the myth of Numa and Egeria:

Hic ubi nocturnæ Numa constituebat amicæ

Nemus et delubra.

Così Giovenal. Nella Sat. 3. Ove pur dice:

In vallem Egeriæ descendimus et speluncas

Dissimiles veris; quanto præstantius esset

Numen aquæ, viridi si margine clauderet umbras

Herba, nec ingenuum violarent marmora tophum

*Tempio di Bacco, or di S.Urbano.

("Here where Numa met his nymph at night, woods and temples. So Juvenal writes in Satire III, where he adds: We descend into the Egeria valley and into its caves, different from the natural ones. How the will of the god would be more present in the waters, if the grass still closed the waves with a green frame and the marble did not violate the native tuff").

We find many traces of the long history of the Nymphaeum in Piranesi's representation. Clearly the 2nd century layout of the rear room remains, with the apse and the various niches, while there is no trace of the central 4th century basin. There is the 16th century intervention of the fountain with the ancient statue of the god Almone in the apse and, from the same period, the ruined remains of the furnishings for the Emperor's

Fig. 4. G.B. Piranesi, Veduta della fonte e delle spelonche d'Egeria fuor della Porta Capena or di S.Seb.no, collected in the Vedute di Roma (1748–1778).

banquet. There is a flat arch window in the east wing, that could refer to the small 17th century tavern. Water gushes out from the innermost niche of the eastern wall and collects in the basin below where a woman washes her clothes.

In the background you can see the church of Sant'Urbano, which still exists today, in a position not consistent with the view of the Nymphaeum.

Although Piranesi used to worsen the conditions of conservation of the archaeological structures represented, to satisfy that typically romantic "taste of ruin", the portrait he makes of the Nymphaeum of Egeria appears quite realistic comparing it with other contemporary engravings.

There is another representation of the Nymphaeum by Piranesi, engraved during his youth, that has smaller sizes (190 × 112 mm). This is the *Spelonca della Ninfa Egeria detta volgarmente la Cafarella*, collected in the series *Varie vedute di Roma* of 1748 (catalogued at n. 108 in Focillon 1918) (Fig. 5). The succession of the two versions is quite clear, beyond the dating of the collections. In the *Spelonca della Ninfa Egeria* we

Fig. 5. G.B. Piranesi, Spelonca della Ninfa Egeria, detta volgarmente la Cafarella, collected in Varie vedute di Roma (1748).

find a less precise and more superficial representation that depends on the reduced sizes of the table but also derives from a less thorough knowledge of the architectural structure by the author.

Comparing the two representations, the difference in scale of the structure is immediately evident: if Piranesi introduces human figures compatible with the real size of the place in the first engraving cited, while following his tendency to magnificence in another way (as we will see in the following chapter), this tendency is amplified to the maximum in the second engraving thanks to the introduction of human figures much smaller than necessary. In addition, the proportions of the architecture are modified, improbably much slenderer in the second engraving mentioned.

In the *Veduta della fonte e delle spelonche d'Egeria*, Piranesi shows a greater familiarity with the place, testified by the great detail with which he represents the various elements, including the lack, still existing, in the upper part of the *opus reticulatum* wall above the apse.

5 Geometric Study

The geometric study concerns the second engraving that Piranesi makes of the Nymphaeum of Egeria, the one entitled *Veduta della fonte e delle spelonche d'Egeria fuor della Porta Capena or di S.Seb.no*.

The desire to understand how Piranesi acted on the representation of the Nymphaeum derives from the different perception one has of the real place. The possibility to access the archaeological site was granted during a survey campaign (Griffo et al. 2019) whose data were very useful for the present study as metric and morphological evidence.

The following geometric analysis is based on the principles of perspective restitution, already theorized in Heinrich Lambert's 1759 treatise *Die Freye Perspektive* (Carpiceci 2012) and it will be dealt with using the current nomenclature of perspective for a question of method clarity.

We proceeded from the two-dimensional analysis of Piranesi's engraving to a three-dimensional analysis supported by the 3D model obtained through the measurements of the survey.

After having scaled the digital image to bring it back to its original sizes (680 × 395 mm) and having necessarily redrawn the main lines of the architectural structures in order to make them more visible, we proceeded with the identification of the vanishing point of the horizontal lines perpendicular to the picture plane, that is the main point (O_0) of the frontal perspective. It i s placed with good approximation as in Fig. 6, and the horizon line (o) passes for it.

It was decided to use the round arches of the niches to define the distance of the projection centre: considering the circumscribed squares relative to each arch, their diagonals identify 45° lines which, according to the perspective rule, converge in a single vanishing point, which is also the overturning of the projection centre (O^*). Since in our case the arches are vertical, this overturned point (O^*_n) belongs to the vanishing line of the vertical planes perpendicular to the picture plane (f'_α). The distance of the projection centre from the picture plane is represented by the distance between its overturning and the main point.

What Piranesi's engraving suggests is a multiplication of the distance of the projection centre (O^*_1, O^*_2, O^*_3) which varies according to the distance of the arches from the picture plane. In particular, the distance of the projection centre increases as the arches approach the picture plane (Fig. 6). The same analysis applied to the arch of the surviving lateral wing of the Nymphaeum, on the left of the observer, would seem to confirm this expedient (O^*_4).

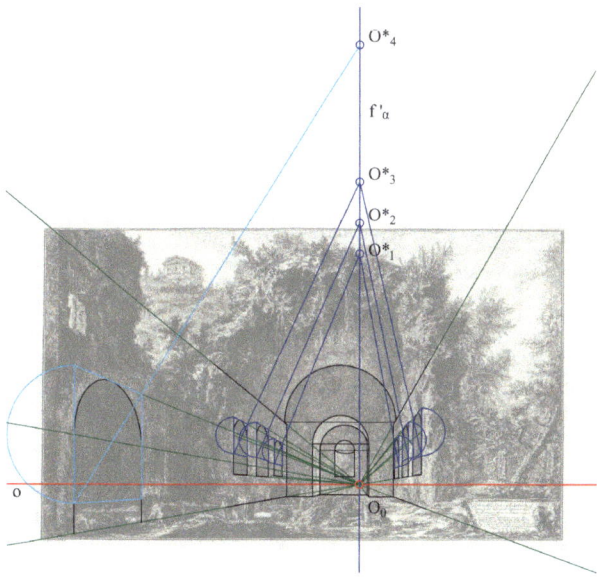

Fig. 6. Identification of the fundamental elements of the perspective space (O_0, o) and the different positions of the projection centre (O^*_n).

Entering into metric issues, it was necessary to find the position of the ground line (f), identified by the measurement of the width of the apse and the *parallelogram rule* (Fig. 7). In this case the size of the apse inserted in the restitution is 350 mm, i.e. in a ratio of 1:10 to the actual size of this apse (about 3.50 m). Following this operation, the ground line is positioned at the lower edge of the etching and this positioning, in such a particular point, could lead to the hypothesis that Piranesi has reasoned on the reduction in scale of its representation. Another observation arising from the previous construction concerns the distance between the horizon line and the ground line, i.e. the height of the projection centre as well as the observer. This measurement is about 80 cm, much less than a human height, but compatible with the point of view of a human sitting on the ground. Perspective, unlike any parallel projection (plan, elevation or axonometric drawing), not only represents the observed object but also its observer: perspective incorporates it. Consequently, the observer becomes module of the perspective space (Migliari 2012). With the expedient of lowering the projection centre, Piranesi intensifies the idea of magnificence that he frequently obtains by inserting human figures in the engravings, smaller than they should be in relation to the architecture. In this etching, as already mentioned, the human figures represented have sizes compatible with the surrounding environment, and magnificence is delegated to the lowering of the projection centre.

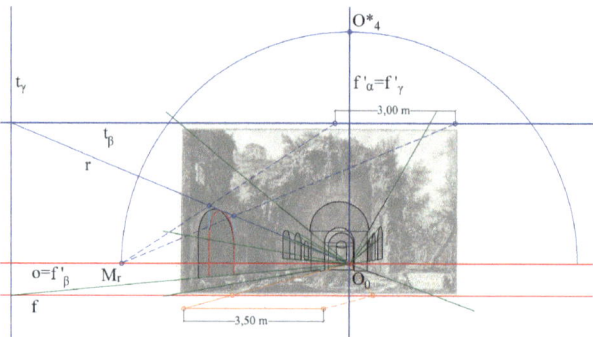

Fig. 7. Identification of the ground line (f) and representation of the arch of the lateral wing according to the survey measurement (in red).

Subsequently, it was verified that the diameter of each arch, measured by the width of the corresponding niche, considering the relative projection centre and therefore the relative measuring point (M_r), was in conformity with the measurements of the survey of the archaeological structure. This circumstance occurs with a small margin of inconsistency, in the order of a few centimeters in real size, on each of the six arches of the niches in the side walls. Such an incongruity is admissible considering the nature of both the object: an archaeological structure with strong deterioration, and the type of representation, an antique print from an etched matrix. The same operation, applied to the east wing arch, led to a much greater discrepancy: of the order of 1 m. In this case, the greatness of the incongruity cannot be explained by small inaccuracies, but by a concrete compositional will of the author. In Fig. 7 it is possible to see the size of this arch, set following the perspective expedient that Piranesi uses in the arches of the side niches, but with the actual measurements of the survey. You can immediately notice how the arch is narrower. An explanation of this expedient can be traced back to an accentuation of the same reason that leads to the expedient of multiplying the distance of the projection centre: presumably to make the details of the architecture more legible, increasing the view of the side walls as they approach the observer.

The next step was to show what the arches would look like when they were viewed from the same projection centre. This operation, for the sake of brevity, was carried out at a two-dimensional level only on the arches of the niches of the wall to the left of the observer and was repeated by taking one of the three identified distances of the projection centre each time (Fig. 8).

Fig. 8. Representation of the niches seen from the same projection centre O_1 (a), O_2 (b), O_3 (c), where the niches consistent with each perspective are highlighted in white.

After investigating Piranesi's perspective expedients in a two-dimensional way, directly on the etching, it was decided to deepen the analysis through three-dimensional modeling.

We wanted to demonstrate the contraction and dilation of the perspective space considering each of the four projection centers (O_1, O_2, O_3, O_4) related to the respective overturned points identified in the two-dimensional analysis of the etching. Taking as a reference the length of the eastern wall, measured on the incision through the four different measuring points corresponding to the relative projection centers, the four 3D models were scaled exclusively along the longitudinal axis, since we know that the variation in the distance of the projection centre from the picture plane only affects the depths of the perspective (Fig. 9). Therefore, there is not a single 3D model and a single perspective setting representing Piranesi's engraving, but the latter is the union of various partial perspectives with different projection centers aligned longitudinally.

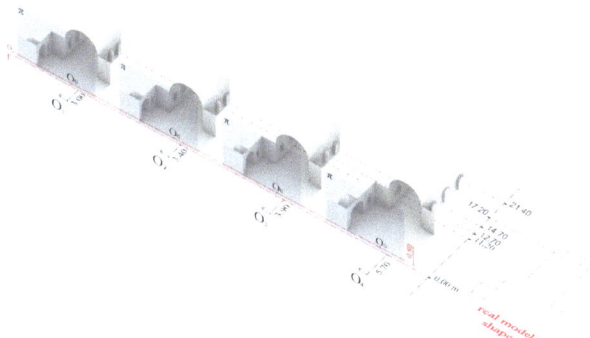

Fig. 9. Scheme representing the variation of the distance of the projection centre (O_1, O_2, O_3, O_4) from the picture plane (π) and the relative contraction and expansion of the perspective space, compared with the actual measurements of the Nymphaeum (in red on the right).

The perspectives obtained through the 3D model with the real measurements (Fig. 10) differ in the foreshortening: the closer the projection centre is to the picture plane, the more visible the side walls are to the detriment of a deformation that prevents the unitary representation of the architecture.

It seems that Piranesi, with this expedient, tries to mediate between the desire to represent the Nymphaeum in its entirety and to describe the side walls in depth. And this mediation would only have been possible by working with several projection centers at different distances from the picture plane.

What in cinematography is called the *vertigo effect*, i.e. an effect achieved by moving the camera forward while zooming out of the scene, or vice versa, and which is experienced in a time sequence, Piranesi synthesizes and freezes it into a single image which is the sum of different perspective.

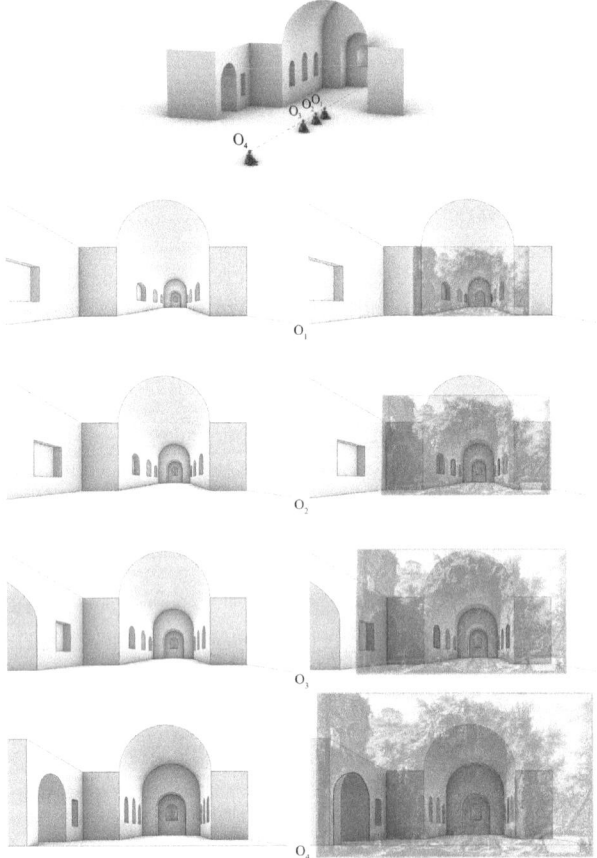

Fig. 10. Above, the schematic representation of the longitudinal displacement of the projection centers (O_1, O_2, O_3, O_4) related to the Nymphaeum model. Next, four perspective views of the relative projection centers (left), compared and superimposed with Piranesi's engraving where the elements consistent with each perspective are highlighted in red (right).

6 Conclusion

The proposed perspective study is only one of the levels of analysis applicable to Piranesi's engravings and is certainly not a priority over other interpretative keys.

Despite the approximations due to the nature of the data used, the author's attempt to force the perspective is evident. The reasons that lead him to the showed expedients are probably to be found in the purpose of his artistic production. As he himself writes in the preface to *Le Antichità Romane*: *[...] e vedendo io, che gli avanzi delle antiche fabbriche di Roma, sparsi in gran parte per gli orti ed altri luoghi coltivati, vengono a diminuirsi di giorno in giorno o per l'ingiuria de' tempi, o per l'avarizia de' possessori, che con barbara licenza li vanno clandestinamente atterrando, per venderne i frantumi all'uso degli edifizi moderni, mi sono avvisato di conservarli col mezzo delle stampe* (Piranesi 1784) ("I decided to conserve, by means of prints, the remains of the ancient buildings of Rome, scattered largely in the cultivated countryside, because I saw that they were diminishing from day to day either through the negligence of the times or the avarice of the owners, who were stealing parts of them in order to sell them for the construction of new buildings").

Piranesi's work is therefore a formidable collection of archaeological documentation on the one hand and of refined art on the other. Both these characteristics must have led him to the stylistic choices adopted, even in their contradiction. Because if on the one hand he emphasizes the details of the ancient structures to the point of representing the degradation as we still see it today (for example the lack on the upper part of the *opus reticulatum* wall of the apse), on the other hand he implements expedients such as falsifying the perspective or positioning the elements in an anomalous way (for example the orientation of the Sant'Urbano represented on the hill behind) for compositional and communication need. This makes his engravings an unreliable instrument metrically and morphologically but does not diminish their importance.

Unfortunately, Piranesi leaves no trace, graphic or literary, of the design of his engravings, so we do not know with what level of geometric consciousness he introduced his expedients. The fact remains that, quoting Panofsky, *one can rightly say that the errors, more or less great, of perspective, indeed even the complete absence of any perspective construction, have nothing to do with artistic value* (Panofsky 2013).

Acknowledgments. The author would like to thank the Appia Antica Regional Park for having authorized the access to the structure; Marika Griffo and Paolo Cimadomo for the previous shared studies on the Nymphaeum of Egeria and for the use of the survey data.

References

Bianconi, G.L.: Elogio storico del cavaliere Giambattista Piranesi celebre antiquario ed incisore di Roma. In: Antologia Romana, vol. 5, no. 34–36, pp. 265–284 (1779)

Carpiceci, M.: Fotografia digitale e architettura: storia, strumenti ed elaborazioni con le odierne attrezzature fotografiche e informatiche, Aracne, Roma (2012)

Castellini, M.: Il Ninfeo di Egeria (II sec. d.C.) e la Grotta degli Animali a Castello. Il ruolo del modello antico attraverso l'analisi dei disegni del GDSU. In: Opus Incertum. Rivista di storia dell'architettura IV, Firenze (2018)

De Cristofaro, A.: Baldassarre Peruzzi, Carlo V e la ninfa Egeria: il riuso rinascimentale del Ninfeo di Egeria nella valle della Caffarella. Horti Hesperidum, Studi di storia del collezionismo e della storiografia artistica **III**(1), 85–138 (2013)

De Cristofaro, A.: Il Ninfeo di Egeria nella Valle della Caffarella a Roma: forma, funzione, cronologia. Orizzonti, Rassegna di Archeologia, vol. XV, pp. 31–49 (2014)

Focillon, H.: Giovanni-Battista Piranesi Essai de Catalogue raisonné de son oeuvre. Henri Laurens Éditeur, Paris (1918)

Gavuzzo-Stewart, S.: Nelle Carceri di G.B. Piranesi. Northern Universities Press, Leeds (1999)

Griffo, M., Cimadomo, P., Menconero, S.: Integrative IRT for documentation and interpretation of archaeological structures. Int. Arch. Photogram. Remote Sens. Spat. Inf. Sci. **XLII-2/W15**, 533–539 (2019)

Legrand, G.: Notice historique sur la vie et les ouvrages de J.B. Piranesi, architecte, peintre et graveur né à Venise en 1720, mort à Rome en 1778. Rédigée sur les notes et les pièces communiquées par ses fils, les Compagnons et les Continuateurs de ses nombreux travaux (1799). In: Erouart, G., Mosser, M., A propos de la Notice historique sur la vie et les ouverages de J.B. Piranesi: origine et fortune d'une biographie. In: Brunel, G. (ed.) Piranèse et les Français, Roma (1978)

Marcos, C.L.: Imaginary Prisons or space as a topic. Disegnare idee immagini, no. 48, pp. 44–55. Gangemi, Roma (2014)

Mariani, G.: Giambattista Piranesi: matrici incise, pp. 1743–1753. Mazzotta, Milano (2010)

Mariani, G.: Giambattista Piranesi: matrici incise, pp. 1756–1757. Mazzotta, Milano (2014)

Mariani, G.: Giambattista Piranesi: matrici incise , pp. 1761–1765. Editalia, Roma (2017)

Migliari, R.: La prospettiva: una conversazione su questioni solo apparentemente banali. In: Carlevaris, L., De Carlo, L., Migliari, R. (eds.) Attualità della geometria descrittiva, pp. 134–137. Gangemi, Roma (2012)

Pane, R.: L'acquaforte di G.B. Piranesi. Ricciardi, Napoli (1938)

Panofsky, E.: La prospettiva come forma simbolica. Abscondita, Milano (2013)

Piranesi, G.B.: Le Antichità Romane. Stamperia Salomoni alla Piazza di Sant'Ignazio, Roma (1784)

Purini, F.: Attualità di Giovanni Battista Piranesi. Librìa, Melfi (2008)

Ovidio: Le Metamorfosi, purgate e corredate da note italiane da Atto Vannucci. Tipografia Aldina, Prato (1848)

Rapp, J.B.: A geometrical analysis of multiple viewpoint perspective in the work of Giovanni Battista Piranesi: an application of geometric restitution of perspective. J. Archit. **13**(6), 701–736 (2008)

Tafuri, M.: La sfera e il labirinto: avanguardie e architettura da Piranesi agli anni '70. Einaudi, Torino (1980)

Vogt-Göknil, U.: Giovanni Battista Piranesi - Carceri. Origo Verlag, Zurich (1958)

Wilton-Ely, J.: Piranesi as designer. Smithsonian Institution, New York (2007)

Yourcenar, M.: La mente nera di Piranesi. Pagine d'arte, Tesserete (2016)

The Project of Maurizio Sacripanti's Science Museum

Francesco Maggio$^{(\boxtimes)}$ and Fabrizio Provenza

University of Palermo, Palermo, Italy
`francesco.maggio@unipa.it`

Abstract. The essay aims to explain, through relation and drawings, the process that led to the critical redrawing o f he New Science Museum in Rome.

In order to understand the 1983 project, it was necessary to establish the key points of Maurizio Sacripanti's design process. The first part is entirely addressed to the biography, specifically analyzing the salient points of his life and considering the artistic and literary influences that at that time gravitated in the Roman context. The second part analyzes Sacripanti's thought in order to understand every facet of his architectural language, together with the study of the most significant projects, in order to draw the necessary conclusions about his modus operandi.

In the final part of the essay, a critical reflection on the project for the New Science Museum is carried out, starting from the exhibition "5 Billions of years ago", up to the elaboration of the drawings came to us, associated with the critical re-drawing in charge that will analyze the composition, volumes and spaces in detail, in order to compare the drawings of 1983 with the realized works, revisiting some aspects to ensure the completeness of the unbuilt project, which represents one of the most ambitious and revolutionary works of Sacripanti.

Keywords: Redrawing · Project · Time

1 Introduction

The very personal architectural thought of Maurizio Sacripanti has contributed with his artistic-architectural language, characterized by extraordinary imaginative components, to the creation of a real avant-garde architecture.

The creative process of the Roman architect originates from a multiple multidisciplinary experience that has pervaded the course of his professional experience; also for this reason the works realized and those merely conceived and remained on paper, seem to anticipate a future to which Sacripanti addresses with his projects with an "antimetodic method, crossed by more than one contradiction".

It is important to point out that between the fifties and sixties the Jungian theories began to spread, changing the face of psychoanalysis and inspiring Fellini's neo-realist film production, which often found itself confronted with Sacripanti in the intellectual context of the time. In that instant an osmosis of notions took place that drastically changed the master's way of designing, who would detach himself from rationalism

L. Agustín-Hernández et al. (Eds.): EGA 2020, SSDI 6, pp. 357–366, 2020.
https://doi.org/10.1007/978-3-030-47983-1_32

to undertake a new and unexplored path of architecture, until now relegated to three dimensions, introducing further points of important importance. Three elements that form Sacredantian architecture have been identified: design, time and science. These three themes serve to understand both the process that the Master went through up to the mature phase of design, and the aspects that led to the completion of the project for the Museum of Science, considered one of the manifestos of his architecture.

2 Maurizio Sacripanti (1916–1996)

Maurizio Sacripanti was born in Rome on August the 8th, 1916. After attending the artistic high school, he enrolled in the Faculty of Architecture of the capital. In addition to the academic teaching that the University offered, Maurizio Sacripanti expanded his knowledge to learn freely, without the restrictions imposed by Fascism, which in those years was trying to condition the minds of men. He graduated in 1943, but the war period did not allow him to practice his profession immediately, and it was only in 1946 that he won, with an ex aequo prize, the national competition for the arrangement of Piazza d'Armi (today Piazza Garibaldi) in Perugia, drawing up the second degree project. The following year he collaborated with Ciro Cicconcelli to design the Albergo della Gioventù and the prefabricated single-family houses in the experimental QT8 district of Milan. In the following decade he realized other projects, among which we can remember the new Auditorium at Borghetto Flaminio in 1950; the San Giovanni Hospital in Empoli, a winning project in 1954 and the Residential Quarter in Verona in 1956, completed four years later. In 1960 he opened his new studio in Piazza del Popolo, moving to the same building where Caffè Canova was located, where art, politics and films by Fellini, Antonioni and Rosi were discussed. Soon the Maestro's studio became one of the major focal points of the artistic-literary activity in Rome. The following year he participated together with Mafai in the competition for the new Peugeot Skyscraper in Buenos Aires, one of his most significant projects.

In 1962 he designed the Italian Pavilion at the Fiera Internazionale di Tolosa per il Mini-stero del Commercio Estero; the Incis district in 1963 for the employees of the Ministry of Foreign Affairs in Rome and the "Cynthia" district of workers' houses in Bagnoli in 1964, while in 1965 he dedicated himself to the new Lyric Theatre in Cagliari, but even this time the ambitious designs of the master did not seem to completely excite the commission, perhaps unable to understand the split with the architecture of the past. One year later he realized the project for the centre for the treatment and re-education of silicotics in Domodossola and participated in the national competition for the new offices of the Chamber of Deputies, a competition with a controversial outcome since no winner was designated. Thus, began what will be identified as the period of unrealized projects, such as the new Museo degli Eremitani in Padua, conceived with Quaroni in 1968, and the Italian Pavilion at the 1970 International Exhibition in Osaka. In 1972 the most significant projects of the last decade were collected in a single volume, forming fragments of an imaginary city conceived by the Roman architect, the "City of the Frontier", a text written by Sacripanti himself in collaboration with Renato Pedio. Despite the countless projects, many of which were not realized, Sacripanti obtained numerous awards, especially through national and international exhibitions. In 1972 he was commissioned to design the Church of Partanna in Trapani, as part of the reconstruction of the areas hit by the earthquake in Belice; this work was also not realized.

In 1983 he drew up the project never realized for the New Science Museum in Via Giulia in Rome and, again in that year, the research centre "Matteo Ricci" in Shanghai, also this one remained on paper because of the Chinese political situation. In 1992 he designed the extension of the Faculty of Political Science, Statistics and Biology of the University of Rome, while in 1994 he dedicated himself to the design of a parish complex in Tor Tre Teste in Rome. In 1995 he left his studio in Piazza del Popolo to move to his home in Via Maresciallo Pildsuski where he continued to work on projects in progress and future projects until his death in 1996, while he was working on the project for the new Auditorium in Rome.

3 The Pure Drawing and the Architecture Drawing

The simple act of drawing can often take on different meanings, depending on the function to which it is associated, and connotations that especially in architecture act as a "bridge" between imagination and reality. It is natural to think about the close link between drawing and architecture in an inseparable pair. Yet, although "concept" and "matter" are linked elements in design, there is a degree of separation between them, a "frontier line" that outlines the boundary between the conception phase and the design phase. It happens, therefore, that for Sacripanti architectural drawing, in addition to taking into account various external factors such as history, the socio-cultural context and other variables to which it must inevitably adapt, clashes with pure drawing, which goes beyond the known graphic conventions because it is a "work in itself capable of giving rise to subversive behaviour", giving rise to a "tension that animates the entire sacrificial drawing" generated by a "prolonged and conflictual transference" (Neri and Thermes 1998, p. 20).

After the publication of the book, the master's architecture achieved a certain success on an international level, first with the international competition for the Peu-geot skyscraper in Buenos Aires and then with the competition for the Chicago Tribune headquarters. But it was in 1959 that Sacripanti reached his design maturity with the international competition for Residential Types in Rome, issued by CECA.

In this project Sacripanti combines the engineering evolution of the pre-fabricated elements with artistic genius to move as far away as possible from the concept of standardization of the architectural artefact and conceives the three-dimensional space with superimpositions, translations and rotations of surfaces in such a way as to offer both concreteness and artistic and visionary value to the project.

Precisely in this example two well-defined and separate areas of design are represented here, one belonging to the sphere of the language used by the architect, the other instead attributed to the sphere of graphic representation. "Both are however interconnected, as if they were two strips of land" (Capiato et al. 2018, p. 5) delimited by the imaginary borderline crossed with the progressive definition of the project.

4 The Time as Design Issue

The neorealist experience that Fellini transmitted to Sacripanti certainly shows a different way of conceiving space. We no longer speak only of the mere static three-dimensional object, which is placed passively regardless of artistic thought, but the architect introduces the fourth dimension. Time, in sacripantian design, can take on various connotations linked to the concept of making architecture and the first way of perceiving it is derived from the changing spatiality of the architectural artefact. Sacripanti wondered how to represent this dynamism and the first step was to reject perspective and the horizon line.

Although perspective can be an "instrument of representation and control of space, it is also the closest to visual perception available to artists and architects" (Capiato et al. 2017, p. 23), for Sacripanti it takes on an obsolete and static meaning.

It is therefore necessary to eliminate the horizon line that "ideally separates the visible world from the unknown, the finite space from the infinite and the above from below" (Capiato et al. 2017, p. 25), to create a continuous space that merges the real dimension with the unconscious. Especially in the central perspective, the loss of the horizon line clears the concept of symmetry, now opening new and more numerous potentially infinite directions, as Sacripanti himself states: "If we look at the sun, it is symmetrical, but if we consider it in another, and perhaps more proper, scale, like a "point" in the "geometry" of the galaxy, the sun is asymmetrical. A tree is undoubtedly round but try dissecting it horizontally and it is no longer round, just like the two parts of a face […] an object of nature is never geometrically perfect" (Giancotti 2015, 61). Sacripantian thought may seem contradictory, because while on the one hand it seems to want to move away from the canonical models of architecture, on the other hand design drawing requires the traditional representative vocabulary, albeit limited, but with a different use that is able to generate new meanings and new interpretations.

Drawing appears almost like a visual puzzle without the possibility of understanding its features, when it is a project conceived for realization, therefore potentially 'realizable'.

A further method of marking time is the use of shadow as an element of design, generated by "archetypes" such as the column, the wall or the hole that constitute the traditional elements of architecture. The reference to archetypes is reminiscent of the Jungian one, which states that they are primitive principles that go beyond cultures, symbols and all forms of expression.

This theme, in architecture, must be addressed with the elements that compose it, since they are "physical nuclei of the mind, suggesting the indissoluble link between architecture and man. In fact, as archetypes, they exist first in the mind, they are forms generated by man's thought" (Capiato et al. 2017, p. 15).

5 The Science Applied to Architecture

For Sacripanti science and art have many analogies: "both respond to a profoundly imaginative stimulus; they enhance new layers of mental functioning; they are driven by tension to freedom and creativity" (Antonucci 2017, p. 65). Scientific progress corresponds to technological progress and with it the foundations are laid for the application of new technologies in the architectural field, for undertaking a path linked to dynamism, transforming architecture into a machine, as in the case of the new Teatro Lirico in Cagliari, one of the extraordinary projects of the Roman architect, in which Sacripanti manages to conceive a "polyorganism" that "goes beyond the limits of a traditional opera house" (Neri and Thermes 1998, p. 60), adopting a technological solution that modifies the surrounding space through the use of blocks placed at the top and bottom, which protrude and retract with electronically controlled guns, thus allowing a spatial modification that consents a possible spatial configuration. This pragmatic view of the scientific approach, by Sacripanti, is given by a certain curiosity derived from the fascination that science can offer, especially astrophysics and the study of black holes in the interaction between time and matter, in correspondence with the horizon of events. Sacripanti was commissioned in 1979 by Giorgio Tecce, former Dean of the Faculty of Science at the University "La Sapienza" of Rome, to design the exhibition "5 miliardi di anni" at the Palazzo delle Esposizioni in Rome, inaugurated on the 29th of May 1981.

This particular event feeds curiosity in the master about the progress of the evolutionary study of stars, galaxies and Cosmology in general, to satisfy his thirst for "knowledge of the Laws of the Universe, because it is Science but the Universe is also the world: poetry, architecture, art, that is the beauty of its own abyss that exists in every heart" (Antonucci 2017, p. 61).

The installation takes the form of a sculpture modelled on the consistency of a black hole, faced through the study of the dimensional trajectories of particles around a Kerr black hole, deciphering its complex contents to compose the idea of his work.

From this derives an innovative experience both in the scientific and artistic-architectural field, since the sculpture takes the dynamic form of a loose knot, inspired by Möbius' tape, capable of involving people in the discovery of the Universe through artistic synthesis.

6 The Science Museum

The Moretta was identified as the ideal area where to locate the project following the demolitions that took place in 1940, indicated in the 1931 General Town Planning Plan, which provided for a renovation program between Ponte Sant'Angelo and Palazzo Farnese, including the district between Via dei Banchi Vecchi and Via Giulia, considered as important streets of the Middle Ages and the Renaissance (Fig. 1).

The museum presents itself as an "anomalous" element with respect to its context, an architecture with an extremely dissonant character that "renounces the pursuit of an own unitary image" but which affirms itself perfectly in its intended role, integrating itself into the complexity of the fabric, capable of "re-evaluating the different environmental and architectural realities that characterize the emptiness of Moretta" (Di Palma 1984, VIII, p. 6).

In the plan it could be associated with a vaguely rectangular figure with a suspended passage, connecting the main body to a square block located further north. Yet the shapes appear to be anything but regular, since first of all the south-west facing elevation, on the Lungotevere, certainly stands out, with an undulating shape that aims to follow the course of the river, thus expressing a dynamism linked to the watercourse emphasized by

Fig. 1. Maurizio Sacripanti. Science Museum in Via Giulia. General plan.

the prefabricated concrete rings that act as sunshades and that, in addition to the rhythm of the elevation, have a different rotation from each other. They seem to be static gears, but apparently in operation, which express the sensation of continuous movement of a machine that "compares itself with the linear dynamism that characterizes the circulation along the Lungotevere" (Di Palma 1984, p. 6).

These rings are closed by glass windows and are arranged along one of the two balconies at the same height, which serve as a link between the two exhibition halls. Four wide wall beams "similar to large arches arranged longitudinally to connect Via Giulia with the Lungotevere, bring to two by two the transversal beams placed to support the horizontal walking and covering structures" (Di Palma 1984, p. 6), thus defining the framework of the project (Fig. 2).

The main entrance can be reached from Via Giulia and once inside it is possible to admire a large space "modelled, at floor level, by the ground sloping slowly down from the Lungotevere towards Via Giulia" (Di Palma 1984, VIII, p. 6) and presumably dedicated to temporary exhibitions, perhaps artists who and inherent to the scientific field (as the master did in the exhibition "Cinque miliardi di anni"); moreover, from the ground floor it is possible to access the lecture hall and the mezzanine floor, which is used as a ticket office and as the museum's administrative and administrative room. The exhibition part, used for machinery and scientific instruments, starts from the "square" located on the first level that follows the course of the staircase on the lower floor and, through an elevated corridor that crosses the entire length of the main body, you can access the various floors that make up the concave part of the exhibition hall, which are designed as a series of "suspended plates" at different heights. In this way, a concave

Fig. 2. Maurizio Sacripanti. Science Museum in Via Giulia. Perspective view in Lungotevere.

and a convex part are generated, connected by several flights of stairs, which allow the visitor a comfortable use, respecting the idea of "bridge" stated in "Frontier City". The spatial articulation of the two exhibition halls "also shapes the course of the roofing level, which is configured as a raised square continuously connected [...] with the exhibition halls below" (Di Palma, VIII 1984, p. 6).

The interior of the building is illuminated not only by the large windows of the elevations, but also by various skylights placed on the roof that guarantee global and dynamic lighting, which changes with the passing of the hours. The different altimetric trend can also be seen in the underground area used for the laboratories and the car park below, creating an overall system that forms "an annular path that can be travelled infinitely, which, in analogy with the passage of time, allows the seasons to flow in a constantly changing spatial fluidity from which ever-changing and stimulating exhibition solutions emerge" (Di Palma 1984, p. 6).

The perspective seems to be absent with the consequent horizon line and wherever you look inside you can perceive a completely different and characteristic space depending on the point of view, thanks also to the considerable contribution that the changing light introduced inside it manages to emphasize the absence of a precise spatiality. Sacripanti puts all his design thinking into practice in a work that pragmatically synthesizes its contents, in fact the museum is the emblematic expression of several themes he deals with; first of all the relationship between space and time is evident and how these two dimensions interact with each other through the archetypes of architecture. It is precisely these elements of design syntax, if combined together in a new key of representation, that can recreate the effect of the fourth dimension: shadows and the absence of the horizon line combine with the material to form apparently static volumes, which instead are iridescent from the observer's point of view (Figs. 3 and 4).

"The Science Museum is the project of a stage: the stage of the universe. A project where there is no perspective.

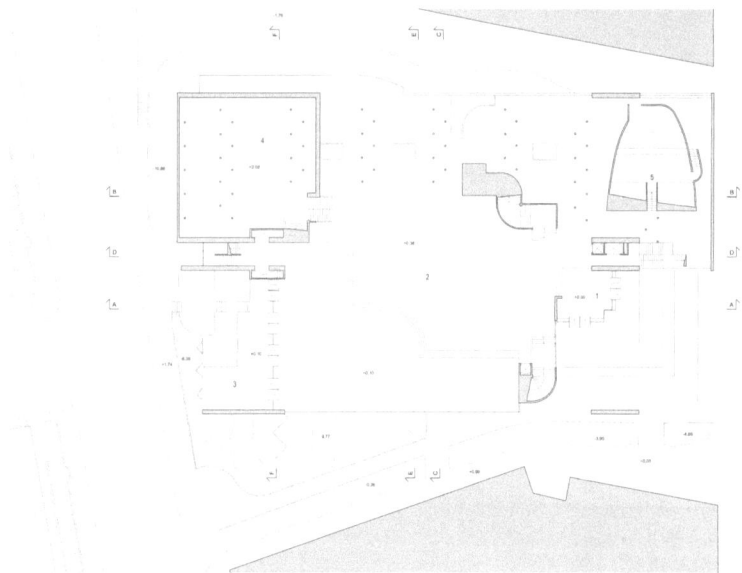

Fig. 3. Maurizio Sacripanti. Science Museum in Via Giulia. Plan at 1, 20 mt.

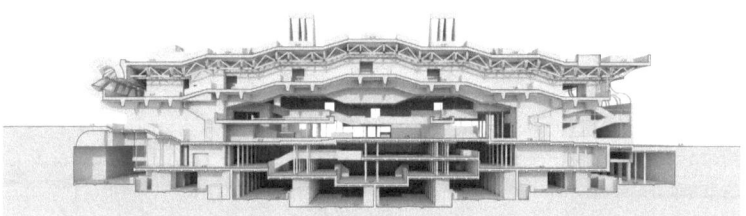

Fig. 4. Maurizio Sacripanti. Science Museum in Via Giulia. Perspective section.

From every point I perceive the genesis" (Serafini 1984, p. 6), as a "microscopic" representation of the sacripantian conception of the infinite "macroscopic" formed by stars, galaxies and black holes. In perfect analogy with scientific theories, which tend to demonstrate physical events through theories, the master uses his own "unconscious and artistic perception" to recreate the inexplicable in the tables in a completely new way of understanding architecture.

The 1983 project continues the path already begun a decade earlier with the publication of the book "Frontier City", in which the most significant projects that embodied Sacripanti's design idea of forming a utopian city characterized by his unrealized projects that, over time, "expanded" the fabric. The Science Museum (Fig. 5), besides being an integral part of this imaginary city, is also of fundamental importance, since it symbolically represents a link, a "bridge" between the real and the unconscious, between architecture and science, overcoming that "borderline" that separated the sphere of the idea from the technical architectural representation, that is, the feasibility of a work that would surely have made time visible and tangible, consequently taking human perception to a higher level simply by admiring the artefact in its entirety (Fig. 6).

Fig. 5. Maurizio Sacripanti. Science Museum in Via Giulia. Bird's-eye view.

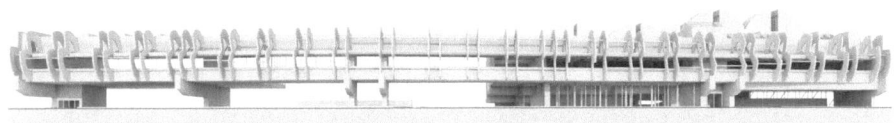

Fig. 6. Maurizio Sacripanti. Science Museum in Via Giulia. Facade.

7   Conclusion

The following essay is mainly intended to tell the design method of Maurizio Sacripanti, addressing a path that is intended to help the understanding of the Science Museum in Via Giulia; however, it was also intended to tell the expanding thought of a master of Italian and international architecture.

Sacripanti's architecture is an almost "paternal" architecture, which tends to visually expose the complexity of modernity with the troubled elaboration of drawings exemplifying a unique way of designing, but which in reality collects a whole series of well known notions: time as a design element, expressed through the absence of perspective, the use of shadow and new engineering solutions to express a Boccionian vision of architecture in constant movement.

All this is told in words and elaborated graphics in the project for the Science Museum, which takes on the characteristics of a sacripantian design-manifesto architecture, since it conveys the multiple arguments of a life based on architectural experimentation. It is unlikely that Maurizio Sacripanti wanted to create an architectural current of his own, but rather he wanted to create his own architecture, far from schemes, and that it was more a mental and artistic exercise than a sterile and infinitely reproducible elaboration.

Credits.   While sharing the topics covered in the essay, paragraphs 1, 3, 5 and 7 were written by Francesco Maggio, paragraphs 2, 4, 6 were written by Fabrizio Provenza.

References

Capiato, E., Cresciani, G., Forlini, F.R., Mancini, M. F., Antonucci, M.: Progettare il mutevole. Nuovi studi su Maurizio Sacripanti, AAA ITALIA, numero speciale (16). http://www.aaa-italia.org/wp-content/uploads/2018/03/AAA-numero-speciale-MAURIZIO-SACRIPANTI.pdf. Accessed 10 Sept 2019

Di Palma, W.: Museo della Scienza, Roma Comune - Progetti per la città. Rivista mensile di informazione e dibattito **VIII**(6), 3–8 (1984)

Giancotti, A., Pedio, R.: Maurizio Sacripanti: altrove. Testo & Immagine, Torino (2000)

Giancotti, A.: Le immagini verranno: Antologia di scritti di Maurizio Sacripanti. Edizioni Nuova Cultura, Roma (2015)

Neri, M.L., Thermes, L.: Maurizio Sacripanti maestro di architettura 1916–1996. Gangemi, Roma (1998)

Purini, F.: Maurizio Sacripanti e il disegno d'architettura. In: Albisinni, P., e De Carlo, L., (ed.), Architettura-disegno-modello. Gangemi, Roma (2011)

Sacripanti, M.: Il disegno puro e il disegno dell'architettura. Fratelli Palombi Editori, Roma (1953)

Sacripanti, M.: Città di frontiera. Bulzoni, Roma (1973)

Graphic Expression of the Transformability of Domestic Space. From Le Corbusier to Andrés Jaque: A Real and Symbolic Evolution

Jose Luis Cabanes Ginés[✉]

Department of Architectural Graphic Expression, Polytechnic University of Valencia, Valencia, Spain
jlcabane@ega.upv.es

Abstract. Coinciding with the centenary of the opening of Staatliches Bauhaus in Weimar, our aim was to succinctly analyse how the appearance of the concept of the rational home of the Modern Movement, which was derived from the Bauhaus, gave rise to the first examples of transformable domestic space. From this first stage we describe the contributions of Le Corbusier in the Weißenhofsiedlung in 1927, Gerrit Rietveld in the Rietveld Schroder House in the same year, and Pierre Chareau in the Maison du Verre in 1932. The simultaneous analysis of the interior mobility elements and their graphic expression, brings us to a second more modern stage, in which both, the elements related to the flexible configuration of domestic space, and the graphic media the architects use, are somewhat more ambitious. For this we describe the work of Rem Koolhaas and the spanish architects Iñaki Carnicero, Vicente Guallart and Andrés Jaque. We thus highlight the real and interesting evolution in the means, and symbolically in their graphic expression, of this essential value in the concept of the modern home, which is part of the contemporary pattern of family unity, and is characterized by the redefinition of the boundaries between the different ways of inhabiting private and public spaces.

Keywords: Domestic space · Graphic expression · Transformability

1 Introduction and Background

April this year saw the 100th anniversary of the opening of the Staatliches Bauhaus in Weimar. The visionary architect Walter Gropius, its first director, published its Founding Manuscript in the form of a four-page booklet, with the strict intention of combining all the artisan disciplines and re-focusing its work from a rational point of view, "the form follows the function", in such a way that the result was a number of products accessible to everybody, for the sake of greater social equality.

The defiance of the traditional standards, which began in the Bauhaus workshops, the authentic epicentre of the new teaching focus, was a radical change in the history of architecture and other fields. Nobody has any doubt that the Bauhaus was the precursor of the Modern Movement in architecture, which from the beginning was also very attentive

L. Agustín-Hernández et al. (Eds.): EGA 2020, SSDI 6, pp. 367–377, 2020.
https://doi.org/10.1007/978-3-030-47983-1_33

to the new concept of home, which not only had a social character but was also committed to modernity.

It is therefore nothing strange that the radical proposal of the semi-detached houses Le Corbusier built in Weißenhofsiedlung in Stuttgart in 1927, breaking with the traditional rigidly compartmentalized domestic space, had a lot to share with the origins of this new dwelling-house culture (Stick 2012).

This new arrangement of the transformable domestic space in this case, was characterised by a clear and adaptable solution of the day/night configuration of the two houses by means of movable boundaries and adaptable furniture. It is also necessary to consider the extraordinary drawings in which it appears, which were also pioneers of the reflection on the conscious use of the means of graphic expression in projects (see Figs. 1 and 2).

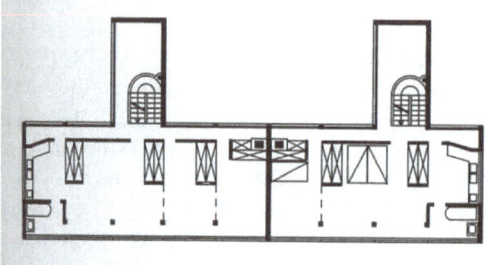

Fig. 1. Modular composition of the dwellings (5 in the large and 4 in the small), supported by the expression of the daytime configuration in the large dwelling, and the night-time in the small. The entire drawing unites content and graphic intention, thus advancing this new focus of architectural graphic expression with very basic means.

Fig. 2. Elements of day/night-time adaptability in the Stuttgart houses: sliding partitions and closet with wardrobe and sliding bed.

In this way, it all gives rise to sparsely explored line of analysis that associates the basic ideals of the Modern Movement's concept of dwelling, as social concern and attention to contemporaneity in its widest sense, together with the evolution of its graphic support, which also exceeds the traditional purpose of the objective and simple representation of the buildings.

2 Discussion

2.1 The Contribution of the Pioneers

Our analysis aims to give a small historical tour of this question, including some other milestones of important examples of domestic architecture with adaptable spaces, projected with the support of drawings, that reflect these intentions by different graphic means. In this regard, we can point out two other examples of the pioneering epoch: the 1924 Rietveld house, and the combined dwelling – doctor's surgery known as the *Maison du Verre,* designed by Pierre Chareau in 1932.

In the Rietveld Schroder House, a brilliant work by the best known representative of the neoplasticism group De Stijl, we can highlight how the architect implemented a system of sliding partitions to form the living room, piano room and bedroom, which occupied three-quarters of the building's floor area. The result is a highly flexible spatial configuration, and thus highly polyvalent, as demanded by the client herself (see Fig. 3).

Fig. 3. Rietveld Schroder House, Utrecht, Holland, 1927. Architect: Gerrit Thomas Rietveld. The rotated axonometric drawing, thus less biased, (foreshortened in the illustration), and the striking use of graphic resources (linear attributes and flat colours), reveal the intention of achieving a very adaptable space, with different day/night-time configurations. The sliding elements are drawn symbolically, with an original lined in their intersection with the ceiling, and their lines contrast with the aspect of the rest of the drawing for their faintness, anticipating the idea of openness. The convergence of the sliding elements in the central part reinforces the prevision of their joint use, impressively modulating the spaciousness of the area.

In contrast with the intention of these architects of adapting domestic spaces to day/night-time configurations, in the combined house/surgery for Dr. Jean Dalsace and

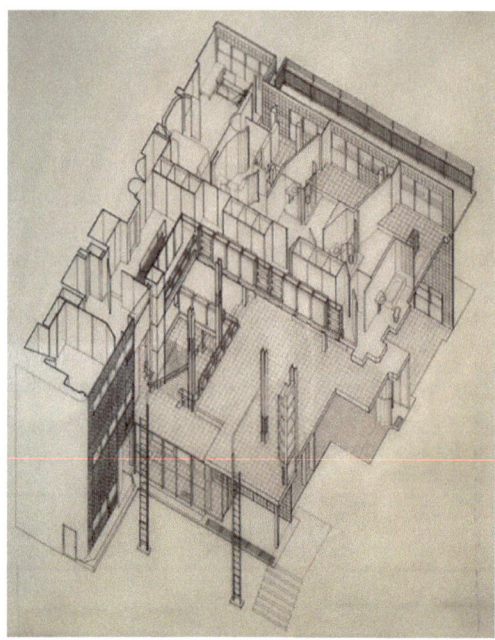

Fig. 4. Axonometry of the Maison de Verre by Pierre Chareau. Drawing by Kenneth Frampton et al., 1985 ("Pierre Chareau: Modern Architecture and Design Exhibition"). Although no known drawings by the author exist, the building's inherent worth drawings of this type are of great interest. The double height between the daytime zone and the second level with bedrooms can be seen, for which, nothing less than one of the most iconic elements is suppressed, i.e. the moulded glass tile façade over the entrance (left foreground). The enormous level of detail suits the exquisite design of the building's elements and shows, among other things, the device thought up by Chareau to control the privacy of the bedrooms: a complex handrail high enough to prevent one from looking down on the ground floor, formed by opaque wooden panels and others of metal in the form shelves, behind which a corridor with wardrobes filters the access to the interior of each room. The retractable stairs can also be seen in the left background.

his wife designed by Pierre Chareau, the aim is to control the space in order to regulate the privacy of certain uses, which always take place on two different levels of the building, by means of very original movable elements.

We can highlight here two spaces with this quality: firstly, Dr. Dalsace's study on the first floor, which contains the daytime zone, and is connected by a private stairs with the surgery on the ground floor. This office can also be integrated with the living area by means of an enormous mechanically operated sliding partition, thus being transformed into a large open space, with views of the front and rear sides of the building (see Fig. 5).

Secondly, we have the retractable staircase that connects the second-floor bedroom with a prívate sitting-room at the back of the first floor, looking onto the interior garden and communicating both spaces. It was designed to be manually operated by a curious folding/sliding system that matches the stairs and the handrails to the base (see Fig. 4).

Also of interest are the diverse transformable furniture designs, considerably more sophisticated than those of L. C. in Stuttgart, such as in the main bathroom, equipped

Fig. 5. Maison de Verre, front view of the living room onto the interior garden. On the ground floor the enormous sliding partition can be seen, that allows the room to be configured with continuity from the front to the back facade, or to close off the doctor's working zone, which communicates with the clinic by means of internal stairs. On the upper floor the modular screen is high enough to prevent a direct view of the lower floor, except in the area of the shelf units in the central position, and in a very limited way, as can be appreciated.

with mechanisms activated by pivots and hinges fitted to the perforated plywood panels on the cupboards, or on glass screens, that compartmentalise the space and provide privacy.

This building is in fact both a complex dwelling place, and doctor's consulting rooms, that strictly interpret the corbusierian ideal of "living machine" by means of elaborate structural, constructive and equipment solutions, and besides, as Kenneth Frampton has pointed out, is able to "*offer by means of the fluidity of the floor plan, the standardization of its components and mobility of its parts*", a novel and demanding control of the relationship between its public and private spaces (Frampton 1969).

2.2   The Contemporary Contribution

However, the real impulse to this binomial of planning and visual intentions derived from the graphic media, arrived when architects became aware of the importance of this strategic feedback relationship in the process of designing buildings (Cabanes and Navarro 2017). Although this approach began with the Bauhaus abstract advance guard, as we have just seen, it did not come to fruition until the 1970s, when the New York vanguard, with Kenneth Frampton and the Five Architects, as its most distinguished advocates, was internationally recognized (see Puebla 2002).

We can here analyse four contemporary examples of this fecund and parallel evolution, between the mutability of internal domestic space, now much beyond the typological supremacy of the rational dwelling of the CIAM, and the symbolism of the drawings on which it was based, and therefore with a certain degree of subjectivity and polysemy, as part of the new creativity.

In first place we put Rem Koolhaas's *Maison á Bordeaux,* built for a married couple. The husband was confined to a wheelchair and this posed a huge challenge to the architect because he demanded "a complex house, as the house will define my world" (Koolhaas 1998). It was thus a combination of eloquent formal features: it was situated on top of a hill, a opaque second floor (with small anti-claustrophobic windows) floating over an open-plan first floor, or the large transversal beams, especially the one over the roof held by a tensing cable at one end, which heightened the anti-gravitational effect of the first floor terrace.

The interior contained a 3.50 × 3 m. elevating platform that "changes each floor and its function. Sometimes it remained flush with the floor and others it floates above it" (Koolhaas 1998), and it could also stop between floors. Around each floor there is a shelf wall that contains all that the client could need, as everything result accessible from a wheelchair, as the platform could be adapted to the appropriate height (see Figs. 6 and 7).

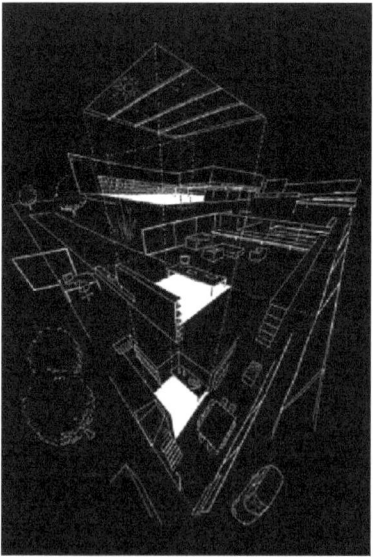

Fig. 6. Schematic drawing of the *Maison á Bordeaux* by architect Rem Koolhaas (OMA), 1998. The choice of a downward conical perspective favours the interpretation of the complex internal space, and the communication of the main floor with the adjacent garden (on the left, with transparency). The white-shaded areas show the situation of the elevator on the building's three levels. The dotted lines symbolize the enormous vertical hollow that pierces the house and the lateral shelf that accompanies it, thus supporting the movement of the *"ejectable room"*.

Fig. 7. Sequence showing the enormous elevating platform furnished with a desk and the changeable spatial effect on each floor level caused by its movement, seen here from the open ground level. In a certain way, it appears to be a much more powerful mechanical device, with a similar approach to the retractable stairs in the *Maison de Verre*.

In the two semi-detached homes designed by Iñaki Carnicero in the mountains north of Madrid, with a program similar to that of Le Corbusier, two interesting questions were considered regarding transformability: one was the functional program, and the other was communication with the immediate surroundings. In the former case, according to the architect, "besides satisfying the client's requirements, we offered the possibility of eventually transforming both houses into one, and so had to consider a wider scenario of adapting to future needs. It was thus a case of re-formulating the commission and designing a single-family home for two families" (Carnicero 2007).

This original idea of re-inventing the commission gave rise to a two-fold modular arrangement of the floor area, with two well defined longitudinal zones, in the rear with stairs and a service area, and in the front with a daytime zone on the ground floor and bedrooms on the first level, with a total of 6 transversal modules, 3 in each house.

We can also highlight the arrangement of the ground floor, with an open plan in the front, and clear access to the outside terrace, dedicated to the daytime common

Fig. 8. Casa *Pitch*, Iñaki Carnicero, 2012. Cross-sectional view of the ground floor (with three modules) and night-time view of dining zone and terrace. The black areas with white lines in the plan outline the rear stairs and service zone, and contrasts with the open space of the other areas. There is a clear boundary between the living areas and the front terrace. The composition in series in both directions, clearly favours the intention of ordering the ambivalent internal organisation, with the option of either one or two dwellings, and coincides with the approach used by L. C. in Weissenhoff, as we have seen, although with a different intention as regards spatial flexibility.

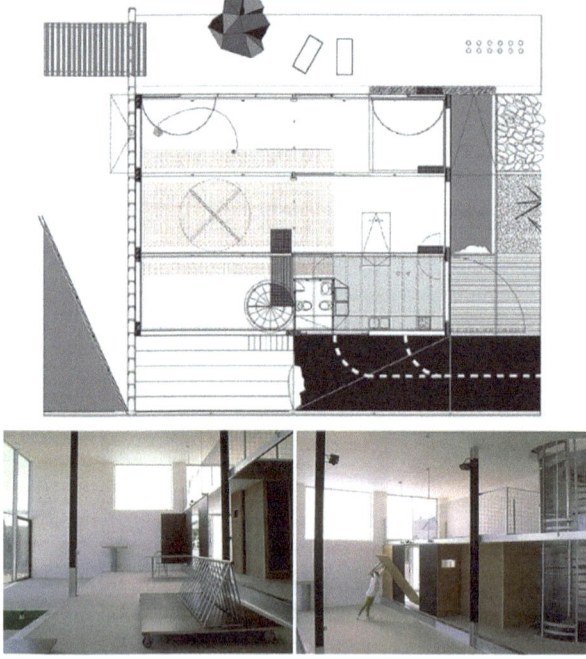

Fig. 9. Metropolitan Loft, Lliria, Valencia (Guallart 1994). In a highly symbolic drawing, with inexplicable suggestions that populate the living-room, an elaborate attention to lines can be appreciated, with different types of stripes, line patterns and shaded areas, which together with other features, express the profusion of mobile elements (pivoting, foldable, etc.) that can transform the open space that invades all the interior (Guallart 1994).

areas. This involved a new approach to domestic transformability: the possibility of total integration with the immediate exterior, together with maximum dilution of the physical internal spatial limits (see Fig. 8).

Also worthy of note are another two examples of contemporary Spanish architects in favour of domestic utopia, as the changeable identification of the individual with the forms associated with contemporary living: the *Metropolitan Loft* by Vicente Guallart (see Fig. 9), which decontextualises this open-plan urban style, and the prototype of the *Rolling House* designed by Andrés Jaque (see Fig. 10), which includes a mutant "intimacy capsule" in another mobile installation, thus multiplying the adaptable effect.

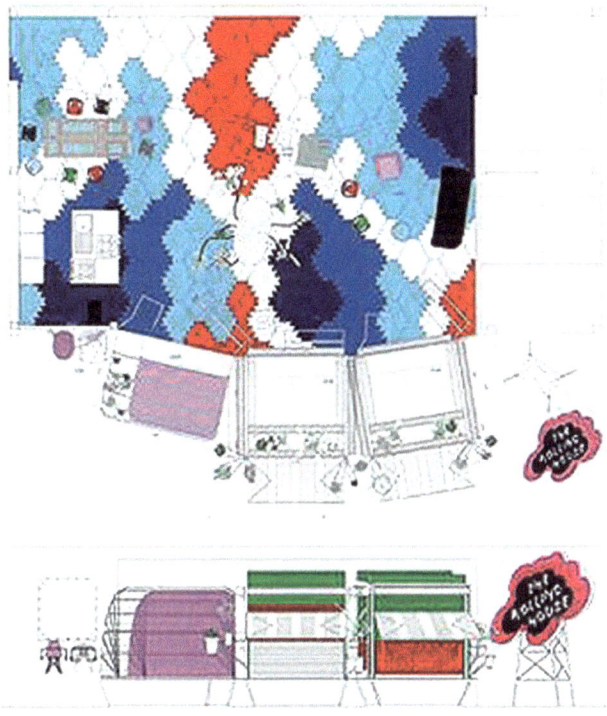

Fig. 10. Rolling *House for a Rolling Society* (Jaque 2009). The projections of the prototype, which include concepts such as an "intimacy capsule", extensions, or shared equipment, are drawn alternating recognisable elements, such as the kitchen and bathroom/shower, together with others symbolic, such as the Hallmark-type greeting card on the outside water tank, or the decorative panels of walls and floor. The result is very eloquent, full of visual values, such as the colour palette, which suggests a new format of changeable rolling domestic space, anticipated, in addition, in its real perception (see Jaque 2009 and Stick 2012).

3   Conclusion

Transformable domestic space, i.e. its ability to assume configurations and even real or visual changeable limits, has been part of the modern concept of a dwelling house from the beginning, as we have seen. To the traditional elements that sustain it, such as curtain walls and adaptable furniture, we can add more recent features, such as elevating platforms, an ambiguous functional program, the physical limits themselves, or the mobility associated with the concept of a rolling home, among others, as we have just seen.

At the same time, the graphic language by which it is expressed, has evolved from the traditional drawings of the early stages, passing through the important turning point of the New York neo-vanguard, to the present time, in which architects, conscious of the role of drawings as "thought-provoking elements" (Alba 2016), aim to anticipate hypothesis on the perception of a building, by means of images that are composed as highly symbolic graphic manifestos, full of suggestions.

Reflecting on this question also implies an approach to the renewed concept of a home, as in the Bauhaus ideals, committed to contemporaneity: a sustainable house framed within the present pattern of the concept of the family, in which the three dualities of its traditional features "work-leisure, public-private, exterior-interior", are again combined. (Amann 2019).

Therefore, "the universalist attitude of the Modern Movement is no longer a viable position in front of the challenges of contemporary society", as regards the present situation, in which "the domestic space-time is increasingly urban and expanding in the city" (Amoroso 2017). In fact, this new identity of private space, physical ambit of the family model of our times, is essentially based on a more intense interaction with the city, through a wide sense of its poly-functionality, and the subsequent adaptability of its elements: the house that assumes new public uses, such as working-from-home, and, at the same time, other private uses, such as restoration or care of clothes, take place in the urban scene. The examples we have treated here clearly represent responses to these challenges.

To conclude, we have tried to analyse and describe how the present architectural avant-garde is able, for the first time, to combine the ideals of domestic space related to spatial transformability, that represent an essential quality of the contemporary concept of private space, with all the values pertaining to conscious graphic expression, full of resources and with highly symbolic contents.

References

Alba, M.I.: La mirada atenta. Aproximaciones a la creación arquitectónica. EGA Rev. de Expresión Gráfica Arquitectónica (27), 88–95 (2016)
Amann, A.: El espacio doméstico, la mujer y la casa. Tesis Doctoral, p. 78. http://www.oa.upm.es/. Accessed 23 June 2019
Amoroso, S.: De género y espacios: hacia una deconstrucción de lo doméstico. Rev. Asparkia (31), 113–130 (2017)
Cabanes, J.L., Navarro, P.: La expresión de los valores básicos del proyecto arquitectónico a través de dibujos esquemáticos. La aportación de la nueva generación de arquitectos españoles. EGA Rev. de Expresión Gráfica Arquitectónica (31), 34–43 (2017)

Carnicero, I.: Pitch House, Madrid, Spain (2007). https://ricastudio.com/portfolio/pitch-house/. Accessed 17 Sept 2015

Frampton, K.: Maison de Verre. Perspecta (Yale Arch. J.) (12), 83 (1969)

Guallart, V.: MetapolitanLoft. Liria, Valencia, Spain (1994). http://www.guallart.com/projects/metapolitan-loft. Accessed 17 Sept 2015

Jaque, A.: House in Never, NeverLand, Ibiza (2009). http://www.andresjaque.net/cargadorproyectos. Accessed 2 4 May 2016

Koolhaas R.: OMA: Casa y Piscina en Burdeos (la Casa Sostenible). El Croquis (131/132), 70–96 (1998)

Puebla, J.: Neovanguardias y representación arquitectónica. 1ª edn. Edicions UPC, Barcelona (2002)

Stick, E.: La arquitectura moderna como experimento: la Weissenhofsiedlung y la relación entre la técnica y la forma. DeArq (10), 102–117 (2012)

Concept and Creation

Thinking with the Hands. The Sketchbooks of the Architects

María Asunción Salgado de la Rosa[(✉)] [ID], Javier Fco. Raposo Grau[ID],
and Belén Butragueño Díaz-Guerra[ID]

IGA Department, E. T. S. Arquitectura, Universidad Politécnica de Madrid, Madrid, Spain
`mariaasuncion.salgado@upm.es`

Abstract. Why do architects draw? The obvious response is that it is the most immediate way to formalize what does not exist. The drawing allows us to represent and analyze any pre-existing element, in addition to encourage experimental processes that lead to the project. The way to addresses these speculative drawings is as personal as the project itself. The key of the narrative is in the genesis. When this type of process is developed on a sketchbook, the order of the pages enables a consecutive reading of the story line. Observed as a whole, the sketchbook work as a procedural manual of the creative thinking for the architect. It is very rare that its content is made public, since it keeps the "guts" of the design thinking. This article makes a review of the most significant sketchbooks in history, focusing on those that display a particularly speculative development. Converted into cult objects, the dissemination of these architects' sketchbooks, has become mainstream, claiming their suitability in all fields of architecture, including professional and academic fields.

Keywords: Drawing · Sketchbook · Architecture · Speculation

1 Introduction. Drawing Is Thinking Architecture

Architects share a common need to draw. Drawing is inseparable from Architecture as it helps to visualize and transmit the imaginary spaces created by architects, making them accessible to others. The drawing is a very powerful communicative tool, but it is also an instrument that requires certain skills and concentration on the making. The act of drawing generates a state of reverie that helps the brain to project all its resources in a single direction, reaching the ideal state of mind for a creative act. In short, thinking architecture involves drawing.

It is important to elaborate on the idea that the drawing triggers the creative process that eventually leads to the project. The hybrid character of the architectural drawing, including artistic resources and stablished technical codes, particularly contributes to experimentation. When combining the technical language with other artistic techniques, we discover a means of expression with endless nuances that are subject to multiple interpretations and contribute to encourage the self-reflection during the process. Juan

L. Agustín-Hernández et al. (Eds.): EGA 2020, SSDI 6, pp. 381–392, 2020.
https://doi.org/10.1007/978-3-030-47983-1_34

Antonio Ramirez states that "the drawing has always been an essential scientific instrument and a means of artistic expression" (Ramírez 1990, p. 11). In any case, it is irrelevant how or what to draw, what is crucial is the development of a process in continuous transformation. Picasso reminded us that "the inspiration exists, but it has to find us working" (Palomo y Triguero 2013, p. 169).

One specific aspect that identifies the graphic process in the project, is the non-linear character. It is a thought that is constantly evolving, which allows the appearance of new paths that will later be taken or discarded. This coming and going characterizes the initial drawing of the architects, making very difficult to verify the time frame of the graphic documentation of a project to establish a parallelism between the thinking of the architect and his graphic work, it is fundamental to work with materials that were dated or developed on a sequential format, such as a sketchbook or a journal. To the authors, the condition of portability of the sketchbook facilitates its use in any circumstance allowing, as well, to review the content. For the observer, the objectual character conferred by the binding, contributes to increase the attractiveness of the sketchbook, as an incunabulum collectible object.

The goal of this article is to show the relationship of the architects with a media such as the sketchbook, understood as a means of expression and reflection, but also as a way to catalog the creative process. Le Corbusier reminds us that "drawing in a sketchbook teaches first to look, then to observe and finally, perhaps to discover... and it is then that inspiration might come" (Farrelly 2014, p. 16). From the notepads, inevitably linked to data collection, to those that unleash the experimental drawing, what we find in their pages is the reflection of the personal evolution of each architect.

2   Travel Notebook as a Germ of the "Diary of an Architect"

Both the sketchbook of architect and the travel notebook, are heirs of the illustrated notebooks developed by William Blake at the end of the 18th Century and William Morris at end of the 19th Century. They were not developed as a formal element to accompany a discourse, but as an introspective experimentation linked to everyday events. In successive decades, many artists share with the architects the interest in the development of these sketchbooks as a means to pour their experiences in an iconographic language. It is throughout the 20th century, when the sketchbooks begin to acquire their own character that go beyond a mere notebook. In their pages, we can find drawings coexisting with accounting records or written reflections. They might include other elements such as a news clipping, leaflets or all kind of receipts. This hybridization is characteristic of the drawing of architecture, conferring a unique character to its pages, related to a particular way of looking and thinking of each author.

All illustrated notebooks share that private and portable condition that turns them into informal spaces for observation and reflection. For that reason, the travel notebooks of the architects are especially popular, as their sketches differ from any other conventional spatial drawing. The versatility of the architectural drawing, allows to shift from a more technical language to a more representative, analytical and even proactive one, transcending the mere representation of architecture. These graphic categories range from the more rigorous survey drafting with diagrams of plans and elevations, to more artistic representations that, beyond their geometry, they show material, sensory and lighting characteristics.

When making notes in a travel sketchbook, we start to build a discourse that integrates the initial states of the project (Sainz 1990, p. 143). All the drawing share "an interest in drawing to learn, to analyze or to record an architecture or a related experience" (Marcos 2014, p. 471).

Many architects and students embrace this vital experience to understand a space that, previously, they only knew through photographs. Fortunately, many travel sketchbooks are kept, that remain as proof of this practice. In the case of Le Corbusier's sketchbooks, they represent the perfect portrait of this process of personal evolution. In their pages, we can appreciate the evolution of his personal interests toward the architecture, corresponding with the construction of the great buildings in Europe, after the First World War. His drawings include portraits developed in his visits to brothels in Paris, as well as written reflections that show his interest for urban development, philosophy and poetry. Le Corbusier himself will declare, "I prefer drawing to talking. Drawing is faster and leaves less room for lies" (Uribe 2015) (Fig. 1).

Fig. 1. Le Corbusier's sketchbook, dimensions 13 × 20.8 cm. Le Corbusier, 1918–19. Source: De Franclieu, F: *Le Corbusier Sketchbooks. Volume 1*, 1914–1948.

Widely disseminated in different publications, the travel sketchbooks of Le Corbusier have been recurrently subject of study, prompting several generations of architects to relive his journeys, but also to become the protagonists of their own initiatory journey. The sketches made by the architect at the beginning of the 1950s in Chandigarh, not only pointed his vision about the future capital of Punjab, but also shared the imprint of a day-to-day landscape symbolized by the sketch of a buffalo, as the paradigm of the domestic animal.

Four decades later, a group of friends, that included the architects Elías Torres, Benedetta Tagliabue and Enric Miralles, travelled to India to visit the architecture of Le Corbusier in Ahmedabad and Chandigarh and the works of Louis Kahn in Dakha. During the trip, the sketches developed by Miralles and Torres, captured a large part of the experiential essence of Le Corbusier, including the buffalo, but this time with a built landscape in the background (Maestre 2016, p. 72) (Fig. 2).

Other recognized architects were previously influenced by Le Corbusier's travel sketchbook. In the late 1950s, a mature Louis Kahn embarks on a journey to the chapel

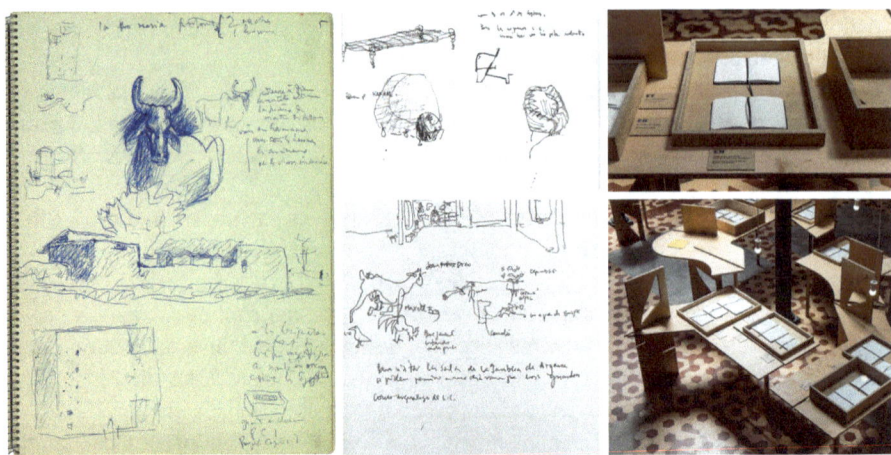

Fig. 2. Sketch of a buffalo and house in Chandigarh. Le Corbusier, 1951. Source: © Fondation Le Corbusier, Paris/Travel sketches in Chandigarh. Enric Miralles and Elias Torres, 1992. Source: Fundació Enric Miralles/Exhibition images "Els colors de Le Corbusier a L' India: el viatge d'Enric Miralles i Elías Torres a la Porte Email, Curator Josep Quetglas, 2019. Source: Fundació Enric Miralles.

Fig. 3. Inner perspective, Ronchamp, France. Louis Kahn, 1958. Source: McCarter 2005/Inner perspective, Ronchamp, France. Norman Foster, 1958. Source: Norman Foster Foundation.

of Notre Dame du Haut in Ronchamp. The same year, a very young Norman Foster performed the same journey searching for an experience that can only be achieved with the direct exploration of the built space (Fig. 3).

The graphic overview of the same building portrayed in the respective sketchbooks of these two architects, reflects coincidences and differences of their graphic "glances" (Bravo de Laguna 2014, p. 151). The initial motivations of each one of them, establish a gradation in their drawings, from the descriptive notes of "the student" Foster, closest to the classic travel notebook, to studies of light that captured the interest of a "mature" Kahn. In both cases, the spirit of their respective sketches on Ronchamp, show more similarities with the travel sketchbooks of Le Corbusier that with the sketches outlined by Renzo Piano almost fifty years later on the same place. To capture the essence of the area of intervention, Piano made a trip to Ronchamp in 2006, developing a schematic data collection, to capture the qualities of the project site, as Le Corbusier did in 1950. His drawings exposed the visual relationships between new and pre-existing conditions, exploring non-material aspects of the architecture. Piano's motivations predetermined his data collection during the journey, according to his design process (Fig. 4).

Fig. 4. Sketchbook for Ronchamp Gatehouse and Monastery, France. Renzo Piano, 2006. Source: RPBW.

Gradually, the travel sketchbook, gives way to a mature sketchbook of the architect, where the memories of the journey are replaced by the persistence of the reflections experienced at specific times. The note-taking gives rise to the experience, to reach the proposition, always from a subjective point of view.

3   The Sketchbook, a Place of Thought

Outside the context of the journey, the sketchbook is understood as an extension of the working strategy, a place to build a line of graphic thinking. The informality of the format, predisposes the author to address a fast drawing, suitable both for taking notes and for the spontaneous draft. The drawings are intuitive, slightly developed, and mainly based on improvisation. It is a private graphic material, hardly accessible, as it hides the heart of the author's process. In the sketchbook, there is room for error and incompleteness, as it is a place for problem-solving. The role of the sketchbook in the design practice of the architect, varies depending on its use. So far, the article has focused on proposals that pursued the construction of a personal thinking through representative and analytical drafts of past architectures. In these sketchbooks, the drawings were intended to understand the reality before us (Fig. 5).

From this point on, we will focus on sketchbooks conceived as an ongoing process of action-reflection. With gestural drawings ranging from improvisation, to the obsessive exploration of one single element, with slight variations, we perceive a graphic introspective, linked to the working process. The drawings are developed as a thinking tool, not with the purpose of communicating. They allow us to observe first-hand the initial phase of the design process, that are rarely shown.

We can find more informal sketchbooks such as those developed by Sou Fujimoto, Alvaro Siza or Peter Märkli, with plenty of gestural drawings whose content is apparently disconnected. There are, as well, more developed sketchbooks from authors such as Scarpa, Tony Fretton, Alberto Ponies, Mark Smout or Steven Holl, that use a relation-based drawing to address the genesis of their projects (Fig. 6).

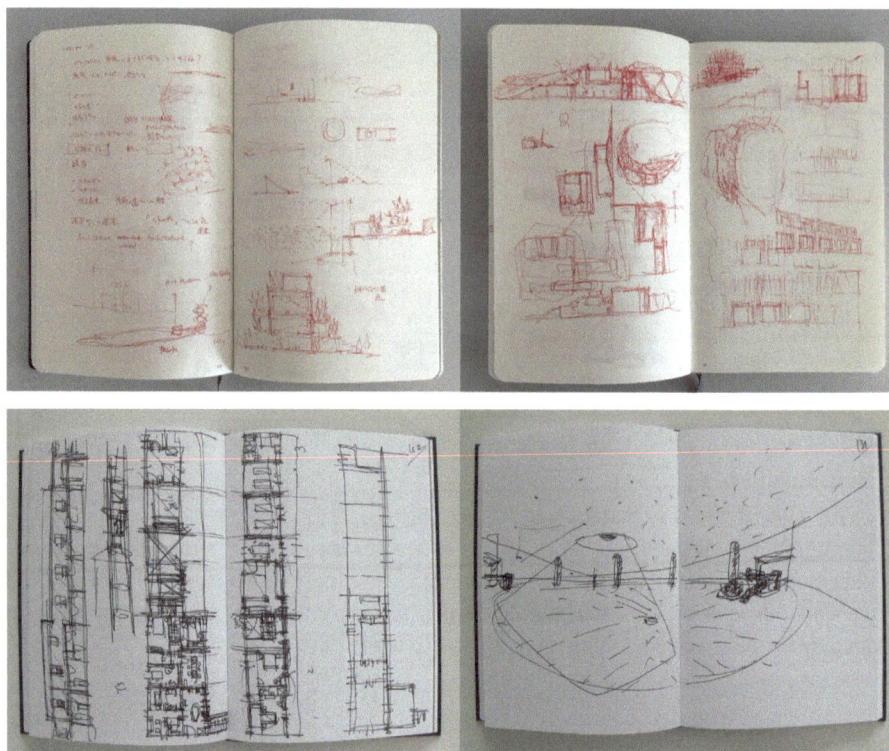

Fig. 5. Sketchbook. Sou Fujimoto, 2012. Source: Sou Fujimoto Sketchbook/Sketchbook. Souto de Moura, 2012. Source: Souto de Moura Sketchbook No. 76.

To illustrate this working methodology associated with the sketchbook, we will delve into the work of Adolfo Natalini and Lebbeus Woods, whose drawings imagine fantastic scenarios and enchanting graphic artifacts.

Superstudio (Natalini's office) presented the project "Continuous Monument" in the summer of 1969, in Graz. The sketchbooks he developed a few months before, include a large part of his graphic creative process. The first sketches show a conception of the project as a belt around Florence, progressively reaching the planetary dimensions aimed by the final project. This sketchbook also contains the genesis of the best-known collages of the group and displays the symbolism of the project. In opposition to the direct language of the completed project, the sketchbook is understood as a dialog between the architect and his environment, an exchange that contributes to put his own speech in order. Its pages reveal not only the project, but the intra-history that surrounds the creative process (Fig. 7).

In architecture, the context surrounding the conception of the idea is a determining factor. In that sense, the sketchbook helps placing the project in his time. In 2001, concurring with the change of millennium, Lebbeus Woods suffered a series of health problems that forced him to reconsider many concepts, according to his own words (Woods 2009). That same year, during the installation of his proposal "The Storm" in

Fig. 6. Sketchbook of Lanzarote Envirographic Architecture project/"Retreating Village" a mobile settlement for a collapsing landscape. Sketches on notebooks. Mark Smout, 2011. Source: Smout Allen

New York, the attack on the Twin Towers took place. In this interlude, he developed a sketchbook that included life-changing experiences, starting with his heart surgery and the end of the world as we knew it. Dated between September 2001 and January 2002, the sketchbook included a series of graphic variations, that recurrently abounded in structural processes of growth. Made in black ink, many drawings connected with his installation proposal for the 2002 Whitney Biennial, and they also included images taken after the destruction of the towers. In his drawings, Woods experimented with what he called "architecture in tension" in which the human being had a limited control of his own nature. Along these pages, Woods unleashed this "active" architecture, alien to the static laws, through graphic exercises that completed a sketchbook which was, unfortunately, his last one (Fig. 8).

The graphic language of these sketchbooks is not intended to communicate, but they convey a lot about the concerns of their authors. The drawings can often be considered ugly, messy or incomplete, but their beauty lies in the overview, allowing the architect to register, record, and build…in short, to think.

Fig. 7. Adolfo Natalini's Sketchbook, development of the Continuous Monument for the 'Grazerzimmer' presented at the Trigon Biennial of 1969, Graz. Adolfo Natalini (Superstudio), 1969. Source: Sketchbook 12 and the Continuous Monument: Adolfo Natalini © Adolfo Natalini

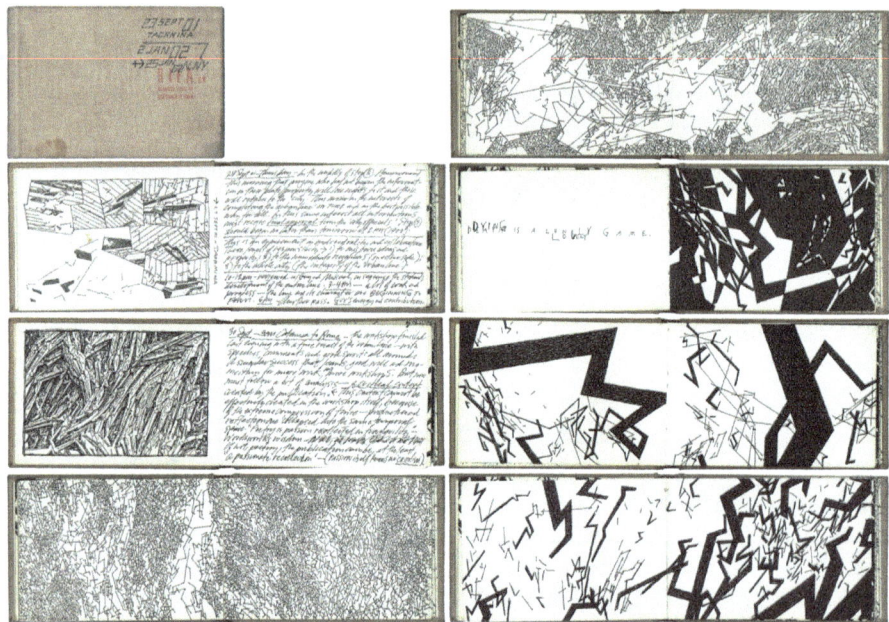

Fig. 8. Sketchbook 01-3 (the last). Lebbeus Woods, 2001–2002. Source: Lebbeus Woods

4 The Architect's Sketchbook as an Artistic Object

Although it is not the norm, certain architects' sketchbooks maintain a consistent line from beginning to end, as they were conceived as a unique piece in itself, regardless the project deployed. Despite sharing format with the artist's notebook, there is no direct parallel between them. While the "livre d'artiste" is a work where the format is a whole, by using the sequential media that characterizes the book as a visual instrument (Crespo 2012), the architect's sketchbook is developed regardless of any a priori, being the result of a process of an intimate and paused reflection.

The architect's sketchbook is rarely conceived in its entirety, as a linear reading document. It is a casual construct that reflects a vital moment of design. In contrast to the artist's book, these sketchbooks are rarely addressed from a single perspective or completed to fill their pages with graphic variations on the same theme. On the contrary, they are a direct reflection of the change and evolution experienced by their authors. However, there are a few exceptions in which the architect, actually, intended a full sketchbook to the experimentation developed on a single process. We will highlight two unique examples belonging to the architects Zaha Hadid and Peter Wilson which, at a given moment, used a complete sketchbook to develop a finite amount of graphical variations.

In the late 1970s, Zaha Hadid produced a series of drawings, inspired by the Russian avant-garde, that helped her explore the spatial concepts and architectural connections developed in the proposal for the residence of the Irish prime minister (Phoenix Park, outside Dublin). Grouped in a sketchbook, these drawings show a graphic research process, with variations that intertwine page to page through die design that connects a geometry with the next, transforming the sketchbook into a set of layers of information. With this technique, Zaha Hadid intended to graphically explore possible solutions to topics related to urban growth in the twenty-first century. Zaha developed drawings that had an apparent geometric simplicity, with a great spatial complexity, attempting to respond to the problem of the densification of the suburbia tissue, by connecting elements, inspired by the suprematist compositions of Malevich and El Lissitzky. The resulting document distilled Zaha's understanding of architecture, as a historical process linked to the cycles of intensification and redevelopment of city life (Fig. 9).

Fig. 9. Study sketchbook for the design of the house of the Prime Minister of Ireland. Dublin (Phoenix Park). Zaha Hadid, 1979–80. Source: Architectural Review. © Zaha Hadid

After completion of the Suzuki House in Tokyo, the customer was impressed by the graphics and gave Peter Wilson a concertina Japanese notebook to encourage him to continue drawing. Captivated by the quality of the sketchbook, between the years 1983 and 1998, Wilson destined its pages to narrate the architectural experience derived from the graphical understanding of two very different urban realities: the Japanese and European.

The characteristic personal touch shown in the project of the Samurai's house (winner of the Shinkenchiku Competition) is applied to the exploration of new means of graphic communication to break with the graphic tradition linked to urban planning projects. According to Wilson, the topics he was facing, required new models of representation, that were able to absorb new concepts related with the "net structures", the "ecology of circuits" or "tactical" places (Wilson 2013, p. 77), among others. This approach led to a lot of criticism from part of his colleagues, who did not understand a graphic discourse that took advantage of the discontinuous spatial experience provided by the Japanese folding-sheet sketchbook (Fig. 10).

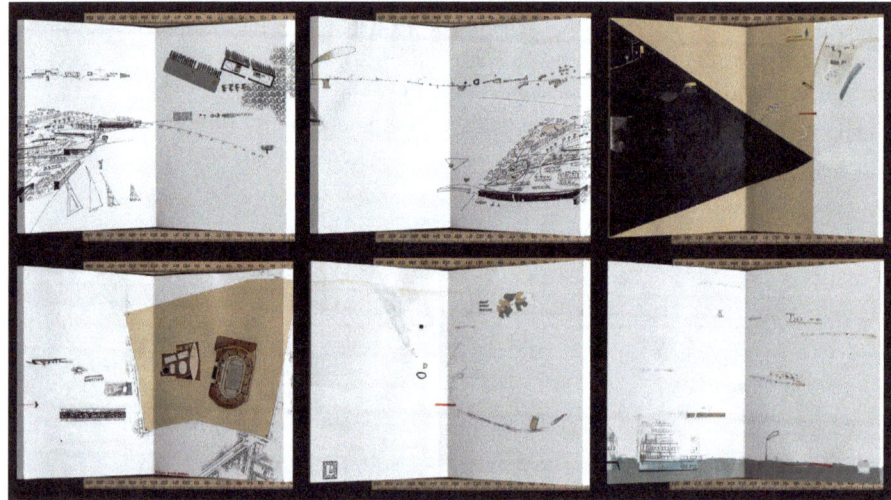

Fig. 10. Eurolandschaft – A Dérive. Peter Wilson, 1998. Source: Peter Wilson.

However, besides the undeniable graphic value, both sketchbooks represent a first-level exercise of architectural criticism, as they graphically explore alternatives to face up to the challenges of their time.

5   Conclusions

The history of architecture is filled with anecdotes related to the emergence of large projects. The authors usually explain that the ideas were outlined on a napkin, a piece of paper or on the sketchbook they always carry with them. Apart from the veracity of these stories, the full reading of the drawings emerged throughout the different phases of a project, are enlightening to understand their authors' thinking.

In these times of immediacy and ephemeral images, the spontaneous graphics processes are fascinating, due to its authenticity. For the observer, is more stimulating to have access to expressive and processual drawings, rather than to merely representative and informative ones. The first ones are rarely shown to the public, as they are considered "work in progress". For the architect, the sketchbooks are useful to retrieve and refer the graphic imprint made in previous stages. They contribute positively to the construction of a personal thinking. The formation of these reflective processes in a sketchbook, helps everyone to establish the chronological sequence that binds all the phases with the project. This would explain the growing interest in this kind of documents, that are not intended to be shown. However, they have become very popular, proliferating publications with this content in facsimile edition or as a compilation agenda-type. This has generated a mimetic stream, where architects and students return to the sketchbook to generate a design experience emerging from the graphic thinking. This taste for the sketchbook, has reopened the debate on the importance of the initial and speculative drawing, a recurring topic in the academic panorama since the penetration of digital media, by the end of the 1990s.

For practicing architects, the sketchbooks are a space of freedom to express their thoughts, released from the impositions of construction; the primary design ideas captured in a graphic act. For students, these graphic diaries are a portable and reviewable tool of self-reflection, that help them to base a critical thinking not only on the observation of other processes, but on the construction of a personal speech through drawing.

For the rest of the public, these sketchbooks have the attraction of inaccessibility, as they were, until recently, an intimate and almost secret aspect of their authors. They are believed to be "moments of inspiration", when in reality, they represent a nonlinear process where exploration paths are opened through drawing. This fact, adding the specific binding format, confer them a certain "incunabulum" character, and has transformed them in an object of desire for the audience, that is eager to acquire a deeper knowledge of the most intimate secrets of the creative process of their authors, awakening, in short, the interest in architecture. For Libeskind, "The architecture is not based on concrete and steel and the elements of earth. It is based on wonder" (Rosenfield 2012).

References

Bravo de Laguna, A.: Kahn, Foster y Piano en Ronchamp. In: Melián García, A. (ed.) XV Congreso Internacional EGA. El dibujo de viaje de los arquitectos, pp. 151–158. Servicio de publicaciones y difusión científica. Las Palmas de Gran Canaria (2014)

Crespo Martín, B.: El libro-arte/libro de artista: tipologías secuenciales, narrativas y estructuras. Anales de Documentación **15**(1), 3–5 (2012)

Farrelly, L., Crowson, N.: Representational Techniques for Architecture, 2nd edn. Bloomsbury Publishing PLC, New York (2014)

Maestre Galindo, C.E.: El fantasma de Miralles. Un viaje a la India de Le Corbusier. Constelaciones, no. 4, pp. 71–84 (2016)

Marcos, C.L.: Viajes interiores. Los viajes del arquitecto. In: XV Congreso Internacional EGA. El dibujo de viaje de los arquitectos, pp. 469–476. Servicio de publicaciones y difusión científica. Las Palmas de Gran Canarias (2014)

Palomo y Triguero, E.: Cita-logía, 1st edn. Punto Rojo Libros S.L., Sevilla (2013)

Rosenfield, K.: TED Talk: Daniel Libeskind's 17 words of architectural inspiration. ArchDaily, 29 April 2012. https://www.archdaily.com/230451/ted-talk-daniel-libeskinds-17-words-of-architectural-inspiration/. Accessed 02 Sept 2019

Ramírez, J.A. in Sainz, J.: The Architectural Drawing. Theory and History of Graphic Language. 1st ed. Editorial Ne-rea, Madrid (1990)

Uribe, B.: Frases: Le Corbusier y el dibujo. Plataforma Arquitectura, 30 July 2015. https://www.plataformaarquitectura.cl/cl/770875/frases-le-corbusier-y-el-dibujo. Accessed 09 May 2019

Wilson, P.: Berlín – Tokio Tierra y otros escritos, 1st edn. Ediciones Asimétricas, Madrid (2013)

Woods, L.: Notebook 01-3 (the last). Lebbeus Woods Blog, 6 October 2009. https://lebbeuswoods.wordpress.com/2009/10/06/notebook-01-3-the-last/. Accessed 11 June 2019

Process Design for Automation

Federico Luis del Blanco García[(⊠)], Ismael García Ríos, and Ana González Uriel

Polytechnic University of Madrid, Madrid, Spain
federicoluis.delblanco@upm.es

Abstract. This paper presents some projects developed for the design of programming tools that allowed to automate repetitive processes. The workshops took place at the Technical School of Architecture in Madrid. The purpose of them was that students and architects developed their own digital tools as a personalized application that could be used in a later phase of design. Using a non-destructible workflow, the different algorithms enabled the possibility to make changes and modifications at any stage during the development of the project. Data analysis, simulations and the production of massive documents under specific conditions are contemplated. Topics include: work with terrains and contour lines from a database provided by NASA; calculation and determination of the slopes on a surface and its extension to a given orography; automated generation of clinometric plans; heights analysis of a terrain, its translation to colors and contours, and the use of scatter tools for the systematic insertion of elements.

Keywords: Procedural design · Automation · Terrains

1 Introduction

Automation is one of the most important factors to speed up the development of a product. It allows the designer to experiment with random combinations, repetitive tasks, parametric design or the exploration of new fields of study (Satti 2007, pp. 607–622) [1].

This paper discusses the utility and efficiency of building in-house digital tools to expand and enhance the capabilities of existing software. With the designed tools, researchers can analyze, make simulations and generate drawings under controlled conditions using digitally generated projects and the designed tools. The proposed workflow propitiates an interdisciplinary approach where architects, computational designers and technical directors could add different values to the final output (Ahlquist 2016) [2].

The film industry is characterized by the high level of specialization of every member in a studio. In this case scenario, a technical director could develop plugins or build in-house tools previously asked by the VFX artists in order to complete different tasks or repetitive work. That scenario is not common in an architecture studio, and the advantages this organization could provide are not few.

During the last decade there has been a resurgence in the field of architecture around computational design, applying computational strategies based on analytical thinking that establishes logical sequences for the design of surfaces, structures or conditioning

L. Agustín-Hernández et al. (Eds.): EGA 2020, SSDI 6, pp. 393–402, 2020.
https://doi.org/10.1007/978-3-030-47983-1_35

Fig. 1. Layered landscape generated by an automated algorithm. Image by Federico del Blanco.

systems. Programming languages gain progressively greater importance, expanding the native capabilities of the software used or even including the development of specific tools that have led to a new approach to architectural design. However, programming is not usually a common skill among architects, so visual programming has become a viable alternative with great acceptance -both in the teaching field and in professional practice- for architects to use visual scripts without the need for textual language. In 2012, ACADIA granted the award of excellence in Innovative Research to David Rutten for his profound impact on architecture and computing when carrying out the development of Grasshopper, a software considered as "pivotal to the transformation of parametric design practices over the past five-plus years." (ACADIA 2012) [3].

2 Case Study: Automation Process to Obtain a Terrain

This manuscript presents a project completed in a workshop to present what kind of objectives can be achieved by developing personal tools which allowed automated processes (Figs. 1, 2 and 3). Attendants were asked to build their own tools using different algorithms to accomplish repetitive tasks.

These tools would be used later for different purposes in a design phase. They could also be framed in a context that contemplates the possibility of transferring technology from the techniques used in the film industry to the field of design and architecture (del Blanco García and García Ríos 2017, pp. 105–118) [4], highlighting the importance of the use of programming languages in the training of students (Rivka and Oxman 2014) [5].

In recent years there has been a growing interest in the field of procedural design that has led to the proliferation of workshops in different Architecture Schools (Spyropoulos, Frazer and Schumacher 2013) [6] that investigate computational systems of calculation

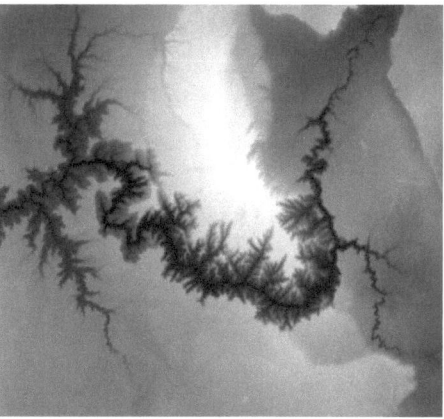

Fig. 2. Greyscale heightmap. It provides the information for the displacement of each point of the map. It can be generated from the three-dimensional surface of the land. Image obtained from the NASA archives.

in search of systemic design applications (Figs. 4, 5 and 6), as well as a notable increase in exhibitions showing different architecture and computational design projects.

The objective of the project designed in the workshops was not to design a specific tool, but for the students to learn how to design their own customized tools. With this in mind, each attendant could add the desired parameters to an existing software, or simply create a new one.

One of the accomplished projects was done using the database provided by NASA. With it is possible to generate a three-dimensional mesh of a terrain located anywhere in the world, as well as project the contours created from a height map (assigning a

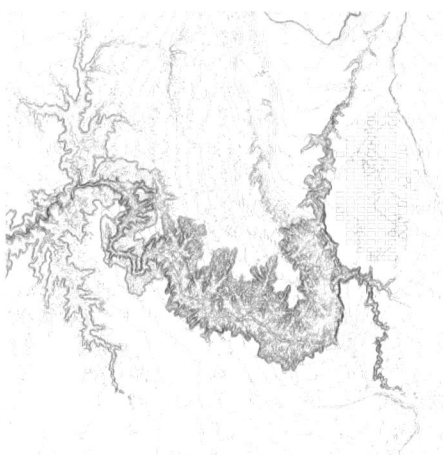

Fig. 3. Contour lines generated automatically from a height map through an automation process. Image by Federico del Blanco.

Fig. 4. Image from Bartlett School of Architecture, Bio Integrated Design Master's in Architecture.

specific height to each pixel of the image). It is a repetitive process that requires a lot of time and needs constant modifications to adapt to the changes that arise in a project. This workflow to generate three-dimensional terrain is not new and is quite frequent in architectural practice. GIS mapping software can achieve these results with less effort and time. However, being able to develop your own tools offers extensive possibilities beyond the restrictions of a specific software (Janssen et al. 2016, pp. 59–68) [7].

Fig. 5. Image from Architectural Association DRL Master's in Architecture.

Fig. 6. Technical School of Architecture of Madrid. MIAU: Postgraduate Program in Advanced Instruments for Project and Computational Design. 2019. Image by Federico del Blanco.

3 Workflow

After developing the digital tools, the process can be automated with different terrains to achieve the same results instantly, allowing homogenization for the study of different cases. Working through a procedural approach allows the terrain to be modified in real time by varying the predefined parameters and experimenting with different combinations. This is an approach to work through non-destructive techniques that allows changes to be implemented at any stage of the project without the need to redo the work done. This is especially relevant when the project is going to be reviewed by members outside the development team (Fig. 7).

The project was developed using scripts and programming, starting with Grasshopper (Piker 2007, stable release 2014) and then using Houdini FX (Janssen and Wee Chen

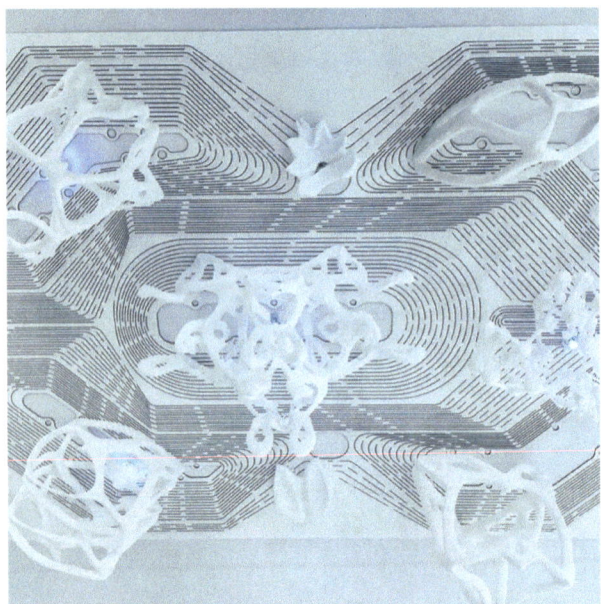

Fig. 7. Automation in Design Exhibition at New York Institute of Technology. 17th April 2019

2011, pp. 801–816) [8]. The results showed a better optimization in the tools developed through Houdini, thanks to its multi-processing capabilities.

Techniques were implemented following three different approaches to develop the tools. Maximum efficiency corresponded to the most complex approach. The first was based on VEX, a high-performance programming language used in Houdini, which provides performance close to the compiled C/C++ code (Houdini documentation of side effects, VEX language reference) [9]. The second was based on VOP (Vex Operators), a visual language based on VEX. The third, which did not require programming knowledge, was based on the use of the standard nodes provided by the software.

Files generated by algorithm sequences are lighter than files produced by polygon mesh geometries or NURBS surfaces. They allow communication between different students, researchers or professionals, facilitating interdisciplinary results. There are many possibilities to obtain similar results using different algorithms (Wassim Jabi 2013) [10], which is especially interesting because it encourages attendees to try alternative strategies.

This workflow allows working with NURBS or polygonal meshes (or exchanging them in the process) without materializing in 3D, making it possible to adapt to different needs. Once the basic algorithm has been established, new variables could be added by adding complexity to the tool or adapting it to different roles (Schumacher 2009, pp. 14–23) [11]. For an additional automation process, scripts could be added that manage the height maps of different databases (Sanchez, 2016, pp. 44–53) [12].

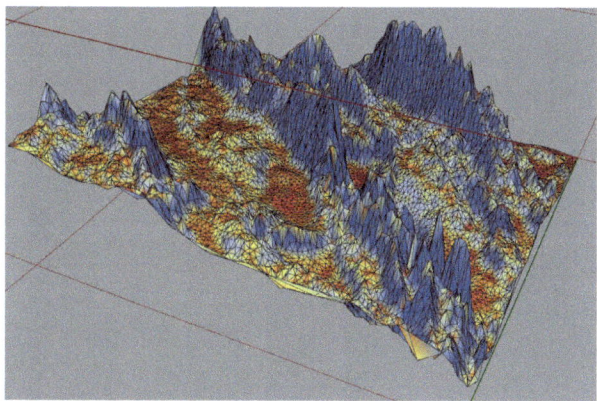

Fig. 8. The designed tool calculates the angle generated by the normal vectors of each point on the surface with the Z axis. Image by Federico del Blanco.

4 Tool Variations

The proposed workflow determines an approach to the development of custom tools that can have different purposes.

Following a similar strategy to the one previously studied, a tool was developed that was able to calculate the slope of the points that shape a surface. If the surface to be analyzed were a terrain, we would automatically obtain a clinometric map.

The clinometric is a map that allows you to classify the slopes of a terrain according to its greater or lesser inclination. With this information, the algorithm that defines this tool could be established in multiple ways. The example shown in Fig. 8 was obtained through three actions applied to a rectangular plane.

The first step was to determine the shape of the terrain in a similar way to the case shown above. A cloud of points is now defined by assigning each one a determined height by an image with an altitude information obtained from the NASA database. To shape the surface, a network was generated from the Delaunay triangulation algorithm. The circumscribed circumference of each triangle (polygons of the surface) must not contain any vertex of another triangle.

Once the terrain surface was defined, a dispersion of points was established so that there was at least one for each vertex of the pre-defined triangles. In each of these points the normal vector of the surface on which it rests is calculated. By calculating the angle generated by the normal vector with the Z axis (vertical unit vector), we obtain the inclination of the surface at each vertex of the triangle.

In this way, we can automate the process to obtain clinometric maps, difficult to obtain through traditional process (Fig. 9).

Using the same terrain as case study, another tool for heights analysis was developed. It is a simpler tool, since it is only necessary to define a color of the different heights at which the points of a surface are located.

Fig. 9. The designed tool generates altimetric maps from a given surface. Image by Federico del Blanco.

Assigning different colors at each height, we would obtain the desired plane. A succession of colors in gradient facilitates the understanding of it. The dispersion points can be grouped by heights, defining the contours of the terrain.

Once the tools were created, they could be reused in any other terrain. These tools are not new and could be obtained with greater simplicity using a GIS software. The purpose of these exercises is to work on the ability to develop automated tools for certain applications.

Scatter tools can be used to add details to the terrain. They could as well be used for analytical or purely aesthetic purposes.

A scatter tool allows the systematic repetition of elements on a surface to fill it. The repetition of several trees and rocks on the terrains shown above could quickly generate the detail needed to perceive the surfaces as forests. Following a similar procedure to the one used in the clinometric plane, the inclination of the elements could be varied in order to avoid unwanted results. The control of the density of the elements would be determined with a greyscale image in which the black colors would correspond to the maximum clogging and the white colors to the empty areas.

Figure 10 shows a more complex case of scattering. The tool designed for this purpose -XGen- allows an exhaustive control of the elements to be repeated. The image shows millions of small-sized prisms located on a previously defined point grid. The height of each prism is defined by a color image, following the same principle of the "heighmaps". When removing the saturation to the image, the upper faces with the maximum height the set of prisms corresponds with the white color and the minimum height with the black. The gray variation determines the different heights. The result is a representation of the natural landscape from the initial image.

5   Simulations

Simulations allow u s t o work by approximation in a different way than most of the times the architects are used to. Instead of directly designing an object, a series of parameters are established. When interacting with each other, they define the object.

Fig. 10. Dispersion of prisms on a surface. The different heights are established following the colors of a base image. Lee Griggs

Simulations can use real parameters such as gravity or wind force, so that we could generate a complete terrain, or at least the details of the surface of the terrain through natural erosion (Quixel 2018) [13].

Houdini or specific programs for the edition of terrains such as Terragen or VUE, have by default an arsenal of tools for this purpose, based on simulations that mimic the agents of nature. However, these tools are primarily focused on the film and video game industries, where they are widely used. The ability to develop our own automated tools opens up a world of possibilities for architecture students.

6   Conclusion

Computational design and analytical thinking are increasingly important in the architecture industry. Students need interdisciplinary training to develop new possibilities and open lines of research related to design and architecture.

The workshops evinced that architecture students had a greater difficulty and less interest in using programming languages. This could be due to the lack of subjects in this area. Visual programming largely resolves this impediment, although it also has more limitations in the design process.

In order to design and implement algorithms that allow automated systems, architecture students need to have training in digital tools and programming languages. An in-depth knowledge of the way in which the most commonly used programs operate and the algorithms in which they are based, allows the user to identify limitations and opportunities of these tools that can be really useful in architectural design.

References

1. Satti, H.M.: Automation of building code analysis: characteristics, relationships, generation of properties and structure of various models. In: Proceedings of the 3rd International Conference on Em'body'ing Virtual Architecture (ASCAAD), pp. 607–622 (2007)
2. Ahlquist, S.: Procedural design. In: Proceedings of the 36th Annual Conference of the Association for Computer Aided Design in Architecture (ACADIA), pp. 11–12 (2016)

3. ACADIA: Awards winners (2012). http://acadia.org/news/K9QGAV
4. del Blanco García, F.L., García Ríos, I.: Technology transfer: from the film industry to architecture. In: Castaño, E., Echeverría, P.E. (eds) Architectural Draughtsmanship, pp. 105–118 Springer (2017). https://doi.org/10.1007/978-3-319-58856-8
5. Oxman, R., Oxman, R.: Theories of the Digital in Architecture. Routledge, Abingdon (2014)
6. Spyropoulos, T., Frazer, J., Schumacher, P.: Adaptive Ecologies: Correlated Systems of Living. Architectural Association Publications, London (2013)
7. Janssen, P., et al.: Parametric Modelling with GIS. In: Proceedings of the 34th Annual Conference of the Association for Computer Aided Design in Architecture (eCAADe), pp. 59–68 (2016)
8. Janssen, P., Wee Chen, K.: Visual dataflow modelling: a comparison of three systems. In: Proceedings of the 14th International Conference on Computer Aided Architectural Design Futures, January 2011, pp. 801–816 (2011)
9. Side Effects Houdini documentation VEX language reference. http://www.sidefx.com/docs/houdini/vex/lang
10. Jabi, W.: Parametric Design for Architecture. Laurence King, London (2013)
11. Schumacher, P.: Parametricism: a new global style for architecture and urban design. AD Archit. Des. **79**(4), 14–23 (2009). https://doi.org/10.1002/ad.912
12. Sanchez, J.: Combinatorial design: non-parametric computational design Strategies. In: Posthuman Frontiers: Data, Designers, and Cognitive Machines, Proceedings of the 36th Annual Conference of the Association for Computer Aided Design in Architecture (ACADIA), pp. 44–53. 978-0-692-77095-5 (2016)
13. Quixel Rebirth Scene (2019). https://www.youtube.com/watch?v=qKVkq3YbwaY

Graphic Imprints, Grids and Diagrams in Architecture

Carlos L. Marcos[1](✉) and Jeff Balmer[2]

[1] University of Alicante, Alicante, Spain
carlos.marcos@ua.es
[2] University of North Carolina, Charlotte, USA
jdbalmer@uncc.edu

Abstract. This text aims to reflect on the imprint that procedures, strategies and graphic instruments may have on design, and as a consequence, to what extent they influence the language of architecture.

Our essay explores the influence of grids and diagrams upon the course of architectural practice. In addition to the possibilities inherent in the *analytical* diagrams popularized by Rowe, we examine the *generative* role that diagrams can play, teasing out the potential structures of order latent in the exploratory stages of the design process. The critical role played by generative diagrams is most abundant in the design strategies of projects from the Renaissance onward, when drawing became the primary tool for conceiving and representing architecture. But the design and construction of pre-Renaissance buildings can also provide clear, if less explicit evidence of generative diagrams: evidence conveyed by the tracery of mediaeval masons, the placement and proportions of cathedrals laid out directly as templates on site; or the architecture of antiquity, with its alignments and optical corrections of walls and peristyles. The prescribed repetition of an array of aligned columns in a dimensional grid or the 'rational magic' of the happy Hippodamian layout are, perhaps, among the brightest moments of architectural reason. We also try to define *transformative diagrams* characteristic of some contemporary architectural practices with topological implications.

Keywords: Grids · Generative diagrams · Transformative diagrams

1 Introduction

The idea of order is integral to any creative act. Works of art are therefore inconceivable without considering a structure that follows an internal logic, a logic that provides it both order and meaning. Be it a musical score, a pictorial composition, a work of sculpture, or merely a good story, all must establish an order – a *logos* – that sponsors the creative work, and attempts to make it intelligible to others.

An author is a demiurge – an inventor of *logos* – who orders the elements of his discipline to artistically shape his creative work. In addition to the technical issues of architectural craft, the ability to engender a work that captures and conveys meaning distinguishes creative work from mere construction.

L. Agustín-Hernández et al. (Eds.): EGA 2020, SSDI 6, pp. 403–417, 2020.
https://doi.org/10.1007/978-3-030-47983-1_36

2 Order

With architecture, the necessity for order is, if possible, even greater than the other arts. In addition to its creative nature, architecture is born from necessity (Milizia 1826); it cannot solely serve a purely expressive purpose: it must also be subject to its utilitarian nature. Buildings cannot be formless structures, lacking order: it would render them unintelligible to their inhabitants. In his 1923 manifesto, *Vers une Architecture*, Le Corbusier stated not only the need for order in the discipline but also the intellectual pleasure that its comprehension could produce in humans: "The regulating line brings forth the sensory mathematics that produces a beneficent perception of order" (Le Corbusier 2007, p. 137).

Beyond its utilitarian obligations, architecture is further constrained by a quality intrinsic to its identity: its scale. Its very size ensures that architecture is bound more fully by gravity, distinguishing it from all other artistic endeavours – even sculpture – due to limits imposed by the resistance of materials. Geometries achieved in a work of sculpture may be unfeasible with larger scale of architecture. Gravity and scale determine much of the physical limits for constructional geometries, and implies a logical, prior sub-set of order for all feasible architectures.

Until the advent of Newtonian mechanics, and subsequent advances in quantifying the strength of materials, builders relied heavily upon structural symmetry for guidance to the structural behaviour of buildings. That is one of the reasons for architecture's typological tradition. Regularly spaced, modular units – of both material elements and the spacing between them – generate three-dimensional grids. These in turn produce rhythmic patterns that make tangible the order upon which they are based. The modular arrangement of beams and columns simultaneously provides for the structural resolution of architecture, systematically channelling forces and loads safely to the ground.

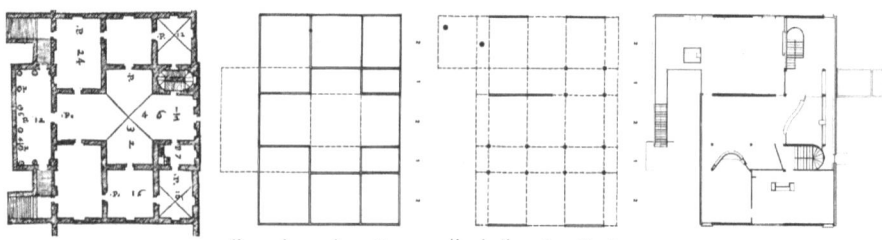

Fig. 1. Le Corbusier (Villa Stein, 1927) and Palladio (Villa Malcontenta, 1550). Comparative analysis of the floor plans and the analytical diagrams of both villas (diagrams first published by Collin Rowe in "The Mathematics of the Ideal Villa: Palladio and Le Corbusier Compared", Architectural Review, 101, Mar. 1947).

Modular tectonics, with its accompanying spatial rhythms, has traditionally helped to define architectural order and make it intelligible. Christopher Alexander has articulated and analysed this sense of order throughout his lifetime, research that rests significantly on the practice of *patterns* (Alexander 2006). Patterns constitute ways of organizing matter in space, and provide an ordered whole that renders it recognizable. For Alexander, it is due to these meaningful arrangements that we come to understand architecture, and beyond that, existence itself. As Mark García (2009, p. 12) has pointed out, Alexander's first approach to patterns in his early *A Pattern Language* extended his influence into other fields ranging from sociology, design and urbanism (Alexander 1977).

Patterns – in the form of grids – were similarly used by Wittkower and Rowe for the analysis of the architecture, revealing their unifying role in the establishment of spatial orders. These were identified graphically through the superimposition of simplified grids over the plans of these buildings (Fig. 1), thus side-stepping the architectural stylistic qualities of each period.

Observing the plan diagrams traced by Rowe (1976) for Villa Stein and Villa Malcontenta, not only do they plainly reveal his comparative analysis, we also recognize how diagrammatic representation can free architecture of its timely or stylistic limitations. These works are pared to their essence, so that nothing further can be removed without eradicating the kernel of the design. This gives evidence of the role of diagrams in relation to plans and to architecture: they are the utmost synthetical notation of a given architectural design. As Carazo and Galván have noted (2017), for this reason alone, diagrams are a vital tool for teaching purposes.

3   Order Systems

For myriad reasons, grids have always been used in architecture (de Molina 2005). In the first place, grids *embody* order. As such, they can also *engender* it. As Ching's *Architecture: Form, Space, & Order* (1979) demonstrates, the organization of architectural space is linked throughout history by what he calls spatial organization systems, and which we refer to as *order systems*. An order system is an *a priori* arrangement of the elements of a composition, one that pre-figures matter while establishing an organization of the material boundaries defining architectural space.

At the same time, ordering systems do not pre-determine design. They provide structure, yet remain open and flexible, a combination of qualities that makes them immensely useful for the design process in architecture. There are unlimited possibilities to produce different architectural designs that arise from otherwise similar linear orders, transcending traditions or styles. Thus, if we compare varied examples of architecture over time with diverse chronology, aesthetics, uses, constructive and structural systems, we can observe striking consistencies. The Egyptian temple complex at Karnak, the Parthenon of Athens, the Basilica Julia, Salisbury Cathedral, the Barcelona pavilion or the Yokohama port terminal: all are examples engendered by the same organizational system. All follow a predominantly linear order – even if most also support a supplementary orthogonal arrangement – wherein matter is ordered along a main axis that can even become a continuous winding gesture. Such is the case with FOA's Yokohama project, which, despite otherwise eschewing the conventions of orthogonality, threads its various flows along an intrinsically linear thrust.

The use of grids as a powerful spatial tool need not be the result of a graphical scheme. Millennia before Alberti first argued for scale drawings as a necessary prelude to the act of building, grids were employed as the initial conceptual framework of 'design', arrayed across the surface of the earth, to literally 'stake out' the project site. For instance,

Egyptian surveyors fashioned ropes with 12 evenly-spaced knots, allowing them to a portable yet precise tool to lay out right angles erected on site (Balmer and Swisher 2019, p. 13). Arranged to form a three-sided figure with $3 \times 4 \times 5$ proportions, Egyptians made use of the special properties of the Pythagorean triangle long before it was given such a name. This practice indicates both: a working knowledge of geometry, and the utility that such an easy and effective geometric tool has proven to be in relation to the setting-out of construction sites ever since.

We know relatively little about graphic representation in antiquity, as the survival of drawings on papyrus or paper are exceedingly rare. There are, however, examples of ancient sculptures in which building plans are represented, suggesting the knowledge and practice of graphic representation based on orthographic projection. Such is the case of the headless statue of Gudea (Fig. 2), as Franco Taboada (2017) has noted.

Fig. 2. Headless sitting statue of Gudea (prince of Lagash), Mesopotmia, 2120 B.C.E (Louvre). (Source, presentation by Franco Taboada UID Conference in 2017). A floorplan is inscribed on a tablet resting on Gudea's lap, a drawing which implies that horizontal sections were known and already in use to represent architecture.

Pompeian frescoes show a knowledge of a rather precise form of perspective in Ancient Rome (though some rules of its construction vary from those first developed centuries later in Florence). Vitruvius' seminal text on architectural theory mentions the term *scenographia,* stating the lines of the side façade are convergent in a point, which some have interpreted as the use of these kind of representations by architects. He also speaks of floorplans and elevations as well as the use of compasses and rulers architects needed to draw them (Vitruvius 1995). In fact, it is known that Vitruvius' original text was accompanied by drawings but, unfortunately, these were lost in the course of time (Rodriguez 1995).

However, despite the most likely professional use of architectural drawing in ancient times, what it is now clear is that architectural representation has been based on projections since the Renaissance, when many of the graphic representation systems of architecture were precisely established, not only perspective. In his letter to Pope Leo

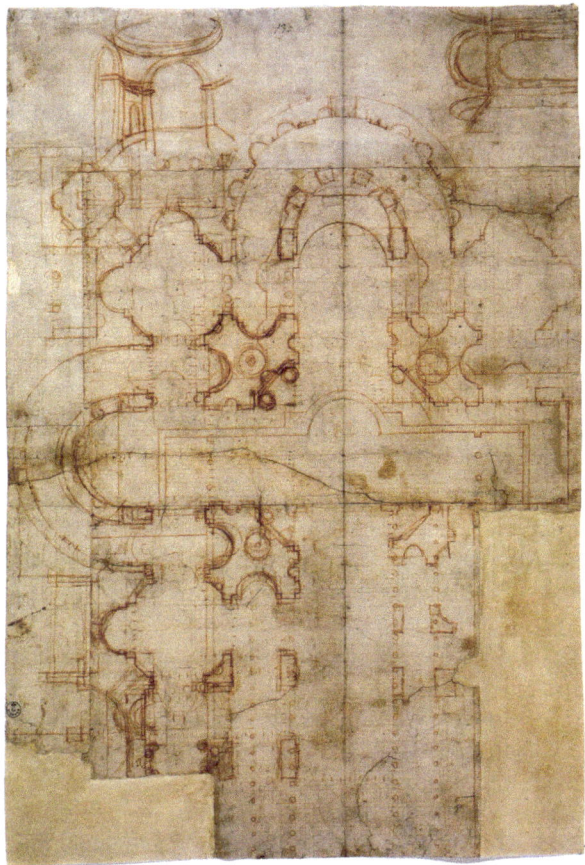

Fig. 3. Bramante. Partial plan of St. Peter's Basilica, Rome, c.1506. (Source: Borsi, F., *Bramante*, Electa, Milan, p. 75, 1989). Bramante's design, drawn over the plan of the earlier Paleochristian basilica, conveys the influence of Imperial Roman vaulted architecture. Bramante's drawing conveys the striking change of scale from Old St. Peter's.

X, dated 1519 (Hart, Hicks, 2006), Raphael's use of the three most characteristic types of orthographic projection in architectural drawing – sections, floors and elevations – was already stated. This system of representation was an attempt to dissect architecture into plans at scale. The unmistakeable influence of Alberti can be traced here, as he already distinguished between the use of the two types of projections suitable for painters or architects, respectively, in his *Re Aedificatoria* (Alberti 1991, p. 95). This intimate relationship between architects and drawings has remained so throughout the centuries.

Once plans could be drawn to represent the design (and anticipate its laying out on site), geometric constructions used to precisely depict architecture in scale naturally led to the use of grids. In Bramante's plan for St. Peter's Basilica (Fig. 3), ancillary construction lines generating a grid can be interpreted as an *appoggiatura* for the drawing itself. Their capacity to embody a certain order can be easily understood, considering the setting of alignments, rhythms and spacing defined by them.

From a strictly compositional point of view, grids are systems of relations between their three constituent elements: lines, nodes and interstices. Lines establish alignments, predominant directions in space, orientation and axiality; in other words, lines define the anisotropic quality of any architectural space. Nodes, the points that are highlighted at the intersection of the grid lines, are given a special status of hierarchical singularity within the grid. Interstices, the spaces in between the lines, are the void areas that engender architectural space itself. Typically, lines and nodes are the geometrical locations where the architectural limits or the structure are placed, while the interstices constitute the architectural space defined by them. The set of relations that the grid defines are equally granted to the elements arranged from them.

4   Grids and Diagrams

Grids and diagrams are, to begin with, abstract notations of architecture. They continue to maintain a certain relationship with cast projections, but they involve a step beyond any form of figuration. That is to say: the basic relation of analogy between matter and space, on the one hand, and the graphic representation with regard to its projective nature, on the other.

Although any type of drawing is a mere shadow on the plane with respect to the richness of architectural space, a section, a floorplan or an elevation are, nevertheless, quite precise and descriptive representations of architecture. Yet, these plans convey relevant architectural information that very significantly cannot be directly inferred through experiencing built architecture. In fact, the findings we can gather from a visit to a building on site and through the observation of its plans is complementary; a critical assessment of an architectural design needs attention to both different realities although based on a same referent (Allepuz 2019).

Grids -considered as schematic and generic order systems-, and diagrams -basically, particularization of grids- make architectural shorthand possible. These graphical notations have the ability to define the basic hierarchy and scale of the different parts with respect to the whole, as well as the topological relations of connectivity of the entire morphological structure. Nevertheless, true architectural diagrams must preserve measurable relationships inherent to scale drawing to represent the geometric essence in a work of architecture. In this sense, we can consider them as synthetic representations of a specific architectural order rather than of architecture itself.

Probably, the first architect who was aware of this synthetic quality of diagrams was Durand (2000 [1805]). He devised this type of graphic notation to free his design method from any particularization associated to historical styles. In doing so, he managed to overcome the limitations of the style and syntax of his time, allowing further development and deepening of its use. Today, Durand's system is generally dismissed as formulaic, based on rigid geometries but this is because we see it with the eyes of our time. Its method, far from being traditional, was quite innovative in its time. He should be attributed the invention of this explicit graphic notation not only in regard to analytical aspects but, above all, as a generative graphic device capable of anticipating and prefiguring architecture.

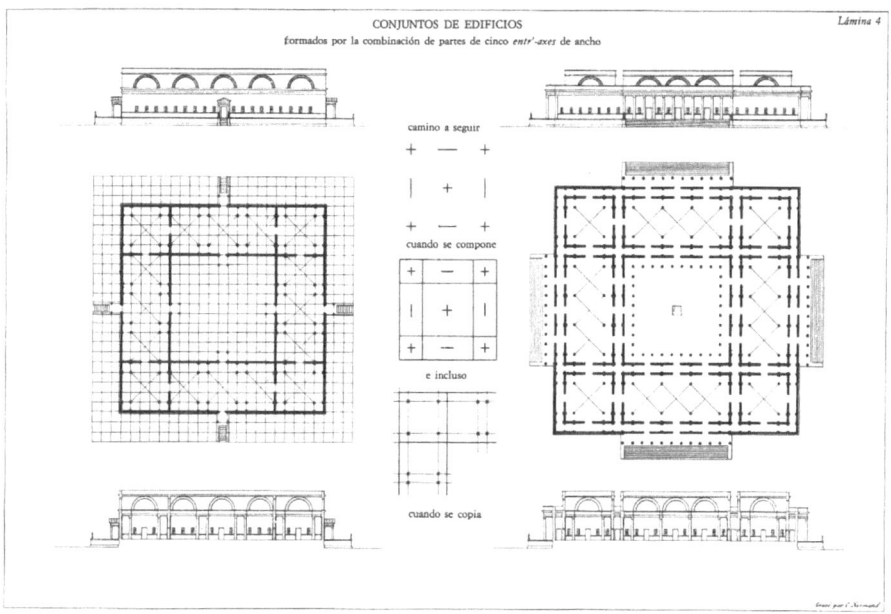

Fig. 4. J.P.N. Durand. Building ensemble. (Source: Cours d'Architecture, Ed. 1821). Note how a same diagram is able to generate two different projects shedding evidence of the generative character of Durand's proposed design methodology.

As it can be seen in Fig. 4, two different design proposals share a same diagram (Madrazo 1994, p. 19). This alone shows the potential of diagrams to generate form in addition to their intrinsic analytical possibilities, regardless of the architectural epoch or the kind of architecture to analyse (Figs. 1 and 5). In Fig. 5, we can observe how diagrams need not be exclusively based on plans but may also be drawn in axonometric projections.

Modernist architecture was also inspired by grids since the dismembering of skeleton and skin in architecture was determined by the International Style through the concentration of vertical loads in columns –*pilotis*–. Even Neoplasticist architecture was inspired by a conspicuous orthogonal plot – although the spacing of the grid is rarely regular. Wright made extensive use of grids, producing some of the subtlest examples of rich geometries in the exploitation of these order systems. Think, for instance, in the design of the Hanna house-Honeycomb (Fig. 6). Even late modernists like Kahn and Breuer continued to make extensive use of them.

One of the most significant examples for the use of diagrams among the early proponents of Modernism was Louis Sullivan. Detailed evidence of his design process illustrates how operational graphic notations can be geared to shape architectural form. Equally significant, Sullivan's system of generative diagramming provided him the means of introducing complexity and harnessing irregularity through them in the design (Sullivan 1990) (Fig. 7).

The incomparable fluidity and richness of Sullivan's ornament is clearly derived from geometric constructions, grids and diagrams, progressively evolving through increasing

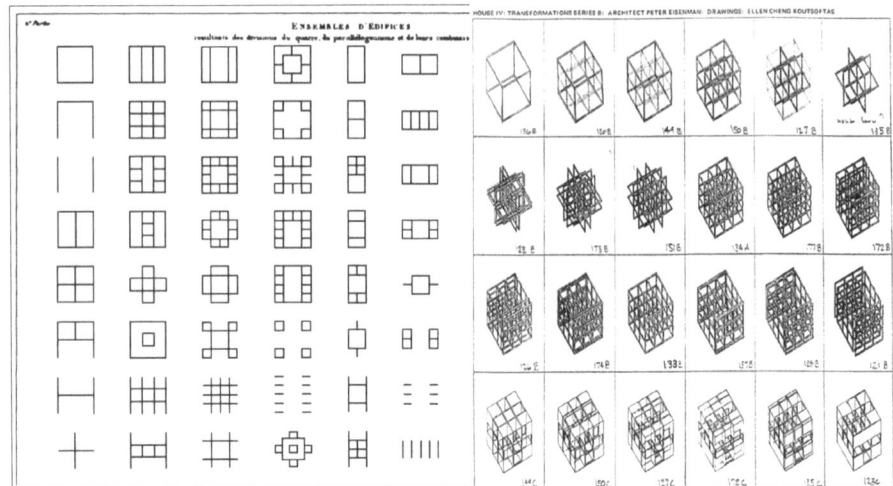

Fig. 5. Left: J.P.N. Durand «Ensembles d'edifices resultants des divisions du quarré...» (diagrams of buildings based on a square plan), (Source: Précis, Ed.1802). Right: Eisenman. Analytical series of axonometric diagrams for House IV, 1971 (courtesy Eisenman Architects). Note the idea of variations developed over a same scheme on both plates. While Durand is proposing a series of possible plan diagrams based on the square, Eisenman is doing the same within a cube which; most of the design process is, however, based on the 9 square grid.

levels of organic complexity, and guiding his design process forward. Without a doubt, Sullivan's generative system of geometric production played a pivotal role in the development of Wright's prolific dominion of form: Serving for several years in his capacity as chief draughtsman to Sullivan, Wright referred to his great mentor as *Lieber Meister* throughout his lifetime. Whereas Sullivan's inspiration for his grammar of ornament is *figuratively* organic, Wright's claim for an organic architecture of his own can be understood as an abstraction of Sullivan's procedures applied to the architectural form, as well as to his geometrized ornamental repertoire.

This generative and evolutionary procedures developed by Sullivan should also be considered in relation to D'Arcy Thompson's *On Growth and Form* (1945). First published in 1917, Thompson's work hinges on a rich array of diagrams produced to explain evolution through the graphical evidence of anatomical transformations. However, Sullivan's approach is a certain blending of geometricism and the use of nature's forms as a source of inspiration. Thompson's *topological or transformative diagrams*, on the other hand, are a graphic depiction of geometric transformations that species have undergone throughout the course of evolution shaping them to make them fitter for survival. To be precise, they are more homeomorphic transformations on a topological space as the changes depicted do indeed deform their anatomies (Fig. 8) through processes of continuity.

What is most remarkable about Thompson's explorations is that a simple distorted Cartesian grid can serve as a guide to follow, interpolate and even to predict these kinds of transformations (Fig. 8). The importance here is the way in which he manages to

Fig. 6. Frank Lloyd Wright. Addition and alterations to Honeycomb House, 1936. (Source: Hanna House Collection. Stanford Libraries) Process drawing of Wright's Hanna Residence, showing the generative influence of its hexagonal grid.

superimpose the cartesian grid over a given anatomy (geometrically defined through drawing projections), just as Bramante's scheme for St. Peter (Fig. 3) is drawn over the plan of the earlier Paleo Christian basilica. The coexistence of two different realities, the cartesian grid and the anatomic depiction in the case of Thompson, or the existing and the proposed basilicas in the case of Bramante, is the only reason for an enhanced reading of potential relationships. It is through this superimposition that Thompson is able not only to graphically verify the validity of biological evolution but, moreover, to predict intermediate evolutionary stages by interpolation to suggest *missing evolutionary links*.

The difference between geometric isomorphisms and topological homeomorphisms rely on the fact that the first keep the proportions of the parts to the whole (symmetry, rotation, translation or homothecy), whereas the latter imply an alteration of these proportions. That is the reason for the common image of the topological space understood as an *elastic* space. These types of transformations can be any kind provided they are continuous, thus, no gluing or breaking of the forms are allowed. Greg Lynn has pointed out that Thompson's deformations of the neutral Cartesian grid superimposed over the anatomies "suggest an alternative to the static morphological transformations of autonomous architectural types" (cit. Carpo 2013, p. 37). In fact, this same procedure

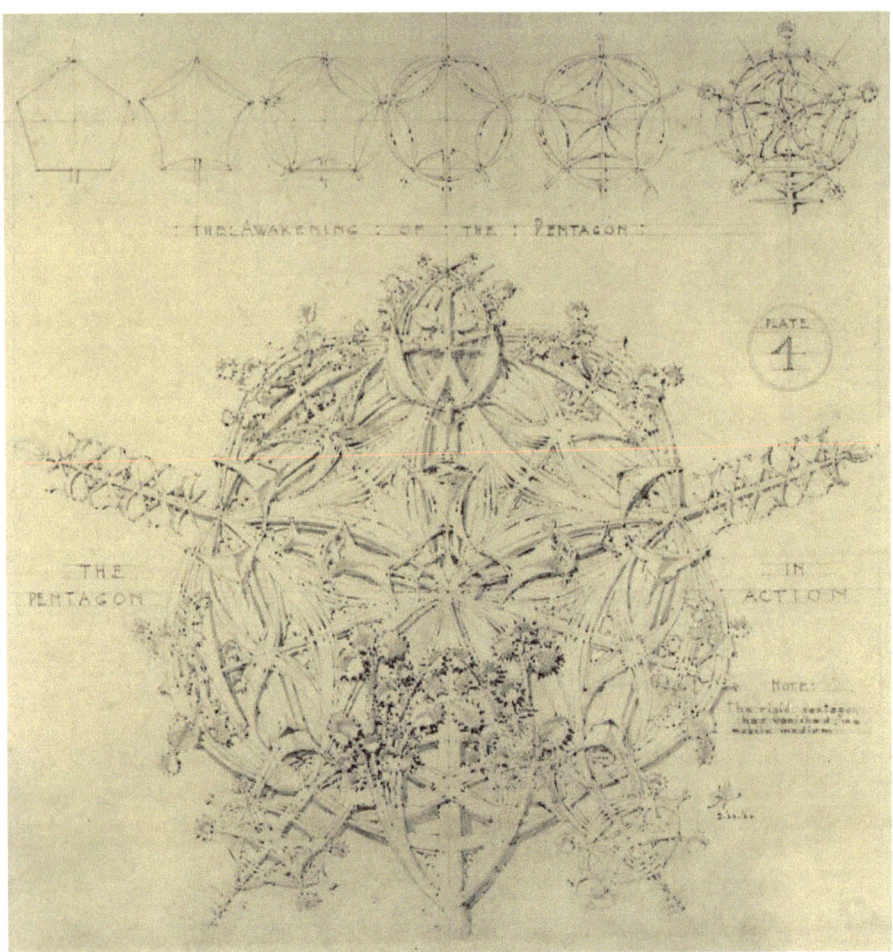

Fig. 7. Louis Sullivan. A pentagonal ornamental arrangement. (Source: A System of Architectural Ornament, 1924, plate 4). Inspired by his observations of geometries found in natural forms, Sullivan's treatise on ornament demonstrates the iterative process of transforming simple geometric shapes towards the full flowering of complexities.

has underwritten algorithms within graphical software, including Photoshop, to attain irregular transformations of a given form.

It is only in the past recent decades, when irregularity has become a hallmark of recent contemporary architecture, when we find a more disruptive strategy in the use of diagrams which can be mirrored in Thompson's procedures. Without any doubt, Eisenman should be credited for the most innovative approach in the use of diagrams and their generative potential (Marcos 2011). Much of his production in the last decades is inspired by this new generative diagrammatic approach; he has written much about his own work and its relation to diagrams (Eisenman 1999).

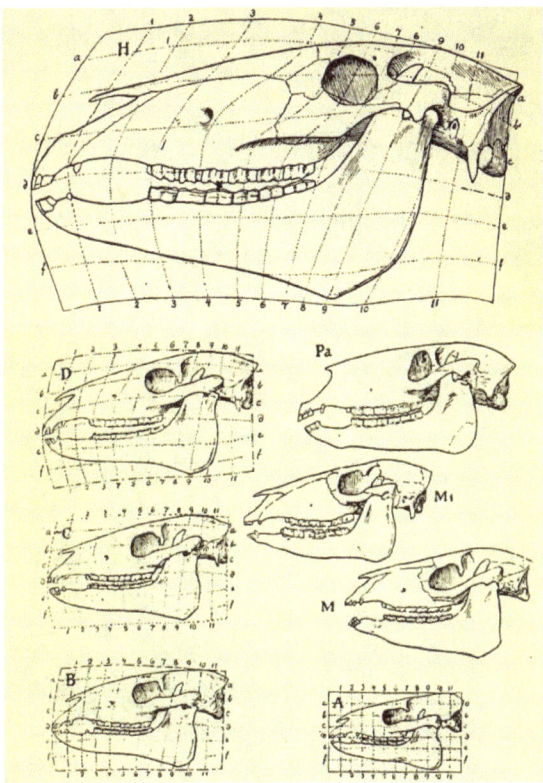

Fig. 8. D'Arcy W. Thompson, 1945, pp. 1078–1079. Evolution of the skulls of different animals through the superimposition of a coordinate system from A (Hyracotherium, Eocene) to H (Horse, Quaternary), and interpolations of imaginary types (B,C,D). The latter to be compared with real extinct species such as M (Mesohippus, Oligocene), Mi (Miohippus), Pa (Parahippus) (Source: On Growth and Form).

Meshed geometries and software have promoted the use of such supple geometries and homeomorphisms. In fact, the diagrammatic strategies followed by Eisenman since the mid '90s could be, in most cases, considered as topological transformations (Figs. 9, 10). It should be noted that, without the previous analytical use of diagrams that Eisenman commenced in 1963 for his dissertation, under the supervision of Colin Rowe (Eisenman 2006), together with the extensive use of this kind of graphic notations for his houses in the '70s, this deep transformation in the use of diagrams might not have been accomplished. His project for the *virtual house* (Fig. 9) is an extraordinary example of the use of *transformative* or *topological diagrams* in architecture.

Such generative and transformative capacities of diagrams to become a potential architecture hinted within them is certainly one of the most extraordinary possibilities of diagrams considered as a design strategy. As Allen has stated, "a diagram is not a thing in itself, but a description of potential relationships among elements, not only an

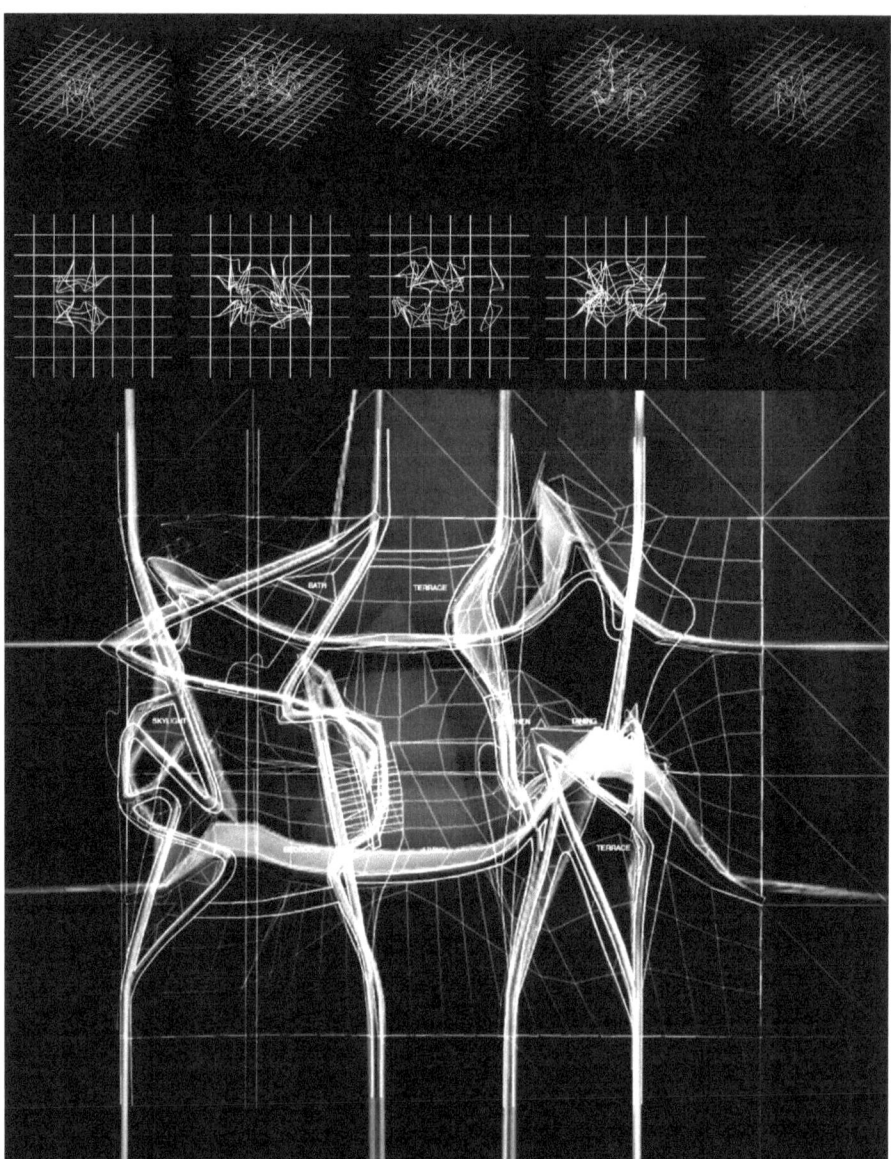

Fig. 9. Peter Eisenman. The Virtual House, ANY competition 1997 (use of generative diagrams). "Use of the notion of the virtual in architecture risks literally materializing the immaterial. Therefore, one needs to address the productive making, or the condition of the virtual within architecture, in order to allow architecture to question traditional ideas of form and space." (courtesy Eisenman Architects).

abstract model of the way things behave in the world but a map of possible worlds" (Allen 1998, p. 16).

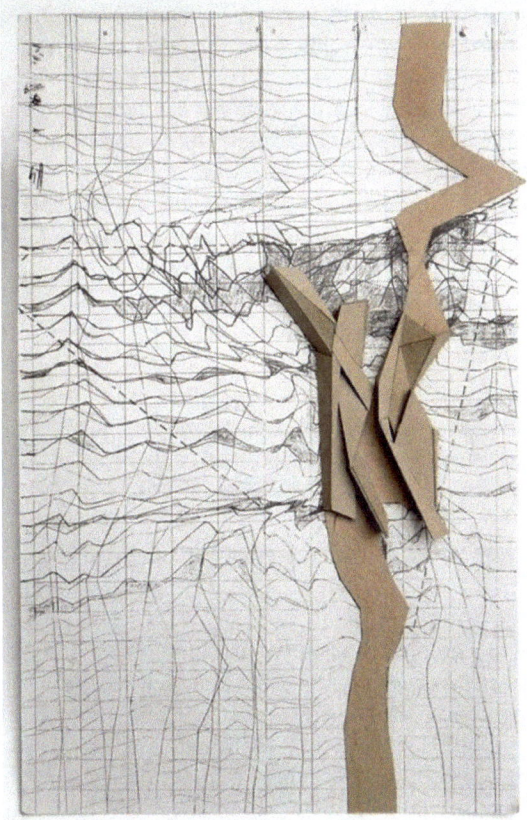

Fig. 10. Peter Eisenman. Study models for L'Huei Library, Geneva, 1997 (use of generative diagrams and physical model extracted from them). Note the different grids and diagrams superimposed -including an alien diagram to site or program which works as an external agent, almost a Deus ex machina (Eisenman 2007, p. 59)-, all of which collaborate in obtaining the final result (courtesy Eisenman Architects).

Nevertheless, most interesting to this research is Vidler's consideration of Eisenman's work as a "late-modern critical and ironic investigation of the Modernist legacy" while attributing the diagrams to be "a device to both recall and supersede its formal canons" (Vidler 2012, p. 55). To a certain extent, this sentence may well serve as a closing statement with regard to the ambivalent analytical and generative potential of diagrams.

5   Conclusions

This research attempts to reveal why diagrams can still be regarded as useful tools in contemporary architectural design as much as they were in disciplinary practice in the past.

The synthetic potential of diagrams to embody architectural order together with the topological relationships between architectural space and the physical limits that build it up make of diagrams a powerful design tool.

Accordingly, they are suitable for professional practice as well as for teaching purposes; they are a graphic compositional resource that students should be familiar with, especially in their first courses.

Surprisingly, the utility of contemporary generative design diagrams is not based on regularity as it was in the past when Durand inaugurated their use. That is another reason for their contemporaneity.

Grids and, consequently diagrams too, do not necessarily have to be regular and arise from the repetition of simple rhythms, nor do the lines that conform them have to be straight.

Early modern and modernist architects have made use of grids and diagrams alike. A possible connection between Sullivan's diagrammatic ornament design methods and Wrights' design practice can be established, but in the case of the latter, it could be also extended to his own architectural *organic* approach, something that needs further research attention.

D'Arcy Thompson devised a diagrammatic method to graphically explain the anatomical evolution of animals making use of *topological diagrams* which allow him to represent morphological transformations.

D'Arcy Thompson's graphic conception of diagrams can be considered *transformative* rather than *analytical* or *generative*. This potential has been embedded in design software and is being used in digitally disruptive contemporary architecture.

If we observe some of Eisenman's projects from the '90s onwards, it can be observed how the intrinsic directionality of the grids -despite its irregular interstices- allows to structure an architecture that arises from the potential of order embodied in these versatile graphic instruments that we can regard architectural diagrams as.

Throughout his long professional practice, Eisenman can be credited as one of the pioneers in the use of the three types of diagrams we have tried to differentiate in this research: *analytical*, *generative* and *transformative*.

To summarize, generative diagrams are neither drawings nor graphic schemes, they are abstract machines for architectural ideation. They possess an ambiguous status in between being and becoming, mediating in a certain way amid form and matter. They prefigure what the design could be but with a degree of openness that makes of them a supple and powerful mechanism to unfold multiple paths in the design process.

Acknowledgements. This paper shows some of the results of the research undertaken at UNCC during a sabbatical leave by prof. Carlos L. Marcos, partially funded by the University of Alicante, Conselleria de Educación, Cultura y Deporte, Generalitat Valenciana, in collaboration with Jeffrey Balmer.

References

Alberti, L.B.: De Re Aedificatoria. Akal, Madrid (1991)
Alexander, C.: A Pattern Language. Oxford University Press, New York (1977)

Alexander, C.: The Nature of Order: An Essay on the Art of Building and the Nature of the Universe. Center for Environmental Structure, Berkeley (2006)

Allepuz, A.: Traces of thought. Graphic processes in autograph drawings of modern architects from its commencement in Borromini until the end of the 20th Century, Ph.D. Dissertation, University of Alicante (2019, unpublished)

Allen, S.: Diagrams matter. In: ANY: Architecture New York No. 23, Diagram Work: at a Mechanics for a Topological Age, pp. 16–19 (1998)

Balmer, J., Swisher, M.T.: Diagramming the Big Idea: Methods for Architectural Composition, 2nd edn. Routledge, New York (2019)

Borsi. F.: Bramante. Electa, Milan (1989)

Lefort, E.C., Desvaux, N.G.: Diagrams: From Isotype to Gif. Notes for a didactic of the graphic analysis of the architecture. Revista EGA **22**(30), 30–41 (2017)

Carpo, M.: The Digital Turn in Architecture 1992-2012. Wiley, Chichester (2013)

Ching, F.D.K.: Architecture: Form, Space and Order. Van Nostrand Reinhold, New York (1979)

de Molina, S.: Materia ritmada: la retícula como sistema de orden. Arquitectura COAM **340**, 106–111 (2005)

Durand, J.N.L.: Precis of Lectures on Architecture with Graphic Portion of Lectures on Architecture by Jean Nicolas Louis Durand. Getty Research Institute, Los Angeles (2000)

Eisenman, P.: Diagram Diaries. Universe, New York (1999)

Eisenman, P.: The formal Basis of Modern Architecture. Lars Müller, cop., Baden (2006)

Eisenman, P.: Processes of the Interstitial, in Written into the Void: Selected Writings, 1990-2004/Peter Eisenman. Yale University Press, New Haven (2007)

Franco Taboada, J.A.: The search for three-dimensionality in professional practice. From the Patesi Gudea to the Elbphilharmonie in Hamburg. In: Cardone, V. (ed.) Territories and Frontiers of Representation, pp. 1551–1557. Gangemi Editore International, Roma (2017)

García, M.: Prologue for a history, theory and future of patterns of architecture and spatial design. Archit. Des. **76**(6), 6–17 (2009)

Hart, V., Hicks, P.: The letter to Leo X by Raphael and Baldassare Castiglione, c.1519. In: Palladio's Rome: A Translation of Andrea Palladio's Two Guidebooks to Rome, pp. 179–192. Yale University Press, New Haven (2006)

Corbusier, L.: Toward an Architecture. Getty, Los Angeles (2007)

Madrazo, L.: Durand and the science of architecture. J. Archit. Educ. **48**(1), 12–24 (1994)

Marcos, C.L.: Being and becoming in diagrams. Traces and protoforms as architectural subtext: from Deleuze to Eisenman. Revista EGA **18**, 102–115 (2011)

Milizia, F.: The Lives of Celebrated Architects, Ancient and Modern. Cresy, E., trans. Taylor Architectural Library, London (1826). Original title: Pasquali, G., dell'Arte di Vedere nelle Belle Arti del Disegno. Venezia (1781)

Rodríguez, D.: Diez libros de arquitectura: Vitruvio y la piel del clasicismo. In: Vitruvio, Los diez libros de la arquitectura. Alianza Forma, Madrid, pp. 11–51 (1995)

Rowe, C.: The mathematics of the ideal villa. In: The Mathematics of the Ideal Villa and Other Essays, pp 1–27. MIT Press, Cambridge (1976)

Sullivan, L.: A System of Architectural Ornament. Rizzoli, New York (1990)

Thompson, D.W.: On Growth and Form. Cambridge University Press, Cambridge (1945)

Vidler, A.: Diagrams of diagrams. Architectural abstraction and modern representation. In: García, M. (ed.) The Diagrams of Architecture. Wiley, Chichester (2012)

Vitruvius, M.: Los diez libros de la arquitectura. Alianza Forma, Madrid (1995)

Too Small to Survive

Clara Maestre-Galindo[(⊠)] [iD]

Department of Architecture and Design, Universidad San Pablo-CEU, CEU Universities,
Madrid, Spain
maestre.eps@ceu.es

Abstract. The sketches of the architectural project of a building which has been demolished acquire a different dimension once the building has disappeared. These drawings, as part of a special "*boite à miracles*", turn into very valuable objects when, after the death of their authors and the disappearance of the buildings, they become part of the graphic legacy of their works. Some of these drawings have been here selected with a view to revealing pertinent information about the authors and their work. Concise and intentional, the thin lines of the drawings will help to understand the character of their authors as well as the initial intentions that guided their proposals.

Keywords: Sota · Fisac · Perpiñá · Demolition · Sketches

1 Introduction and Background

During the period between the end of the twentieth century and the beginning of the twenty-first century several interests arose that brought about the demolition of certain buildings. But, thankfully, at least the interests in making the drawings that devised them disappear are virtually inexistent. It may be easy to understand that, while the drawings themselves belong to their authors, this is not the case with the actual real estate property. Some of the key elements that have led to the disappearance of these buildings and, consequently, the merits that have turned their architectural sketches and plans into documents of incalculable value, will be discussed here.

This study is illustrated by way of three examples of architectural constructions from Madrid. These constructions, unfortunately no longer existing, are the following: the Arvesú House (1955–1987), the Jorba Laboratories (1965–1999) and the headquarters of Banco de Valladolid in the Centro Colón building (1967–2018), works by the architects Alejandro de la Sota, Miguel Fisac and Antonio Perpiñá together with Luís Iglesias, respectively.

2 Discussion

The Arvesú House, owned by Angel Arvesú, was designed by Alejandro de la Sota in 1955. It occupied a plot of land in Madrid located at number 14 Doctor Arce Street.

L. Agustín-Hernández et al. (Eds.): EGA 2020, SSDI 6, pp. 418–428, 2020.
https://doi.org/10.1007/978-3-030-47983-1_37

The complex built with the intention of housing the new headquarters of the Jorba Pharmaceutical Laboratories was conceived by Miguel Fisac in 1965. It was located on a plot of land at the entrance to Madrid from Barcelona, owned by the Catalan businessman José María Jorba. The Banco de Valladolid building, which was part of the group of buildings known as Centro Colón (formerly Edificio Génova), was designed by Antonio Perpiñá and Luís Iglesias for the company Génova, S.A. in 1969. It was located to the north of the plot delimited by Paseo de Recoletos Boulevard, Génova Street and Marqués de la Ensenada Street (Fig. 1).

Fig. 1. Arvesú House. Jorba Laboratories. Centro Colón building.

2.1   Context of the Demolitions

Some of the characteristics common to all of them are laid out below. They were built in the mid-twentieth century, during the course of the twelve years that coincided with the great economic boom that took place in Spain during the sixties. The young architect de la Sota was 42 when he planned the Arvesú House, the same age Fisac was when designing the laboratories. Somewhat older than the latter, Perpiñá was 51 years old when he designed the Centro Colón building. It is worthwhile to pause momentarily to focus on particular dates because they are significant.

The demolition of the Arvesú House took place in 1987, curiously shortly after Alejandro de la Sota was recognized and awarded for his work, receiving, as he did, the Gold Medal for Merit in Fine Arts from the Ministry of Culture in 1985, and only a few months before he received the Gold Medal from the Higher Council of the Colleges of Architects of Spain in 1988. Coinciding with that year, 1987, the Ministry of Public Works commissioned de la Sota to remodel the building of the Arcades of Paseo de la Castellana Street in Madrid, which had been selected to host a large monographic exhibition on the architect Mies van der Rohe. Then only a few months later, in December 1988, the first major monographic exhibition on de la Sota's work was opened in the exhibition hall of the building of the Arcades that he had just remodelled.

The Centro Colón building was designed as one of the first multifunctional complexes, combining commercial, administrative and residential uses (Fig. 2). Anticipating at the time what is now considered customary, residential use was then intended for highly mobile residents. The Centro Colón shared commercial premises, offices and car parks within its three buildings, as well as a large part of its general facilities. It was made up of three buildings of different sizes that housed the different uses: administrative in the area of Paseo de Recoletos, commercial on the corner with Génova Street and residential around Marqués de la Ensenada Street. The one we are now studying, on the corner, was the enclave of Banco de Valladolid bank, one of the Rumasa companies dismantled in 1983. "Soon the premises were acquired by Barclays Bank, since the British financiers wanted to witness firsthand the economic revolution that was taking place in Spain... and what better place to behold it than this building in Madrid that had: on the left, Paseo de la Castellana Street; opposite, Plaza del Descubrimiento Square, and on the right, Paseo de Recoletos Boulevard with so many cutting edge evocations" (Azorín and Gea 1990, p. 105). It seems an unfortunate circumstance that the building has recently been almost completely demolished just one hundred years after the birth of the architect behind the project, Antonio Perpiñá.

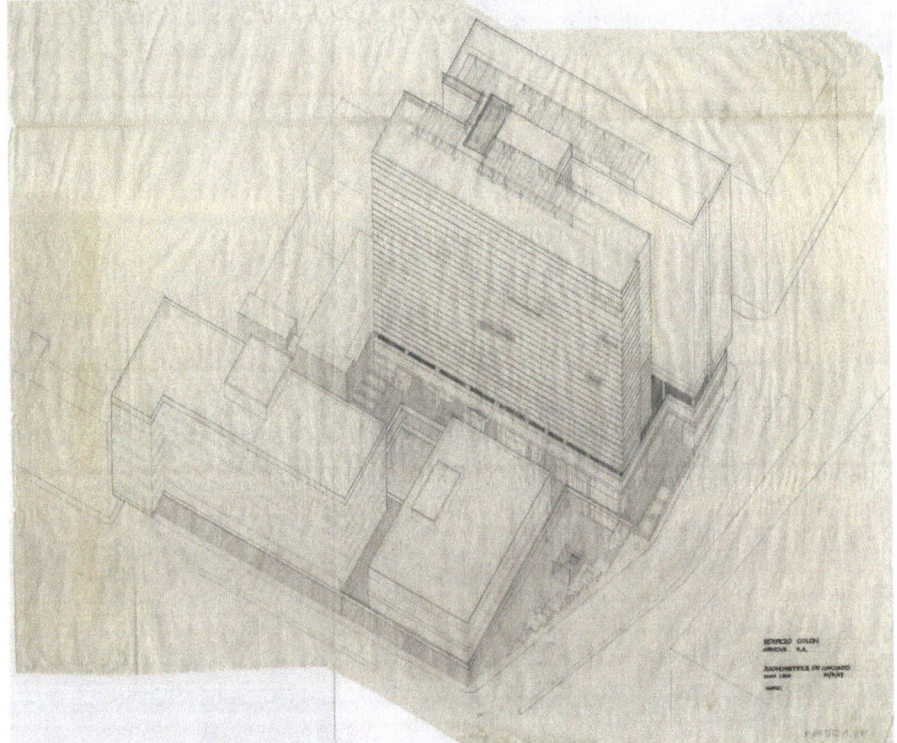

Fig. 2. Antonio Perpiñá. Centro Colón building, Madrid. Axonometry of the set. 1967. Source: Fonds and Legacy. COAM Foundation

There is also a certain irony in the proximity of the dates between the demolition of the Jorba Laboratories in 1999 and the awarding of two of the most prestigious prizes that may be given to a Spanish architect, namely the Gold Medal for Architecture received by Miguel Fisac from the Superior Council of Colleges of Architects of

Spain, in 1994, and the Spanish National Architecture Award granted by the Ministry of Development, in 2002. Last July marked the twentieth anniversary of the disappearance of the Laboratories.

There are similarities between these three buildings that help to understand why they were demolished. The promotion of all of them was entirely private. Ángel Arvesú, in the case of the house located on Doctor Arce Street, and then José María Jorba, businessman and owner of both Jorba Laboratories, and the entity Génova S.A., promoter of the set of buildings that formed the Centro Colón. None of the buildings was included in the Catalogue of Protected Buildings, drawn up by Madrid City Council as part of the General Town Planning Regulations. Subsequently all of them were acquired again by private developers. Grupo Lar purchased the Jorba Laboratories in 1999, the building that had been owned by Barclays Bank was acquired last year by the British investment fund CBRE and the plot where the Arvesú House was located is currently occupied by an office building.

All demolitions were carried out quickly, one could almost say surreptitiously. This was the case with the Jorba building, whose demolition was carried out without prior notice (Buey 2019). The Jorba Laboratories were demolished and the Barclays building dismantled capitalizing on the summertime. Regarding the Arvesú House, De la Sota Rius recalls, that "the owner had gone to see my father to request a renovation, when he learned that someone was going to declare the house a protected heritage site, so realizing that this was going to be a problem for him, he got rid of the house on Good Friday" (Alemany 2017). A change of use of the plot at number 14 Doctor Arce Street was needed in order to house the new office building. However, such a change of use was not necessary for the corner of the Colón building, since it already had the commercial and administrative uses. This also applied to the plot at the entrance to Madrid, which was also zoned for administrative use. The building that replaced the Jorba Laboratories significantly increased its volume and, consequently, the built surface area. Doctor Arce's office building doubled the floor area and tripled the total built area. The dismantled building of Colón will see its height increased to almost double. It is more than a foregone conclusion that the expansion of the buildability and redefinition of uses has allowed and will allow new investors to make huge profits on all the operations carried out there.

2.2   The Authors

From de la Sota's "Recuerdos y experiencias" (1989) it can be discerned that he always liked to talk about architecture as entertainment and that the thrill of architecture made him smile and laugh while life did not. This concise sentence reveals some aspects of the architect's personality. Therefore, it is not untoward when observing his face to find a serious look together with a slight smile. His son recalled: "In the case of my father, Alejandro de la Sota, life and work walked hand in hand, so together in fact that it is still impossible to separate them" (De la Sota Rius 2018). Regarding Miguel Fisac, Fernández-Galliano affirmed in the year of his death, that he was "an architect of fertile technical imagination, as gifted with craft skills as with mechanical inventiveness" (Fernández-Galiano 2006). According to his wife, Ana María Badell, Fisac had a very lively genius that he gradually transformed until he stopped getting impatient at the end of his life (Fisac 2007). Within the trio of studied authors Antonio Perpiñá is, perhaps, the least famous of all. Perpiñá was well known for his urban work and not so much in the field of building, "but the truth is that to build the Ministry [of Industry and Commerce] he won a competition involving people of the highest order, as Francisco Javier Sáenz de Oíza" commented recently Monje (cit. Plaza 2018). Referring to Perpiñá, Díaz de Tuesta expounds that he was part of a group of architects from the 1950s who were not the best known and who did not want to be. "These secondary architects built almost socially, putting aside authorship. Their concerns were more prosaic: to solve problems" (Díaz de Tuesta 2010).

Despite notable differences between them, one may find not a few coincidences. They worked intensely in Madrid, although they were not from Madrid. Alejandro de la Sota was from Pontevedra, Miguel Fisac from Daimiel and Antonio Perpiñá from Girona. Their talents allowed them to make outstanding works of architecture in a fruitful period in the capital city during the fifties and sixties. Throughout their work it is easy to foresee their wandering gazes and the consequent repercussion on their work. De la Sota had travelled to Berlin at the end of 1956, paying special attention to the work of Erich Mendelsohn (Fig. 3).

Fig. 3. Mendelsohn: stairs of the Metalworkers' Union building, Berlin. Source: Ofhouses Sota: Stairs of the Arvesú House, Madrid. Source: Alejandro de la Sota Foundation.

Fisac's trips to the United States and the Far East during the 1950s are well known (Fig. 4). Perpiñá also travelled to different destinations during the same decade, as Spain's representative on the Urbanism Commission of the International Union of Architects. He himself points out that "at one time, every year, the conference would be held in a different city, once in Paris, then in London, Prague, Warsaw, Vienna, Washington, thus linking up with the town planners of each site and exchanging opinions on the economic and political issues and so on that were everywhere, which provided him with a global vision" (Nasarre 1992, p. 90). It can therefore be assumed that all these journeys contributed to broadening his cultural references in general and his architectural references in particular (Fig. 5).

Fig. 4. Miguel Fisac: Lever House, New York, 1950. Source: [Cat. 124] inv. MF 102

2.3 Their Drawings

Pep Llinás remarks that of Alejandro de la Sota's buildings "there remains intelligence and sensitivity, culture and humour and the virtuosity of a magician, the more virtuous the less architect" (Llinás 1989, p. 11). But what is left after the disappearance of the works? Such is the case of the Arvesú House, of which only its drawings remain, more valuable now than ever. Observing the sketch made by de la Sota for the stairs of the Arvesú House, it is possible to imagine the "Mendelsohn" effect described by Baldellou (Fig. 6). In addition to "their fondness for music -with Bach as intermediary- their devotion to a feminine ideal -Louise, Sara-, their elitism and religious beliefs, and their sensitive intuition", he explains that both had "a way of trapping ideas obsessively through the drawing of 'the idea'" (Baldellou 1999, p. 13). This idea can be fully appreciated drawn as it is in a small sketch whereby the closed aspect of the house towards the outside is shown (Fig. 7 left). "To live quietly inside at home, with your back facing the world" de la Sota himself explained in 1955 (Sota Martínez 1985, p. 34). In his document of the project he went on to explain that "if the architect had completely forgotten about the

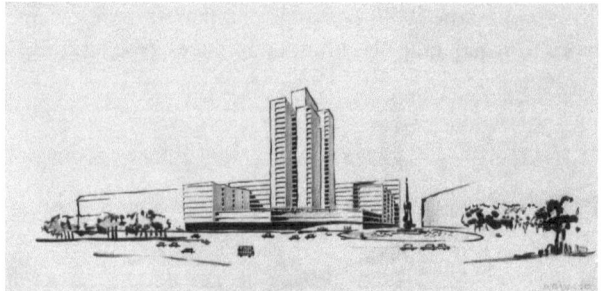

Fig. 5. Antonio Perpiñá and Luís Iglesias. Sketch for the Centro Colón building, Madrid. Source: Fonds and Legacy. COAM Foundation.

owner, (…) not a single window would have been put in what is known as the façade in all the houses". In spite of the fact that openings to the outside were finally incorporated, the idea of a closed façade would remain unalterable in the architect's mind. It is thus explained that in the second sketch the original idea lives on. It can be thought that the second sketch for the façade was made after the construction of the house, since there are some subtle differences between the two (Fig. 7 right).

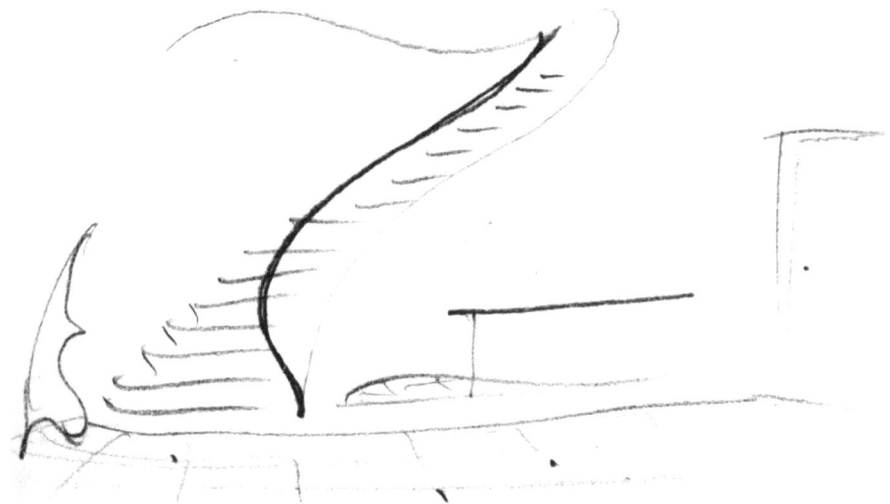

Fig. 6. Alejandro de la Sota. Sketch for the stairs of the Arvesú House in Doctor Arce Street, Madrid, 1955. Source: Alejandro de la Sota Foundation.

The first was drawn with graphite, using the technique to play with the pressure of the stroke on the paper. In a more intense way, the lateral entrance in the structure protruding from the façade stands out, shaded. The representation of the windows, however, has been reduced to a minimum, converted into small specks on the paper. The soft outline of the envelope allows us to guess that its author could have refined it later. The second

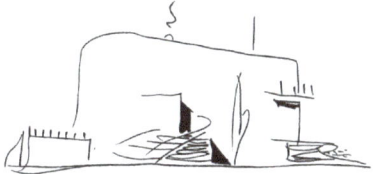

Fig. 7. Alejandro de la Sota. Sketches for the exterior of the Arvesú House. Source: Alejandro de la Sota Foundation.

drawing has been made with a continuous line, possibly with a felt-tip pen, a technique that does not enable the nuances of the previous one. Unlike the first one, it offers up more information on the relationship between the building and the street. It shows the access to the dwelling slightly raised from the street and incorporates new elements such as the exterior fence, a tree on the façade (whose existence we know from the archive photographs), and a projecting balcony on the upper floor. Curiously enough, the windows have all completely disappeared.

The Jorba Laboratories watercolour has a distinct value since, unlike the rest of the selected sketches, it was made at the time of the demolition of the building. On many occasions Fisac would speak of his great difficulties in drawing as well as his lack of innate ability, despite the enthusiasm he displayed in doing so. It is logical to think that the constructed building is undoubtedly more interesting than its drawing (Fig. 8). The demanding technique and the soft tones of the watercolour are not capable of transferring the power of the shape or texture of the material. We can imagine that this was probably not the author's intention and that Fisac painted this watercolour to keep alive the memory of the building that was not meant to be preserved. It is not a sketch, nor a previous idea, nor does it express the intentions of the project. It is the farewell to a work that would inevitably disappear.

Fig. 8. Miguel Fisac. Jorba Laboratories, Madrid, 1999. Source: [Cat. 3989] inv. MF 403

Fig. 9. Antonio Perpiñá. Elevation sketch for the Centro Colón building, Madrid. 1967. Source: Fonds and Legacy. COAM Foundation.

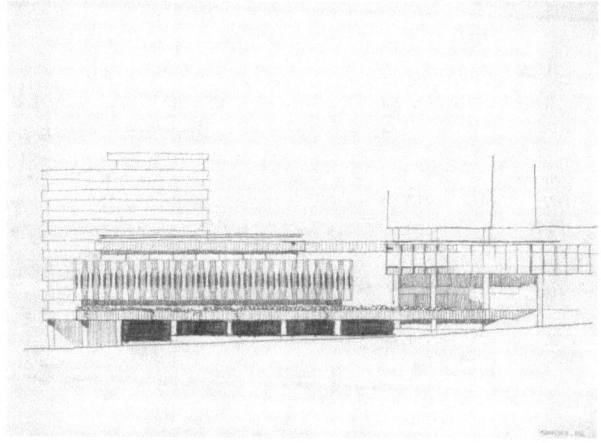

Fig. 10. Antonio Perpiñá. Elevation sketch for the Centro Colón buildings, Madrid. Source: Fonds and Legacy. COAM Foundation.

There are also noteworthy differences with the drawings of the Centro Colón building. Perpiñá, unlike Fisac, was an extraordinarily capable draughtsman. It can be seen in the two selected sketches. In them it is observed, not only the rigorous proportion control, but the remarkable dexterity in portraying the depth of the façades by means of the play of light and shade, the textures of the different materials, as well as the relation with the street within the neighbouring environment. The first of the sketches represents the elevation of the set towards Génova Street (Fig. 9). Perpiñá is able to express the existing dialogue between the three volumes by making use a very rapid graphite line and using the study of the shading. The drawing serves to demonstrate that all the volumes were part of a single project and that they were conceived together. It is evident from the relief of the façade of the office building, which extends to the edges of the residential building. It can also be seen in the proportions of the horizontal bands of the old Barclays building, aligned with those existing on the mezzanine of the apartment building. The second sketch depicts the elevation of the old bank headquarters to the Génova Street, without forgetting its relationship with the other two buildings. This second sketch shows a more detailed study that allows a greater adjustment between the smallest building and the other two higher buildings. The larger scale of the drawing allows him to incorporate a new graphic variable of texture as well as to play with conceptual differences between the foreground and the background (Fig. 10).

3 Conclusion

The demolition of the Arvesú House was as controversial and disputed as the demolition of the Jorba Laboratories. Within a cloud of secrecy, the Colón building has recently vanished. The way in which these buildings have disappeared is directly linked to the character of the architects who designed them. De la Sota's is ironic, as he states when he learns that the Casa Arvesú was to be demolished on Good Friday, "of course, that's why they say it's Glory Friday" (Alemany 2017), whereas Fisac's is churlish "what was not acceptable was to arrive in Madrid and that the first thing one found was a strange thing, which was generally pleasant" (Fisac 2007). Fortunately, Perpiñá did not get to witness the demolition of two of the three buildings that made up the Centro Colón, from which only the residential building endures, probably because this one is not *too small*.

But even greater is the evidence of personality in their drawings. Alejandro de la Sota's sketches reinforce the perseverance of his way of being, "everything had to be clear in the head long before drawing the first line" (Sota Rius 2018). Looking at his drawings, one can perceive that the initial intention has lived on, above and beyond the construction itself. Having the house been demolished, the idea of the blind façade seems to have prevailed over the building. Miguel Fisac's watercolour expresses the great vitality and passion of his character. The discretion and natural elegance of Antonio Perpiñá is manifested, tangible, in the outlines of his sketches for the Centro Colón building. At present, only the apartment building remains, which is perhaps more difficult to understand once the other two have disappeared, but magnificent when studied inside the Centro Colón alongside the project's sketches and drawings. For various reasons, but with common characteristics, these drawings are part of a prized legacy of our graphical architectonic heritage.

Acknowledgments. María Loureiro Sumay. Translation.

References

Alemany, L.: El propietario derriba la emblemática Casa Guzmán de Alejandro de la Sota. El Mundo (2017)
Azorín, F., Gea, I.: La Castellana, escenario de poder: Del Palacio de Linares a la Torre Picasso, 1ª edn. La Librería, Madrid (1990)

Baldellou, M.Á.: La forma continua: Sobre el efecto Mendelsohn. Revista Arquitectura COAM **318**, 08–13 (1999)

Buey, M.: ¿Por qué no se salvó 'La Pagoda de Fisac' del derribo? El País (2019)

Díaz de Tuesta, M.J.: La brillantez de los otros arquitectos. El País (2010)

Fernández-Galiano, L.: Muere Fisac, un referente de la elegancia. El País (2006)

Fisac, M.: Miguel Fisac, Apuntes y Viajes, 1ª edn. Scriptum, Madrid (2007)

Llinás, J.: Nada por aquí, nada por allá. Alejandro de la Sota. Pronaos, Madrid (1989)

Nasarre y de Goicoechea, F.: Antonio Perpiñá Sebriá: Historia del Urbanismo Contemporáneo Español. Revista Urbanismo COAM **16**, 88–92 (1992)

Plaza, A.: La fachada del edificio Barclays de Colón que nadie quiso salvar. Eldiario.es (2018)

de la Sota Rius, J.: Semblanza de De la Sota. Diario de Pontevedra (2018)

de la Sota Martínez, A.: Vivienda unifamiliar Sr. Arvesú en la calle Dr. Arce. Alejandro de la Sota, 1ª edn. Pronaos, Madrid (1989)

History of Medicine and Planimetric Analysis

Fernando Vilaplana Villajos[✉]

University of Seville, Seville, Spain
fvilaplana@us.es

Abstract. The adequate analysis of several historical floor plans, and the production of up-to-date floor plans in the case study we present, contributed in a substantial way and led to novel conclusions in an academic field in which it is not common for researchers from the field of Architecture to be involved in; the History of Medicine. We are not referring to the part of this discipline that deals specifically with the history of hospital architecture, in which, evidently, there is much that architects can contribute to, although that was the area in which this research was initially developed. The contributions made in this field supposed a substantial change in the previous hypotheses on the formal evolution of the hospital analysed, being this of such magnitude that it transcended the architectural, and even affected the previous theories about the treatment of the disease that said hospital dealt with. Therefore, it was necessary to make a critical analysis of them, and to set out in the research conclusions the review of settled and widespread ideas about the treatment and social consideration of leprosy over the centuries. Hypothesis that has finally been positively considered amongst researchers of the History of Medicine. It is therefore another case in which the positive effects produced by the synergies between different fields of research are confirmed, and support the claims for the need to have research teams that use graphic documentation, with researchers who adequately handle its language.

Keywords: Planimetry · Medicine · History

1 Introduction

The architectural evolution of hospitals throughout history occupy a prominent chapter in the manuals of the history of architecture, and of course in the history of medicine. Approaching the study of this architectural typology implies assuming that the concept of hospitals have evolved over the centuries and at the same time, how could it be otherwise, the shape of the buildings that host them. Within the study of this type of architecture, lazarets or leprosariums have been little studied, for different reasons, including the significant decrease in the incidence of this disease at the end of the Middle Ages.

In the case of the San Lázaro Hospital in Seville, we have news of its existence since the mid-thirteenth century. Its origin and operation being similar to that of other lazarets of the late Middle Ages. However, unlike most leprosariums, it continued to function during modern and contemporary times, continuing to be used as such, even until the beginning of the last century.

L. Agustín-Hernández et al. (Eds.): EGA 2020, SSDI 6, pp. 429–436, 2020.
https://doi.org/10.1007/978-3-030-47983-1_38

A series of circumstances derived from its use, location and cultural significance, led it to be practically unknown by society in general and hardly studied at the university level. As for its architectural evolution, this was hardly decipherable. Especially after the important reforms implemented at the end of the 19th century, and, above all, those that, during the 20th century, facilitated its conversion into a contemporary hospital; for this centre is still in use, and is comparable to any other medical centre today.

Typological studies on leper hospitals are scarce, particularly in modern and contemporary times. For this reason, the contribution the findings and analysis of the graphic documentation of San Lázaro have been relevant, both in the field of architecture and in medicine. Contributions that would not have been possible if the tools provided by the knowledge of work techniques and graphic analysis had not been used during the investigation.

2 Preliminary Studies

The previous university studies were carried out after the transformations performed throughout the 20th century, what greatly hindered the task of deciphering the formal evolution of the Hospital. On the other hand, researchers did not have any historical floor plan of the hospital and based their studies on the documentation of the hospital's archive, the reading of the ancient authors of the sixteenth and eighteenth centuries, and the contribution of nineteenth-century historians of Seville.

These complications were joined by the inevitable prejudice with which all researches, including myself, begin research of any study object related to leprosy. The medieval image of a stigmatizing disease, which socially annulled those who suffered from it, and, therefore, leprosariums as infectious places where the sick were little less than abandoned to their fate, has been assumed by society, and by most of the researchers, throughout the twentieth century.

However, the earlier the texts consulted the more the image they convey of the hospital differ from this preconceived idea. The authors of the XVI speak of a sumptuous and solemn hospital, and refer to the abundant privileges granted by the crown to the institution. In the seventeenth century it is described as a comfortable dwelling and in the eighteenth there is mention again that the house is accommodated and that it is in a very delicious place.

Like other previous investigations, the one I describe here, was impervious, initially, to this information. Existing prejudices about leprosy, and, what is supposed to be a lazaret, prevailed over what the texts were contributing. It had to be the direct and abundant exposure of data that the graphic documentation provided, which finally redirected the course of the study.

3 Three Floor Plans

Assuming fully the doctrine of Camilo Boito, summarized in his sentence: The monument as a document, the direct analysis of the hospital has been a fundamental part of this investigation, and the main tool used for this analysis, the execution of an analytical survey. This planimetry of the building was carried out by taking on-site data, photographs that were later rectified by ASRix, as well as georeferencing with GPS. This analysis of the building led us to date the construction of several elements between the nineteenth and twentieth centuries, including the galleries of the access courtyard that, until then, had been dated in the Middle Ages and which spatial and constructive data, revealed by the drawing, clearly induced us to rethink its construction to a closer date.

Properly dating this part of the building became a key element for the study of the hospital. The strategic situation of the courtyard, its appearance in Neomudéjar style (as we now know), and its location next to the church of the group, this one truly Mudejar and therefore medieval, led to the hypothesis of locating in this place a supposed Mudejar cloister; considered as the foundational space of the hospital. An assumption that was the base of subsequent studies and did not allow the formal evolution of the hospital to be adequately developed.

3.1   The 1890 Floor Plan (Fig. 1)

Starting from the assumption, established during the analytical survey, of the incorrect dating of the access courtyard, the search for historical plans of the hospital that could confirm it began. In this work, the Provincial Council Archive of Seville was fundamental, as abundant of the hospital's more recent floor plans were preserved here, and thanks to this research work, the first known floor plan of the hospital in the XIX century was located and catalogued.

Dated in 1890, it represents on sheet paper, 65×93 cm, the ground floor of the hospital, and it indicates the surface area of the building; 5689.08 square meters and the scale of the plan 0 m 005 m (1: 200). This floor plan reaches levels of precision comparable to current standards, and already uses the metric system. Being therefore the first floor plan of San Lázaro that we could consider in accordance with the current standards o f drawing.

It is probably a survey made after the important works carried out in the hospital in the 1860s, and leaves no doubt; at the end of the 19th century there were none of the galleries that currently exist in the hospital access area. Moreover, there was not even a patio in this place. The access was through an open space, which we could not call a patio, as it bordered on one of its sides with a large garden.

From a more general point of view, it shows us a hospital in which aspects such as ventilation or the adequate treatment of sewage have already been considered; key elements of the hospital typology initiated with the hygienist current of the 19th century. It is also the first document in which infirmaries are seen represented, which constitutes a milestone in the typological evolution, especially in hospitals for lepers, as this type of patient used to reside in individual rooms.

3.2   Search of the 18th Century Floor Plan (Fig. 2)

While it is true that the first of the floor plans we have described was a fortunate finding, the existence of the second floor plan we present was known. It appeared in a 1993 publication, the Historical Dictionary of the streets of Seville, and it showed the image of an old floor plan of the San Lazaro hospital, but unfortunately, probably due to a

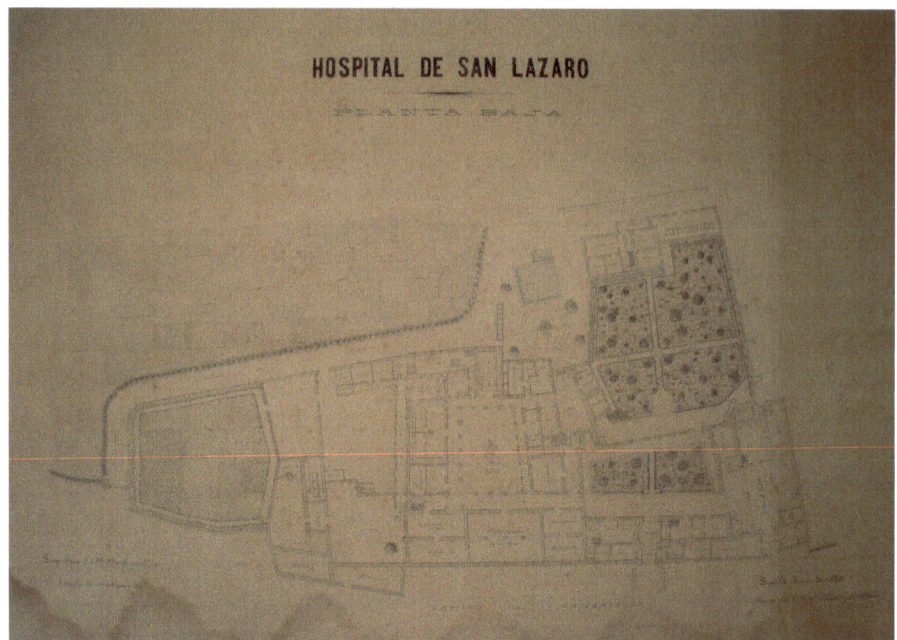

Fig. 1. Ground floor of the San Lazaro Hospital. Archivo de la Diputación Provincial de Sevilla. Mapas, Planos y Dibujos 394.

printing error, its bibliographic reference was wrong. It was not dated, and it was very difficult to extract information from it, given the size and resolution with which it was reproduced. In fact, the work that contained it, of a generalist nature, did not analyse it, and it was solely used as a "fill-in" image.

The importance of bringing this floor plan to light and analysing it prompted its search through different archives and libraries in the city. Finally, it was possible to locate it in a Military Archive in Seville. It is a manuscript on paper, 55.5 × 84.0 cm, made with ink and gouache, in good condition. The floor plan includes graphic scale measured in rods and feet and, an ample legend. The quality and geometric precision of this floor plan, which has been possible to date around 1760–65, is more than acceptable, even from current parameters, and makes it a very useful tool for studying the formal evolution of the building, as it had never been used in previous studies.

This floor plan gives us especially relevant information about the building, as it shows its structure prior to the important reform of the 19th century, and is the first graphic document on which the shape of the building can be seen prior to that date. On the other hand, the comparative analysis of this floor plan, with other graphic documents, signed by an outstanding military engineer working in Seville at that time, has made it possible to attribute the authorship of the floor plan and link the hospital with the well-known relationship of military engineers and illustrated hospital architecture.

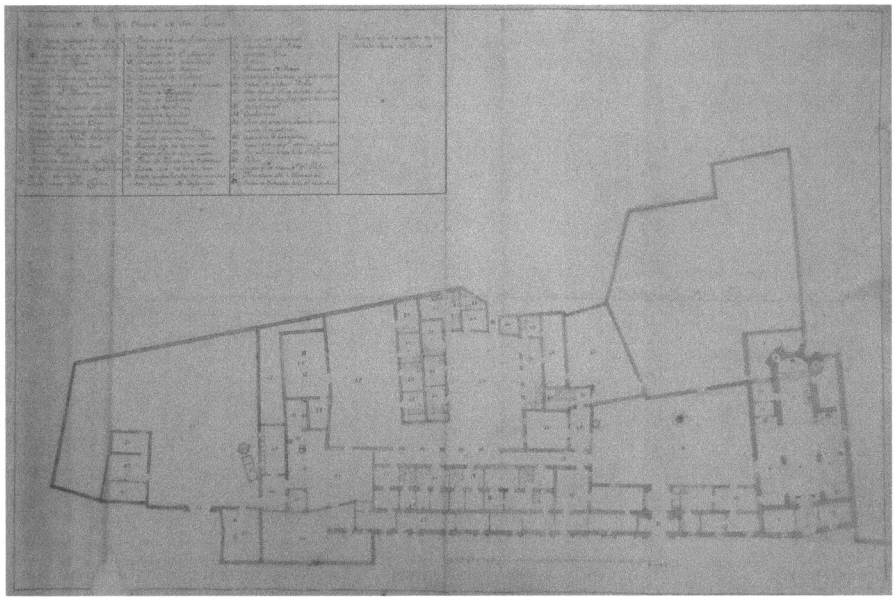

Fig. 2. Floor plan of the San Lazaro Hospital. Archivo Intermedio Militar Sur 8/24

This link was unknown until this investigation and through it; the relationship between the hospital and the Academy of Medicine of Seville has also been strengthened. This medical institution considered, at such an early date, that leprosy had a cure, as we know to be true today. As a result, this floor plan formed part of an initiative that emerged here and promoted the reform of the hospital into a healing centre.

3.3   Finding of the 1685 Floor Plan (Fig. 3)

The search for graphic documentation exceeded the local level, and an interesting floor plan dated in 1685 was discovered in the Simancas General Archive. This was part of a Royal Visit, a kind of audit performed by the crown. The supervision of the Visit was entrusted to two canons of the Cathedral, and the person in charge of the survey was a prominent member of Sevillian architecture who became the Master architect of the city.

Made using the technique of scratching with a punch and later inked, on sheet paper, 43.5 × 86.5 cm. It is an "orthogonalized" hospital floor plan. We can observe several errors within the work resulting from the artist's problems to relate the dimensions of the exterior and interior of the building.

We were therefore faced with an undiscovered floor plan that gave us unpublished and relevant information about a time when information about the hospital were very scarce. For the proper analysis of this floor plan, whose geometric definition, as we have said, was not very precise, we carried out a transcription. This work was based on the data provided both by the current survey and by the eighteenth century floor plan, closer in time to the floor plan that was intended to be transcribed. In this transcript, attempts

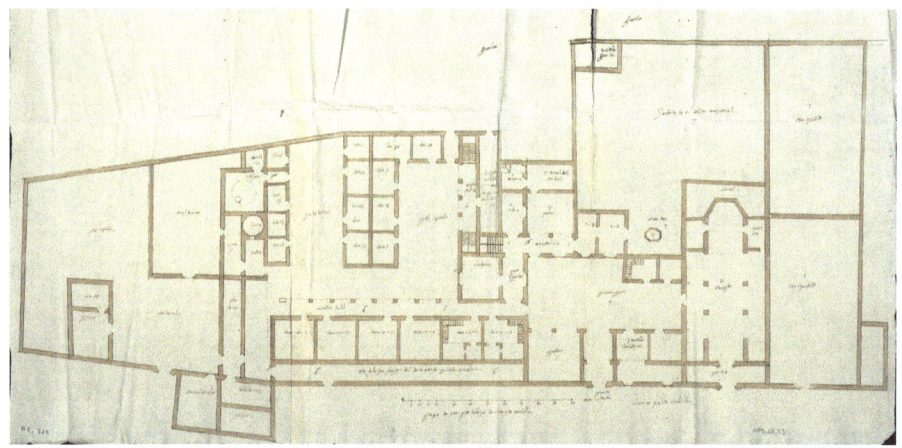

Fig. 3. Floor plan of the San Lazaro Hospital. Archivo General de Simancas. Mapas, Pliegos y Documentos. 68,73

were made to reverse the errors detected and it was possible to obtain a hypothetical floor plan of the building at the end of the 17th century.

Fig. 4. Floor plan of the San Lazaro Hospital in the 17th century. Drawing and hypothesis by the author.

This floor plan (Fig. 4) not only shows the shape of the building in the seventeenth century, it reflected the works carried out in the sixteenth century; an unknown Renaissance-centred intervention performed on the hospital and that this investigation brought to light.

4   A Review

We have tried to convey to the reader the thought that guided the research on San Lazaro, which, as we have stated, was mainly graphic. During this study, the use of a survey as an analytical tool of historical plans, and the transcription of this same plan, facilitated the historical rereading of the San Lázaro Hospital. As a lazaret, it has also represented an important advance in the study of a specific hospital model, leprosarium, barely studied to date.

The initial approach to the object of study, by carrying out an analytical survey of the building, aroused the suspicions that the assumed chronology of the hospital was wrong. Assumption that was confirmed by the finding of the 1890 floor plan. In this way, the result of the works of the 1860s could also be specified, which, although known, and their importance acknowledged, had never been formally defined.

He finding and analysis of the eighteenth century floor plan, not only allowed us to observe the form of the hospital to that date, it also brought to light an important intervention of an enlightened character. A reform, which even considered that leprosy had a cure and the floor plan that formed part of the architectural intervention necessary for the application of this novel concept about the disease and lazarettes.

The 1685 floor plan shows a hospital on which a complete Renaissance reform had been carried out, which is, in itself, an unprecedented discovery in this type of hospitals, but even more, the building that it shows us is much more than a simple hospice to shelter sick poor people. In this floor plan, large patios, galleries and, above all, a porticoed façade that related the Lazaretto to the road and, more importantly, to the city are observed. That is, completely different to the idea of an isolated medieval lazaretto that commonly suggests the mere mention of leprosy.

This information allowed the review of some texts that had been consulted and that now made sense, such as the Gussow/Tracy hypothesis. A hypothesis that affirmed that for medical/social reasons, during the 19th century, facing a reactivation of leprosy that doctors of the time were not able to fight, a so-called leprophobia was encouraged as a method that favoured the isolation of the sick. This idea, which is still alive today, has permeated most of the research and studies carried out on any object of analysis related to leprosy during the twentieth century.

The study of this institution has proved to be significant for the history of medicine. The San Lázaro Hospital in Seville is the leprosarium with the most extensive history known, which has its roots in the Middle Ages and which remained in use as a leper hospital until the early twentieth century. This unique circumstance therefore makes it a paradigmatic institution in which to study not just the formal evolution of this hospital typology. Its long history makes it possible to understand the changes that have occurred in the treatment and social consideration of this disease and where we can confirm the hypothesis that the current concept of Leprosy as a stigmatized disease is a social construction born in the 19th century that has lasted to this day.

5   Conclusions

Graphic documentation is a basic element in historical research, beyond the strictly architectural or the field of heritage.

Knowledge of graphic language can substantially improve the results of this type of research.

The elaboration of current building surveys and the transcription of historical plans are tools that enrich this type of research.

It is necessary to include in multidisciplinary teams that treat this type of documentation, trained researchers to analyse the graphic documentation adequately.

Acknowledgments. I would like to express my gratitude to the archives consulted and especially to their staff for their invaluable help.

References

1. Aires Mateus, M.: Hablar de proyectos es hablar de dibujos. In: Proceedings of the XI International Congress of Graphic Expression in Architecture. D.E.G.A, Seville (2007)
2. Arévalo Rodríguez, F.: El análisis documental y el levantamiento como metodología de investigación e n arquitecturas desaparecidas. EGA **20**, 134–143 (2012)
3. Carmona García, J.I.: Enfermedad y sociedad en los primeros tiempos modernos. University of Seville (2005)
4. Contreras, F., Miquel, R.: Historia de la lepra en España. Gráficas Hergón, Madrid (1973)
5. Echeverría, E., Celi, F., Casa, F.: El dibujo como herramienta de investigación. EGA **25**(25), 180–191 (2015)
6. Espigares Rooney, B.: Corografías contemporáneas de granada. La investigación a través del dibujo y sus resultados inesperados. EGA **23**, 70–79 (2014)
7. Giménez Muñoz, M.: Los establecimientos benéficos más relevantes de Sevilla hasta 1849. Diputación Provincial de Sevilla (2008)
8. Gussow, Z., Tracy, G.: Stigma and the leprosy phenomenon: the social history of a disease in the nineteenth and twentieth centuries. Bull. History Med. **44**(5), 425–449 (1970)
9. Laín Entralgo, P.: Historia universal de la medicina **3** (1972)
10. Leistikow, D.: Edificios hospitalarios en Europa durante diez siglos. Boehringer Sohn, Ingelheim (1967)
11. López Piñero, J.M.: La Medicina en la historia, Madrid (2002)
12. Pevsner, N.: Historia de las tipologías arquitectónicas, Barcelona (1979)
13. Rosen, G.: The hospital. Historical sociology of a community institution. In: The Hospital in Modern Society. Free Press of Glencoe, London (1963)
14. Sigerist, H.E.: Civilization and Disease. Cornell University Press, Nueva York (1941)
15. Thompson, J., Goldin, G.: The Hospital: A Social and Architectural History. Yale University Press, New Haven (1975)

Sketch for the Mural of the Aránzazu Basilica, by López-Villaseñor

Antonio Álvaro-Tordesillas$^{(\boxtimes)}$ ⓘ, Noelia Galvan-Desvaux ⓘ, and Marta Alonso-Rodriguez ⓘ

School of Architecture, University o fValladolid, Valladolid, Spain
tordesillas@arq.uva.es

Abstract. In 1961 the second contest for the decoration of the apse of Aránzazu was summoned since the winner of the first had died prematurely. The young painter Manuel López-Villaseñor won a second prize. Villaseñor was an educated painter between Renaissance academicism and contemporary informalism that he lived. Painter of great security in the drawing, of firm strokes that always strengthen the silhouettes and lend their stylized figures gravity and autonomy. And painter capable of essentializing the architecture that composes his landscapes from a geometrization of forms, especially influenced by Piero della Francesca. This text briefly narrates his journey as a muralist and explains the sketch he presented to the event; which he ordered around a cross, the panegyric of the Virgin of Aránzazu.

Keywords: Aránzazu · Mural painting · Apse contest · López-Villaseñor

1 The Decoration Contest of the Basilica of Aránzazu

In mid-1952, when the building of the Basilica of Aránzazu was being completed, a contest was called for deciding who would make its pictorial decoration. On the jury were the architects Francisco Javier Sáenz de Oiza and Luis Laorga –authors of the architectural project– and also the architect Secundino Zuazo, the sculptor Jorge Oteiza, and the painter Vázquez Díaz. Ten proposals were submitted, and the winner was Carlos Pascual de Lara.

In 1954, the Bishop of San Sebastián dictated that the decoration of the Sanctuary did not conform to the established canons of Sacred Art; what was corroborated and prohibited by the Pontifical Central Commission of Rome in the summer of the following year [1].

But in addition, Lara would die with thirty-six years, in 1958. The work of another artist who fit that 'established' canon had to be chosen. So in 1961, a national contest for the termination of the apse was convened, which was started and was incomplete. Forty-two proposals of all trends were presented, since the call included that 'the conditions of the contest do not require brushwork, but open possibilities for sculpture, wrought iron, mosaic, artistic games of lights, to very varied mixed solutions' [2]: Lucio Muñoz, winner of the contest, Eusebio Sempere, Rafael Aburto, Manuel Hernández Mompó,

L. Agustín-Hernández et al. (Eds.): EGA 2020, SSDI 6, pp. 437–444, 2020.
https://doi.org/10.1007/978-3-030-47983-1_39

José Luis Sánchez, Julián Ugarte, Susana Polac, Ramón de Vargas, etc. And a young man, but already recognized, Manuel López-Villaseñor, who obtained one of the five second-prizes [3].

2   Manuel López-Villaseñor

Villaseñor was an academic of the Royal Academy of Fine Arts of San Fernando since 1956, and this year he obtained the chair of Mural Painting and Pictorial Techniques at the School of San Fernando, where he would reach 'as a teacher the same credit as a painter' [4]. And to his credit, there were prizes of the First Hispano-American Biennial (1951), Gold Medal in the National Exhibition of Fine Arts (1952), First Prize in the International Exhibition of Agrigento (1952), Gold Mill in the Regional Exhibition of Valdepeñas (1955) or First Medal in the Exhibition of Sacred Art of Zaragoza (1958), among others.

But before that, he had gone to Italy in 1949, to the Academy of Spain in Rome as a Penetration in Painting, where he spent four years. There he got to know first-hand, both the Quattrocento painting and the contemporary Italian abstract painting.

Of the Renaissance, Giotto, Ucello had caused a sensation in him but the Pompeii frescoes and those of Piero della Francesca awakened their vocation towards the mural painting. The vocation is something that was inside and it was in Italy where it sprang up: 'the decision was immediate. I dedicated myself to the study of mural painting with a lot of interest and intensity'.

Of the abstracts, Sironi, Morandi, Carrá, Campigli… took a reporter to inform. The assimilation of new artistic languages such as Cubism and Expressionism, within an academic education based on the teachings of the great masters of Spanish and Italian painting, created a style, an unquestionable personal language. To him, also a person with a strong personality. 'Being figurative, it is technically abstract: being abstract, it contains figurative elements (…) Neither abstract, then, nor figurative. The opening of his vital retina is formally flooded with all those means that tend to realize the pictorial or human emotion' [5].

On his return, in the Fifty, Villaseñor was among the young artists involved in the discreet modernity that combined tradition and current, influenced by Italian modernity, together with Capuleto or Pascual de Lara himself [6].

3   The Murals of Zaragoza and Ciudad Real

Between 1954 and 55 he had painted part of the mural of the Deputation de Zaragoza. His Roman stay, still recent, was reflected in the atmosphere that reigned the ensemble: the costumes, the hairstyles, the characters, the landscapes, the formal essence of things. In addition, the scene chosen for the mural was that of fifteenth-century Italy (Fig. 1). 'The representation of the horse that in a marked foreshortening shows its rump to the spectator and that Villaseñor takes from a detail of the *Exaltation of the Holy Cross* of Piero della Francesca, or the slight perspective that applies in the representation of the scene and that recalls a again to the work of Della Francesca' [7].

Fig. 1. Mural of the Deputation of Zaragoza, 1954–55. Source: [8].

But there is also a certain informal influence, of Cubism, which is evident in the volume and the strong geometrizing character of the hard folds of the robes, the profiles of the landscape and the characters.

Other murals would follow this: the famous, award-winning and missing mural for the *Cabo de San Roque Transatlantic* (1954), the *Commerce Mural at the School of High Mercantile Studies of Barcelona* (1961), the *Mural of the Ulta Research Institute*, in Zaragoza (1966). But now he had just signed the mural of the Deputation of his native Ciudad Real (1960). Monumental, with historical scenes of the province and the harshness of country life; Schematic figures, linear, static, essence of them... In the centre, a bishop with a red robe of pyramidal composition protects motherhood. Flanking him, medieval warrior monks and Masters of the Military Orders in grey, white and black tones. On the right wall, he presents scenes from La Mancha life: ploughing man, peasant family, agriculture, industry, mining... all in brown and earth tones (Fig. 2). 'One group centres the other wall with a markedly dramatic sense of the town, the fields, the industry, the mining of my town, Puertollano, an iron town and deep geological and vital galleries, towns of human floods that ended up finding their sediment in those lands' [5]. This group is completed with the representation of a goat at the feet of an old man, motive perhaps took from an earlier painting of 1959, *Goats* (Fig. 3), a recurring motif in many of his paintings, and used to in the sketch of the mural presented to the Aránzazu contest.

In both murals, it is noteworthy the way he analyses and then paints the architecture. Influenced by Della Francesca, the analysis of the forms of that allows him to rescue his geometric essence. From its relationship with informalism, this geometric essentialization of forms sometimes becomes radicalized. In these years, and in later years, when he painted the sketch of the Aránzazu mural, his canvas painting was characterized by these geometric essentializations of Architecture. His pictures of cities are an obvious sample of this (Figs. 4 and 5).

Fig. 2. Mural of the Deputation of Ciudad Real, 1960. Source: [8].

4 The Sketch of Aránzazu

When Villaseñor appeared for the contest, he chose as his main motive the historical theme of the sanctuary, as did the great majority of the contestants, including Pascal Lara himself. The traditional plan of the panegyric of the Virgin of Aránzazu was to first make a dreadful description of the centuries that preceded its manifestation: the account of civil strife and the persistent drought that ravaged the area; so that, 'on such a dark background (…) the blessed silhouette of the Virgin emerges and radiant' [9].

Centred, in the lower part of the apse, Villaseñor painted on his knees and with his arms raised towards the Virgin the shepherd boy who had found his image hidden among a bush of thorns. Next to him, two goats of his flock; Cubists, bony (here the goats appear) (Fig. 6). Above this imposing and sharp hawthorn fly white angels that sing the Glory of God. This setting determines the base of a cross that orders the composition of the apse. In both arms, groups of pilgrims gather to pray before the Virgin. And in the upper part of the cross, there is a representation of the basilica and the Franciscan-Canaan congregation, guardian of this sanctuary since the early 16th century.

Around, in the four holes that the cross leaves free, Villaseñor narrates other scenes of the history of the place. So first, on the left, there are some sailors with their ships that are invoked under the protection of the Virgin, as did Juan Sebastián Elcano, Miguel López de Legazpi and Antonio de Oquendo. On the right, there are more pilgrims and what seems a memory of the fires that ravaged the sanctuary, two first fortuitous and a third, caused by the liberals in the middle of the first Carlist war. In the lower part Villaseñor paints, on the left, a medieval scene of the fight of sides, between the families of the Gamboa and the Oñaz, and the drought that ravaged the place; while on the right, the scene now represents the embrace of both families, ending the con-shop and the end also to the drought when the Virgin appeared. The Virgin ends with drought and confrontation and returns peace and tranquillity to the Basque people. 'In gratitude, they build a basilica and periodically go on pilgrimage to their sanctuary' [10]. There are also some Mercedarian and Dominican friars, painted in white, guardians of the sanctuary

Fig. 3. Goats, 1959. Source: López-Villaseñor Museum, Ciudad Real.

Fig. 4. Moon over Plasencia, 1957. Source: [11].

at the beginning; also a recurring reason that reminds us of those medieval monks of previous murals.

Thus, the image that looks like a messy set of scenes, is not done because it is around a large cross that distributes the space and inside which is the Virgin's dressing room.

Compositionally, Villaseñor plays in this mural in addition to colour. The dressing room of the Virgin appears white; even more with the black border that surrounds it. The rest with grey colours, earth tones of great sobriety. The set is of a canonical compositional clarity. Villaseñor 'composes harmonizing forms, balancing volumes, looking for symmetric weights' [11].

Fig. 5. Mántua Channel (prop. Rudolf Müller). Source: [11].

Fig. 6. Manuel López-Villaseñor. Sketch of the mural for the apse of Our Lady of Aránzazu (second prize). Oñate (Guipúzcoa), 1961. Source: [10].

5 Conclusion

Although they can be rescued from the text, it is convenient to summarize as a conclusion the characteristics that define the way of seeing and embody the reality of Villaseñor on a canvas. Characteristics that result after studying the work of the painter Villaseñor, after studying his retirement years in Rome, his travels in Italy, reading his comments, interviewing people who knew him, with the López-Villaseñor Museum in Ciudad Real, with the Basilica in Aránzazu. And on the other, after analysing and comparing his work with the references he himself denoted as his: first the Zuloaga, Solana, Palencia and Vázquez Díaz; then the Giotto, Ucello, and Piero della Francesca; and finally the contemporaries Sironi, Morandi, Carrá, Campigli… All of them created an unquestionable personal language.

His paintings use a strong geometrization of reality that rescues the formal essence of things, of characters, landscapes, volume and folds of costumes, but especially of architectures. Geometrization exercised in his (sub) conscious by Della Francesca and the informalism that he had to live. Facts that make him analyze the shape of the architectures that he later paints on his murals highlighting, above all, his original geometric component; even forcing it.

The sketch of Aránzazu shows these characteristics apprehended and tested in previous murals; schematic, linear, static, essential figures…

And the colour, as its other quality. Villaseñor's painting based on a palette of earthy, grey colours… that distinguish it. And that he uses in Aránzazu, with the particularity of being able to highlight the dressing room of the Virgin, contrasting black and white in a unique way.

Compositionally, Villaseñor is clear and orderly, as in all his murals, using the expressed geometrization of the figures in his paintings to even divide the space of the mural. In Aránzazu, the composition, as we have seen, is ordered according to a cross that divides the wall and tells the panegyric of the Virgin. Also noteworthy is the recurrent use of some figures that seem to be essential in their murals: medieval monks, goats, essential architectures, the people…

Villaseñor composes harmonizing forms, balancing volumes, looking for symmetric weights; as well as if he were an architect composing his project.

References

1. Fernández Cobián, E.: El espacio sagrado en la arquitectura española contemporánea. Colegio oficial de arquitectos de Galicia, Santiago de Compostela (2005)
2. Sitio web de Arantzazu. https://www.arantzazu.org/index.php/es/32-basilica/abside/59-el-abside-historia-de-su-decoracion. Accedido el 19 Oct 2019
3. Eraso Iturrioz, M.: Oteiza y vanguardia. Ondare. Cuadernos de artes plásticas y monumentales **6**, 297–336 (1989)
4. Martínez Cerezo, A.: Diccionario de artistas españoles. Martínez Cerezo, Santander (1999)
5. Ponce, F.: Villaseñor. Servicio de publicaciones del Ministerio de Educación y Ciencia, Madrid (1971)
6. Cabañas Bravo, M.: La Primera Bienal Hispanoamericana de Arte: arte, política y polémica en un certamen internacional de los años cincuenta. Universidad Complutense de Madrid, Madrid (1991)

7. Grau Tello, M.L.: La pintura mural en la esfera pública de Zaragoza (1950–1997). Universidad de Zaragoza, Zaragoza (2012)
8. Villaseñor, Z.A.: Biografía y catálogo razonado de su obra. Servicio de Publicaciones de la Junta de Comunidades de Castilla la Mancha, Toledo (1990)
9. Lizarralde, A.: Historia de la Virgen y del Santuario de Aránzazu. Aránzazu, Oñate (1950)
10. Monforte García, I.: Aránzazu, arquitectura para una vanguardia. Diputación Foral de Guipúzcoa, Donostia-San Sebastián (1994)
11. Camón Aznar, J., Villaseñor, L.M.: Dirección General de Bellas Artes, Madrid (1958)

Space Without Substance: The Drawing of the Material in the Work of Modern Architects in Japan - Kazuo Shinohara, Toyo Ito, Kazuyo Sejima (SANAA)

Ángel Allepuz Pedreño[✉]

Department of Graphic Expression, Theory and Projects, University of Alicante, Alicante, Spain
allepuz@ua.es

Abstract. Japan has developed its own idea of modernity outside of the Western model. To talk about contemporary Japanese architecture is like talking about domestic architecture. Based on this architecture with popular origin, Kazuo Shinohara will set a critical differentiation from the idea of space that prevailed in modern western architecture. We owe Shinohara the denial of the idea of space understood as an architectural substance and raw material, overall, not comparable to the idea of emptiness typical of Japanese culture. This would explain, in our opinion, not only the conception of the relationship between structure, form and space, but also a way of drawing architecture in Japan. In our judgement, the way in which specific contemporary Japanese architects draw, such as Toyo Ito, Kazuyo Sejima or Atelier Bow-Wow, has its origin in the way of drawing of Shinohara, whose teaching, based on his conception of the idea of space and its relation to matter, extends up today.

Keywords: Kazuo Shinohara · Architectural drawing · Architectural space

1 Introduction

Japan remained completely self-contained to all types of trade, apart from the Netherlands and China, for more than two hundred years, occupying almost the entire Edo period (1615-1868). Japan underwent a process of modernization that included the abolition of feudalism and strong industrialization after the forced opening of the country to Commodore Perry's threat to bomb Japan from Tokyo Bay in 1853. Based on the international character and ecumenical vocation of the Modern Movement it does not seem right to maintain an eurocentric vision of these relations. It has been a round trip. Japan was the place of early attention of modern architects. F.LL. Wright builds the Imperial Hotel in Tokyo between 1915 and 1923. This matter involved another european teacher: R.M. Schindler, who was trained in the small group of disciples of Otto Wagner, collaborator of Adolf Loos and fellow of Richard Neutra (March and Sheine 1995). His training as an engineer at the Imperial Technische Hochschule made him be able to join

© The Editor(s) (if applicable) and The Author(s), under exclusive license
to Springer Nature Switzerland AG 2020
L. Agustín-Hernández et al. (Eds.): EGA 2020, SSDI 6, pp. 445–457, 2020.
https://doi.org/10.1007/978-3-030-47983-1_40

the office to develop the hotel execution project. It began on January 1, 1918 and ended definitively in 1922. In the course of this period he was practically in front of the office, since Wright traveled, too, frequently to Japan. From Wright's office comes Antonin Raymond, who in 1923 finishes his own house, considered the first modern house in Japan (Frampton 1998 [1993], p. 262). Early, too, was the long stay of Bruno Taut, who studied in detail and devotion the Japanese popular dwellings - the houses of the peasants and fishermen - (Fig. 2). He also made some drawings, few. Taut said: "The Japanese house is completely designed for summer. Actually, it seems to be just a summer house" (2007 [1936], p. 95).

In addition to the aforementioned Wright, Schindler and Taut, the trips of other Western architects to Japan such as Neutra, Gropius, Le Corbusier, Perriand, Fisac, Scarpa or Holl, are well studied today (Masson 2017). This current also occurred in reverse. Le Corbusier's office received two Japanese architects in the decade of the twenties (Kunio Maekawa 1928-30 and Junzo Sakakura 1929-36) who returned to Japan bringing back with them the Corbuserian ideas. Maekawa joined to the Raymond's office in Japan, and was, in turn, master of Kenzo Tange, and this one, of Arata Isozaki (Kostof 1998, p. 1269).

At the same time, Jun'ichirō Tanizaki writes his *In Praise of the Shadow* (2003 [1933]), a plea for the virtues of the aesthetics and traditional ways of life of domestic Japan, not palatial, which has already come into conflict with the devastating influences of the foreign architecture with which it does not find a possibility of conciliation. In this intellectual environment, the Japanese popular house - not the palaces, nor the temples will be the center of the theorizing reflexion of Japanese architecture.

The search for a solution to the confrontation between tradition and an imported modernity will mark, and, in our opinion, still marks the line of debate. There is a transit from the mimetic acceptance up to the mixture of the modern formal repertoire with the traditional Japanese constructive elements and, finally, has given rise to the development of an own architectural thought. The reconstruction of the country after World War II was interpreted as a kind of colonialism imposed by the foreign administration, whose basic budgets did not fit with the Japanese way of living. Tadao Ando believes that between 1955 and 1960 there were many attempts to reconcile modern vocabulary with the aesthetics of traditional Japanese architecture. They failed because of the lack of understanding of the character of the wood - traditional construction material - whose mock eaves, lattices and terraces were a sign of the complete disconnection with the new forms. At the beginning, the influence was tangential and did not affect the way in which the Japanese made their constructions, but the picture was going to change with the post-war situation. Tadao Ando says: *"I suspect that no Westerner can understand the difference between the antiquated life that the Japanese people led and the one which was introduced in the country during the interwar period"* (Ando 1985, p. 138). This matter has interest to us, because the denial of the idea of space in traditional Japanese architecture is deeply rooted in the configuration of the popular way of life.

2   The Idea of Space in the Written and Constructed Work of Kazuo Shinohara

The critical review of the fundamentals of the idea of space in modern architecture in Japan was initiated by the architect Kazuo Shinohara (Frampton 1998 [1993], pp. 287–288). The architect had completed his doctoral thesis entitled Study of the spatial composition of Japanese architecture (Massip-Bosch 2011, p. 7). This architect devotes the same attention and care to both what he writes and what he builds and photographs. His work is short; but its influence will be very great in an environment, that of Japanese architecture, where architects hardly recognize a teaching or even a mutual influence. It has its starting point in 1964 - the first year that Japanese architects are allowed to travel outside the islands - with the text entitled The Japanese Conception of Space (2011 [1964]) and continues in 1967 with the so-called Theory of Residential Architecture (2011 [1967]). This revisionist attitude spread among a group of young Japanese architects.

After World War II, the hope of making compatible the traditional way o f perceiving Japanese space and the evolved forms that modern architecture had achieved was maintained; but soon one realizes that, with exceptions, that Japanese historical sensibility about the nature had been excluded. The cause comes from afar, and according to the criteria of the architect Shinohara, it is because concepts such as "Space" do not maintain any relationship with Japanese daily life; there may be an idea of "emptiness", but no concept of space arose (inside traditional Japanese architecture). The deep study of tradition was not a refuge for shelter, but a starting point to build a critical and genuine review o f he architecture that remains to be done.

Shinohara observes how Giedion devised a method to identify each historical period with a way of treating and managing space and, consequently, useful to discern what the spatial manifestation of modern architecture should be. Shinohara focuses on the correlation Giedion establishes between mathematical advances and the emergence of the corresponding spatial manifestation. Shinohara wrote: «what we have learned here is that in Western architecture a philosophy supporting it definitely existed and that this developed hand in hand with progress in the spatial conception related to the sciences» (2011 [1964], p. 244). If this biunivocal relationship between theory and space is true, it cannot be given in Japan, because such a notion of "space" has not developed. Tanizaki, aware of the dependence between science and artistic production, had already anticipated in 1933 that the lack of aesthetic accommodation that was observed in Japan for Western products derived from the fact that Western scientific foundations do not match well with the "Japanese spirit" (pp. 19–22).

We should not overlook that Shinohara never abandoned his point of view as a scientist. He wrote: «I specialized in mathematics before studying architecture. Therefore, for me, thinking about mathematics is almost the same as thinking about architecture. They are two sides of the same coin» (Shinohara and Obrist 2014, p. 3). Shinohara doubts that in Japan the idea of space as a "raw material" of modern architecture could be culturally assimilated. This idea dominates the analysis, criticism and architectural design in the West in the modern period in a hegemonic way and constitutes one of the paradigms of modernity. In Japan, it cannot be in this way, by the reason that such a conception of "space" as a substantive idea, has not historically existed, and, consequently, cannot be translated.

We know from David Stewart's article Kazuo Shinohara's three spaces of architecture and his first and second styles (p. 25) that Shinohara was, in 1976, a precocious reader of Gilles Deleuze. For this year, Deleuze had already published some of his non-academic works of philosophy: *Difference and Repetition* (2002 [1968]), *The Logic of Sensation* (1989 [1967]), and, in collaboration with Félix Guattari: *Anti-Oedipus* (1985 [1972]) and *Rhizome*) (1976). The influence of Deleuze's work that Stewart makes known to us is credited by the fact that Shinohara commented on Deleuze's ideas in his own writings. On our side, we observe similarities in some of the ideas presented here by Shinohara. Deleuze had begun his critics on the "deep," the dismantling of "the essential," the denial of "the universal," and the dissolution of "the substance." The critique of the idea of space as a substance is worthwhile as an example: According to Shinohara, the Japanese do not recognize the substantial condition of space; they identify, at most, a certain idea of emptiness, of vacuity. Shinohara says (2011 [1964], p. 245):

Katsura Villa, Kinkaku-ji and all the other beautiful and elegant buildings considered to represent Japanese perfection in architecture do not contain any "space" as such. Such beauty as exists is a beauty stemming from the "non-existence of space". This is no mere wordplay, since "space" here refers to that concept of generated space in the West, which envisions a substantial entity as opposed to space as simple void of cosmic emptiness.

The spatialist movement of the criticism and theory of Western architecture of the first half of the twentieth century (Frankl, Giedion, Zevi and Moretti) establishes a predomain of space - as a defined vacuum - over the shape as envelope, and this over the bearing structure. The structure is subjugated to the configurable shape of an identifiable space without interfering with it. Inside the canonical modernity (Le Corbusier or Mies) the delimiting shape is released from the supporting structure -free plan with dividing planes and non-structural walls. The space-envelope relationship is worked from the modulation of the degree of enclosure and the search for a fluidity between the interior and the exterior. What seems clear is that there is no interference between the idea of space, confinement and structure. The space-shape (envelope or delimiter) and shape-structure (support or support of the form) relationships in the Western tradition are maintained as a constant (from the Pantheon, through the Gothic cathedrals and ending in Nervi, Candela or Pérez Piñero). On the other hand, a "logic of emptiness" gives an air of transience to contemporary architecture that is produced from Japan (Fig. 9).

In Japan, however, the relationship between "space" and structure is not hierarchical, the lack of conception of a substantive and unitary idea of space makes that the presence of structural elements that interfere with space is not considered as a design failure. (Figure 1 and 9). The organization of the space takes place by setting physical references, setting guidelines for the arrangement of matter—pillars, beams and enclosure walls—. The plot line does not consist in making them disappear, hiding the structure, eliminating supports and increased spans, since the presence is necessary to measure and recognize the limits of the void. The materiality of the structural elements tends to vanish, without being absent. Arata Isozaki, an architect somewhat younger than Shinohara, has internalized this idea and expresses it this way:

Fig. 1. Interior view of *Tanikawa* House (1972-74) Kazuo Shinohara. (2011, p. 133).

No time and no space. But we have "ma": between objects and objects. In-between space, sound and sound, there are silences apart, pauses. This is called "ma". Space is important; in-between space is more important. Everywhere we have using "ma". You have a concept of Time: cronos plus "ma", and Space: void plus "ma" (The GAA Foundation 2017, pp. 1:20–2:05).

3   The Coeval Architectural Drawings of Shinohara with His Writings on the Japanese Conception of Space

Following the chronology established by Stewart (2011), and for the purposes of this study, we will limit ourselves to the drawings of the earliest Shinohara houses known as *First Style* (1954-66) that ends with the *house in White* (Fig. 3 and 5). This period culminates with the publication of the two texts previously referenced and that mark the distance with the idea of Giedion space based on the mathematical formation of Shinohara. We will extend the study to the *third style* (1974-82), which coincides with the reading of Deleuze's work and we exemplify in the house for the poet Tanikawa (Fig. 1 and 7) of which he writes about his "desire to discover an antispace" (2011, p. 132). In this second period it will be the work of a French philosopher that allows Shinohara to reconcile the space-shape-structure disjunction.

Stewart states that, in the opinion of many, Japanese construction is rudimentary; but in ours one it is the structural character of the Japanese constructive practices what deposits the content in the immanent matter. For Shinohara, Japanese constructive practices are essentially structural. All the attention of the designer and drawer is focused on the presence of the exiguous matter. (Figure 7). The negation of the spatial condition explains not only the conception of the relationship between structure, form and space, but also the way in which the architecture is drawn. The study of the architectural documents produced for the construction of the project has its correlation with the way in

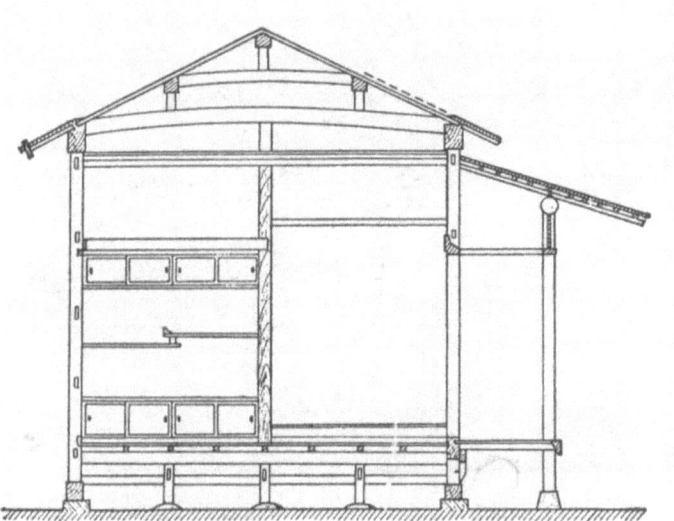

Fig. 2. "Cross-section through a small house, niche, *tokonoma* and veranda, from *Japanese Design*". Reproduced by Bruno Taut. (2007 [1936], p. 50).

which it is represented. Constructions in Japan lack the tectonic character they present in the tradition of Mediterranean cultures that, from Egypt to Rome, have marked the canonical way of building. In Shinohara we find a precise correlation between his thinking and the mode of representation. Among the available drawings of the *House in White* (1964) (Fig. 6) we found a paradigmatic one made in September 1965 at 1/50 scale. It is composed with a detailed elevation, where the traditional construction elements of the wooden eave, blind panels, and carpentry carefully reproduce the material texture and the precise relative position within the plane. The window that is placed on the inclined plane of the roof presents the foreshortening of the oblique projection. In contrast, the plan, which is represented on the same scale, lacks any indicator of materiality, modulation or depth. It has no symbolic elements, neither graphic nor extra graphic. There is no environment nor are there hidden and projected edges, although they would be of great help to interpret it - as we can see in the following example (Fig. 6).

On the opposite, all the graphic intensity will be found in the constructive actions, since the basic sections do not follow the rules of geometric section-projection (Fig. 6). Sections are large scale drawings (Fig. 7). More than constructive details stacked in relative position, we will always find complete sections of the building fully resolved in its constructive and dimensional aspects. Shinohara finds no satisfaction in making conventional sections where a distinction is made between the contour, the interior and the surroundings. You do not need to draw sections that mark the distinction between the "interior space" and the exterior by means of the interposition of a line marked in thickness that turns a continuous mass residue still to be built: a *Poché* (Castellanos Gómez 2010). The section must have sufficient scale to show precisely what is happening "within it", since a new universe of emptiness and matter unfolds on a smaller

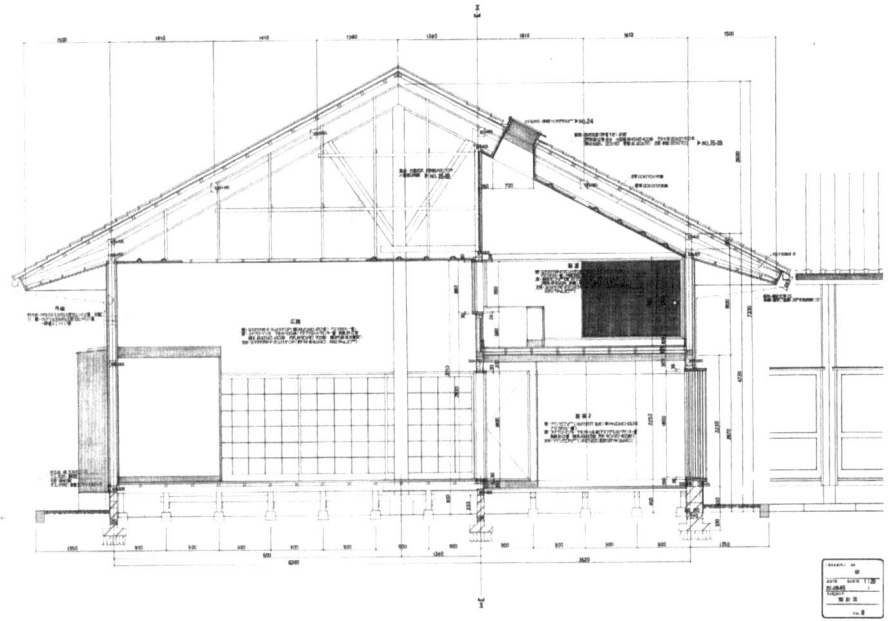

Fig. 3. Cross-section *House in White* (1964) Kazuo Shinohara. (Shinohara et al., 2011, p. 79).

scale, whose placement sequenced in time and whose disposition must be studied by the architect.

This will be a drawing mode that accompanies Shinohara's graphic production during his career, you just compare the reproduced drawings as Fig. 3 and Fig. 7.

4   Vestige of Shinohara's Thought and Drawing in Contemporary Japanese Architecture

Japanese architects are refractory to the idea of being grouped in schools. They generally do not recognize direct and more difficult influences from a nearby architect. Shinohara is an exception, because his influence has been expressly recognized by Toyo Ito (1941-) (Frampton 1998 [1993], p. 345) and the architects of the younger generations such as Kazuyo Sejima (1956-) and Ryue Nishizawa (1966-) (Fig. 9). Sejima, recalls the deep impression produced by the vision of some photographs of a house of the architect Shinohara (Harvard University. Graduate School of Design 2019a, b). In our judgement there is an implicit recognition in the work of Atelier Bow-Wow created by Yoshiharu Tsukamoto (1965-) and Momoyo Kaijima (1967-) (Donati 2019) and another implicit that is evidenced in the comparison between the drawings of one and the other, while a graphic code unification that persists after fifty years becomes highly visible (Fig. 7 y Fig. 10).

Toyo Ito (1997, p. 26) by placing 13 tubes on the floor of the *Sendai media library* denies the possibility of existing in the "main space". It defines its pillars as "large trees", within which the "useful spaces" are inserted, which, in turn, contain voids trapped within

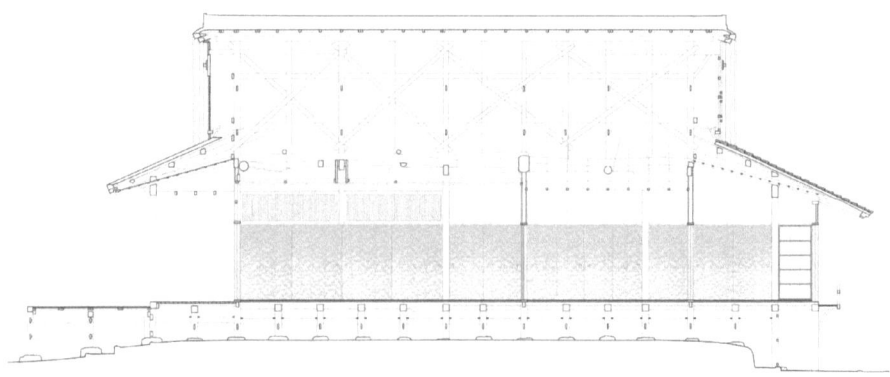

Fig. 4. Longitudinal section of *Katsura Imperial Villa,* Japan (1616-1660) Unknown author.

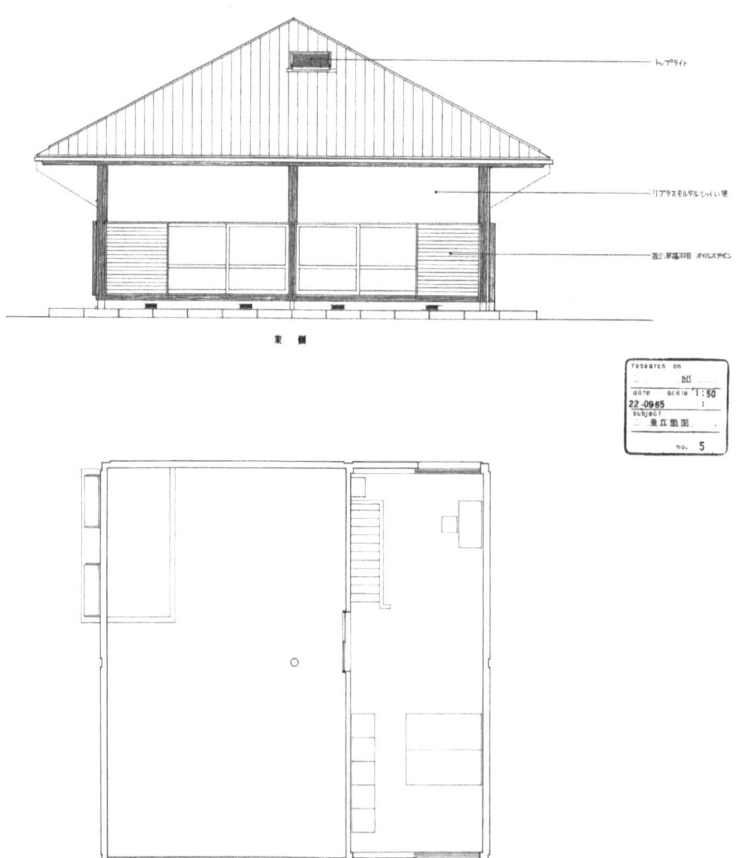

Fig. 5. Elevation and loft plan of *House in white* (1964-66) Kazuo Shinohara. (2011, p. 79).

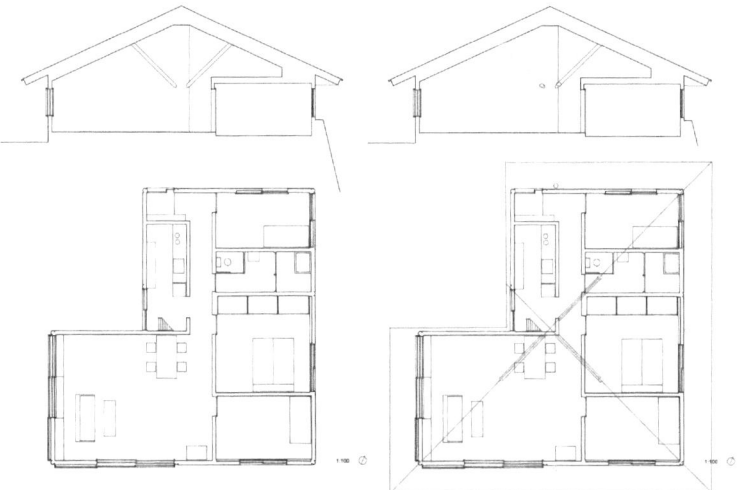

Fig. 6. Ground plan and section of the *Hanayama South House* (1966-68) left. Kazuo Shinohara (2011, p. 71) Right view. plan redrew by the author.

the interweave of its bar-branches through which energy flows, information, the air and the users (Fig. 8). Drawings that refer to the "column-trees" of the *Tanikawa house* (Fig. 1) and *Hanayama south house* (Fig. 6) and that recalls the latticework of beams that support the roof of the *Katsura palace* (Fig. 4). The action will happen between the objects, the objects are not important: matter what is in between.

From an examination of the drawings of the sections of the *Katsura Palace* (Fig. 4), the house for the poet Tanikawa (Fig. 1) and the *Nora House* of Atelier Bow-Wow (Fig. 10), we can identify some characteristics common. The graphics used is concentrated in a few graphic variables: the intense use of the nuances that the line in black ink, in its different styles and very few thicknesses, masterfully solve the variety of nuances with very limited resources. The sharpness in the definition of the structural and constructive elements clearly distinguished between the undifferentiated intermediate spaces is appreciated. There are no spots, scratched surfaces or dark areas that configure the edge in which a portion of interior space is limited. It is, on the contrary, an organization of lines that precisely define concrete objects, with a known constructive purpose and with scrupulous respect for emptiness, the silence that lies between them. It is difficult to interpret the medium and large-scale project plans, as it happens with the media library, or the SANAA works, where the structure is diluted in a scattered set of thin columns, the enclosures are constructed with the minimum glass section and the *Pochés* are hollowed out and are able to be walked, as is the case in the Glass Pavilion of the Toledo Museum of Art (Ohio). Only a large jump in the scale between drawings allows a partial understanding of the building (Fig. 9). Sejima declares its preference for working with models; even at 1/1 scale (Sejima et al. 2004, p. 12).

Others, like Tadao Ando (1941-), who does not recognize the influence of Shinohara, as far as we know, makes the following reflections:

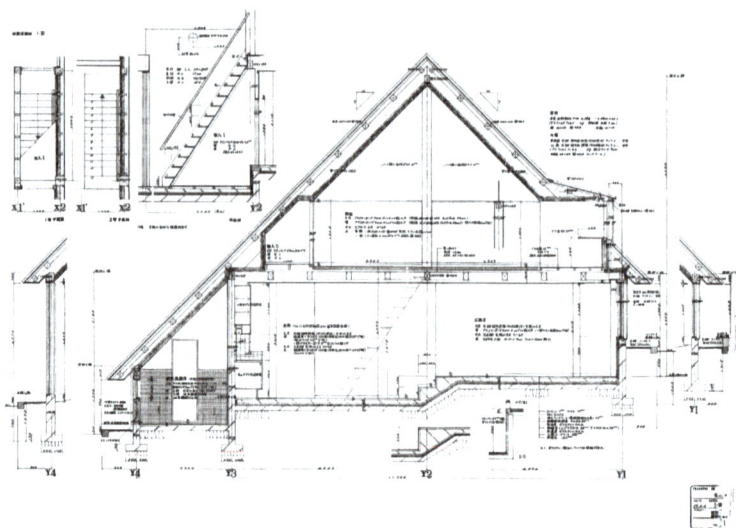

Fig. 7. Constructive section of the *Tanikawa House* (1972-74) Kazuo Shinohara. (2011, p. 144).

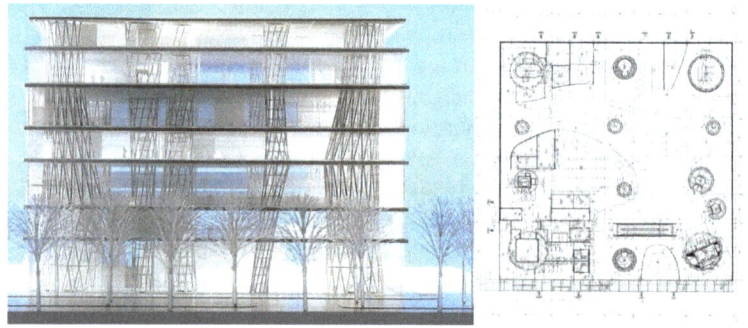

Fig. 8. Model and plan o fthe *Mediateca of Sendai* (2000) Toyo Ito. (1997, p. 29).

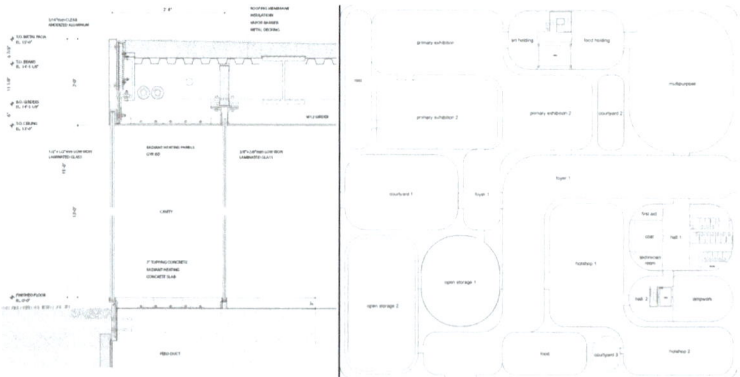

Fig. 9. Constructive section of the glass screen and ground plan, *Museum of Toledo*, Ohio (2014) SANAA (2004, pp. 114, 120).

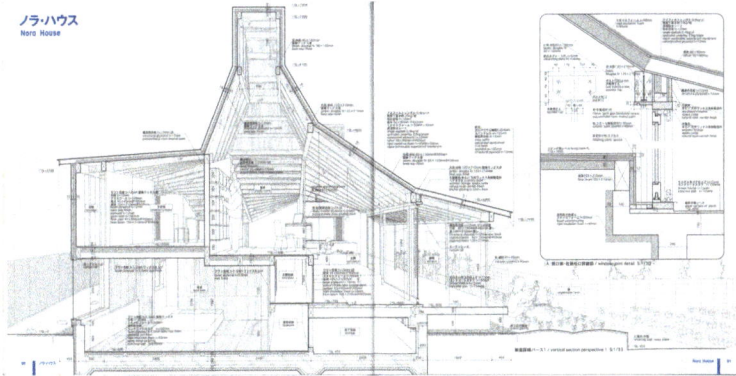

Fig. 10. Constructive section of the *Nora house* (2006) Atelier Bow-Wow (Tsukamoto et al. 2007, pp. 90–91).

Japanese architecture has a marked horizontal guideline, and its spaces are not only non-geometric but also irregular. You could say that it is a formless architecture, in which architecture and nature are integrated, producing a floating space. For me, the Pantheon and Piranesi are representative of the western architectural space as opposed to the Japanese architectural space. After long considerations, I decided that the object of my work should be the integration of these opposite spatial conceptions into a unified and transcendent architecture (Ando 1990, p. 6).

There is no doubt that Ando seeks a reconciliation where Shinohara had detected a caesura. Is the same dialectic; in fact, it defines the two concepts of space by an opposition. Ando could already travel to Europe (1965), where he comes into contact with the work of Le Corbusier; but, in addition, he experiences the net sensation of spatial plenitude produced by the effect of entering the Pantheon. Ando also recognizes the source of inspiration for his work in traditional domestic architecture, specifically in Sukiya's betrayal due to the impulse he gave to traditional constructions to find a formal definition consistent with the presence of human life,—vibrant and vigorous in direct relationship with nature—with the light and the shadow. Ando explains that: "Sukiya, at the lower end of the scale, is nothing more than a tea house; in the upper one, it can be presented as a group of premises, as the Katsura Palace shows" (Ando 1985, p. 140).

5   Conclusion

Shinohara carried out a critical review of the idea of space understood as a raw material of modern architecture in the 60s. Shinohara denies the existence of an idea of space in the Western-homologous Japanese tradition based on a mathematical model. The lack of substantiation of space in the popular peasant architecture (origin and source of all Japanese architecture) was sublimated in the idea of a certain non-substantial emptiness because of the fluidity and transitory characteristic of popular architecture. Shinohara consecrates space as the void that remains between objects, far from the substance space defined by the material boundary conditions. Japanese constructions lack clear bounds, both in their mobile perimeter closure, and in the covering that protects them, composed of a framework of structural elements. The "structure does not define space" nor "is in the middle of space" but the space that flows "between the structure". We think that this way of understanding space has been transmitted to Ito, SANAA or Ishigami. The difficulty found in the understanding of a plan or section of a medium-scale building comes from the absence of apparent clear limits caused by the lack of material in the enclosures and the absence of resonance with the structural elements. Space is something almost impossible to represent; but even more when there is no mass to section. The lack of materiality makes very difficult its graphic representation in accordance with the usual Western model, and has unleashed a particular way of drawing.

References

Ando, T.: Desde una autoconfinada arquitectura moderna hacia la universalidad. In: Frampton, K. (ed.) Tadao Ando. Edificios, Proyectos, Escritos. Gustavo Gili, Barcelona (1985)

Ando, T.: Composición espacial y naturaleza. El Croquis **44**, 5–6 (1990)

Castellanos Gómez, R.: Poché o la representación del residuo. EGA. Expresión Gráfica Arquitectónica **15**, 170–181 (2010)

Deleuze, G.: Lógica del sentido, 1ª edn. Paidós Ibérica, Barcelona (1989[1967])

Deleuze, G.: Diferencia y repetición. Amorrortu, Buenos Aíres (2002[1968])

Deleuze, G., Guattari, F.: Rizoma. Pre-textos, Valencia (1976)

Deleuze, G., Guattari, F.: El Antiedipo. Capitalismo y esquizofrenia, 1ª edn. Paidós Ibérica, Barcelona (1985[1972])

Donati, G.: Architectural Ethnography (2019). https://www.nan-ban.com/en/chronicles/architectural-ethnography. Accessed agosto 2019

Frampton, K.: Historia crítica de la arquitectura moderna, 9ª edn. Gustavo Gili, Barcelona (1998[1993])

Harvard University, Graduate School of Design "Reflecting on Shinohara": Kazuyo Sejima and Seng Kuan in Conversation (2019a). https://www.gsd.harvard.edu/event/reflecting-on-shinohara-kazuyo-sejima-and-seng-kuan-in-conversation/. Accessed 01 Sept 2019

Harvard University, Graduate School of Design, The House is a Work of Art: Kazuyo Sejima on her fascination with "Shinohara's way" (2019b). https://www.gsd.harvard.edu/2019/09/the-house-is-art-kazuyo-sejima-on-her-fascination-with-shinoharas-way/. Accessed 15 Sept 2019

Ito, T.: Mediateca en Sendai. 2G Toyo Ito. Seccion 1997 **2**, 26–33 (1997)

Kostof, S.: Historia de la arquitectura, 1ª edn. Alianza Editorial, Madrid (1998)

March, L., Sheine, J.: RM Schindler: Composition and Construction, 1ª edn. Academy Press, Londres (1995)

Massip-Bosch, E.: Kazuo Shinohara: beyond styles, beyond domesticity. 2G Kazuo Shinohara. Casas **58–59**, 4–18 (2011)

Masson, A.: Viajes de arquitectos occidentales a Japón. UPM, ETSAM, Madrid (2017)

SANAA, Sejima, K., Nishizawa, R.: Ocean of air. El Croquis (2004)

Sejima, K., Díaz Moreno, C., García Grinda, E.: Liquid playgrounds [fragments from a conversation]. El Croquis, III-IV **121–122**, 9–26 (2004)

Shinohara, K.: The Japanese conception of space. 2G Kazuo Shinohara Casas **58–59**, 242–245 (2011[1964])

Shinohara, K.: Theory of residential architecture. 2G Kazuo Shinohara Casas **58–59**, 246–259 (2011[1967])

Shinohara, K., Massip-Bosch, E., Stewart, D.B., Okuyama, S.-I.: 2G N.58/59 Kazuo Shinohara. Gustavo Gili, Barcelona (2011)

Shinohara, K., Obrist, H.-U.: Hans-Ulrich Obrist en conversación con Kazuo Shinohara. Quaderns d'arquitectura i urbanisme **266** (2014)

Stewart, D.B.: Kazuo Shinohara three spaces of architecture and his first and second styles. Kazuo Shinohara. Casas Houses **58–59**, 19–34 (2011)

Tanizaki, J.: In Praise of Shadows. Leetes Island Books, Chicago (2003[1933])

Taut, B.: La casa y la vida japonesas. Fundación Caja de Arquitectos, Barcelona (2007[1936])

The GAA Foundation: Arata Isozaki—Time Space Existence (2017). https://www.youtube.com/watch?v=E54K8wACQRc. Accessed 02 July 2019

Tsukamoto, Y., Kaijima, M. (Atelier Bow-Wow): Graphic Anatomy Atelier Bow-Wow. ToTo Publishing, Tokio (2007)

Emotional Laboratory: Drawing and Creative Experience as a Collective Visual Narrative

Mara Sánchez Llorens[✉] and Iván Pajares Sánchez

Polytechnic University of Madrid, Madrid, Spain
mariadelmar.sanchez@upm.es

Abstract. This communication focuses on answering three questions: What do we mean by emotional, creative expression? How to learn to research from visual storytelling? How to do it collectively? We respond to these issues through the analysis of the pedagogical experience entitled *Emotional Laboratory*. We have asked architecture apprentices to turn an investigation into a visual narration. They have made it in teams and use drawings and other free experiences as creative tools.

Students need to understand reality as a human construction that they work. Apprentices discover and communicate the meanings of that reality, and both, in turn, interact through their mutual understanding.

They discover the collective intelligence in this reciprocal work and their co-workers. As they must reveal their graphic group research, presenting the process repeats every week, they develop their public oratory and increase their critical capacity.

Emotional Laboratory brings every apprentice develops and shares the ideas around a topic visually raised in the course so that they are validated, rejected, or combined by the rest of the group. In addition to the visual experience, the apprentice begins in collaborative work. This cooperative learning takes place among people coming from different contexts, so the members of every team feedback and generate a coherent and shared visual narration.

Keywords: Emotional Laboratory · Collective intelligence · Narrative creation

1 Introduction. Emotional Lab

"[The experiment] wants to become the expression of creativity with a free will that, without trying to deny rational values, can aspire to be fruitful under an emotional conception (Goeritz 1953)."

This communication focuses on answering three questions: What do we mean by emotional, creative expression? How to learn to investigate from the visual narrative? How to do it collectively?

We respond to these issues through the analysis of our recent teaching experience entitled *Emotional Laboratory*, and that consists of raising in the classroom research that the apprentice architects must turn into a visual project, by groups, drawing and other creative experiences.

© The Editor(s) (if applicable) and The Author(s), under exclusive license
to Springer Nature Switzerland AG 2020
L. Agustín-Hernández et al. (Eds.): EGA 2020, SSDI 6, pp. 458–468, 2020.
https://doi.org/10.1007/978-3-030-47983-1_41

1.1   A Laboratory in the Atelier

The *Emotional Laboratory* is a teaching experience based, in its theoretical approach, on an educational research methodology called "critical communicative research" The research becomes a ploy for the apprentice to participate in an absolute studied reality around a raised topic of which there is a group-based subjectivity and communicate it visually.

This form of research tries to understand reality to interpret it, but also to explain it since its ultimate purpose is to transform it. "Critical communicative research has taken its first steps for more than a decade, and its applications in the social sphere result in success (Bizquerra 2004)".

The following reflexions must be considered in the field in question, that of architecture:

Why Propose Group Creative Processes in Architecture Ateliers? In learning in Architectural Schools, it is common to consider that creativity originates in individual processes. Spaces of creative diversity, however, favour the development of collective intelligence. By reversing the reflection, we can affirm that without collective processes, the creative experience is impoverished, and therefore the group creative processes are more enriching, although sophisticated.

The experience analysed in these lines is born from a group process since the beginning. First, we propose a topic to research in the class, and then the working groups reflect on the reality bounded from the action and try to explain it make-do artefacts and drawings that seek, in turn, new approximations. That is, they speculate from team practice.

The examples of the processes set out in this communication are from the course 2018-19, *A World of Migrants: The Power of Movement*. One of the groups focuses its gaze on the lost paradises and stops at the relevance of plastic on the migratory routes of individual animals altered by the barriers that generate this material in the landscape. The words disappear from the narrative when the members of the group manipulate a plastic object to turn it into an artefact that triggers, in turn, a composite drawing that maps the presence of plastics on the limited routes (Fig. 1).

The eye directs the hand because collective action precedes the word; the group does not speak; the word comes later. Action generates more action than reflexions.

Drawing and Creative Experience. For Jerome Bruner, knowledge and human creation are divided into two modes, the paradigm and the narrative. The latter is more focused on the humans being, on their experiences, desires, and needs. An example of this form of approximation is the series on the daily routines of architect Clorindo Testa from Buenos Aires (1923-2013) and is a primary reference for several groups in our classroom (Fig. 2).

In the specific example, we discussed in the previous section, the lost climatic havens, and plastics, the narrative creation lies. The group's experience shows that plastic barriers are holding back ancestral migrations of animals and that human interventions in the landscape must favour or at least not disadvantage them.

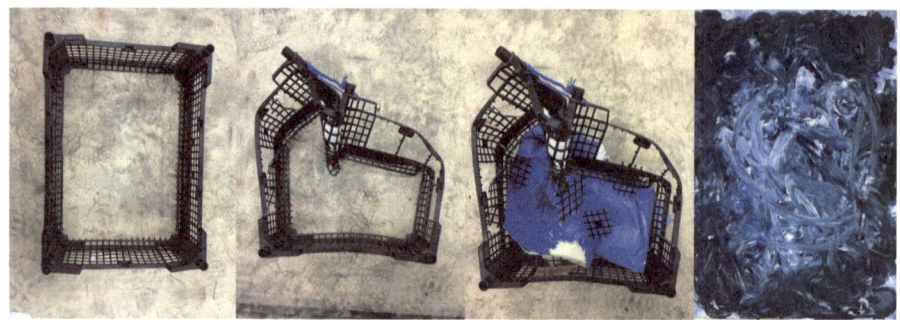

Fig. 1. Artefact and drawing, group 8. The work *Barriers* is a cartography of landscapes transformed by humans by artificial barriers of various kinds and plastic materiality. Source: Authors

Fig. 2. Clorindo Testa, *Inhabit, Move around, Work, Entertain*, 1952-1974. Source: Glusberg 1983

We conceive of this mode of viewing as an experimental inquiry that turns our atelier into a laboratory. Architectural creative processes, understood as a way of collective visual declaration of reality, trigger a narrative-critical process.

Fifty-seven apprentices are integrated into the atelier-laboratory and are organized into fourteen groups of three or four members to carry out the experience described in this communication. The first group presentation of the research in the process takes place in the classroom on March 5, 2019, followed by other collective presentations on April 12 and 26, 10, and 17 of May.

The training of the architect requires the communication skills we acquire through practice. In the proposed exercise, the apprentices share these skills with the rest of the School in a public event.

The Open Day of the School holds on 24 May, and the last presentation of the experience is delayed to this date and becomes a public event open to all those attending the Open Day. The fourteen members participate and present their research at a time of ten minutes per team. We invite four artists/architects, and they dialogue with all the teams and spectators.

1.2   A Case Study *Emotional Laboratory*

"The detonator is to get them to talk, to vote ideas, without judgment, without restriction or anything, because that is where I think the big things come up-throw the idea you want, Say what you like and go! (…) At a first moment, that would be for the detonator, to make them talk, to present ideas, to make anything they come up with sound like what it sounds like (Valbuena 2018)".

The issue raised in the classroom *A World of Migrants: The Power of Movement* is connecting the first-year apprentices with their daily lives. First, they express their reflections in the initial drawings, without judging; then there are other ideas that they also express in drawings and other formats.

As with the day-to-day life by Clorindo Testa, some groups to feel part of a migrant world, empathize with other migrants (Fig. 3), or to be so themselves by becoming urban migrants wandering in their daily lives through the spaces that make up their routine (Fig. 4).

Fig. 3. Drawing and photography of light in motion, group 3. The work *The Forgotten* fits a group of migrants from their place of origin to their destination, reconstructing the atmosphere of the spaces travelled. Source: Authors

Concern for climate change, loneliness, tourism-phobia, or gentrification are the orientations of group work. These approaches derive from the artefacts, maps, videos, installations, and performances that occur in the atelier. The members of each group narrate at one time and graphically, their conclusions about the relationships and simultaneities existing between the reflected reality, space, and emotions.

The work of the groups visualizes subjectivities. These subjectivities result from an egalitarian dialogue of all individualities. They not only speak but communicate visually. The instrumental dimension lies in the graphic and experimental process, and the group gives meaning to research that does not deny rational values and bears fruit under an emotional conception.

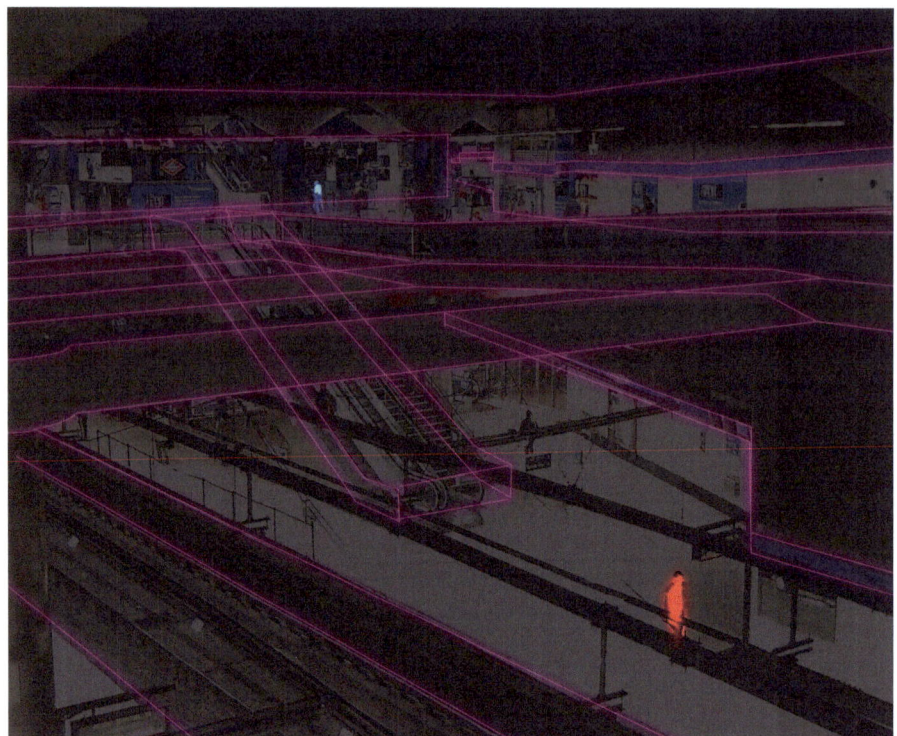

Fig. 4. Manipulated photography, Group 1. The work *Converged Trips* follows the same routine of two persons who, without knowing each other, use the subway of the city in which the course takes place daily. It delves into the invisible networks that they create in these daily lives and the public spaces in which they develop. Source: Authors

The recent internal surveys that we conducted in our groups verify that the apprentices have experienced and highly value the collective intelligence that appears in this form of exercise, much more abundant in its way than a sum of individualities. It is the apprentices themselves who discover this idea of collective intelligence in their collective work.

2   Shared Drawings and Creative Experiences

"Collective problem solving is not a significant part of our education. Virtually all evaluations and reviews are about individual performance. And, in the few exceptions in which group problem solving is involved, it is a homogeneous group of colleagues, all from the same country, who make up the whole team (Buxton 2007)."

This teaching experience aspires to exercise collective intelligence in the classroom, from a humanistic approach close to the theories of Wilhelm Dilthey and tries to make it visible to the apprentices and members of the groups. This collective intelligence is called cultural intelligence by some authors (Bizquerra 2004).

The apprentice needs to understand the reality in which he works as a human construct whose meanings he communicatively constructs through interaction in mutual understanding. In this reciprocal work, he discovers in his companions the collective intelligence and by presenting it repeatedly to them he develops his public oratory and increases his critical capacity.

2.1 An Iterative Shared Process

The teaching experience analysed here approximates the rational to the intuitive: it is an open process. Although this statement may seem recurrent, what is important about the *Emotional Laboratory* is the creative process it triggers, a process that is iterative and shared.

Repeated Critical Ability or Self-criticism. The experience takes place in nine weeks with two weekly sessions (Thursday and Friday) of three hours each. In the first sessions, the maestros present an auxiliary tool that is a rubric in which logic and subjectivity converge. According to this rubric, maestros lead the first constructive criticisms. From the second session of presentations, the trainees lead the discussions after the presentations accompanied by the maestros.

The maestros elaborate the rubric according to specific indicators of Kasei engineering applied to sociology and served the apprentices to elaborate their analysis of the evaluable discourse later.

Kansei Engineering is a method to translate feelings and impressions into product parameters since the 1970s. Professor Mitsuo Nagamachi designed them. The syllable "Kan" in Japanese, means sense and the syllable "sei" sensitivity, "Kansei" means something like "the sensibility perceived through the senses."

This rubric tool has the pedagogical value of being a reference for the apprentice to be critical of their work, critical of their work in the group, and critical of the work of others. They do critique through action. That is, the apprentices criticize their work with more work; hence the process is iterative.

Shared Shares and Their Instrumental Dimension. The apprentice elaborates and shares his ideas around the issue raised so that they are validated, rejected or supplemented by the rest of the team: a collaborative work is initiated between people coming from different contexts that feedback, draw and experience in a shared way.

The sequence of sessions in the classroom simultaneously: they work individually, a session on Thursdays, and they work on the group, the session on Fridays. They work outside the atelier, too (Fig. 5) (Sclater 2003). The students prepare the exhibitions that they share weekly with the rest of the groups. In the last session of the course, they share their final and group presentation to an external audience in a public presentation of a playful character that performs in open (Fig. 6).

2.2 We Base the Approach to Our Area of Knowledge from the Art World

The creative experience shapes the images that arise in the eye of our mind, our dreams and our daily life (experiences, desires, needs). An *Emotional Laboratory* is a form of praxis, through which the knowledge of we can mature in wisdom by the empathy that is born later with group work. This teaching experience offers a fresh air path for all

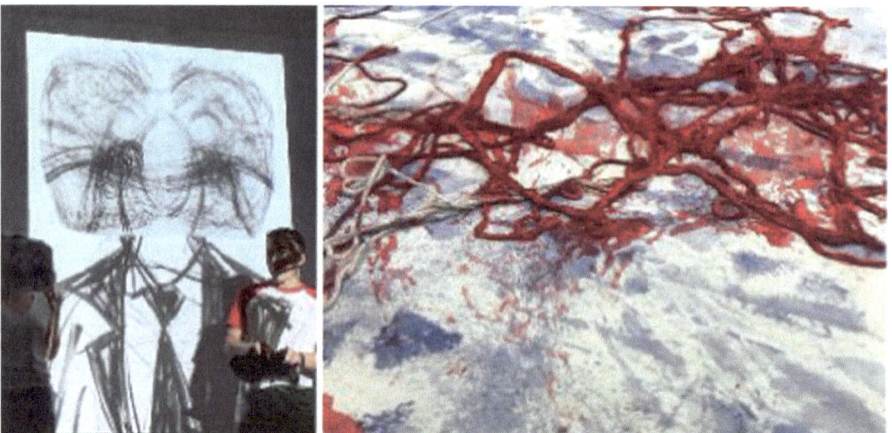

Fig. 5. Presentation, group 9. The work *2173. Lost climate paradises* that explore, map and establish connections between certain common spaces for the members of the group in the present and their idea of how these spaces will be in the year 2173. Source: Authors

Fig. 6. Poster *Urban Infidelities*, group 11. The open day is announced using posters prepared by the groups distributed by the School in which this experience is developed and disseminated on social networks. Source: Authors

members of each group to explore artistic creation from this spirit of self-discovery and discovery in the group.

In the triggers raised (references of the art-world), the students discover their personal history, recognizes processes, patterns, and themes in life, identify and liberates memories, combines the diary and the creation of new images, practice the ability of active imagination and connects it with others.

Interwoven with all this is the author's creative process and his journey to connect his emotions through creation (Allen 1995).

The graphic heritage of the Italian-Argentine architect Clorindo Testa (Glusberg 1983) features an immense mosaic developed over twenty years (1952-1974) and composed of 107 drawings (70 × 70 cm) in which he imagined the daily routine of a boy who lived in the Buenos Aires town of Avellaneda to show that, over time, in contemporary cities, some problems were aggravated rather than eradicated.

With this reference and this way of connecting graphic processes and architectonic production, apprentices produce their images (Fig. 7).

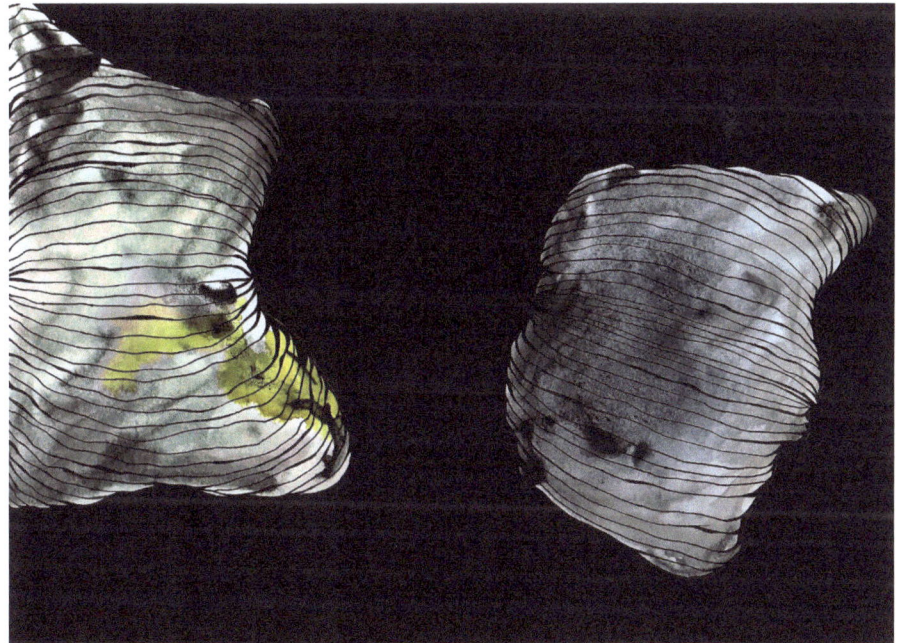

Fig. 7. Mapping the flooding process of the Kiribati Islands, Group 13. The work *Exitum* explores the idea that people without tradition is a people without any origin, it generates an atlas of geographies that will be transformed with the increase of the sea level and the anthropological consequences that the migrations of the inhabitants of those geographies will trigger. Source: Authors

This approach, which initially starts from triggers of the art world, as we have already stated, helps students to approach the issue posed by provoking, that will accept, reject or end up in discussion groups through other drawings and other creative experiences that are a critical form of graphic communication. The trainees investigate the proposed reality and together investigate themselves, narrating to themselves, relying on images.

Our goal is to enhance the emotional and to do so collectively, in mixed groups. We propose a flexible work scheme in its technique that, with the interaction of the members of the teams, triggers the search and the appearance or chance encounter with new techniques, which also makes the experience instrumental research. This process has only one condition: to think with the generated images. This process of thinking with the generated images is, in fact, iterative.

"I have always said that paintings generate ideas, and not that ideas generate images; that is, I try to get the works to generate thought. [...] It is necessary to think before and after but not during the moment of creation because sometimes thought paralyzes (Barceló 2011)."

The drawings, the artefacts, the installations, and the performances show us atmospheres that directly enliven our senses through landscapes that sometimes vanish. They are radio-graphed in other cases, they investigate the deepest structures of their architecture, revealing them sometimes, other times they become defined and floating fields of colour that interact (Fig. 8) but, above all, generate thought. They made everything with colour and graphics, with texts and images superimposed. The universe liberated by these drawings and other creative experiences is suggestive and very tempting.

Fig. 8. Models and drawings, groups 9, 10 and 14. Source: Authors

The *Emotional Lab* comprehension is not the same now. It is a footprint that shares the past and present of the groups and alerts them to individualities, collections, and processes loaded with emotions.

3   Collective Graphic Narratives. Conclusions

"Now some might argue that the same purpose could be achieved simply by having a project website, where images could be shared among the community. Although cheaper and well understood, a grounded approach to the conventional network is far inferior, as it does not have the property of visible persistence or provide the same sense of background consciousness. With it, one must take explicit action to look at the work. It does not come for free completely by being in the studio (Buxton 2007)."

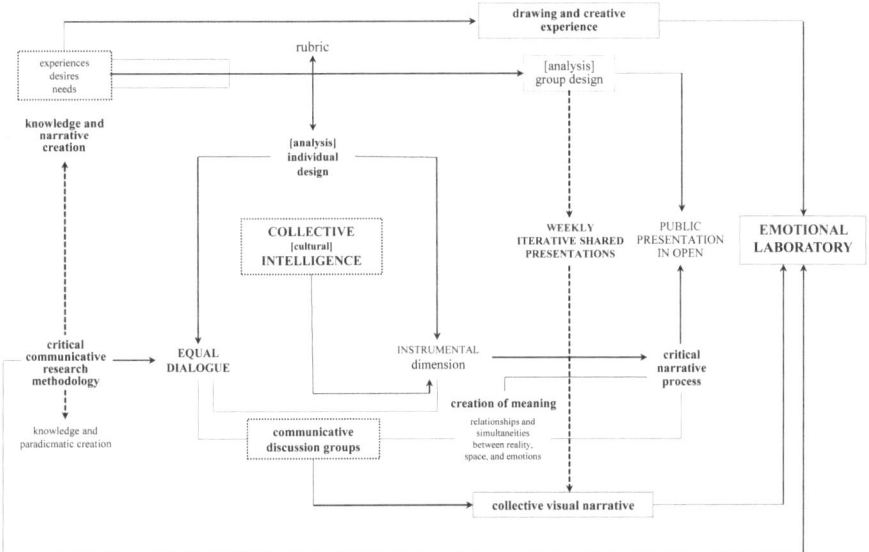

Fig. 9. The process of the *Emotional Laboratory*. Source: Authors of this article

The presence on the web, in our case, is not enough. The presence of work, pedagogically, is what generates thought in-depth, thought, and narration. This narrative has enthusiasm apprentices that are part of a group. They tell their approach to the topic proposed through their own experiences and needs. Empathy is born by sharing successive presentations.

"Things have their life (…) further than the ingenuity of nature, and even beyond the miracle and magic (García-Márquez 2007)."

The topic for the research proposed in the atelier *A World of Migrants: The Power of Movement* is the detonator of works entitled: *2173, Lost Climate Paradises, Barriers, Tourism and the Ephemeral Beauty, Creatio apud devastate, Exitum*. Testa's work was the criticism of the Italian-Argentinean architect towards the modern city through the routines drawn on a real scale in a suffocating urban environment. Apprentices are critical of the proposed topic by communicating how they are searching for new urban, climatic, and work. The atelier becomes a space of shared ideas born of a living collective intelligence that learners experience and value positively.

The visual project shown has that pedagogical value plus that results from the immediate comparison with the graphic research heritage of Clorindo Testa, which is one of the references shown by the maestros, as we have already mentioned.

The final work combines the imposing scale of the panoramic views and the intimate portrait of the life of apprentices. Both are a criticism of the subject of the exercise. Both scales coexist with the overlap of techniques and highlight the group character of the work. The experience becomes a form of a collective visual narration that is the result of a process of drawings and other creative experiences (Fig. 9).

The reader can now deduce the answers to the three initial questions. What is meant by emotional, creative expression? How to learn to investigate from the graphic narrative? How to do it collectively? Our answer is open, the topic proposed in the classroom is the ploy: the apprentices themselves are the researches. Finally, they draw narrations of what their intersubjective objectivity is, we mean, the atelier becomes a space of shared ideas. Experience is the result of collective intelligence that apprentices experience and value. A new area of work arises in which drawing and creative experience are a form of collective graphic co-communication: An *Emotional Laboratory*.

References

Goeritz, M.: El Eco In: Website of the Experimental Museum El Eco. http://eleco.unam.mx/eleco/manifiesto-de-la-arquitectura-emocional-1953. Accessed 14 Oct 2019

Allen, P.: Art is A Way of Knowing, 1ª edn. Shambhala Publications, Boston (1995)

Barceló, M. In: Desirée, M.: Creatividad y emoción. Universidad Complutense de Madrid. Ph.D. thesis (2011)

Bizquerra, R.: Metodología de la investigación educativa, 3ª edn. La Muralla, Madrid (2004)

Bruner, J.: Actos de significado, 2ª edn. Alianza Ensayo, Madrid (2006)

Buxton, B.: Sketching User Experiences: Getting the Design Right and the Right Design. Morgan Kaufmann Publishers, San Francisco (2007)

García-Márquez, G., Fuentes, C.: "Cien años de soledad" y un homenaje, 1ª edn. Fondo de Cultura Económica, Ciudad de México (2007)

Glusberg, J.: Clorindo Testa. Pintor y arquitecto. Ediciones Summa, Buenos Aires (1983)

Sclater, D.: The arts and narrative research—art as inquiry: an epilogue. Qual. Inq. **9**(4), 621–624 (2003)

Valbuena, W.: Prefigurar, co-crear, entretejer. Diseño, creatividad, interculturalidad. Arte, Individuo y Sociedad, **31**(1), 111–129 (2018)

Places of Memory
The Conceptual Drawing and Its Symbolical Structure

Angelique Trachana$^{(\boxtimes)}$ (iD) and Ioana Georgiana Şerbănoiu (iD)

Universidad Politécnica de Madrid, Madrid, Spain
angelique.trachana@upm.es

Abstract. In the beginning was the Drawing. During the process of architectural graphic ideation, the atmospheres and the meaning that the work transmits once materialized are revealed. When it comes to memorials and museums dedicated to the tragedies of the last century – the objects of this analysis – their symbolical structure is articulated within the first traces. The symbol is a type of language and the narrative that is, therefore, graphically constructed, is intelligible, although it leaves the receivers some degree of freedom for their final interpretation. The abstract architectural form does not obey the laws of a narrative that has a closed structure of significance. As Croce (1967) would affirm, its ability to convey emotions is directly proportional to its degree of freedom and its symbolism and it doesn't correspond to the apprehension of a motionless concept. In the conceptual drawing, the idea as an image is a carrier of symbolic potential, a horizon of very general and abstract ideas that is part of the operational gestures. The images are linked to the places that are revealed as the generating powers of the projects in these first strokes.

Keywords: Architecture · Drawing · Symbol

1 Introduction

Pierre Nora (2008, p. 111) developed the "places of memory" notion from a historiographic perspective. The places of memory are "all significant units, of either material or ideal order, from which the will of the humans or the effect of time has created a symbolic element of the memorial patrimony of a community". The term "places of memory" is usually referred in relation to the material and symbolical appropriation of space. In the last century and especially after the Second World War, there has been a resurgence of the memorial architecture and, at the same time, a shift of paradigm in the way in which the places of memory were designed. From wars to the Holocaust, from terrorism to violent conflicts and all the victims that these events caused, all of them have been the object of commemoration during the last century, through the construction of symbols and containers of their history. Objectified in the form of monuments, memorials and museum, memory requires space and architectural configuration.

L. Agustín-Hernández et al. (Eds.): EGA 2020, SSDI 6, pp. 469–483, 2020.
https://doi.org/10.1007/978-3-030-47983-1_42

1.1   Objectives

This paper studies the form that is born as the final result of the drawing; it is about relating the shape and content of this particular architectural heritage, exploring the drawings of its genesis. The hypothesis of the work is that the significance underlies within the first strokes. In this sense, we propose to reveal and discuss the foundation that organizes significance through graphic expression.

As Benedetto Croce (1967, p. 40) proved, it is within the unwavering relationship between form and significance where this communicative capacity, that the work possesses as an articulated whole, lies. Each configuration responds to both rationality and emotion, to the axiological references of the society and to those of the author, who filters into the project, through his sensitivity, very disperse variables and intentions. In the architectural ideation there is clearly no valuative neutrality and it is within it where the production of meaning underlies. While it is true that modern architecture, as an abstract art, is endowed with meanings from the outside and is polysemic, it is under this perspective from which a double analytical problem of remarkable importance emerges: intentionality and the way in which the social actors will appropriate and interpret these spaces-places of memory. Like all symbolic forms, architecture is not a mere reproduction of a limited reality, belonging to the field of "possibilities and not to the field of realities" (Cassirer 1976).

It is, therefore, a question of detecting, in the initial strokes, the underlying motives to the way in which the event is intended to be commemorated, to try and decipher, in the conceptual drawing, the meaning that will be lived later as a bodily experience in the built work.

1.2   The Conceptual Drawing. the State of the Art

Boudon (1993) distinguishes between the conceptual and the representational drawing. Both moments possess a communicative power but, whereas the second moment is conventionally communicative – given that it is drawn in order to be presented while the object has already been configurated – the generative drawing is, fundamentally, self-communication. The idea as an image "is the bearer of symbolic potential, a horizon of very general and abstract ideas that we face by mobilizing, above all, our desire – perhaps our desire to be" (Brea 2010, p. 9).

For the science of embodied cognition, the drawing is a reflexive tool that allows us to download cognitive information to the environment, the paper being a medium where, through the hand, we activate image schemes. The environment is part of the cognitive system. Cognitive activity does not only come from the mind, but it is a mixture of the mind and the environmental situation in which we find ourselves. The activity of the mind is based on mechanisms that interact with the environment. Drawing helps us to convey ideas that are difficult to express with words, it helps us visualize what we cannot imagine. But the relationship that exists between drawing and imagining a space with certain characteristics is not just visualizing. This visualization is very vague, so there is a need to touch, smell, move, establish relationships with the whole body as a perceptual system and also relate ideas or experiences through tracing. The traces on paper are analog body gesture.

According to John Dewey (1933), a process of creation of meaning also implies reflective thinking that establishes a conversation with the situation in question; this allows generating an understanding of an idea or experience, as well as establishing relationships between different ideas or experiences (Clark 1998). Therefore, drawing as a reflective practice, generates meanings through action. The tools and the materials are also a part of this action. The interaction between the body and the external situation, through tools and means creates adaptive structures in the world (Schön 1983).

Drawing – as a vehicle for cognitive and sensitive discharge – processes information that is not always conscious and that can be difficult for the mind to handle, information that is imagined. Through drawing, tracing and gesturing, experiences and sensations are revived. At the same time, these can generate new experiences that, one can speculate, are transmitted intersubjectively and, therefore, can affect the others.

Modern abstract architecture denotes values associated with subjectivity and freedom of expression, appealing to an open structure of meaning. The idea can be inspired by other existing ideas that were brought to memory or it could very well be a product of the imagination. The images are linked to the places that are revealed as the generating power behind the concept of the project from its initial strokes. According to Norberg-Schulz (2001), the places are the concretization of the "existential space".

"The foremost skill of the architect is, likewise, to turn the multi-dimensional essence of the design task into embodied and lived sensations and images; eventually the entire personality and body of the designer becomes the site of the design task, and the task is lived rather than understood. Architectural ideas arise 'biologically' from unconceptualized and lived existential knowledge rather than from mere analyses and intellect" (Pallasmaa 2014, p. 12)

Thus, Pallasmaa reinforces the prevalence of action against the mental work in the process of imagining a space. The mind of the designer generates an image scheme that stems from his own experiences and interactions with his environment. Unlike visual mental images, image schemes are more abstract and consist of dynamic spatial patterns that underlie spatial relationships and the movement found in real concrete images. Image schemes are multimodal and have a strong relationship with the metaphorical approaches to abstract concepts (Gibbs 2010). In fact, much of our concepts are based on our bodily actions and perceptual interactions with the world.

While drawing – imagining an architectural space – vivid sensorial and emotional image schemes are relived. According to the neurologist Frank O. Wilson, the drawing implies spatial and temporal – therefore bodily – relations, insofar that, while drawing, we are "touching" the lines with the hand and the strokes activate image coupling schemes with the world that we generate throughout our lives (O'Donnell 2011) – symbolic constructions in constant interaction with the subjects charged with positive or negative affectivity (Pallasmaa 2014).

A symbolic dimension rooted in the subconscious – as studied by Freud and Jung – goes beyond the projective action. For Ernst Cassirer (1976) the symbol is the language. In the neo-Kantian conception of the symbol elaborated by Cassirer, the judgment, the verbal discourse and the formal representation precedes and predisposes the knowledge of the symbolic forms. Language manages to reproduce the true structure of what exists.

The narrative that is graphically constructed connects – through a language that is, at the same time, rational, emotional and symbolic – managing to be intelligible even when leaving the recipients a degree of freedom in their final interpretation. The Vietnam Veterans Memorial in Washington, the Jewish Museum in Berlin, the Memorial to the Murdered Jews of Europe in Berlin, the Yad Vashem Complex in Jerusalem, Ground Zero in New York, all of these examples create an environment which, in a discreet manner, guides the visitors to certain conclusions; however, these conclusions are not necessarily imposed, but rather suggested.

2   Places of Memory and Their Conceptual Drawings

2.1   The Bodily Gesture

One of the first projects of places of memory that adopted an abstract and minimalist architecture as a form of communication was the Vietnam Veterans Memorial in Washington DC, designed by Maya Lin, in 1982. The memorial consists of the iconic landmark that is The Wall in which the names of the fallen and the missing are engraved. Its symbolic efficacy gravitates on a crack in the ground, a large pit that acts as a wound in the earth and symbolizes the severity of the loss. The designer imagined herself "taking a knife and cutting into the earth, opening it up, an initial violence and initial pain that in time would heal" (Klein 2015).

The trace, in its true essence, as a conceptual abstraction, is generated by a physical gesture – function of the vitality of the hand in motion – consequent with the energy put to work and completely identified with its execution (Seguí 2000). The abstract annotation code that transmits sensitive and conceptual information is the Architecture Degree Zero. The neutrality of the conceptual drawing, the "architecture degree zero" designates the drawing as a movement in which a succession of empty signs takes on meaning through its articulation sustained in the creative moment.

Lin's decision was preventing her design from being influenced by either one of the narratives regarding the politics or the controversies of the Vietnam War. Her work, according to Sturken (1991, p. 126), is a serene exercise of contemporary art, made in a vacuum without any knowledge of its subject; thus, the main discourse is replaced by a multitude of cultural discourses on commemoration and healing (Figs. 1 and 2).

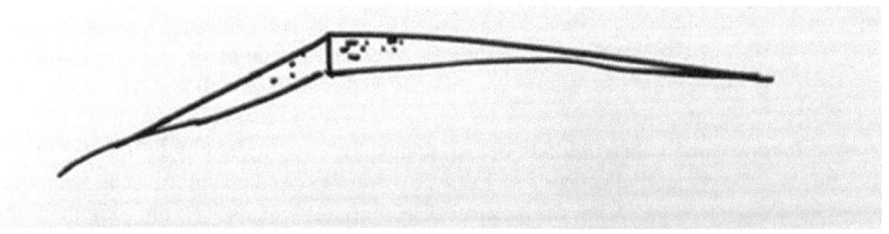

Fig. 1. Maya Lin. The first strokes for the Vietnam Veterans Memorial. Washington. Source: https://www.liveauctioneers.com/en-gb/item/60948493_maya-lin-signed-drawing-of-the-vietnam-memorial.

Fig. 2. Maya Lin. Vietnam Veterans Memorial. Washington, 1982. Source: https://www.flickr. com/photos/archirazzi/3422831315/in/set-72157648051611569.

2.2 Performativity

The Holocaust memorials and museums continued in the same direction with the intention to provoke a Holocaust experience rather than to represent ideas or values.

The Jewish Museum in Berlin, the work of Daniel Libeskind, built between 1989 and 1999 is an example of a container that integrates the horrors of the Holocaust and offers visitors a sensation of terror, loss and helplessness. The architect underlines the symbolic dimension of the museum's design process. The plan of the new building arises from a performative act based on a star of David which has been broken and distorted (Figs. 3 and 4).

Fig. 3. Daniel Libeskind. Conceptual drawing of the Jewish Museum. Berlin. Source: https:// veredes.es/blog/en/el-espacio-de-la-ausencia-marcelo-gardinetti/.

The museum organizes a space-time of the visitors' experience, so that they are likely to empathize with the victims, the absent. The physical construction becomes a receptacle for the two variables of space and time to manifest themselves in the face of the complex test that is mastering the representation of tragedy and terror, while letting the world know what happened.

2.3   Visual Metaphor

The Memorial to the Murdered Jews of Europe, in Berlin, designed by Peter Eisenman and built between 2003 and 2004, is a work that can be categorized anywhere between a sculptural piece and a public space (Eisenman and Rauterberg 2005). As the architect himself said, "this is not a sacred space" (Åhr 2008, p. 286); however, due to its intellectualism and cold materiality, it does not cease to evoke loneliness in a Jewish cemetery.

Fig. 4. Daniel Libeskind. Jewish Museum. Berlin, 1999. Source: https://www.enlacejudio.com/2013/11/10/daniel-libeskind/.

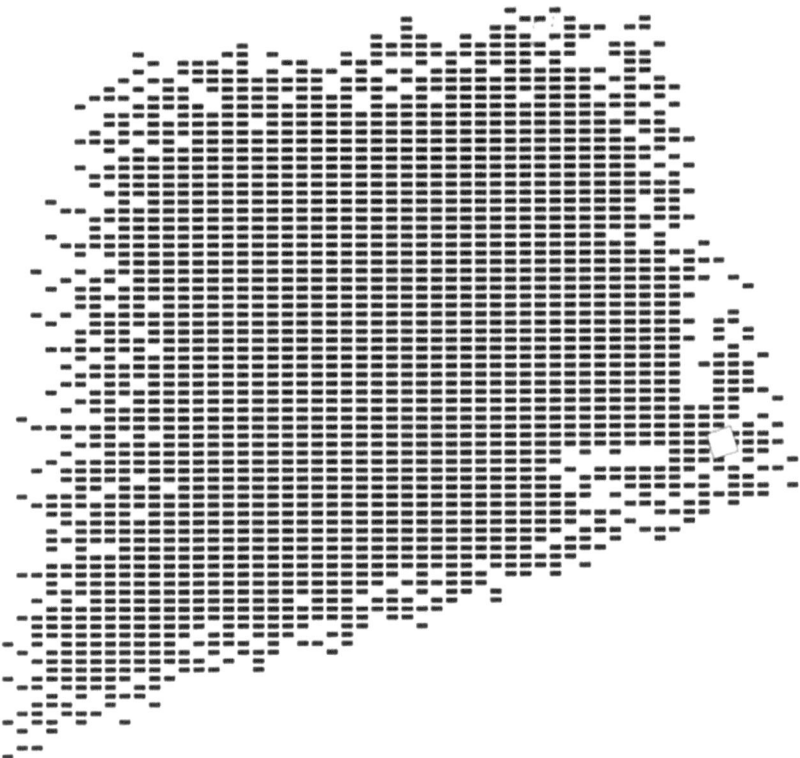

Fig. 5. Peter Eisenman. Scheme for the Memorial to the Murdered Jews of Europe. Berlin. Source: http://www.ayp.fapyd.unr.edu.ar/index.php/ayp/article/view/87.

In the same way as the Jewish Museum, this memorial recreates the *locus eremus* as a deep emotional impact of the absence and the emptiness (Figs. 5 and 6).

It does not possess any ability to communicate with the place other than imposing its own presence. It is a self-referenced form that responds through a purely visual metaphor. It alludes to the mere occupation of the land, imposed by the colonizers, through its homogenous checkerboard structure. Eisenman imposes its indefinite plot over 19.000 square meters, consisting of 2.711 concrete steles, 2.38 m long and 0.95 m wide, varying in height from 0.2 to 4.8 m, while deforming the mesh in a wave-shaped warp.

In and of themselves these steles are not – nor they intend to be – a particular sign. The homogeneity of the plot can be read as a metaphor for silence, which is the determining factor in the perception of the memorial: the deprivation or diminution of one of the senses in order to emphasize and sharpen the feelings that visitors are expected to be experience. Due to the ubiquitous gray of the concrete, a state confusion is created; this is another characteristic of the memorial that is revealed, quite paradoxically, given that sight is the privileged sense in its perception. According to Dekel (2009), the cartesian planimetry refers to the obsession of the Nazi regime with the order. However, its varied, waved-like elevation, compared to the sensation of enclosure, manages to rise and impose itself as the memory – a diffuse, indeterminate and silent memory, that almost doesn't want to manifest and express the discomfort of the German people in the face of history.

2.4 Symbolic Movement

The Holocaust Museum, the main piece of the Yad Vashem commemorative ensemble in Jerusalem, work of Moshe Safdie, in 2005, is an excavation with only one apparent prism-skylight. The preliminary inspiration, as the architect admits, can be found in the underground cities of Cappadocia and in the Beit Guvrin caves. It can suggest a tomb and a ritual path. The museum itself is a tour that leads the visitors in a way which completely blocks skipping any chapter, thus avoiding shortening the route and loosing anything from the history that develops in front of their eyes (Murphy 2011, p. 97). The different rooms are accessed from a zenithal illuminated corridor which leads towards

Fig. 6. Peter Eisenman. Memorial to the Murdered Jews of Europe. Berlin, 2004. Source: https://www.pinterest.es/pin/515310382354818804/.

the exterior, always in sight, full of syncopates in the form of physical barriers, which guide the visitor to a chronological foray of the Holocaust. This movement seem to be symbolic of the inexorable path of sacrifice of the Jews and makes the visitors to empathize upon the lack of choice of the path and, finally, upon the life of the Jewish people toward the Holocaust. The drawing is eloquent (Figs. 7 and 8).

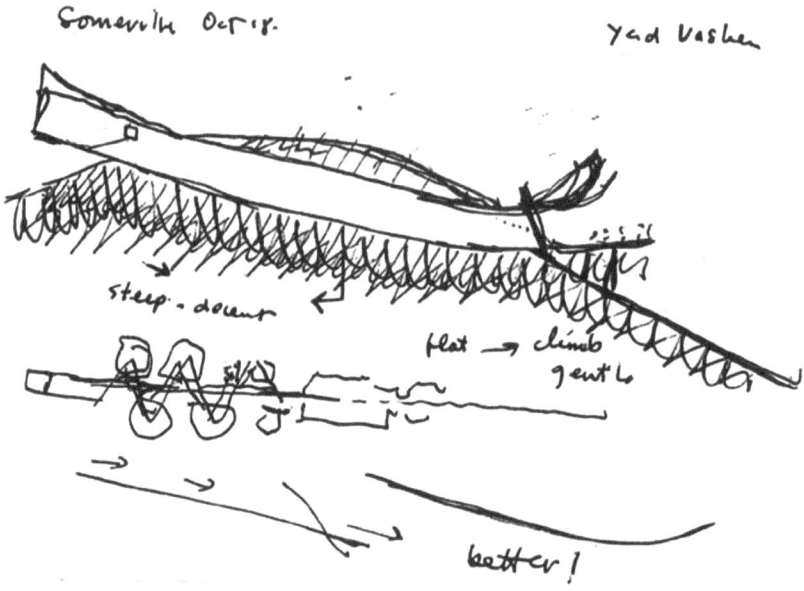

Fig. 7. Moshe Safdie. Conceptual drawing of the Yad Vashem Holocaust Museum. Jerusalem. Source: https://architizer.com/projects/yad-vashem-holocaust-memorial-museum/.

2.5 The Representation of Absence

The 9/11 Memorial in New York, the work of Michal Arad inaugurated in 2011, highlights the absence of the Twin Towers, the emptiness that they left, preserving their footprints. The two black holes in an initial sketch unequivocally symbolize the void, the negative shape. The metaphors of emptiness, the absence and the deconstruction as codes, reflect an aesthetical strategy of negativity; "the formalised negativity of the deconstructivist countermemorial is, in fact, the positive form under which a contemporary, pluralistic, civic sacred has come to appear" (Lê 2013, p. 456). "A countermemorial is simply that kind of memorial that explicitly acknowledges antimonumentalism" (Lê 2013, p. 461) by discussing the ethical risks of the monuments (Figs. 9 and 10).

3 Places of Experience

The pieces presented here – highly subjectively configurated – are not confronted with the history but with the present itself. They are not landscapes of remembrance but

Fig. 8. Moshe Safdie. Yad Vashem Holocaust Museum. Jerusalem, 2005. Source: https://www.yadvashem.org/es/museum/holocaust-history-museum/architecture.html.

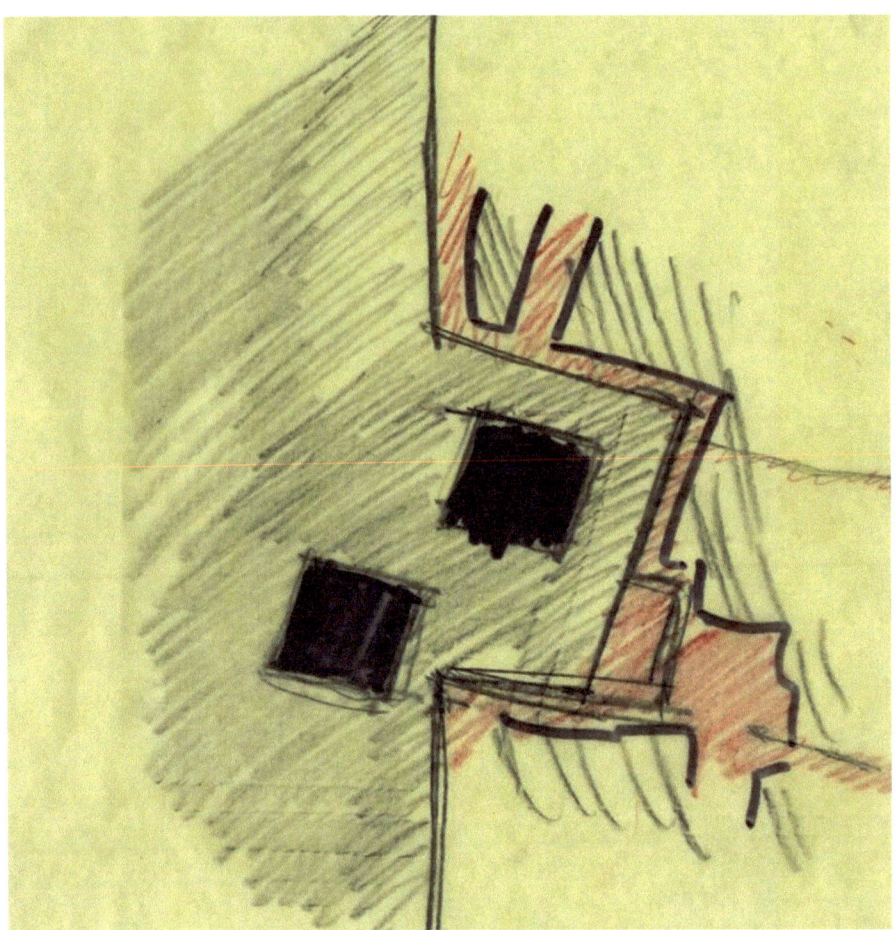

Fig. 9. Michael Arad. Conceptual drawing for the 9/11 Memorial. New York. Source: https://www.nytimes.com/2011/09/04/arts/design/how-the-911-memorial-changed-its-architect-michael-arad.html.

places of experience, which can be accessed in the present moment. The metaphorical images thus become the own image of the present. The preliminary drawings retain the immediacy of a physical experience, the same with which we approach the built space and, in this way, also the history and ourselves. An abstract consciousness in the present, here and now, involves us in a vague notion of the creative moment that enables the work to overcome time and change our attitudes.

The poetics of the abstract art, based on the liberty and immediacy of expression, while the process of its formalization (the autonomy of the form) erases history and stylistic if linguistic conventions (commitments with the collectivity), establishes a compromise with the present and the individual- subject of the action. An "amodal" architecture now burdens the author with all the moral responsibility of the form. The purpose of the created forms is not as much as to bring to our understanding the significance of which they are a link but to create a particular perception of the object (Trachana 2013).

Fig. 10. Michael Arad. 9/11 Memorial. New York, 2011. Source: https://www.911memorial.org/blog/exploring-memorials-sustainable-design-earth-day.

The identifiable elements establish a corporeal identity as a sense of continuity with the experience of ourselves, a continuity that includes values, beliefs and a sense of belonging to something supraindividual, to something that is beyond ourselves, but in any case is a complex experience that includes the memory, the self-image, the experience of time, the emotions and the values. Now, the drawing is not a mean of transmission, but it is an end of significance. Acquiring a total significance of the forming practice, that is an immediate significance, the form is silenced, but at the same moment a new identify is being created (linguistical), a new institution of the architecture (Trachana 2013).

As we well know, architecture as a language, presents clear inabilities to transmit information beyond certain thresholds. Faced with the theory of communication that is governed by mechanisms of rigid structure, architecture does not obey the laws of a closed structure. The symbolic structures of architecture allow a certain degree of liberty to the interpretation. In addition, the capacity of architecture to transmit emotions seems directly proportional to its degree of liberty in respect to the conventions and codifications. The more degrees of liberty it enables, the more symbolic power it reaches. An "open work" reaches a high level of possible suggestions. The metaphor is being used almost as an equivalent to the symbol.

Effectively, the language of abstract architecture has an enormous potential of significance and interpretation. It begets open and neutral works, with ability to induce in atmosphere, so that each individual experiments its own experience and relives its own memories. The processes of significance and symbolic structures present themselves as eclectic and more complex than some simple theories which can be found in the linear schemes that human mind creates according to the law of minimal effort (Hegel 1971, p. 104).

The re-presentation of a significance through sensitive signs establishes an analogy between language and architecture which lies in that overlap in its semantic structure of emotional and propositional aspects (symbolic process) (Sánchez-Rojas Fenoll and Vera Botí 1985, p. 34). But architecture encloses an intention that is beyond the drawing, it configures itself as a total sign, its significance being given in its unity (and not in the succession of its signs) and includes a reflection upon the human and social use of the form.

4   Conclusion

Observing these phenomena of unobjectionable complexity certainly requires different types of analysis, methodologies and disciplines. A semantic analysis of the phenomenon-object-drawing involves observing its operational laws among a series of abstract signs that can be handled independently of the object itself. These laws constitute a pre-structure that explains the phenomenon. According to phenomenology this pre-structure belongs to the symbolic system in which architecture assumes the role that gives it its historical becoming and the symbol is one of the ways that leads us to an objective view of things.

Facing the semiotic studies, in which the information of the visual forms refers to their ability to signify within a cultural context where their meanings, coined by tradition and normativity, can be captured in a global way, the analysis of conceptual drawings of the architecture of the memorials arises here from the need to explain the generative, subjective but, as it has been intended to demonstrate, no less transcendental moment. As it was suggested in the previous analyses, inside the formal systems in question there exists an overlap of symbolic structures transferred from the funeral art (the cemetery or the tomb) and the sacred places that are inscribed within a great symbolical tradition or familiarization. Inexorably, the subjectivity is crossed by archetypes. Due to the archetypal character of the human perception and human society, according to Jung (1981), its sacred and mythical essence is latent in the perception of these places. As Mircea Eliade (1974, p. 12) points out, the symbolic thinking "is substantial to the human being: it precedes the language and the discursive reason. [...] Images, symbols, myths, are not irresponsible creations of the psyche; they respond to a need that fills a role: they leave naked the most secret modalities of the human being."

The symbolic in art and architecture as Croce (1967) states, is not corresponding to the apprehension of a motionless concept, but to the continuous formation of a judgment. The work, then, won't be a simple representation, but a representation of the judgment. The methodologies of possible interpretations (formalist, iconological, semiologic, structuralist, materialist, sociological, technical, etc.) are gathered within the entire study of symbolic structure.

References

Åhr, J.: Memory and mourning in Berlin: on Peter Eisenman's Holocaust-Mahnmal (2005). Mod. Jud. **28**(3), 283–305 (2008). https://doi.org/10.1093/mj/kjn015

Boudon, P.: El dibujo en la concepción arquitectónica: manual de representación gráfica. Ed. Limusa, México (1993)

Brea, J.L.: Las tres eras de la imagen. Imagen-materia, film, e-image. Ediciones Akal, Madrid (2010)

Cassirer, E.: Filosofía de las Formas Simbólicas. Ed. F.C.E., México (1976)

Clark, A., Chalmers, D.: The extended mind. Analysis **58**(1), 7–19 (1998). https://doi.org/10.1093/analys/58.1.7

Croce, B.: Breviario de Estética. Espasa Calpe, Madrid (1967)

Dekel, I.: Ways of looking: observation and transformation at the Holocaust Memorial, Berlin. Mem. Stud. **2**(1), 71–86 (2009). https://doi.org/10.1177/1750698008097396

Dewey, J.: How We Think: A Restatement of the Relation of Reflective Thinking to the Educative Process. D.C. Heath, Lexington (1933)

Eisenman, P., Rauterberg, H.: Holocaust memorial Berlin: Eisenman Architects. Lars Müller, Baden (2005)

Eliade, M.: Imágenes y símbolos. Ensayos sobre el simbolismo mágico religioso. 2ª edn. Ed. Taurus, Madrid (1974)

Gibbs, R.W.: Embodiment and Cognitive Science, p. 91. Cambridge University Press, Cambridge (2010)

Hegel, G.W.F.: Introducción a la estética, p. 104. Ediciones Península, Barcelona (1971)

Jung, C.G.: The Archetypes and the Collective Unconscious. Princeton University Press (1981)

Klein, C.: The remarkable story of Maya Lin's Vietnam Veterans Memorial. (2015). https://www.biography.com/news/maya-lin-vietnam-veterans-memorial. Accessed 10 July 2019

Lê, D.: On «reflecting absence»: negativity and the sacred at ground zero. Lit. Theol. **27**(4), 452–471 (2013). https://doi.org/10.1093/litthe/frt040

Murphy, D. (ed.): Yad Vashem. Moshe Safdie – The Architecture of Memory. Yad Vashem Publications, Jerusalén (2011)

Nora, P.: Les lieux de mémoire. Ediciones Trilce, Montevideo (2008)

Norberg-Schulz, C.: Genius loci. G. Gili, Barcelona (2001)

O'Donnell, T.: Sketchbook: Conceptual Drawings from the World's Most Influential Designers, p. 176. Rockport Publishers, Beverly (2011)

Pallasmaa, J.: La Mano Que Piensa: Sabiduría Existencial y Corporal En La Arquitectura, p. 12. Editorial Gustavo Gili, Barcelona (2014)

Fenoll, S.-R., del Carmen, M., Vera Botí, A.: Lenguaje y símbolo: la Arquitectura en la encrucijada. Imafronte **1**, 23–42 (1985)

Schön, D.A.: The Reflective Practitioner: How Professionals Think in Action. Basic Books, New York (1983)

Seguí de la Riva, J.: Dibujar proyectar. Instituto Juan de Herrera, Madrid (2000)

Sturken, M.: The wall, the screen, and the image: the Vietnam veterans memorial. Representations **35**, 118–142 (1991). Special Issue: Monumental Histories

Trachana, A.: El grado cero de la arquitectura. En: EGA. Expresión Gráfica Arquitectónica **18**(22), 142–153 (2013)

On the History of the Image in Cinema: Translucent Shadows and Segundo de Chomón

José Mª Gentil-Baldrich[✉]

University of Seville, Seville, Spain
jmgentil@us.es

Abstract. The Aragonese filmmaker Segundo de Chomón directed the film *Une excursion incohérente* in 1909. In my opinion, this film is the one where, for the first time, translucent shadows are used as elements of a plot in cinematography. This work relates the film to its origin in shadow shows, to the legendary expressionist film *The Cabinet of Dr. Caligari*, the later influence of these images in the cinema, and its widespread use in North American *film noir* up to the film *Psycho* by Alfred Hitchcock. Without claiming that the numerous examples presented are all directly derived from Chomón's film, this does not prevent him from being recognised as the pioneer of a technique that would later have extensive influence.

Keywords: Shadows · Translucent · Cinema · Chomón · Caligari

1 The Acknowledgement of an Aragonese Man

It is very revealing that, in the origins of Spanish cinematography, there are various Aragonese leading players who, despite being acknowledged in specific studies, have not reached the place that would belong to them as part of the specific patrimony of the graphic image. At the inception of the new technique, Eduardo Jimeno Correas (1870–1947) is one of the creators who has been acknowledged. He, in 1897, shot and broadcast Salida de misa de doce del Pilar de Zaragoza, which is considered the pioneering film in Spanish history. Another creator is Segundo de Chomón, a very interesting character with an important place in the origins of cinematography, who has been subject of study, even exhibitions, for some time (Tharrats 1988; Sánchez Vidal 1992; Lorenzo 2011). I want to re-assert the value of some iconic aspects of his production, which, in my opinion, are not sufficiently highlighted and related to a theme specific to Graphic Expression: the shadows.

Segundo de Chomón was born in Teruel in 1871 and died in Paris, still young, in 1929, and his figure is truly representative of the Aragonese diaspora. He left his native land around 1895 and went to Paris, where he came into contact with the newly appeared cinematographic technique. He was a volunteer for the Spanish army in the Cuban War and, after the disaster of 1898, he returned to the French capital, where he specialised in certain technical devices. These devices were always his field of greatest notoriety. After a stay in Barcelona—between 1902 and 1906, when he developed a series of productions

L. Agustín-Hernández et al. (Eds.): EGA 2020, SSDI 6, pp. 484–498, 2020.
https://doi.org/10.1007/978-3-030-47983-1_43

and shootings that, finally, did not achieve the desired success—, he returned to Paris, hired by Pathé Frères, at first as a technical assistant of their productions and, finally, as scriptwriter and director of his own films. The most notable of these were *La maison hantée* (*The House of Ghosts*) in 1907 and *Hôtel électrique* (The Electric Hotel) in 1908, both characterised by their fantastic and innovative nature. However, the one that is of most interest regarding image effects is *Une excursion incohérente*, filmed and premièred in 1909, and whose convoluted and surreal plot faithfully corresponds to its title. The film was forgotten for a long time and its authorship has even been badly assigned in some databases. It has only recently been recognised as an influence on the subsequent surrealist movement, even in Buñuel's *Un chien andalou* who was, precisely, also Aragonese[1].

2   Shadows in Cinema

In the history of cinema, the pioneering use of the shadow as a protagonist element in dramatic discourse is usually attributed to the works that emerged from the German Expressionist school in the second decade of the twentieth century. Even though we recognise this significance, we usually forget the important precedent set by our subject of study, who used it as an important element of the plot in his aforementioned 1909 film, and of which we can see clear influences in other later works (Gentil 2019). Since Chomón was an expert in the special effects technique—probably the best of his time—, he did not hesitate to use them extensively in *Une excursión incohérente*, a film with a length of little more than eight minutes, which was his last individual work of that French stage. In it, we can cite, as outstanding examples of the effects used, the appearance of images of worms and other filth in food—some of which have been linked to similar occurrences in some shots by Spielberg for Indiana Jones—and the use of cartoons, which he applied in the construction of a bridge and other compositions, although this curious use is not generally recognised[2]. And, above all, we must highlight the new use he gave to shadows.

The use of projected shadows was not very original in shows of that time, and Chomón had already done it the previous year in another film featuring shadow play (Fig. 1)[3]. What was new at the time was the use of shadows with real characters and the fact that they acted as significant protagonists within a plot, as is the case with the shadows of this film. There is no doubt that this use was influenced by the Parisian shadow theatres of the time, especially the sessions held in Montmartre's *Le Chat Noir*. Produced through

---

[1] Another Aragonese protagonist on the subject of cinematographic shadows was Florián Rey (1894–1962) in *La aldea maldita* (1930), but these were not translucent shadows. See (Gentil 2019).

[2] Although Emile Cohl (1857–1938) had made an animated short film the previous year—*Fantasmagorie*, two-minutes-long in its most extended version—, this cannot be compared with the sequence by Chomón. When Lotte Reiniger (1899–1981) and her *The Adventures of Prince Achmed* (*Die Abenteuer des Prinzen Achmed*, 1926) are mentioned, no one cites Chomón as an antecedent.

[3] Chomón had used them in *Les ombres chinoises* (1908), but for these he used cartoons and not characters.

the projection of silhouettes cut out on a screen, some were enormously successful, such as *L'Epopée* drawn by Caran d'Ache, which premièred in 1886 and captured the Napoleonic epic narrative at a time of the necessary recovery of national identity after the defeat in the Franco-Prussian war. Chomón himself, upon his arrival in Paris, had the opportunity to see this technique when it was still active at this cabaret, then located on the Boulevard de Clichy. This is where he performed his last shadow show, in January 1897, as the technique was already in decline as a visual spectacle precisely because of competition from the emerging cinema. Curiously, *Le Chat Noir* was also home to a markedly iconoclastic artistic group, meeting under the name of *Salon des Arts Incohérents*, a term that we see present both in the title and, above all, in the surreal situations of the film we are dealing with (Lope Salvador 2013; Mendieta Rodriguez 2017).

Fig. 1. Segundo de Chomón. *Les ombres chinoises* (1908). Museo Nazionale del Cinema (Turin)

However, as far as this work is concerned, what should be highlighted most about *Une excursión incohérente* are the scenes where the translucent shadows that interest us are used—for the first time, in my opinion. These are developed in various sequences, which are also mixed with animation, in which the protagonist couple retires to bed and the wife—who sleeps in a separate bed—goes behind a translucent curtain that separates the beds. There, as she gets ready, her projected shadow appears. Initially, this shadow reflects a real domestic action, but later, as they have fallen asleep, the male protagonist participates in a dream in which the shadows projected against the light on the canvas—as if on a virtual screen—give way to an episode that, because it is oneiric by nature, is difficult for us to explain (Fig. 2)[4].

Fig. 2. Segundo de Chomón. *Une excursión incohérente* (1909).

The film was released by the then-powerful Pathé Frères, from mid-1909 throughout Europe[5] and, from February 1910—under the title *A Panicky Picnic*—, it was presented in New York and Chicago[6]. The visual impact that his approach must have generated is unquestionable, and we can appreciate its influence in other subsequent films. Here, we will see some significant examples related to our theme[7].

---

[4] Tharrats (1988: 33) refers to *transparent backgrounds* but does not mention this film. Lope Salvador (2013: 687) does study it, but without discussing the shadows, and notes a similar use of the curtain in Frank Capra's *It Happened One Night* (1934).

[5] The release date figures as November 1909, but in Finland it is listed as September 1909, which is contradictory.

[6] It also appeared, in some English version, under the title *The Traveller's Nightmare.*

[7] The film is available for viewing on various sites. For example: https://www.youtube.com/watch?v=TzgcQ8IZBWg.

3 Dr. Caligari

In 1920, a legendary film in Expressionist cinema was released: *The Cabinet of Dr. Caligari* by Robert Wiene. This film is often credited with the emergence of shadows in cinema, in such a generalised way that references to it have become clichés for the critics: the well-known shadow cast by the doctor—played by actor Werner Kraus—, amplified on a wall, is paradigmatic (Stoichita 1999: 154; Michalski 2009: 54; Zunzunegui 2009: 71]. However, that image is deceptive, because it was used only as an advertising strategy for the film and does not appear as a frame in it. Although the film takes place in dark environments and it was filmed entirely with a stationary camera and in shady avant-garde décor—with the exception of the madhouse, it is difficult to find orthogonal lines in the film—, there is only one sequence in the film with shadows acting as protagonists, which is precisely what characterised the method filmed by Chomón, as we can see. In fact, the moment in which Cesare—the sleepwalker used by Caligari for his crimes (min. 24:33)—murders Alan is conveyed with shadows that Wiene casts behind the protagonist, pretending to appear on a back wall; however, in reality, they were made against a backdrop over a curtain. In addition, at the beginning of the scene, Alan is asleep in his bed, as is the case with the protagonist of the 1909 film we are discussing. Undoubtedly, Wiene gave the procedure a more technically refined character—as more than ten years had already passed—, but the similarities with the sequence of *Une excursión incohérente* suggest the origin of Wiene's shot. The technical factor of the scene has been credited to the German director and, in a recent study (Glover 2017); it has been hailed as an authentic moment of genius. This method has even recently been included in some theatrical performances of the work, with images of translucent shadows for the staging of the moment mentioned, which some authors have come to take as images of the original film itself[8]. But Chomón has never been cited as a precedent (Fig. 3).

Fig. 3. Robert Wiene. *Das Cabinet des Dr. Caligari* (1920).

---

[8] Staging of *The Cabinet of Dr Caligari* in London, 2013, at the Arcola Theatre, by the theatre company Simple 8 in an adaptation by Sebastián Armesto and Dudley Hinton. The same method has been used in other shows and theatrical performances.

When discussing the importance of shadows in cinema, it is evident that we cannot fail to note their appearance in the later *Nosferatu*, by Murnau (1922), a film where they reached a widely-recognised prominence: those of the vampire projected on the wall of the staircase, or projected on defenceless or sleeping beings—scenes that would extensively influence later cinematography. On their own, they deserve to be the particular focus of a study that is outside of our scope here, as translucent shadows were not used in *Nosferatu*, and that is why we do not discuss the shadows in this film. The appearance of these projected shadows in cinema later helped films to nuance or blur compromising or imaginary scenes, such as dreams, or terrifying or erotic situations, where the shadows came to adopt, discreetly or exaggeratedly, tacitly or expressly, a life of their own. And, regarding the specific use of translucent shadows, which are the ones that concern us, we can see their presence in other examples.

4   Other Examples

In 1924, the film *Greed* by Erich Von Stroheim was released in the United States, after a long and tempestuous process of shooting, production and editing. In one of its sequences, an act that gathers the protagonist's shadows, reflected on a canvas—which previously, as in Chomón's film, contrasts with the light behind her—, is projected as part of a long and complex plot (Fig. 4). Although we may think that their connection is dubious, we cannot forget that the première of *Une excursión incohérente* coincides with the date of Von Stroheim's trip—or rather, his flight—through Europe on his way to the United States. In any case, there is no doubt that he saw the film in his new home at the beginning of the following year[9].

Another image that leaves less doubts when establishing a connection is the one that appears in 1923 in *La Roue* (Fig. 5) by Abel Gance, in which the protagonist, for a specific intimate scene, goes behind a curtain where her shadow is projected while she cleans herself. Gance knew Chomón well, both because of his previous Parisian work and because of the work he had done in Italy before the Great War. And, significantly, he had him as a collaborator at least since 1925, in his last French period, with a special role in his *Napoleon*, which premièred in 1927. However, the French director—who was very arrogant—did not include his participation in the film's credits[10].

---

[9] Von Stroheim arrived in New York from Bremen or Cherbourg on 26 November 1909 and Chomón's film was shown in Europe precisely during his trip. The influence of this image from *Greed* can be seen literally reproduced in the painting *Outre Songe* (1978) by the Paris-based Argentine surrealist painter Leonor Fini (1907–1996). The sequence of shadows has been removed from some of the different versions of this film.

[10] The original title of the film was also the director's name: *Napoleón vu par Abel Gance*. It was, curiously, a magnified cinematographic version of Caran D'Ache's L'Epopée, which we have mentioned regarding shadow theatres.

Fig. 4. Erich Von Stroheim. *Greed* (1924).

The same usage is present in a 1927 American film, *Fashions for Women* by Dorothy Arzner, one of the few female directors active in the then explosive universe of Hollywood. The film, which must have been considered markedly frivolous, is currently lost and we only have references to its images from the photographs used for its promotion. One of them portrays actress Clara Bow's naked body behind a translucent decoration, conveniently blurred to avoid censorship (Fig. 5). But while in this film—or in *Greed* or *La Roue*—the shadows give nuance to scenes that could be compromising when portrayed directly, there are others in which the protagonists—as a variant of the method—give us access to situations through translucent shadows.

In 1915, The Cheat, directed by Cecil B. de Mille, premièred; this is, possibly, one of the first indirect uses of Chomón's method. In it, the protagonists listen to a conversation between other characters—or so it appears, because it is a silent film—in front of their projected shadows on a translucent partition, as in Chomón's film (Fig. 6). This physical element acquires symbolic importance because, at the end of the film, when the evil protagonist—who, during the distribution of the film, began as Japanese and ended up as Burmese, under pressure from American socio-political groups—is wounded while shadowed on the other side of the partition, and that virtual screen has to be destroyed in order to gain access to him. A similar use of the translucent partition wall was made in

Fig. 5. 1. Abel Gance. *La Roue* (1923)/2. Dorothy Arzner, *Fashions for Women* (1927).

1924, in the avant-garde film L'Inhumaine by Marcel L'Herbier, a director so influenced by The Cheat that he himself would expressly acknowledge it—see image—in the credits of Forfaiture (1937), a French remake of the American film where, in this case, the Asian protagonist was now Mongolian.

Fig. 6. 1. Cecil B. de Mille. *The Cheat* (1915)/2. Arthur Robinson. *Schatten* (1923).

We see a similar method again in a German film from 1923, with the meaningful title of *Schatten*[11], by the American director Arthur Robinson, who was then active in Germany (Fig. 6). In it, shadow play is featured, in a main and autonomous way, in a jealousy plot with a certain erotic component. In this film, in addition to the shadows cast on the wall (derived from the original shadow theatre), which were the plot's own and dominated the film, translucent shadows also appear in a situation very similar to *The Cheat*. Such double use of shadows is also featured in Howard Hawks' *Scarface* from 1932. In it, the first murder is portrayed through a translucent shadow and, at the end, the well-known Valentine's Day massacre in Chicago is conveyed through the shadows projected on the wall of the garage where it takes place (Fig. 7). On both occasions, shadows are used to avoid the problems caused by images that were in conflict with the then demanding American censorship.

The kind of shadows we discuss, used in various films, started changing their dramatic function from the 1920s onwards[12]. In George Fitzmaurice's *Raffles* (1930), the thief observes how the policeman's shadow passes through a wall hanging (Fig. 7); in *The Testament of Dr. Mabuse* (1933)—very interesting for its architectural images—by Fritz Lang, the leader of the criminal network maintains his anonymity by speaking behind a curtain that, like in *The Cheat*, is pierced at the end of the film. Lang himself would use this type of shadow again in 1944, already in his American period, in *Ministry of Fear*; in 1945 *Scarlet Street* and in *House by the River* in 1950. Above all, he would use them in 1956, in *While the City Sleeps*, where the beautiful protagonist does particularly suggestive gymnastic exercises behind a translucent curtain, within a double plot structure (Fig. 8). By then, the shadows were used very frequently, and by the best directors, in American *film noir* (Fig. 8)[13].

---

[11] The full title is *Schatten. Eine nächtliche Halluzination.*

[12] For example: Josef Von Sternberg, uses it in *Underworld* (1927); *Der Blaue Engel* (1930) and *Shanghai Express* (1932).

[13] They appear in *Rebecca* de Alfred Hitchcock in 1940; *Citizen Kane* by Orson Welles (1941); *The Maltese Falcon* by John Huston (1941); *This Gun for Hire* by Frank Tuttle (1942); *The Great Flamarion* by Anthony Mann (1945); *The Dark Corner* by Henry Hathaway (1946); *The Big Sleep* by Howard Hawks (1946); *A Double Life* de George Cukor (1947); *Johnny O'Clock* by Robert Rossen (1947); *He Walked by Night* by Anthony Mann (1948); *The Enforcer* by Raoul Walsh (1951).

Fig. 7. 1. Howard Hawks. *Scarface* (1932)/2. George Fitzmaurice. *Raffles* (1930).

Fig. 8. 1. Fritz Lang. *While the City Sleeps* (1956)/2. Howard Hawks, *The Big Sleep* (1946)

A novel use of translucent shadows was made in 1943, on the shower curtain in Mark Robson's *The Seventh Victim* (Fig. 9), and the same use would be given to them in the justly famous scene of Hitchcock's *Psycho* in 1960 (Fig. 10), which has recently been the subject of somewhat exaggerated reinterpretations (Fig. 10)[14]. Ironically, it will be used—in the spirit of the film—by Jacques Tati in *Mon Oncle* (1958); even David Lynch, in *The Elephant Man* (1980), makes it so that, when the researcher shows the deformed being that is the object of his studies before the scientific community, the spectator initially sees him through a translucent shadow to maintain anticipation.

Fig. 9. Mark Robson. *The Seventh Victim* (1943)

---

[14] For example: *78/52* by Alexandre O. Philippe (known in Spain as *78/52*), 2017. The translucent shadows in the shower was also used in *Niagara* by Henry Hathaway (1953), but with a different meaning.

Fig. 10. 1. Alfred Hitchcock. *Psycho* (1960)/2. Alexandre O. Philippe *78/52.* (2017)

5   Final Considerations

It does not seem as though Chomón used again the same method with which he closed that period of his career. In 1912, he left for Italy, where he culminated his international career by collaborating with Giovanni Pastrone, both as assistant director, cameraman and special effects manager, as well as a project manager in Italian cinematography. His work in the great production *Cabiria* (1914), especially the simulation of Archimedes' burning of the Roman fleet in Syracuse, is among the outstanding moments of the use of special effects in the history of cinema. It is interesting that many of Chomón's claims come, precisely, from Italian scholars, eager to recognise his unjustly forgotten historical cinematography, and to acknowledge the creators who gave it a boom until the First World War (Nosenzo 2007). The Aragonese filmmaker has been credited with the invention of the travelling or tracking shot, the camera movement that he practised in the shots of some of his shoots and which was non-existent in, for example, *The Cabinet of Dr. Caligari*. However, Segundo de Chomón was careless with his intellectual property, and so the "carrello", as it was called in Italy, was patented by Pastrone himself as if it had been his idea. He also developed several technical gadgets for the colouring of films and the stop motion technique—or shooting frame by frame—, indispensable for his special effects. One of the last ones he made for Spanish cinematography was in 1927, the magnificent oneiric act—also a dream, but now without translucent shadow—of a very young Conchita Piquer in *El Negro que tenía el alma blanca* by Benito Perojo, which was filmed precisely in Paris during his last stay in the French capital before he died.

Finally, the purpose of this work is not to state that all those translucent shadows that we have seen used in cinema—and the many that we did not mention or that have yet to be studied—are directly derived from the sequence of the film *Une excursión incohérente*, by Chomón, that we have discussed. The most probable scenario is that the majority of those who used this technique did not even know about the existence of our Chomón—who received very little recognition—, especially when his memory faded with the passage of time and the diffusion of cinema meant that, in the commercial development of the new medium, some directors repeated what others did, as it is easy to confirm. But we must praise the originality of this iconic method, which has, in my opinion, never been recognised and has always held great interest for the history of the cinematographic image and, by extension, of the Graphic Expression itself.

References

Gentil Baldrich, J.M.: "Elogio de la sombra en el cine: notas a Junichiro Tanizaki". en: *Acca 17. Análisis Y Comunicación Contemporánea de la Arquitectura*, Seville, Departamento de Expresión Gráfica Arquitectónica, 2019, pp. 18–35 (2019)

Glover, J.: *How has German expressionism laid the foundations for future films and is it still relevant in modern filmmaking and what influences has it had throughout film from its origins to modern time?* M.A. thesis, Canterbury Christ Church University (2017). http://create.canterbury.ac.uk/17052/. Accessed 22 May 2019

Lope Salvador, V.: "Incoherencia y pesadilla en el cine primitivo". en: *Actas del 2º Congreso Nacional sobre Metodología de la Investigación en Comunicación*, Universidad de Valladolid, pp. 645–706 (2013)

Lorenzo, R.: *Segundo de Chomón. El alquimista impasible y su cine de fantasía*. LABoral Centro de Arte y Creación Industrial, Gijón. Exhibition, 25/2/11 al 23/5/11 (2011)

Mendieta Rodríguez, E.: "La tensión dramática en el cine de Segundo de Chomón: El Hotel Eléctrico y los grandes hallazgos técnicos". en: *Cuadernos de Aleph,* vol. 9, pp. 120–140 (2017)

Michalski, S.: "Sombras de soledad. Sombras de amenaza". en: AA. VV. *La Sombra*, Madrid, Thyssen Bornemisza, pp. 52–59 (2009)

Nosenzo, S.: *Manuale técnico per visionari. Segundo de Chomón in Italia, 1912–1925*. Turín, Biblioteca Fert (2007)

Sánchez Vidal, A.: *El cine de Segundo de Chomón*, Zaragoza, Caja de Ahorros de la Inmaculada de Aragón (1992)

Stoichita, V.I.: *Breve historia de la sombra*, Madrid, Siruela (1999)

Tharrats, J.G.: *Los 500 films de Segundo de Chomón*. Universidad de Zaragoza, Zaragoza (1988)

Zunzunegui, S.: "Sonata de espectros". en: AA. VV. *La Sombra,* Madrid, Thyssen Bornemisza, pp. 70–81 (2009)

The Influence of Dance in Lawrence Halprin Spatial Drawings

María Aguilar Alejandre$^{(\boxtimes)}$

Departamento de Ingeniería del Diseño, Escuela Politécnica Superior, Universidad de Sevilla,
Seville, Spain
maraguilar@us.es

Abstract. Lawrence Halprin (1916–2009) is an American architect who is internationally well-known for having developed a prolific work especially in the field of landscaping and public spaces. His architectural legacy is directly influenced by his activity as a researcher in the field of interdisciplinarity, a work he exercised throughout his life in tandem with the avant-garde dancer Anna Halprin (1920).

The works and dance proposals of Anna Halprin as well as the projects that they developed together, determined his work and ways of doing and looking. Among these influences, there is one that stands out and that has to do with his way of graphing space and what happened in it. The architect firmly believed that the architectural graphic representation could include aspects such as movement. This paper aims to present part of these graphic works by Lawrence Halprin and to point out, in a critical way, how the experience of dance is present in them.

Keywords: Dance · Architecture · Movement · Notation · Graphic ideation

1 Lawrence and Anna Halprin, a Multidisciplinary Life Project

Lawrence Halprin (1916–2009) was an American landscape architect and urban planner based on the west coast of the USA, specifically in the surroundings of San Francisco although his projects extend beyond these limits. Due to his architectural education at Harvard, he is introduced to the theories and dictates of the emigrated teachers of the Bauhaus, where he establishes a very special relationship with Walter Gropius [1]. The approach of the popular director of the Bauhaus leaves a great impact on Lawrence Halprin, especially regarding the relationship between disciplines. According to Gropius, art and architecture must be merged, an issue that the American architect will put into practice in the first person.

In San Francisco, Lawrence Halprin daily routine as an architect is going to be immersed in a panorama of a strong countercultural atmosphere of which he is fully involved. In addition to his well-known projects in the public space such as *Freeway Park* (Fig. 1) (Washington) or *Franklin Delano Roosevelt Memorial* (Maryland), Lawrence Halprin also did an important job as researcher, disseminator and activist. This task was most of the times in tandem with the avant-garde dancer Anna Halprin [2].

L. Agustín-Hernández et al. (Eds.): EGA 2020, SSDI 6, pp. 499–507, 2020.
https://doi.org/10.1007/978-3-030-47983-1_44

Fig. 1. Freeway park, Lawrence Halprin. Washington, 1976 (opening date). Image: Freeway Park Association.

Lawrence Halprin and Anna Schuman met in 1939 being both students at the University of Wiscosin. A year later they got married and from then on, the profession of one and another will radically influence the work of each other. Lawrence Halprin is very attracted to the study of body movement through drawing as well as the role that the moving body plays in space [3], while his partner is interested in how architecture determines spaces and how these spaces define the movement of bodies.

In 1941, thanks to a scholarship granted to Lawrence Halprin, the couple moved to New York so that he could study landscape urbanism at Harvard. The experience is a revelation for both of them because they attend together (Anna Halprin as a listener) the art and design classes taught by the exiled masters of the Bauhaus from whom they learn, among other ideas, that art is not fragmented into isolated areas but the different branches of art complement and interconnect with each other. Moreover, the artistic disciplines are integrated and diluted within a whole of greater importance. Architecture and dance, the disciplines of the Halprin couple, are also included in this approach. It is in this way that they have developed the merge of these two fields, dance and architecture, in common works such as the workshops *Experiments in the Environment* [4] or the method *The RSVP Cycles* [5].

The *Experiments in the Environment* are constituted by a set of activities aimed at dancers and architects so that both professions inform each other. Several editions are developed during the summer season, the one of 1966, placed in The Sea Ranch (California) is especially important due to the presence and collaboration of the architect Charles Moore. The kind of activities carried out in these workshops were: blindfold walks to refine body perception, spatial appropriation exercises, structural knowledge work with the body, etc. Lawrence Halprin used to invite his office colleagues to join these workshops so that they could get to know these common territories of dance and architecture from a new point of view (Figs. 2 and 3).

Fig. 2. 1966 Summer Workshop. The Sea Ranch, California. Architectural Archives, School of Design, University of Pennsylvania.

Fig. 3. 1966 Summer Workshop. The Sea Ranch, California. Sketch by Lawrence Halprin. Architectural Archives, School of Design, University of Pennsylvania.

Throughout his life, Lawrence Halprin, worked individually and jointly with Anna Halprin from whom he learnt her way of understanding and creating with the body in the space. Very especially, Lawrence Halprin was interested in dance from the point of view of graphic representation, how it could provide new levels of information mainly from the introduction of body movement in the architectural and urban drawings.

2 Lawrence Halprin: Architecture, Dance and Graphic Representation

The first documentary sources found on the influence of dance on the architect Lawrence Halprin are not directly linked to the issue of graphic representation but to the concern that the space designed by architects and urban planners should be a suitable place for the body. Space and body constitute an indissoluble couple in both fields, dance and architecture, however, in this latter discipline the body is not yet as present as in dance.

Fig. 4. Untitled. This is my beloved series. Lawrence Halprin. 1946. E. Cella Art &Architecture

Fig. 5. Anna Halprin, The Branch, the Halprins' dance deck, Kentfield, California, 1957. Photograph by Warner Jepson. Museum of Performance and Design.

In this sense, Lawrence Halprin wrote in 1949 a first article in which under the title *The choreography of gardens* [6] underlines how gardens should be a support for the moving body.

Lawrence Halprin's early graphic attempts in relation to the world of dance begin with some simple collaborations with Anna Halprin and her ensemble. He will start by drawing her and her dancers as they dance, some first sketches in the traditional way that will be of great help to him for the subsequent construction of the outdoor dance platform generally known as *The Deck* [7]. This is one of his smallest but more transcended works and in which drawing has a key role. The freehand drawing of dancing people always entails the added difficulty of the representation of the movement and, in addition, the approach of the drawing of the body in relation to its associated space (Figs. 4 and 5).

As he works with Anna Halprin on the creation and recording of her choreographies, Lawrence Halprin meets the world of *notations*. These are a kind of scores for the record of body movement that, although they do not have the usual use that musical scores, may have awakened in numerous architects and theorists an important interest since they show parallels with the representation of the architectural space and the inclusion of movement in it [8].

Throughout the history of dance several notation systems are known, however, by Halprin's time, the one that had transcended the most was *labannotation* [9], although it never became common use even among dancers and choreographers. This notation system is based on a series of symbols that include the direction of movement, the part of the body with which it is performed, its duration, its quality and its purpose. This results in a series of attractive black and white drawings of very difficult translation if you do not have a specific training in the writing system.

Once Lawrence Halprin has known this system, he decides to take a step further and creates what he calls *motations* in order to, according to the author himself, introduce the concept of movement into the word. Lawrence Halprin firmly believes that one of the elements to work with in the design of buildings and cities is movement and, for this reason, it is crucial that it is represented. The *motations,* unlike the other notations, were intended to be a record of movement in the architectural and urban space, and therefore, it focuses not only on representing the movements of the human body but also on giving data about its environment. This is because Halprin's intention is that *motations* would not only serve to be used by choreographers and dancers but could also be used by architects, designers or artists of any other discipline who might consider them useful (Fig. 6).

The *motations* are organized, such as labannotation, based on two vertical and parallel stripes to be read in a complementary way. In the strip on the left, or horizontal axis, different schemes are made in plan as a frame of the state of the objects or people in motion. It is mainly useful to see the path of the movement. Small sketches are inserted on the right stripe or vertical axis as elevations in which it can be read simultaneously how this movement occurs. Somehow, it might be seen like an attempt of movement's ortho projection system. This system is complemented by a set of 26 symbols representing concepts such as person, car, acceleration, etc. [10] (Fig. 7).

Among Lawrence Halprin's main interests was to find a new way to graphically represent the architectural and urban space so that it overcome the geometric definition

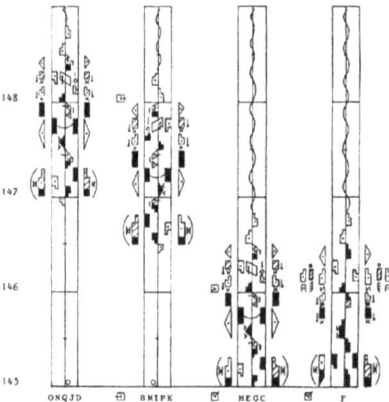

Fig. 6. A piece of a labannotation score.

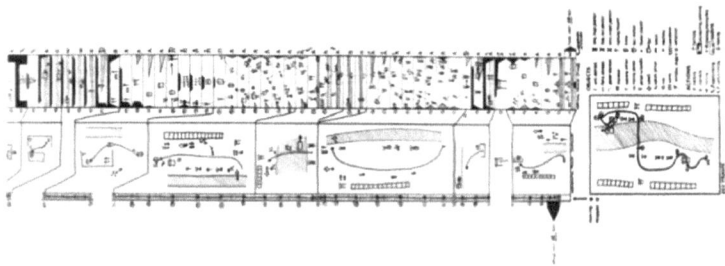

Fig. 7. Journey in Collide Corner, Lawrence Halprin. 1974. Source: Harris, A.: Choreographing the Space. Architectural Studies Integrative Projects 62,12 (2014)

of its elements and incorporate the action of bodies into that space. An intention very similar to that is the one that Bernard Tschumi would have years later and which is very clearly reflected in his *Manhattan Transcripts* [11]. Unlike Tschumi, who stays on a more artistic and intentional plane, Lawrence Halprin wants to develop an objective method applicable to different situations and by different professionals. His initial approach is far from any subjectivity. And although the method is tested on many occasions (Dancer's workshop, Freeway trip, Student Union Plaza, etc.) Lawrence Halprin himself recognizes the difficulty and patience to have for its learning [10], it resembles the time required to start reading musical scores.

Lawrence Halprin's *motations* could be summed up in a mixture of a set of plans to explain the space and a few notations to describe the movements that are happening in this space. Somehow, Lawrence Halprin thought architecture would be humanized if the architectural representations included the human being. In fact, it might be suggested that Lawrence Halprin himself took a step further on labannotation by being able to enter space in a more intuitive way. It is easier to read the space on a plan than through a set of symbols as was the case with Laban's system.

3 Architecture, Dance and Representation. Other Approaches

Lawrence Halprin is not the only architect who has looked at dance [12] in order to make transfers to the field of architecture, although it could be said that he has been one of the pioneers in doing so, and one of those who has shown the most dedication in this regard. Among the architectural professionals who have looked at dance because of the representation of the moving body the figure of Bernard Tschumi could be highlighted.

Bernard Tschumi also finds in the notation of dance a certain suggestion to enrich the architectural graphic representation, however, he departs from Lawrence Halprin as long as his intention is more related to a way of introducing life, activity and meaning in architectural drawings. The goal was to collect some of the action that happens in that architecture and not only its formal definition and material characteristics. For Bernard Tschumi, there was no architecture without an event, so it was important to bring the event to the representation of architecture.

Probably, Bernard Tschumi's most paradigmatic work in relation to the use of notation is *The Manhattan Transcripts,* which are neither real projects nor mere fantasies, just an attempt to transcribe a certain interpretation of reality. The main idea of the *transcripts* is to transcribe things that the representation of conventional architecture does not, for example, the complex relationships between use and space. It is also, therefore, a critique of the representation of architecture from which Tschumi intends to expand its boundaries.

The *transcripts* take place in Manhattan, a real space where real things happen. Its structure is divided into three worlds: that of architectural objects represented by plants, sections or volumetrics, that of movements, for which notations and traces are used, and that of events, for which newspaper news or photographs are chosen so that they document some action happened in that space. As Tschumi suggests in the publication of the same name, the choreographic notation is brought to analyze the movement since the notation suggests an energy that sculpts a volume, or, conversely, how an enveloping space is generated within a rigid space (Fig. 8).

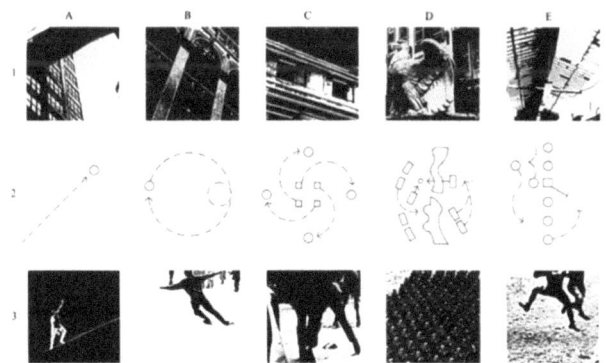

Fig. 8. The *Manhattan Transcripts*. Bernard Tschumi. 1976–1981. © Bernard Tschumi Architects.

In a lecture given by Tschumi himself at the Architectural Association in 1982, he attributes to the notation another utility. This is the ability to eliminate preconceived ideas or images about bodily actions in a way that what prevails is its spatial effect, that of movements in space. In this regard, nowadays certain Laban supporters continue to defend the use of notation for the registration of choreographies against audiovisual media as it is the only way to communicate the essence of choreography and not only a visual image. In other words, if we were talking about music, it would be as if the musicians learned to play the pieces from recordings instead of playing the scores.

It is interesting how architecture has come particularly close to the means of representation of dance, especially when in the field of dance its use has not spread too much. In fact, most of the choreographers and dancers do not use them even to register or to create their choreographies. According to Paul Virilo, the interest of architecture in these drawings elaborated by choreographers is probably the fact that they are intimate drawings in which the energies related to the body become visible. Indeed, Paul Virilo himself acknowledges having begun to study *labannotation* to see if he could somehow complement the plans and the sections since they do not include time or the feeling of volume. To Halprin, Tschumi or Virilo, we can add the architect Sony Devabhaktuni [13] who has studied the notations and drawings of the choreographer Merce Cunningham from a spatial point of view with the intention of revealing new approaches to architecture from the dance notations.

In both the case of Bernard Tschumi and Lawrence Halprin, their use of notations does not become systematized in a way that can be used as a language of representation of any project or work. It gives the impression that the use of notation offers them an appropriation of the perceived or created space more than a real language to put into practice. However, even if this has not happened, both acknowledge that their work would not be the same without their studies on notations. Tschumi states that neither *La Villette* nor *Le Fresnoy* could have existed without the *transcripts* which are included in the graphical documentation of each architectural project. Just like Lawrence Halprin whose notations for the motion study of Anna's workshops appear as graphic and project material with the same treatment as other architectural works in his monographies.

4 Conclusions

For much of his professional life, Lawrence Halprin tested with his team a series of experiences about the representation of space and what was happening in it that were clearly influenced by the discipline of dance with which he lived so closely. In some, the incorporation of the space-body wisdom of dance introduces what it could be called small advances, while, in others, an overly academic look at dance arises somewhat more debatable results. That academic look focuses on notation systems, an issue that has also been addressed in architecture by Bernard Tschumi or Paul Virilo. Lawrence Halprin focused primarily on the *labannotation* that was neither commonly used by Anna Halprin herself, and which departs to some extent from other more enriching experiences such as *Experiments in the Environment,* workshops made up of architects and dancers in which the two exchanged experiences through different activities and where there was also a presence of the drawing even if it did not have a main role.

In any case, although the *motations* did not extend as a system for different reasons they did serve as a personal challenge through which a more human architecture based on drawing could be considered. The use of *motations*, and his knowledge of the world of dance, allows Lawrence Halprin to better understand the perceived or created space [13], this goes beyond the traditional plans and elevations by promoting a new reading on the project. It should not be forgotten that the proposal of the *motations* must always be read as a complement to the planimetry and not as a substitute for it, according to the writings of Lawrence Halprin himself [3].

It could be said that Lawrence Halprin's approach to *motations* paved the way for experiences that are now being studied accompanied by new motion-recording technologies such as mappings or *space syntax* which manage to incorporate movement in a more advanced and effective way. Their initial motivation, in any case, is shared. Therefore, it is important to emphasize that although today we are closer to a more complete and reliable representation in the architectural and urban space, this representation should never be understood as a substitute for the experience of the body in space but as a complement that helps us to understand it. This question is paramount and is very present in Lawrence Halprin who tried to transfer his knowledge of dance, and therefore of body and space, not only to representation but also to architecture, understanding that the disciplinary fields are diffuse and, thanks to this, more enriching.

References

1. Ross, J.: Experience as Dance, 1ª edn. University of California Press, Berkeley (2007)
2. Blancafort, J., Reus, P.: Pioneros de la participación colectiva en los procesos de planificación urbana. Legado Halprin. Arquitectura, ciudad y entorno **28**, 57–76 (2015)
3. Wasserman, J.: A world in motion. Landsc. J. **1–2**, 33–53 (2012)
4. Halprin, A.: Collected Writings and Others: San Francisco Dancer's Workshop (1974)
5. Halprin, L.: The RSVP Cycles. George Braziller, New York (1969)
6. Halprin, L.: The choreography of gardens. Impulse Dance Mag. **30**, 30–34 (1949)
7. Halprin, L.: Dance deck in the woods: use of the deck. Impulse Dance Mag. **24**, 21–27 (1954)
8. Virilio, P.: Gravitational space. Traces of Dance Choreographers' Drawings and Notations, pp. 35–60. Éditions Dis Voir, Paris (1994)
9. Kiplin Brown, A.: Dance notation for beginners: Labanotation, Benesh Movement Notation. Dance Books, London (1984)
10. Halprin, L.: Motation. Prog. Archit. **46**, 126–133 (1965)
11. Migayrou, F.: Bernard Tschumi: Architecture: concept and notation. Centro Georges Pompidou, Paris (2014)
12. Gálvez Pérez, A.: Materia Activa, la danza como un campo de experimentación para una arquitectura de raíz fenomenológica. Universidad Politécnica de Madrid, PhD Dissertation (2012)
13. Devabhaktuni, S.: Merceinspace. AA Files **62**, 32–63 (2011)

Parametric Processes in Graphical Architectural Thought

Francisco González-Quintial$^{(\boxtimes)}$ and Enrique Sancho-Pereg

ETSA-AGET, University of the Basque Country, Leioa, Spain
`francisco.gonzalez@ehu.eus`

Abstract. The idea of parametric systems in architecture is usually connected with relatively new digital processes and the production of architectural forms inseparable from the use of IT tools in both Computer Assisted Design and processes of calculation and computation. However, the use of parametric systems in design and construction processes can be identified much earlier than the arrival of computers in everyday graphic architectural design work.

This paper identifies the factors that define these processes and how they are more closely linked to thought, fundamentally graphical thought, than the tools which may in given circumstances be used in architectural design and creation. It argues that through these parametric processes we can read diverse architectural forms which, though separated in time, share a common geometric and generative foundations. This cannot be correctly understood without the generative process that gives rise to these architectural forms having previously been identified.

Keywords: Graphical thought · Parametric architecture · Computational design

1 Introduction

We can understand a parametric process as one in which, through a set of modifiable variables, we can obtain correspondingly different results. The structure or algorithm that processes variables is held constant and this is what generates solutions.

Generative design, parametric and algorithmic architecture are terms that we automatically associate with "cutting edge" tendencies in architecture. This association can to some extent be understood and is based on the proliferation of these types of computational design architectural processes, which today have an intrinsic connection with the utilization of IT technologies, both hardware and software. However, if we sufficiently disassociate the terms "computation" and computerization," [1] this relationship can be seen as purely instrumental and the result of circumstance.

In mathematics, computer science and related disciplines, an algorithm (from Greek and Latin, dixit algorithmus and this in turn from the Persian mathematician Al-Juarismi) is a pre-written set of well-defined, ordered and finite instructions or rules that allows an activity to be carried out by successive steps that do not generate doubts for those who are carry out said activity. Given an initial state and an entry of data, following the steps established successively, a final state is reached and a solution is obtained. The

L. Agustín-Hernández et al. (Eds.): EGA 2020, SSDI 6, pp. 508–516, 2020.
https://doi.org/10.1007/978-3-030-47983-1_45

development of an algorithm and its writing or programming is nothing other than the construction of a logical chain of processes, previously established in a rational way and organized by means of a tool, be it analogical or digital in nature.

If we focus purely of the concept of parametrization, we can identify it as a process which involves making it possible to include different values within different logical chains. Any parametric process will yield results in function of the values introduced which in certain cases will be useful and in others will only facilitate the identification and ruling out of invalid solutions.

We will attempt to demonstrate how these systems can be identified in the history of architecture and the development of graphic process which have sustained it. These processes, as happens with any other system of any type linked with human activity, have adapted to the tools available at any given time. Through the rewriting and reinterpretation of graphical processes used in the past, we can interrelate them and obtain new architectural forms based on the same logical chains, which can also be called algorithms.

2   Pre-digital Parametric Processes

Just as the definition of algorithms comes from a pre-digital age, we can identify parametric design processes prior to the incorporation of computers into the area of architectural graphics. It is before the Albertinian definition of drawing as the correct and distinctive tool of architecture [2] and prior to the establishment of this way of representing ideas which, through a set of projections, freezes a design process in time, that graphical and constructive processes are structured more clearly as a process which can be identified as algorithmic or parametric. Within the work of Alberti himself a process of digitalization and parameterization can be identified in his search for reliable structures for the gathering of data and the transfer of information. This is understood as a system of codification, and even as an alternative system of codification alternative to graphical representation [3], even from within the renaissance itself.

Attempting to focus the problem, we can understand the Gothic system for covering large spaces as a purely constructive process. As it is a modular system, it can be adapted to the dimensional requirements of any specific space. The graphical representation of this process creates, through the introduction of different numerical values or parameters defined in any system of units, the final design for the buildable space. The apparent overall complexity is reduced to a logical progression along an optimized central logical backbone process into which values, in some cases very simple, are fed in to produce a final result. This is an adaptable, flexible system which aims to provide a simple answer to complicated design problems. It was able to work in this way because graphic the process had not yet been standardized as purely architectural. It also works as a codified system in a series of graphical systems which, through the introduction of a series of values or parameters, produce flexible designs, apt not only for covering quadrangular spaces but also any intermediate space, perimeter or external to the final structure, including possible singularities. From this viewpoint we can interpret Villard de Honnecourt's notebook as a compilation of a series of parametric definitions, a possible pre-digital algorithm [4] (Fig. 1).

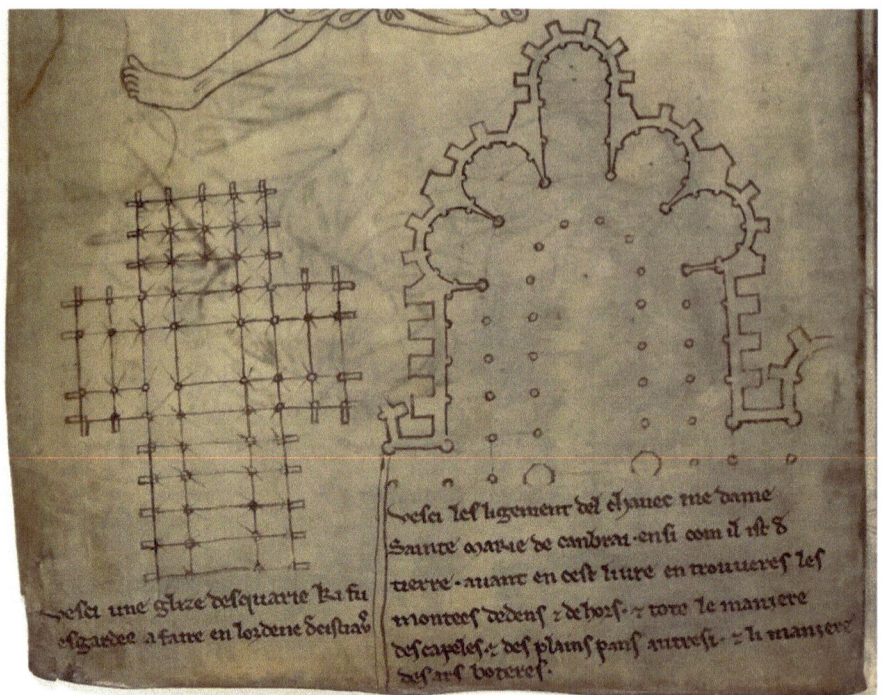

Fig. 1. Villard de Honnecourt, *Album de dessins et croquis*. Page 14v. Detail of the plan for a Cisterian church and coral floor of the Cambrai Cathedral. Source: https://gallica.bnf.fr/ark:/12148/btv1b10509412z/f30.image.r=honnecourt (accessed March 2019)

Generically, we can formulate that the system of sketching gothic vaults is based on the adaptation of a module with sufficient geometric flexibility to evolve from the groin vault, where the intersections of two straight cylinders produced a pair of elliptic curves for cylinders of identical generatrix, which produced for surfaces with zero or parabolic curvature [5]. This plotting is rigid and if not based on a square floorplan, the system ceases to have geometric validity as the basis of a simple plotting system. The sail vault, with a spherical surface and positive or elliptical curvature, in which all of the sections are circular, is more flexible with respect to the floorplan and moreover the surface that is generated is perfectly spherical, as are the circular guide ribs of the side arches which support it. But the system evolves as far as the ribbed vault where all of the ribs are circular arcs and in certain cases of the same radius, adaptable to any quadrilateral floor. This enables a solution for covering transitional spaces, of almost any floor geometry, gaining independence from both the rigid imposition of a square grid and the regularity of the surface which closes the vault itself, which acquires a geometry not easily identifiable as being derived from a primitive geometry, and comes to be labeled "irregular fill" [6].

If we concentrate specifically on the plotting of the tierceron vault proposed by Hernán Ruiz (Fig. 2), the sketching of the plotting of this vault can be easily understood, as is also true in the generic case of a rib vault, as a parametric system where, starting with a governing quadrilateral floorplan, all of the dimensions necessary to carry out the complete construction of the vault over the space can be obtained. All this is demonstrated through a rewriting of the physical plotting system on paper as an algorithm, in this case digital, through a CAD programming tool such as *Grasshopper*, [7] a plugin for *Rhinoceros*, [8]. In this way it can be demonstrated that the system used to design in a physical space is not significantly different on paper or as a digitally programmed definition. The difference lies basically in the tools used in the drafting process, while the process, or algorithm which generates the geometry (Fig. 3) remains identical. With respect to the directrix curves, a simple translation of the rib section is enough to obtain a complete geometric construction of the vault. We see that these graphical mechanisms are found to be much closer to accepted hypothesis for the physical and graphical construction of these vaults, than might be the pre-digital plotting systems proposed in the XIX century as the ideal for this type of element.

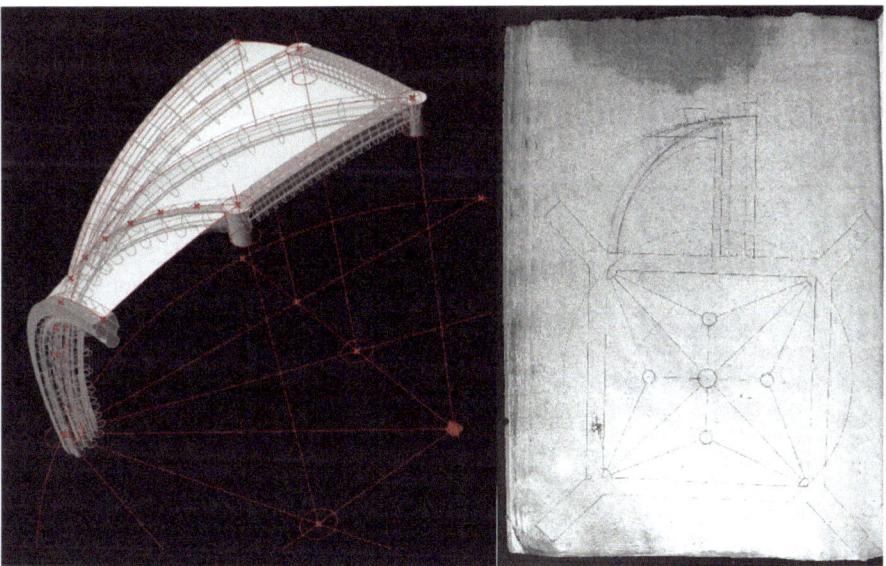

Fig. 2. Right. Hernán Ruiz, plan of a tierceron vault [6]. Left. A model obtained from parametric definition by means of the transcription of the process of designing a tiercion vault based on the definition by Hernán Ruiz. Source: Author

3   Graphical Connectivity of Pre-digital Parametric Processes

Beyond the fact that the previously described processes can be identified as parametric processes in themselves, the importance of the question is not especially one of definition, but instead lies in the possibilities represented by the system as such and the potential which it offers as a space for the graphical exploration of architecture.

The catenary systems that were employed profusely at the end of the XIX and beginning of the XX Century in Catalan Modernism, recognized most often but not exclusively

Fig. 3. From right to left. Evolution of the rib vault from the groin vault. The ability to choose a height independent of the side arches results in a flexible system that can be adapted to any floorplan. Source: Author

in the architecture of Antonio Gaudí, have already been identified as design methods [9]. Certainly, the potential that these processes represent can only be explored through their reinterpretation and rewriting through processes similar to the one described above for the quadripartite tierceron vault. This is, in fact, made even clearer through an analysis of the catenary systems used by Gaudí himself to achieve these forms. Nowadays, we automatically associate the computational processes known as *form-finding* with a system that bases form on structural optimization. Via means very similar to those well-known from illustrations of "Gaudian" construction in which physical catenary models are used, we use virtual space and physical simulation in virtual graphical constructions to the same end. These developments are based on pre-digital procedures formulated and documented more than a century ago, not only in physical forms, but also through the use of static graphic processes. (Figure 4) Once again, the shift occurs more in the medium used rather than in the foundation. The difference is quantitative and the possibilities offered by this change in the system of representation are apparently limitless.

Other lesser-known plotting techniques, such as obtaining catenaries on site, can also be understood as analogue or pre-digital parametric processes, where the result of the generative algorithm can be understood as the form acquired by the chain when subject to variable loads and anchor points, these being the initial data inputs in said process. In addition, because of its plotting and construction located on site, it is closely related to the previously mentioned systems for the plotting of designs of constructions in stone. All these forms can now be simulated in a virtual space through algorithmic tools for generating forms that introduce into their logical chains aspects of physical behavior conditioning the resulting surfaces.

Through the use of computational parametric definitions constructed on the basis of certain pre-digital processes, it is possible to interrelate geometries without apparent connection. This does not mean that the process of generating these geometries was

Fig. 4. The generation of antifunicular surfaces using the form-finding tool tools in the *Kangaroo* plugin for *Grasshopper* [10]. Analog generation of catenary surfaces in the roof of Casa Milá. Antonio Gaudí. (Source: Author)

consciously understood at the time as a generative process as such, but it does seem reasonable to maintain that, from a contemporary perspective, these processes are articulated in exactly the same way as those that are currently commonly accepted as such, that is to say that the procedure can be understood structured in the same way and it is the medium, digital or analogical, that varies.

We can clearly identify this type of process in the work of Felix Candela. In his shells we find a profusion of apparently indeterminate forms which, after a deeper analysis from an exclusively geometric point of view, leaving aside intimately related and not less important questions related to form such as structure and construction, are based on a reduced repertoire of basic forms that follow diverse generative processes that interrelate them offering different geometric results.

In this geometric language, the profusion in the use of the *hypar,* warped quadrilaterals, sections of the surface of a hyperbolic paraboloid, which can be understood in multiple ways, stands out. Taking into account its generative mechanism, it is defined as a ruled surface but also as a transition surface and as a translation surface as well [11]. Being a surface of negative or hyperbolic double curvature and in spite of being ruled not developable, it can be discretized into flat quadrilateral elements and therefore semi-discrete surfaces or developable strips. [12].

This geometric versatility allows us to get an idea of the adaptability of this surface in the field of construction of architectural forms, which, added to the intrinsically resistant nature of this geometry, lets us understand the profusion of its use, especially in the case of *Candela*, without forgetting Gaudi's use of the form.

The hypar has a clearly parametric reading, easily identifiable in the generative process of the roof of the Medalla de la Virgen de la Milagrosa Church [13]. By means of the variation of the perimeter that defines the surface, a formal repertoire is obtained that allows the transformation of the horizontal covering, which has a rectangular floorplan, in the semi-section of the nave of the Church (Fig. 5).

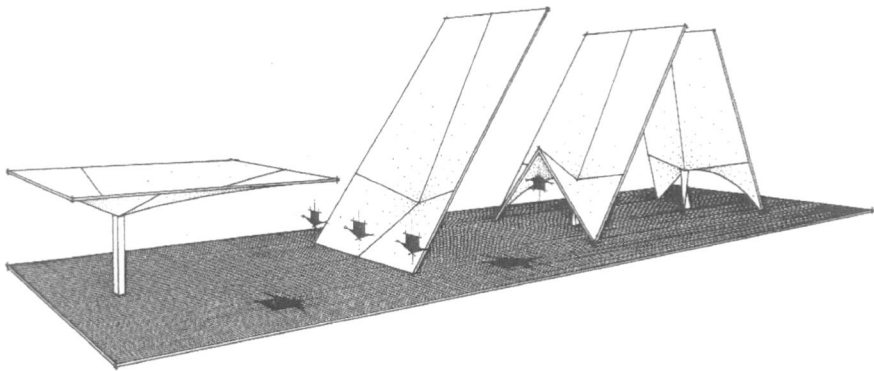

Fig. 5. Derivation of the shape of the roof of the Medalla de la Virgen de la Milagrosa Church. (Faber, 1963) Page 103. Explanatory sequence of the generative process of the section of the nave of the Medalla Milagrosa Church. Narvarte, Felix Candela 1954-55

Taking as an example the Central Hall of the Bolsa (Stock-Exchange) Building in Mexico City, (Fig. 6) described as the orthogonal intersection of two hyperbolic paraboloids on a rectangular floor plan, it can be understood from the starting point of a parametric definition of the quadripartite vault, with flat ribs and severies of positive or elliptical Gaussian double curvature, the negative or hyperbolic surface that constitutes the shotcrete shell of the *Candela* can be obtained by means of modification of the starting parameters (Fig. 7).

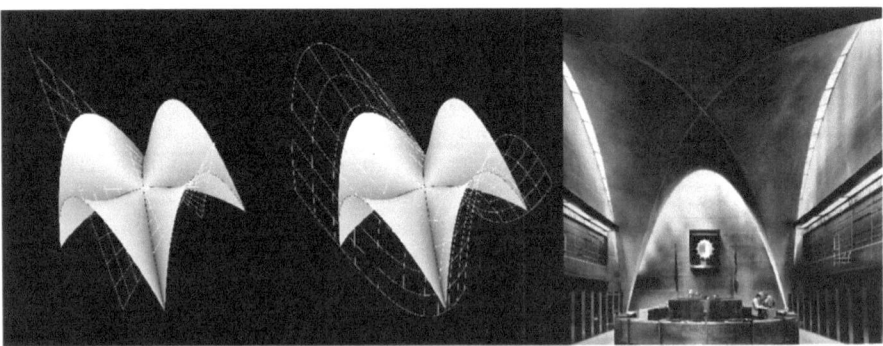

Fig. 6. Intersection of hyperbolic paraboloids, on the left, generation as ruled surfaces, in the center, generation of the paraboloid as a translation surface and therefore developable, of a generatix parabola along another directrix parabola. Mexico Stock Exchange Building. Groin vault with hyperbolic surfaces. López Carmona, Fernando, De la Mora Enrique, Candela, Félix. Source: Félix & Dorothy Candela Archive, Princeton University

As in the case of ribbed vaults, in which a radical change is produced by the adaptation of a flexible plotting system for pointed arches from the rigid plotting of semicircular arches, the appearance of the *spline* curve and later the *NURBS* curve mean a total change in the possibility of reproducing and generating any type of geometry.

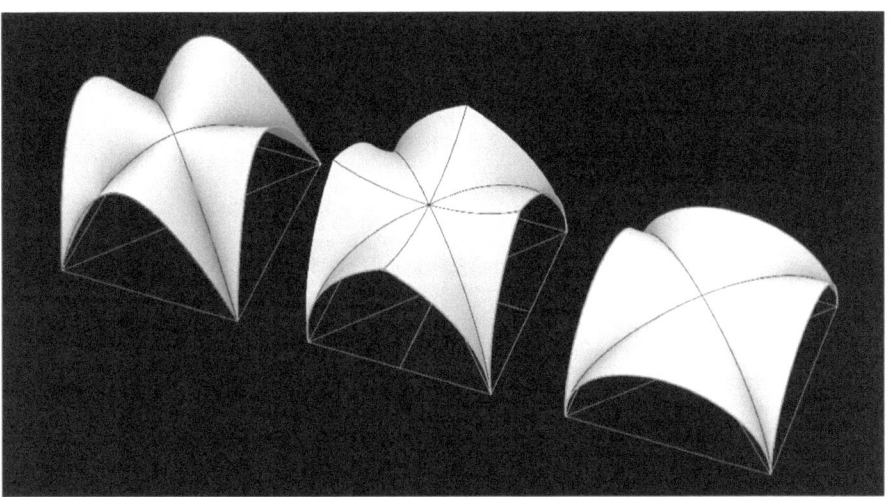

Fig. 7. From right to left. Evolution of the hyperbolic vault from the groin vault. The position of the height of the key below the height of the side arches makes the cylindrical surface hyperbolic. Source: Author.

With the same mathematical or geometrical definition of the curve, which came from Bezier's work and Casteljau's algorithm, [14] we can represent any conic section simply by varying the conditions of the points in the control polygon that determines the curve. In this way we can transform circular edged directrices into parabolic edged ones. A cylindrical groin vault is transformed into a hyperbolic-section vault by a vertical movement of the key point.

The surfaces that compose the vaults modify their curvature, going from being zero for the cylindrical surface of groin vaults, to being positive or elliptical for the quasi-spherical surface of ribbed vaults, and finally becoming a negative or hyperbolic surfaces when the position of the key is below the height of the side arches. This transformation not only has a geometrical but also an eminently structural significance.

4   Conclusion

The analysis above cases and the rewriting of the processes that generate them through the use of digital parametric tools demonstrates, first of all, how the graphic development of certain constructive forms can be understood as a timeless mechanism linked, as could not be otherwise, to the tools available at the moment in which they were generated.

Secondly, we can affirm that the tool manifests itself as the backbone of the creative process. In the case of parametric and generative design tools, these tools facilitate the materialization of an extension of formal possibilities since they not only cover the possibility of obtaining shapes that in certain cases are predictable, but also allow the generation of solutions which, as they depend on values and data chains of an undetermined extension, produce solutions which are similarly indeterminate and practically infinite in number.

References

1. Terzidis, K.: Algorithmic Architecture, 1st edn edn. Architectural Press, Oxford (2006)
2. Carpo, M.: The Alphabet and the Algorithm. The MIT Press, Cambridge (2011)
3. Carpo, M.: The Second Digital Turn: Design Beyond Intelligence. The MIT Press, Cambridge (2017)
4. https://gallica.bnf.fr/ark:/12148/btv1b10509412z/f30.image.r=honnecourt. (acceso marzo 2019)
5. Fitchen, J.: The Construction of Gothic Cathedrals; A Study of Medieval Vault Erection. Clarendon Press, Oxford (1961)
6. Rabasa Díaz, E.: Forma y construcción en piedra: de la cantería medieval a la estereotomía del siglo XIX. Akal Ediciones, Madrid (2000)
7. https://www.grasshopper3d.com/
8. https://www.rhino3d.com/es/
9. Jormakka, K., Schürer, O., Kuhlmann, D., Shürer, O., Kuhlmann, D.: Basics Design Methods. Birkhäuser, Basel (2008)
10. https://www.food4rhino.com/app/kangaroo-physics
11. Schober, H., Schaffert, C.: Transparent Shells: Form, Topology, Structure. Ernst & Sohn, Berlin (2016)
12. González Quintial, F., Sánchez Parandiet, A.: Método de adaptación de formas de doble curvatura mediante superficies desarrollables: tesis doctoral, S.N. (2012)
13. Faber, C.: Candela, the Shell Builder. Reinhold Pub, Corp., New York (1963)
14. Pottmann, H., Bentley, D.: Architectural Geometry. Bentley Institute Press, Cricklewood (2010)

The Model as Expression, Analysis and Synthesis of the Creative Process in Architectural Design

Javier Fco. Raposo Grau$^{(\boxtimes)}$ ⓘ, María Asunción Salgado de la Rosa ⓘ, and Belén Butragueño Díaz-Guerra ⓘ

IGA Department, E. T. S. Arquitectura, Universidad Politécnica de Madrid, Madrid, Spain
javierfrancisco.raposo@upm.es

Abstract. Through the action of drawing, we can imagine, explore, recognize, analyze, propose, register. With strokes, spots, gestures, lines, and different graphical resources such as drawing, assembling, photomontage, collage, model. We explore, analyze and communicate experiences. We can state that the drawing and the model maintain the leadership as the main language and a thinking tool among architects, as they respond to all those actions, procedures and tools. Drawing means tracing, describing, showing, revealing something that was hidden or silent. Drawing is a process that combines different procedures, associated with diverse graphic media, as it is the model. The model allows the expression, analysis and synthesis during the creative process of the architectural project.

According to the dictionary, "modeling" is *making a prototype of a publication that will be printed; and also, a scale model of a construction.* The term "model" is defined as *a theoretical scheme, usually in the mathematical form, of a system or a complex reality, as the economic developments of a country, which is developed to facilitate the understanding and the study of its behavior.* This definition encompasses important concepts linked to the actions of drawing and architectural project. We draw, we model and we design since we are children. We use the model as a mechanism to learn, to apprehend, to understand the exteriority, and to lay hold of our interiority in the creative processes. The present article will make a trip around the model as a means and procedure for architectural creation.

Keywords: Model · Drawing · Design

1 Introduction. Drawing with Models

The article addresses, from a teaching and professional perspective, the use of the model in the processes of architectural creation. The frame is a graphic environment of hybridization of languages, associated with different analog and digital procedures (Chiarella and Redondo 2011, 184) the goal is to facilitate the understanding of complex architectural developments, relying on hybrid graphic procedures, using the model as a procedure of experimentation to explore and encourage self-reflection in the creative process in architecture.

L. Agustín-Hernández et al. (Eds.): EGA 2020, SSDI 6, pp. 517–529, 2020.
https://doi.org/10.1007/978-3-030-47983-1_46

Miguel Angel de la Cova, paraphrases Arthur Drexler in his preface at *The architecture of the École des Beaux-Arts*, entitled *Engineer's architecture: Truth and its consequences*, wondering what the role of the model should be in architecture. It could represent a faithful copy of the defined building, or an expression of its perception, or it could be exclusively linked to the interests of the architect, distancing itself from reality. There is an open debate between art and technique in the discipline of architecture, and consequently, between the means of production, analysis and social communication, compared to what it means to work with models, as they are usually connected to the physical and representative condition (De la Cova 2016, 13). We think with our hands, in a permanent and continuous "action of making" that becomes an "action of thinking".

We define the artistic creation as a searching for something that is only found when done. "A doing that invents the form of doing" (Pareyson 1988, 225). The talent returns to the hand, that has recorded in its palm the labyrinthine lines, as the neuronal network of the brain (Sennet 2009). Again, brain and hand actively seek each other. There is a parallelism between the development of models and the architectural drawings, in the different moments of the creative process (Fig. 1). The hybrid character of the drawing brings together graphic resources and adjusted codes, that encourages experimentation.

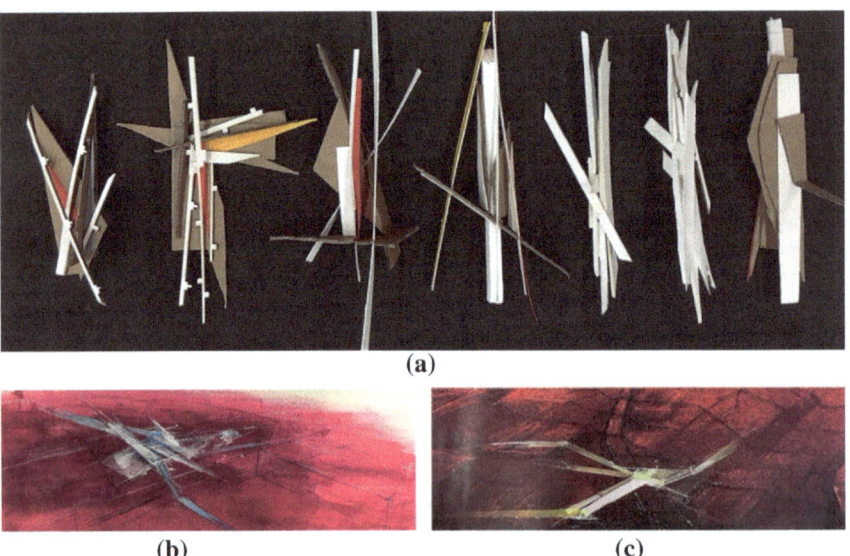

(a)

(b)   (c)

Fig. 1. a. Student Pedro Trueba Padial. b–c. Student Melanie Wailder Heisecke. Polytechnic University of Madrid (UPM). School of Architecture (ETSAM). Department of Architectural Graphic Ideation (DIGA). Subject: Drawing, Analysis and Ideation 2 (DAI 2). Course 2018–19. Professors Raposo, Salgado, Butragueño

We can find models that work as a thinking tool in the initial moments of exploration and architectural expression (Carazo 2011, 35); others represent analytical and diagrammatic models (Carazo and Galván 2014, 63); and some others, work as models of representation and communication of architecture (Carazo 2011, 32). The model is a procedure of simulation and miniaturization, to imagine, design and build meaningful architectural entities, that work as a mediator between thinking and acting. The miniaturization is a mechanism of activation of the interior space, strengthening the dynamic qualities of design. Certain studies on the experience of art from the new understanding of the vision, show an active process that allows the brain to be selective and make decisions during the creative processes of the artist (Zeki 2005).

It would be restrictive to classify the models exclusively based on one condition: the imaginary, the processual, or the communicative. Throughout the history of architecture, the different means of expression have made possible the use of the procedures and modeling tools for different purposes, such as seductive strategies, city planning, constructive and structural tests, specific expressions in competitions (Carazo 2014), and even the restitution of lost architectures (Úbeda 2011, 158–159). Representative and communicative models are the ones that had a wider dissemination throughout history, with significant milestones that have been subject of successful exhibitions, such as *The ten models* of the *Modern Architecture Exhibition* of 1932 (Montes and Alonso 2018, 36).

Today we are facing a paradigm shift in the designing mechanisms, due to the incorporation of new graphical tools linked to the digital transformation and Industry 4.0. Undoubtedly, it brings a significant impact on the creative processes linked to the discipline of architecture (Fig. 2).

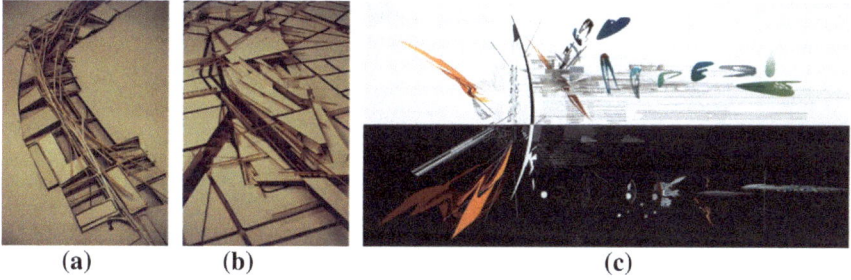

(a)	**(b)**	**(c)**

Fig. 2. a–b. Eliyahu Keller. Rothschild Boulevard Study, 2012. c. Zaha Hadid. The World (89°), 1984

The incorporation of computer media, during the past three decades, has allowed the use of computer as a tool of construction and graphic transformation, leading to the development of extremely realistic 3D drawings or renders. These systems can generate a large number of conclusive images of the project, that mainly focus on the communicative aspect. This evolution has led to the discredit of the analog procedures versus the digital ones, abandoning certain means such as drawings, collages and models, in the different phases of the creative process in architecture: initial, intermediate and communicative phases.

In recent years, the development of ICTs and the insertion of digital vs. analog, have questioned the model as a mechanism of faithful representation of architecture, becoming more important for other different purposes. The hybridization of the analog and digital models opens a spectrum of interesting possibilities. Using these procedures of graphical hybridization between drawings and models, we could reinforce the value of the model as a thinking procedure, linked to the initial moments of the creative processes. The goal is to transmit the qualities of the architectural project, over its appearance and final formalization, in an expressive and intuitive manner (Cabas 2017, 253–254).

The neurologist and writer Oliver Sacks (Llopis 2013 in Sacks 1990, 211), explains that children begin immediately to explore the world by looking, feeling, touching, smelling… the same way animals do from birth. He adds that this feeling does not happen in isolation, but combined with movement, emotion and action. Together, movement and sensation become the predecessors of meaning. Sacks considers that we learn as the result of a complex coordination of manual actions, linked to the vision and manipulation of objects. We can identify the models as real objects, that enable the development of the concepts.

2   The Model as a Process of Exploration and Experimentation of Dynamic Qualities in the Architectural Design

For Dollens [2002, 11] "maquettes are a primary mode of communication […] they constitute a game of communicative forms, […] and transmit ideas and emotions; we learnt as children to invest in fantasy through models and games, to visualize other worlds and link these worlds through metaphors and fables […]". Throughout every architectural imaginary process, the expression and experimentation act as engines of the vital activity (Fig. 3). As a reaction to stimuli, expressions of movement on different materials arise, leaving the trace of the qualities and dynamics of the future project as operational speculations of the creative process (Colombo 2017, 83).

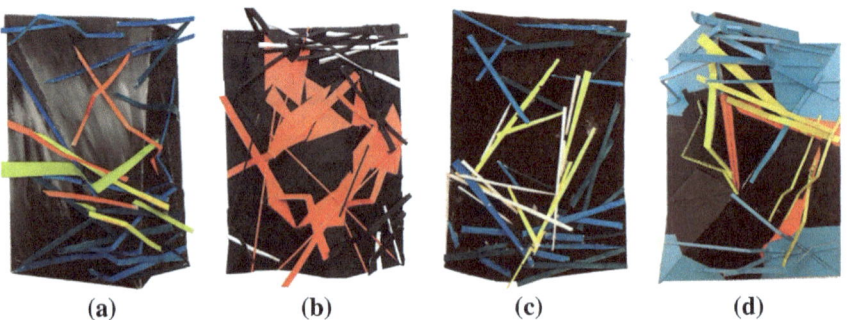

(a)   **(b)**   **(c)**   **(d)**

Fig. 3. a–d. Student Celia Asenjo López. Intuitive models. First approaches. UPM. ETSAM. DIGA. Subject DAI 2. Course 2018–19. Professors: Raposo, Salgado, Butragueño

On the topic of the beginnings and the initial traces of the process, the architect Enric Miralles states that the most important aspect "is the art of launching the thinking, the path to invent and represent things", inducing the importance of the beginning and the laborious process of explaining the project from outside the project (Miralles 1996). The creative and productive processes of Enric Miralles and their relationship with the models are widely known. We can talk about different type of models: sketch, process and communication, at diverse scales of approximation, linked to the territory and the projected buildings. Both the hands and the different materials of models' development are very important, and constitute an irreplaceable working material in the creative processes, as well as in the processes of communication with partners and customers.

In the process of the student's learning, it is necessary to train their sight to see things differently, and to stimulate and strengthen their capacity for observation and instrumentation, expanding the boundaries of understanding the reality under study (Bosh 2000, 7). Through these mechanisms of stimulation, we will be able to overcome the appearance of things and keep the essences and invariants that remain in a territorial model.

Gehry uses the models in all phases of the development of the projects. The project of the Guggenheim Museum in Bilbao means a declaration of intentions about the relationship between drawing and thinking (Fig. 4). "This is how I think, I just move the pen. I think about what I'm doing, but I'm not looking at my hands." From his remarks, we appreciate a parallel with the automatic writing. He states that "the drawing is a tool. The model is too. All of them are tools. The building is what has a meaning, the completed building" (Van Bruggen 1997).

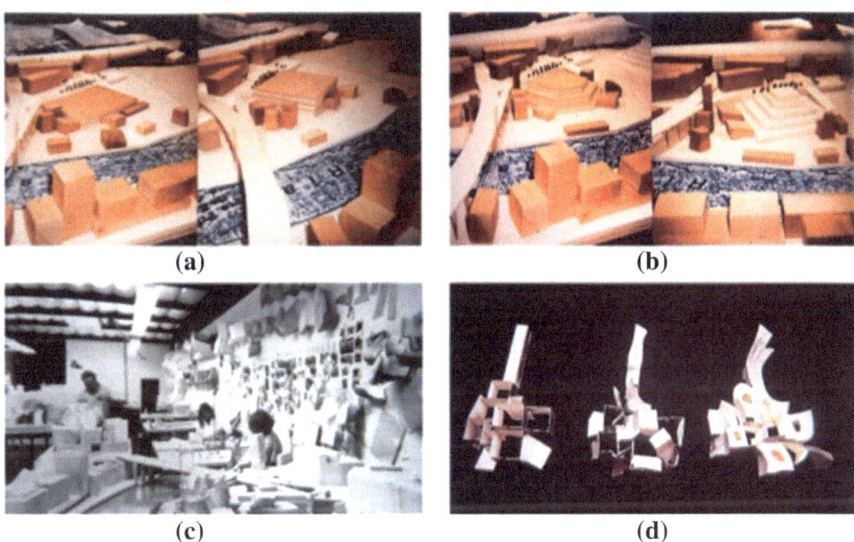

(a)   (b)

(c)   (d)

Fig. 4. a–b. Preliminary models of the project, developed by collaborators (Edwin Chang). Guggenheim Bilbao. July, 9th, 1991. c. Gehry's Office. d. Atrium models

If the "thinking processes" are preceded by the "action processes", the result is merely a product of thinking, is dead and frozen before it is born, not responding to an evolutionary and transformative process of the executive action. In that case, we would be dealing with models that represent closed architectural products, semblance and formal models. The ultimate goal would be the contemplation, and not the experience and internalization of the qualities and dynamics of the projected space, nor would respond to an imaginary act. In this kind of processes, there is room for intuitive, germinal models that give feedback to the creative process.

These other models are a manifestation of the dynamics of the project, its qualities, and ultimately the essences of the designed architecture (Fig. 5).

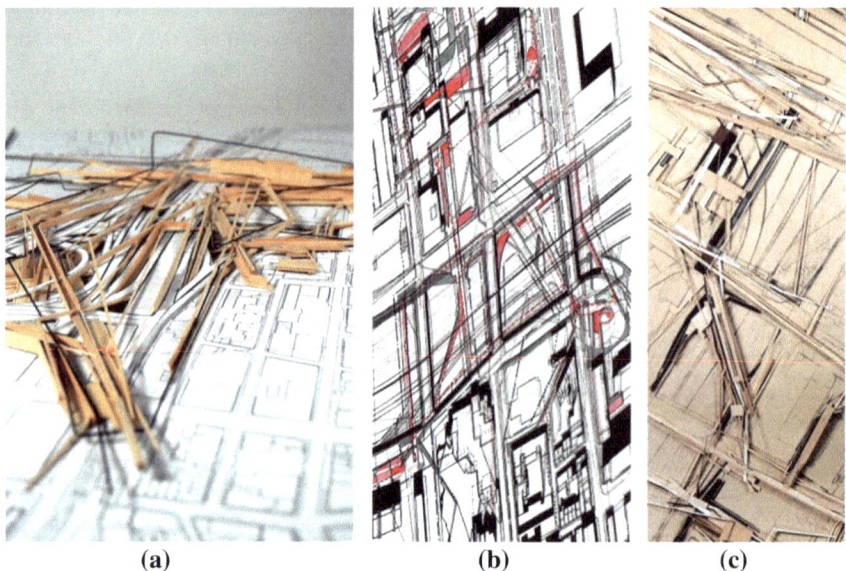

(a)   (b)   (c)

Fig. 5. a–c. Luke Douglas Ericson. Anachronous Trajectories, Master Architecture, 2015

The architectural graphic language and its different modalities and procedures, such as the models, are knowledge tools to explore and analyze situations and propose the construction of the environment with a new spatial order. This language has incorporated heterogeneous aspects linked to the functional, cultural, symbolic, and communicative aspects (Gagel 1997). These aspects are always linked to the different moments of the creative process.

The model plays a very important role in the teaching strategies, but it is necessary to guide the approximations in 3 or 4 structured and progressive phases. This approach provides the students with the skills and understanding of the instrumental procedures applied.

The first approach, with an imaginary character, the intervention is understood as an interpretative and comprehensive process of the site and its architecture. It is a territorial approach of the proposed area.

The second approach addresses the study of a specific area, reinforcing the imaginary aspect of the architectural production, to reach a complete understanding of a smaller and approachable area with greater precision, from a modeling perspective. With the third approach, the student will develop the abilities to process, transform and manipulate the working area or building, operationally and analytically, introducing the students in the architectural design.

The Architectural Project is addressed from the intuitive experience of ideation and architectural imagination, in the first phase (all project is born from an imaginary and intentional interpretation of reality). The origin of the architectural form as the interpretation of the established order (Fig. 6).

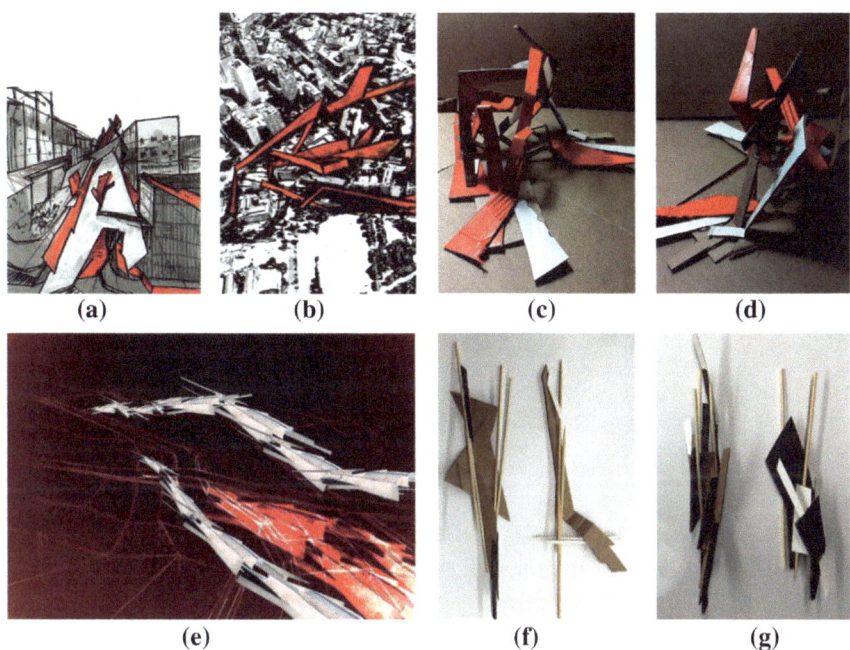

(a)   (b)   (c)   (d)

(e)   (f)   (g)

Fig. 6. a–d. Student Kattalin Bárcena Ochandiano. e–g. Student Alejandro Ruiz de la Puente. Analytic and geometric processes of experimentation with models and drawings. UPM. ETSAM. DIGA. Subject DAI 2. Course 2018–19. Professors: Raposo, Salgado, Butragueño

The actions taken are related to the sensory experience. There is a strong relationship between the capacity to imagine and the ability to abstract the complexity of the object of study (Puebla 2006). The students operate individually with the comprehensive tools acquired in previous years (Fig. 7). The drawing, collage and the model, are essential comprehensive tools for the students. Perception, expression and internalization are linked through the movement and the action of drawing. (Armstrong, Stokoe and Wilcox 1995).

The students need to find their own procedures of action and experimentation in creative processes and understand the creative approach as a game. Some authors, such as Schiller have even considered the playful momentum as the foundation of the artistic impulse (Schiller 1990).

Huizinga considers the game as a source of enchantment and fascination for all human beings, "Play casts a spell over us; it is "enchanting", "captivating". It is invested with the noblest qualities we are capable of perceiving in things: rhythm and harmony." (Huizinga 1987, 23).

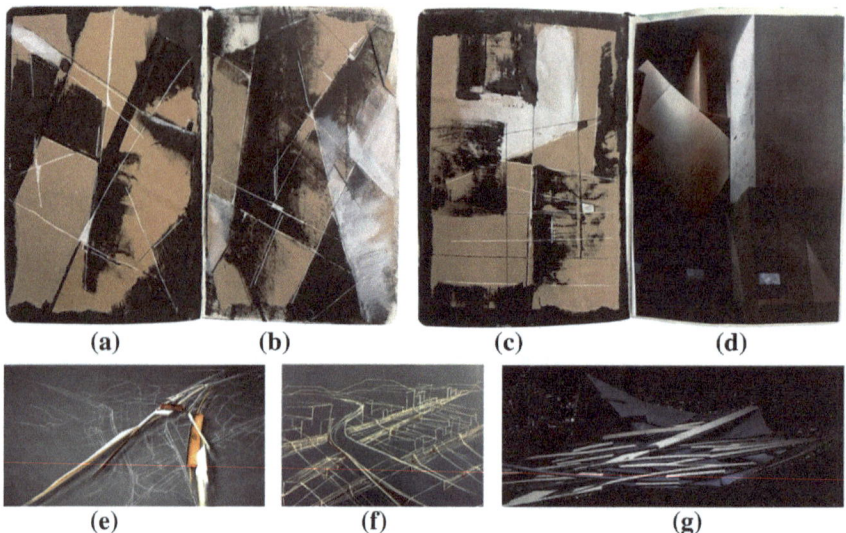

Fig. 7. a–d. Student María Mendoza Sanz. e–g. Student Marta Vacas de Miguel. Analytic and geometric processes of experimentation with models and drawings. UPM. ETSAM. DIGA. Course 2018–19. Professors: Raposo, Salgado, Butragueño

3   The Model as a Geometric and Analytic Process of the Aesthetic Qualities in Architectural Design

The model is not unique. It responds to a process of expressions that progressively acquire complexity, and gain a greater significance in the analytical stage. In that stage, the spatial qualities will be geometrically arranged and will become measurable elements, even when the models keep their expressiveness and freshness, showing a strong relationship between imagination and abstraction (Puebla 2006). We rely on models that reflect those qualities within geometrically controlled experiences.

The rigor and control of the architectural elements and their relationships are fundamental to ensure the existing ties from the generating ideas to their final configuration. This is a phase of production and architectural transformation, where the architectural form is under construction, validating quantity (versus quality), proportion and spatial metric (Fig. 8).

We are addressing the creative processes in a sequential manner, but there is a certain contradiction, given that the operating system is discontinued by multiple interferences that encourage an open, speculative and tentative process, with a non-linear quality, extremely useful in the teaching practice.

These models connect us with the spatiality and the architectural experience. Even though they have a high degree of abstraction, they introduce geometry as the sustaining element of reality.

The drawing and the architectural model require a deep knowledge of geometry. The drafting of the models in the analytical stages intend to accurately address the problems. The intentionality and scale of the models have the goal to be as effective as possible, in

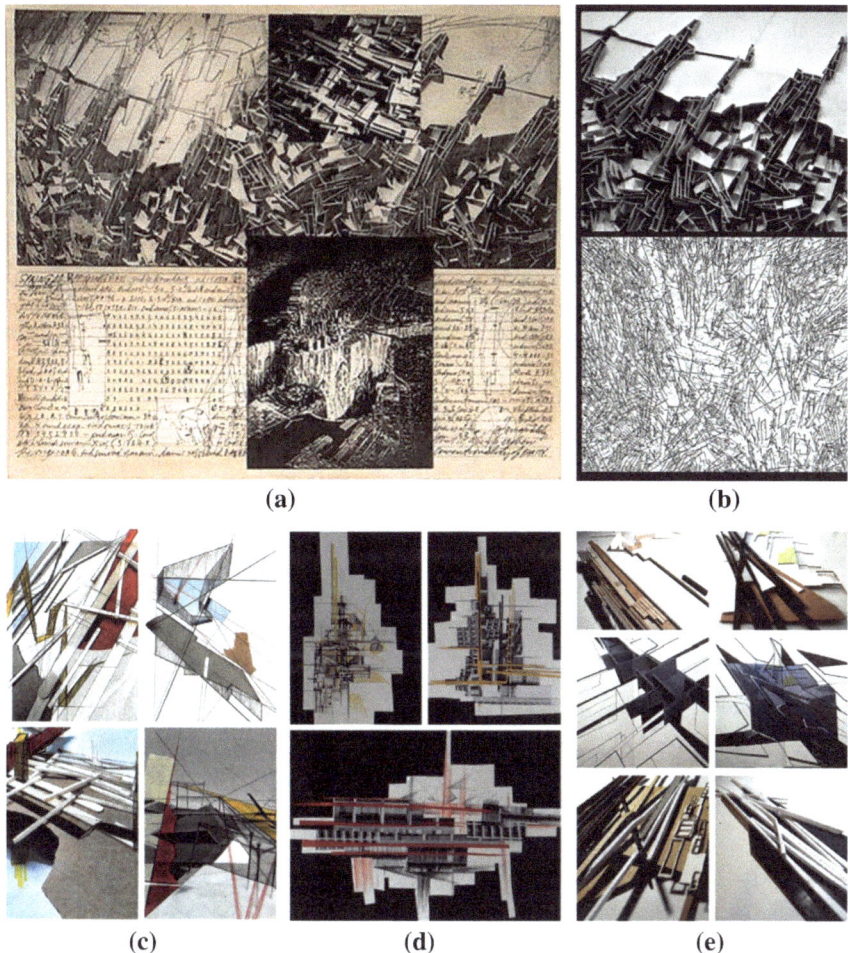

Fig. 8. a–b. Lebbeus Woods. Terrain/Tectonic landscapes, 1999. c. Student Patricia Romero Díaz. d. Student Pilar Jiménez Rubio. e. Student Sofía Prieto Gaitero. The origin of the architectural form, Imaginary phase. Urban area subject of study. UPM. ETSAM. DIGA. Subject DAI 2. Course 2017–18. Professors: Raposo, Salgado, Butragueño.

relation to the precise moment of the creative process, and their shaping may vary if the model is intended as a document for transmission of the processes or for transmission of the results.

As we move forward, we give more importance to the serial processes. Progressively, the triggers are replaced by the personal drawings of the students, that are used again as referents in a cyclical manner (Fig. 9).

In the professional field and in the teaching practice it is important to give way to criticism, as a complement to the development of the models, to establish solid criteria and continue with the analysis, to achieve a coherent solution, as a consequence of the analysis of the problems detected. The literality progressively disappears and the

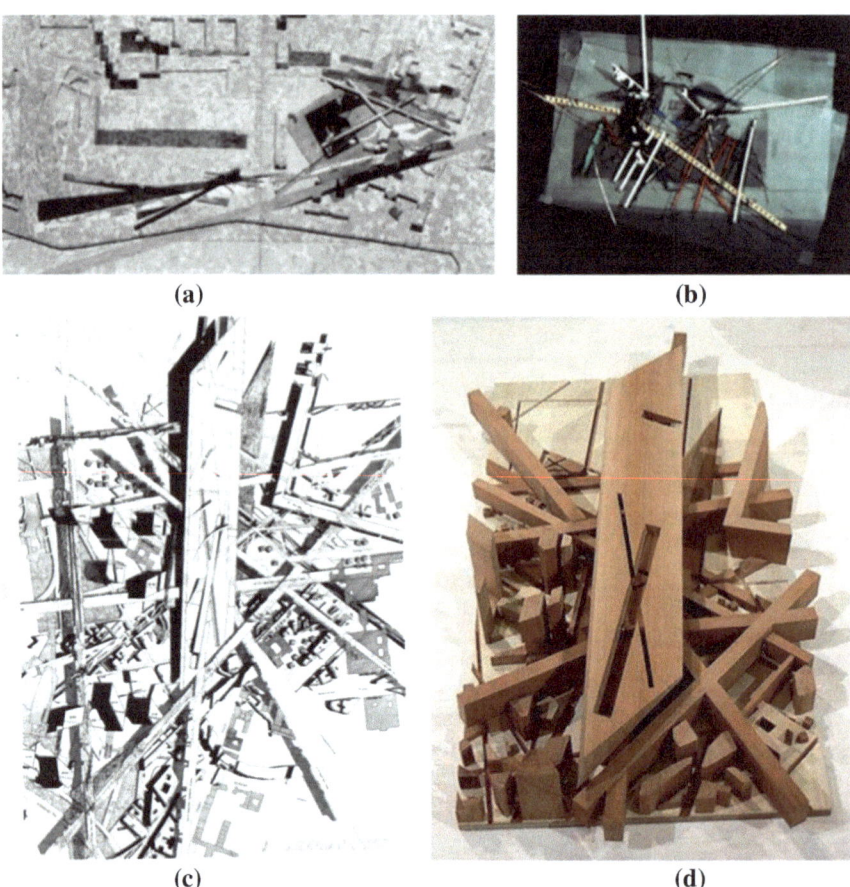

(a)

(b)

(c)

(d)

Fig. 9. a–b. Daniel Libeskind. Edge City, 1987. c–d. Daniel Libeskind. Potsdamer Platz Berlin, Project 1991, Realstadt Exhibicon Opening

discourse complies with greater maturity. The drawing and the model go hand in a close relationship, and each of them performs the task in every moment of the process.

In this phase, the students will develop skills that allow them to build the analytical processes of transformation, evolution and operational handling of the model, on its way to the design of the architectural form (quantity, proportion and spatial metric). The goal is to quantify and measure the spatial qualities explored in the previous imaginary phase and to operate with increasing geometric rigor when approaching the final solution.

4   The Model as Synthesis and Communication in Architectural Design

As architects, we design with the intention to build, and the synthesis of the design process should conclude with the materialization of the design. The model should cover the communicative phase of design, by closing the multiplicity of the initial records corresponding to experimentation and the analytics of the process. We are not referring to a scale model as a strict representation of the design, but to a model that values the essential aspects in terms of dynamics and not in terms of forms. In any case, the option to materialize as planned has always been a resource used to produce architectural models.

The size of the models may vary, although they are generally handy small-size models (Campo 2013), far from the representation. This operation eliminates the superfluous, representing a synthesis of the architectural idea in its pure state (Carazo 2016).

In this phase, the models refer to a language of interpretation and conceptualization, covering the expression, representation and interpretation, under a universal graphic code, backed up by a solid geometric construct.

According to Constant Nieuwenhuys, we can refer to the model as a resource for speculation and imagination. The specific task of the creative figures at that time, was to present a new and exciting reality, rather than representing the unsatisfactory reality that was about to disappear (Fig. 10).

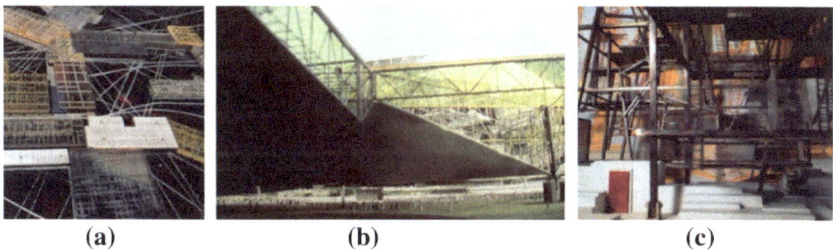

| (a) | (b) | (c) |

Fig. 10. a–c. Constant Nieuwenhuys. New Babylon. Yellow sector, 1969

This critical approach makes a revision of the analytical interventions developed in the first and second phases. Based on that, the students will draft a conclusive interpretation that meet the conceptual proposals that will materialize in conceptual models when working on formal processes, or in formal models when we are moving toward the communication of the architectural proposal. There is more interest in the spatial qualities of the proposal, that in its formal definition.

This is the most exciting phase for professionals and students, as it culminates the creative process but, in the second case, it is not usually most successful in terms of results.

5   Conclusion

When introducing the article, we stated that the drawing (considered as a general procedure of the architects to produce knowledge) and the model (subject of this study) maintain the leadership as creating and thinking tools of the architectural creation. The different procedures are part of a game. Everything is a game. As children, we played building models, and we still do models with a playful purpose. As adults we must integrate the game in the artistic expression. Building scale models is a game, an entertaining and carefree wandering, when not forcing a formalized, closed, dead, or empty solution. The focus is in the meaning, the essence, the provocation, the design as an artistic creation and as a game.

This article presents the different procedures of the architectural creation linked to the development of the model, from the teaching and professional perspective, validating the experimentation and action versus results, and establishing relationships between the model and a graphical environment of hybridization of languages (analogue and digital procedures).

As a general consideration, we can say that the models should not be directed to the production of comparable results in any of the creative phases, but rather, they should lead to a continuous experimentation of possible routes of action and significance, opting for the learning of the theory based on the active experience of its implementation. For that purpose, it is necessary to make a personal effort, supported by a relentless pursuit in the development of models, translating the findings into educational and professional procedures, and avoiding less fertile actions. This approach must always seek more meaningful connections between the drawing (graphics settings and analytical categories of the process) and the project (stages and attentions in the architectural creation) with different procedures, especially when developing models.

Given the need to unify the graphical environment, and the hybridization of languages, with both analog and digital techniques, it would be important to incorporate other procedures such as photography (as a substitute for representation), reprographic reproduction, different manual and digital techniques and the stimulation in the elaboration of models, which as we have seen, are a highly productive resource for the architectural creation.

References

Armstrong, D., Stokoe, W., Wilcox, S.: Gesture and the Nature of Language. Oxford: University Press (1995)

Bosh, E.: Prólogo. In: Berger, J. (ed.) Modos de ver. Gustavo Gili, Barcelona (2000)

Cabas, M.: La maqueta: herramienta esencial en el proceso de diseño de Richard Meier. Rev. EGA 22(29), 248–255 (2017)

Campo, A.: An idea in the palm of a hand. DOMUS (972), 10–11 (2013)

Carazo, E.: Maqueta o modelo digital. La pervivencia de un sistema. Rev. EGA (17), 30–41 (2011)

Carazo, E.: La maqueta como realidad y como representación. Breve recorrido por la maqueta de arquitectura en los 25 años de EGA. Rev. EGA 22(34), 158–171 (2018)

Carazo, E.: El componente lúdico de la maqueta de arquitectura. Notas para una explicación de su pervivencia en el tiempo. En: Echeverría, E., Castaño, E. (eds.) XVI Congreso Internacional EGA, vol. 1, pp. 609–616. Fundación General de la Universidad de Alcalá, Alcalá de Henares (Madrid) (2016)

Carazo, E., Galván, N.: Aprendiendo con maquetas. Pequeñas maquetas para el análisis de arquitectura. Rev. EGA (24), 62–71 (2014)

Chiarella, M., Redondo, E.: Unfolding architecture. Cuatro estudios de caso para una nueva propuesta pedagógica del proyecto arquitectónico a partir de geometrías plegadas generadas por medios digitales y tradicionales. Rev. EGA (18), 184–191 (2011)

Colombo, S.: Los modelos de New Babylon: del urbanismo unitario al modelo digital. Rev. EGA 22(31), 80–89 (2017)

De la Cova, M.A.: Vida de las maquetas: entre la representación y la simulación. Rev. Proyecto, Progreso y Arquitectura (Año VII) (15), 112–125 (2016). Universidad de Sevilla, Sevilla

Dollens, D.: De lo digital a lo analógico. Gustavo Gili, Barcelona (2002)

Esquinas, D.J., Zaragoza, I.: Enric Miralles y las Maquetas: Pensamientos ocultos entrecruzados y otras intuiciones. Rev. Proyecto, Progreso y Arquitectura (Año VII) (15), 112–125 (2016). Universidad de Sevilla, Sevilla

Gagel, C.: Literacy and technology: reflections and insights for technological literacy. J. Ind. Teach. Educ. **34**(3), 6–34 (1997)

Huizinga, J.: Homo ludens. Alianza Editorial, Madrid (1987)

Llopis, J.: El boceto arquitectónico en la era digital. Rev. Arquiteturarevista **9**(2), 143–152 (2013)

Miralles, E.: Mélanges. Rev. Archit. d'Aujourd'Hui **1996**(312), 68–81 (1996). Archipress & Associés, Paris

Montes, C., Alonso, M.: Las diez maquetas de la modern architecture exhibition, 1932. Rev. EGA **23**(32), 36–47 (2018)

Pareyson, L.: Estetica, teorĺa della formatività. Bompiani, Milán (1988)

Puebla, J.: Sobre la innovación expresiva del proyecto contemporáneo. Rev. EGA (11), 132–141 (2006)

Sacks, O.: Neurology and the soul. New York Rev. **37**(18), 44–50 (1990)

Schiller, F.: Cartas sobre la educación estética del hombre. Anthropos, Barcelona (1990)

Sennett, R.: El artesano. Anagrama, Barcelona (2009)

Úbeda, M.: El valor de restitución de la maqueta. Una maqueta de barro para reconstruir Arg-e-Bam. Rev. EGA **16**(18), 158–169 (2011)

Van Bruggen, C.: El Museo Guggenheim de Bilbao. Publicaciones del Museo Guggenheim, New York (1997)

Zeki, S.: Visión interior: Una investigación sobre el arte y el cerebro. A. Machado, Madrid (2005)

Influences of the Architectonic Heritage in Mackintosh's Glasgow School of Art

Gonzalo Sotelo Calvillo[1][(✉)] and María Teresa Raventós Viñas[2]

[1] ETSAM, Universidad Politécnica de Madrid, Madrid, Spain
gonzalo.sotelo@upm.es
[2] EPS, Universidad San Pablo-CEU, Alcorcón, Spain
traventos@ceu.es

Abstract. The most direct knowledge of the architectural heritage during the training of architects is produced through the drawings contained in their travel sketchbooks. This communication aims to analyze the influences that these teachings operated on Charles Rennie Mackintosh during the design process of the Glasgow School of Art.

Our objective is to transcend the formal approaches that have been made in the preceding publications to attempt to reveal the mechanisms inherited from the study of vernacular buildings in the methodology that Mackintosh used in this project.

To support the hypothesis, this essay not only analyzes the previous writings, but also the original sources of the graphic material developed by Mackintosh archived in the Glasgow School of Art, the University of Glasgow and the National Library of Ireland in Dublin.

This article proposes a circular route that begins with the education of Mackintosh in Glasgow, continues with the medieval architectures drawn in Britain and Italy, and culminates with the proposal for the educational building back to his Scottish hometown.

Keywords: Architectonical drawing · Travel · Design process

1 Introduction

This paper proposes to trace the teachings that the Scottish architect Charles Rennie Mackintosh (Glasgow, 1868—London, 1928) extracted from the study of the architectural heritage to articulate a new language. Therefore, this study analyzes the influences, collected in his travel sketchbooks, present in the design methodology of his most iconic work, the Glasgow School of Art (1896–1909).

The collection of these documents is due to the special relevance in this investigation, as they constitute not only a record of the buildings that Mackintosh visited, but they can also reveal the characteristics that aroused their interest. The design of the Glasgow School of Art is chosen for two main reasons. It configures a pioneer architecture, precedent of modern language, which unfolds at the threshold of the twentieth

© The Editor(s) (if applicable) and The Author(s), under exclusive license
to Springer Nature Switzerland AG 2020
L. Agustín-Hernández et al. (Eds.): EGA 2020, SSDI 6, pp. 530–541, 2020.
https://doi.org/10.1007/978-3-030-47983-1_47

century. Likewise, the development of this project covers more than a decade, so its extensions illustrate the evolution of the new compositional and graphic lexicon proposed by Mackintosh. An author who also stood out as a precise artist, as his various designs showed.

Hence, Mackintosh drawn a graphic journey that starts with his training in Glasgow, continues with the drawings of British vernacular buildings, reaches the medieval and Renaissance Italian architectures, and ends when he returned to design the new building for the aforementioned educational building in his hometown.

2 Methodology

As a method of analysis, this essay examined the previous studies referred to Mackintosh, especially those dedicated to his travel sketchbooks and the Glasgow School of Art in particular. To try to transcend the merely formal approaches, this communication studied the original sources of graphic material developed by Mackintosh, archived in the Glasgow School of Art, the University of Glasgow and the National Library of Ireland in Dublin.

Therefore, this paper analyzes the relationships between the drawings by Mackintosh, where he studied the constructions of historical heritage, and the designs of his proposal for the Glasgow School of Art, which illustrate its various phases. Consequently, we articulate a comparison that tries to reveal the existing connections between the teachings and influences received from the study of historical architecture and the different production stages of this Scottish architect.

3 Educational Precedent

At the end of the 19th century, Glasgow was Britain's second and Europe's sixth largest city with more than 650,000 inhabitants, whose relevance laid in heavy industry. It was an example of a Victorian city built of sandstone and stylistically conservative. Glasgow had overcome the austerity of Greek neoclassicism to embrace a frivolous imitation of the Nordic Renaissance.

In this city, Mackintosh's education was forged between practices in architectural studies and night painting and drawing classes at the Glasgow School of Art. He combined these teachings with his work, first as an apprentice with the architect John Hutchinson between 1884 and 1889, and finally as a draftsman in the office run by John Keppie and John Honeyman, during the next fourteen. Equally, Honeyman was a restorer accustomed to work on ecclesiastical buildings that included the Glasgow Cathedral, built next to where Mackintosh was brought up. A unornamented building dated from the 12th and 13th centuries, most appealed to Augustus Pugin, and which presents the unique particularity in Great Britain of its double-storied east façade, which Mackintosh detailed in his watercolor in 1890 (Fig. 1).

Beyond the affinities that James Macaulay (in Buchanan 1989, p. 142) found between the cross section of this Cathedral and the west wing of the Glasgow School of Art, it is clear that this temple served as a starting point for the proposal that Mackintosh presented in 1903 for the Liverpool Cathedral competition (Fig. 2). This project illustrates a latent

Fig. 1. Charles Rennie Mackintosh. Glasgow Cathedral at Sunset. 1890. Watercolor on paper, 39.4 × 28.3 cm. Source: Glasgow School of Art

drive that guided him to a morphological language where he produced a rereading of Gothic art.

Mackintosh was educated under the influence of Arts & Crafts design principles and the theories defended by Morris and Ruskin. Therefore, he inherited a spirit of respect for medieval architectures that led him to visit buildings that conform the historical heritage. Ruskin proposed a return to the production processes of the late Middle Ages and held

Fig. 2. Charles Rennie Mackintosh. Competition design for Liverpool Cathedral. South elevation. 1902. Pencil on paper, 61.1 × 74.8 cm. Source: Glasgow School of Art

an unhistorical point of view that involved a return to solutions of the twelfth century, becoming a champion of the neo-Gothic revival.

Among his influences, we also highlight the writings of W. R. Lethaby, especially Architecture, Mysticism and Myth (1892), where Mackintosh found confirmation of the thought he was cementing. Lethaby emphasized the existence of composition order in the old buildings, where he found forms and materials appropriate for their function (Billcliffe 1977, p. 10). This is the reason Mackintosh began to study vernacular architectures, but without trying to imitate them through a historical mimesis, because he tried to keep up affinities with the past, but not being dependent on it.

Additionally, Mackintosh had traveled through various towns in Scotland and England, filling in sketchbooks with drawings of the buildings that caught his attention, but he longed to be able to make a trip of greater depth. His dream came true when he was the winner of the Alexander Thomson Travelling Studentship in 1890, which allowed him to visit the main Italian towns between April 5 and July 7 1891, which completed his education.

4 Travel Notes

For a Victorian architect, the sketchbooks achieved both an educational and a practical function. However, Billcliffe (1977, p. 9) considered that Mackintosh performed them with a double purpose: to serve as thinking loud, and to record what he had visited. Mackintosh called these architectural sketches 'jottings' (Grogan 2002, p. vii) and assigned them a private role, as they were conceived as a record of the elements that had interested him until it constituted a reference book that could be incorporated into future projects. These notes formed a historical vocabulary that should not interfere with the new architectural language that he wanted to create.

Drawing teachers such as Alexander McGibbon or Francis Newbery influenced Mackintosh. Newbery encouraged their students to abandon the mechanical copy and to trace more personal and analytical drawings, and demonstrated his mastery in his strokes, the product of a selective and intentional gaze.

His travel sketchbooks are a reflection of his concerns oriented by his education and serve as a means of approaching his most reserved facets, so they could be translated as "a kind of unexpected autobiography" (Maestre 2015, p. 291).

In his first travel sketchbooks, the young Mackintosh explored the possibilities he found within historical architecture to find his own style from a collection of vernacular British constructions where castles (Fig. 6) and religious buildings predominate, such as the Merriott Church in Somerset (Fig. 3).

Fig. 3. Charles Rennie Mackintosh. Merriott Church, Somerset. 1895. Pencil on paper, 36.2 × 12.7 cm. Source: University of Glasgow Art Collections

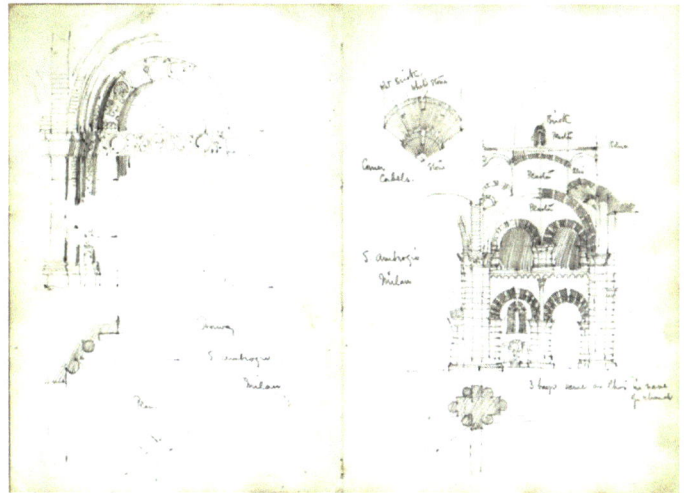

Fig. 4. Charles Rennie Mackintosh. Sant'Ambrogio, Milan. 1891. Pencil on paper, 23 × 16 cm. Source: Glasgow School of Art

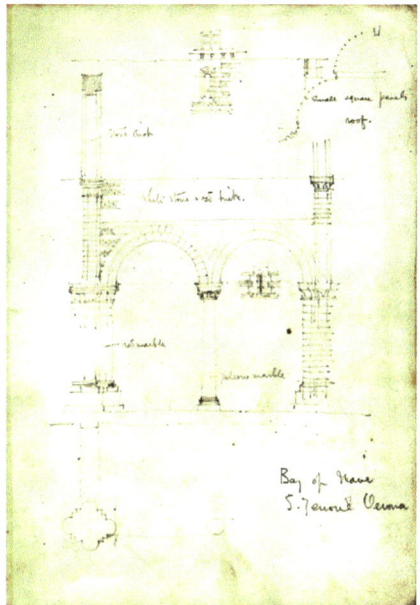

Fig. 5. Charles Rennie Mackintosh. Bay of nave of San Zeno Maggiore, Verona. 1891. Pencil on paper, 23 × 16 cm. Source: Glasgow School of Art

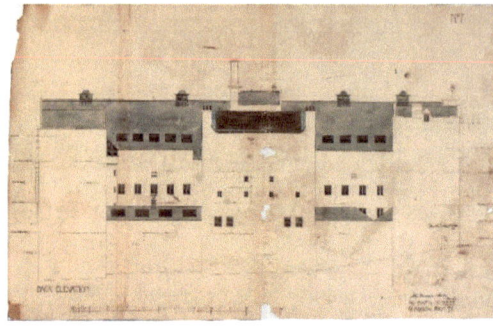

Fig. 6. Charles Rennie Mackintosh. The Castle, Holy Island (above). 1901. Pencil on paper, 20.3 × 26 cm. Glasgow School of Art, Glasgow. 1897. South Elevation. Ink on paper, 56.7 × 88.9 cm. Source: Glasgow School of Art

During his trip to Italy Mackintosh showed a closer vision to Ruskin than to Alexander Thompson, so he was especially interested in the Byzantine or Romanesque constructions instead of the Renaissance ones, such as the Basilica of Sant'Ambrogio in Milan (Fig. 4).

Likewise, Mackintosh wrote down in his sketchbooks references on the arches of the Monastery of San Zeno (Fig. 5), a building that he described as a beautiful example of Italian Gothic in brick and stone work (Billcliffe 1977, p. 9). These drawings seem to shape sketches of the constructive solutions that Mackintosh resumed several years later in the polished design for the vaulted spaces of the Glasgow School of Art. Although Ruskin was considered one of Mackintosh's favorite authors, this circumstance did not prevent him from disagreeing with the principles contained in writings such as The Stones of Venice (1853), or Ruskin's dislike for Perpendicular and Elizabethan Gothic styles, and post fifteenth century Italian architecture, especially the work of Palladio (Robertson 1990, p. 21).

These sketches served to lay down the skills he used during the next decade, although he did not draw the main historical buildings. With the expansion of photographic reproductions, Mackintosh could have become familiar with the most iconic constructions from publications. This is the reason Grogan uses (2002, p. 51) to justify the absence of graphic records of the Duomo in Florence or the Campanile at Pisa in his travel sketchbooks.

Billcliffe (1977, p. 10) underlined that in Mackintosh's drawings there are no sketches of modern architecture, although he noted that he was aware of contemporary constructions, likewise other authors (Pevsner 1998, Buchanan 1989) have established connections between elements of his architecture and the designs of Voysey or MacLaren.

5 The Glasgow School of Art

The project submitted by the studio where Mackintosh worked was the winner in 1896 of the competition for the design of the new Glasgow School of Art. The conditions of the competition detailed the submit should be illustrated by a plan of each roof, three elevations—north, east and south—, a longitudinal section and two cross sections in outline only, specifying that these drawings must be sent without shading, etchings or perspectives (Buchanan 1989, p. 205). In addition to the program of spaces, the project was conditioned by an uneven site and the impossibility of having any lights in the lower floors of the south elevation, due to its proximity to the adjacent buildings. This last conditioner forced Mackintosh to draw a composition for this façade that seems to come from his sketches of the Holy Island Castle (Fig. 6).

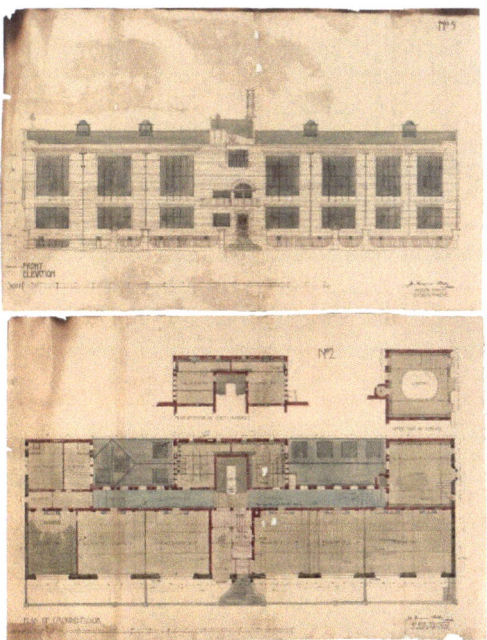

Fig. 7. Charles Rennie Mackintosh. Glasgow School of Art, Glasgow. 1897. Front Elevation (above). Ink on paper, 52.2 × 86 cm. Plan of Ground Floor. Ink on paper, 56.7 × 86 cm. Source: Glasgow School of Art

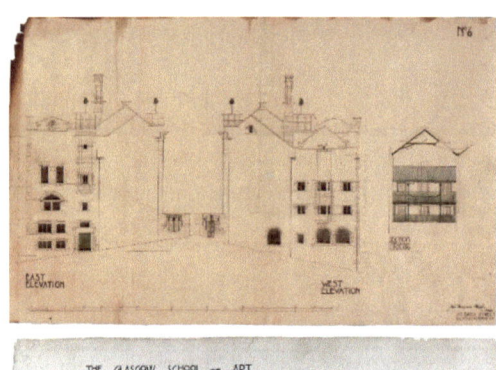

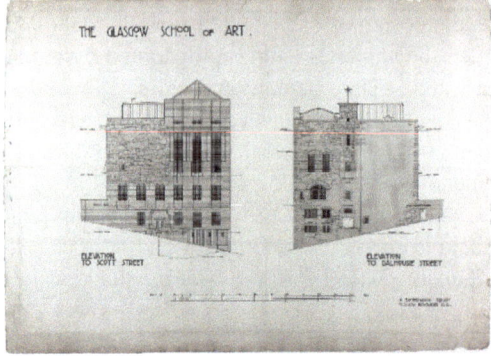

Fig. 8. Charles Rennie Mackintosh. Glasgow School of Art, Glasgow. 1897, 1910. East and west elevations (above). Ink on paper, 56.7 × 93.3 cm. East and West Elevations in the 1910 version. Ink on paper, 68.6 × 96.2 cm. Source: Glasgow School of Art

Macaulay (1993) argued that, although the functional structure of the Honeyman and Keppie study is not known, Mackintosh was allowed to develop the design concept and project details. Thus, Mackintosh could be the author of the winning proposal for the new School of Art. He evolved a building that not only introduces a formal simplicity that anticipates rationalism, but also integrates a symbolic layer of Gaelic and Celtic references, as well as an influence of Japanese engravings on the structural solutions for the library and the western tower. A combination that makes his architecture approaches Pre-Functionalism.

With a heterodox composition, Mackintosh articulated a serene and massive main façade, where he opened large glazed surfaces to allow an illumination of classrooms and workshops from north, as a response to the needs of the occupants and the functionality of the building. The spaces are orchestrated by various qualities of light, taking advantage of the uneven roof sections and the orographic qualities of the site superimposing the slope of the land on the cross section.

In the north elevation (Fig. 7) he presents a superb entrance that underlines its asymmetry with the vertical direction of a fortress-like polygonal tower, where the echoes of the sketches Mackintosh draw in 1895 of the Merriott Church in Somerset (Fig. 3) can be found (Buchanan 1989, p. 81); likewise, the precedent of the design of the windows can come from the drawings of houses in Lyme Regis and Chipping Campden (Billcliffe

1977, pp. 10, 32, 44). The steel bars of vegetable motifs that adorn the carpentry of the classrooms assist the cleaning of the windows, so they show a blend of functionality and decoration (Pevsner 1998, p. 14). This elevation composition conforms a careful balance between simplicity and monumentality to this unique building, arranged on a severe rectangular order that structures the various spaces on the plan.

When approaching the extension of the building to the west between 1906 and 1909, Mackintosh was already a partner of the architect's office where he worked, which allowed him more autonomy that is evidenced in the drawings for his new proposal by showing features of greater maturity and security. Features that are exemplified in the new solution for this elevation, which abandons the medievalist scheme of balanced asymmetry similar to the east façade towards compositions closer to rationalism (Fig. 8). However, there are authors like Macaulay (1993) who find a great similarity between the solution proposed by Mackintosh for this west elevation and the Bristol Central Library designed in 1906 by Charles Holden.

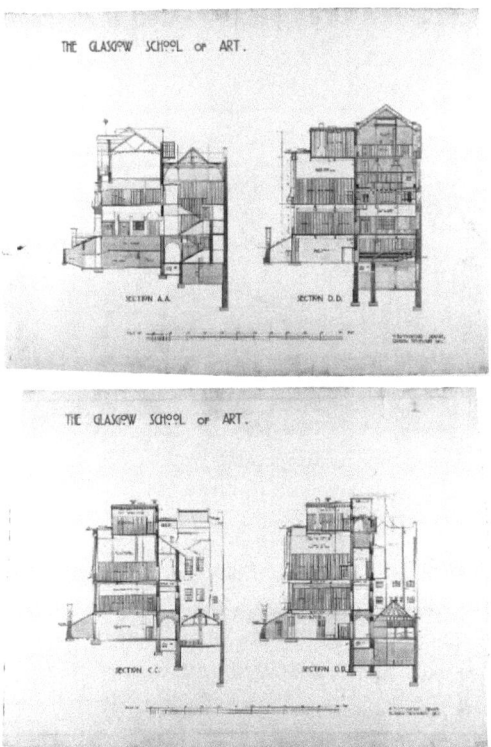

Fig. 9. Charles Rennie Mackintosh. Glasgow School of Art, Glasgow. 1910. Cross Sections. Ink on paper, 69.2 × 92.7 cm. Source: Glasgow School of Art

Mackintosh did not forget the lessons learned from the study of architectural heritage, nor did he hide the inheritance in his work, because "behind every style of architecture there is an earlier style in which the germ of every form is to be found" (Robertson

1990, p. 203). Although many studies have found direct formal relationships between contemporary architectures and the Glasgow School of Art facades, there are interior rooms that seem to show a debt with much older constructions. In this way, the gravid entrance hall forms a dark vaulted space reminiscent of Piranesi prisons. On the other hand, the series of brick arches that make up the groin vaults of the loggia on the second floor, seem to translate the lessons obtained from the Lombard Romanesque he drew when travelling to Italy, as seen in the cross sections of the project (Fig. 9).

6   Conclusions

After having placed it in its historical, geographical and formative context, Mackintosh's graphic production is revealed as a suggestive object of study. I n his way, the teachings that the Scottish architect extracted from the analysis of the classical architectures he visited during his travels armed with sketchbooks that accompanied him throughout his career until generating a vital record could be traced and valued. Mackintosh seems to rely heavily on ancient sources and does not shy away from the heritage of vernacular architecture.

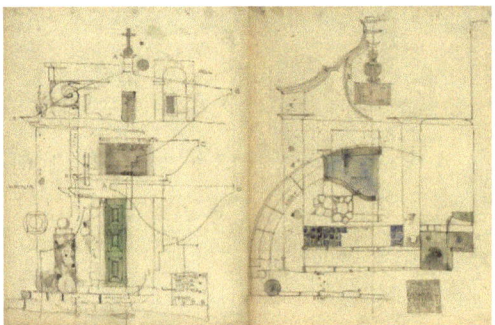

Fig. 10. Charles Rennie Mackintosh. Shrine and Well in Sintra, Portugal. 1908. Pencil and watercolor on paper, 26 × 20.3 cm. Source: University of Glasgow Art Collections

During the design process of the Glasgow School of Art, we have found a complexity of overlapping references that transcends the formalist imitation to generate a new constructive language that approximates Mackintosh to rationalism. It also demonstrates a formal evolution in the decade that separates Scottish picturesqueness in the east elevation, to solutions that denote more mature and sophisticated compositions in the west elevation or the library drawings.

These references are juxtaposed with a holistic principle that integrates each part within a harmonic ensemble. A methodology that is also displayed in the expertise displayed in the drawings that illustrate his trips in the early twentieth century, which are more difficult to decipher as they evolve with constant overlaps and changing scales, as seen in the Holy Island sketches (Fig. 6) or in Sintra drawings (Fig. 10). Therefore, both in his travel drawings and in the Glasgow School of Art proposal, Mackintosh operated an orderly overlap of the components of the architectural heritage that aroused his interest. This Scottish architect considered shameless to pretend to live in another century by rejecting the revival, since he defended that the work of an architect cannot be limited to the references of tradition and the past without a purpose.

References

Billcliffe, R.: Architectural Sketches & Flower Drawings by Charles Rennie Mackintosh. Academy Editions, London (1977)

Buchanan, W. (ed.): Mackintosh's Masterwork: The Glasgow School of Art. Ernst & Sohn, Berlin (1989)

Grogan, E.: Beginnings: Charles Rennie Mackintosh's Early Sketches. Architectural Press, Oxford (2002)

Macaulay, J.: Glasgow School o f Art: Charles Rennie Mackintosh. Phaidon, London (1993)

Galindo, C.E.M.: Cuadernos de viaje. El apunte íntimo y personal del arquitecto. Doctoral Thesis. Universidad CEU San Pablo, Madrid (2015)

Pevsner, N.: Charles R. Mackintosh. Canal Éditions, Paris (1998)

Robertson, P. (ed.): Charles Rennie Mackintosh: The Architectural Papers. MIT Press, Cambridge (1990)

Organising Complexity: A Reflection on Parametric Design

Ángel J. Fernández-Álvarez[✉]

Departamento de Expresión Gráfica Arquitectónica, Universidad de A Coruña, A Coruña, Galicia, Spain
angel.fernandez.alvarez@udc.es

Abstract. This paper reflects on the role of parametric design in architecture carried out with digital tools.

The digital revolution has profoundly altered the architectural discourse by introducing debates on theory and design that are based on ideas arising from the intersection between art, science and technology.

Parametric design has become a powerful tool for organising complexity and this creates the need for critical reflection on its current and future influence on architectural projects.

The mechanisms for bringing together the information processing power of computers and the intuitive skills of designers poses new challenges that require the inclusion of computational thinking in the training of future professionals.

A redefinition of the relationships between architecture, the digital technologies and production and manufacturing techniques is needed so that we can have intelligent design thinking that allows us to properly organise the complexity of the activity involved in the architectural project.

Keywords: Digital architecture · Parametric design · Complexity

1 Introduction

Architectural discourse has been profoundly altered by the so-called "digital revolution" and simultaneously we are seeing new debates arising in relation to theory and design that are based on ideas coming from the intersection between art, science and technology.

The success and growth of parametric and algorithmic design techniques and strategies (Terzidis 2006, Sakamoto and Ferré 2008; Jabi 2013; Aiello 2014) has led to an identity crisis for the discipline and new considerations regarding the authorship of architectural works.

This situation provokes the need for critical reflection on the role of parametric design in the practice of architecture carried out with digital tools.

L. Agustín-Hernández et al. (Eds.): EGA 2020, SSDI 6, pp. 542–549, 2020.
https://doi.org/10.1007/978-3-030-47983-1_48

2   Towards Generative and Relational Design

As a design tool or strategy, parametricism defines relationships between elements by assigning values in order to organise and control complexity, so its underlying principles are connectivity and interrelation (Dunn 2012). It is a relational design methodology based on the consideration of systems instead of objects. The designer in some way becomes an "editor" of relations between the different elements and later selects the results obtained based on different criteria that can be linked to aesthetics, functionality, finance, interaction, etc.

The parametric approach was one of the first operational concepts in the field of computer-aided design. In 1963, Ivan Sutherland proposed the first graphic user interface in his famous doctoral thesis: Sketchpad: A Man-machine Graphical Communications System. This system allowed him to draw with the computer and, at the same time, apply changes to the design parametrically. Sutherland himself pointed out the possibility of establishing a kind of "conversation" between the user and the computer through the use of graphic information, as opposed to communication through written instructions as had been carried out previously.

The use of code languages and scripting techniques (Reas et al. 2010) has become a very important project tool that uses construction modelling, geometric programming, structural optimisation, environmental simulation, genetic algorithms and digital manu-facturing techniques (García Alvarado 2013). The potential of software and the power of hardware make it possible to explore design alternatives very rapidly, while progress in digital manufacturing tools allow us to switch from conventional artisanal and industrial models to a new digital model in which variations can be made at no extra cost.

What Patrik Schumacher in his 2008 manifesto calls a "sense of organised complexi-ty" is in addition to the strong emphasis on differentiation. The complexity of the project is addressed in a similar way to the strategies developed by natural systems, the final form being the result of the interaction between forces according to pre-established laws (Schumacher 2008). So, in contrast to the modern mechanistic concept of space, para-metricism considers the notion of "field" with the dynamic vision of a changing reality based on trends, flows and gradients in which variation and deformation are regarded as organised information structures.

The multidisciplinary nature inherent to parametric design adds great flexibility to the design process and transforms it into a collective and collaborative task that calls into question the role of authorship in the architectural design process. The designer goes from being the creator/generator of the form to becoming the editor/programmer of the processes and systems, with the task of defining some initial starting conditions and creatively selecting an appropriate end result (Fernández-Álvarez 2014).

One key aspect is precisely the need to define existing relationships through the use of formal notations that require the designer to have some experience, but that also offer the advantage of quickly being able to explore new solutions with great freedom and a certain amount of randomness.

In this context, the parametric tool may be seen more as a production system than a representational construct and, although some consider it to be a true style (Schumacher 2009), others consider it simply to be a design methodology. In reality, Schumacher provides a style idea as a "design and research program" following methodological criteria taken from Imre Lakatos' theories of the philosophy of science and also strongly taking into account the communicative dimension of architecture.

In contrast to the widespread idea of programmed automation, parametric design implies an intentionality, a user-defined logic that transforms into a conscious digital design. This logic lies in the ability to achieve a suitable definition of the problem through an abstract diagram and its correct mathematical description. The resolution of complex projects is where the parametric tool allows the integration of multiple variables and the realisation of successive iterations that allow us to obtain versions that evolve towards the most suitable solution, in a shorter time and without losing the previous modifications.

In contrast to the primacy of visualisation in conventional design methods, parametricism involves the proper definition of a system of relations, the parametric software being responsible for the graphic results of each proposal. This situation, in which the planner becomes the designer of systems and processes, had already been foreseen in the era of the pioneers of digital in architecture, such as Gordon Pask when he proposed the idea that "design is control of control" (Pask 1969).

Finally, we must also consider the capacity offered by the new tools to mediate with the tectonic, establishing a link between information and matter. The processes involved in searching for form are structured around three principles (Oxman and Oxman 2014): the differentiation processes characteristic of natural systems, [in]formed or integrated tectonics and continuity from the design phase through to the production phase by including the logic of the material in the parametric approach.

3   Towards Design Intelligence

The possibility of reusing code modules, the concept of "open work" that allows for collective and participatory knowledge strategies, a conceptual change from the object to the process and the exploration of the unexpected, caused by the introduction of randomness, make parametric design a powerful tool for organising complexity with a huge current and future influence on architecture projects and the configuration of an intelligent design thinking.

Immersed in what some call a "post-digital" era or what Mario Carpo has de-scribed as the "Second Digital Turn", there is a need for critical reflection on the phenomenon that goes beyond the mere academic review of technological tools, always in continuous evolution. These have become true tools for "thinking" rather than tools for "making" (Carpo 2017).

From the viewpoint of theoretical criticism, it is also necessary to highlight the importance of cultural approaches to the problem from a cross-cutting and multidisciplinary perspective in order to take into consideration the social, ethical, political and philosophical implications that any technology involves. The popularisation of parametric generative systems (Agkathidis 2016) allows us to contemplate digital design, process and production technologies not only as tools but also as ways of thinking. Computational thinking (Wing 2006), intelligent digital design or design intelligence become emerging ideas, offering new ways of seeing, thinking and doing architecture.

In terms of existing examples, we can think about the introduction of cybernetic theory into architecture during the 1960s and 1970s, when we had the curious situation of advanced digital thinking but without the technological infrastructure required to implement it, proving once again that ideas are what are truly innovative and disruptive, rather than technologies.

The dissemination and democratisation of parametric design and BIM methodologies have contributed to the development of a growing trend towards the consideration of what is known as "computational design", consisting of the development of a certain mental model that allows us to organise thinking in architectural design processes developed with digital tools. According to the definition by Jeannette Wing (2011), "computational thinking" consists of "the thought processes involved in formulating problems and their solutions so that the solutions are represented in a form that can be effectively carried out by an information-processing agent".

This involves developing skills that allow us to harness the power of computing for the study, analysis and resolution of complex problems. To do this, it is necessary to rely on the profound knowledge of the underlying principles behind the different tools rather than on practical training in the use of the different commands for the software programs. As stated by Senske (2011), the transfer of knowledge is one of the hallmarks of this type of approach to design problems, as it promotes the application of that knowledge outside of its own learning contexts.

Senske deems it a priority to consider this transfer as an objective in the educational programmes of schools of architecture and for this he suggests three basic conditions in the digital training of future professionals: taking different contexts of using software into consideration, promoting self-discovery (perfectly compatible with the well-known phenomenon of self-learning linked to these technologies) and introducing meta-cognitive approaches that allow for an active learning experience. This is intended to highlight the computational aspects of design so that students understand the internal processes that underlie the creation of forms through parametric strategies. To do this, the writing of code is enhanced through scripting languages such as Processing or Python and the generation of compositional rules following the logic of algorithmic design. Programming skills allow designers to interact with other media, thereby expanding the repertoire of operational tools that can be used in a project.

The ultimate goal is to achieve a conscious and flexible use of digital design, overcoming the limitations of simply giving training in the routine use of different programs. The aim is to ensure that the concepts of computational thinking can be applied to any type of software and their different technological evolutions and new versions. This same issue had already been raised by Stan Allen (2005) when he advocated "a relaxed, pragmatic, inventive and direct approach" to digital technologies, emphasising the need to promote the user's digital "astuteness" that, when "moving beyond the logics of visualisation", would allow new design potentials to be found.

4   Towards a New Concept of Authorship

The simplifying term "digital revolution" is now referred to as more of a "digital turn", an expression that suggests the idea of introducing digital technology into architectural design processes that began at the beginning of the 1990s and still continues today. In the initial phase, which we can call the phase of the "pioneers", the possibility of an electronic space as an alternative to the physical space was even considered, in which bits would take the place of the traditional materials with which, until that time, architecture had been built (bits not bricks).

As outlined above, the emergence of a new generation of 3D modelling software along with the progressive evolution of hardware power has led to significant changes in the way we use computer tools to design and produce architecture. For Dunn (2012), the role of the computer branches off in two directions. First, it has the function of improving productivity by becoming an advanced tool for designing complex forms and a powerful interface for the proper visualisation of the design processes. Second, it offers the chance to manipulate and work in the very core of the act of devising through programming using scripts and algorithms.

Terzidis (2006) distinguishes between "computerization" (which relates to the first option) and "computation" (more linked to the second). It is in this second area where we see more than the mere digital interpretation of the ideas in the planner's mind, by transferring the designer's intentions to the algorithmic process but taking advantage of the machine's capabilities to explore and offer a set of alternative new solutions.

It should be noted that the key lies in obtaining a generic, open and parametric notation that faithfully reflects the conditions of the process. However, this working method raises the possibility that ownership may be shared by the different agents participating in the design process: architects, owners, builders, manufacturers, users, etc. To this we need to add the potential possibilities for the materialisation of the design provided by the new 3D manufacturing technologies, together with the introduction of concepts taken from the theory of systems and the sciences of complexity (energy considerations, fractals, indetermination, chaos theory, fuzzy logic, emergency, non-linearity, etc.), which adapt perfectly to the way computers work.

It is also worth highlighting the impact of what could be called a symbolised "participatory turn" on the advances of Web 2.0 and 3.0 (semantic web), which allows us to go beyond the limited zoning of traditional spaces. The new technological developments facilitate hybridisations and interactions between physical and virtual spaces with user participation in the configuration of "augmented" spaces. These open up the possibility of introducing human experience and communication into a real-time performative design process.

The philosophy of free software and the possibilities of collaborative working offered by the Internet also open up the path towards a hypothetical open source architecture, with designs that can be freely downloaded and edited. This basic idea is reflected in Alastair Parvin's WikiHouse project, whose objective is the democratisation of building at the same time as, through simplification, also pursuing sustainability principles.

The utopian idea of architecture with a Creative Commons licence is based on the creation of libraries of digital models that can be downloaded, manufactured and assembled in a simple and low-cost way, adapting to the needs of the users. This concept of "participatory authorship", somehow implicit in the development of the BIM methodology and in the implementation of parametric strategies, calls into question the authorship model that has been a feature of the discipline since the Renaissance era. This is at a time when the dominant trend is moving towards the adoption of collaborative strategies.

As stated by Carpo, we are entering a "post-digital" or a "second digital turn" phase characterised by the consolidation of the use of digital tools in design practice, by a moving of the boundaries between previously very clearly defined disciplines and that now see their boundaries becoming more blurred, promoting multidisciplinary hybridisations and actions that become characteristic elements of the new situation.

In these trends we can see the influence of what is happening in the field of artistic experimentation, with a much more dynamic and innovative model of reflection and research that provides innovative and disruptive ideas and cutting-edge architectural concepts. Concepts therefore emerge that characterise post-digital aesthetics such as, for example, the "bastardisation" of technology advocated by John Richards, researcher in digital electronic music. This consists of "forcing a system in to a state in which it was never intended, or appropriating something for a use other than what it was initially designed for" (Richards 2006).

Along with experience from the world of art, it is worth highlighting the influence of the DIY (Do It Yourself) postulates that characterise the most advanced and experimental architecture and that support a relaxed and free use of software based on the reuse of code and also of physical objects in a contemporary recovery of the concept of "ready-made".

It is this notion of "digital DIY", linked to the idea of heterogeneity, which in some way characterises the "post-digital" concept. Permanence, change, integration and separation coexist simultaneously with the valuing of uncertainty, indeterminacy and ambiguity, aspects that can only be properly described with the communicative metaphors from the abstract logic underlying the digital world. This gives rise to such interesting suggestions as experimentation with a so-called "aesthetics of error". This takes advantage of technological failures which it uses as inputs for exploring new design possibilities, freely and without prejudice expanding the conventional functions and uses of the software (Cascone 2000).

The concept of authorship becomes a key aspect in the theoretical reflection on digital architecture. Mass customisation, which contrasts with the standardisation advocated by modernity, and digital variability at no extra cost link the current situation to the artisanal model that existed prior to the Industrial Revolution, suggesting a new paradigm of "digital craftsmanship". This links to the ideas of the Maker Movement based on the participation and democratisation of production and design (Anderson 2012). This situation may lead to resistance in the discipline of architecture, characterised as it is by a "strong" authorship concept and with a strict sense of control of the processes. Distributed authorship, made possible by the new design tools and strategies, represents a break from the tradition of separating the design and the material realisation of the project, and therefore with the concept of individual authorship developed in the era of Renaissance humanism (Carpo 2011).

For Carpo, a new type of "authorship" is created which he classifies as "generic" (Carpo 2009). This connects, for example, with the way of working in the great works of medieval Gothic architecture. However, this apparent pre-modern optimism is marred by a negative vision of the resistance that may appear to these new forms of "diffuse" authorship, calling into question the very definition of the discipline's professional framework that has existed since it first appeared in the 15th century.

5 Conclusion

All of the foregoing means that we are seeing a disciplinary crisis of unpredictable consequences as a result of the idea, frightening from the perspective of the humanist paradigm, that the end product of the design process is no longer preconceived in the mind of the designer but is instead obtained through a complex collaboration mechanism involving the information processing power of computers and the naive and intuitive skill of the designer.

This threat to the traditional idea of authorship leaves us in a difficult place, half-way between disruption and nostalgia (Picon 2019), which requires urgent theoretical and critical reflection and new approaches to the design of academic courses so that they include computational thinking in the training of designers.

The new teaching plans should introduce strategies and principles from computer science in order to consciously apply them to the analysis and resolution of complex design problems while promoting the transfer of knowledge between various contexts (Senske 2011).

Together with parametric and generative design, the constant advances and developments in the fields of Artificial Intelligence, Machine Learning and Big Data, and the consequent emergence of applications linked to architectural design, will lead to new and exciting challenges that must be faced with realism and determination. The final objective is an ambitious redefinition of the relationship existing between architecture, the digital technologies and production and manufacturing techniques. This is required in order to create intelligent design thinking that allows us to properly organise the complexity of the activity involved in the architectural project.

References

Agkathidis, A.: Generative Design: Form-Finding Techniques in Architecture. Laurence King Publishing, London (2016)

Aiello, C. (ed.): eVolo 06: Digital and Parametric Architecture. eVolo Publishing Company, New York (2014)

Allen, S.: The digital complex. In: LOG, vol. 5, pp. 93–99. Anyone Corporation, New York (2005). (Spring/Summer). Spanish version: "El complejo digital: diez años después", in Ortega, Ll. (ed.) La digitalización toma el mando. Editorial Gustavo Gili, Barcelona, pp. 159–168 (2009)

Anderson, C.: Makers: The New Industrial Revolution. Crown Business, New York (2012)

Carpo, M.: Revolución 2.0. El fin de la autoría humanista. Rev. Arquitectura Viva **124**, 19–25 (2009)

Carpo, M.: The Alphabet and the Algorithm, p. X. The MIT Press, Cambridge (2011)

Carpo, M.: The Second Digital Turn: Design Beyond Intelligence. The MIT Press, Cambridge (2017)

Cascone, K.: The aesthetics of failure: post-digital tendencies in contemporary computer music. Comput. Music J. **24**(4), 12–13 (2000). Winter 2000

Dunn, N.: Proyecto y construcción digital en arquitectura. Blume, Barcelona (2012)

Fernández Álvarez, A.J.: Cabalgando la nube. Información y representación arquitectónica en la era post-digital. EGE: Rev. de Expresión Gráfica en la Edificación **8**, 95–105 (2014)

García Alvarado, R., Lyon Gottlieb, A.: Diseño paramétrico en Arquitectura; métodos, técnicas y aplicaciones. ARQUISUR Rev. Año **3**(3), 17–27 (2013)

Jabi, W.: Parametric Design for Architecture. Laurence King, London (2013)

Oxman, R., Oxman, R.: Theories of the Digital in Architecture. Routledge, New York (2014)

Pask, G.: The architectural relevance of cybernetics. In: Architectural Design, vol. 7(6), pp. 494–496. Wiley, London (1969). Spanish version: "La significación arquitectónica de la cibernética", in Ortega, Ll. (ed.): La digitalización toma el mando. Editorial Gustavo Gili, Barcelona, pp.15–28 (2009)

Picon, A.: Digital fabrication, between disruption and nostalgia. In: Ahrens, Ch., Sprecher, A. (eds.) Instabilities and Potentialities: Notes on the Nature of Knowledge in Digital Architecture, pp. 221–236. Routledge, London (2019)

Reas, C., McWilliams, C.: LUST: FORM + CODE in Design, Art, and Architecture. Princeton Architectural Press, New York (2010)

Richards, J.: 32 kg: performance systems for a post-digital age. In: Proceedings of the 2006 International Conference on New Interfaces for Musical Expression (NIME06), Paris, p. 283 (2006)

Sakamoto, T., Ferré, A. (eds.): From Control to Design: Parametric/Algorithmic Architecture. Actar-D, Barcelona (2008)

Schumacher, P.: Parametricism as Style-Parametricist Manifesto, presented and discussed in the Dark Side Club1, London 2008. 11th Architecture Biennale, Venice (2008). https://www.patrikschumacher.com/Texts/Parametricism as Style.htm. Accessed 14 Jan 2020

Schumacher, P.: Parametricism: a new global style for architecture and urban design. In: AD, Architectural Design, "Digital Cities", July/August 2009, vol. 79(4), pp. 14–23. Wiley, London (2009)

Senske, N.: A curriculum for integrating computational thinking. In: Parametricism SPC: Conference Proceedings of the 2011 Regional Conference of the Association for Computer Aided Design in Architecture. ACADIA, 10–11 March 2011, pp. 91–98. Lincoln, NE (2011)

Terzidis, K.: Algorithmic Architecture. Routledge, New York (2006)

Wing, J.M.: Computational thinking. Commun. ACM **49**(3), 33–35 (2006)

Wing, J.M.: Research notebook: computational thinking - what and why? In: The Link Magazine, Issue 6.0/Spring 2011, pp. 20–23. Carnegie Melon School of Computer Science, Pittsburgh (2011). https://www.cs.cmu.edu/sites/default/files/11-399_The_Link_Newsletter-3.pdf. Accessed 14 Jan 2020

Objects as Phenomenological Provocation in the Graphic Ideation-Abstraction Process

Josemaría Manzano-Jurado[1]([⊠]) [iD] and Santiago Porras Álvarez[2] [iD]

[1] Department of Architectural Graphic Expression and in the Engineering,
University of Granada, Granada, Spain
jmanzano@ugr.es

[2] Department of Architecture, College of Engineering, Korea University, Seoul, South Korea

Abstract. The concept of ideation-abstraction in the project, has created a new perspective in the architectural creation process. It is based on phenomenal principles, where the haptic, acoustic, olfactive and taste,... establish an alternative way of "looking".

Experiments carried out recently, have been based on pictorial, literary and musical provocations and have allowed, through graphic ideation, the establishment of new architectural creation strategies, starting from the drawings generated in the creative process.

The conclusion of these experiments is that the graphic medium developed from external provocations is a great precursor to the ideas that will later be transformed into architectural artifacts, Ideas, that through conceptual manipulation, will become part of the architect's formal imagery he or she will retrieve and use in his/her design work.

The innovation of the present experiment is the use of a palette of provocations based on quotidian objects: Quotidian objects stir ideational provocations. This work is based on everyday objects, objects that are easily recognisable from our own archive of images and shows, again, that the process of project learning becomes attractive and prolific.

This strategy brings the incorporation of new ideational material, where the tactile becomes the protagonist and generator of sensitive experiences aimed at the creation of architectural artifacts as proto-ideas.

From the object as provocation to the abstract object that becomes the architectural model, we can establish an intimate relationship between the phenomenal and intuitional to the specifics of the architectural form.

Keywords: Ideation-abstraction · Objects · Phenomenology

1 Origin

The research into the concept of ideation and its link to creative work, allows us to enter creative processes that have changed the paradigm relating to the graphic generation of architectural forms based on phenomenal stimulations [1].

L. Agustín-Hernández et al. (Eds.): EGA 2020, SSDI 6, pp. 550–559, 2020.
https://doi.org/10.1007/978-3-030-47983-1_49

The muses do not exist. However, what really does exist, and is worth considering, is a set of specific procedures that provoke previously unconsidered ideas.

Architecture is, unquestionably, built work. It is meaningful because its essence consists in giving material shape to an idea. If not, it would simply be a project, a document, a graphic code. The vital experience of form and space, the haptic, thermal, acoustic, chromatic, functional (and so on) perceptions are also part of what we consider today as architecture [2].

However, everything has its origin in the intuitive process, through which we are able to retrieve information and creative capacity from non rational worlds, taking an ontological positioning regarding man and the phenomena that affect him [3].

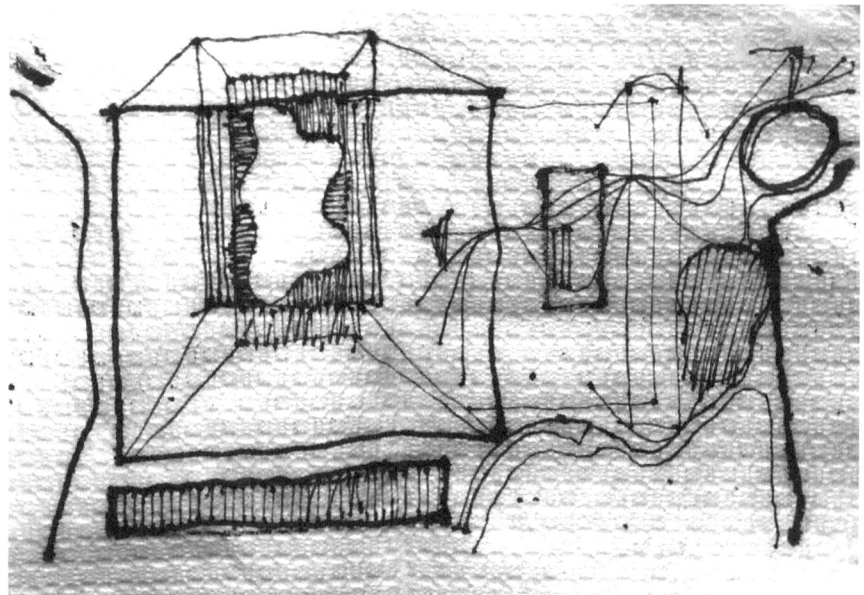

Fig. 1. Ideational Sketch by Manzano-Jurado, J. Source: His own production, included in his Doctoral Dissertation.

That creative drawing is more than just lines on paper is beyond discussion; it is the result of a process that starts in the interpretative capacity of intuition and ends in the precise configuration of forms and spaces that have the final objective of satisfying the needs of human activity (see Figs. 1 and 2).

Even though neurobiology is a science apparently unlinked to architectural ideation, and therefore to artistic creation, there is a growing body of research focusing on the mechanisms that, within our brain, produce artistic acts. The new theories on brain plasticity, that is to say, its ability to adapt to individual experiences, to specialise, open new perspectives in the knowledge of the act of artistic creation, simple but simultaneously complex and intimate.

This research, related to creativity within the field of neurobiology, has allowed us to understand that our brain and our own organic cognitive channels are able to provide us with a non objective idea of reality, discovering its inner mechanisms, and giving special prominence to intuition and the spontaneous process of the generation of ideas [4].

The latest theories on the brain's decision making mechanisms (of course, artistic and creative decisions included) demonstrate that we take decisions before being conscious of even having made them. That is to say, our brain has decided at a given moment prior to our being conscious of taking the decision. This fact is undoubtedly interesting in the field of artistic creation and its automatic connotations [5].

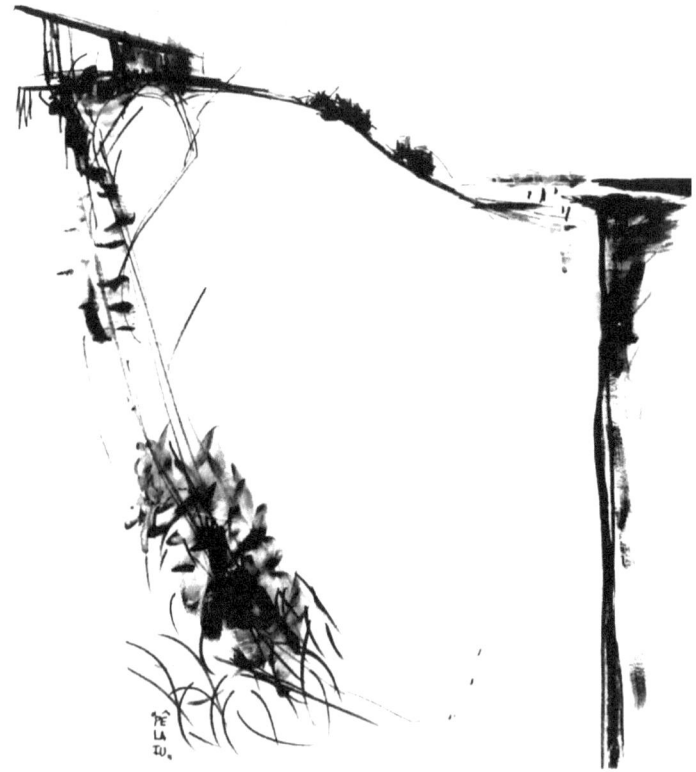

Fig. 2. Ideational sketch by Norwegian Architect Sverre Fhen for the Maritime Museum of Bygdoy, Oslo. Source: Nasjonalmusseet Arkitectur. Oslo.

Experience, as the fundamental source of our perceptual library, is the basis of phenomenal thinking. This new positioning against reality brings a different vision of the architectural space: inhabited space versus geometrical space [6].

What, in essence, phenomenology intends is for us to take consciousness of the fact that our experience of reality, the given, is our most accurate knowledge. Phenomenon, that is to say, all that can be experimented, show themselves to consciousness, not to thought. Thus providing a category of truthfulness to the experience of reality and, for that purpose, also becomes a tool of knowledge.

Intuition is the knowledge from experience in which the known object becomes present to us, differently to considering it as a conceptual object. This positioning against reality makes us consider architecture as the response to individual needs in our everyday life and our relation with all we perceive. It is understanding architecture through the perception of architectural artifacts, considered as phenomena, through the emotions and sensations they transmit to us and which are perceived by our senses.

This is not a random process, rather, it follows lines comparable to scientific processes. It is a field of research on the creation of ideas, where the graphic medium is extremely important. Among all stimuli we receive visual stimulus provides us with enormous amounts of information (see Fig. 3 and 4). However, we have already established that from the phenomenal and fundamental aspect, they are not unique [7].

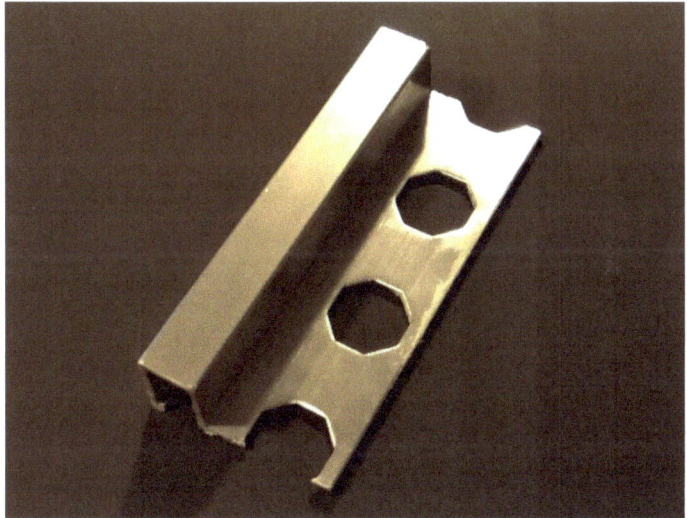

Fig. 3. Manzano-Jurado, J. Object: Aluminium form. 2019. Source: Author's work.

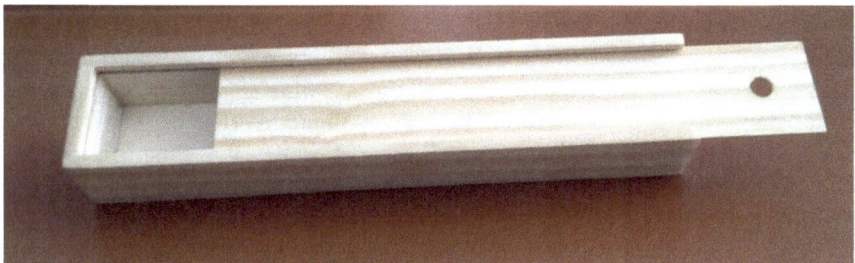

Fig. 4. Manzano-Jurado, J. Object: Wooden box. 2019. Source: Author's work.

The vision to which we will refer is not just contemplative; rather it needs an important effort of imagination. From the stimuli we receive, through the analysis of objects, we obtain elements and parts that constitute a generative idea (see Fig. 5). It is a way of seeing in which the details are part of the process, rather than one global vision of what is in front of us.

Previous research has focused on ideational possibilities of provocations from the pictorial, literary, musical etc. discovering new graphic procedures in the development of architectural objects, artifacts and new forms, but is this enough?

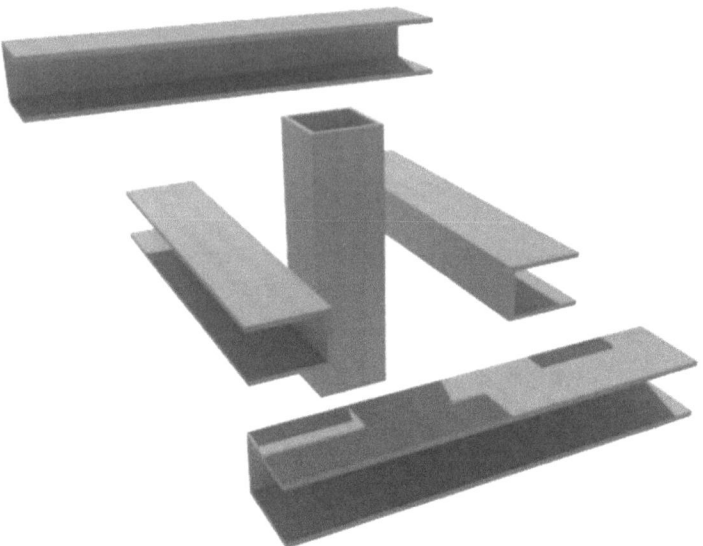

Fig. 5. Barragán-López Cebreros. Manipulation of forms from object. 2019. Source: Academic exercise.

2 Experience

Experimenting within the university environment is particularly difficult: even though it is a space for simulation, the absence of the motivational component of professional work, somewhat lessens the intensity.

Trying to harmonise innovative and heterodox experiments with an extremely regulated teaching system, under a rigid Architectural Studies curriculum, is rather a complex task. The experimentation with second year students of Architecture, in spite of the handicap of their disciplinary inexperience, allows us to combine fresh ideas without the commercial pollution inherent in professional work.

The initial hypothesis asks whether quotidian objects can be interpreted from the abstract and ideational perspective, through graphical manipulation, to contribute to the creation of new architectural forms.

In previous sessions, we had worked using experiences based on evocations from music, literature and painting, with encouraging results. The final production of the student and in general of the creative designer is intense and diverse. As the spectrum of formal solutions opens and widens with the incorporation of the abstraction with a great intuitive effort.

Rather than producing a set of rules, the intention is to create an instrumental base of architectural ideation, through the proposal of a process of image creation that stimulates and generates ideas and conceptual architectural images. A graphic procedure based on phenomenal perception, with the support of strategies that facilitate the analysis of forms of objects of everyday use that are easily recognisable from our archive of images (see Fig. 6).

The transformation of the object's concept through graphic medium explores new possibilities aimed at the creation of shapes that develop into architectural forms.

 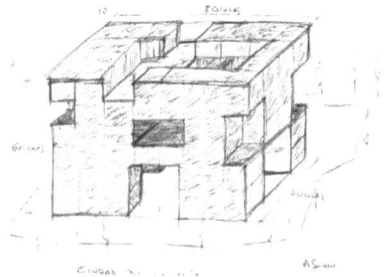

Fig. 6. Marín-Sánchez-Vico. Object and drawing derived from it. 2019. Source: Academic exercise.

We start from the fact that everyday objects provoke plastic suggestions. The consequences derived from their analysis and their formal decomposition helps generate new architectural forms through the mediation of graphic support.

It consists, therefore, of seeing objects through imagination; a look from the thought, stressing that the apparent is visible. Needless to say, our intuitive capacity to outline elements that are not initially perceptible has a strong prominence [8].

Curiosity also manipulates our way of seeing and focusing in specific traits that help transform simple objects, fragments, shades, into new creations.

The initial stages of the exercise were quite disconcerting. It is very difficult to extract the abstract component from everyday objects, which already have a strong emotive and utilitarian sense. They are familiar objects where their functional character is prominent. The drawings of these objects are the main element that helps to detach the utilitarian component and unveil their abstract and conceptual facets. The drawing transforms, that is to say, goes beyond the form and transcends it. From this attitude, the creative process changes both qualitatively and quantitively. The student creator conceives the object and observes it from a different aspect. Intuition and imagination take over the creative process.

From a graphic point of view, through this experience, the student exercises his imaginative capacity, from the random or accidental, discovering forms initially unnoticed, but that become relevant thanks to a more focused concentration.

For this experiment, the objects chosen as an initial reference were very different. Three of them were particularly relevant, due to their strongly evocative character, and due also to the brilliant results achieved by the students after the process of abstraction. The first was a prismatic three layered 'sandwich' type construction material, with two dense layers and one central porous and cavernous layer of cork. The second was an aluminium extruded profile and the third a wooden box with a sliding cover.

Each of these objects are made up of very different shapes and materials. The creative process that followed ended with three building proposals, also with three very different programmes: One, a building block with a large complex for Musical and Scenic Performances. Another, a house in a forest and finally, an Artistic and Academic Centre.

3   Epilogue

Schools of Architecture are breeding grounds where the controlled and tutored experimentation of innovative creational processes yield promising results. The creative possibilities of architectural project making are far from exhausted, as the ideation and abstraction process opens new ways and expectations. The most relevant objective of imagination is arriving at the creation of pre-forms as architectural artifacts, that establish the formal keys for the consequent development of the idea from a wide spectrum of phenomenal perceptive provocations, trying to stimulate our creative capacity as widely and intensely as possible.

To quote Javier Seguí de la Riva, *"Imagination: an inner event with the appearance of new situations, possibilities, personal roles or sequences of behaviour, projected on a mental screen as a visual figuration, auditive or composed of diverse sensorial components (olfactive + tactile + etc.) starting from complex associations which integrate past events (states of memory) with schemes of currently happening events"* [9].

The drawing communicates not only the meaning of the object it represents but also holds other references to the creative process. The reference objects are starting points to generate architectural forms (see Fig. 7).

This drawing must be able to contain and express the initial intuitive process, and the future intentions, as well as the necessary characteristics of the objects, the ideated architectural artifacts that, afterwards, must be represented and experimented phenomenally. Creating a sketch is exploring sensations, is thinking perceptively. By exercising our graphic imaginative capacity, from the random or the incidental or from the proposed

Fig. 7. Abellán-Del Rio-Plaza. Final Process Model. Artistic and Academic Center. 2019. Source: Academic exercise.

Fig. 8. Barragán-López Cebreros. Final Process Model. House in a Forest. 2019. Source: Academic exercise.

objects, we discover forms that initially remain unnoticed but that, with a special way of looking, stand out and become prominent.

Fig. 9. Marín-Sánchez-Vico. Maqueta. Scenic and Musical Complex. 2019. Source: Academic exercise.

The creative provocation based experiments encourage the ideation effort. The academic experimentation, considered as a space of simulation, started from the provocations of a musical, literary, pictorial and video graphic nature, demonstrate the creative and expressive possibilities of the drawings created from the situations phenomenally stimulated. The result is an endless series of images and proto-ideas (see Fig. 8).

In our case, the experimentation with simple objects, fundamentally the haptic experiences, and the cultural references extracted from that, create the breeding ground for new ideas that, through their graphic transformation, direct the author towards recognisable architectural elements, generated following the normal design process.

The step from object as a provocation, to object as abstract (the architectural model), allows us to establish an intimate relationship between the phenomenal and intuitive, with the specific of the architectural form (see Fig. 9).

Regarding the creative reality of the architect, we can summarise by saying that the graphic ideation-abstraction applied to everyday objects can also be considered a preliminary stage of the formalised architectural idea. This process, by means of graphic tools, preferably free hand drawings, and their latter formalization as object-architectural models, embodies a phenomenal interpretation of the subjective, intuitional, interferences (sensations such as smell, light, taste, humidity…) and influences (visions, images, memories).

The final results are extremely encouraging. The process running from the abstract object to the architectural formalisation is slow but very fruitful, resulting in excellent proposals, suggestive and rich in nuance (see Fig. 10).

Fig. 10. Barragán-López Cebreros. Render of a House in a Forest. 2019. Source: Academic exercise.

References

1. Pallasmaa, J.: The Thinking Hand: Existential and Embodied Wisdom in Architecture. Wiley, Hoboken (2009). Translated into Spanish as "La mano que piensa. Sabiduría existencial y corporal en la arquitectura". Editorial Gustavo Gili, Barcelona (2012)

2. Pallasmaa, J.: The Eyes of the Skin: Architecture and the Senses. Wiley, Hoboken (2012). Translated into Spanish as "Los ojos de la piel. La arquitectura y los sentidos". Editorial Gustavo Gili, Barcelona (2014)

3. Husserl, E.: Der Encyclopedia Britannica Artikel from Edmund Husserl's Phänomenologische Psychologie. Die Krisis des europäischen menschentums und die philosophie. Die philosophie als menschheitliche selbstbesinnung, selbstverwiirklichung der vernunft. (1925). Translated into Spanish as: "Invitación a la fenomenología". Ediciones Paidós Ibérica, Barcelona (2008)

4. De la Gándara, J.: Psico-neuro-biología de la creatividad artística. Arte y Psiquiatría Cuadernos de Psiquiatría Comunitaria **8**(1), 44 (2008)

5. Libet, B.: Mind Time: The Temporal Factor in Consciousness. Harvard University Press, Harvard (2005)

6. Bachelard, G.: La Poétique de L'espace. Les Presses universitaires de France, Paris (1957). Translated into Spanish as: "La poética del espacio". Fondo de la Cultura Económica, México (1986)

7. Holl, S.: Questions of Perception: Phenomenology of Architecture. A+U Publication, Tokyo (1994). Translated into Spanish as: "Cuestiones de percepción. Fenomenología de la arquitectura". Editorial Gustavo Gili, Barcelona (2011)

8. Jenny, P.: Zeichnen im Kopf: An der Quelle Ihrer Bilder (Schule des Sehens) (2004). Translated into Spanish as: "La mirada creativa". Editorial Gustavo Gili, Barcelona (2013)

9. de la Riva, J.S.: Anotaciones acerca del dibujo en la arquitectura. Rev. EGA (1), 73 (1993)

Critical Thinking and Graphic Action

Belén Butragueño Díaz-Guerra[(⊠)] ⓘ, Javier Fco. Raposo Grauⓘ,
and María Asunción Salgado de la Rosaⓘ

IGA Department, E. T. S. Arquitectura, Universidad Politécnica de Madrid, Madrid, Spain
b.butragueno@upm.es

Abstract. The action of drawing is inherent to the human being. Drawing is our primal way of expression, even prior to the speech. Drawing is, therefore, a powerful means of communication and beyond that, it is a fundamental tool of discovery and exploration of the world around us.

Drawing allows us not only to discover the world, but to analyze it and apprehend it, to decompose it and even to create new realities out of those pieces. Finally, through drawing, we can communicate these new emerging realities and express them in the most appropriate way, closing the loop from drawing as an expression of the individual to an expression of the individual's creation. This circle can be easily replicated in the teaching of architectural drawing, drawing to design. In this article, we analyze the process from the original drawing of pure expression, to the drawing of knowledge and learning, the analytical and interpretive drawing and, finally, the propositional drawing as a creative and thinking tool, closing the circle with the drawing of creative communication. The entire process leads us to conclude that the drawing is a form of language, that requires the previous learning of a code. Once acquired it, there is a possibility of decomposing it and constantly reinterpreting it.

Keywords: Drawing · Analysis · Exploration · Thinking · Communication

1 Introduction

As mentioned, drawing is, an innate action we do from childhood, as it is a primal form of expression and a form of knowledge and analysis of the world that surrounds us (Sennet 2009). Each drawing has a specific purpose and, therefore, a corresponding graphic language. However, in the collective imagination there is a pre-determined idea of what a "good drawing" means, usually equated with a mimetic representation of reality. When a child performs his first strokes, the adults around him can't help but asking him to "tag" the drawing. They usually tend to ask questions such as "what is this?", and "what are you drawing?". From rationality, we need to find a precise meaning to the drawing and we expect an accurate representation. In addition, the child is usually complimented when he is following the limits of the drawing or they draw in a neat and clean manner. In short, it is considered valuable obtaining a recognized outcome, true to reality, and a flawless execution. In most cases, such requirements cause in the child a disconnection from the drawing, especially when there is a lack of natural skills, as

L. Agustín-Hernández et al. (Eds.): EGA 2020, SSDI 6, pp. 560–570, 2020.
https://doi.org/10.1007/978-3-030-47983-1_50

the playful and imaginative component disappears, and the drawing is perceived as an obligation to give a precise and predetermined response. (Berger 2011).

A different strategy is to encourage the expressiveness and the creative freedom, not requiring a specific significance or technical rigor. This methodology stimulates the uninhibited stroke, the simple expression through the use of color and full paper occupancy… thus, we find results that connect with the excitement and passion. It is when the drawing becomes a thinking mechanism at all stages of life (Rojas 2012) (Fig. 1).

Fig. 1. Guided graphic process versus expressive graphic process in childhood.

As first-year drawing professors at Bachelor's Degree in Architecture, in multiple occasions we face an initial blockage of our students that state: "I don't know how to draw" or "I don't have spatial vision". These preconceptions come from the traditional assimilation of the spatial ability and graphic skills, exclusively with figurative or representative virtuosity. Without denying the beauty, pregnancy and the value of the representative drawing in itself, we must not forget that the drawing presents multiple aspects and forms of expression, depending on the aim of the action of "drawing". In this article, we focus on the graphic processes linked to architecture, considering drawing as a tool of exploration, expression, analysis, interpretation, proposition and communication.

Under this premise, our teaching proposal is supported by open, speculative and tentative mechanisms, linked to the critical learning and not merely to an instrumental learning. We develop experimental, dynamic and collaborative pedagogic strategies, to encourage the students' acquisition of the necessary graphic skills to adopt a personal language and a critical thinking. Throughout this text, we will deepen in the various actions of drawing, associated with the different phases of the creative process in architecture.

2   Drawing as an Action of Exploration and Knowledge

The first challenge we face as professors on speculative subjects, is the need to disassemble the structure of certainties and realities apprehended by the students, as they come from an excessively scheduled and determined educational system. We find talented students that feel more comfortable when required measurable results. They don't realize that the learning occurs mainly along the way and not so much in the finishing line. This phenomenon can be extended to society at large, anchored in the pursuit of certainty, feeling comforted by what is recognizable and precise and crumbling in the face of indeterminacy and uncertainty. However, it is well known that creativity and imagination are fed primarily by imprecision and indetermination, and require of open and speculative procedures to emerge (Claxon 1999). In the case of the learning of "drawing", this paradox is synthesized in the fact that we are asking students educated in figuration and representation, to "get out of the limits" and embrace abstraction. Well aware of the complexity of this cognitive challenge, our priority as teachers is to involve students in this elaborated process of "deconstruction" and to share with them the difficulties at all times, establishing a relationship of mutual trust, which is critical to the success of any open and speculative pedagogy. It is assumed that this "trust" has a critical nature, based on a permanent dialog and communication between students and teachers. A fundamental task is creating an enabling environment to work and exchange experiences, allowing students to act freely, fostering a climate of cooperation, open dialog and collective trust. One of the pedagogical strategies that we use, is the critical session, where students participate in the comments of their teachers and peers about their own work and the others, contributing to the formation of a critical thinking. In this first phase, we address the architectural project from the knowledge of the early stages of the creative process, ideation-architectural imagination. The main objective is exposing the students to a nonspecific drawing, focused on values of expressiveness in relation to external references and their own interiority, from personal proposals that may gain a certain capacity of suggestion and evocation (Fig. 2).

Fig. 2. Expressive drawing processes. DAI 1. Course 2017–2018, ETSAM, UPM.

We start with a variety of triggers to provoke the action of "drawing", in order to break the blockade of the blank sheet (Berger 1997). The stimuli are diverse and range from images, texts, videos… even the physical space. Each of them is directed toward specific objectives and capabilities to develop in each moment, and they are linked respectively with the "gestural expression", the "developed expression" and the "synthetic expression". We resort to fast, and alterable graphic techniques (charcoal, pastels, acrylics, collage, photography, computer procedures, etc.), that encourage the immediacy and essentiality of the drawings and enable the action of "erasing", the rectification and the attempt (Fig. 3).

Fig. 3. Expressive drawing processes. DAI 2. Course 2018–2019, ETSAM, UPM.

This way, we enhance the expressive qualities of the reference models with the mechanisms of graphic abstraction. It provides a free and unrestricted exercise for the student to delve into the playful character of the drawing and find their own graphic language to investigate and define their personal journeys. We search for movement and the essentiality of the actions, in order to obtain the maximum visual meaning with the fewest number of graphic elements. In short, in this phase we define the graphism as an autonomous language, devoid of significance, favoring the processes of natural abstraction, based on the generic expression, from the gesture to the sensation.

3 Drawing as an Action of Analysis

Along this phase, we address the "architectural design" through the development of processes of production-architectural transformation. We delve into the idea of the project starting with a certain control of the creative phases, that will ultimately lead to the construction of an architectural form (quantity, proportion and spatial metric), affecting the relationship between form and geometry (Fig. 4).

Fig. 4. Analytic Drawing Processes. DAI 1. Course 2018–2019, ETSAM, UPM.

As previously mentioned, the learning of architectural drawing requires the acquisition of a specific graphic language. As with any other language, this implies the learning of certain common codes that allow the transmission of knowledge and the expression of ideas, in a way that is intelligible to the interlocutor, who should have knowledge of the code. In addition, the proficiency and knowledge of the codes, will encourage an ulterior manipulation and reinterpretation, to promote the generation of a distinctive language linked to a personal creative and productive discourse. Such language should be, equally, recognizable and assimilated by the receiver; namely, it should be guided by the principles of the code while its ultimate expression is customized.

By using the graphic code, we encourage the analysis of the expressive drawings corresponding to the previous phase, from a reflective and interpretive prism. This progressive graphic analysis draws conclusions that promote the development of the complete graphic process. In this phase, the action of graphic thinking is ignited by the act of drawing, which we define as "thinking by drawing" or "drawing by thinking" (Marina 1994) (Fig. 5).

The main advantage of this type of drawing is to introduce the concept of "spatiality" in architecture and emphasize its cognitive dimension. Students grasp the architectural space from a technical and formal perspective, which will allow them to develop processes of manipulation in subsequent phases.

It forces the students to gain a high degree of abstraction and the introduction of geometry as a nurturing element of reality. The architectural drawing as the "action of analysis", requires a known and recognizable geometric support, both in its generation as in its properties, which is based on the use of well-known mathematical relationships, that define the code (Raposo 2010).

At this stage, the drawing is manifested as the language of visuality, as a visual mediator. The representation is added to the expression, in the analytical sense of the term, giving rise to a "developed expression", based on the visual abstraction. We use perception versus gesture and action versus sensation.

Fig. 5. Processes of analytic drawing. DAI 2. Course 2018–2019, ETSAM, UPM.

One of the most utilized methodologies is working with serial processes, where triggers are progressively replaced by images developed by students, meaning their own and personal references. This is a key moment for the group, regarding the flourish of talent. They start to emerge talents that remained "dormant", arising as a result of the acquisition of certain graphic skills and a critical thinking. It encourages the self-esteem and self-confidence in their personal approaches (Butragueño, Raposo, Salgado 2017) (Fig. 6).

Fig. 6. Processes of analytic drawing. Course 2018–2019, ETSAM, UPM.

4   Drawing as an Action of Thinking

This is one of the most decisive actions of the creative process. It addresses the architectural project through the development of processes of synthesis, materialized in three phases: imagination-transformation-conceptualization of architectural form. The result is a structured analytical expression, focused on the concept, through an intellective abstraction. Undoubtedly, the ultimate purpose of drawing in architecture is to catalyze the architectural thinking. Thinking and acting emerge at the same time and they are interdependent (Bellardi 2014). Our pedagogical proposal aims to foster the emergence of architectural thinking and to promote the acquisition of the required visual culture in a rapid and effective manner.

For this purpose, we rely on strategies that allow to establish a link between the thinking dynamics and the strategies of drawing as active and operational mechanisms of design. At this point of the process, the collective has a fundamental importance, as the group evolves as long as the emerging individualities are put at its service, through dynamic and collaborative pedagogical experiences (Fig. 7).

Fig. 7. Image of collective working dynamics. Course 2018–2019, ETSAM, UPM.

In this last step, the design is an interpretative language for the conceptualization of the idea, covering expression, representation, and interpretation of the architectural concept, from a purely propositional perspective. This proactive requirement makes more evident than in previous phases the need for a universal graphic code, based on a solid geometric construct, allowing an analysis and a precise interpretation of drawing.

In this regard, the plan and the architectural section play a special instrumental role, as they provide a qualitative and global understanding of the project and the interrelationship between the different layers of information, enabling decision-making at different scales. In this phase, the ultimate goal is the development of a personal architectural proposal, where each student has the opportunity to freely define a programmatic development, the scale and location, within the area defined during the semester. The key is to stimulate a students' investigation on the most appropriate ways of representation for the analysis and the understanding of the project, which subsequently enable its use as an instrument of criticism and transformation (Fig. 8).

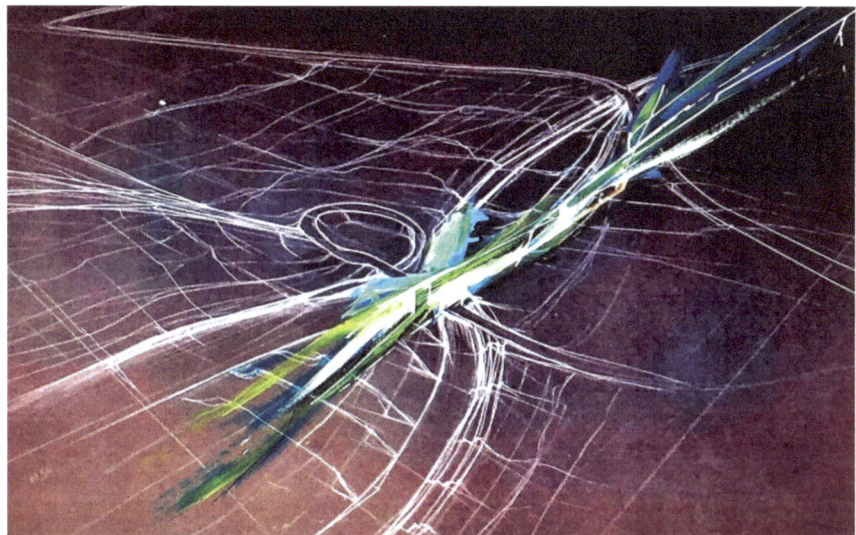

Fig. 8. Processes of interpretative drawing. Course 2018–2019, ETSAM, UPM.

On the other hand, there is a voluntary distance from the references that have triggered the process at this stage, understanding the architectural drawing fundamentally as an "interpretative action", once they have acquired sufficient skills and have the proper graphic tools.

5   Drawing as an Action of Communication

The action of "drawing to communicate" is one of the fundamental aspects of the architectural drawing. A transcendental part of the architect's work is the successful transmission of the processes developed and the results obtained, to the diverse actors involved. This requires the use of very different codes, covering practically all the graphic phases referred in the text: expressive, analytical, interpretive and proactive drawing. In addition, a growing number of authors, including ourselves, defend the presence of the communicative aspect of drawing as a creative agent from the beginning of the process. We insist on the transformative value of the drawing, that accompanies the project along the process, not being exclusively its ultimate tool of expression.

The transmission of the processes requires a necessary rationalization, that favors the emergence of alternative routes and the breakdown of linear thinking. It is equally fundamental the introduction of external stimuli, that apparently are alien to the processes, leading to the reconsideration and enrichment of the strategies followed. In the relationship described between the drawing and the action of thinking, we must necessarily incorporate the playful component, as well as the fate and the discontinuity (Allen and Pearson 2016) (Fig. 9).

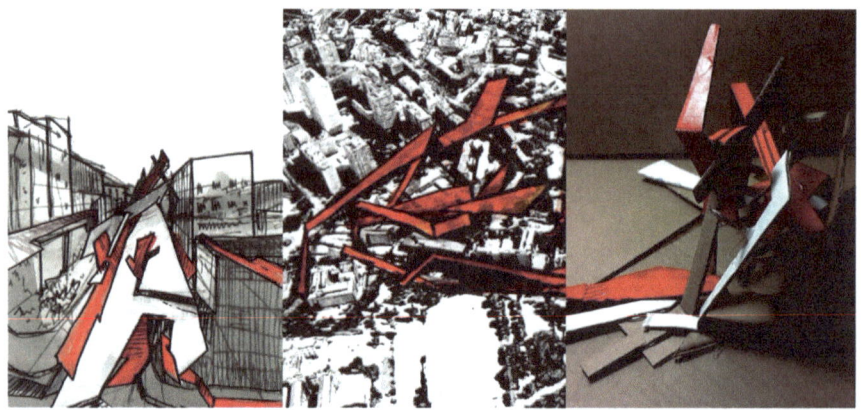

Fig. 9. Processes of communicative drawing. Course 2018–2019, ETSAM, UPM.

6 Conclusion

Along the article, we have covered the different actions related to the "drawing". It allowed us to verify the implicit idea of this research: the creativity and the graphic learning are necessarily developed by prioritizing an exhaustive drawing process versus the results. The dynamics of the collective work between teachers and students favor the detection of talent in the different phases of learning. It is well known that talent is more related with an attitude towards work than with an alleged aptitude. Great masters defend that the talent must be worked each day. In the words of Picasso, "the inspiration exists, but it has to find you working" (Palomo and Triguero 2013, p.169). In other words, the talent is not innate, but it can be cultivated, by creating an appropriate atmosphere in the room, to learn without complexes, so the talent can evolve in a natural way.

In our opinion, enhancing the flourish of talent means to make a commitment with innovation, creativity and excellence, in the benefit of the collective. The teaching strategy is successful to the extent that each student manages to find a personal form of expression, promoting his imagination and the evocation of new realities. Therefore, the evolution is measurable only in relation to the departure point of each student. This type of dynamics requires a high degree of flexibility and effort on the part of the teachers, in relation to the teaching dynamics developed in the classroom, favoring the exploration of different personal paths and proposing activities that stimulate their individual capabilities, without neglecting the collective ones.

The teachers delve into the areas of personal interest of students, as a response to the processes raised in the class' dynamics, validating the permanent exchange of ideas. The main interest of our pedagogical proposal lies in the students' exposure to multiple graphic techniques and the different actions of drawing raised, and not so much in the collection of virtuous results (Fig. 10).

Fig. 10. Processes of communicative drawing. Course 2018–2019, ETSAM, UPM.

As noted, this is a development in continuity, that prioritizes the processes versus the results, requiring a maximum commitment of the students. In parallel, it promotes the acquisition of a solid criterion, as well as communicative abilities, both graphic and expressive, in order to achieve the maximum significance and synthesis of the creative processes developed. In conclusion, our pedagogical proposal sustains every act of drawing in four major areas: action, actors, referential framework and means of production. The priority "action" is the teaching of "drawing to design", through the stages described in this text. The fundamental "actor" is the room, understood as the community formed by teachers and students. The room becomes a catalyst for ideas that arise from permanently questioning the individual and collective proposals. The personal interests of students are enabled, fostering diversity and developing a personalized process of self-discovery for each student. In this context, the validation of the processes occurs in relation to the evolution of each individual from his starting point and not compared to the collective. At the beginning of the year, the burden of motivating teaching strategies lies in the teacher, as could not be otherwise. However, in subsequent phases, the students take the initiative and define the topics of interest. The "referential framework" is based on the promotion of a proactive enrichment of the collective and critical thinking. The aim is the acquisition of a transversal knowledge that connect directly to the needs of society, which is the role of the architect. The "means of production" is intimately related to the necessary knowledge of the graphic language, already referred in the text. The mastery of the codes o f language, allows us to experience beyond the known techniques and make use of instrumental tools related to analog-technological means of hybridization, that facilitates the resolution of problems from open perspectives. From our Teaching Unit, we believe that this educational path through the different actions of the drawing, gives students the ability to make progress in learning to design. It provides them with all the conceptual and technical skills that are necessary at the time of initiating a process of creative learning, which ultimate goal is the architectural project.

References

Sennett, R.: El artesano. (ed.) Anagrama, Barcelona (2009)

Berger, J.: Sobre el Dibujo. (ed.) Gustavo Gili, Barcelona (2011)

Rojas, E.M.P.: La evolución del dibujo infantil. Una mirada desde el contexto sociocultural merideño. Educere (2012). http://www.redalyc.org/articulo.oa?id=35623538016

Claxon, G.: Aprender: El reto del aprendizaje continuo. Paidós Transiciones, Barcelona (1999)

Berger, J.: Algunos Pasos Hacia Una Pequeña Teoría de Lo Visible. Ardora Ediciones, Madrid (1997)

Marina, J.A.: Teoría de La Inteligencia Creadora. Anagrama, Barcelona (1994)

Raposo, J.F.: Identificación de los procesos gráficos del "dibujar" y del "proyectar arquitectónico, como procesos metodológicos de investigación científica arquitectónica. Rev. EGA **15**, 102–111 (2010)

Butragueño, B., Raposo, J.F., Salgado, M.A.: Aprendizaje líquido…desde la incertidumbre. Rev. JIDA **5**, 100–115 (2017). textos de arquitectura, docencia e investigación

Bellardi, P.: Why Architects Still Draw: Two Lectures on Architectural Drawing. The MIT Press, Cambridge (2014)

Allen, L., Pearson, L.: Drawing Futures: Speculations in Contemporary Drawing for Art and Architecture. UCL Press, University College, London (2016)

Palomo y Triguero, E.: Cita-logía. 1ª ed. Punto Rojo Libros, S.L., Sevilla (2013)

About Sert's Drawing for Harvard Planning, a Reference for Pedagogical Innovation?

Jesus Esquinas-Dessy[1](✉), Isabel Zaragoza[1], and Paula Esquinas[2]

[1] Universitat Politècnica de Catalunya, Barcelona, Spain
jesus.esquinas@upc.edu
[2] Area Metropolitana de Barcelona, Barcelona, Spain

Abstract. The aim of this communication is to show some reflections on the graphic representation used by Sert to "make understand" the suitability of the urban planning proposal that he presented to Harvard University to guide its growth in the middle of the 20th century, in order to evaluate how that graphic language is the result of a rational and universal culture especially disciplined, as well as how the given context transformed a generic visual rationality into a more specifically rooted in the urban conditions of the site, in the human activity of its users and, in all the objectives of the assignment.

Recently, Sert's work is contemplated as blurred among the dense architectural culture of the Modern Movement, and his experience in urban planning is not perceived as a continuous evolution of his functionalist vision of the city. However, there are aspects in his multifaceted experience as an architect that make understand his work under a humanist and existentialist point of view with a singular graphic language.

Our work brings to light the narrative nature of the set of graphic material presented throughout the Harvard University Planning document, focused mainly on the impactful capacity of the graphic resources used. The research achieves an original and reasoned interpretation of some graphic strategies used by Sert to persuade a diverse audience with his thoughts, and where artistic references are fundamental.

Keywords: Architectural representation · Sert · Modern movement · Urban planning · Teaching innovation

1 Introduction

The current disruptive change of technologies is identifiable as a new age of divided representation [1], similar to that at the beginning of the Renaissance when the statement of perspective laws meant a stimulating cultural change. Simultaneously, the new graphic resource caused an innovative look to the historical legacy of ancient Greece and Rome.

Related to this interest in looking carefully historical legacies, the work of Josep Lluis Sert, as a committed "activist" of the modern movement and conceived with rigorous ideological reflections, is a useful theoretical contribution as a timeless reference for the architectural practice.

© The Editor(s) (if applicable) and The Author(s), under exclusive license
to Springer Nature Switzerland AG 2020
L. Agustín-Hernández et al. (Eds.): EGA 2020, SSDI 6, pp. 571–584, 2020.
https://doi.org/10.1007/978-3-030-47983-1_51

Recently, the figure of Sert is seen as "blurred and misunderstood in contemporary architectural culture" [2] and his experience in urban planning, more linked to theoretical diffusion than to professional practice, is interpreted characterized by a continuous evolution of his initial functionalist vision of the city, to another more humanist and existentialist. However, his significance as an urban planner is completely justified, by his ideological activity in the influential CIAM (Fig. 1), by his academic responsibility at Harvard, or by his ties with the art world. Therefore, it is enough to intuit that his work contains a valuable graphic language.

 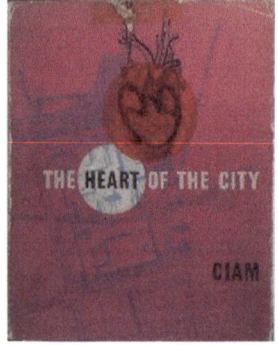

Fig. 1. Cover of the renowned illustrator Herbert Bayer, trained at the Bauhaus and collaborator of Sert, in CIAM publications. Left: Can our Cities Survive? 1942, written by Sert. Right: CIAM 8, The Heart of the City, 1952, co-edited by Rogers, Sert, Tyrwhitt.

Harvard Planning is considered urbanistically as a paradigmatic implementation of university in the city [3]. It is a work of maturity, when it already surpassed an abstract interpretation of an orthodox functionalism of the city, he had evolved into a new humanism and existentialism and, where intentional artistic references were glimpsed in his representation [4].

The examination of the original material of Sert's work for this planning [5] reveals a rational "story" intended to "persuade", in which the graphic representation has a key role and, whose successful result supports it as a useful model of graphic narrative, and therefore, necessary in the pedagogy of architectural representation.

This look is possible thanks to specialized archives, such as the Special Collections of the French Loeb Library of the Harvard GSD, which carefully guard these original graphic documents and which, as with unbuilt architectures, pedagogically allow reinterpretation of the thoughts that configured them [6].

New technologies allow us to observe and, therefore, draw and teach reality in a different way [7], and that possibility, adequately encouraged by past timeless experiences, as in the stimulating Renaissance, is an open door to innovative creativity every once more insistently demanded.

2 A Humanistic Planning for Harvard

Sert arrived to the United States in 1941, at his 40, and together with Wiener founded the office Town Planning Associates, TPA, dedicated to architecture and urban planning, which was especially focused on Latin America. This experience that lasted until 1958, accompanied by a continuous and progressive self-criticism, bequeathed a whole series of "functional" city plans, which illustrated the International Congresses with concrete examples of that "modern" urbanism.

Sert's planning for Latin American cities show the controversies and new formal directions that modern urbanism took as its social significance changed throughout the decade [8]. Although they were almost never implemented, they allowed Sert to develop a graphic language of great synthetic capacity and unavoidably attractive [8] (Fig. 2). The shadow of the teacher Le Corbusier and his artist friends Miró, Leger or Picasso is not far…

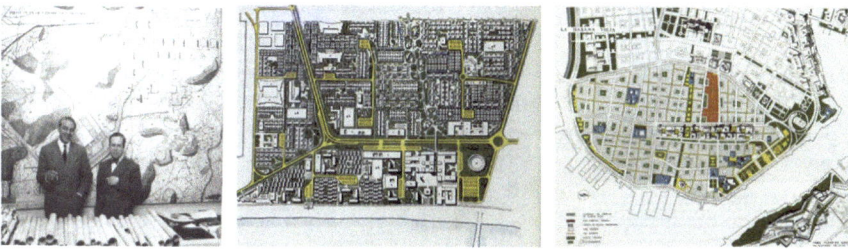

Fig. 2. Sert's urban planning for Latin America. Left: Wiener and Sert presenting the Cidade dos motores. Source: [8] p. 168. Center: Chimbote. Source: [4] p. 131. Right: La Habana. Source: [4] p. 184.

Thus, endorsed as an experienced and self-critical urbanist [9], when he was about to finish the 30-year historical cycle of CIAM, in 1953, he was named dean of the Harvard Graduate School of Design and, at the same time, architecture studies director of that school replacing Gropius. In this academic context, Sert has the opportunity to apply his "functionalism" to Cambridge's historical complex in the United States. It is 1957 when the president of Harvard University orders him the advising of a planning office for the entire Campus. The same year, he chairs the Cambridge City Planning Commission where he establishes a program to stop the growth of the university at the expense of the city [10].

The urban proposal [re]presented in a document, which has been disseminated iconically through only "one of two" images (Fig. 3), which synthesize urban ideas of a document where there is more.

The first image is published, just ending his responsibility at Harvard, in a book that as a catalog collects the work done by Sert until then [11] (Fig. 3 left). Harvard Planning is represented as an exciting "green" environment, typical of rationalist urbanism.

Twelve years later, a paperback book on Sert's work [12] shows a second image. (Fig. 3 right). It is a single plan; the university plan is similar, although here, it incorporates the drawing of the roofs of the planned buildings, already constructed.

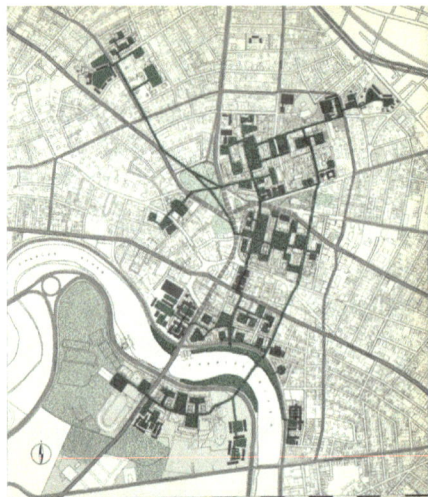

Fig. 3. "Iconic" plans for dissemination of Harvard Planning. Left: book written by Sert and his collaborator Bestlund [11]; university community floor plan, highlighting a system of open spaces connected by pedestrian paths and represented with green frames linked by continuous thick lines of pale gray color. Right: book written by Freixa, [12] former collaborator of Sert; floor plan without the previous green hatches. It is a proud representation of "achievements." Sharpening the look, we see that the buildings of Le Corbusier (7), the married residence (2), the science center (4) are drawn. In addition, this is enhanced by a numbered list of buildings and, coded by shadow cast that reveal their great heights.

They are, in a way, graphic records of a "before and after." And, it is confirmed that the intention of the "iconic" drawings evolves over time. In fact, the second drawing, although referring to Harvard planning is not part of the original document, it is a "redrawing" focused on recording the human scale of the city defended by Sert.

Much later, Rovira, without a professional relationship with Sert, chooses the latter in black and white as the only representation of this planning for his books [4, 13]. He insists on valuing this work with the "achievements" of the proposed urbanity.

And surely, this "achievement" of vertical density injection is the most significant urbanistically and graphically! Therefore, pedagogically, narrated basically in graphic form.

However, in reality, the urban proposal is part of a planning document of future facilities for the university, which is guided by the conviction of effectively integrating university and city.

3   A Graphic Narrative

A document is bound, with 60 folios printed on both sides, composed globally and rationally [14], with text on the left pages and plans on the right, so that open, effortlessly relates the descriptions of the text with the information of the plans. In a way, it resembled the presentation formula of the architecture illustrated magazines that were successful in those times: large images, accompanied by supporting texts and headlines.

Sequentially, those ideas that according to Sert configure the functional city were presented. They are illustrated with diversity of registers and graphic intensities, until reaching a result of modern buildings that are understood to be absolutely integrated in an urban fabric of a city that respects and dialogues with the traditional physiognomy and, with the "modern" myth of living in nature. A certain "vintage" alchemy strange in the initial purist utopias of functionalism.

In a more focused way of seeing over the whole than on each component separately, intentions are revealed that go beyond simple technical, or aesthetic criteria, or sensitivity of each of the drawings or the overall of them. A clear storytelling intention is interpreted, to tell what is proposed and why it is proposed. The sequence of graphic records presented in a way that is full of connections confirms this statement.

Only by looking at the covers of the sections (Fig. 4) in which the document is divided, without imagining much, is the continuous growth of the city perceived as the origin, and finally, new buildings that respond to that "assumed" need for growth. It is not a new narrative, but in this case, the fact is that the answer is implemented soon. And this it is significant, especially in the world of urban planning.

Through concatenations of images, and limited graphic resources, a future attractiveness of modern homes is dreamed. At the end, some technical representations, in plan accompanied by perspectives, have just set in the reader the hypnotizing materiality of the dream (Fig. 5).

The question arises: does a university planning document have the ultimate goal of building new homes with modern comforts? Perhaps yes, but it is very likely that this was not the main assignment. Surely, it has more than one explanation. The most possible is that it is a way to approximate the proposal to a bewitching future for the general public, whether citizens or university students. Anyone dreams and imagines enjoying a home equipped with modern features, dressed in blue "sky".

In fact, the story shows that "the ultimate goal of the city is housing" [15]. Specifically, the last drawing of the document, graphically describes a new residence for students. Singular ending for an inventory of a university. However, the few implementations of previous functional planning warned about the need to look for new representation strategies. The innovative aerial views did not seem effective enough.

The surprise arises! Although the need to implant density in height is obvious and, this could be controversial, unlike previous plans, he avoids drawing explicitly in elevation or perspective, so any new high-rise building. Sert has known the problems of the city for years and knows how to respond to them. In addition, he also has known for years how to spread ideas, or even how to influence public opinion through the media [16]. And, on that was Sert!

He decides to transmit the ideas of the university integrated growth together with the city, through an attractive graphic format -comfortable and easy- for a large audience, such as an illustrated magazine, where the visual component is basic! He discards more technical formats aimed only at expert readers. He relies on his experience in publications of great diffusion so that the graphic presentation of his ideas embellishes both the university community and the group of citizens, whether experts or laymen. He knew

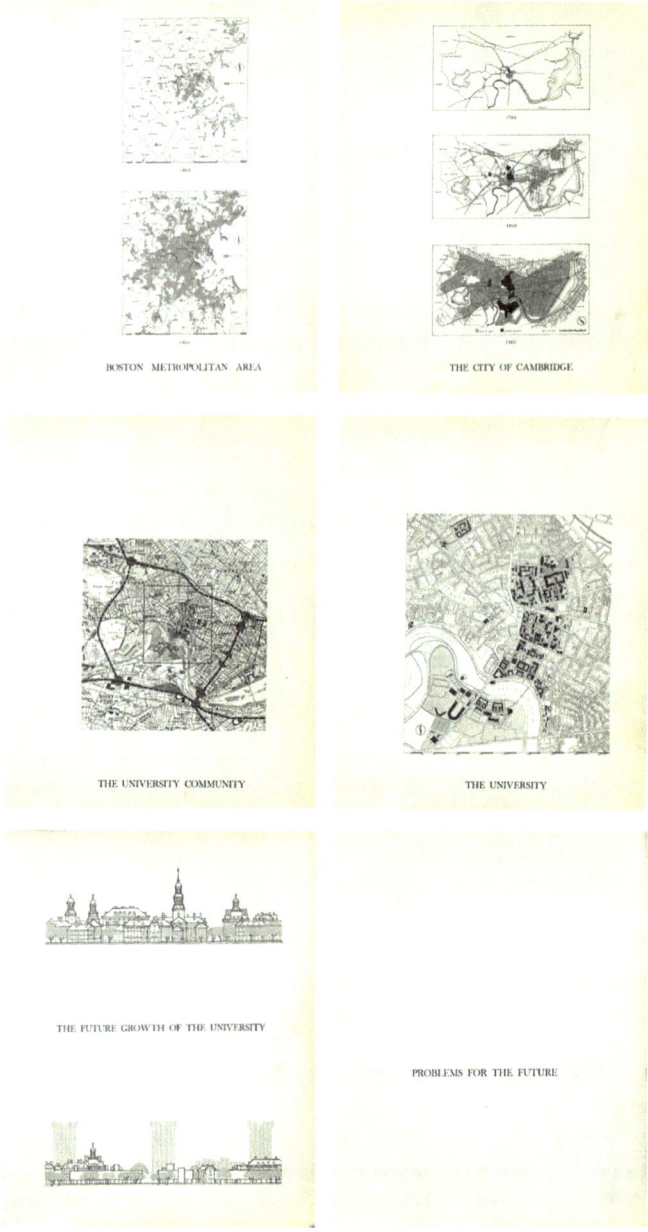

Fig. 4. Sequence of "covers" singled out by the white predominance. They synthetically narrate an origin and an ultimate goal. Up-left: initial cover p. 1-1. Down-right: final cover. p. 7-1. Source: An inventory for planning, HUPO (1960) [5].

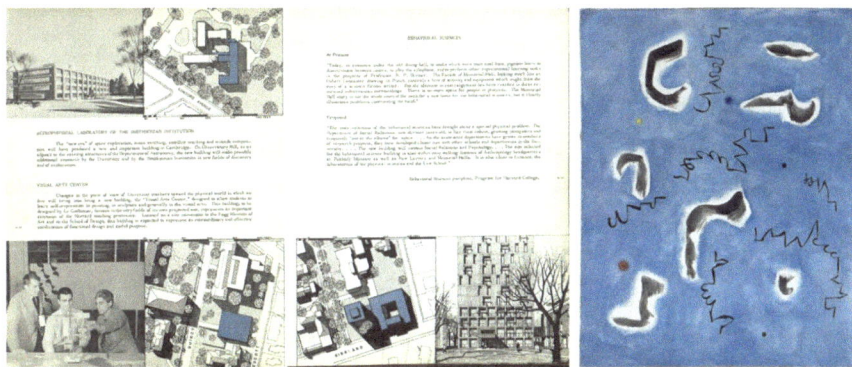

Fig. 5. Result of modern buildings described graphically with an artistic expressiveness. Left: drawings of the inventory of new buildings. Source: An inventory for planning, HUPO (1960) [5] p. 6-10, p. 6-11. Right: Joan Miró, Oiseaux dans l'espace (1946). Source: www.museoreinasofia.es

that visual language was memorized more than linguistic; and more, those images with a certain abstraction that still maintain clear links with their referent. In other words, images with a certain ease of reading accompanied by a slight effort to understand the message.

4 Before and After

In any graphic register, ordering the elements is considered a key in the intention of "understanding" the creative process of the designer, and "making it understood" to the public. The order and structure make the story readable. In this sense, Sert provides it to the general public, carefully orders his planning arguments and intentionally manages the types of images and graphic techniques.

He understands that the final goal of the work is the identification of projects series of new buildings for the university that solve the growth needs of the city, but at the same time, he avoids raising misgivings in the rest of the citizens. Thus, in these two senses, the document is structured in two parts: the first about urban reasoning that ensures the harmonious growth of the two communities and, a second part, with a graphic inventory of future buildings (Fig. 5).

First, he outlines complicities between city and university, emerging common topics. Thus, an old engraving (Fig. 6 up) appeals to the pride given to both communities by the historical singularity of the university, the oldest of the British colonies in America.

Following, in a first part, he develops the argument sequentially on different scales: metropolitan area, city, university community and, finally, the university. It makes readable that the metropolitan event conditions the evolution of the university.

The common approach focuses on the graphic story of "before and after". Through this technique, progressively and sometimes repetitively, he highlights the improvements of his proposal on the problems underlined as pressing from a "functional" perspective (Fig. 7 up). It begins with urban growth, and ends with residential distributions, in between, evaluates public transport, land uses, traffic, public facilities, urban renewal

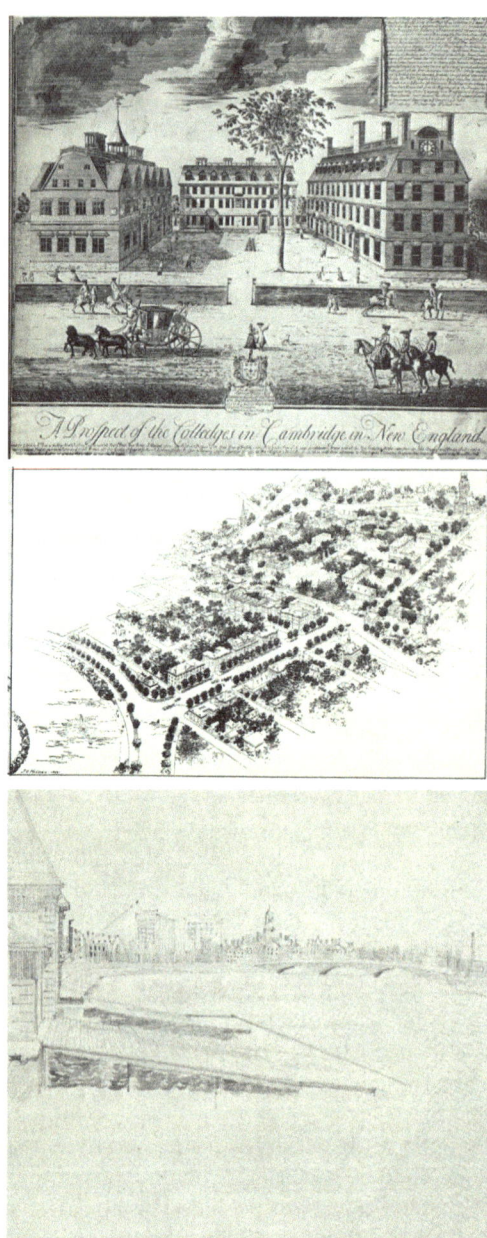

Fig. 6. The "romantic connection" in the story. Up: engraving: view of Harvard in 1743 pag.VI. Center: Study to dignify the approach to Harvard, aerial view (1902) p. 4-4b. Down: Walk along the Charles River in Cambridge (1960) p. 4-16. Source: An inventory for planning, HUPO (1960) [5].

projects, parking, and the network of open spaces and pedestrian paths. The covers of each section graphically synthesize the argument (Fig. 4).

The romantic graphic sequence that connects the story and the open spaces is especially relevant as introduction to the new building list (Fig. 5).

In the **second part**, a brief introduction of plans with general ideas gives way to the concretion of future buildings. Drawings of diagrams, tables and statistics precede plans, perspective views or schemes of each building. It is an end of sweet promises, "desirable" by all (Fig. 5) and, more, seeing that it improves the functioning of the city.

The document prolongs the end with an epilogue of future works and some back covers. It is now the turn of realistic aerial photographs (Fig. 7 down), large and, in black and white. They insist on visualizing university buildings corseted by a continuous residential urban fabric as the proposal origin.

But, in addition, **creative strategies** based on the expressive emphasis on the narrative as a whole are observed. There is a clear connection between intentions and images that discard a pure exercise of banality. A completely graphic language of lines, colors, patterns, etc., individually or related to the rest of the illustrations, weaves the consistency and final coherence of the whole document.

It is not just the willingness of expression but it is the creation and construction of a narrative that pretends to persuade, and in that sense, arguments are expressed in a way that is sufficiently impressive.

It acts on parts of the graphic element, on its totality, or on some of its qualities. Narrative compositions are generated by **repetitions** of elements, or permutations, or substitutions. Thus, for example, the same gray base of the urban fabric is used in the drawings of each of the areas, maintaining same scales, orientations and surface area, illuminating only selected facts. Likewise, the usual practice in functionalist urbanism of repeating colors as coding to represent similar facts is confirmed.

It does not seem coincidental that the plans referring to cars are emphasized with **red lines** on the gray of the urban plot. It is immediately related to the vitality of blood circulation. However, it is also possible to relate them to the bloody traces of blades that dissect the pale gray of the residential skin (Fig. 8 up-left), an accentuated perception when seeing the red dots indicating the traffic congestions. Its forms record blood drops. (Fig. 8 up-right).

Perhaps the most significantly expressive is the cover of the university growth. Halfway through the document, synthetically, the "before and after" of that space shared by both communities are drawn. Their comparison makes perceive some physiognomies that do not differ significantly. The new tall bodies are drawn abstractly with a texture similar to that of trees (Fig. 9 up).

Although this blur of skyscrapers seems like a graphic attempt to appease its visual impact, it can also be explained as a manifesto for reducing apparent volumes, as an articulation of the city's large scale in relation to the existing landscape. In short, a friendly fit of the new buildings in the city is perceived. The graphic idea evokes that drawing of the master showing the attractive profile of minarets in Istanbul as a pattern alternating high and low buildings (Fig. 9 down).

Being convinced that "if the information presented could find its public, the ordinary citizen would understand the validity and feasibility of rationalist proposals" [17], he

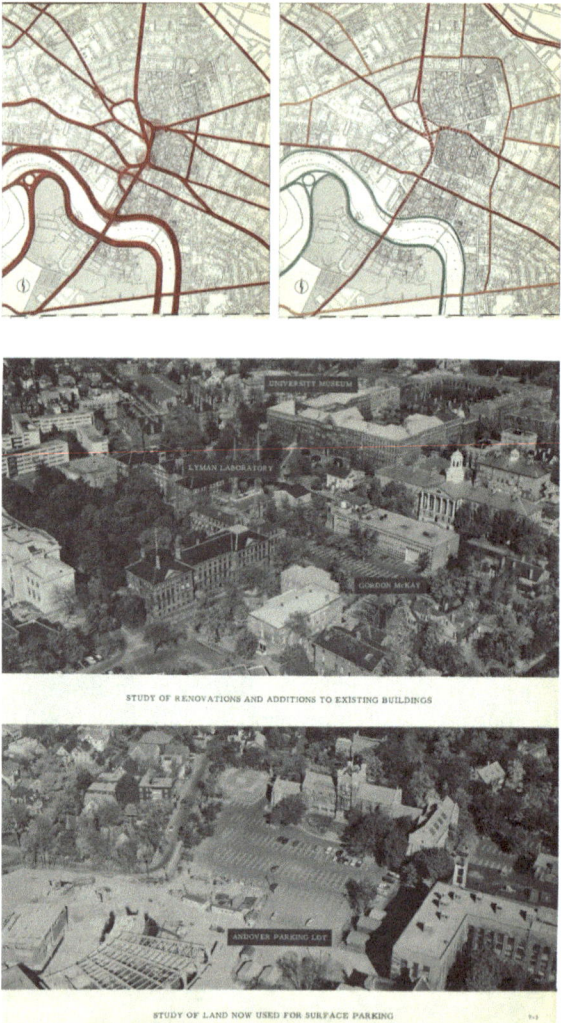

Fig. 7. "Before and after". Up-left: Existing traffic flow p. 3-7. Up-right: Proposal of road classification p. 3-9. Down: Photographies of places for future planning work p. 7-3. Source: An inventory for planning, HUPO (1960) [5].

materializes a printed document accessible to the general public. The coherent ordering between graphic and conceptual messages allows the reader to memorize a synthetic mental map as a result of an intentional sum of concatenated synthetic messages.

This is a proposal for the implementation of public facilities argued from a planning "story", and therefore capable of being told to the general public. "The architect has

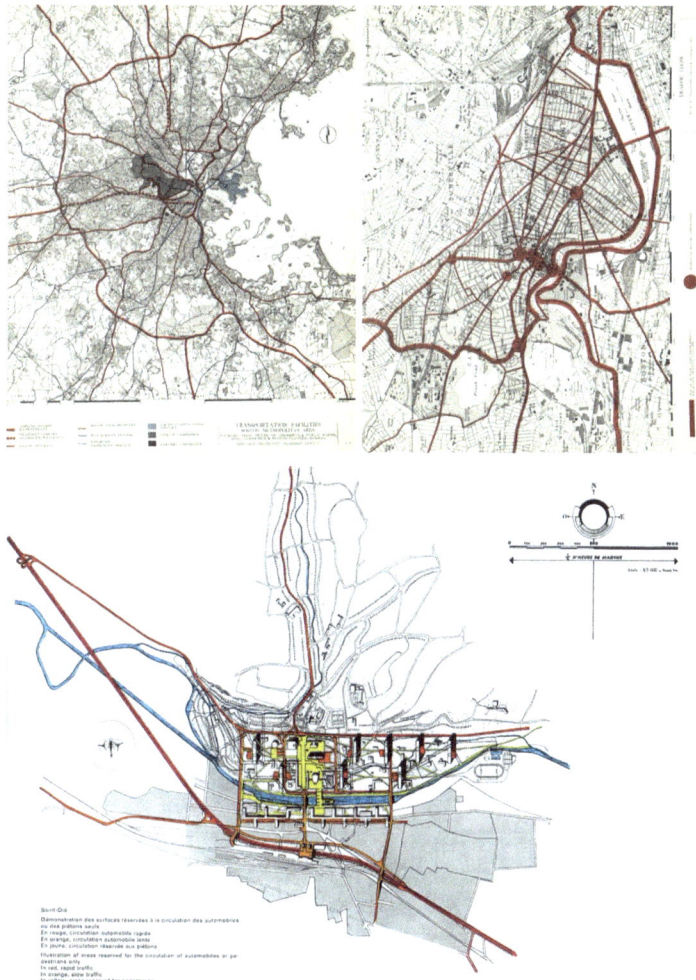

Fig. 8. Expressive emphasis by the rhythmic transmission of the same idea: color, repetition. Up-left: roads in the metropolitan area. Source: An inventory for planning, HUPO (1960) [5] p. 1-5. Up-right: road and congestion in the city of Cambridge. Source: An inventory for planning, HUPO (1960) [5] p. 2-7. Down: Planning of Saint Dié de Le Corbusier (1945). Source: Le Corbusier, Complete Works n°4 p. 136

approached them, being able to tell, explain, give a reason for what he does" and, therefore, "becomes more easily understood by who has to assume it" [18]. The construction of the planned buildings, which keep its functionality today, demonstrates the value of this urban theory, but, above all, the value of persuasion of the document.

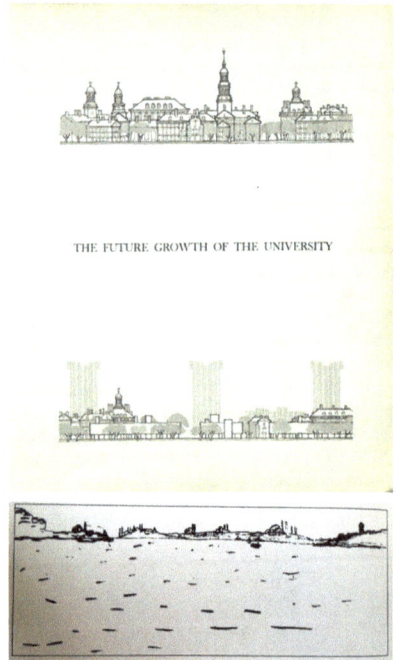

Fig. 9. Up: Cover, the future growth of the city. Source: An inventory for planning, HUPO (1960) [5] p. 6-1. Down: view of Istanbul from the Marmara Sea, at the entrance of the Bosphorus, by Le Corbusier (July 1911), source: Voyage d'Orient Carnets AFLC 6128

Fig. 10. Artistic expression and architectural representation. Left: **Cui Ruzho** 红梅 – 镜 心– 设 色 纸 本,. Right: Harvard University Facilities Plan; Source: An inventory for planning, HUPO (1960) [5].

5   Conclusion

Thus, a careful look at the original inventory for planning at Harvard University unveil intentional graphic strategies. It is a material prepared to persuade, and all, through visual resources that rely on communicating a permeability, sensuality and respect among all the communities involved: the metropolitan area, the city and the university. The value of the narrative form, and that of its graphic quality, are key to achieving the predetermined objective of persuasion.

At a time when the constant technological advances guide towards representation systems with greater and faster graphic resources, it is opportune, as in a new Renaissance, to reimburse with a new and deep critical sense those graphic patrimonies of the history of Architecture, which as in the case of the legacy of Sert's graphic work in Harvard Planning, it contributes to show us, in addition to his urban thinking, a timeless ability to "persuade", concretized through a "shocking" graphic representation and an ordered sequence of images, and where the artistic contribution is essential (Fig. 10) and useful, to make "seductively" perceive those new ideals that will shape the utopian construction of the "modern" city.

Time has shown that, despite the daring proposal, the university and the city have been configured in their sense, and this makes their "graphic narration" as a persuasive reference of success.

It is a way of doing usually forgotten, or diffused in the teaching of Architectural Representation which is mostly focused on working on one or several isolated images, disconnected from a larger set of drawings and abandoning all narrative ideas. This form of restriction limits the achievable power with a graphic story, while preventing modeling a more intense mental image of memories and meanings of the transmitted arguments.

It consists on a mode of drawing, "of revealing and hiding" that not only "is understood" in the individual value of a drawing by itself, but in the value of a set of drawings, all of which are understood as part of a unit. In somehow, it reminds it to the way of organizing documentation in an exhibition, or "how to tell a story" [19], or even more distant, to represent perceptions embedded in memories and meanings of a city [20].

The study of this document helps to show graphic narrative resources that improve the ability to communicate ideas more persuasively. Therefore, seeing the importance that this currently has, it is evident the interest of this type of resources for architects, and even more, for those who have to teach them.

Acknowledgements. All images from An Inventory for planning, HUPO (1960) are Courtesy of Frances Loeb Library. Harvard University Graduate School of Design. The authors would like to thanks to Ines Zalduendo and Ann Whiteside for their collaboration. This investigation was supported by the Research Group ADR&M (UPC BarcelonaTech) and the Consortium for Advanced Studies Abroad (CASA). Isabel Zaragoza, one of the authors, is a Serra Húnter Fellow.

References

1. Vesely, D.: Architecture in the Age of Divided Representation: The Question of Creativity in the Shadow o fProduction. MIT Press, Cambridge (2004)

2. Mumford, E., Mostafavi, M.: The writings of Josep Lluis Sert, p. 13. Yale University Press, New Haven (2015)
3. Roca, E.: Campus y Ciudad, la experiencia del Barcelona Knowledge Campus. Rev. Iberoamericana de Urbanismo (5), 97–105 (2011)
4. Rovira, J.M.: José Luis Sert: 1901–1983 (English ed.). Milano: Electa Architecture. Phaidon Press (2003)
5. Harvard University Planning Office: Harvard University, an inventory for planning. Harvard University Planning Office, Cambridge (1960)
6. Millan-Gomez, A.: Polémicas berlinesas. Representaciones en la obra de Mies van der Rohe, años veinte. EGA Rev. de Expresión Gráfica Arquitectónica (13) (2008)
7. Amado, A., Fraga, F.: El dibujante digital. Dibujo a mano alzada sobre tabletas digitales. EGA: Rev. De Expresión Gráfica Arquitectónica (25) (2015)
8. Mumford, E., Sarkis, H., Turan, N., Hyde, T., Sekler, E., McAtte, C.: Josep Lluís Sert: the architect of urban design, 1953–1969. Yale University Press, New Haven/Harvard University Graduate School of Design, Cambridge (2008)
9. Rubert, M.: Ciudades en América Latina. En Sert Arquitecto en Nueva York. Museu d'Art Contemporani de Barcelona – Actar, Barcelona (1997)
10. Freixa, J.M.: Josep Ll. Sert (1ª ed. Catala/portugues. Ed. Obras y proyectos) Gustavo Gili, Barcelona (1979)
11. Bastlund, K., Sert, J., Giedion, S.: José Luis Sert: architecture, city planning, urban design. F.A. Praeger, New York/Verlag für Architektur (Artemis), Zurich (1967)
12. Freixa, J.: Josep Ll. Sert (1.a ed. español/ingles. ed., Obras y proyectos). Gustavo Gili, Barcelona (1989)
13. Rovira, J.M.: Sert half a century of architecture: 1928–1979, complete work. Fundació Miró, Barcelona (2005)
14. Granell, E., Redondo, E.: Los dibujos de proyecto de J.L. Sert para la fundación Joan Miro de Barcelona. EGA: Rev. de Expresión Gráfica Arquitectónica **22**(30), 224–237 (2017)
15. Le Corbusier: El regalo de los Técnicos (1952)
16. Mendelson, J.: Sert y la imagen, creating a public for modern architecture: sert's use of image from GATCPAC to the heart of the city. DC Papers (13–14), 130–139 (2005)
17. Sert, J.L.: Can Our Cities Survive? An ABC of Urban Problems, Their Analysis, Their Solutions. Harvard University Press, Cambridge/H. Milford, Oxford University Press, London (1994)
18. Moneo, R.: La construcción del relato Nueva sede del BBVA Madrid, España. Arquitectura Viva (186), 26–35 (2016)
19. Marquez, G.G.: Como se cuenta un cuento. E.I.C.T.V. Ollero & Ramos, Madrid (1995)
20. Lynch, K.: The Image of the City. MIT Press, Boston (1960)

Correalism and *Biotecnhique* in the Drawings of Frederick Kiesler

Starlight Vattano[⊠]

Free University o fBozen, Bozen, Italy
starlight.vattano@unibz.it

Abstract. Starting from the reflections carried out by Friederick Kiesler on theatre as place of experimentation of the architectural matter, the issue deals with the deepening on the relationship between the drawing and the theatral space of the kieslerian vision, through the redrawing of the *Universal Theatre* of the 1961, architectural synthesis of his *correalist* and *biotechnique* poetry in which the interaction between actor-spectator and stage-auditorium is focused, already launched in Vienna around 1920 with the projects of the *Endless Theatre* and the *Raumbuhne*. The aim is to produce new images and new places of one of the most emblematic works of Kiesler, tracing again the set-design constructions and the studies of the possible movements made by the actors in the scene for the definition of the theatral space. The redrawing of Kiesler's work raises some material questions on the existence that the architect faced looking at the project as an abstraction process, from the nature to the machine, passing through an architecture in progress. The graphic production of Kiesler testifies the coexistence of two principles characterizing his work: the concept of space-time in architecture and the continuous construction rendered through the tension of the shell. The digital elaborations produced following the graphic study aim at metabolizing these two principles, framing the interaction in the place of the drawing.

Keywords: Correalism · Biotechnique · Graphic interpretation · Redrawing · Set design

1 Introduction

The graphic study aims to trace the key points of Kiesler's theatrical experimentation, who elaborated a *correalist*[1] and *biotechnical*[2] vision of architecture, recognizing in the theatre a space for the experimentation of his philosophy of the *Endless*: dynamic set designs, projections of light and galactic structures in shell, crystallized in a "whole"

---

[1] Frederick Kiesler published the *Manifesto of Correalism* in 1939, in response to modernism, addressing the problem of the house designed as a place of regeneration of vital forces.

[2] By biotechnology, or *biotechnique*, Kiesler means man's ability to influence life in specific directions. The concept was borrowed from the term coined by Patrick Geddes *biotechnics*. For further information on the concepts of *correalism* and *biotechnique* see: Kiesler F., "On Correalism and Biotechnique", in *Architectural Record*, settembre 1939.

© The Editor(s) (if applicable) and The Author(s), under exclusive license
to Springer Nature Switzerland AG 2020
L. Agustín-Hernández et al. (Eds.): EGA 2020, SSDI 6, pp. 585–595, 2020.
https://doi.org/10.1007/978-3-030-47983-1_52

scene. The scenic director of the productions of the *Julliard School of Music* in New York and co-founder of the *International Theatre Arts Institute* in Brooklyn, Frederick Kiesler emigrated in 1925, unknown from his Habsburg homeland, dedicating himself to the practice of art in various forms, from the figurative arts to the theatre, up to architecture. In a monographic issue of the series directed by Bruno Zevi, *Gli Architetti*, dedicated to Kiesler, Maria Bottero observes that was probably his image of the integrity of the human being with the continuity of everything to generate "multiple disciplinary encroachments" (Bottero 1999, p. 6).

In relation to the reading of space perceived by the integrity of the human body, as early as 1950 the architecture critic Douglas Haskell had already recognized Kiesler's invention of a modality of material modelling, of a *second modern 'order'* (Philips 2017, p. 2), which contrasted the modern building with box structures in steel and glass, in favor of more advanced technologies for the creation of organic and naturalized spaces; the architect, in fact, used to make up for the scarcity of economic resources with his figurative, holistic and multi-material sensitivity of the scene.

The galactic dimension of sculptural, figurative and scenographic formalization will be declined to the architectural scale between 1950 and 1960. The structural morphologies based on the relationship between vacuum and tension are flanked by the spatial research that also finds correspondence in the musical field, in that silence of Cage's sound composition, an impression of magnetic forces that in the vacuum generate the "nothing-between-part" (Bottero 1999, p. 9) (Fig. 1).

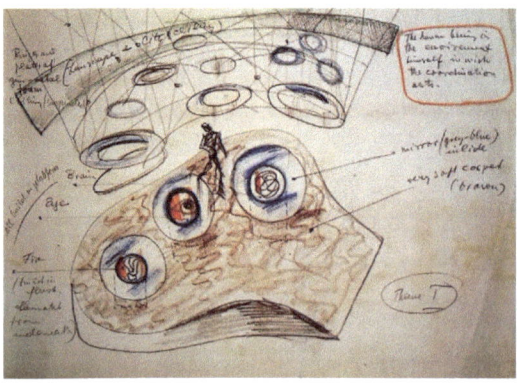

Fig. 1. F. Kiesler, sketches for the *Vision Machine*, 1959–1962. Source: Frederick e Lillian Kiesler archive, Vienna.

Around 1990 the research of the intellectual and technological project began to turn to the continuity of the envelope that echoes the Kieslerian *Endless*; in fact, also in the philosophical field, Gilles Deleuze proposes a reflection on the plastic fluidity of architecture, on the possibilities of a moldable organism through the bending of material-energy that, in continuous form, pursues its development in the absence of vacuum (Deleuze 1990).

2 Shapes and Body in Kieslerian Theatre

As an amateur actor and a great theatre lover, Kiesler recounts his space through sculpture and cognitive schemes on the machines of vision. As shown by the diary recorded in 1961, concerning the project of a musical work proposed by Edgar Varèse, the set designer and architect will move more and more towards the automation of life in the theatre, arriving at the hypothesis of a total replacement of engineers with the machine. The graphic themes tackled in Kieslerian theatre are influenced by the Futurist, Constructivist and De Stijl artistic avant-garde, polarizing the gaze on the cyclicality of repetition and the circular scheme. At the centre of this speculation on the continuity of space, Kiesler places the formation of the human personality, replicating the industrial production of consumer goods with the creative realization of the self, a philosophy of daily, technological and sensory continuity that takes shape through *correalism* and *biotechnique*. Preliminary studies and scenographic sketches complete the reflection on the poetic home-theatre that aimed at shaping a new society through the creation of new ways of living based on the trinomialism *correalism-endless-galaxy*.

But the symbolic and significant value of the word intertwine with those of the gesture, defining the educational and therapeutic purpose of improvisation in that theatrical experience theorized by the doctor Jacob Levy Moreno who in 1921 gave his *Steigreiftheater* (theatre of improvisation) a circular stage.

For Kiesler, this interweaving of theatre and life corresponds to the insertion of the stage into the space visualized and constructed by the actors and spectators. The innovation consists in the alteration of the spectator's perceptual mechanisms through a synchronization of scene changes with the dynamics of the action, thus recurring to the expedient of light and image projections and introducing mobile screens that refer to the temporariness of life (Fig. 2).

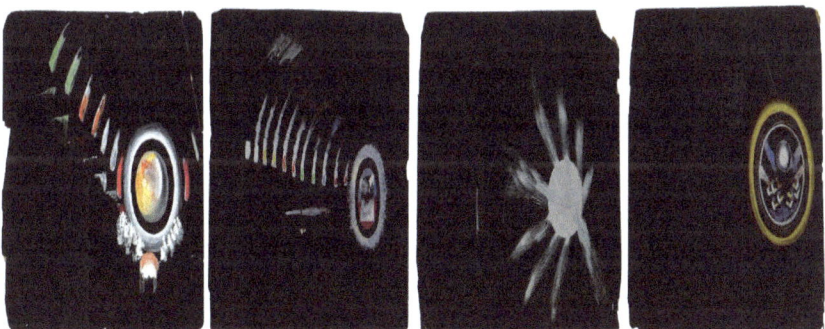

Fig. 2. F. Kiesler, studies on the light and images projections for the *Film Guild Cinema*, 1928–1929. Source: Frederick e Lillian Kiesler archive, Vienna.

The ideogrammatic transparency of his drawings also responds to the reflection on the incorporeality of the self, symbolized by the architectural tensionism and the dematerialisation of spaces.

His vision of a future theatrical space coincided with the transformation undergone by the human body within the theoretical speculation on the mixture of machine-humanity. In fact, Kiesler was strongly influenced by the androids and living machines of Karel Čapek who in his work *R.U.R.* announced the obsessive unification of the technological device in everyday life.

In dealing with the relationship between the animate and the inanimate, with the aim of proposing new forms of observation on the moving image, Kiesler measured himself by experimenting with futuristic environments that provided his architectural and tonal visualizations with the dimension of temporal depth necessary to immerse himself in the drama of the cybernetic advent.

His vision of architecture drew on the fields of biology, psychology and aesthetics to rethink the remodulation of the space occupied through the curvature of surfaces, the bending of scenographic concavities and the bending of volumes that would physically envelop the spectator inside the theatrical belly, projecting him into the scenic action. The correalist diagram proposed by Kiesler considered these ways of declining space within a total environment, structuring it on the basis of three subsets: the human environment, the natural environment and the technological environment. The three environments interact with each other converging in that dimension of human heritage which, according to Kiesler, gives shape to the performative space (Fig. 3).

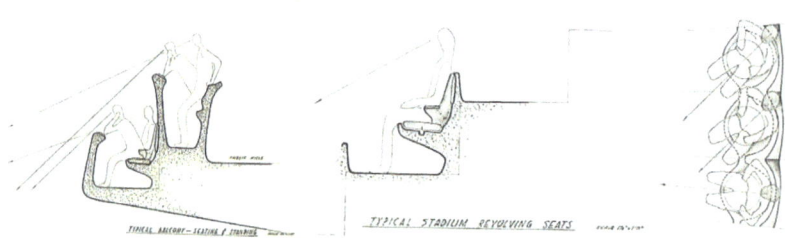

Fig. 3. F. Kiesler, from the left, section studies on the seats of the balcony and auditorium. Source: Frederick e Lillian Kiesler archive, Vienna.

The physical and psychological worlds open to the multidimensionality of space by establishing a system of life cycles and establishing a new connection between individual and collective, between subjectivity and environment. The scenic devices designed by Kiesler, unlike what happened in the theatre of the 19th century, were conceived to provide a technological dynamism capable of animating future puppets and robots. It was a subversion of the scene confronted with the new aims of science and the pulsating need to visualize social, temporal and material experience through the enormous media potential.

Its theatrical envelope coincided with the illusory *mise-en-scène* narrating the infinite space through the fusion of kinematics on stage with spectators and artists. Kiesler, in fact, tried his hand at adopting optical and haptic techniques to explore the filmic environment as well, stimulating the perceptive qualities of the viewer who was thus assimilated/developed by his/her immediate surroundings.

This is how he thought of space as Étienne-Jules Marey had developed his chrono-photography, a dimension in which to experiment with the dynamic possibilities of the human body by scanning them over time as if they were a multiplicity of frames on a two-dimensional support to describe movement. The surface plunged into the temporal dimension, the scenic space into the kinaesthetic one.

The maximum lightness of the structure operates in the act of the material curvature, the space of the theatrical world bends to the language of the symbol: the tensegral structures weave linear units along forces wefts. Thus *continuity* and *change* become two modes of internal-external interpenetration, memory-future.

In the organization of the actions on stage, Kiesler thinks of a figure, a real coordinator, for the synchronization of musical time and dynamic space. In this way, the act of drawing in Kiesler's *action living* is part of the same path beaten by the artists on stage: it deepens the materials, crosses kinetic circuits and reaches the definition of a stereophonic and multidimensional space. In 1955 the preparatory drawings for the theatre-tent for the *Ellenville Music Festival* embodied the intentions of the architect-actor who connected two stages at the extremes of a large auditorium designed with two proscenium spaces and a walkway that followed the entire theatre ring perimeter. An additional central corridor allows the actors to pass from one stage to another within the main stage or to walk behind the spectators. The aim is to envelop the audience by means of the physical, sound and visual space, precisely for this reason he feels the need to free himself from the fixed elements of the stage in favor of mobile systems that can be easily assembled and disassembled.

As it happens in the theatrical spaces proposed by the Russian school based on the teaching of Stanislawskij-Mejerchol'd, in the theatre of cruelty of Artaud in France, or in the section of the Bauhaus enhanced by Schlemmer, Kiesler also intervenes in the redefinition of the scenic space by establishing a new relationship between the actors and the audience.

3 Redrawing of the *Universal Theatre*, 1961

The graphic study developed for the *Universal Theatre* project is based on a first phase of iconographic investigation of Kiesler's sketches on spaces, sets, mobile devices and stage movements.

The analysis concerns the identification of the spatial metrics adopted during the design phase in which the signs of spaces interpenetration, displacements, acoustics and light projections establish a graphic code to trace the drawing-project integration. Since the first sketches it is possible to recognize what would have been Kiesler's final idea: the unification of a large ovoid shell, firmly fixed to the ground through a curvilinear base, with a massive turreted volume at the back characterized by a more regular section in the studies of 1959 and strongly helical in the study drawings of 1962 (Fig. 4).

The tensional force through which Kiesler expresses his principle of the infinite is embodied in the *Universal Theatre* project, a container conceived as the sum of "enveloped shells", which builds the space «for infinite variations of sound, light and stage action» (Bottero 1999, p. 20). The design of the theatre, which Kiesler called "white elephant" (Bottero 1996, p. 80) provided for a capacity of 10,000 people fully enveloped

Fig. 4. F. Kiesler, sketches of the *Universal Theatre* project, 1959–1962. Source: Harvard Theatre Collection, Cambridge.

by a double shell designed to contain the heating and cooling system, together with a series of ramps, elevators and platforms, a "continuous" structure that would favour total integration between the audience and the actors at any point of the space (Fig. 5).

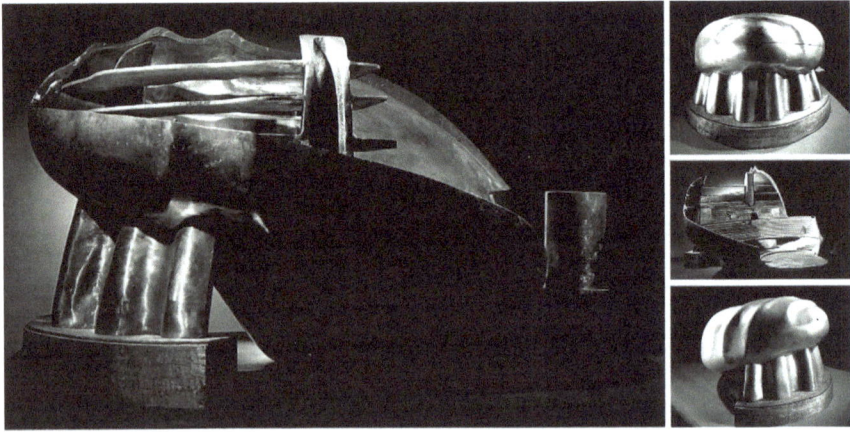

Fig. 5. Photos of the original model, 1961. Source: Frederick e Lillian Kiesler archive, Vienna.

We could speak of a Kiesler's approach to the cinema dimension in which we proceed with the enhancement of the viewer's proprioception completely immersed within a film, in the absence of the curtain, the proscenium and the stage.

His dynamic scenography involves the projection of images in place of painted scenes accompanied by the use of symbolic forms and abstract elements to create different configurations. The *Universal Theatre* is the latest project for Kiesler's theatre and therefore represents a synthesis of the architectural values already expressed in the *Endless*, the *Performing Arts Center* in Brooklyn and the *Woodstock Theatre* (Fig. 6).

It is a competition project launched by the Ford Foundation for the creation of a new prototype of American theatre for large cities; Kiesler will work nine months on the

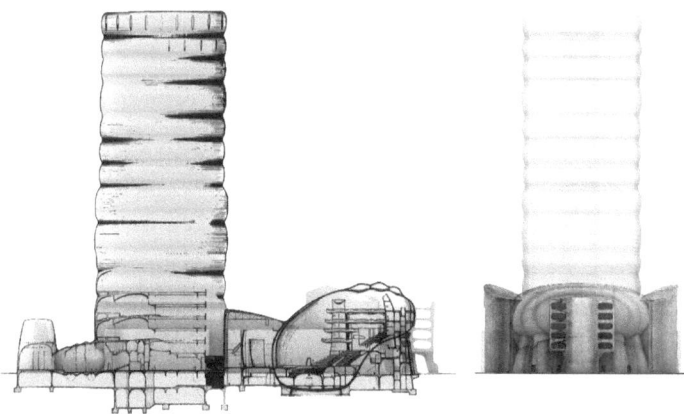

Fig. 6. Redrawing of the theatre with the original section and façade. In the images of the digital model have been inserted the forebody at the entrance (section) and the central tower (façade) not included in the original drawings.

body of drawings and the plaster and wood carving of the complex of buildings designed as a service centre for the show.

The large "white elephant" consisted of an amalgamation of volumes differentiated by size and shape, a large 1,600-seat auditorium defined by a stage with peripheral spaces that could be transformed into an arena. The sinuosity of the envelope follows the morphology of a shell that builds a fluid space in which lights and projections evoke the philosophy of Kiesler's infinite. The large ceiling-screen is a reinforced concrete shell designed without pillars or beams to contain infinite sound, light and scenic variations. The three projection towers, once the machinery for suspending and lifting traditional scenes has been eliminated, perform the function of vertical support and disengagement of the balconies on the different levels (Fig. 7).

The control booth designed to amplify the possibilities of projection in different directions is placed in the center, while two other booths are placed among the spectators to resolve the issue of the projection of films or slides both on the bottom and on the cyclorama. The buildings system is accessed through a curved envelope, treated as a large entrance curtain in which four openings are made. There is no information on the function of the forebody placed in axis with the central volume which is not indicated in the plan (Fig. 8).

Kiesler inserts a central tower of thirty floors, thus interrupting the horizontality of the concatenation of volumes, designed to house a series of small theatres or audition rooms for an audience of 120–130 people, television studios, radio stations, publishing offices, spaces for film and record production or for the staging of exhibitions. The large shell with double envelope and the central tower are joined by a connection volume that contains two curved screens probably designed in response to the possibilities of projection and image alteration. Both in the plan and in the section it is possible to read two concentric rings in correspondence of the stage that suggest a mobile scene designed

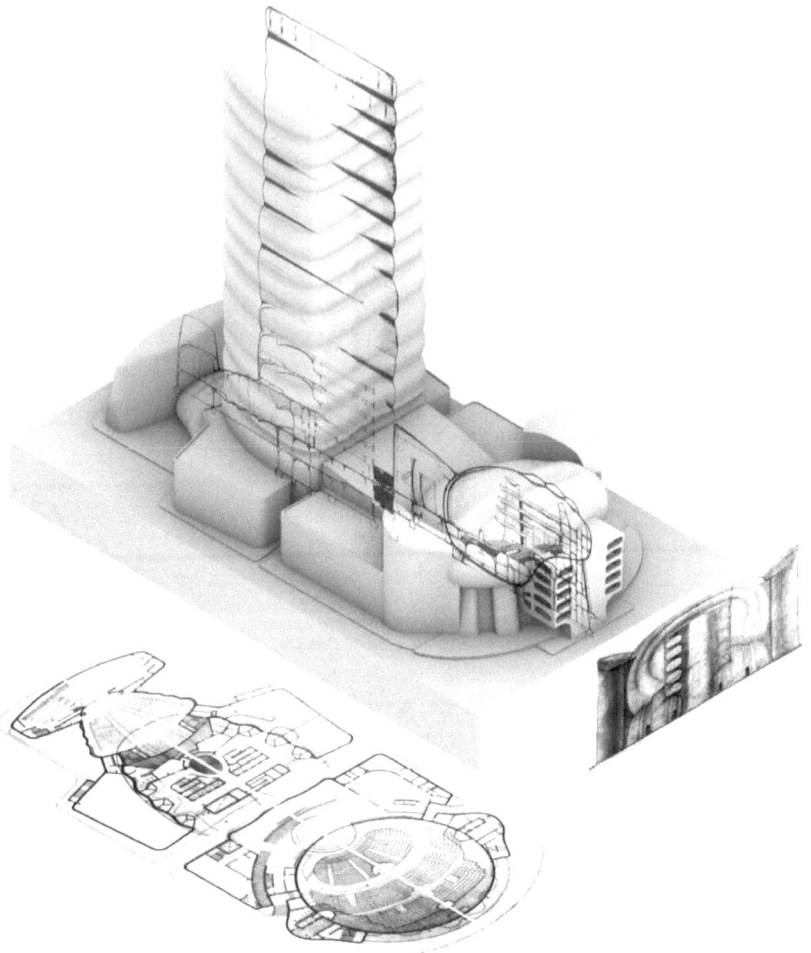

Fig. 7. Digital model of the theatre with insertion of the original drawings.

to incarnish that philosophy of the *continuum* from which Kiesler was inspired in his life-theatre project. It is probably no coincidence that both bodies used as auditoriums have been modelled as if they were a primitive shelter or a uterus that refers to the origin of living, to a primordial place in which the scenic action develops.

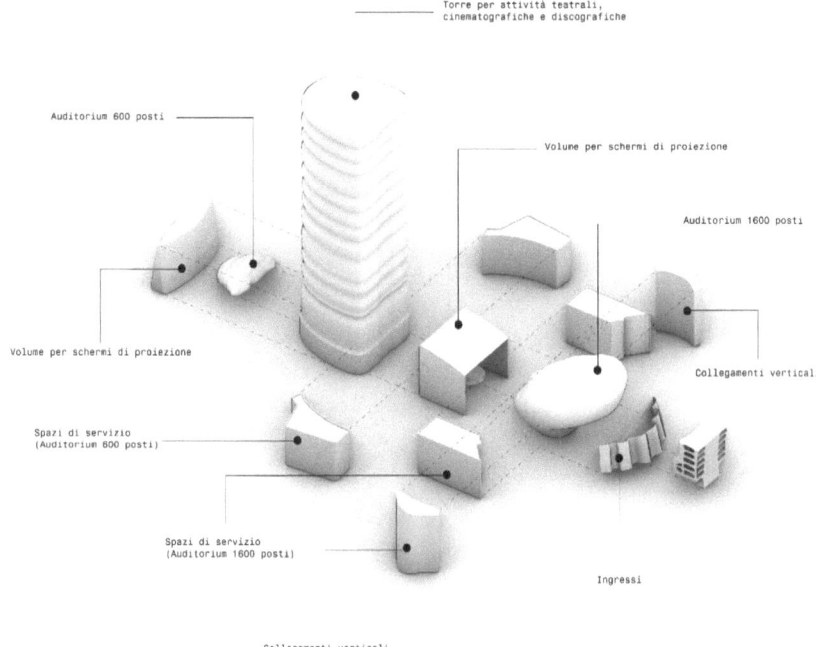

Fig. 8. Exploded digital model with functional identification of volumes.

4   The Correalist Sign by Kiesler: Conclusions

The theatre contains the founding character of Kiesler's work, in fact, between 1921 and 1923 he created the innovative set design of R.U.R.[3] which, through its kinetic images, attracted the attention of many avant-garde artists. The theatre for Kiesler represents the place of architecture-nature, body-shell of living. The sound and theatrical space are conceived as stereophonic and multidimensional actions in relation to the movements of the actors and the projection of the images on mobile surfaces, as shown by the sketches for the total-stereum theatre conceived with Edgar Varèse in 1961 or the circular stage of the *Raumbuhne*, created for the Festival of Music and Theatre, which balances Kiesler's architectural vision on the modification of the actors-spectators' system: the stage is placed inside the space for the audience, abandoning the two-dimensionality of the nineteenth-century backdrop in favor of a three-dimensional and mechanized scenography[4] (Fig. 9).

On the project hypothesis Kiesler would say: «I did not draw a set design but a theatre for sets, complete with ceiling, seats and stages» (Kiesler 1966, pp. 427–428). In this continuity that links everyday life with technology and education through the

---

[3] The set design was made for the play taken from the utopian science-fiction drama *R.U.R.* (*Rossumovi univerzální roboti*, The universal robots of Rossum) written by Karel Čapek.

[4] In the same years similar experiences were carried out in Russia, as demonstrated by the sceno-graphic work of director Mejerchol'd for Vladimir Majakovskij's *Mystery Bouffe*, in 1921, which completely renounced the backstage with a stage invading the audience.

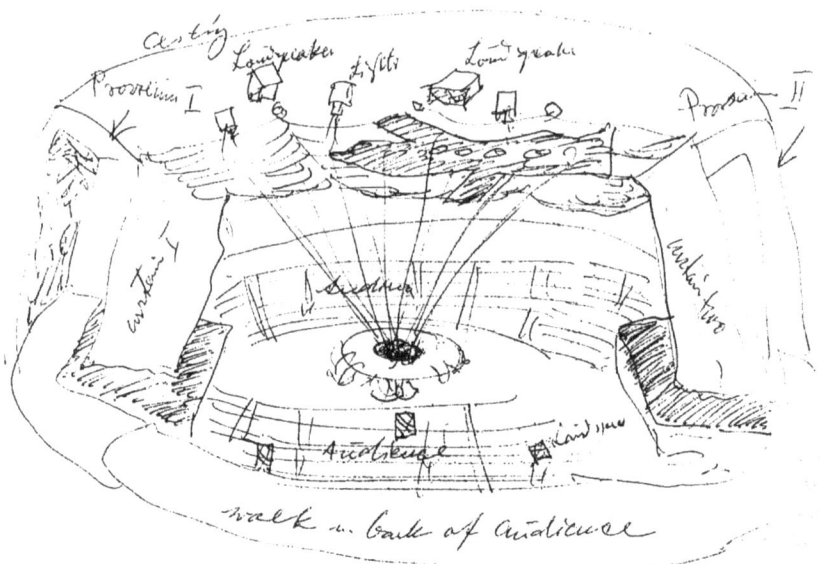

Fig. 9. F. Kiesler, sketches of the tent-theatre designed for the *Ellenville Music Festival*, 1955. Source: Frederick e Lillian Kiesler Archive, Vienna.

manipulation of art and philosophy of the body in space Kiesler affirms his being-present by saying: «I deny that there is a dualism between love and work or between life and dream [...]. The horizons of imagination are as real as an open wound in our body. There is no division at the root. The split is only at the top. This rupture of vital events is due only to the insufficiency of our senses and especially of the brain. When I feel, I know. When I feel, I am. If I think, I am divided» (Bottero, 1999, p. 18).

A continuity of being interpreted both in Kieslerian philosophy and architecture. In this dialogue between the theoretical and the concrete, the operation of redesign explores the two-dimensionality of design traces by returning three-dimensional orbital spaces in the digital dimension. The project sketches and mono-material models created by Kiesler make it possible to read the sculptural apparatus of the architectural object. This information is integrated through the redrawing elaborated starting from a ground floor plan, a façade at the entrances and a longitudinal section that cuts through the sequence of volumes, from the 1,600-seat auditorium to the space for the projection screens, expanding the architectural and spatial knowledge of the theatre. In fact, the new images integrate with the original drawings, the axonometric correspondence between the digital model, the plan and the project section place in space the sequence of volumes just mentioned in the two-dimensional representations and distinctly legible in the digital model. The creation of three-dimensional drawings also makes possible to broaden the experience of abstraction and synthesis of architectural content through the possibilities that the exploded axonometric scheme offers in identifying the functional organization of internal and external spaces. The continuity of the world, matter and life constitute the representation of Kieslerian infinity through form, symbol and time. The elaboration of new images aims at the digital reconstruction of the *Universal Theatre* in which the paradox of this existential *continuum* is configured: the primordial permanence and the transient of the ephemeral (Fig. 10).

Fig. 10. Digital model of the *Universal Theatre*.

References

Bottero, M.: Gli architetti, Frederick Kiesler. L'infinito Come Progetto, Testo & Immagine, Torino (1999)

Bottero, M.: Frederick Kiesler: Arte, Architettura, Ambiente. Mondadori Electa, Milano (1996)

Deleuze, G.: La Piega. Leibniz e il Barocco. Einaudi, Torino (1990)

Kiesler, F.: Inside the Endless House. Simon and Schuster, New York (1966)

Phillips, S.J.: Elastic Architecture: Frederick Kiesler and Design Research in the First Age of Robotic Culture. MIT Press, Cambridge (2017)

Millimeter Towers in a Lost Space. The Vision of Saul Steinberg

Luis García Gil[(✉)] and Javier Fco. Raposo Grau

IGA Department, E. T. S. Arquitectura, Universidad Politécnica de Madrid, Madrid, Spain
luis.garciag@upm.es

Abstract. Saul Steinberg, one of the most innovative artists of the 20th century, revolutionized the world of illustration, becoming an indisputable reference in modern graphic narration.

This cartoonist and architect was born in 1914 in Râmnicu Sărat (Romania). In 1933 he began his life's journey, moving to fascist Italy to enroll in the Faculty of Architecture of the Royal Polytechnic of Milan, earning a doctorate in architecture in 1940. During these years he began to publish his drawings in magazines such as Bertoldo. In 1941 Saul Steinberg began a difficult flight to the United States of America, where he would take up his final residence. In 1942, The New Yorker magazine began to publish his drawings, starting a collaboration that would last until the end of the artist's life in 1999.

The aim of the paper I am presenting is to relate Saul Steinberg's critical vision with the work of two emblematic European artists in 18th and early 19th century: Giovanni Battista Piranesi and Francisco de Goya. Three artists who, beyond their formal innovation, have bequeathed us a graphic heritage capable of narrating time and anticipating future eras.

Keywords: Steinberg · Piranesi · Goya

1 Introduction

Madrid, June 1, 2019. National Library of Spain. Exhibition by Giovanni Battista Piranesi [1].

The rooms are in semi-darkness, a dim light forces one to concentrate on Piranesi's engravings. My initial search leads me to the engravings of *Le Carceri d'Invenzione* (1749–1761). They are in the second room, extraordinary as always, just as I remember them. The question that already inspired me is still up in the air: when or why does an artistic creation, a graphic, a representation, become the voice of an era or acquire immortality in order to transcend its time? Perhaps the answer to this question has to do with this insistent idea that leads me to relate Piranessi, with Goya and with Saul Steinberg: their capacity to symbolize not only a world of their own, but the spirit of their whole era, its antecedents and its most fearsome consequences. Giovanni Battista Piranesi and Francisco de Goya are the two authors I have selected to accompany Saul Steinberg in my comparative research because the exceptionality of their hand and their

L. Agustín-Hernández et al. (Eds.): EGA 2020, SSDI 6, pp. 596–606, 2020.
https://doi.org/10.1007/978-3-030-47983-1_53

vision has transcended their time, and their works are part of the graphic heritage of Humanity.

The baroque and pre-romantic vision in *Le Carceri d'Invenzione* by Piranesi somehow opens the way to the illustrated and pessimistic vision of the series of prints *Desastres de la Guerra* or *Los Disparates* by Goya [2]. And we could say that both artists anticipate the great convulsions and changes of the 20th century.

From these glances at the past, the philosopher Walter Bejamin [3] reflects on how the 19th century is not aware of the destructive energies of technology, and thus, after the splendour of the change of century, 'the nightmares of the 20th century', as the philosopher called them, were lurking; two world wars that massacred entire generations of young Europeans and put an end to the illusions of progress as conceived by the generations that preceded them. This trace is collected in what Bejamin called narcotic historicism; a complex conceptualization through which he explains how the dreams of the 19th century, with its ideals of freedom, equality and progress, with the strengthening of bourgeois domination and its rigid structures and hierarchies ended up generating the elements of their own destruction, giving way to totalitarianisms and their darkest consequences.

It is these images, these traces and their reflections in the post-war period, that shape an existentialist and dehumanizing vision of reality, where the everyday being becomes a hero and the protagonist of the social story: those ordinary men and women that Saul Steinberg s o acutely and brilliantly synthesizes in his illustrations, through his recurrent themes about the universal human being and his existentialist vision applied to a mutilated and wounded society after surviving two world wars.

Not always a generation has in its memory a Piranesi or a Goya or a Steinberg.

Highlighting these milestones in the timeline and collecting their graphic and symbolic heritage, I would like to reflect on the significance of some drawings by Saul Steinberg by comparing them to Piranesi and Goya.

2 The Vision of Saul Steinberg, Appearances, Solitudes and Masks

Saul Steinberg was one of the main renovators of graphic humour in the 20th century and a great *flâneur* [4] of his time. Piranesi and Goya were also two great observers and storytellers of their respective epochs.

Their portraits propose rhetorical and deeply penetrating glances, glances from the 18th, 19th and 20th centuries, linked to the figuration and representation of their time.

Piranesi looks forwards, but from the sidelines, surrounded by an ancient taste, cladding his image with references to culture, perhaps to his training as an architect. He looks down on us from above, reminding us that we are not the chosen ones.

Goya shows a frontal, self-absorbed, solitary, critical, enlightened look, perhaps meditating on the failure of the human being and his inability to prevent war, showing an essence of the human being that is controversial, trapped, capable of dreaming great utopias and on the way losing reason for them (Fig. 1).

Saul Steinberg, longing for past centuries, looks at himself with the mestizo language of the 20th century, as subtly represented in his singular self-portrait-mask, he rejects

Fig. 1. Felice Polanzani. Portrait of G.B Piranesi. Rome. 1750. Source: Le Antichitá Roma..., Volume I. 1756, etching, INVENT/19044. Biblioteca Nacional de España. Madrid [1]. Francisco de Goya y Lucientes. Self-portrait. Madrid. 1795. Source: Museo del Prado, Madrid [2].

the frontal look, looking at the past, at that area of the paper that represents the origins, perhaps nostalgic for what he left there, his lost future.

In Roland Barthes' words, the portraits could be understood as traces or reflections;

"... All of Steinberg's characters look like someone I know (I'm sure) but whose name I never remember. Who could it be? Where did I see that face, that appearance? I struggle with my memory, as in a dream whose precision (and not vagueness, contrary to what is believed) is an enigma to me. Steinberg does not give me an outline; it is again and again a subtle and penetrating face that seems to emerge from my life. In Steinberg's world, in short, nothing but identities, of which I try to find their proper names..." [5].

Saul Steinberg photographs himself, draws himself, creates a collage, with simple lines reminiscent of children's games ... a six and a four, the face of your portrait ... which marginally reminds us of James Joyce's word games, ...

... Co Co Coroc. Cluc Cluc Cluc. The black one is our hen. She lays eggs for us. When she lays the egg, she's very happy. Coroc. Cluc Cluc Cluc. ... [6].

In his 1951 drawing, he represents the universal human being of the 20th century, as an irony of an era. If there were portrait catalogues this would be *"The Portrait"*. It contains the exact number of details so that the person it represents cannot be confused with any other and at the same time by exchanging the pieces, it can be anyone. It is a geography that has its own climate, almost a joke, every furrow is impregnated with the character's DNA, it can be you or it can be me, every stain is in the exact place so that the observer does not get confused (Fig. 2).

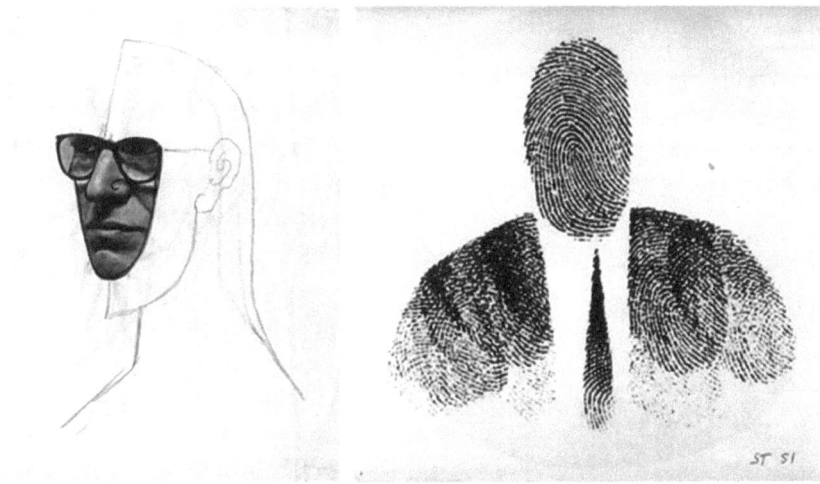

Fig. 2. Saul Steinberg. Untitled, Self-Portrait. New York. 1965. Source: Private collection. Untitled, Self-Portrait. New York. 1951. Source; Saul Steinberg Foundation. New York [12].

It is the pattern of Saul Steinberg. If someone wants to get their best portrait, the artist just needs their fingerprints. Of course, with a suit and tie. Avoid laughing. You could be Saul Steinberg. All you need is paper, ink and the pencil is you (or your fingerprint).

The work of these singular artists, a fight against dehumanization that without a doubt could already be intuited in the 18th century, grows throughout the 19th century and culminates in the 20th, where Steinberg's drawing represents it in its universal un-named portraits. Steinberg's abstraction in which "HE" is the character.

3   The Millimeter-Tower

Saúl Steinberg, a traveller, an observer, a cartoonist of everything that happened on the edge of the road, places his attentions on the boundary, on the border, on a personal place where from the distance, it is possible to look towards the center. And from his border viewpoint, in 1950, he proposes drawings of the city-metropolis, of the buildings that make it up with a critical and subtly.

Graph paper towers or furniture transfigured in buildings, as new symbols of an unknown audacious corporation, will occupy and possess the space. They are drawings in conflict between concept and form.

Saul Steinberg seems to review and criticize the homogenization of the International Style versus the American Art Deco vision, the own versus the alien. In his words this thought is intuited:

"... As soon as I arrived in New York, one of the things that immediately fascinated me was the great influence of Cubism on American architecture. And Art Deco was nothing but the decadence of the influence of Cubism, the Cubism that has become decorative: Chrysler Building, Empire State Building, the jukeboxes, the cafes, the shops, the women's dresses, their hairstyles, the ties, everything was made of Cubist elements..." [7].

This series of thoughts or graphic ideas were published in "The Passport", [13]. Steinberg presents some illustrations that are precursors of experimental proposals that are about to arrive. Saul Steinberg puts on print his critical opinion on the city-metropolis (Fig. 3).

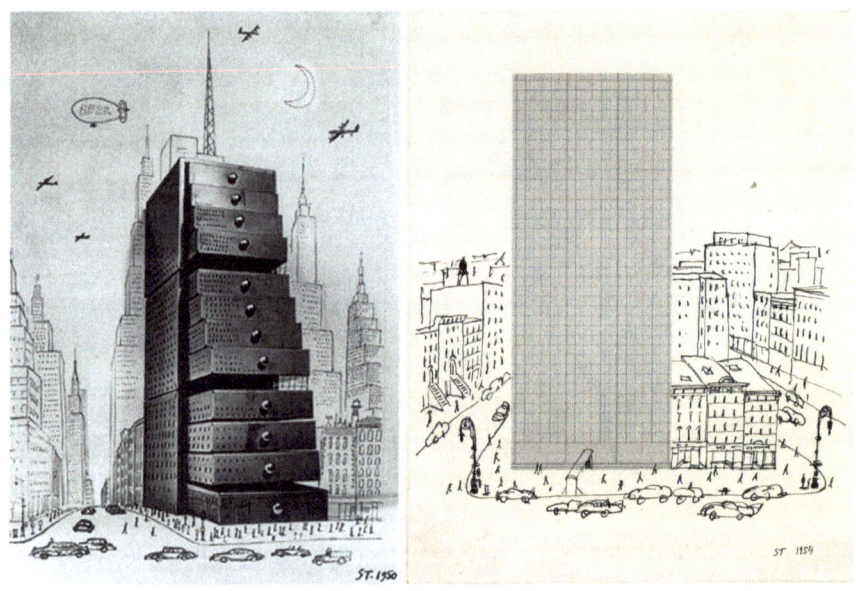

Fig. 3. Saul Steinberg. Drawer furniture. Urban landscape. New York. 1950. Fountain: Saul Steinberg Foundation. New York. Saul Steinberg. Graph paper architecture. Urban landscape. New York. 1954. Source: Leon and Michaela Constantier's collection. New York [12].

He is ahead of "Pop-Art", "Archigram", "Postmodernity", "Deconstruction". Saul Steinberg gives us a glimpse of the future and does not stop reflecting on the contemporary city.

Separated by two centuries, Steinberg's vision is comparable to the one developed by Piranesi in his views and reconstructions of Imperial Rome. In Piranesi's engravings *Le Antichità Romane* (1756), his influence, more than as an architect, would be remarkable as a theoretician of architecture, facing the clichés of his time that considered the artistic and architectural superiority of Greece was imposed over Rome; since Roman art was supposed to be no more than a mere extension of Greek art. Contrary to the postulates of *Winckelmann, Dumont, Leroy or Ramsay* [8], Piranesi tried to prove the exact opposite: the superiority of Roman art and architecture. To this end, Piranesi linked Roman architecture not with Greek architecture, but with Etruscan and Egyptian architecture, giving it its own originality (Fig. 4).

Fig. 4. Giovanni Battista Piranesi Secondo Frontispiece. Rome. 1757. Source: Volume II, published in Lettere di Giustificazione scritte a Milford Charlemont... 1757 etching, ER/4812. Giovanni Battista Piranesi. Carceri d'invenzione VII. Rome. 1764. Etching, INVENT/45323. Source: Biblioteca Nacional de España. Madrid [1].

Both artists, Piranesi and Steinberg, trained as architects, and although they are not remembered as such, their works have had a great influence on Piranesi's and Steinberg's contemporaries.

Their reflections on architecture and the city have made generations of architects, town planners, writers and thinkers meditate. Their graphic testimonies have shaped criteria of analysis and ideation of the architectures that we have seen imagined, designed or built in these last three centuries.

4 The Lost Spaces

In 1740 Piranesi leaves Venice, he needs to be at the epicentre of the Old Empire. In another context, in 1933, but also on a path of personal quest, Saul Steinberg moves to Milan to begin his training as an architect.

Their architect's souls will resonate within them. And their works will reflect this. Their cities and spaces could be placed at the confines of the world, and somehow their graphic works have become a kind of graphic narrative, universal scenery, an inner look, a singular vision of their times.

Whether they are wooden cities, engravings, line drawings, collages or ironic or critical visions, both artists proposed their intimate version of the city-metropolis, of impossible spaces and perhaps of their true architectural dreams, "a city on the edge". Cities, prisons, buildings, spaces for the far edges of the world, metaphors of their true works with which the future could be urbanized, art for the places where the world seems to end and where everything can be invented again.

Between 1745 and 1760, the editions of *Le Carceri d'Invenzione* - dated 1742 - are documented and in 1761 the second edition of this series of engravings is published.

Piranesi revised and reworked previous images and added two more engravings to the fourteen of his first version. This new version, composed of sixteen engravings, shows a darker, more distressing and claustrophobic vision of Piranesi. A certain disconsolate sleep and a disturbing presence of an excessively twisted ideation can be inferred from their observation.

Piranesi opens the door to surrealism, to marginal paths, to spatial labyrinths, to the confusion of reason, he seems to be a tortured mind. They are labyrinthine architectures that, almost two centuries later, will be recreated in some of Saul Steinberg's drawings. Piranesi's influences on the generations that will succeed him will be very noticeable, his technique and his graphic language, which is no longer Baroque, is entering the Romantic language and will give way to Goya, Bacon or Escher and many other artists and architects (Fig. 5).

Fig. 5. Saul Steinberg Excavation. New York. 1951. Saul Steinberg. Untitled. New York. 1951. Source: Saul Steinberg Foundation. New York [12].

In these drawings from 1952, Steinberg experiments with photographic image and drawing, decontextualizing objects and places to turn them into impossible cities, inhabited by tiny beings. He plays with the very architectural concept of scale and the fission of languages.

In Harol Rosenberg's words about Steinberg's work, he states; ...*His work is very current and yet it has a halo of ancient style...* [9].

In the drawing entitled, *"Canal Street"* dated 1988, he presents an urban image of the metropolis where the agglomeration is no longer human but architectural. The chaotic appearance of massive human shadows and cars lets us see the layout of the street. As we can see in Steinberg's drawing, there is an underlying order, a symmetry, a horizon, a grid plan urban layout of the capitalist and speculative city, where the architecture stirs restlessly (Fig. 6).

In these late Steinberg drawings, it seems that the buildings mutate into aggressive and violent beings and that the citizens lose even their shadow or their identity.

Fig. 6. Saul Steinberg Canal Street. New York. 1988. Source: Valencian Institute of Modern Art. Valencia. Saul Steinberg. Canal Street Frontispiece. New York. 1990. Source: Saul Steinberg Foundation. New York [12].

These drawings of the last metropolis sound and speak gloomy, threatening, Steinberg's light and shaky line is now firm and hard, the totem of civilization is disturbing, the *"sorcerer"* of the 20th century seems to be closing a cycle of creative exploration [10].

The multidirectional story of Steinberg's drawings is not that of the single-subject. The authorship is collective, public and generational. The historical evolution of a changing society, that decade after decade has been dismantled and redrawn by Steinberg.

Piranesi's *Le Carceri d'invenzione* and Steinberg's urban drawings place the human figure on a secondary plane, highlighting the insignificance of the human being and his loneliness in the face of desolation. Both artists describe dehumanized and dark spaces and architectures.

5   The Dreams of Reason

The 19th century opens with a new look that can perhaps be synthesized in one of Francisco de Goya's most boisterous paintings, *"La pradera de San Isidro"* [2], gentle visions of a world illuminated by Reason and Enlightenment, happy and calm people, a festive spirit to decorate the palace halls, to satisfy the blind vision of some kings who are about to lose their heads. In many ways Goya's work, and especially the series *"Caprichos"* and *"Disparates"* [2], exemplify a world in crisis, a world in process of change. His aesthetic proposals anticipate modern sensibility and the shift towards an art dominated by subjectivity and creative freedom. Goya can be considered a philosophical artist and a master of satire, with a genius for invention, which he used to show his distrust of human beings and the symbols of a society that was about to disappear (Fig. 7).

For Saul Steinberg the end of the Second World War was a period full of hope and longing for the future. Immersed in the post-war American spirit, liberal capitalism seems to be synonymous with opulence and life can become a kind time with the only condition of not looking beyond and remain in a present that awakens.

Fig. 7. Francisco de Goya y Lucientes *"La pradera de San Isidro"*. Madrid. 1788. Francisco de Goya y Lucientes. *"La romería de San Isidro"*. La Quinta del Sordo, Madrid. 1795. Fountain: Museo del Prado, Madrid [2].

Steinberg's proposals for the U.S. Pavilion at the 1958 World's Fair in Brussels illustrate this point of view.

Beings of kraft paper, line drawing cities, clichés and commonplaces of a carefree time, which make Steinberg think and wonder about something ironically new, anticipating, perhaps unconsciously and with apparent carefreeness, the imminent struggles for civil rights... *How do you draw black people?* [11] (Fig. 8).

Goya picked up the Piranesian spirit, when the Spanish War of Independence (Spain 1808–1814) changed his life, and his drawings and paintings show a very real vision of his existence.

Goya is meridian and leaves no doubt, ...*the dream of reason produces monsters*... [2] (Goya. 1797–1799), the infernal vision of a new *La pradera de San Isidro* (Goya. 1819–1823) is germinating, dark and gloomy, a time without hope. The darkness and the lack of culture overcome the light and the renewal proposals of The Enlightenment (Fig. 9).

The events that marked the second half of the 20th century in the United States, led Steinberg to live his own *Le Carceri d'invenzione,* and to find the hell that Americans suffered with the disappointment brought by the Vietnam War; the Counter-Culture; the Hippy Movement.

Fig. 8. Saul Steinberg. Small Town, The Americans. Brussels. 1958. Source: Royal Museum of Fine Arts of Belgium. Brussels [12].

Fig. 9. Saul Steinberg. The City. Cover of The New Yorker. New York. 1973. Saul Steinberg. Bleecker Street. Cover of The New Yorker. New York. 1970–1971. Source: Saul Steinberg Foundation. New York. Saul Steinberg. Untitled. New York. 1990. Source: Saul Steinberg's papers. Beinecke Manuscript and Rare Book Library. Yale University [12].

All this is reflected in Steinberg's work and becomes darkness, irrational dangers, psychedelia and, in short, the fear of loss and loneliness.

This parallel journey of these three image creators, who have contributed a graphic heritage for humanity, can be closed with the critical, disturbing and dark vision that Saul Steinberg shows us in some of his latest works.

A city with a certain dark gaze, inhabited by masked crowds, violent-looking beings who, like in a Goya coven, would like to distance the spirit from the human.

A small world facing chaos, in the best spirit of Walter Benjamin's thought and his dialectical reflections on how the dreams of the 19th century gave way to the nightmares of the 20th century.

6   Conclusion

With their drawings, Piranesi, Goya and Steinberg have left us their thoughts on the semantics of art, investigating and proposing reconfigurations of the graphic language of their historical moments, contributing, each in their own time, a conceptual revision of the signs and styles of their societies, their cities and their architectures. Their virtuous drawings showed us, with irony, critical vision and innovative creativity, their way of seeing the world they lived in.

Saul Steinberg defined drawing as *"...a way of reasoning on paper..."*, the act of drawing was a lifelong commitment and he removed the carefully forged masks of 20th century civilization.

"...You can't draw well if you lie ..."

And it could be said that these three unique artists do not lie and are part of the graphic heritage of humanity.

References

1. Piranesi, G.B.: Monographic Exhibition of Engravings. Biblioteca Nacional de España, Madrid (2019)
2. Museo del Prado. https://museodelprado.museumsmadrid.org/. Accessed 12 Jan 2020
3. Cappannini, C.: Remembrances and tensions in Walter Benjamin's fingerprint images. Research work developed at the Institute of History of Ar-Gentine and American Art of the Faculty of Fine Arts. Theme, The Dialectic Image in Walter Benjamin's Theory. Projections to the Teaching of Aesthetics. Rio de la Plata (2014). https://es.calameo.com/read/00065810471104ce87a16. Accessed 12 Jan 2020
4. Benjamin, W.: Book of Passages, p. 421, Akal edn. Edited by Rolf Tiedemann, Madrid (1982–2005)
5. Barthes, R.: All except you. Saul Steinberg. The Eiffel Tower, p. 117. Ediciones Paidós Ibérica, S.A. Barcelona (2001)
6. Joyce, J.: Ulysses. Critical edition Garland, New York, prologue and translation by José María Valverde, p. 331. Editorial Lumen, Barcelona (1984)
7. Steinberg, S., Buzzi, A.: Reflections and Shadows, p. 44. Half a Cow, Valencia (2012)
8. Piranesi, G.B.: Monographic Exhibition of Engravings, p. 377. Biblioteca Nacional de España, Madrid (2019)
9. Rosenberg, H., Steinberg, S.: Catalogue of the Retrospective Exhibition on the Work of Saul Steinberg, p. 10. Whitney Museum of American Art, New York (1978)
10. Rosenberg, H., Steinberg, S.: Catalogue of the Retrospective Exhibition on the Work of Saul Steinberg, p. 22. Whitney Museum of American Art, New York (1978)
11. Steinberg, S., Buzzi, A.: Letters to Aldo Buzzi, 1945–1999, p. 36. Half a Cow, Valence (2012)
12. Saul Steinberg Foundation website. http://saulsteinbergfoundation.org/. (Illustrations). Accessed 12 Jan 2020
13. Steinberg, S.: The Passport. Harper & Brothers, New York (1954). Revised ed., with an introduction by John Hollander. Vintage Books, New York (1979)

The Graphic Heritage of the Utopia of "La Fonction Oblique"

María Pura Moreno Moreno[(✉)]

Universidad Politécnica de Cartagena, Cartagena, Spain
mpura.moreno@upct.es

Abstract. This paper analyses the graphic heritage of the Theory of "La Fonction Oblique", created by the French Group 'Architecture Principe' in the 60s, established as one of the utopias critical of modernity. The stable order of architectural space, both vertical and horizontal, was modified with the inclination of its envelopes, generating, in the opinion of its predecessors Claude Parent and Paul Virilio, a dynamism far from merely static living. The sense of vision ceased to be a protagonist in spatial perception and the rest of the senses were also included, leading to a new type of imbalance which, from a historical perspective, had only been seen in what were known as "archaeologies of the oblique." But this was seen in few constructed examples. However, its graphics paved the way for lines of research that would go from the domestic to the urban scale. The analysis of the drawings published in the magazine of the same name, used for the introduction of this theory in architectural debate, will act to highlight the convergence there was between possible ideation, theory and practice. Its formal vocabulary will be analysed according to characteristics such as its monochromatism, its different points of view from aerial or infinite perspectives, or the dramatism of its sometimes a-scalar condition. All analysis will direct the discourse towards conclusions referring to an architecture which is detached from an only static and visual compression giving priority to all of the senses, as a whole.

Keywords: Fonction Oblique · Claude Parent · Paul Virilio · Graphism · Architecture

1 Introduction

During the second half of the 20th century, space utopias, represented by all kinds of visual resources such as drawings, collages, photomontages and models, questioned a certain uniformity in the architecture of the modern movement, which had put identity aside, in favour of rationalization.

Criticism arose from various academic sources in the form of architectural, tectonic and urbanistic approaches whose ambitious scale forced their proposals to be limited to the graphic field, with little built experimentation.

Some of these approaches were disseminated in publications aimed at inaugurating new disciplinary debates, while others are still waiting to be discovered in personal archives; the rescue of which remains a pending task.

L. Agustín-Hernández et al. (Eds.): EGA 2020, SSDI 6, pp. 607–619, 2020.
https://doi.org/10.1007/978-3-030-47983-1_54

The analysis of this graphical heritage constitutes a line of research that allows us to determine the importance of what went from something that was imagined to be the genesis for the architecture of the following generations.

The certification of the intellectual force predicted by these more or less utopian approaches, is demonstrated in a drawing that managed to foresee a future marked by the acceleration associated with liquid modernity in our time (Bauman 2014, 35–48).

This paper heritage modelled a present that became aware of the experience accumulated during the past, to project it towards a better future, both socially and spatially. A future that, although it was then impossible to build, was at least worth both imagining and representing by means illustrations.

That projection of a new long-term spatiality was tried out by collectors such as the Japanese Metabolists, like Kisho Kirokawa, the situationists such as the Archigram and Superstudio groups, or by the persons Yona Friedman, Cedric Price, Hans Hollein and Frei Otto.

Within the same time frame, but in the socio-political context of the French prole-gomenous of May '68, the Architecture Principe Group was founded in 1963, by the architect Claude Parent (1923–2016), the philosopher Paul Virilio (1932–2018), with an occasional contribution by the painter Michel Carrada and the sculptor Morice Lipsi (1898–1986).

Its theoretical discourse, monopolized by the idea of La Fonction Oblique, exposed a challenging thought of the Cartesian order, through a conception of space, which was later classified as one of the "Utopias of air, concrete and paper" (Cohen 2014, 11–192).

Its objective was to face a society in crisis that needed to be rethought with renovating architectural and urban programs.

"…The state of crisis that can clearly be seen in all human activities… the enormous contradiction of values and disciplines indicate the proximity of an unprecedented event. Historically, we have observed numerous transformations of societies, but we have never witnessed the mutation of man himself…" (Virilio 1966a, b).

The difficulty of putting it into practice, especially on a territorial scale, made its more radical experimentation be reflected mainly in drawings, and it was only specifically radicalized in three constructed projects: the Venice Biennale Pavilion of 1970, the Église de Sainte Bernadette in Nevers (1963–1966) and the Thomson-Houston Research Center (1964–1971). Other buildings that could be subsequently associated with the theory, such as the Tinqueux supermarkets (1967–1969), Sens (1967–1979), Ris Orangis (1967–1971) or D'Épernay-Pierry (1968–1979) also took from it the inclination of walls and envelopes, and limited movement to inclined ramps. However, their exterior morphology emphasized ideas complementary to La Fonction Oblique, such as the fracture, the collision of the masses or the formal resemblance of the artifice of architecture with the natural failures of the earth's geography and topography more intensively.

This article will analyse the drawings that accompanied the texts of La Fonction Oblique to demonstrate how their graphic diversity in aerial perspectives outside human vision - with bird's - eye view approaches, their axonometric schemes, or their conic perspectives with points of distant escape that directed the arch-tectonic to an infinite linear development, exerted a mechanism of thought and reflection for the Theory's own development (Fig. 1).

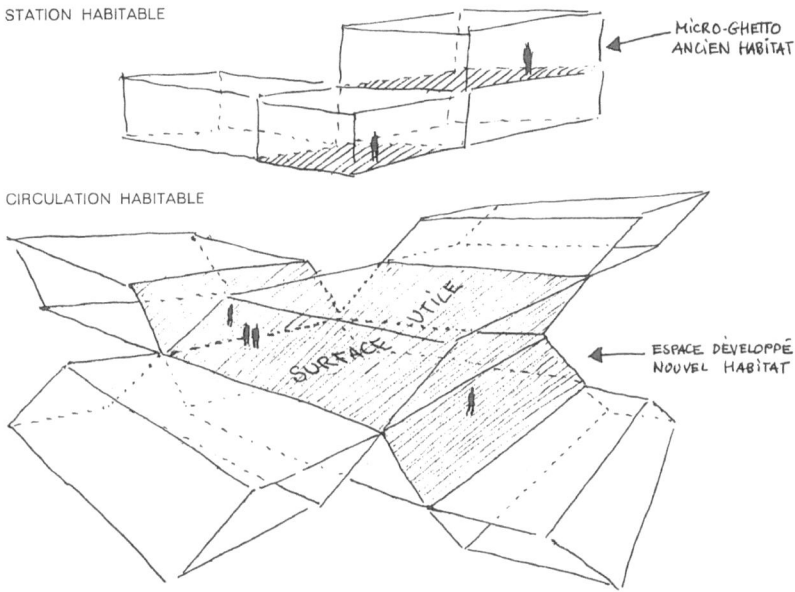

Fig. 1. Station y Circulation Habitable. *Architecture Principe* n°5, Aout, 1966.

The assessment of both the dramatism of the sketches, which were always in the colour gamut of grays or the contrast of black and white, and of their own completion, which combined the quick completion and the virtuosity of drawings of very defined urban pieces, will endorse its contribution to the understanding of the oblique hypothesis (Fig. 2 and Fig. 3).

Fig. 2. Les Turbines. *Architecture Principe* n°1 Février, 1966

In the same way as it was for Pallasmaa (2008, p. 63), for whom the architecture of the past allowed us to imagine the lives of the past, now the graphics of this hypothesis of Living in Oblique focused on a future that anticipated other ways of living. This is

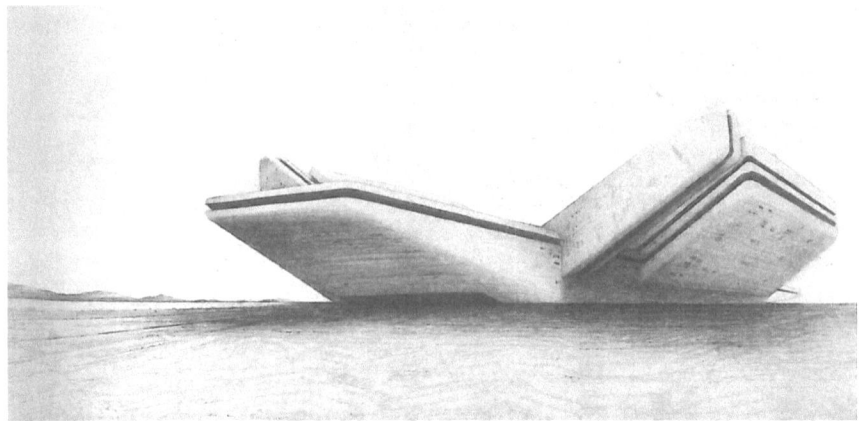

Fig. 3. Nautacité. Architecture Principe n°1, Févreir, 1996.

explained by later constructions such as the Oslo Opera (2007) by the firm Snøhetta, the Museum in Neuhaus (2006) by Odile Decq or the Passenger Terminal of Yokohama (1995) by Foreing Office Architects, FOA, as well as the materialization of new geometries as complex as the Moebius ring or the Klein bottle (de Dalmau 1992, 184). All of these were made possible thanks to the joint development of ideas suggested in the field of drawings and graphs, combined with the necessary progress of technology. Together, they have been responsible for the fusion of the interior-exterior binomial or the principles of spatial fluidity for the configuration of a new urban and intimate dwelling, always linked to the action of movement. This combination of inhabiting combined with more movement, and flow had previously been criticized for the dissociation between the two prior to the proposed idea of La Fonction Oblique.

> *"The principle of fluidity requires abandoning the interior-exterior duality, the essence of the ghetto of our inherited cities, to write about space in terms of surfaces, surfaces of inferior, superior, enveloping, network layers to describe our actions in terms of trajectories, paths and displacements."* (Parent 2001, 45–46).

The methodology used in the analysis of the drawings will include themes, lines, points of view or scales of representation. Its variety of format and finish will provide guidelines for a classification by drawing techniques that link the trinomial formed by ideation, underlying theory and future practice.

2   Architecture Principe: La Fonction Oblique

The recognition of the need to expose their ideas as an indispensable part of the elaboration of the theory, prompted the members of the Architecture Principe group, mainly Claude Parent and Paul Virilio, to publish a magazine of the same name, as a permanent manifesto, which would become the tool used for its diffusion throughout 1966, (Johnston 1996, 12).

The coordinated purpose of texts and graphs or drawings combined two objectives at the same time. The first was the attempt to establish the favourable principles and circumstances for La Fonction Oblique in architecture. And the second was a direct denunciation of the urban planning of the time that was concentrated either in the horizontal expansion of the vast city, or in its spreading through a vertical architecture in the new metropolises.

This duality between the imaginative proposition and the criticism of what was established was cited in articles, drawings and photographs by way of the titles of the nine issues of the magazine: "La Fonction Oblique" (February 1966), "Le Troisième Ordre Urbain" (March, 1966), "Le potentialisme" (April, 1966), "Nevers: chantier" (May–June, 1966), "Habitable Circulation" (July, 1966), "La Cité Médiate" (Agost, 1966), "Bunker: archéologie" (September–October, 1966), "Pouvoir et Imagination" (November, 1966), and Charleville (December, 1966).

Its retrospective approach to the history of architecture focused the group's attention on the association of each past civilization with a specific spatial definition that referred to "…a system of geometric references in which society develops; a system underlying its social and political organization; to their economic development, to their philosophical and religious conceptions…" Along the same lines, the theory developed its own classification into three successive forms of crystallization of space.

A first one, of horizontal order, in which the conquest of the land and the territory played a major role.

A second, of vertical order, associated with the conquest of space, with a social hierarchy that was strongly criticized for the rupture brought about in social relationships, by separating the actions of inhabiting and movement. In parallel to this complaint, and from an architectural or urban planning point of view, this vertical order was considered erroneous because although it contemplated the low surface occupation of the tower plant, it did not warn of the extra surface area needed for the equipment, car parks and traffic? In this sense, the benefit to urban development in the form of towers, as opposed to residential, horizontal multiplication, as claimed by Le Corbusier, was considered by Parent, to be absolutely false (Parent 1970, 17).

The last and third urban order, which the new theory represented, was that of the obliqueness of the domestic, private enveloped walls of the homes to be lived in. As a result, they represented both the geological world and the law of fluidity, acoustics or hydraulics which, although they had been seen in the world of engineering, they had not yet been experienced in the architectonic field (Fullaondo 2006, 78–80).

The generalizable concept, from the oblique on a territorial and urban scale to the size of domestic and intimate enclosures, was frequently seen to be the intentions of the text and graphics published in the nine issues of the Architecture Principe magazine. When it ceased to be published, in December 1966, it meant that the release of a number done by Lipsi and Michel Carrada reflecting the relationship of La Fonction Oblique with the arts, was not released (Parent 1981, 48).

The events of 1968 divided the group, given the divergence of opinions regarding the integration o f theory into the political debate of the moment. Its dissolution was due to Parent's refusal to place it outside the strict architectural or urban planning spheres, in favour of integrating it opportunistically into the socio-political debate of the time along with the ideas of many of his contemporaries.

Such a lack of cohesion, among the members of the group, did not prevent Parent from publishing a book in 1970 called Vivre à l'oblique which summarized the most significant theoretical development. It was a contemporary text that spoke o f the Venice Biennale Pavilion project, where the theory's postulates were experienced on a very small scale, the carrying out of which was driven by Parent thanks to the participation of other artists with whom he did not hesitate to work, given the refusal of former Architecture Principe members to do so.

This text insisted on the social and sensory advantages of the inclined plane, through architectural and urban theorems that were accompanied by thick stoke graphics and eventual completion (Fig. 4); and where the theory was explained with easy-to-read diagrams using only the simplest of tools – the black ink marker (Parent 1981, 48).

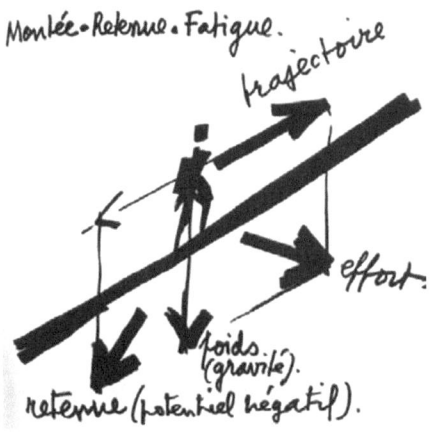

Fig. 4. Parent, Claude, Vivre à l'oblique, L'Aventure Urbaine, Paris, (1970).

3   Architecture Principe Magazine

3.1   A Thematic Classification

Paul Virilio presented the theory of La Fonction Oblique in the first of the nine issues of Architecture Principe (February–December 1963) on the grounds of refuting the classic orders of the city's historical approach, and accepting as inevitable "…the end of the vertical as an axis of elevation, the end of what was horizontal as a permanent plane, to the advantage of the oblique axis and the inclined plane that meets all the necessary conditions for the creation of a new urban order and that also allows a total reinvention of architectural vocabulary…" (Virilio 1966a, b).

The group's historical judgment of both the concentric juxtaposition of the ancient models of cities, such as Rome, and the vertical superposition of the modern metropolis, as in, for example, New York, dissociated the two key functions of the life of the individual in society: those of motion and living (Parent 2009a, b, 14). Under this premise it was considered that the permanence of the ground plane in a horizontal position, throughout civilizations, had been an obstacle to developing renewed ways of inhabiting. In the face of such analysis, the new theory advocated to modify, first graphically and then constructively, the angle 0° of all the envelopes mainly that of the ground.

In his opinion the change should be radical.

"...The abandonment of the notion of height will be of capital importance: and it will finally allow the materialization of the reality of space-time. Therefore, the original principles of this art of space that is architecture will be radically modified, its objectives and its power will be singularly expanded..." (Virilio 1966a, b).

The experimentation of space with the whole body was specified in the reinvention of an architectural language and vocabulary that was depicted in the magazine in two ways:

The first gave prominence to the virtuosity of various types of pencil and ink drawings that emphasized this conceptual speculation (Fig. 5).

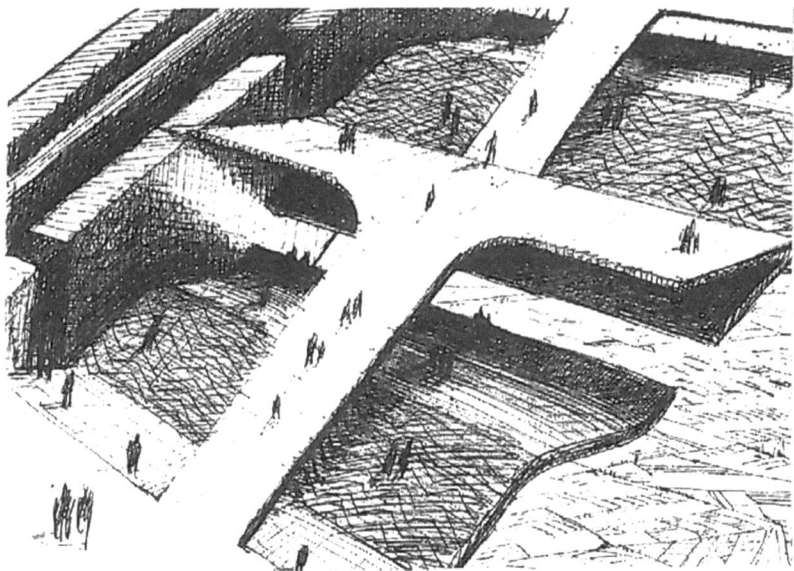

Fig. 5. Habitat sur plan incline. Site de derivation. Côte surplomb. *Architecture Principe* n°3 Avril, 1966

And the second, somewhat less speculative, in the May–June issue, published a set of photographs of the Église de Sainte Bernadette de Nevers, while in the September and October issue, there were images of the Atlantic Wall Bunkers, which Virilio had investigated prior to the setting up of the group, and whose morphology and materiality were reflective of the new oblique architecture.

Drawing thus became a means of expression and a useful mechanism that brought together the tasks of communicating, expressing and representing time.

Before becoming an architect, Parent had been a publicist and illustrator, and confessed that, since he was a child, he and his father, an aeronautical engineer, were always drawing realities or fantasies. This vital information has recently been corroborated by his daughter Chloé Parent in a text entitled "The hand that draws," where she said:

> *"He was totally absorbed in the interaction between his imagination, his hand and the paper. No one could interrupt it. Sometimes he shaded meticulously, filling in the surfaces with his pencil, a task that, being more automatic, allowed him to think about other things… He however, did not put the pencil down. If he did, the rhythm and the quality of the drawing could be altered. He drew in silence, without music or conversation"* (Parent 2019, 183).

Using drawings to show volumes and spaces on paper provided the key to questioning historical formalities such as the never-ending situation of the horizontal ground plane, and to give substance to a theory that, without these drawings, future built architectures would not have been nurtured so directly and immediately.

The aspects developed in the Architecture Principe magazine can be classified by topics that included a territorial scale; L'abandon des villes, Le troisième ordre urbain, Dominer le site, Habitable Circulation, Structure, Civilization, Les cités immiates, La I cited Médiate, Infraestruture or Infra-Urbanisme, or in more specific imaginary projects that drew up new cities such as, Les Turbines, Nautacité, Les Vagues and Les Crateres (Fig. 6).

All of them rejected the two Euclidean approach and they were depicted by aerial points of view with infinite leak lines coming out of its high horizon lines which delimited an architecture and urbanism of inclined planes; generated by the so-called "inclisitio" (Fig. 6) or habitable inclined slat conformed by surface and sub-fiction (Parent 2009a, b, 35–36) (Fig. 7).

The second theme, more linked to phenomenological and sensitive aspects, appeared in reviews such as La Fonction Oblique, L'Homme Déraciné, Experimentation, Fluidité, Civilization or Habitable Circulation. The way in which it was shown using drawings was more direct as schematic sections or axonometric perspectives referring to space discourse were specified.

The last issue, from December 1966, was the only one in which the project was presented (but never came to be), of the Palacio de Congresos de Charleville, demonstrating with it that all the virtual reality of aerial perspectives could really be built when included in a more technical drawing of plants, elevations and sections. Even images of a wooden model that homothetically reflected the projected reality could be used.

3.2 A Graphic Classification

On envisaging the set of drawings with which La Fonction Oblique was expressed, we are invited to remember what the three languages of the architect are, in order of specificity: the natural, the graphic and the architectural (Sainz 1990, 21). Having been warned of

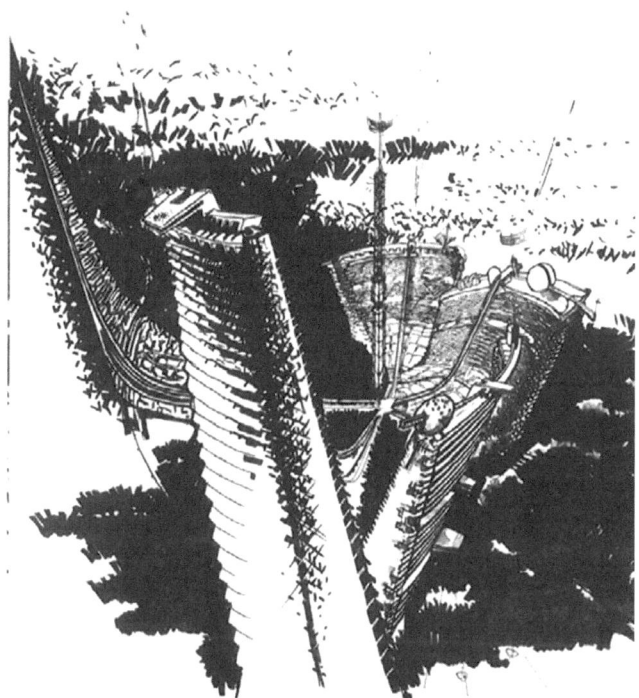

Fig. 6. Les Turbines. Architecture Principe n°1, Févreir, 1996.

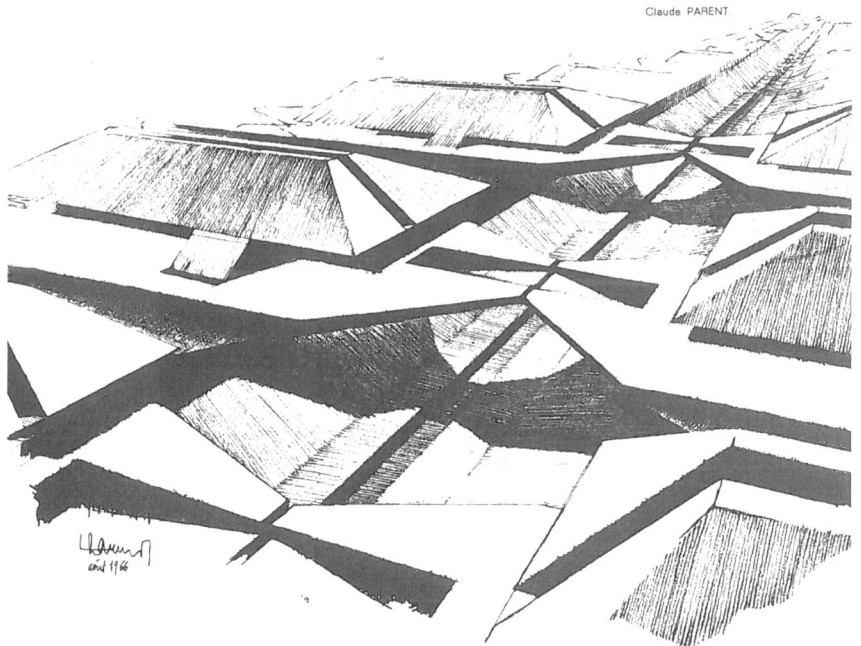

Fig. 7. Espaces habitable. *Architecture* Principe n°5. Juillet 1996.

this particular lexical pathway, it is not surprising to assume that, in the exposition of the theory, natural language was directly substituted by language related to drawing.

What was drawn was not a tangible reality that was simply copied, but an imagined and utopian material that referred to a spatial absence. This characteristic of architecture and unreal urban situation was represented, in many cases, from visions outside the human perspective and so therefore, the horizon line at the drawer's eye level was not taken into account. It did not matter if this line was placed either far above or far below the real view contemplated from a normal topographic situation. The absence in certain drawings of the human figure prevented the mental association with the true scale of the proposals, forcing the reader to make a more imaginative effort. Its appearance in others meant that dimensions could be specified. And the presence of many of them allowed them to be compared with buildings designed by civil engineering which was governed by hydraulic gravity, and therefore were as plausible as the architectures of Sant'Elia (Fig. 8).

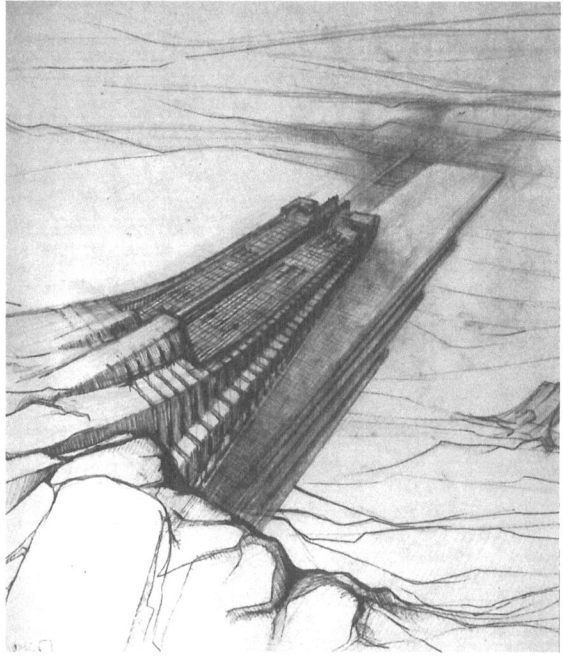

Fig. 8. La Vague. Architecture Principe n°2. Mars 1996.

The logical path between the three cited languages - natural, graphic and architectural - firstly stimulated the sight, to be followed by the activation of the image that filtered the mental information, which was finally expressed on paper by combining specific tools such as the pencil and marker as well as techniques such as shading, contrasting and outlining. This transition was reduced in the proposals of La Fonction Oblique to just the last two phases as they are utopian realities that were intended to suggest new realities rather than portray them; where the first visual stimulation was ignored, in favour only of the imagination. Or as Parent stated, in favour of the purely mental.

> *"...The objective vision is always partial, the knowledge of an architecture can only be mental... This mental universe suggests more than it determines. It is more a sleight of hand than a logical expression..."* (Parent 2001, 20).

Several classifications can be established by contemplating the graphics of the Architecture Principe review:

A first could focus solely on the tool of its completion. That is, in differentiating the drawings done in pencil, which used the entire chromatic range of the grays of coal, compared to those which took on only the dramatic contrast between white and black, or the full and the empty. The latter were mostly done with black ink markers of different thicknesses.

In all of them, this monochromatism reflected the importance the light-shadow duality had. A contrast of light that emphasized the role given to lighting in architecture by Parent, which he pointed out, for example, in his article "Sur le sentier de la lumière", when he recalled the use of the oil lamp, while visiting the architecture of his childhood home, which added mystery to the discovery and sensory manipulation of the different parts of the house.

A memory that claimed the liturgy of spatial/three-dimensional reading through light and its contrast with shadow, which was, in turn, supplemented by the evocation of the reflective characteristic of lacquered materials turned into mirrors which he named 'Tanizaki Jumochiró', poetic light in the text "Praise of the shadow" (Parent 1991, 112) (Fig. 9).

Fig. 9. Nautacité. Architecture Principe n°1. Févriers, 1996.

This monochromatic intent, in addition to evoking the light and shadow binomial through its dramatic extent, that went from greater to lesser, projected the idea of an architecture built primarily in reinforced concrete, massive and cryptic. A ductile and malleable material was taken to be the main reference due to the analysis carried out by Virilio of the constructions of the Bunkers of the Atlantic Wall that were described in the number titled Bunker: archaeology" (September–October, 1966).

A second illustrative classification could be that which takes into consideration the points of view chosen for the presentation of the new space proposals. It would be necessary to distinguish, in the first place, the drawings carried out from an aerial point of view in a downward direction towards an urban reality that should be expressed from the position of superiority of the bird's eye view. Secondly, there would be the drawings that, with very low horizon lines, visualize material realities with ascending views towards the sky, even reflected in a ground plane whose aim it was to be reflective. In the latter case, these views expressed the idea that inclined cities took up more volume in their spatial development in all three dimensions, but in contrast, they needed less support surface in the ground plane (Parent 1970, 57). The development of movement was not reduced to the horizontal plane linked to the territory, as in the schemes of Le Corbusier, but it moved out through the vertical space, always directed through the inclined planes.

Finally, in a third classification, we would include the explanatory diagrams chosen as axonometric representation, which, as a result are not representative of a real visualization by human eye, but instead try to be as explicit as possible to achieve spatial understanding (Fig. 10).

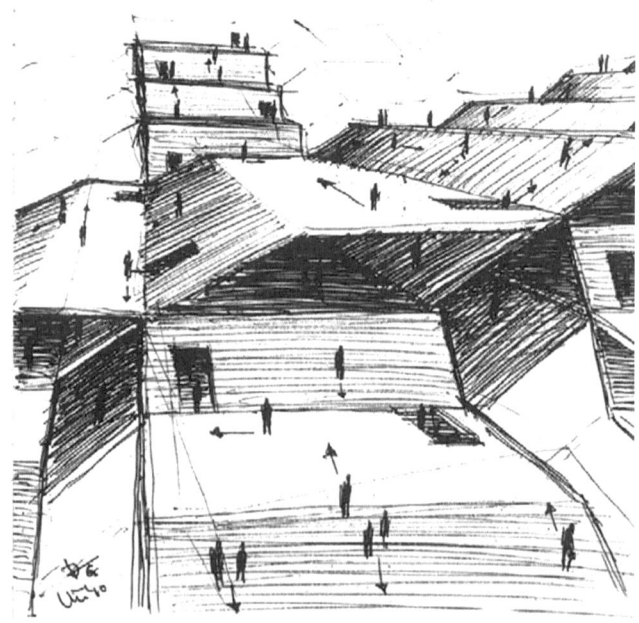

Fig. 10. Civilitation. Architecture Principe n°5. Juillet 1966.

The combination of all these drawings left no doubt about what the theory was trying to accomplish from whatever view it was seen - neither urban, nor architectural, nor spatial.

4 Conclusion

The analysis of the graphic heritage developed in the preparation of the La Fonction Oblique obliges us to eventually reflect on the importance of paper architecture in later architectures.

The synthesis of its graphism is currently self-evaluating itself, questioning the materiality of real space in the era of the untimely and immaterial emergence of virtual space. Its doubts about the validity of Euclidean geometry, as a constant rule, were projected into a synthetic graph that anticipated future fractal geometries, chaos theory or gravity questioning. All the drawings, regardless of their completion and point of view, jointly confirmed the idea that by making the language of graphics evolve, in the first instance, the constructed space could be transformed, and in that way that would achieve, in a near future, an architecture whose sensitive perception would involve all the senses by unifying the actions of inhabiting and moving in the space-time binomial.

This analysis aims to show the importance of the heritage of drawings of this theory in order to show how difficult it is to find the balance between the forecast of a reasonable future and the limits of utopia (Fullaondo 1967, 9), always from an ethical and social point of view... aspects that should preside over any architectural act.

References

Bauman, Z., et al.: Arte, Líquido. Ediciones Sequitur, Madrid (2014)

Cohen, J.L.: L'architecture au XXe siècle. Modernité et continuité. Hazan, Paris (2014)

del Dalmau, D.F.B.: La invención de La Fonction Oblique. Ph.D. disertación, Departamento de Proyectos de ETSAM, Universidad Politécnica de Madrid (2012)

Johnston, P.: The Function of the Oblique: The Architecture of Claude Parent and Paul Virilio, 1963–1969. Aa Publications, Londres (1996)

Fullaondo, J.D.: Toujours l'oblique, en Claude Parent Vue par...50 témoignages du monde entire, Ed. Groupe Moniteur, Départament Architecture, Paris (2006)

Parent, C.: Vivre à l'oblique, L'Aventure Urbaine, Paris (1970)

Parent, C.: Errer dans illusion. Les architectures hérétiques S.L, Paris (2001)

Parent, C.: Sur le sentier de la lumière. L'Architecture de Aujourd'hui (274), 112 (1991)

Parent, C.: Entrelacs de l'oblique, Collection Architecture « Les hommes » , Éditions du Moniteur, Paris (1981)

Pallasmaa, J.: Los ojos de la piel. La arquitectura y los sentidos. Gustavo Gilio, S.L., Barcelona (2018)

Parent, C.: Vivir en lo oblicuo. Gustavo Gili, S.L., Barcelona (2009a)

Parent, C.: The hand that draws. In: Claude Parent Visionary Architect, pp. 183–185. Rizzoli, New York, (2009b)

Sainz, J.: El dibujo de arquitectura. Nerea, Madrid (1990)

Virilio, P.: Avertissement, Architecture Principe, n°1. Recogido en Architecture principe: 1966 et 1996, 8 Février 1966. Les Editions de l'imprimeur, Besançon (1996a)

Virilio, P.: La Fonction Oblique, Architecture Principe, n°1. Recogido en Architecture principe: 1966 et 1996, 8 Février 1966. Les Editions de l'imprimeur, Besançon (1996b)

Degli edificj antichi e moderni di Roma. Vedute in contorno, 1817. Notes on an Graphic-Architectural Experimentation by Giovanni Battista Cipriani

Martino Pavignano(✉)

Department of Architecture and Design, Politecnico d iTorino, Turin, Italy
martino.pavignano@polito.it

Abstract. The contribution proposes a critical analysis of some architectural engravings by Giovanni Battista Cipriani, Sienese architect, draftsman and engraver, working in Rome between the 1780s and 1830s. In particular, I related, by means of engravings as visual language, Cipriani's graphic production '*in contorno*' (outline drawing) - representing classical and modern Roman architecture, published in the 1810s - to the figurative experiences by the French painter Bénigne Gagneraux and the English sculptor John Flaxman, both in Rome around 1792. A 'formal' link is therefore hypothesized, highlighting Cipriani's originality from the point of view of the architectural representation for the communication of artifacts and his deviation from the communicative paradigm promoted in Rome in that period, still strongly influenced by the striking visual outcomes of Giovani Battista Piranesi.

Cipriani emerges as an experimenter of a communicative practice drawing its origin from the consolidated practice of architectural drawing and from the debate on the supremacy of drawing between the arts. In this sense, Cipriani can be highlighted as an ideal anticipator of the much more famous works by Paul-Marie Letarouilly, as well as a simplifier of the visual code aimed at popularizing architectures as an extreme graphic-visual remodeling of the cultural soul of neoclassicism.

Keywords: Giovanni Battista Cipriani · *Vedute in contorno* · Outline · Bénigne Gagneraux · John Flaxman

1 Introduction

Drawing, intended as the graphic expression of an intellectual practice of representation which involves the mind and the hand, has always been one of the main linguistic tools of the visual communication of Architecture and, since the earliest times, many efforts have been made to make its results reproducible, at most for widen the diffusion of the messages conveyed [1]. Over time, we witnessed the multiplication of opportunities that allowed us to «arouse those *stimuli* and those communicative opportunities that

L. Agustín-Hernández et al. (Eds.): EGA 2020, SSDI 6, pp. 620–632, 2020.
https://doi.org/10.1007/978-3-030-47983-1_55

extended the practice of representation [and therefore of Drawing] to all human activities, especially those relating to the use of image, both as a qualified objective of aesthetic communication/information and as a design tool for all sorts of artefacts» [2]. This primary led to the birth of the various techniques of reproduction of graphic messages by means of printing (xylography, etching, lithography, etc.) [3].

The research exposes the results of part of my doctoral thesis related to the analysis of the collection of prints *Degli edificj antichi e moderni di Roma. Vedute in contorno*, by the architect, draftsman and engraver Giovanni Battista Cipriani (hereinafter GBC), published in Rome, starting from 1817 (Fig. 1a, b).

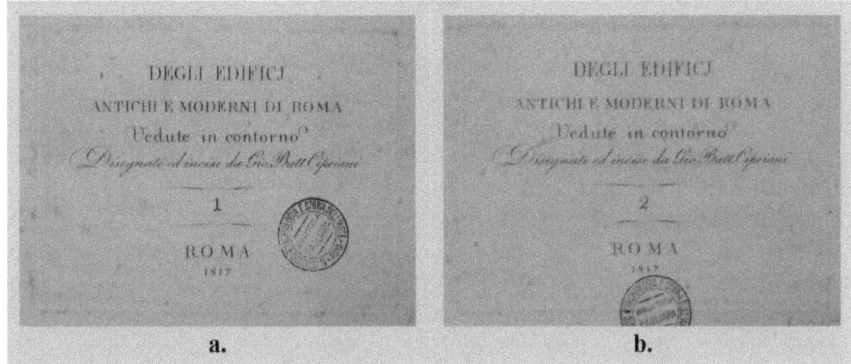

Fig. 1. Frontispieces of the two volumes of *Vedute in contorno*: a. vol. 1, f. 1r; b. vol. 2, f. 2r.

This set of small *vedute* is an example of technical and formal innovation in architectural representation addressed to the communication of buildings and landscapes, in the light of the possible influences (also by means of concordant dates and places) of the artists Bénigne Gagneraux and John Flaxman on the subject of representation of subjects in pure outline (*trait*, outline).

2 Giovanni Battista Cipriani (1766–1839)

Born in Siena, 1766, GBC devoted himself to fine arts and started his studies at Giuseppe Silini's atelier (Sienese sculptor and architect). In the early 1780s, he moved to Rome in order to continue his studies with the architect Giuseppe Palazzi, by winning the prize in memory of Marcello Biringucci and Giulio Mancini [4]. GBC's sketchbooks preserved at BiASA confirm his presence in Rome at least starting from 1784. From 1790 he was «half scholar» of the philosopher Leonardo De Vegni [5]. In the same period, he met the architect Giandomenico Navone, with whom he elaborated the *Nuovo Metodo*, published in Rome, 1794 [6]. It is possible that GBC studied engraving and etching at Raffaello Morghen's atelier, together with his brother Galgano or, at least, that the gained indirect experience of these techniques [7]. GBC was also a member of the architect and art historian Francesco Milizia's cultural club, gathering many intellectuals and artists [4]. GBC worked with Milizia for drafting the first full set of illustrations for *I principj di architettura civile*, published in Rome by Salomoni with the title *Indice delle figure relativi ai principj di architettura civile di Francesco Milizia* in 1800 [8]. As far as is known, GBC never took part in the composition or re-signification of a built artefact; on the other hand he dedicated his career to applications of what we could now name as indirect survey, not in the sense of a set of operations mediated by an instrument other than the 'meter' for direct survey, but as a result of careful critical observation of other authors' representations of architectural artefacts. In the same time, he worked on a continuous comparison between his perceptual survey and such graphical sources [9].

During his career, he frequently put himself at the service of that branch of Architecture dedicated to the study, representation and communication, also for explicit educational purposes. In this regard, it is necessary to say that in the last decades of the 18th century, a new debate on training tools was developed in Rome – whose cultural environment was dominated by studies and comparisons on the fine arts dealing to drawing [10] – in order to provide architects and architectural students with up-to-date didactic tools.

Among his works, often reproducing that of other famous architects and draftmen, for example Antoine Desgodetz and Julien-David Le Roy [4], the originality of the aforementioned *Vedute in contorno* have to be highlighted. Furthermore, it should not be forgotten that GBC was used to publish his original books almost always in «*commodo sesto*» («comfortable sesto») [11]. This was for two reasons: to be able to «easily study» on such volumes and to offer the reader affordable products. Within this paper I do not enter within the graphic production of GBC's notebooks and manuscripts, that were already the subject of some studies by Bentivoglio [12, 13], Debenedetti [4, 14], Pasquali [5], Olschki [15] and that I analyzed in my doctoral thesis [16].

3 Cultural Context and Visual Preludes

GBC's graphic production is very vast. Here I propose a critical analysis of the volumes *Degli edificj antichi e moderni di Roma*. This collection is characterized by the absolute clarity and linearity of many of the representations produced by mean of etching, as well as by their particular, almost experimental visual outcome. For such reasons, GBC can be defined as an experimenter in the field of graphics for architecture, configurator of new visual hypotheses on the use of the sign as formalized graphic trace.

Here, the analysis of the graphic value of the linear component of the drawing is required, by framing it in its theoretical debate of the second half of the 18th century. Winckelmann in his *Gedanken über die Nachahmung der griechischen Werke in der Malerei und Bildhauerkunst* (1755) established a completely new and original relationship between History and Nature (in a strict relation to Art), by linking the concept of 'beauty' to History and not just to an Idea. In 1786, in his text *Principi del disegno* [18], Raffaello Morghen underlined the importance of drawing, to be understood as the highest and purest expression (also visual) among the Arts, specifying how it was easier (for him) to understand Nature by means of other artists' representation of it, rather than by Nature itself [19].

Without forgetting the graphic typology of the illustrations of Winckelmann's work – mostly illustrated with linear and outline drawings – a few years later, in 1792, the

German writer Karl Philipp Moritz argued that «as it happens in our thought, the 'cultivated' spirit loves order, light, clarity, so, in the Art, what is well ordered, what is easy to observe and to understood without effort, must necessarily come first over what is intricate, enveloped, bulky» [19]. In other words, and with the aim of recovering the original purity, he intended to elevate the Antiquity to be a model of a clear and linear thought [19].

Furthermore, in the 1790s, artists and engravers working on new editions of classical literature, such as Hesiod, Omero, Aeschylus, began to reconsider the pre-classical art in order to create new illustrations for these works. This debate was mainly developed in the artistic, non-architectural field; in a sense, it followed the best known legend on the origin of painting, reported by Pliny the Elder (in the first century BC) [20] and by Athenagóras (in the second century BC) [21]. In this myth the concept of outline drawing is already exposes, since painting originated by the act of tracing the projected shadow of a man on a wall; in this way the Corinthian girl Diboutades obtained the *silhouette* of her lover [22], who was leaving for the war, in order to preserve his visual memory [23].

3.1 Bénigne Gagneraux (1756–1795)

Born in Dijon in 1756, Gagneraux was trained as a painter within François Devosge's *Académie de peinture et sculpture*. With the sculptor Alexandre Renaud, Gagneraux was the first winner of the scholarship banned in 1776, thus having the possibility of moving to Rome (where he had already been in 1774) to continue his studies as a *pensionnaire* of Burgundy. Just like that of his colleagues in the *Académie Royale*, Gagneraux's Roman training included visits to churches, monuments and important art collections (Albani, Borghese, Farnese, Ludovisi, Mattei, etc.) while drawing, copying and verifying the ancient and the modern masters (Raffaello, Carracci, etc.) [24].

At the same time, he devoted himself to numerous activities, including the graphic reproduction of ancient statuary elements for Francesco Piranesi (in 1781) [25]. In 1784, thanks to him, Gagneraux became the official 'art supplier' of Gustav III, King of Sweden [26]. In this way, the French artist found his maximum professional gratification, becoming the reference man for the Swedish nobles who were about to visit Rome. In addition, he had several meetings with Pope Pius VI, who appreciated his production and commissioned him some works and copies. Because of the revolution against French power, he was forced to leave Rome in early 1793 and moved to Florence, where he remained until his death [25].

3.2 John Flaxman (1755–1826)

Born in York in 1755, John Flaxman was one of the sons of a craftsman specialized in the manufacture of plaster casts. Since when he was a child, he was interested in the possibilities that clay and wax offered to the plastic modeling and started his study with his father, soon proving to be an artist of great talent [27]. Between 1770 and 1775, Flaxman attended the Royal Academy School of Art and worked as designer of Etruscan-style ceramics at Josiah Wedgwood Workshop [11]. This last experience was fundamental for its artistic education, since the concept of outline drawing passed from the reproduction of figures copied from the Etruscan and Greek potteries [28].

Between 1787 and 1794 he undertook his *Grand Tour* in Italy, where he met William Young Ottley. Here he traveled all around the Peninsula and visited Florence, Rome, Naples, getting into contact with many of the most significant examples of Italian art, from Cimabue to Raffaello, up to the classical antiquities of Ercolano and Pompei, developing a series of notebooks rich in sketches reproducing pieces of art, buildings and views in general [29].

In 1794 he returned in his homeland, where he was appointed an associate member of the Royal Academy in 1797, then an academic in 1800 and a professor of sculpture in 1810. At the same time, he worked for numerous private clients [27]. Since his return he no longer devoted himself to outline drawing, with the exceptions of the figuration of the poem by John Milton in 1810 and of the poems of Hesiod, in 1817. He died in London in 1826.

4   The Importance of Gagneraux's *Trait* and Flaxman's Outline

In this context, the interest for Gagneraux's and Flaxman's works is absorbed by the production of printed representations, thus engravings. In fact, thanks to the study of ancient vascular painting, Flaxman and Gagneraux realized the non-indispensability of color for a non-mimetic reproduction of reality. In this sense, the pure outline and the «paratactic rhythm» of those vascular representations were able to replace the rigorously perspective reconstruction of Renaissance-style representations, thus simplifying the composition of the images, still maintaining their naturalistic characteristics [19].

The French painter published his *Dix-huit estampes au trait* in Rome in 1792 [31, 32]. The interest in this collection is not dictated by the subjects represented, but by the graphic characteristic of the engravings, here understood as a true formal connotative language. Such language proclaims the superiority of the line over all the other elements of the image created by the artist, even on the drawing itself. Being aware of the fact that the «reduction of a figure to an outline drawing, without shading or modeling [...] is a procedure as old as the artistic practice» and that the practice of engraving does not lack examples of «linear drawings» formalized by simple, demonstrative and sufficient traits at the same time, this typology of representation is defined as «in direct function of the economy of the (graphic) medium» [23]. It is therefore important to remember that the *trait* drawing of Gagneraux does not constitute an absolute novelty in the panorama of representation, having already been widely used, for example in the tables of the 16th century architectural treatises [32], among all the *De Architettura* by Cesare Cesariano and *The Four Books of Architecture* by Andrea Palladio.

The eighteen prints, of different sizes [33], illustrate single episodes and show a synthetic use of the line: the *trait* surrenders its meaning from simple graphic expedient to the true essence of the representation. The approach to the line restores an analytical procedure that «contravened any speculative tension» [32]. Gagneraux was not the first interpreter of this formulation of communication: as early as 1778 the sculptor Jean Jacques Lamarie claimed that «drawing is all in the stroke» [30].

In Fig. 2 it is possible to notice how the attention for the use of the *trait* (outline) appears both in the figures that animate the representations, as in the context, which is not specified (Fig. 2a), or a landscape (Fig. 2b) or a pseudo-architectural one (Fig. 2c).

It is precisely this last case (one of the largest engravings of the series) that becomes the ideal link between the use of the outline in the previous or contemporary architectural treatises and the specific use, or rather the graphic expression, which GBC will interpret few years later.

Fig. 2. Engravings from the *Dix-huit estampes au trait*, Musée des Beaux Artes, Dijon: a. *Portrait des deux filles de M. Le Comte de Sellon. De Genève*, 115 × 157 mm [33]; b. *La peinture*, 132 × 178 mm [33]; c. *Hébé*, 196 × 156 mm [33].

The peculiarity of these representations is also reflected in many original drawings by Gagneraux, that confirm his graphic intention, describing a possible transformation over time, from a sign still looking at the natural context (Fig. 3a), (Fig. 3b) to a sign probative of the simple physicality of a *predella* (Fig. 3c).

Concerning Flaxman's work, here I focus on the production of collections of engraved prints which might have provided a visual and cultural pretext to GBC for the realization of his *Vedute in contorno*.

In 1793 Flaxman, helped by the engraver Tommaso Piroli, published the first illustrated volume in Rome: *The Iliad of Homer*; *The Odissey of Homer*, (Fig. 4a, b). In the same year (or perhaps in 1802) Flaxman illustrated Dante Alighieri's *Divina Commedia*, (Fig. 4c) still published in Rome.

With this type of representations he had set himself the goal of eliminating visual irregularities due to the figuration of the incidence of light and the effects of the texturing of the drawing, bringing his graphic vocabulary to the basic level of a rudimentary language formed by the pure sign/outline drawn on monochrome paper [23]. It is therefore clear that the intention to simplify one's visual research shines through the printing of

Fig. 3. Drawings by Bénigne Gagneraux: a. Ante 1783, *Adam et Eve pleurant sur le corps d'Abel*, pen and black ink on paper, 350 × 260 mm, Musée des Beaux Artes, Dijon [33]; b. Ante 1782, *Le festin des dieux champêtre*, pen and black ink on paper, 521 × 697 mm, Gabinetto Disegni e Stampe degli Uffizzi, Firenze [33]; c. 1787 circa, *La cérémonie du jeudi-saint*, black ink and carcoal on paper, 360 × 530 mm, Musée des Beaux Artes, Dijon [33].

«atmospheric quality» and «analyticity of chiaroscuro» which reflect all the «tension» and attention «to volumetric restitution» which will be subsequently completely overcome by the search for pure outline [10]. With this techno-graphic expedient Flaxman was able to place the reader, for example in the case of the *Divina Commmedia*, on a level that was now ethereal, now infernal compared to the representations of the same subjects that had happened up to then.

Fig. 4. Engraving by Tommaso Piroli on John Flaxman's drawings from: *The Iliad of Homer; The Odissey of Homer*: a. 1793, *Thetis finds Achilles mouring over the corpse of Patroclus* [23]; b. *Council of Jupiter, Minerva and Mercury* [34]; c. *Divina Commedia di Dante Alighieri*: 1793, *Circle of Angels around the sun* [23]; d. *Compositions from the tragedies of Aeschylus: Oath of the Sevens against Thebes* [23].

In 1795 he published the *Compositions from the tragedies of Aeschylus* in London (Fig. 4d). Still in 1817 he illustrated the *Compositions from the Works and Days, and Theogony of Hesiod*, engraved by his friend William Blake [23].

Flaxman's representations were «witnesses of the same fervent search for elemental purity and […] of the same mannered and exquisite results» [23], regarding the rediscovery of the classical canons in Greek art, likewise of architecture, the figurative arts of the late eighteenth century. As Piera Tordella states: the remarkable «diagrammatic reduction of form», being the symptom of Enlightenment rationalism, allowed Flaxman to operate a definitive process of «simplification and synthesis» [10], reaching the more important visual formalization of the most known «attempt to reduce the visual arts to a minimum vocabulary» [11]. The clever dosage of the thickness and the intensity of the outline, sometimes combined with the schematic shadows shown on the scene, allowed the flat Flaxmanian figures to create an intricate overlapping of parallel planes, [23], thus allowing the Artist to produce a style that is absolutely not mimetic, aimed at reducing the vision of objects to punctual information.

At this point is clear that Flaxman succeeded in the extremely synthetic intent of composing the image in all its forms by using line as a «visual device of modernity

[…] (highlighting) contours, shape, chiaroscuro, composition, movement, etc.» [35]. Precisely those images with a pure outline, configured as «linear abstractions of shapes and volumes» were able to define, at the end of the 18th century, a new type of «visual text sometimes sharply re-edited, often intensely accepted, even by artists of other movements» [10] (like Henri Fuseli, Jaques-Luis David, William Blake). In this case, this anti-mimetic specification, in my opinion, will meet the interest of GBC. Furthermore, the two artists, like architects [35], were the proponents of the main theoretical and practical speculation of the neoclassical idea on the superiority of drawing and line over all other type of languages in the field of visual communication, particularly in the field of printed graphics and the representation of the human figure.

5   Giovanni Battista Cipriani: *Degli edificj antichi e moderni di Roma. Vedute in contorno*

This collection of *vedute* of the city of Rome consists (in the specimen I consulted at the BiASA of Rome) of two volumes, each containing 62 prints plus the frontispiece, for a total of 124 *vedute* of the dimensions of approximately 110×90 mm.

Within GBC's vast graphic production, these volumes can be considered as breaking elements in the consolidated language of representation and communication. In fact, these views translate Flaxman's communicative theory into paper by representing the architectural subjects in a outline, even if not in a consciously elaboration. This abstraction leaves room for the interpretation of the graphic sign, allowing the author to make an extremely significant eidetic synthesis.

GBC applies this methodology of representation to architectural subjects, for example the *Anfiteatro Flavio* (Fig. 5a), as well as to subjects more related to landscape thems, such as the *Circo di Caracalla* (or *Circo di Massenzio*) (Fig. 5b).

Fig. 5. GBC, examples of *disegno in contorno* (outline drawing): a. *Colosseo*, vol. 1, f. 5r; b. *Circo di Caracalla*, vol. 1, f. 23r.

His experimentation starts from *vedute* characterized by a light, almost air, graphics [14]. In such occasions, however, the attempt to simplify the graphical 'act' does risk falling into a semantic contradiction of superabundant signs describing buildings, for example in the case of the view of the *Basilica di S. Pietro* (Fig. 6a), see the colonnade. This kind of experimentation arrives at representations where the difficulty of rendering the infesting vegetation on ancient ruins becomes a pretext to bring up the pure volumes of the building masses, as in the case of the *Bagni di Paolo Emilio* (Fig. 6b).

Fig. 6. GBC, examples of *disegno in contorno* (outline drawing): a. *Basilica di S. Pietro in Vaticano ed Obelisco Egizio*, vol. 1, f. 61r; b. *Bagni di Paolo Emilio*, vol. 1, f.18r.

Due to the previous considerations, I believe that GBC obtains the best application, or rather the most convincing and strong communicative results, in the representation of architectural interiors.

Here, I bring for example the views of the hall of the Pantheon (Fig. 7a) and of the *Tempio di Claudio* (Fig. 7b). In both cases, it is evident how GBC manages to make the architectural space of the two interiors clearly legible without using chiaroscuro and contrasts but using only lines. Obviously, these views are quite rigid, but, in their small dimensions, fully embody the severe taste of neoclassicism based on Winkelmann's aesthetic theories. And, again, the view of the *Tempio di Claudio* takes this concept to its extremes by placing at the center of the picture not the intercolumniation, but the column; this vertical element is the base of the rigorous symmetry of the image produced GBC.

Therefore, this collection of *vedute* is a very interesting expressive declination of what the city of Rome was able to arouse in each single interpreter of its views, also thanks to the innumerable stratifications of its urban facies. Moreover, the *vedute* acted the role of possible forerunners of the – now somehow obsolete – postcards.

The Author himself, within the preparatory manuscript of the textual descriptions to be attached to the *vedute*, provides us with another possible hint for their interpretation. In fact, he reveals his innovative communicative intentions regarding pure outline drawing by declaring his rigorous inclination to 'simplify' the reading – or the visual comprehension – of such images, while taking care of the 'economy' of the cultural and material production process. Still, he highlighted that all the views represented the as built as surveyed by him [36].

Fig. 7. GBC, examples of *disegno in contorno* (outline drawing): a. view of the hall of the *Pantheon*, vol. 2, f. 4r; b. view o f he interior of the *Tempio d iClaudio*, vol. 2, f. 42r.

6 Conclusion

Thanks to the analysis here presented, it is clear how GBC implemented a continuous mediation between the aesthetic of Giovanni Battista Piranesi' archaeological curiosity and of the Karl Friedrich Schinkel's neoclassical rigor. Moreover, sometimes he mediated these with visual expressions with other that, somehow, anticipate the Romanticism which its dawn int that years. This anti-mimetic visual specific, in my opinion, met the interest of GBC, whom, on the other hand, elevated the vision as a probative instrument of the goodness of his work and, consequently, of the visual and communicative result of his *vedute in contorno*.

Here, GBC emerges as an experimenter of a communicative practice that found its origin both from the consolidated practice of architectural Drawing and from the debate on its supremacy over the other Arts. Still, he arises the ideal anticipator of the much more famous works by Paul-Marie Letarouilly, as well as a simplifier of the code of visual communication aimed at popularizing the architectural fact as an extreme graphic-visual reworking of the cultural prerogative of neoclassicism.

References

1. Carpo, M.: L'architettura dell'età della stampa. Oralità, scrittura, libro stampato e riproduzione meccanica dell'immagine nella storia delle teorie architettoniche. Jaka Book, Milano (1998)
2. De Rubertis, R.: Verso quale rappresentazione? Diségno (2), 23–32 (2018). https://doi.org/10.26375/disegno.2.2018.5
3. Hind, A.: La storia dell'incisione dal XV secolo al 1914. Allemandi, Torino (1998)
4. Debenedetti, E.: Giovanni Battista Cipriani. Studi sul Settecento romano **22**, 235–236 (2006)
5. Pasquali, S.: Fortuna di G. B. Montano del tardo Settecento: un taccuino di disegni di Giovan Battista Cipriani. Il disegno di architettura 25–26, 18–23 (2002)
6. Amadei, E.: Tre architetti romani dei secoli XVIII-XIX. Capitolium, XXXV (10), 18–22 (1960)
7. Cipriani, G.B.: Libraccio, o miscellanea di memorie spettanti alle belle arti di Giò. Batt.a Cipriani Sanese. Manoscritto con disegni, BNCR VE 1207 (1801)

8. Cipriani, G.B.: Indice delle figure relative ai Principij di architettura civile di Francesco Milizia. Salomoni, Roma (1800)
9. De Bernardi, A.: Forma, Spazio, Percezione. Giardini, Pisa (1979)
10. Tordella, P.G.: Il disegno nell'Europa del Settecento. Ragioni teoriche ragioni critiche. Olschki, Firenze (2012)
11. Cipriani, G.B.: Monumenti di fabbriche antiche estratti dai disegni dei più celebri autori da Gio. Battista Cipriani Sanese. Tomo I. s.e., Roma (1796)
12. Bentivoglio, E.: Alla ricerca del disegno smarrito: «Lettera» da Roma. Il disegno di architettura, 1–3 (1989)
13. Bentivoglio, E.: Il "Libraccio" di Giovanni Battista Cipriani. In: Architettura nella storia. Studi in onore di Alfonso Gambardella, vol. 1, pp. 368–373. Skira, Milano (2007)
14. Debenedetti, E.: I Taccuini di Giovanni Battista Cipriani. Studi sul Settecento Romano **31**, 207–236 (2015)
15. Olschki, C.: Giovan Battista Cipriani. Quaderni di Studi Romani **11**, 7–20 (1940)
16. Pavignano, M.: Rappresentare l'architettura. Il viaggio ideale di Giovanni Battista Cipriani tra disegni, libri e stampe. Ph.D. dissertation in Architectural and Landscape Heritage. Politecnico di Torino, Scuola di Dottorato, 11 July 2019 (2019)
17. Marotta, A.: Un linguaggio trasversale: il segno come traccia grafica. In: EGRAFIA 2012, pp. 457–460. Urbanismo y Diseno de la Universidad Nacional de Cordoba, Cordoba (2012)
18. Morghen, R.: Principi del disegno. Pagliarini, Roma (1786)
19. Fragonara, M.: Incisione a contorno e l'idea del bello. Appunti sull'incisione neoclassica. Rassegna di studi e notizie, vol. XXVI, pp. 71–96 (2002)
20. Plinio il Vecchio: Storia naturale, vol. XXXV (2014)
21. Athenagóras: The apologetiks of the learned Athenian philosopher Athenagóras, Londra (1914)
22. De Rosa, A.: Geometrie dell'ombra. Storia e simbolismo della teoria delle ombre. CittaStudiEdizioni, Milano (1997)
23. Rosenblum, R.: Transformation in Late Eighteenth Century Art, 3rd edn. Princeton University Press, Princeton (1974)
24. Laveissière, S.: Bénigne Ganeraux primo Prix de Rome degli stati di Borgona. In: Bénigne Ganeraux (1756–1795) un pittore francese nella Roma di Pio VI, Roma, Galleria Borghese, Aprile – Giugno 1983, catalogo della mostra, pp. 20–25. Accademia di Francia a Roma – De Luca Editore, Roma (1983)
25. Sandström, B.: Bénigne Gagneraux e la Svezia. In: Bénigne Ganeraux (1756–1795) un pittore francese nella Roma di Pio VI, Roma, Galleria Borghese, Aprile – Giugno 1983, catalogo della mostra, pp. 31–40. Accademia di Francia a Roma – De Luca Editore, Roma (1983a)
26. Hoffmann, P.: Pio VI e Roma – cultura, arte e società. In: Bénigne Ganeraux (1756–1795) un pittore francese nella Roma di Pio VI, Roma, Galleria Borghese, Aprile – Giugno 1983, catalogo della mostra, pp. 26–30. Accademia di Francia a Roma – De Luca Editore, Roma (1983)
27. Thomas, J.: John Flaxman, R.A. (1755–1826). J. Roy. Soc. Arts **104**(4966), 43–66 (1955). http://www.jstor.org/stable/41368419. Accessed 18 Mar 2018
28. Whinney, M.: Flaxman and the eighteenth century. A commemorative lecture. J. Warburg Courtauld Inst. **19**(3/4), 269–272 (1956). http://www.jstor.org/stable/750298. Accessed 18 Mar 2018
29. Brigstocke, H.: Refocusing the Grand Tour: John Flaxman and the reappraisal of early Italian painting and sculpture, 1787-94. In: Brigstocke, H., Marchand, E., Wright, A.E. (eds.) John Flaxman and William Young Ottley in Italy, pp. 3–24. The Walpole Society, Londra (2010)
30. Sandström, B.: Gagneraux e l'antichità. In: Bénigne Ganeraux (1756–1795) un pittore francese nella Roma di Pio VI, Roma, Galleria Borghese, Aprile – Giugno 1983, catalogo della mostra, pp. 47–51. Accademia di Francia a Roma – De Luca Editore, Roma (1983)

31. Laveissière, S.: Il tratto. In: Bénigne Ganeraux (1756–1795) un pittore francese nella Roma di Pio VI, Roma, Galleria Borghese, Aprile – Giugno 1983, catalogo della mostra, pp. 55–56. Accademia di Francia a Roma – De Luca Editore, Roma (1983)

32. Rossi Pinelli, O.: Il secolo della razione e delle rivoluzioni. UTET, Torino (2000)

33. Laveissière, S.: Catalogo. Il tratto. In: Bénigne Ganeraux (1756–1795) un pittore francese nella Roma di Pio VI, Roma, Galleria Borghese, Aprile – Giugno 1983, catalogo della mostra, pp. 67–188. Accademia di Francia a Roma – De Luca Editore, Roma (1983)

34. Morris, B.: Flaxman's illustrations to homer as a design source for glass decoration in the 1870 s. Burlington Mag. **129**(1010), 318–321 (1987). http://www.jstor.org/stable/882964. Accessed 21 Mar 2018

35. Leone, F.: L'officina neoclassica: anelito alla sintesi, ricerca dell'archetipo. In: Leone, F., Mazzocca F. (eds.) L'officina neoclassica. Dall'Accademia de' Pensieri all'Accademia d'Italia, pp. 18–53. SilvanaEditoriale, Milano (2009)

36. Cipriani, G.B.: Spiegazione delle Vedute di Roma antiche e moderne Disegnate ed incise in contorno da Gio. Batt. Cipriani. Manoscritto, BAN 1660/1 (1816)

Sequences, Tracks and Footprints: Graphic Lessons Gathered from Swarm and Spider Web Robotics

Jose Carrasco Hortal[(✉)], Salvador Serrano Salazar, and Francesc Morales Menárguez

Department of Graphic Methods, Theory and Design, University of Alicante, Alicante, Spain
jose.carrasco@ua.es

Abstract. The development of automation in construction is opening up a whole field of research on how to optimise movements and manoeuvres.

Furthermore, important studies have also been conducted into the emulation of spider webs and the advances of swarm robotics. The latter consists of a series of identical robots that collaborate when performing tasks, covering a broad spectrum of situations without the need to specifically pre-programme each robot.

Graphics included in this paper demonstrate the potential of small robotic units with a certain autonomy. Some research and design cases created by Tomas Saraceno, Maria Yablonina and Gramazio & Kohler are referenced for discussion.

The methodology first translates, on a new digital graphic, how a spider of the Theridion sisyphium species builds its web through different stages sequenced by Benjamin and Zschokke [1]; second, it explains how a Delaunay triangulation that undergoes subsequent relaxation helps to obtain a plausible arrangement of tense strings that can then be manually replicated. At a real performance level, the study describes how a group of students from the University of Alicante emulates the filament-throwing techniques of the aforementioned arachnid species during a design workshop completed in May 2019.

Keywords: Graphic sequences · Swarm robotics · Spider-based design · Virtual-physical

1 Context

"By the means of Telescopes, there is nothing so far distant but may represented to our view; and by the help of Microscopes, there is nothing so small, as to escape our inquiry. Hence there is a new visible World discovered to the understanding" [2].

In emerging fields such as virtual reality, augmented reality or component design, a single simulated image or graph is increasingly becoming insufficient to explain a final product. In any case, this fact helps us to understand that architecture, and the socio-cultural heritage it encapsulates, needs to reconsider the discipline's methodologies.

L. Agustín-Hernández et al. (Eds.): EGA 2020, SSDI 6, pp. 633–643, 2020.
https://doi.org/10.1007/978-3-030-47983-1_56

1.1 Drawing the Living Being (Capturing Phases)

Tools such as the Kinect sensor, which probes people by recognising the human skeleton and then following its movements, or eye-tracking devices that translate the observer's eye movement trajectories into chromatic maps, are examples derived from the seminal studies conducted by Muybridge and Marey, who used chronophotography to obtain animal movement trajectories [3]. They even incorporated pneumatic and electrical sensors [4]. In the late nineteenth century, Muybridge succeeded in demonstrating that any racehorse will, at some point, just for an instant, have all its limbs in the air when galloping. This result was achieved via a system of high-speed captures, using white sheets in the background attached to one another with vertical lines (a timeline).

Almost a century later, Bernard Tschumi introduced the spectator to visible synthesis of space and human action based on a series of photographic drawings and diagrams, using codes learned from ballet choreographies. "Architecture is not simply about space and form, but also about event, action, and what happens in space" (see Fig. 1). Tschumi was attempting to incorporate elements usually absent from architectural or landscape representations. He was referring to relationships between spaces and uses, between type and programme, between objects and events. Relevant to this study, in his initial approach to his Manhattan Transcripts, he introduces a notation mode based on three-panel graphs to portray any experience (known as "chronotope" in language theory): the first component is composed of photographs showing the actions; the second consists of planimetric maps revealing changing architectural manifestations; and the third are diagrams presenting the movements of main actors [5].

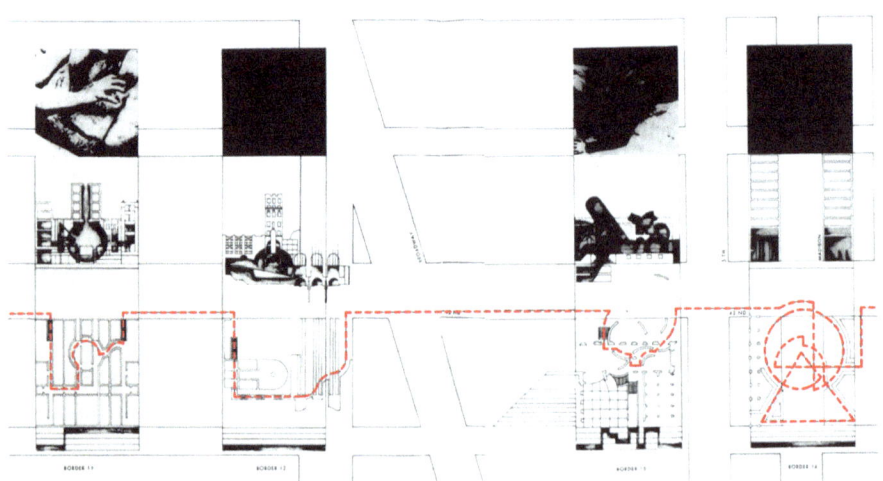

Fig. 1. Bernard Tschumi. The Manhattan Transcripts 1976–1981.

1.2 Drawing the Robot (Capturing Phases in Automatons)

Moveable Generative Units. John and Julia Frazer pioneered the architectural use of individually operated robotic units, with their project "The Universal Constructor" [6]. In this project, they illustrated the functioning of a self-organised environment of small cells able to communicate with the neighbour ones. Project's definition referred to John Von Neuman's research on self-replicating machines. They succeeded in designing devices called cellular automata, in which the communication through cells included instructions on how to proceed with evolutionary disposition. Each unit explores its sides, looking for the presence of neighbouring units, using processing times based on LED lighting. In this case, the light draws and reports the process. The information flow is equivalent to what Neuman developed with his "instruction string". Frazer turns the visitor into a necessary actor who participates in the information process. This auto-generation path was also explored by cellular automata such as "Game of Life" (John Conway) or the making of the Universal Computer Turing machine, thanks to which any complex problem could be oriented to solutions.

Automata Functioning as Swarms. In more recent times, other researchers have been able to create and use robots to weave thread systems over a range of different supports [7]. In a similar way to Saraceno, Yablonina based her research on observing how a spider builds its web [2]. In Yablonina's experiment, a spider of the *Tegenaria Atrica* species was placed inside an clear box and its movements were recorded using infrared cameras that captured images every 30 s, and a laser scanner that scrutinised the clear box space in a succession of horizontal slices. The experiment resulted in defining the spider's web path and obtaining a digital file with a three-dimensional mesh so as to analyse the thread's orientation. In addition, geared robotic devices created protocols to sew strings (Fig. 2).

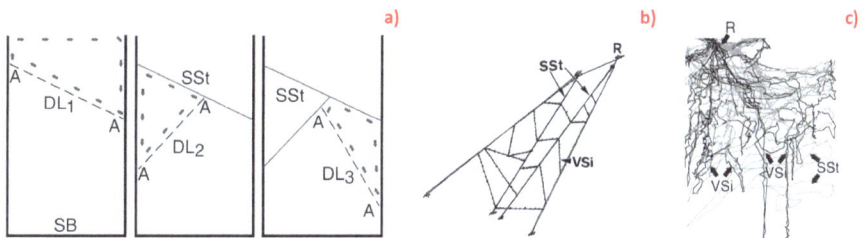

Fig. 2. Graphic formats in Benjamin and Zschokke: movements (a); stages (b); result (c).

In another model, Yablonina created a series of robots specialised in different structure-weaving tasks. Some moved along vertical surfaces to which they adhered thanks to a fan that creates a vacuum between the surface and the robot itself; others moved by means of a carriage-like system through pre-installed ropes. Both types of robots carried string coils to produce fastenings to points on the wall. Using an electromagnet system, robots can pass the coil from one to another so that, all together, the automata can weave on different planes. The result was a three-dimensional tense configuration made up of intertwining strings (see Fig. 3).

Parallel to Yablonina's climbing species, Fabio Gramazio and Matthias Kohler explored self-construction using drones, which present a greater capacity to access any point in 3D space, beyond that allowed by a real room's surfaces or a virtual prismatic

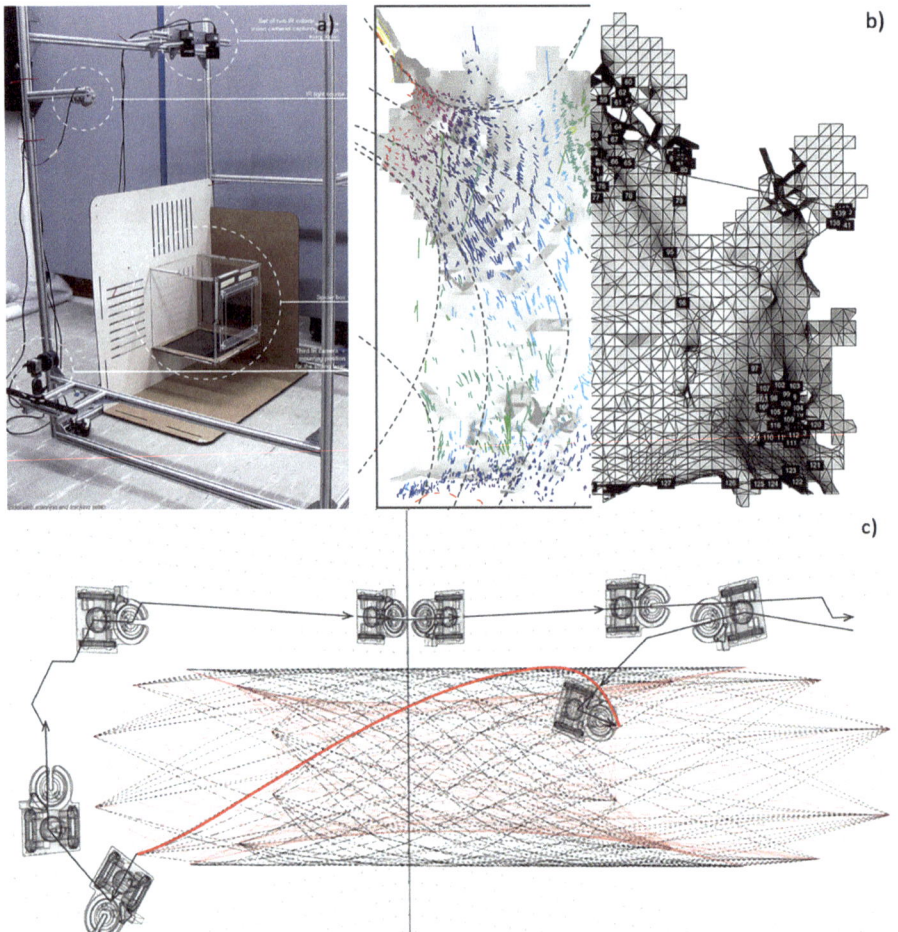

Fig. 3. Yablonina. Spider Web, Stuttgart University, in 2013 (a, b); spider-robots passing the string to each other (c).

universe [8]. In this case, they used quadcopters loaded with string coils that progressively extend, creating fastenings and knots between the strings themselves (see Fig. 4). The contribution at the code level is that this swarm of drones needs to regulate three parameters: the flight path that each drone follows; the force they apply to the string when tied; and the drone's orientation when dragging the string. This enables "constructive primitives" to be obtained that, when concatenated, define each drone's flight path to build a structure designed beforehand. The interesting point is the fact that the interaction between the drones and their environment generates a family of visualisations and graphs, which is related exclusively to coordinated movements to avoid collisions.

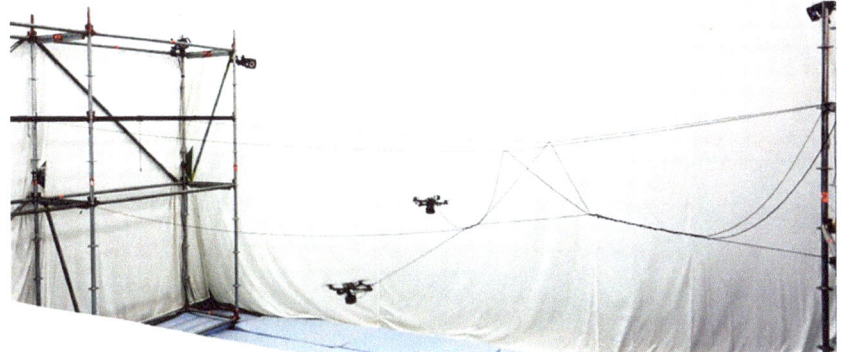

Fig. 4. Gramazio and Kohler. Tense strings placed by a swarm of drones.

1.3 Recording via Stereophotography and Sheet Laser Scanner

Working in an aeronautical hangar in Frankfurt, Tomás Saraceno and his team succeeded in replicating, at a scale magnified 30 times, the shape of a three-dimensional web created in a transparent plastic clear box by the black widow (*Latrodectus mactans*) spider. To do this, he used resources such as stereophotogrammetry or sheet laser, a new tomographic method (slice-based images) implemented in collaboration with the Darmstadt TU [9].

Spider webs, besides representing for many scholars a visualisation model of the origin of the universe, prove challenging for most scans: firstly due to the invisibility of the filaments, as their diameter nears a nanometre; and secondly, because of non-flat joining between more than two threads. In this case, stereo-photographs of the laser-lit slices (110 pairs of stereo-photographs) were taken, creating in each a sort of constellation, as if it were "a photographic image of a slightly starry night".

A script then attempted, unsuccessfully, to replenish the thread section corresponding to the unscanned space, that is, in the interstitial space between thin slices. The paradox, after using such cutting-edge technological resources, is that the problem had to be solved by four architects, adding lines, one by one, in a 3D space.

Before starting the installation of the spider web in Bonames (Frankfurt am Main), orthographic projections (specular images between them) were placed on the walls, floor and ceiling of the room.

Vertical nylon threads were used as master lines, tensed between floor and ceiling. The key to the procedure was that they hosted the web's initial knot positions, before losing their shape. Formally, they were small elastic ropes that would then become knots in irregular polygons [2] y [9] (see Fig. 5).

2 Methodology

Based on the studies referenced, this article begins by describing how one of the procedures, that of the silk thread web described by Benjamin and Zschokke, was parameterised using a generative algorithm to make it operable within a three-dimensional

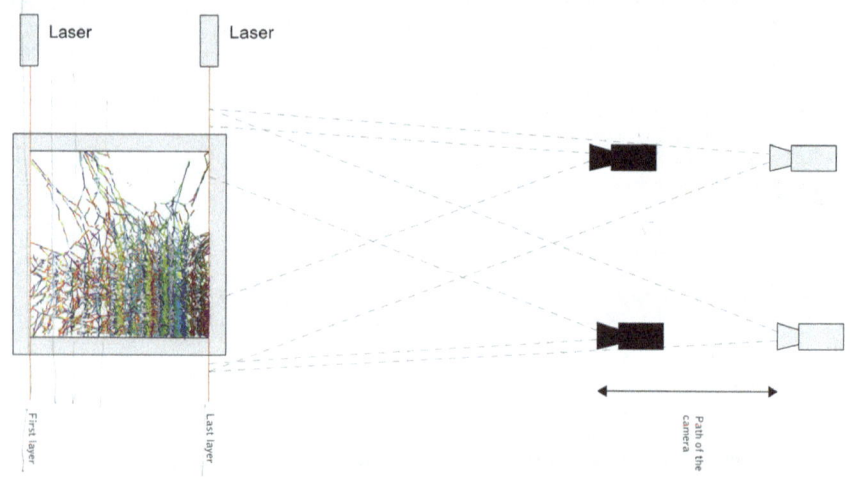

Fig. 5. Saraceno. Camera and laser placement map

universe of a prismatic envelope. It then goes on to show how another parametric version of a three-dimensional thread web is produced for a students' installation in a courtyard at the University of Alicante.

The method is based, firstly, on the translation of Benjamin and Zschokke's rules into an algorithm that generates a spider web–of the Theridion Sisyphium species–in a three-dimensional modelling environment, simulating the natural process. Secondly, the method reduces the system to another algorithm based on Delaunay triangulation, which then undergoes relaxation to facilitate the installation of a real string system in a few hours of assembly.

3   Results

3.1   Running the Virtual

To translate Benjamin and Zschokke's rules into an algorithm, the Grasshopper visual programming environment was used within the Rhinoceros software. Using these digital tools, we built an algorithm that reproduces the spider's action when weaving the web within a virtual urn-universe. Benjamin and Zschokke observed three different types of thread in the web of the *Theridion Sisyphium*: the fastening thread, which functions as a guide and is not definitive; the structural threads, which act as a general web support; and the viscous thread, which aims to capture prey.

In a first iteration, the virtual spider creates a fastening thread. The ends of this latter thread rest on the walls of the delimited virtual urn, and the virtual spider moves along it, adding more silk, thus reinforcing it and turning it into structural thread. At each successive iteration of the algorithm, the virtual spider adds a new fastening thread that connects an existing one to one of the urn walls, repeating the previous process and resulting in a new structural thread. Once there are a sufficient number of structural threads, the virtual spider begins to create viscous threads. These threads do not rest on the urn walls: their two ends rest on previously created structural threads.

The process is stochastic, because threads and wall positions to place new threads are randomly selected. This condition derives from the observation made by Benjamin and Zschokke that there is apparently no pattern of preference when the spider adds threads.

At each algorithm iteration, that is, each time a thread was added to the fabric, a relaxation process was executed using the Kangaroo plugin, adopting a shape that matches with the structure's state of minimum energy (see Fig. 6).

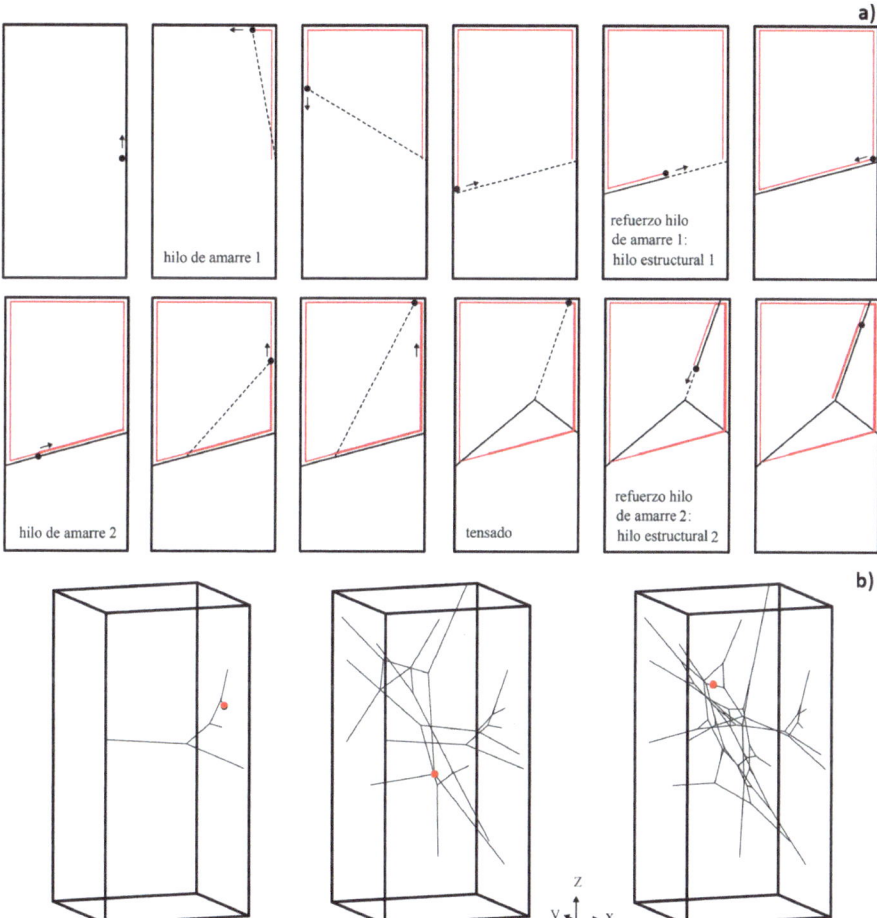

Fig. 6. Stages of structure set-up in 2D (a) and 3D (b). Author: Serrano, 2019.

3.2 Building the Physical

Once the algorithm had been defined, the next step was to physically apply what had been learned: a tensed structure was made and assembled in May 2019 in which devices designed by students of the University of Alicante were incorporated using point loads.

At this stage, an alternative model was explored to approximate the virtual generation of three-dimensional spider webs based on Delaunay triangulation. This new algorithm begins with a cloud of dots located within the clear box-universe, distinguishing the fastening points–allowing the fabric to be attached to the support–from loading points, where the devices were placed for point-in-time loading.

Delaunay triangulation was generated from these points. The algorithm then removes threads that go from one fastening point to another from the structure, as these threads do not help to support the loads. Finally, a relaxation process was carried out on the resulting wire mesh, so the loading points acquired a determined position within the urn-universe (see Fig. 7).

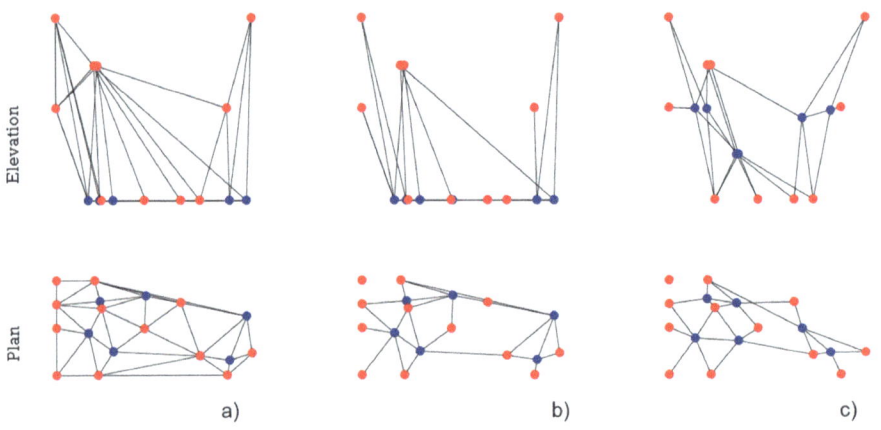

Fig. 7. Delaunay triangulation on 2D cloud of dots on base surface (a); removal of strings with no load (b); relaxation of the whole (result, c).

The urn-universe was, in this case, a courtyard measuring approximately 4.50 m (width) × 8.50 m (length) × 8.50 m (height) in the Polytechnic building of the University of Alicante. The behaviours of the collaborative robot of Maria Yablonina's installations (or of the Theridion Sisyphium spider that walks on its own thread, throws new filaments, goes back in search of new positions, segregates gluey silk for fastenins, etc.) were somewhat reproduced by the team of students-teachers in their attempts to throw ropes, add sliding knots, tie ends to gravel bags, design knots allowing for six passing ropes, etc. (see Fig. 8). Some of the contents tested consisted of using a spatially dimensioned universe, geometry controlled using analogue references and stereographic projections, and a parametric method to obtain node positions.

To build the structure, almost 200 m of jute rope, 6 mm in diameter, were used. Printed knots were made in 3D, and bags full of stones were placed as counterweights. Student's productions were laid out, hanging from each knot and contributing to the final

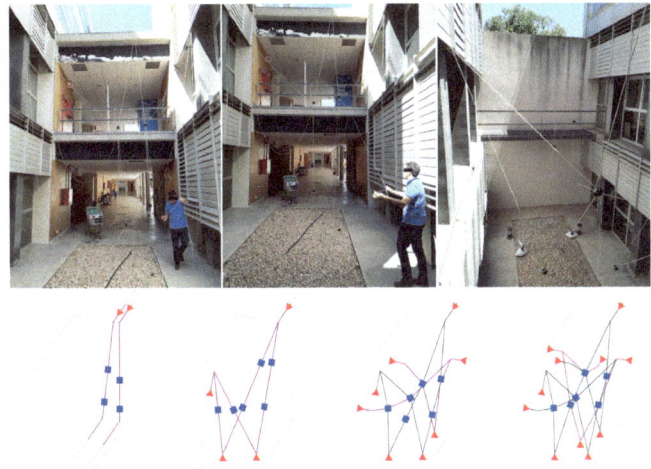

Fig. 8. Installation "La ruta de la seda". Assembly stages, UA. May 2019.

tightening. They generated interactions with observers using social media data to make certain decisions regarding the design itself (Data Driven Design).

4 Discussion and Conclusion

What can we learn from how spiders weave? How can this knowledge contribute to new automation methods in construction thanks to advances in robotics? The use in architecture of robotic units, which work individually with nearby inputs generating a higher-order global behaviour, foreshadows a new construction paradigm: that of a shift towards the process rather than remaining focused on the object. Research efforts should focus, therefore, on how to recognise behaviours and draw the itineraries of these particle systems.

"Architecture is everywhere and cannot be viewed essentially as the science of constructing houses, cities, etc. I think that the aims and interactions between disciplines must be continuously re-invented for each specific context. …We have to try to activate a process of re-actualization in relation to ever-changing contexts (to become) capable of imagining more elastic and dynamic rules". [2]. Kastner, Saraceno and other researchers use the term "ecology principle" to denote the framework that connects academic disciplines currently operating separately (engineering, social sciences and humanities) and which give meaning to future societies, learning how nature self-regulates through rules of cooperation and survival. In addition, thinking about sequences or stages gives value to other types of graphic depictions that go beyond the usual planimetric maps or deployments using cylindrical projections. It is something linked to digital production and its direct robot execution. (see Fig. 9).

The above supports the recognition of space-time as an inseparable cathegory within the creative process, and not only in architecture and urban design. In fact, it is hugely useful, since it constitutes a value that can be approached from diverse disciplines, leading

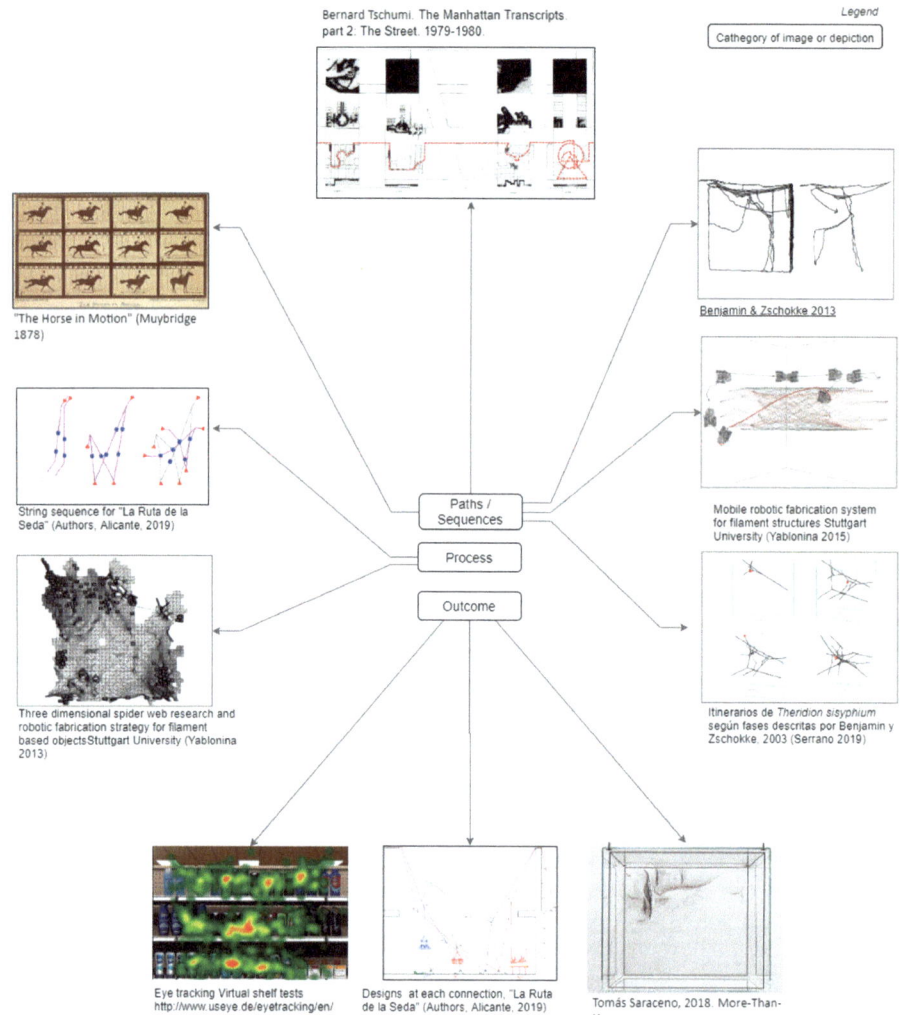

Fig. 9. References and productions (diagram: authors).

to confluences. For example, *chronotope* is a term from the Theory of Language, created by Bakhtin. The term gives value to the components that provide a space-time meaning to the story: "*In the literary artistic chronotope, spatial and temporal indicators are fused into one carefully thought-out, concrete whole. Time, as it were, thickens, takes on flesh, become artistically visible; likewise, space becomes charged and responsive to the movements of time, plot and history. The intersection of axes and fusion of indicators characterizes the artistic chronotope*" [10]. Terms such as nature or landscape (scape) and any of their variants (literary, visual, tactile, sound, olfactory, gustatory…) imply a time for experience and the possibility of apprehending its elements [11]. Its recording or spectrum eventually requires a set of parallel tracks that can host contents digitally.

And the term *narrative,* applicable to all these variants, implies the need for a structure to explain the process in all its complexity [12].

Comparing the works of Yablonina (artificial spiders) and the work of Gramazio & Kohler (drons), the conclusion is that in both installations, points and itineraries were controlled within the space. In the first case by 2D movements, in the second by 3D movements. Comparing how real spiders build their webs and how "La Ruta de la Seda" installation" was mounted at the University of Alicante, both cases use threads that gradually increase their shape complexity. Control lines, tensed vertical cables, delimit the position of Saraceno web's knots, while a Delaunay on the ground defines the first distribution of points in the courtyard of the University of Alicante. They work taking into consideration gravity, continuous fastenings, ground adhesion or anchors and passing knots as if, at each iteration, the whole was achieving a new distribution of the balance; the parallelepipedal clear box in the real spider experiments is comparable to the courtyard space in the university building.

To conclude, this work shows the technical possibilities of a methodology that includes multiple stages, each of which requires a precise and exquisite exercise of creativity, definition, labeling, depiction, etc. taken to extremes and within the grasp of students in a learning exercise lasting a single term.

References

1. Benjamin, S.P., Zschokke, S.: Webs of theridiid spiders: construction, structure and evolution. Biol. J. Linn. Soc. **78**(3), 293–305 (2003)
2. Kastner, J.: The Spinner and the Web. Tomas Saraceno in twenty jumps. In: Saraceno, T., Arrhenius, S. (eds.) 14 Billions (Working Title), pp 74–85. Skira, Milano (2011)
3. Brookman, P., Braun, M.: Helios Eadweard Muybridge in a Time of Change. Steidl, Göttingen (2010)
4. Lynn, G.: Animate Form. Princeton Architectural Press, New York (1999)
5. Tsumi, B.: Architecture and Disjunction. The MIT Press, Cambridge (1996). http://www.tschumi.com/projects/18/
6. Frazer, J.: An evolutionary architecture. University of Minnesota Architectural Association, Minnesota (1995)
7. Yablonina, M., Menges, A.: Towards the development of fabrication machine species for filament materials. In: Robotic Fabrication in Architecture, Art and Design, pp. 152–166. Springer, Cham (2018)
8. Augugliaro, F., Schoellig, A., D'Andrea, R.: Generation of collision-free trajectories for a quadrocopter fleet. A sequential convex programming approach. In: IEEE/RSJ International Conference on Intelligent Robots and Systems, Vilamoura, pp. 1917–1922 (2012)
9. Arrhenius, S.: Intergalactic spiders. In: Saraceno, T., Arrhenius, S. (eds.) 14 Billions (Working Title), pp. 8–18. Skira, Milano (2011)
10. Bemong, N., et al.: Bakhtin's Theory of the Literary Chronotope: Reflections, Applications, Perspectives. Academia Press, Gent (2010)
11. Wankhede, K., Wahurwagh, A.: The sensory experience and perception of Urban spaces. Int. J. Emerg. Technol. **7**(1), 741–744 (2016)
12. Weinstock, M.: The Architecture of Emergence. The Evolution of Form in Nature and Civilization. Wiley (2010)

Apple by Foster, Reinventing Patrimony

Carmen Escoda Pastor[✉], Abdulhadi Jawad I. Alhelal, and Josep Fort Mir

ETSAB, Universidad Politécnica de Cataluña, Barcelona, Spain
carmen.escoda@upc.edu

Abstract. It is interesting to see how, in a context where identity and virtuality have ever greater significance, digital corporations use architecture and its representation as a defining element of their own entity as trademark.

These big dotcom companies are affecting architecture, through work on emblematic buildings, converting them into icons that show their economic power and the questions that define them.

The case of Apple is paradigmatic, and is linked to both the participation of Steve Jobs in the creation process of the Apple identity as well as the whole set of designs that constitute their products. When the company considered the creation of a network of stores, they were understood and developed as another Apple product. Taking the example of Foster, we see how his approach and that of the company, were not only not contradictory, but are in fact complementary.

This paper focuses on Apple stores in historical buildings, with "the brand" Foster, in which technology was used as a form of expression in the restoration or, shall we say, in the neo-use. Buildings that take on a new identity, while their historical value is revalued.

Keywords: Apple · Foster · Tradition · Modernity · Neo-use

1 Introduction

"As an architect, you design for the present, with an awareness of the past, for a future which is essentially unknown" (Foster 2017, p. 10).

The growth of digital corporations in recent years today forms the third industrial revolution, which we can also call the "digital revolution", where some of the main digital corporations such as Apple have become an essential centre of the world. This corporation is moving to link its identity to historical, natural and cultural values, in conjunction with other values such as technology, the social work environment and entertainment, thus strengthening its corporate identity.

The vision of Steve Jobs in industrial design is one of the factors that has set Apple apart from other digital corporations since it began. The role that Jobs played in industrial design provided a special vision in terms of combining art and technology when a balance is achieved between the electronic development of the product and the industrial design, with the aim of creating products characterised by their clarity of shape and elegance. He also attempted to transfer this balance to the physical sites of his offices, to their architecture, highlighting the identity of the company, enabling the public to see it and feel it through the company's products and architecture.

L. Agustín-Hernández et al. (Eds.): EGA 2020, SSDI 6, pp. 644–655, 2020.
https://doi.org/10.1007/978-3-030-47983-1_57

1.1 Background

The concept of corporate architectonic identity is not a new idea, because the use of architectonic design to express a particular idea has a historical dimension. After the industrial revolution, new influencing powers arose, such as the factories that had significant economic power in Europe. This influence was extended to various aspects of social, political and cultural life.

AEG Turbine Factory (Fig. 1), designed by Peter Behrens, is a clear example of corporate architectonic identity. Built in 1909 in Berlin and called the "Cathedral of Work" by Le Corbusier (Bletter 1996), it is a clear example of what we call a design strategy. It has architectonic characteristics that reflect the values of identity adopted by AEG. At that time, AEG was considered one of the most important international companies, so the director of the company, Walter Rathenau, was aware of the importance of design to compete in the global markets (Putnam 1988). The interest in design was not limited to the products, it also extended to everything related to the company, from the logo to the buildings, for which Behrens was named the architect of the Company in 1907.

Fig. 1. Newsroom Electrolux, AEG Turbine factory, Berlin, 2012, https://newsroom.electrolux. com/de/2012/05/23/happy-birthday-aeg-deutsche-traditionsmarke-wird-125-jah

2 The Apple Product

Starting with the idea of the "Apple product" the initial store model incorporated the same criteria as the brand's other products. The initial cubic model, developed under the influence of Steve Jobs and subsequently patented as an industrial product, establishes the basis for the development of the stores built later. Adopting the square as the base incorporates a symbolic value into the proposal that reminds us of the Garden of Eden and its square floor plan. It is a mythological element that has been taken as a reference for countless new architectonic typologies throughout the centuries, whether consciously or unconsciously. The cube and the bitten apple greatly reinforce this link with the origin, Eden, deeply rooted in the collective subconscious of current majority cultures, beyond religions and personal beliefs (Fig. 2).

Fig. 2. James Hancock, Drawing of the Apple store on 5th Ave., New York, 2019, where the cube is shown with 15 panels of glass. https://www.archdaily.com/359403/all-the-buildings-in-new-york-drawn-by-hand/51674388b3fc4bf57e000050-all-the-buildings-in-new-york-drawn-by-hand-image?ad_source=myarchdaily&ad_medium=bookmark

The store built in New York constituted a new iconic component for the city. I t manages to stand out despite being relatively small, as is surrounded by skyscrapers. A determining factor is without doubt the association the public makes between the object and the brand. It is not simply "going" to an Apple store, but rather literally "entering" into an Apple product, which is quite different. In Apple stores, everything is designed in this sense. The object interacts with the environment. The massive use of glass, its transparency, enables it to integrate into the site at the same time as to incorporate the activity of the store into the context. Both identities respect and complement each other, that of the brand and that of the surroundings.

It is very illustrative how, despite successive stores being located in historical buildings and neighbourhoods, they always managed to maintain their identity as Apple products. The same strategy as for any Apple product is applied. Like the fact that in any Apple computer or telephone only apps or programs that meet the conditions established by the Apple programmers can be installed, remember that it is a closed-code system, all the buildings where Apple stores are located meet certain conditions that, in reality, are quite specific.

If the building already has iconic connotations, it is ideal, as Apple can make use of them to link them to the brand. It is preferable if the building has large windows or its location allows an extension with a glazed pavilion, so the interior and exterior are as communicated as possible, at least on the floor where it is accessed. The renders (Fig. 3) of the Apple store in San Francisco show this direct communication between exterior and interior space, through the long-glazed façade.

Fig. 3. Transparency as a communication strategy. Foster and Partners, Apple Union Square, San Francisco, 2014, https://9to5mac.com/2014/03/11/san-francisco-gives-final-approval-for-brand-new-union-square-apple-retail-store/. https://www.fosterandpartners.com/projects/apple-union-square-san-francisco/#gallery

Reusing historical buildings is one of the strategies that Apple uses in its search for values that may have social meaning and reflect the concept of art and technology, expressing the vision and identity of the company. By maintaining the original characteristics of the building fully or partly and combining innovative architectonic aspects in its design and construction, Apple wants to transmit to people that the company itself has become part of that historical value of the building.

Following this criterion, they borrow the most emblematic elements of the building, maintaining its original external façade, preserving spaces or redesigning them with a modern language. The techniques and materials selected for the first of the brand's stores were used to generate this modern-historical image, with significant presence of structural glass. These reflect the idea of the art-design dichotomy through the expressive use of technology.

2.1 Apple in Union Square, San Francisco. Foster + Partners

The second generation of Apple stores preserves the intention of combining art and technology, but adds new strategies that reinforce the presence of the company on a local scale. This store is a reference model for all the others, at world level, containing new characteristics while still showing the earlier values of the company.

The interaction of the company with the community is a value that characterises this stage and enables it to drive more educational, cultural and entertainment activities within the stores. Thus, social values linked to Apple's corporate identity are reinforced. It also improves the relationship of the company with the surrounding area, reinforcing sustainability.

"We are not just evolving our store design, but its purpose and greater role in the community as we educate and entertain visitors and serve our network of local entrepreneurs" (Apple Inc. 2016).

The team responsible for the architectonic design in the second generation of Apple stores was the British architecture company Foster + Partners, the Apple teams led by Jonathan Ive, Apple's art director, and their senior vice president of retail and online stores, Angela Ahrendts.

The store is in Union Square, one of San Francisco's most important squares. With a height of approximately 13 m, the access sliding doors give the store two important characteristics: the connection between interior and exterior, and the ventilation of the interior.

Due to their size, the doors provide a welcoming sensation, making it seem more like a public space than a store, especially with the doors completely open. "It all starts with the storefront - taking transparency to a whole new level where the building blends the inside and outside, breaking down barriers and making it more egalitarian and accessible" (Apple Inc. 2016).

With the doors open, the store benefits from the temperate climate of San Francisco. Natural ventilation is one of the sustainable solutions the company worked on for this generation of stores, linking the corporate identity of Apple to nature and sustainability.

On the north side, between the store and the hotel Grand Hyatt, there is a rectangular square and a fountain, (Figs. 4 and 5), which Apple completely renovated. It was a collaboration between the team of Apple designers and the Foster and Partners studio.

The company considered the square a gift for San Francisco, showing its respect for the community. The square is connected to the store by glass doors, but Foster uses other elements that link the interior and exterior, such as the green wall covered with Ficus Repens, a plant that is part of the city's environment.

On the stairways that lead to the square from street there is a historical fountain (Fig. 4), patrimony of the city of San Francisco, made by the sculptor Ruth Asawa and built in 1979. Apple preserved it during the renovation and converted it into one of the references for the store. In addition, it constitutes a meeting point where people can make use of the free services of both the store and the square. This historical fountain maintains the desire to recover the historical values, very present in the first generation of stores when Apple, for example, made use of some historical elements in its Opera store in Paris.

"This is an incredible site on Union Square and a chance to create a new public plaza" (Howarth 2016).

Fig. 4. Integration of the historical fountain with the building.

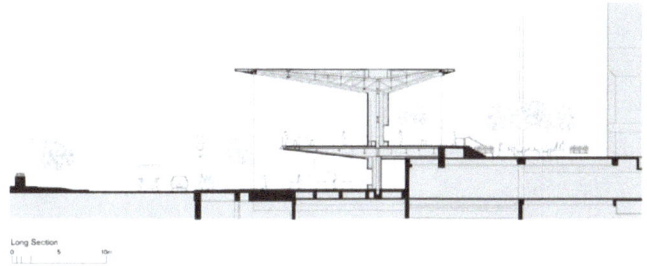

Fig. 5. Foster and Partners, Apple Union Square, San Francisco, 2016. https://www.fosterandpartners.com/projects/apple-union-square-san-francisco/#drawings

2.2 Apple on Regent Street. London. Foster + Partners

Apple Regent Street Store is one of Apple's flagship stores. It was the first one to open in Europe, in 2004, in London. The historical building on Regent Street (Fig. 6) meant a chance for Apple to link its brand to the history of this city, by reusing the building and restoring its historical façade. That gave the company the opportunity to improve its image, combining art and technology. The historical value of this building and the mosaics on the façade express aesthetic values in line with Apple's artistic values.

"Regent Street looked quite different when we opened in 2004" (Apple Inc. 2016) (Fig. 7) .

Fig. 6. Bedford Lemere & Co, Regent Street, London, 1910, https://www.architecture.com/image-library/ribapix/image-information/poster/231237-regent-street-london/

Fig. 7. Foster and partners, the Apple Regent Street store, London, 2016, https://www.fosterandpartners.com/projects/apple-regent-street-london/

Apple used the image of the historical façade, from 1898, and its location to launch a message to the public showing how Apple had become part of an important street in the city. Preserving the façade showed its commitment to the community, reinforcing the company's values. The façade is catalogued as Grade II and contains Portland stone,

Carrara marble and hand-cut Venetian smalti glass tiles. This was all respected and restored.

In parallel with the modern interior design, Foster used natural materials such as wood, stone and terrazzo that were in keeping with the existing materials in the historical building, and showing respect for the local historical monuments. According to Jonathan Ive, "By choosing materials sympathetic to the historic nature of the building, we were able to modernise the space while remaining authentic to its surrounds" (Apple Inc. 2016). Jonathan's statement indicates that the company wanted to show its respect for the historical building, and that it was not a mere coincidence or the architect's unilateral decision.

When the Apple Regent Street store opened for the first time in 2004, the stairs used were the glass stairs that Apple had installed in previous stores, but in the 2016 update, the new stairs were made of stone, bringing the proposal closer to the historical nature of the building, an aspect that was not so present in the first version of the Regent Street store.

Another of the characteristics of the Regent Street store was a large central hall, 7.2 m high, open to the avenue. It was designed as a kind of exterior space surrounded by interactive screens and trees, giving the sensation of being in a village square, a flexible and welcoming space. The ceiling has the same characteristics as the Apple Union Square store. The natural light comes from the glass in the arches of the façade that opens onto Regent Street.

Stefan Behling, one of the architects at the Foster + Partners studio, stated in an interview:

"Think about how much retail actually happens on devices. People will increasingly buy online. The role of Apple in a community is to get people together and make the experience fun: see amazing stuff together, meet other people. It's a humanistic approach instead of a sales approach" (Kwak 2016).

2.3 Apple in Washington, Carnegie Library. Foster + Partners

What sets this store apart from other Apple stores is the fact it shows how far Apple can go reusing historical buildings. This is due to the significance of this historical building and its size, which makes it Apple's largest restoration project to date.

The Carnegie Library (Fig. 8) was the first public library in the city. It was called Beaux-Arts and opened in 1903. It was designed by Ackerman & Ross and was conceived as centre for learning and discovery for the public and that was what it was for more than 70 years, until the building was abandoned before Apple announced it would reuse it. Like other Apple stores in historical buildings, Foster was able to adapt its historical nature giving it back its original grandeur, to bring new life to this great cultural icon. He did this by preserving the historical façades, recovering the original style of the interior spaces, the skylight and the large patio, and restoring the unmistakable details from the early XX Century (Figures 9 and 10).

With the usual Apple furniture, tables, chairs and wooden trees, as well as interactive screens in the walls and displays of Apple products, a contrast between old and modern is created, which not only contradicts the company's vision, but also supports it, combining art and technology.

Fig. 8. Streets of Washington, Carnegie Library, Washington D.C., 2019, https://www.flickriver.com/groups/43753617@N00/pool/

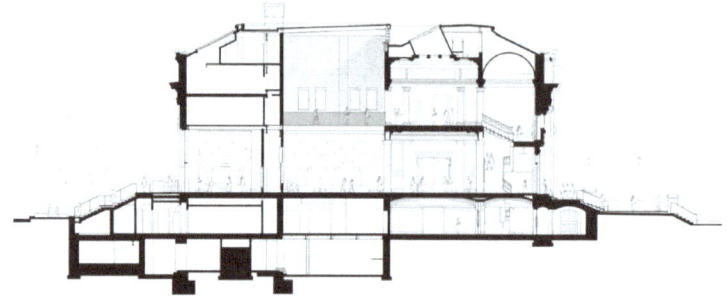

Fig. 9. Cross-section through the skylight and central patio. Foster and partners, Apple Carnegie Library Section, Washington DC, 2019. https://www.fosterandpartners.com/projects/apple-carnegie-library-washington-dc/#drawings

"I love the synergy between old and new, the juxtaposition of the historic fabric and contemporary design. It is the layers of history which create the rich tapestry of urban life. In its 'new' phase of life, Apple Carnegie Library will be a way for us to share our ideas and excitement about the products we create, while giving people a sense of community and encouraging and nurturing creativity" (Foster + Partners 2019).

Apple, once more, uses the historical value not only on a physical level of the building, but even through the memory of the original library as a public learning centre at the service of the local community. The Library was founded by Andrew Carnegie as a public learning centre, and Apple now takes that specific vision and converts it into one of its main characteristics. Thus, the objective was not only to preserve the building physically, but also preserve the intention with which it was constructed.

In that regard, one of the characteristics of the store is the part of the building where historical photographs of the original library are on display on the ground floor. Dedicating part of the building to speaking about the building itself was something new

Fig. 10. Before and after. The original loans desk of the library, which was inaugurated in 1903, and the skylight have been converted into an interior patio and meeting point called the Forum. https://www.apple.com/es/newsroom/2019/05/apple-carnegie-library-opens-saturday-in-washington-dc/

for Apple. That showed how far their vision could go to make their store a place for culture, playing an active role in the community.

3   Conclusion

Studying the strategy and design process behind Apple stores reveals that, in reality, it is a different way of understanding the concept of restoration. This concept, to a certain extent, assumes the wish to restore something that belonged to the architecture, but that had been lost. Using a metaphor, it is as if restoration, traditionally, wanted the building to say the same as it used to, though in another language. In the case of Apple, however, efforts focused on making the building say something else, changing the content of its message.

There is a consensus about the relationship between Apple's identity and its architectonic identity, that is reflected in various characteristics of the case studies of the three Apple stores. Through contemporary architectonic elements, such as glass doors and ceiling panels in the San Francisco store (Fig. 3), and also through the use of modern techniques in building stone staircases in the Regent Street store in London. All of this is an expression of the combination between aesthetic values and contemporary technology.

The use of solar panels, the promotion of natural ventilation and the reuse of historical buildings is a trend that reflects social responsibility, that reflects values of sustainability. Through the three case studies of Apple stores, Apple's role in the preservation of patrimony is clear. The social and entertainment environment on which Apple focusses in its stores, more than just showing their products, is defining a new future for the concept of the retail stores and their role in local communities, at least in the next ten years.

If we take the Carnegie Library store, for example, we will see that the company does not functionally need this massive building merely to physically display their products. Nevertheless, by adopting the same functional vision of the building as a public space for learning, Apple has committed to part of the history of that library and its future. We can see that the future of Apple stores in historical buildings will focus on preserving the purposes of those buildings. In addition, it will not only preserve the building physically, it will also attempt to partially preserve its historical function.

In this sense "the brand" is complemented with the architecture. And Apple has chosen the "Foster brand". Some time ago Foster stopped being an individual architect and became a brand. Inspired by both historical buildings and scientific progress, his projects reconcile tradition and modernity, the ability to transform and technological innovation.

His use of technology as a form of expression and the desire for atemporality in his proposals have become, paradoxically, what gives him his identity as a brand. And it is just that which enables him to adopt Apple's criteria as, in some ways, they are the same. "Apple by Foster" stores are able to attract the interest of the followers of both, culminating in an idea that began with the graphical representation of the initial store showing that, at least on this occasion, a picture is worth more than a thousand words.

References

Apple Inc. Apple Regent Street to reopen with new design, newsroom (2016). https://www.apple.com/newsroom/2016/10/apple-regent-street-to-reopen-with-new-design/

Apple Inc. Apple union square highlights new design elements, community programs, newsroom (2016). https://www.apple.com/newsroom/2016/05/19Apple-Union-Square-Highlights-New-Design-Elements-Community-Programs/

Bletter, R.H.: The Modern Functional Building (Texts & Documents). Getty Research Institute, Santa Monica (1996)

Foster + Partners. Apple Carnegie Library breathes new life into a much-loved city icon, foster and partners (2019). https://www.fosterandpartners.com/news/archive/2019/05/apple-carnegie-library-breathes-new-life-into-a-much-loved-city-icon/

Foster + Partners. Apple Union Square, San Francisco, Foster and Partners (2016). https://www.fosterandpartners.com/projects/apple-union-square-san-francisco/

Foster, N.: Norman Foster: futuros comunes. Madrid. Fundación Telefónica, October 2017–February 2018. https://espacio.fundaciontelefonica.com/wp-content/uploads/2017/06/norman_foster-.pdf. Accessed 25 Aug 2019

Howarth, D.: Foster + Partners unveils Apple Union Square store in San Francisco, dezeen (2016). https://www.dezeen.com/2016/05/21/foster-partners-jonathan-ive-apple-union-square-store-san-francisco/. Accessed 11 Oct 2019

Kwak, C.: Jony Ive collaborates with Foster + Partners on a Smart New Apple Store in San Francisco, architectural digest (2016). https://www.architecturaldigest.com/story/jony-ive-foster-and-partners-open-san-francisco-apple-store. Accessed 11 Oct 2019

Putnam, T.: The theory of machine design in the second industrial age. J. Des. Hist. **1**(1), 25–34 (1988)

The Extension of the National Museum in Helsinki. Graphic Strategies of Contemporary Design Competitions in Heritage Contexts

Hector Mendoza Ramirez[(✉)] [iD] and Mara Gabriela Partida Muñoz[iD]

ETSAB Barcelona School of Architecture, Universitat Politècnica de Catalunya,
Barcelona, Spain
hector.mendoza@upc.edu

Abstract. The graphic material of the five finalists of the design competition for the extension of the National Museum in Helsinki is carefully observed. The Finnish competition is a context conducive to contemporary creation, not only because the rules of the competition make an explicit call to innovation, looking for an up-to-date image that complements the heritage complex, but for the culture of the competition existing in Nordic countries. This tradition prioritizes an open, anonymous and transparent competition process, different from others carried out for interventions of equivalent relevance, such as the extension of the Prado Museum in Madrid or the Rijksmuseum in Amsterdam that prior to the presentation of design proposals, the participants are selected by their curriculum.

The presentation rules and requirements in the Finnish competitions result in varied architectural proposals, some of those with fresh and groundbreaking ideas. It is of key interest for our field of study, to observe and learn from projects that coincide with representations enhanced by a clear "graphic discourse" (Crespo et al. 2012) in which architectural heritage and avant-garde proposals are nuanced, or manipulated, to make the dialogue evident between the elements of the different historical periods.

Keywords: Architectural competition · Architectural representation · Architectural heritage · Finnish architecture

"Far from threatening the permanence of the past, creativity exercised with rationality and knowledge underlines and expands our heritage legacy" (Trillo 2006).

1 Introduction

Uusi Kansallinen (New National) is the name of the open international architecture competition, developed in two phases for the extension of the National Museum in Helsinki. The tradition of the architectural competition in Finnish culture has been recognized as an open, anonymous and transparent process. Although a restricted competition could guarantee the quality of the proposals by the simple fact of preselecting those participants that meet the requirements set by the calling entity, an open competition process allows "the exploration of unexpected roads or the emergency of new talents" (Rojo 2012). In

L. Agustín-Hernández et al. (Eds.): EGA 2020, SSDI 6, pp. 656–666, 2020.
https://doi.org/10.1007/978-3-030-47983-1_58

the context that concerns us, we can mention the cases of the Central Library in 2018 by ALA or the Steven Holl Kiasma Museum in 1998, both buildings a few meters away from the National Museum and were selected in extremely attended competitions with 544 and 515 participants respectively.

The urban environment that concerns us is characterized by the concatenation of unique buildings from different periods on one side and another of the important Mannerheimintie Avenue. Among the relevant buildings near the site of the competition, we can mention the Olympic Stadium of Lindegren and Jäntti of 1934, the Finlandia Hall of Alvar Aalto of 1967, the Central Library and the Kiasma mentioned above, the Parliament by Siren of 1931, the Helsinki Music Center of LPR architects of 2011 and the new Amos Rex museum of JKKM architects of 2018. (See Fig. 1).

Fig. 1. Helsinki ortho-photo highlighting Mannerheimintie Avenue and the immediate context where the National Museum is located next to several unique buildings. City map and Approach.

The specific location of the National Museum, unlike the buildings on Mannerheimintie Avenue, that are accessible from the street through large green areas, is located within a plot delimited by a perimeter wall. In patrimonial terms, the competition rules indicate that the constructions and open spaces within that walled plot of the National Museum are designated with the "ark label" of historical, cultural or architectural value. This label indicates that demolition, alteration or addition of constructions that may compromise the value or historical-cultural style of facades and roofs is not allowed. The Finnish Heritage agency offered to the competitors an appendix dedicated to the explanation of the value observed in the architecture style called National Romantic of the Museum, work of the architect firm Gesellius-Lindgren-Saarinen from 1902–1904.

In terms of intervention possibilities, the brief suggested areas where it was acceptable to place new buildings, above surface and underground, as well as the spaces that could be intervened inside the heritage building. With that information, participants could focus on how to intervene and not so much in where to intervene. Even so, as it will be explained later, some variations in the placement generate interesting relationships between architectural heritage and contemporary views.

2   Graphic Dialogue Between Heritage and Contemporary Intervention

Recalling that sequence of those charcoal perspectives of the competition for the skyscraper in the Friedrichstrasse, in which Mies Van der Rohe insisted again and again to clarify the geometry, material and atmospheric relationship of its new architecture with the context where it was inserted, this communication focuses on the graphic material of the finalist proposals of the competition in Helsinki. It is of key interest to observe the strategies of representation of a context with patrimonial elements in front of the proposals called to be innovative and the new image of an avant-garde institution. From those drawings by Mies Van der Rohe, it is worth highlighting the search for the precise point of view, having used own photographs different from those offered by the organizers of the competition, or the insertion of that new and foreign architecture between recognizable elements in the foreground and in the background. The controlled exaggeration of the lateral vanishing points of the architectural object has also been highlighted and, above all, the manipulation of the built context that was initially represented with photographs and would end up being abstracted as a great shadow in preparation for the new construction emerging as a light and transparent object. (see Fig. 2).

Fig. 2. Sequence of photomontages and subsequent charcoal drawing of the Friedrichstrasse by Mies Van der Rohe. Source: Gastón, C.: Mies: Competitions in the Friedrichstrasse. Proyecto, Progresso, Arquitectura. 7. 54–67. (Gastón 2012)

It should be mentioned that, within the National Museum competition brief, creativity was encouraged. Several points of the brief explained that the potential for future development of the proposals was more important than the details or perfect construction. A contemporary vision was sought with special attention in the placement of the new construction in relation to the current museum, the walled park and its respective presence with the urban landscape and surrounding buildings.

The organizers of the competition, anticipating that they would receive a large number of proposals, defined the requirements and contents of the entries. Among the presentation requirements, it was necessary to include a situation plan showing the shadows cast at 45° and the realization of a "photo-infography" (Taboada 2011), understood as the superposition of the proposal on a photography; all participants using the same base photograph. The base photograph was an aerial view on a clear summer day, with a large visual field that shows in the foreground the walled park where the annexed volume will be placed, followed by the current museum on Mannerheimintie Avenue that is presented as a backbone that links a sequence of architectural landmarks from different eras, such as the Finlandia Hall, the Central Library, the Kiasma museum and the historic city center as a background.

3 Finalists, Design Proposals

For the development of this communication, although the whole global of the graphic presentations is analyzed, we focus on the material that the competition rules marked as mandatory, especially the photo-infography of the 5 finalist proposals.

Atlas. Under the motto *Atlas* a pavilion with a circular plan is presented. It occupies the quadrant closest to the current museum. The infography shows a floating plate that is inserted without modifying the original atmosphere of the photograph, meaning that the heritage building or the landscape is not altered for the montage; the material reality of the preexistent elements is preserved in the image. The communication of the idea is effective in representing the new object in an abstract and alien way, approaching just enough to the built or natural heritage, suggesting the intentional introduction of light to the spaces underground. The plate is represented elevated just enough, in order to show its presence on Mannerheimintie Avenue above the wall. It is a construction with a very light presence, which from this point of view, does not disrupt the perception of the heritage building.

On the one hand, the representation of the plate, in this view, avoids defining its materiality or color and manages to separate from the current museum due to the contrast of its appearance. However, at ground level, it does suggest material integration, through a radial agreement between the landscaped area and the paved area under cover. This rendering work, together with small details, such as the shadow that the old building throws on the pavilion, or the vegetation that is slightly placed in front of it, makes the foreign object perceived integrated, subtle and kind in the natural context that is also part of the heritage. (See Fig. 3).

In floor plan and sections instead, the heritage construction is drawn in a simplified way. The spatial geometry is defined, the visual limits of the spaces are outlined. The scale of the drawing does not allow to elaborate the structure or the internal layers of the enclosures, leaving the inner space of the walls and roofs blank. This way of drawing is coherent in continuity with the new underground spaces of the proposal and the geometry of the circular pavilion. The use of color or the definition of materiality can be seen only in outdoor spaces, including the museum's courtyard, gardens and the city as a background, that is, the material reality of the pre-existence. (See Fig. 4).

Lände. The proposal titled *Lände* stands out visually since it is the only, out of 179 entries, that uses black background in its representation. It is a decision that graphically supports the contrast work developed to the spatial and material level. Drawings mainly

Fig. 3. Motto: Atlas. Photo-Inphography and vision from Mannerheimintie 2019. Source: Uusi Kansallinen Arkkitehtuurikulpailu Image Bank

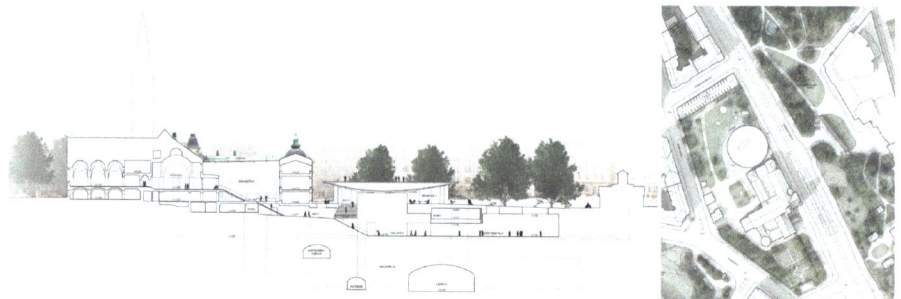

Fig. 4. Motto: Atlas. Floor situation and longitudinal section 2019. Source: Uusi Kansallinen Arkkitehtuurikulpailu Image Bank

reinforce the heritage wall structure by representing it solid, while the new architecture that is intuited transparent, light and organic; new architecture is drawn with the prominence and graphic weight similar to that given to the elements of the park, pavements and vegetation.

The photo-infography, coherent in emphasizing the contrast mentioned above, chooses to modify the atmosphere of the original photo, intensely darkening the context, especially the open spaces of the site, in order to show the powerful light that emerges from the underground and from the courtyards. This light finds echoes of integration in the context by leaving the singular buildings illuminated along Mannerheimintie Avenue. (See Fig. 5).

The spatial relationship between the world on the surface and the world underground occurs through the skylights that emerge as crystal cylindroids inhabited with reflective bubbles, like floating drops of mercury that contribute to depositing light inside the basements. The surrounding geometry of the underground vestibule is irregular, and organic, as the result of a series of geometric subtraction operations.

Despite looking like a proposal that breaks and formally moves away from heritage building, within the presentation, it establishes a series of instances that able the observer to recognize situations or experiences similar to those in the existing premises. For

Fig. 5. Motto: Lände. Photo-infography and Ground Floor-access 2019. Source: Uusi Kansallinen Arkkitehtuurikulpailu Image Bank.

example, the board displays in one corner of the panel, the photograph of one of the most representative spaces inside the old museum, the dome and its lantern, together with a visualization of the interior of the new construction. In those linked images, the new spatiality is suggested as an echo of the existing, a recognizable and familiar environment; despite the boldness of the proposed geometry. In the same way, in the perspectives of the outer space, the recognizable texture of the natural stone is shown in the foreground, looking for resonance in the materiality of the plinth facade of the existing building, and between these two planes the glass cylindroids are placed as if they were even temporary or removable. (See Fig. 6).

Fig. 6. Motto: Lände. National Museum skylight. Interior and exterior visualizations. 2019. Source: Uusi Kansallinen Arkkitehtuurikulpailu Image Bank

Asuuri. *Asuuri* is one of the most discreet proposals, even though its volume is the largest one, especially in height. The pavilion is placed as a bar construction, repeating the width of the current museum, seeking to consolidate a facade towards Mannerheimintie.

In the frontal perspective, as well as in the elevation from the street, the park wall is observed as the enclosure of the proposal itself, where the main entrance takes the visitors directly to the interior of the new construction. Next to this access, there is an urban window that allows a visual relationship, from the sidewalk, with the different interiors of the new building. This window looks out over the new double-height lobby, and the large staircases that descend to the second basement.

The photo-infography accurately shows the materiality of the proposal that mimics the zinc used on the roof of the existing museum and the Finland Hall. It also appropriates the same color and textures of the original stones on its facades but with a simple and contemporary expression. This aerial point of view has the advantage of having the park in the foreground, enhancing the virtue of being the proposal that maintains a large park area in continuity with the landscaped area of the original museum. However, if you look closely in comparison to other proposals, Asuuri is the project that mostly modifies the urban image of Mannerheimintie Avenue by getting rid of the leafy trees that were on that edge of the park. This fact could represent a critical point in any other context, but in Finland, if the proposal justifies it, it would not have to be considered an ecological drama to move or remove some trees. In the competition rules, it is specified that this area of the park can be modified, and it is already assumed that it would be necessary to remove some of the trees. (See Fig. 7).

Both, at the level of architectural design, and at the level of presentation of the idea, Aasuri is a project that seeks simplicity and clarity, avoiding the excess of information that would saturate the effective communication of the idea. The presentation panels turn out to be the cleanest and most aligned of the five finalists. The use of color is limited to the three perspectives of controlled size in the upper part of the presentation boards (aerial, frontal and interior) and to the vertical projections (elevations and sections) of those points where it is important to clarify the material dialogue that is mentioned above, and the great presence of trees that would not be altered.

Fig. 7. Motto: Asuuri. Photo-infography. Elevation and frontal view of Mannerheimintie street. Source: Uusi Kansallinen Arkkitehtuurikulpailu Image Bank

Happio and Kolme-Pihaa. *Happio* and *Kolme-Pihaa* are two proposals whose placement on the site coincides when configuring two clearly differentiated exterior spaces: a park area that is left as intact as possible, at least at the surface level, and a paved area in front of the existing museum.

Happio places a white carpet as an abstract base, without clearly defining its texture or material layout. The proposal suggests a platform or plinth where the built elements of the plot concur exclusively, leaving the gardened areas outside the plinth.

In the photo-infography you can see how the new volume, with a width similar to the bays of the heritage building, gravitates over the limits of the carpet, generating a porched space that identifies the new entrance to the museum. The irregular geometry of this element has been worked in such a way that it refines its edges to reduce the expression of the facades towards the square and towards the garden, seeking to be perceived as a construction more similar to a pergola or thick roof, and not so much to a volume with a height that would admit two floors inside. It is a pavilion with enough personality to identify access to the underground interior of the new museum, and with a light presence, a wooden facade that conceptually links it to the park and that should not harm the entire perception of the heritage site. (See Fig. 8).

However, the best perspective of this project is from the height of a pedestrian point of view, together with the elevation from the park. These drawings show the following sequence: patrimonial building-square-pavilion-park and vice versa. The new construction shows its unique geometry, suspended and permeable, a spatial threshold, which welcomes the descending path to the underground rooms, the ascent to the restaurant on the first floor and formalizes the transition between park and plaza.

Kolme-Pihaa is the proposal that with a single gesture manages to accumulate most of the qualities that have been mentioned in the other projects. In the aerial view you can see a thin cover with abstract geometry in the form of a boomerang or stylized "L" shape. Such a figure, already in abstract, generates space in its concave part and suggests limits in the convex part. The concave part, drawn with a wide curve, embraces the south-east and south-west limits of the park that is left virtually free of construction on its entire surface. The convex part of the roof faces two completely different environments. On the one hand, we have Tölö Street, where the roof looks out over the perimeter wall of the museum grounds. This allows that from the street, the new construction is announced and can be perceived discreetly. The pavilion appropriates the perimeter wall of the park to turn it into its facade. The other side of the convex part, is presented within the site to configure a controlled outdoor space together with the main building, a welcoming environment, where, like Happio, coexistence occurs between the architectures of two different periods.

In plan, the new roof maintains a prudential distance with the patrimonial construction, allowing the existence of a vertebral corridor between both. This corridor joins the two new pedestrian entrances, in Mannerheimintie and Tölö respectively. In the central part of the corridor, the pre-existing construction and the proposal coexist opposite to each other. This relationship generates a superficial tension that is controlled from the aerial perspective despite the fact that the new structure sharpens its end at that specific point. It is interesting to note that the height of the new construction equals the upper limit of the dark stone basement of the old museum. In this perspective, that basement becomes the background of the sharp end of the new structure that, fortunately, due to the solar incidence, receives the shadow that the existing construction throws. The shadowed area graphically blurs the space in which the vertebrate corridor could be perceived narrow. Spatially, this point is resolved by preventing the glazed boundary of the interior of the pavilion from reaching the pointed end, bending much earlier to allow an outer space under the pergola. (See Fig. 9).

Fig. 8. Motto: Happio. Photo-infography and front view of the new entry square. Source: Uusi Kansallinen Arkkitehtuurikulpailu Image Bank

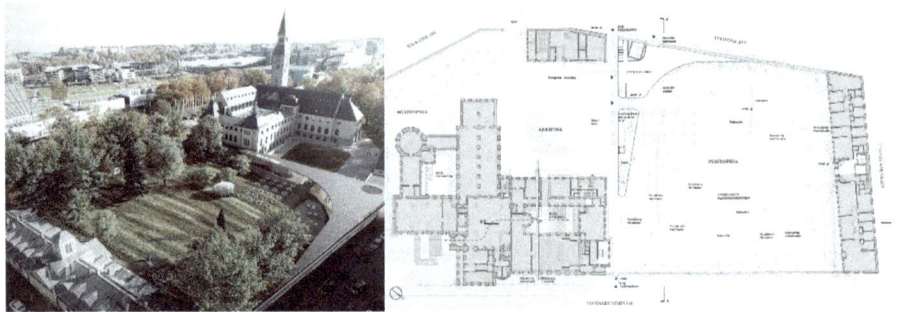

Fig. 9. Motto: Kolme-Pihaa. Photo-infography and Ground Floor- Access. Source: Uusi Kansallinen Arkkitehtuurikulpailu Image Bank

About the height of the pavilion, we can add that it rises just enough to try to deposit natural light into the underground interior spaces by means of glazed enclosures. We can observe a glazed roofed area that interrupts the opaque materiality with a horizontal skylight. In the pedestrian perspective, from the park, the transparency of the facades and the shallow depth of the new construction allow to show the roof as the only solid element of the intervention, which subtly embraces the outer space. The cover from this point of view is reduced to its minimum expression: a line. This vision brings together the main qualities of the proposal, positively redundant in the geometry that embraces the park, the transparency of the enclosures and the fineness and lightness of the roof that allows for a new dialogue between outdoor spaces, heritage buildings and contemporary architecture. (See Fig. 10).

Fig. 10. Motto: Kolme-Pihaa. Visualization from the park. Source: Uusi Kansallinen Arkkitehtu-urikulpailu Image Bank

4 Conclusion

The proposals, different from each other, suggest attractive ways to resolve the relationship with heritage architecture. The graphic strategies are coherent in that each proposal insists on a form of dialogue or relationship with what exists: park and current museum. And, in this sense, they could be distinguished in three types of approaches:

1. Abstraction and Contrast
2. Continuity
3. Emerging spatiality

Abstraction and Contrast. The first approach would include both, the proposal with the motto Atlas, and Lände. Both graphic proposals try to differentiate the materiality of the heritage construction, well defined in the aerial view in the case of Atlas, and in the perspectives in the case of Lände. In both, the proposed architecture is not defined constructively or materially, it is not their interest to do so in the competition phase. They follow the indication that was made in the brief, suggesting new spaces before defining a perfect constructive detail. In that sense, they enjoy some formal and graphic freedom, with lines suggest transparent surfaces and pure geometries that seem not to be affected by gravity, or the weight of materiality that do characterize the heritage spaces of the pre-existing ones.

Continuity. The word continuity understood as the union between different parts that manage to form a unit or a whole that develops over time, is consistent with Asuuri. In all drawings and representation systems, they describe in detail the geometry and material of the heritage building. It does not take shortcuts or synthesize a graphic language that detracts from the construction of earlier times, nor does it avoid defining the geometry, materiality or chromatisms of contemporary architecture that is added to the heritage environment.

Emerging Spatiality. The strength of the proposals with the slogans Happio and Kolme-Pihaa, coincides in the possibility of creating new spaces, or spaces with innovative features in the environment. The three previous finalist proposals, as well as the majority of the participants of the competition, sought to continue or respond geometrically or materially, with greater or lesser courage, their relationship with the heritage building, focusing on affecting or modifying it as little as possible, both built and natural. But this last approach, which includes Happio and Kolme Pihaa, for the position they offer in the placement of the entrance pavilions, inhabiting the boundary between the new entrance square, and the well-defined area of the park, they manage to give a new character to the exterior spaces and generate a dialogue between equals: Heritage Building - Natural Landscape - Contemporary Architecture.

Although, for this communication we focus on photo-infographies, as it was the mandatory and constant material in all proposals, it should be noted that in addition to being a tool capable of superimposing and calibrating the value of history, place and strategies in a clear and forceful way, it was not the only material studied. Each proposal completed and suggested new visions that reinforced its own graphic discourse, such as the use of elevations or perspectives from Mannerheimintie Avenue in the case of Atlas and Lände, which were the ones that most modified the urban image, or the underground perspectives of Asuuri, that with its innovative geometry make winks and calls to the interior spatiality of the current museum. But above all, we highlight the perspectives at the pedestrian level in the Happio and Kolme Pihaa proposals, which allow the observer to be placed in these new exterior spaces that are generated thanks to the successful placement of the entrance pavilions, which give unprecedented values to the existing architecture without leaving behind the proposed contemporary architecture.

References

Crespo, I., Font, J., Martínez, F.: Composición y discurso gráfico en los concursos. En: Úbeda, M., Grijalba, B. (eds.) Concursos de arquitectura: 14 Congreso Internacional de Expresión Gráfica Arquitectónica. Oporto, 31 mayo a 2 junio 2012. Universidad de Valladolid, Valladolid (2012)

Resano, D.: Entre lo cristalino y lo transparente. El rascacielos en la Friedrichstrasse de Mies van der Rohe. EGA Expresión Gráfica Arquitectónica 22(31), 132–139 (2017)

Rojo, J.: De jurados y arquitectos: ideas sobre los concursos. Proyecto, Progreso, Arquitectura. Universidad de Sevilla 7, 26–38 (2012)

Taboada, J.: Sobre perspectiva, fotografía e infografía. Apuntes para una fenomenología de la representación. EGA Expresión Gráfica Arquitectónica 16(17), 54–65 (2011)

Trillo, J.: La nueva arquitectura y la ciudad histórica. En: Ruesga, J. (ed.) In vitro. El concurso de arquitectura en la ciudad histórica, pp. 31–39. FIDAS/COAS, Sevilla (2006)

Gastón, C.: Mies: Concursos en la Friedrichstrasse. Proyecto, Progreso, Arquitectura 7, 54–67 (2012)

Competition Brief. Image Bank. Uusi Kansallinen Arkkitehtuurikulpailu 2019. https://www.uusikansallinen.fi/en/. accedido June 2019

Reflections in the Space of Luca Cambiaso and Franco Albini

Cristina Càndito$^{(\boxtimes)}$ (iD) and Valter Scelsi (iD)

Department Architecture and Design, University of Genoa, Genoa, Italy
`cristina.candito@unige.it`

Abstract. The monographic exhibition which Genoa dedicated to Luca Cambiaso in 2007 opened with the famous *Self-portrait with the portrait of his Father Giovanni* (ca. 1570), where the figure in natural size of Luca can be seen in a room in penumbra, in front of the painting he is making. The personages depicted present gestures, positions and attitudes which may refer not only to other episodes of Luca's work and to the echo they may have had on his later pictorial experience, but also to various hypotheses concerning conception and dating. In 2004, the painting found its place in the Gallery of Palazzo Bianco in Genoa. Now that it has become part of the museographic setting designed by Franco Albini according to the cultural project of Caterina Marcenaro (1949), the painting has created in combination with it a dizzying apparatus. The room on the courtyard floor provokes more than one reflection in the onlooker concerning the properties of space, codes regulating its use and perception and memory of things. The location of the canvas is decisive in producing an extension into the real environment of the space depicted in it, theatre both of the transmitting capacities of the work and of our possibility for establishing meaningful relations with it.

Keywords: Mirrors · Perspective · Museography

1 Introduction

The traces which the two protagonists of the artistic life of Genoa have left come together inside Palazzo Bianco, which since 2004 has been hosting one of the versions of Luca Cambiaso's *Self-portrait with the portrait of his Father Giovanni* (ca. 1570) in the context of the setting made by Franco Albini (in 1949). The main studies on the famous sixteenth century Ligurian painter (Magnani 1995; Boccardo et al. 2007; Magnani and Fiore 2015) have made it possible to rewrite the history of the painting and place it in its context within the work of the painter.

In this paper, we dispute the dating of the work, which would have as its *ante quem* term the date of Giovanni's death. In our hypothesis, based on the composition of the painting and its possible derivation from a previous portrait, this term would no longer provide certainty for dating it.

We will find an assonance between the work and Albini's setting, based on previous studies (Marcenaro 1950; Bucci and Rossari 2005), capable of bringing out what is significant in the works on exhibit, which remains effective even though such works were not contemplated in the original layout.

L. Agustín-Hernández et al. (Eds.): EGA 2020, SSDI 6, pp. 667–677, 2020.
https://doi.org/10.1007/978-3-030-47983-1_59

2 Representation of Luca Cambiaso

The importance of the role of Luca Cambiaso (1527–1585) in Genoese painting appears from the vast extent of his production, ranging from paintings on canvas to large frescoes, and in how he was able to bring a local school of painting up to date, even exporting its principles abroad, when he moved his activity to Spain, starting in 1583 (Magnani 1995; Boccardo et al. 2007).

His *Self-portrait with the portrait of his Father Giovanni* (ca. 1570, oil on canvas, cm. 104 × 97) represents the painter in the act of portraying his own father and master, Giovanni Cambiaso (1475–1579). The interior can be identified as that of the painter's study, also thanks to the easel in the painting (Fig. 1).

The scene includes:

1. The painter, turning around by three-quarters, looking at the observer, the brush in his left hand resting on the painting;
2. The portrait of his father, looking at the observer in the same way;
3. An interior with a tall cupboard and some items hanging from the wall.

Fig. 1. Comparison between the two versions of Luca Cambiaso, *Self-portrait with the portrait of his Father*. On the left: Palazzo Bianco, Genoa. On the right: Galleria degli Uffizi, Florence

The items fit only in part into the visual frame on the top right, above the painter's head: there appear the legs of a dummy, between two medallions, which were perhaps part of the equipment of the painter's study.

There is also a Florentine version (*Self-Portrait while Painting his Father*, ca. 1570, oil on canvas, cm. 86.5 × 71) which has no setting and may be a copy (Berti 1979).

2.1 Reflections and Copies

To understand the meaning of the work, it may be of interest to explore the interest in optics which Cambiaso expresses in various ways in his work, for example in his study drawings for major paintings, where "cubic shaped" figures often appear (Magnani and Fiore 2015), treated as geometric solids, so they can be foreshortened and correctly placed in space. In the drawing made for the fresco of the vault of Palazzo Grimaldi, *Odysseus Slaughters Penelope's Suitors* (Nationalmuseum, Stockholm, ca. 1565) (Fig. 2), the representation from a front perspective of a refined classic architecture is no mere frame; it is actually a three-dimensional space in which to fit schematic human figures. This kind of preparatory drawing allows one to prefigure - without needing to anatomically detail the bodies - the spatial consistency of all the elements of the composition, which include not only the architecture and the figures, but also the shadows they cast.

Fig. 2. The cubic form of Luca Cambiaso. *Odysseus Slaughters Penelope's Suitors*, study drawing for the fresco of Palazzo Grimaldi, Genoa (Nationalmuseum, Stockholm, 1565 ca.).

A similar interest in optics can be seen in certain paintings, where the lighting effects are emphasised by the choice of a nocturnal setting. Especially, the *Virgin of the Candle* (Palazzo Bianco, Genoa, 1570–1575) (Fig. 3) is able to provide surprising depth effects despite the poor lighting provided by a candle in an interior populated by few but significant elements (a cradle, a hanging basket, an opening), which provide what is necessary to place the figures in space, just as in the self-portrait we are examining.

As happened with some painters of Cambiaso's times, relationship with space, its perspective representation and the appearance of lighting were studied in depth and extended, also thanks to the use of tools, such as flat mirrors, which had only recently begun to be produced.

The introduction of flat mirrors had a profound influence on sixteenth century art, as can be seen in the shift from images deformed by convex mirrors - for example

Fig. 3. Luca Cambiaso's Nocturnals. *Virgin of the Candle* (Palazzo Bianco, Genoa, 1570–1575).

Parmigianino's *Self Portrait* (Kunsthistorisches Museum, Wien, 1524) - to others which could provide a more objective image of reality, as Albrecht Dürer precociously showed in his *Self-Portrait with Fur Coat* (Alte Pinakothek, Munchen, 1500). As is well known, Dürer was the first scholar to publish a text on mathematics in the German language (*Underweysung der Messung...*, 1525), which also contained an original treatise on perspective, the outcome of his previous sojourns in Italy. Dürer showed special interest in the theme of self-portrayal, where the identification of the subject and observer took on various forms: he illustrated different stages of his own life with drawings and paintings which portrayed him, but in the version of 1500 we mentioned, he represented himself in a Christological image, and the proud comparison was emphasised by the words, "*I, Albrecht Dürer from Nurnberg, at the age of 28, with everlasting colours have created myself in my own image*".

2.2 Hypotheses and Comparisons

Luca Cambiaso's *Self-Portrait with Portrait of his Father* therefore falls into a tradition which can illustrate not only the techniques employed by artists, but also how they were able to provoke metaphorical inspirations. We should observe how the painting can be spatially interpreted in at least two main ways (Fig. 4):

A) The scene is seen by an observer placed behind the painter.
B) The scene shows what is reflected in a mirror placed behind the painter.

We can see how the hypothesis A comes up against evident contradictions, including the fact that Luca Cambiaso would be using his left hand; though he was actually ambidextrous (Boccardo et al. 2007, p. 206), this would have been unusual in the sixteenth century because of its negative connotations.

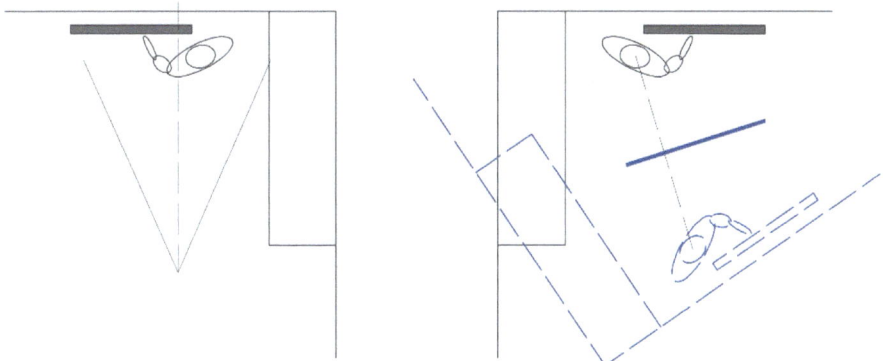

Fig. 4. Planimetric reconstructions A and B.

Both hypotheses, however, are incompatible with the presence of the subject being portrayed: the canvas, supported by the easel, is placed almost against the wall, in an unnatural position for a painter, who normally places canvas and subject to include both in his visual field.

Fig. 5. Luca Cambiaso, *Self-Portrait* (Accademia di San Luca, Rome, ca. 1570).

Therefore, we may suppose the portrait to be the reproduction of two previous paintings: a self-portrait of Luca and a portrait of his father. In the space of the painting itself and in Luca's production, we can identify other evidence which seems to confirm this hypothesis. In fact, there exists a self-portrait of Luca Cambiaso, kept today at the Accademia di San Luca in Rome (ca. 1570), which seems to coincide with the one shown in the painting we are discussing (Fig. 5). The subject of the painting would therefore not actually be two living models or their reflections, but a depiction of two previous paintings, and in hypothesis A of Fig. 4, we would only need to replace Cambiaso's figure with a previous self-portrait of his. The "mirror phase", then, would be transferred in the reproduction of the original self-portrait.

Fig. 6. Luca Cambiaso, *Christ before Caiphas* (Accademia Ligustica, Genova).

We have to admit no documented portraits of the father Giovanni exist. However, in Luca's production, there appears a face which seems to have been a preparation, at least for the frowning expression of a mature subject, whose appearance resembles that of Giovanni. We are speaking of the *Christ before Caiphas* (Accademia Ligustica, Genoa) (Fig. 6), this too, like the *Virgin of the Candle*, with a nocturnal and essential setting. The fact that it belongs to the "nocturnal" thread of his work allows us to place the painting in the early 1570s (Sommariva G. sheet no. 57, in Boccardo et al. 2007, p. 318) and the posture, physiognomy and severe expression of Caiphas bring to mind the portrait of Giovanni Cambiaso contained in Luca's painting.

Putting aside any undefinable consideration concerning the interpretation of the severity of the father's expression, let us concentrate on how this hypothesis opens up the possibility of dating the painting differently, which may be placed after the death of Giovanni (1579), allowing us to attribute to it a widespread allegorical meaning, also through the symbolic image of the mirror: the vanity of the world facing death. The theme can also be associated with that of the "three ages of mankind", where Giovanni could represent old age, Luca maturity and the spectator, childhood. The severe expression of Giovanni could therefore be admonitory, but it could also be tied to his role as master in the art of painting - an art under the protection of the apostle who gave Luca his name. However things may be, it seems to provide a conscious reflection on the role of the painter and the meaning of the representation at a moment when the artist, having achieved success, was seeking new reasons and stimulation for his creative activity, which would continue in Spain.

We can also mention at least two paintings (Fig. 7) which in some way echo the composition of Cambiaso's painting. The *Self-portrait* (Uffizi, Florence, 1646) by the Austrian Johannes Gumpp (born in 1626) sends us back the gaze of the painter through two images of himself, generated by the mirror and by the self-portrait on easel inserted into the paintings. In this case, it is an optical play where the painter takes on his correct posture, looking at the painting and the subject, that is the mirror, which reflects his image.

The *Self-Portrait with a Portrait on an Easel* (Fogg Art Museum, Cambridge, Massachusetts, 1623–1624) of Nicolas Régnier (1591–1667) presents a similar composition, with the painter however facing the observer. In this example, however, the painter is correctly using his right hand and leaves no room for interpretations concerning use of a mirror.

Fig. 7. On the left: Johannes Gumpp, *Self-portrait* (Uffizi, Florence, 1646). On the right: Nicolas Régnier, *Self-Portrait with a Portrait on an Easel* (Fogg Art Museum, Cambridge, Massachusetts, 1623–1624).

It is not yet possible to establish with certainty how Cambiaso's *Self-Portrait with Portrait of his Father Giovanni* was made; but his enigmatic composition seems to provide an opportunity to enrich with meaning the stages of evolution of one of the sixteenth century artists who was best introduced into the figurative culture of his time, and who was a carrier of original cues tied to perspective drawing and luministic representation.

3   The First Room

The documented record first mentions Luca Cambiaso's Genoese painting only in the mid-nineteenth century. It was in Genoa, in the picture collection of the *palazzo* which then belonged to the brothers Anton Maria, Bendinelli, Vincenzo and Francesco Spinola, today number five of via Garibaldi (Alizeri 1846–1847). From here it was moved on temporary loan to the gallery of Palazzo Bianco in 1892, before returning to the collection of the *marchesi* Spinola, around 1910. In the twentieth century, it was hosted in the Guala collection (Suida-Manning and Suida 1958, p. 101, Fig. 351), then in Lugano, in a private collection. After being moved to the Canesso Gallery in Paris in 2003, in the following year it was bought by the Art Foundation of Compagnia di San Paolo, which gave it on gratuitous loan to the gallery of Palazzo Bianco. In the original, nineteenth century museographic layout of Palazzo Bianco, it was kept in what was then room eight, probably the room where the works of Bernardo Strozzi are currently on exhibit. In its present position, in the first room of the museum, it has taken the place of three works put by Albini in the same environment (a late-sixteenth century Flemish tapestry, a fifteenth-century Saint Sebastian of the Tolmezzo school, and fragments of English alabaster of the same century), therefore there are no original relations, one could say relations of direct knowledge, between the project of the work and that of the hosting space: Cambiaso did not paint the picture for Palazzo Bianco, nor did Albini design the setting to host Cambiaso's canvas.

However, the setting of the canvas (Fig. 8) plays a decisive role in producing a place able to extend the space depicted in it into the real environment, theatre both of the transmitting capacities of the work and of our possibility for establishing meaningful relations with it. All of this, even though the work of Cambiaso and that of Albini were born in absence of each other. If we take a closer look, the traces of the work of Albini inside Palazzo Bianco merge with those of others who contributed to the current image of the museum.

The vicissitudes of the palazzo hosting the collection, as is well known, lie at the heart of modern museography (Fontanarossa 2015). Last in the series of Strada Nuova, Palazzo Bianco (Spesso 2011) was erected on the structures of the previous, sixteenth century Palazzo of Luca Grimaldi, in the early eighteenth century, on a design by the architect Giacomo Viano. In 1884 the *marchesa* Maria Brignole Sale De Ferrari donated it to the Municipality of Genoa to make a public collection (1892). During World War II, the palazzo was seriously damaged by bombing, and in the immediate postwar, restored by the Civil Engineering Office under the supervision of Carlo Ceschi for the Superintendency for Monuments of Liguria. Starting in 1949, reordering of Palazzo Bianco saw the renowned collaboration between the curator Caterina Marcenaro (1906–1976) and the architect Franco Albini (1905–1977) from Milan. Actually, the urgent need for postwar rebuilding largely conditioned the decisions and timing of the recovery project, and the building was already completely rebuilt when Albini stepped in. When, in 1949, Mayor Gelasio Adamoli appointed the commission to reorder the collections of the *palazzo*, its members were Orlando Grosso, Director of the Office of Fine Arts of the Municipality of Genoa, Antonio Morassi, Superintendent of Galleries and Works of Art of Liguria, and Mario Labò, President of the Ligustica Academy of Fine Arts, as well as Franco Albini as external consultant and Caterina Marcenaro, who officially succeeded Grosso in April 1950. So, if it is true that the traces of Albini's work blend in today with those of all who have contributed to the present image of the museum, it also seems possible to say that what critics have considered to be one of the theoretical

Fig. 8. Palazzo Bianco (Genova), court and room 1 (photo by Gian Luca Porcile).

reasons for Albini's layout - the intention to bring out to the full the expressive capacities of the individual works of art inside the museum - is clearly enacted in the first room.

The absence of other paintings near Cambiaso's canvas fits in to a surprising degree with what Albini seems to have intended, asking for each work to be given its own empty space, capable of being "almost a zone of influence of its pictorial space." (Albini 2006, p. 73). Indeed, the current solitary condition of Cambiaso's canvas actually exceeds this advice. Located on the back wall of a rectangular room, the first in the route suggested to the public, it takes over the use of the whole space, making it available to the visitor, inviting him to come close and, in exchange, revealing to him the plot of an intimate tale. Critics, almost always unaware of the actual starting conditions of the hosting architecture, have insisted on the consistency of the Albini-Marcenaro museographic project with the ideas upheld in those years by Giulio Carlo Argan and, before, by Lionello Venturi. Argan warned that the monumentality of historic buildings hosting important collections - when the unity of monument and museum is not in itself a document - would end up by hindering the development of the museum according to a rational scientific and museographic plan (Argan 1949).

In the canvas, surrounded by a sturdy gilded frame, the figure in natural size of Luca appears against a background of warm colours, inside a room basking in the half-light. Some items on the wall (the dismounted parts of a painter's dummy) suggest that that was his study, a familiar, comforting environment.

Observing the painting, we feel as if we were witnessing memory being handed down, because this portrait of Cambiaso, like every other portrait, contains the promise, or the acknowledgement, of a farewell. This way, what we might call the *ekphrastic dimension* of emptiness, that is the mesh of relations between the work of art and the dimension in which it was placed and which demands the need for a void, or what we perceive to be such, because it is only in and thanks to the void that the meaning (actually only one of the infinite possible meanings) of the painting which we are gazing at fulfils itself.

It is in the void of the room that dynamic phenomena, due to the perceptive surprise, can be sparked off. The capacity the picture has of telling a narrative on successive, increasingly internal levels, is what sparks this off.

The father stares at the person looking at the painting, binding himself to the onlooker in what appears as an entirely mutual relationship, yet it is not an exclusive one. In the instant in which Giovanni's eyes seize the viewer, they deprive him of his freedom to move undisturbed, they control him on behalf of his son, who is engaged in painting. Actually, so far it is the spectator who feels extraneous, if, in a truly mutual relationship, he imagines the father looking at his *real son* painting the canvas, and the *painted son* looking at someone or something outside the scene. It is the discovery of what Luca is looking at that brings the spectator back into the game. Luca (probably) is staring into a mirror where he reflects himself in order to copy himself, and in so doing, portrays himself with the brush in his left hand. Since the relationship between subject and reflected image is a direct relationship, with no mediation (Eco 1985), it is the discovery of the possible presence of the mirror in which Luca is seeking his own image which permits us, spectators, to enter virtually into the painted space. We know how to use the mirror because we know there is no other man in the mirror, no one to whom to refer right and left. The mirror is a neutral instrument. The mirror comes onto the stage, for Cambiaso, through a message which he delivers to us on a strictly cultural plane, the meaning of which survives through the ages and reaches our times: inappropriate use of the left hand. If the painter portrays himself in the act of painting with the "devil's hand" it is because he cannot do otherwise reflecting himself in the mirror, or else, more probably, because he wishes to communicate to us the presence of a mirror. He is not observing his painting subject, Giovanni, as the composition and pose seem to suggest - rather, he is busy observing himself. Luca's message is clearly addressed now to the third vertex of the triangle, the spectator, and the means of transmission is the mirror, a tool which does not *translate,* but *records.* Because, as Eco suggests, if a mirror names, it names only one person at a time and if reflected images were similar to words, they would be personal pronouns, where if "I" say "I", it means "me", but if another person says it, it will mean that other. Cambiaso's *implicit mirror*, after all, appears in its force when it shows us the full capacity, before a portrait, to do away even with our instinct to entrust ourselves to the *spectator's physiognomics*, the tendency to judge the nature of subjects from their somatic features, the structure of their bodies, the shape of their eyes, the fold of a smile. We do not evaluate Luca and Giovanni, we do not even observe whether they resemble each other as much as a father and son should.

The volume of the first room was not designed by Albini, nor were its floor or ceiling. The layout he designed for the three works has disappeared, and today, in their place, there stands a painting the architect never even saw. However, today in the first room, the room of Luca Cambiaso's self-portrait, Argan's precepts are reflected: the work is independent of the architecture, solitary in its undecorated environment, one can concentrate on it avoiding architectural distractions. Albini's work in Genoese museography has made these natural consequences.

In the end, the first room - with the painted figures of Luca and Giovanni, the glass door on the court drawn by Albini, the opening letting out onto the hanging garden, the floor made of slate and white Carrara marble - is a place where dissonances and reflections inevitably share in one general atmosphere.

4 Conclusion

The mysteries contained in a sixteenth century portrait/self-portrait today transmit, according to our codes, a message about the artist's relationship to the social and personal significance of his job. But in our culture, as in every other, there exists, between the use of ordinary codes and reflections on art, the direct experience of art through senses. The location of the painting inside a renowned twentieth century museographic setting, thus is not something neutral, instead it is an opportunity to remind us how designing space and its representation merely return us an idea of fleeting, broken, unstable and imperfect time: the deceptively immobile territory of continuous reflections.

Acknowledgements. The contribution was drawn up in collaboration. Paragraph 2 was written by Cristina Càndito, while paragraph 3 was written by Valter Scelsi.

References

Albini, F.: Le funzioni e l'architettura del museo: alcune esperienze. In: Bucci, F., Irace, F. (a cura di) Zero Gravity. Costruire le modernità, Electa, Milano (2006)

Alizeri, F.: Guida artistica per la città di Genova. Editore Libario, Genova (1846–1847)

Argan, G.C.: Il Museo come scuola. Comunità **3**, 64–66 (1949)

Berti, L.: Catalogo generale degli Uffizi. Centro Di, Firenze (1979)

Boccardo, P., Boggero, F., Di Fabio, C., Magnani, L.: Luca Cambiaso, un maestro del Cinquecento europeo, Catalogo mostra, Genova, Palazzo Ducale e Palazzo Rosso, 3 marzo-8 luglio 2007. Silvana Editoriale, Milano (2007)

Bucci, F., Rossari, A.: I musei e gli allestimenti di Franco Albini. Electa, Milano (2005)

Eco, U.: Sugli specchi e altri saggi. Bompiani, Milano (1985)

Fontanarossa, R.: La capostipite di sé. Una donna alla guida dei musei. Caterina Marcenaro a Genova 1948–71. Etgraphiae, Roma (2015)

Magnani, L.: Luca Cambiaso: da Genova all'Escorial. Sagep, Genova (1995)

Magnani, L., Fiore, V.: Luca Cambiaso: dalla ricerca storica al virtuale. Genova University Press, Genova (2015)

Marcenaro, C.: Introduzione al Catalogo provvisorio della Galleria di Palazzo Bianco. Direzione delle Belle Arti, Genova (1950)

Spesso, M.: Caterina Marcenaro, musei a Genova 1948-1971. Edizioni ETS, Pisa (2011)

Suida-Manning, B., Suida, W.: Luca Cambiaso, la vita e le opere. Ceschina, Milano (1958)

Earthen Construction. Graphic Mediation in Spontaneous Architecture

Alice Palmieri[✉]

Department of Architecture and Industrial Design, University of Campania 'Luigi Vanvitelli',
Aversa, Italy
alice.palmieri@unicampania.it

Abstract. This paper proposes a reflection on spontaneous architecture, a theme dealt with in detail by Bernard Rudofsky, who in the exhibition *"Architecture without architects"* is critical of the author's architecture that places more emphasis on the creator at the expense of collective logics that have formed the basis of architectural thought. Considering that the evolution of building has not been shaped by architects, but by certain social, political and economic contexts, means reading the history of architecture not on the basis of a scientific production carried out in individual graphic elaborations, but by communitarian reflections and awareness. In this sense, architecture is free from aprioristic thought, and not constrained by a tension that starts from the final form, but generated by the constructive process that progressively occupies the space defining a place. Two fundamental questions are connected to the mediation of design in the project conception (and its possible lack): the first concerns the authorship of the work that is the need -found by Vasari for the first time- to identify an architect and his cultural context in order to attribute meaning and value to a building. The second theme, on the other hand, is closely linked to the constructive reasons and processes that generate a type of participatory architecture, sometimes without overall planning and graphic design representation. This is the case of earthen architecture, its evolutions and the approach to this technique in the contemporary culture.

Keywords: Earthen architecture · Graphic mediation · Spontaneous architecture

1 Introduction

As result of an exhibition held in 1964 at MOMA in New York, curator Bernard Rudofsky publishes a book entitled *"Architecture without architects"*. It is a catalogue of spontaneous architectures in the world, buildings without authors and empirically realized, almost without the mediation of project design. In the press release of the exhibition we read a statement by the architect Pietro Belluschi, who is enthusiastic seeing the iconographic material and declares: *"Somehow for the first time in my long career as architect I had an exciting vision of architecture as a manifestation of the human spirit beyond styles and fashions and, more importantly, beyond the narrow straits of our Greco-Roman tradition"* (p.r. MOMA 1964). A peculiar point of view is therefore highlighted, according to which the so called 'architectural tradition', the one referred to

L. Agustín-Hernández et al. (Eds.): EGA 2020, SSDI 6, pp. 678–688, 2020.
https://doi.org/10.1007/978-3-030-47983-1_60

in the usual debate, is not "the" tradition but "one" tradition, that is, the Greco-Roman version, which has the undisputed *vate* in Vitruvius (Corrado 2017). Our cultural codes allow us to contextualize the architectures of the time starting from the examples of ancient Greece, where the elements and styles that are the foundations of Western culture are first identified. And even so, there are many expressive languages that are free from these references in which, more than form, it is technique that defines the creative process that generates spaces born from experimentation and the search for possibilities for action: spontaneous architecture (Fig. 1).

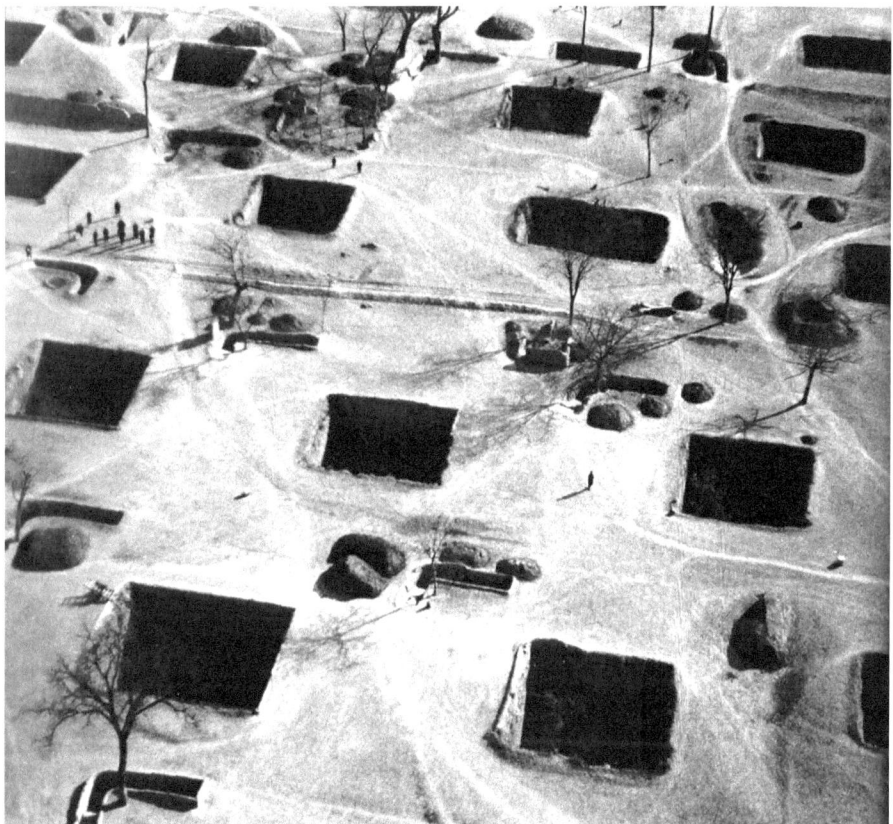

Fig. 1. Photo of a settlement at Tungkwan (Honnan) of a rigorous, almost abstract design, evidence of applied geometry despite the absence of an overall graphic planning. Source: Rudofsky 1964, pp. 16.

"It is so little known that we do not yet have a term to define it. In the absence of any label, we will call it vernacular, anonymous, spontaneous, indigenous, rural architecture, as the case may be" (Rudofsky 1964, preface). Not infrequently, in his publication, Rudofsky is provocative (and sometimes controversial) about the architecture of Western countries, which often appears self-referential and full of "architectural prejudices". The critics are moved towards the consideration that a building without an author is less

respectable and valuable than the representative works of architectural history, which emphasize the work of the individual creator and generate noble buildings. There is, however, a diametrically opposed key, which considers precious the buildings that express the initiative of a community. From here we begin to define the concepts, now belonging to contemporary culture, of collective or participatory architecture *"not produced by a few intellectuals or specialists, but by the spontaneous and constant activity of an entire participating people of a same culture and operating on the basis of a common experience"* (Rudofsky 1964, cit. Pietro Belluschi).

2   The Issue of Authorship

A first critical reading of this premise invites us to reflect on the theme of authorship, that is the identification of architecture with its creator. In fact, faced with the discovery of a new contemporary work, the first question we are accustomed to asking ourselves concerns the designer, almost as if we want to identify someone "responsible". Actually, this question underlies our need to contextualize a building through its author. Yet history is full of extraordinary examples without known architects, such as the immense heritage of Gothic constructions and the amazing cathedrals built in open-air factories that were almost "cities within cities". The functioning of the medieval yard was a complex reality that we know thanks to written sources and representations of the various craftsmen engaged in simultaneous work. Due to its size, technology, tools used and economic effort, the construction of a cathedral is an extraordinary event that involves the identification of increasingly specialized professional figures and more advanced masonry techniques. The cathedral building worksite therefore becomes a laboratory for the exchange of skills and knowledge, a place of new experimentation. The direction of this great machine was entrusted to the person whom the sources of the time indicate with the generic term *magister* (or *aedificator* or *artifex*). This figure is identified with the expert master builder in which the manual approach prevails over the intellectual, theoretical and design attitude. This technician, who we might consider the architect of the cathedral, is trained directly on site, through the oral transmission of practical knowledge, often in the form of "trade secrets". It is not a coincidence that the fascination of gothic churches is given by the correspondence between aesthetics and technological innovations, which through continuous empirical experiences, come to realize elements such as the cross vault and the pointed arch. The integration of these two systems will allow the construction of point-shaped architectural structures, without loading the perimeter masonry, permitting the installation of large windows with colored glass that constitute a recurring element of the Gothic language. These considerations reaffirm the role of the architect, not as a designer, but as a builder, deeply aware of the management of construction processes. Some of these master builders also knew how to read and write, and almost all of them had knowledge of geometry and proportions, so later becoming able to draw up scale projects. But in Gothic architecture, the question of the author remains more or less unsolved.

"So many important buildings were therefore built in Italy, and beyond, of which I could not find the architects" so Vasari for first tries to investigate the origin of the artworks with the intention of linking the designers and their cultural backgrounds with

their creations. The publication *"The lives of the most excellent painters, sculptors, and architects"* (Vasari 2015) represents in this sense the turning point marking the genesis of a radical change in artistic production that could no longer be considered independent from its creator and this (almost obsessive) cult of authorship still persists today. However, the history of human habitation is very rich of anonymous and participatory architectures, born from community knowledge and needs. And this is precisely the central pivot of Rudofsky's work: he aims to build a convincing thesis in favor of architecture without an author, understood as an instrument of sustainable design that has existed for thousands of years in cultures around the world (Ratti 2014, pp. 20).

3 Earthen Architecture

In the same (and opposite) way of Vasari, Rudofsky tries to give prestige to the image of those non-authorial architectures, often devoid of graphic mediation, demonstrating the validity of planning "from below". *"My project is openly polemical, as it compares the serenity of architecture in so-called underdeveloped countries with the degradation of industrialized countries"* (Rudofsky 1964).

The difference between these two approaches to architecture is basically given by the role of the graphic design in a cognitive process that generates "a priori" the image of a place. The project, intended as the conception of a built space, is defined through a detailed technical representation, while a realization without the mediation of the drawing of the form, works instead concentrated "on" and "in" the construction process, following the nature of the material. For this reason, in the context of this research a fecund area rich in expressions and visual languages is that of raw earth constructions. From a graphic point of view it is interesting to note how the techniques to be applied are often carefully documented in manual form, leaving a gap in the representation of the shape that is being configured. For raw earth constructions, the most common techniques used are *adobe* (bricks made with a mixture of clay, sand and sun-dried straw) and *pisè*. This last system has undergone various contaminations, even in the contemporary world, and for this reason it has manifested multiple declinations. Originally the realization of a *pisé* wall was completely handmade, without the aid of any support. Later it is defined as a masonry construction technique based on the working of wet earth compacted with special tools, inside wooden formworks. The wall is made in layers, beating the clay with large-headed ligneous instruments, pointed or wedge-shaped, in order to make it compact and accelerate its drying. Once the earth has hardened, the formwork is dismantled. This "technological" introduction, which uses formworks to define the masonry, maintains the layered construction process, but forces the morphology of the architecture. In fact, while a process linked to the installation of wooden planks implies a right angle plan, the original manual procedure, invited to draw with the rammed earth soft and curvilinear forms more manageable empirically. Furthermore, these primordial construction methods (which still exist in developing countries) involve a continuous renewal of buildings that cyclically need to be regenerated with the contribution of new earth. This last process, similar to what Rudofsky calls "from below", is characterized by the incessant mutability of a shape whose realization is interrupted when the word 'end' opposes it. Forms are born from the nature of the material chosen, from their processing possibilities and grow in the hands of those who make them, without having a design to complete, a predefined expectation and an image established in the beginning. It is therefore a *modus operandi* that highlights the present, the discovery and the sense of surprise and occupies the space according to a procedure that is similar to the concept of 'natural growth' (Campos 2018) (Fig. 2).

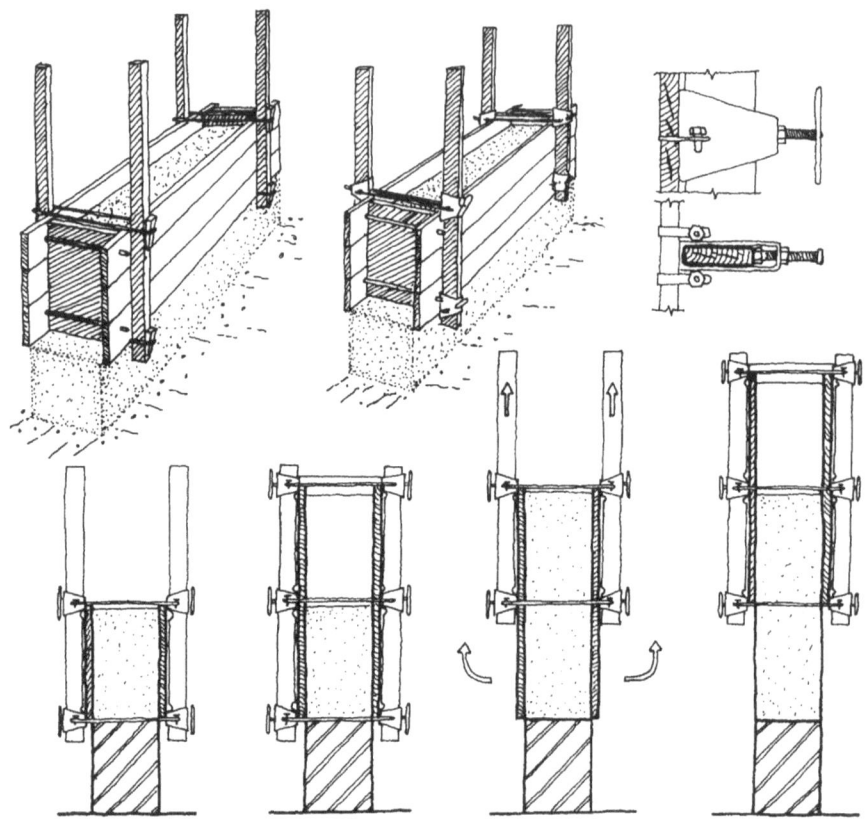

Fig. 2. Construction of a pisé wall using formwork. Source: Minke 2006, pp. 53

4   Gestures and Processes

As already highlighted, much of the graphic documentation relating to earthen architecture does not reproduce its compositional aspects, but represents the construction techniques that define a production process made of gestures that are repeated in order to define a replicable executive dynamic.

The traditional techniques most commonly used to build earthen constructions can be divided into two main types: the first approach uses rammed earth to obtain monolithic elements capable of performing a load-bearing function, such as *pisè*; while the second method uses the natural material to produce blocks of various shapes and sizes called *adobe*. In both cases, it is interesting to note that the representation of these techniques reproduces movements: the steps and phases of work narrate a succession of human actions that generate the construction process in an anthropocentric key, which cannot disregard the illustration of manual skills and physical involvement in the procedures of realization. In this sense, there is an interesting documentation that since the 18th century has been involved in the graphic design of technical manuals, in which the methods, phases and instruments used are illustrated. And even if sometimes the contents were repeated in a similar way, the language of drawing was modified coherently with its time and used easily interpretable codes and conventions; at the same time, the manner of representation narrates the graphic sensitivity and interpretative reading of the designers in different historical moments (Figs. 3 and 4).

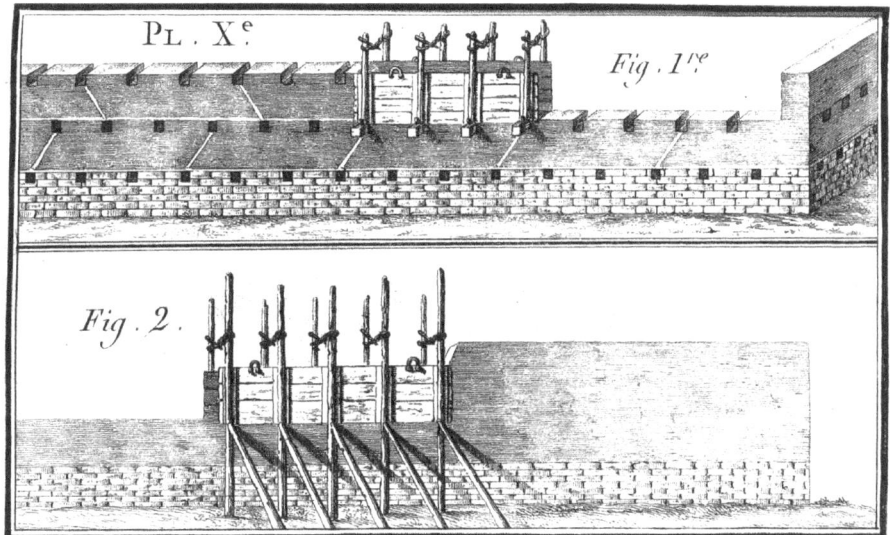

Fig. 3. Cointeraux, 1793. *École d'architecture rurale. Premier cahier. Seconde édition. Planche X.* Source: Conservatoire numérique des Arts & Métiers

Among the first testimonies of earthen architecture, there are dwellings dating back to the Bronze Age and found in what was southern Gaul, where a building technique in use between the Greeks and the Carthaginians had been imported. Even in ancient Rome the construction technique of *pisé* was known, to which the Roman builders preferred stone or fired bricks. In Europe, a revival towards the use of earth as a building material is due to the work of Professor François Cointeraux, who in 1791 published a scientific treatise entitled *"Traité sur la construction des manufactures et maison de campagne"*. The great editorial and scientific success of this publication is due to the exhaustive treatment of the subject, which has been widely discussed, proposing a new way of building walls in raw earth that also encompasses the art of making the tools necessary for its realization. The treatise is therefore accompanied by graphic plates illustrating the

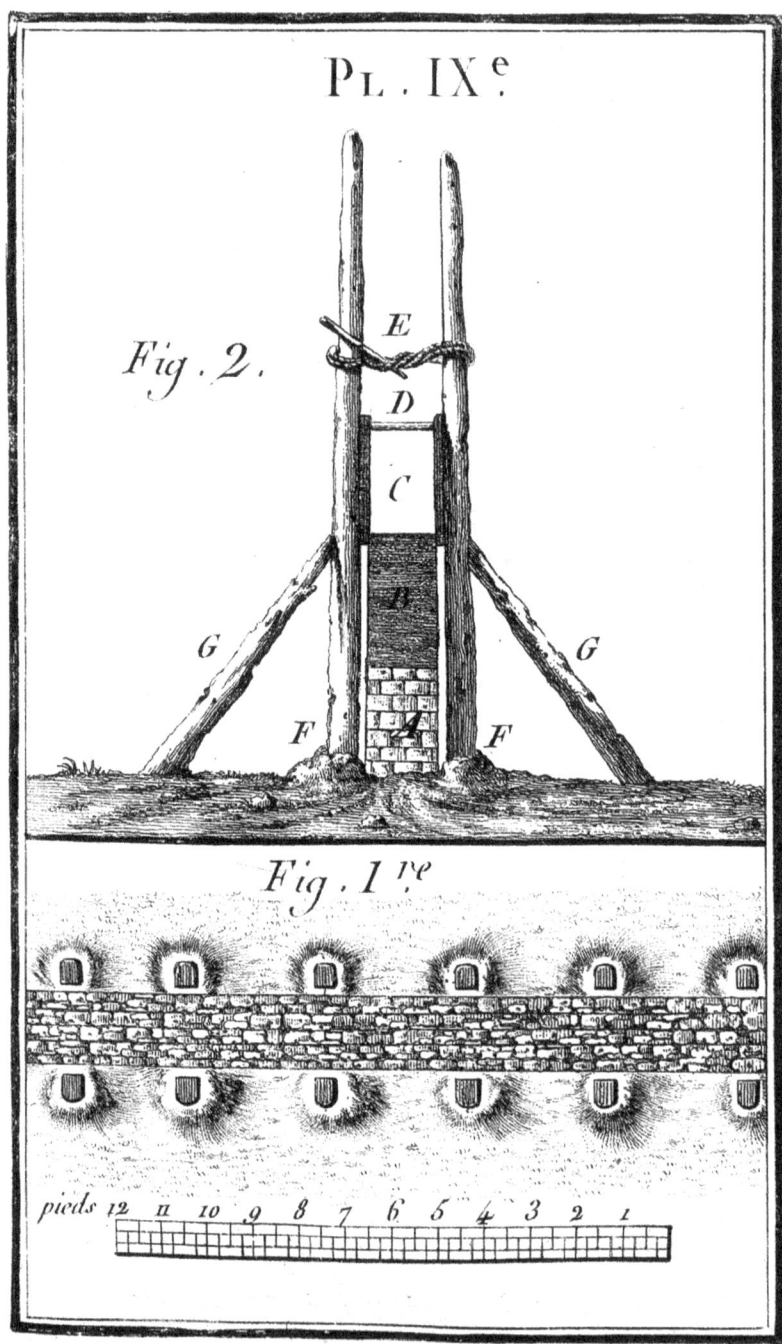

Fig. 4. Cointeraux, 1793. *École d'architecture rurale. Premier cahier. Seconde édition. Planche IX.* Source: Conservatoire numérique des Arts & Métiers

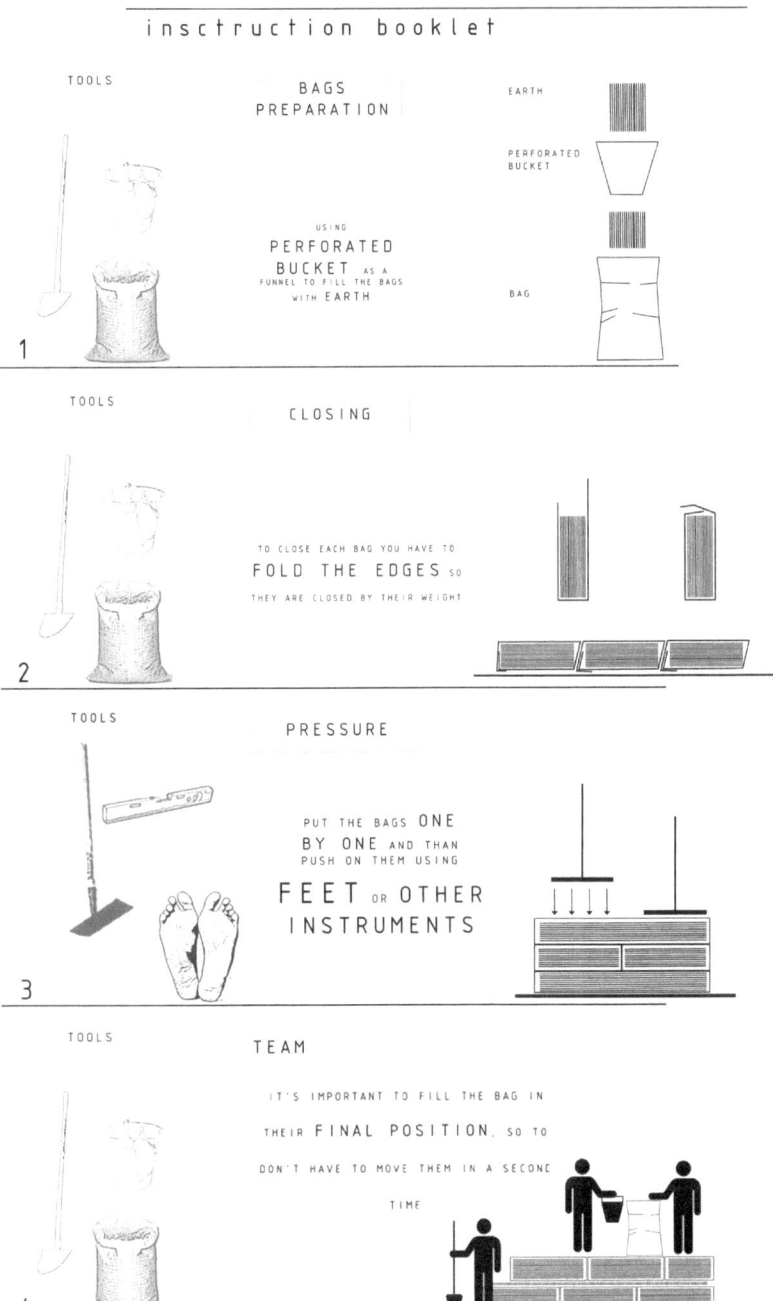

Fig. 5. Summary table - ARCò, Al Khan Al Ahmar Primary School, Instruction Manual

construction techniques in *adobe*, *torchis* and *pisé*, and equipment necessary for working raw earth, straw, wood and everything necessary for good construction.

At the contemporary time, the need to graphically represent construction techniques in an efficient and immediate way sometime makes use of pictograms and therefore a direct visual communication, not necessarily technical, which becomes self-explanatory. Especially in contexts where architect and builder have different languages, cultures and generally different codes, this kind of image shortens distances and is able to communicate building processes even not common. This is the case, for example, of the graphic project conceived by the Milanese team called Arcò - Architecture and Cooperation, for the construction of the children's centre on the Gaza Strip, in the village of Um al Nasser. In 2011, for the realization of this structure, a system of "*earthbags*" was planned: a simple procedure that requires attention during installation, especially for the connection between the different rows, which is carried out by non-specialized local personnel and which combines an innovative component (given by the bags) with a traditional technique of architecture in rammed earth (the *pisè*). Through a succession of pictograms representing the tools, the operational gestures and the people involved, the entire production process is declined, replicable and "appropriated" by the local population, regardless of the compositional and dimensional characteristics of the building that is being realized. Also in this case the iconographic language overcomes territorial and cultural distances and through a communication that uses different methods of representation (from symbols to photographic images) not only describes the process of realization, but indirectly tells something about a context, an historical moment and the need to understand each other beyond any location. The narration of the earthbags installation procedure reveals that there are few rudimentary tools available, that

Fig. 6. Children's Land_Um al Nasser - setting up earthbags wall ©ARCò

in order to re-propose the pisè technique the weight of man is necessary and that to ascend in altitude the collaboration of more workers is requested. For this reason, the story behind the pictograms is the richness of a representation that in its conciseness allows us to intuit the social environment in which architecture wants to take shape (Figs. 5, 6 and 7).

Fig. 7. Construction of a rammed earth wall, Benin. Source: photo Thierry Joffroy, https://archeorient.hypotheses.org/4562

5 Conclusion

We can distinguish *"two phases of the project operation: one connected to the project as a document and history of the formation of an architectural image, the second to the organization of this image in the project according to a series of notations essentially aimed at communicating the project itself according to its correct execution"* (Gregotti 2008, pp. 20).

Vittorio Gregotti separates two steps in the genesis of the architectural image, two moments that are not temporally successive, but which influence each other in the formation of the project. What we have said so far with regard to earthen architecture takes this dialectic to the extreme, almost to the point of cancelling it out, because in the reflections that are being proposed, the prevailing thesis is that the ultimate image of the construction is not generated by the conception, but by the material itself. Choosing to build with earth implies a pre-emptive visual language, made of colours and textures inherent in the material and which prove to be dominant with respect to many other formal aspects. Consequently, the graphic mediation is led to describe the themes related to the production process of architecture, since the "constructive demands" of the material itself guide the design choices, almost as if it were a developing organism. This motivation justifies the lack of graphic elaborations relating to the ultimate form of ancient earthen architecture or the examples used by Rudofsky. It is a building in progress with which architects confront themselves to seek to define a model of behaviour that from the simplicity of the renewed construction techniques produces new forms of art (Fathy 1986)

References

Campos, C., Cirafici, A., Fiorentino, C.C.: Del acontecimiento al indumento: traducciones intersemióticas y diseño contemporáneo. In: Proceedings CIMODE 2018: 4° Congresso Internacional de Moda e Design (2018)

Corrado, M.: Quando l'architettura non la facevano gli architetti (2017). http://larchitetto-nella-foresta-design.blogautore.repubblica.it/

Fathy, H.: Costruire con la gente. Jaca Book, Milano (1986)

Gregotti, V.: Il territorio dell'architettura. Feltrinelli, Milano (2008)

Minke, G.: Building with Earth. Birkhäuser – Publishers for Architecture, Basel (2006)

Ratti, C.: Architettura Open Source. Einaudi, Torino (2014)

Rudofsky, B.: Architecture Without Architects: A Short Introduction to Non-Pedigreed Architecture. Doubleday, New York (1964)

Vasari, G.: Le vite dei più eccellenti pittori, scultori e architetti, Edizioni la Biblioteca Digitale, Firenze (2015)

http://www.ar-co.org/it/progetti/realizzati/terra/index.php

Geometry and Art in Decorative Panels of Salerno Cathedral's Floor

Stefano Chiarenza[1] (ID) and Barbara Messina[2(✉)] (ID)

[1] San Raffaele Roma Open University, Rome, Italy
stefano.chiarenza@uniroma5.it
[2] University of Salerno, Fisciano, SA, Italy
bmessina@unisa.it

Abstract. The paper dwells on an interesting Norman-era artistic episode that is in the San Matteo cathedral, in Salerno. Specifically, it is the mosaic floor of the church, stylistically attributable to the "Cosmatesque" art that, between the XII and XIII century, spread in the regions of central-southern Italy culturally influenced by the Byzantine world. After deepening the stylistic implications of the Cosmati, attention is paid to the geometries of the inlaid floor drawing, that define the aesthetics of the multiform motifs made by artisans of different cultural backgrounds. Leaving aside the intellectual, psychological or religious reasons that inspired the realization of the floor decorations, the paper traces and analyses, through considerations of a mainly graphic nature, the logic underlying the drawing of the inlaid marble pavement. The vast repertoire of shapes is therefore examined and, with the help of drawing, the harmonic synthesis between geometry and mathematics in the various modular compositions is highlighted. The aim is to identify the compositional structure and the geometric rules of the decorative patterns that define its overall image.

Keywords: Cosmatesque style · Geometric drawing · Arab-Norman art

1 Introduction

In the context of the Southern Italy artistic heritage of the Middle Ages, the Arab-Byzantine-inspired decorative repertoire plays a particularly significant role. It is a heritage that, mixing styles and figurative traditions of oriental taste with local languages and construction techniques, becomes an emblem of an artistic practice that, precisely in Norman times, unites the regions of the Mediterranean basin. What distinguishes this repertoire is the strong geometric imprint, the founding matrix that guides and regulates the figurative composition of the suggestive ornamental systems.

Starting from the numerous interdisciplinary studies on the rich artistic heritage produced in this era, the paper intends to outline the graphic logic underlying the decorative motifs that characterize it, with the aim of defining compositional rules and geometric relationships at the basis of the structuring the final image. In particular, attention is paid to the inlaid floor of Cosmatesque style preserved in the transept and choir of the

L. Agustín-Hernández et al. (Eds.): EGA 2020, SSDI 6, pp. 689–700, 2020.
https://doi.org/10.1007/978-3-030-47983-1_61

Salerno cathedral, emblematic of the cultural syncretism typical of this region in that time. Thanks to a critical-interpretative reading, made possible also by specific graphic investigations, we intend to highlight the great compositional and decorative quality of this floor, strictly dependent on the rigorous geometry at the base of its ornamental motifs.

2 Cosmatesque Stylistic Implications in the Decorative Apparatuses of Southern Italy

Southern Italy, in the Norman era, knew a particularly favourable artistic moment: the numerous commercial exchanges and the intense contacts, also cultural, with Byzantium and with the Islamic world, in fact, determined the flourishing of a stylistic language of refined beauty that is based precisely on the deep integration between figurative canons and building systems of different cultures. Looking with specific interest at the decorative apparatuses, it should be underlined how, in Norman Italy, the interaction between the Byzantine and the Arab world translates into an absolutely original ornamental repertoire. The propensity for the drawing of geometric patterns, evidently inspired by Islamic art, coexists for example with materials and construction techniques–such as *opus sectile* and mosaic decorations–in vogue by the Byzantine art. The result are artefacts that, thanks to the mixed-use of forms and elements, express the Hegalian idea of "*una totalità di differenze essenziali, che non solo si palesano come differenze e opposizioni, ma mostrano, nella loro totalità, unità e connessione*" (Hegel 1972, p. 158).

With reference to the figurative suggestions, the contribution of Cosmatesque production, by marble workers active in Rome and Lazio between the twelfth and fourteenth centuries (Chiavoni et al. 2017), is also significant. The Cosmati–patronymic with whom the creators of this artistic language were known–will influence artists and stonecutters from Southern Italy, becoming for them a model to be emulated. The recurrence of some aspects and some distinctive signs characterizes their works, so much so that, beyond the personal way of expression of each master, we can speak of a real stylistic identity. While not wanting to achieve a rigid schematization, which could appear reductive for the purpose of a deep understanding of the entire Cosmatesque artistic production, it must first be emphasized that it focuses, above all, on the decoration of flat surfaces: from the floors to the walls of liturgical furnishings, such as *plutei* and ambo, from entablatures to the frames of various architectural elements, also including, in some cases, structures with a three-dimensional development, such as the columns shafts or the candelabra for Easter candles. It should be noted that "*anche quando fecero opere d'architettura essi restarono sostanzialmente lapicidi, cioè non affrontarono mai un vero problema costruttivo: concentrarono il loro interesse non sullo spazio, ma sulla superficie e restarono quindi decoratori*" (Matthiae 1980, p. 837).

In figurative terms, what makes the Cosmatesque style recognizable is, above all, the polychromy of the decorations obtained thanks to mosaic tiles, or stone inlays of various kinds, which, skilfully combined, give life to elegant geometric patterns (Fig. 1). The coloristic taste is therefore accompanied by the search for figurative techniques and suggestions directly deriving from the classical or Paleo-Christian world, of which the artistic approach in the conception of the decorations is especially proposed.

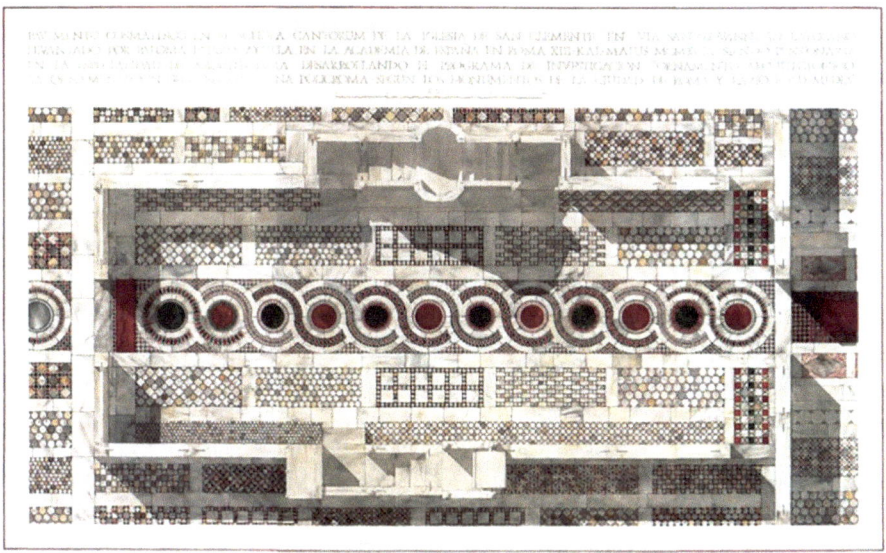

Fig. 1. Rome. Basilica of San Clemente, XII century. Watercolour drawing of the Cosmatesque floor. (Source: Pajares-Ayuela 2001).

The deeply retrospective taste associated with the "*carattere aulico connesso alla rarità e alla qualità dei materiali impiegati [...] contraddistinguono in maniera specifica le opere cosmatesche dai linguaggi 'volgari' del Romanico europeo, apparentandole piuttosto ad altre espressioni, legate in qualche modo al mondo bizantino. Questo atteggiamento, che vedeva nello spoglio o nella imitazione dell'Antico il mezzo e il fine dell'operato artistico e insieme il segno di una recuperata romanità, appare una costante delle botteghe marmorarie fino almeno a tutti gli anni sessanta del Duecento ed emerge con orgoglio anche nella insistente autocelebrazione presente nelle firme dei maestri sin dalla fase iniziale di attività*" (Bassan 1994). Concerning the graphic stylistic elements, certainly the artistic production of Cosmatesque derivation offers consolidated models, formal solutions referable to usual patterns which, adapting each time to the surface to be decorated, allow their full ornamentation. A large repertoire–even 133 patterns catalogued in the 1950s on Roman production alone (Piazzesi et al. 1954)–whose fil rouge is the use of compositional principles conformed on common rules. The ornaments' drawing always develops from square or rectangular panels, or from basic modules–usually circles–delimited by white marble bands. These, framing the dense polychrome motifs obtained with the mosaic technique or in *opus sectile*, "*ne costituiscono, con la loro zona di riposo, la ritmica scansione*" (Matthiae 1980, p. 838).

A solid geometric structure is the basis of the internal development of the entire composition, which always starts with elementary figures such as the triangle, the square, the rhombus, the hexagon, the octagon or the circle, often derived from each other. Thus, "*un rombo [può essere] creato da due triangoli equilateri, un triangolo generato dalla divisione di un quadrato lungo una delle sue diagonali o un rettangolo formato dall'unione di due quadrati*" (Moran and Williams 2007).

In other cases, the geometric drawing at the base of the decoration, when more complex, is obtained by the juxtaposition of elementary figures (Fig. 2). *Vesica piscis* (or ogive) motifs are, for example, derived from the intersection of two congruent circles, each with its own centre on the other; star motifs from the superposition of two squares or two triangles (rotated to each other by 45°, in the first case, and 180°, in the second); *ad quadratum* or *ad triangulum* compositions are still obtained by combining two squares (rotated by 45°) or two triangles (rotated by 180°) which, inscribed in each other, define a gradual reduction of the figurative module. An ornamental system, therefore, the Cosmatesque one, which makes explicit the need to translate the strict mathematical-geometric structure, that regulates the composition, into a graphic sign. The rigor of the formal configuration gives aesthetic strength to the decorative patterns, which are thus charged, at the same time, with symbolic as well as aesthetic values (Cigola 1993).

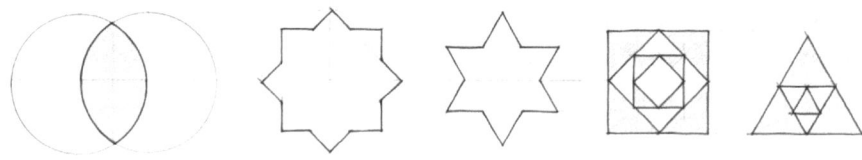

Fig. 2. Compositional patterns at the basis of the drawing of some geometric decorative motifs in the Cosmatesque style. (Drawings by the authors).

3   The Salerno Cathedral's Floor: Geometry of the Inlaid Drawings in the Choir and Transept

The floor of the Salerno cathedral, built between 1080 and 1085 at the behest of Archbishop Alfano I–former monk of Montecassino and friend of Abbot Desiderio–is attributable to the production of Cosmatesque influence. Indeed, *"l'arte della Campania si trovava all'epoca sotto la forte influenza dell'abbazia di Montecassino, il cui abate Desiderio, sostenitore dei Normanni, era nello stesso tempo in buoni rapporti anche con Bisanzio"* (Zarnecki 1997). It should be noted, in this regard, that the abbot had called inlaid masters and mosaicists from Constantinople to decorate his abbey and, at the same time, instruct and train local artisans, in a sort of construction yard-school. This explains the strong ascendant that, on the one hand, the marble workers operating in Cassino, on the other hand the Byzantine art had on the Salerno workers who, precisely, were inspired by these models. The Salerno cathedral, dedicated to San Matteo, was therefore designed according to a basilica plan, now in use in central-southern Italy: the reference compositional scheme was, precisely, that of the Montecassino abbey, which revisits the traditional model of the Christian basilica in the light of the new Carolingian era trends. The church's space–preceded by an external quadriporticus–has in fact a longitudinal body, divided into three naves, and is closed by a triple apse transept. The transept, used as a presbytery, is finally preceded, in the last two bays of the central nave, by the *schola cantorum*–a space reserved for the singers who sang the psalms, then followed by the faithful.

Just in the transept and in the choir there is the splendid floor: its surface (Fig. 3), of about 800 m^2, shows an elegant inlay drawing that recalls, in many ways, the Cosmati floors. Even without going into historical-critical issues, which would require extensive study (Carucci 1983), it should be noted that the dating of the floor, as well as the attribution to specific workers, are open questions. The most accredited thesis refers to the execution of the floor by two different craftsmen groups, that may have operated simultaneously or, more likely, at different times (Braca 2003, pp. 135–148).

Fig. 3. Salerno. Cathedral of San Matteo. Transept floor, XI century. (Source: Carucci 1983).

This hypothesis derives from the iconographic comparison of the floor's decorative patterns. The pavement is entirely designed as a sequence of rectangular or square panels, bordered by a dividing band in white marble: the panels present common characters in the choir and in part of the transept–precisely in its central and northern portion–while they clearly differ in the southern part of the latter. The choir area (Fig. 4), as well as the transept's central and northern areas, show, in fact, a certain regularity in the drawing's articulation, which adapts, although with some exceptions, to the spatiality of the building. Here the floor shows an alignment, both longitudinal and transversal. Regularity is interrupted, however, suddenly in the transept's southern area, which seems to respond to an autonomous design criterion.

To further underline the hypothesized difference in execution are the decorative motifs, which seem to confirm the different cultural contexts in which they would have been conceived. If, in fact, almost all the panels, regardless of their location, propose the circle as the fulcrum of the ornamental motifs (few, and mostly peripheral, are those

Fig. 4. Salerno. Cathedral of San Matteo. Choir floor, XI century. The *guilloche* motif (Source: Carucci 1983).

in which this figure does not appear), the configurative composition shows completely different results in the two previously identified areas.

The choir, and the transept area homogeneous to this, have geometric patterns of Cosmatesque's derivation: here there are rhomboid motifs, guilloche motifs (exclusively in the choir)–that is a linear sequence of circumferences connected by a band that envelops them–or quincunx motifs (Fig. 5)–composition of five circumferences, one central and four angular, connected by a frame tangent to them and contained within a square. The decorations in the southern area of the transept, on the other hand, have an evident oriental taste: in fact, if some panels with more basic drawing are excluded, the patterns generally include only circles of variable radius, connected with frames mutually tangent to them.

In this regard, Braca notes that "*i due riquadri del lato meridionale formati da sole circonferenze intrecciate fra di loro non trovano riscontro nel panorama regionale dei primi decenni del XII secolo*", making their appearance "*solo in pavimenti bizantini, come in Santa Sofia a Nicea e nel monastero di Ivirone sul monte Athos*" (Braca 2003, p. 139). Byzantine art would not, however, be the only model to which the linguistic novelties of the Salerno floor can be traced. In it, in fact, the drawing's sinuosity seems to give marble the softness of fabric: the intertwining of the lines "*finora impiegato per i macromodelli curvilinei, qui coinvolge anche i micro-modelli, precorrendo le originali soluzioni di matrice islamica che più tardi caratterizzeranno la produzione siciliana*". This suggests "*il sopraggiungere di una nuova cultura architettonica e figurativa la quale, piuttosto che essere ascrivibile alla tradizione bizantina o islamica, appartiene al filone innovativo di una koinè mediterranea, recepita in Italia meridionale per intermediazione del regno normanno*" (Longo 2010, p. 182).

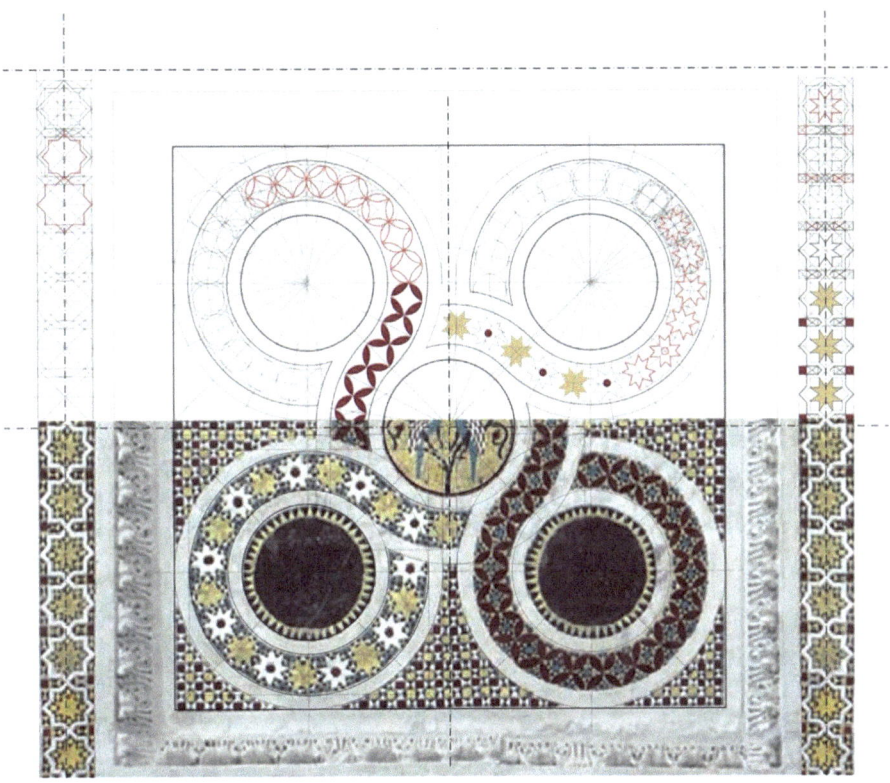

Fig. 5. Salerno. Cathedral of San Matteo. Ambone D'Ajello. Graphic study of flat pattern with linear development (translation or rotation) consisting of a quincunx motif present in a decorative panel. (Drawings by the authors).

4   Geometric Reading of Decorative Motifs

Some considerations can be made about the mathematical-geometric structure underlying the graphic composition of the flooring.

Numerous doubts persist on the correspondence of the current flooring to the original conception (Severino 2011; Braca 2003). However, in terms of overall design, the presence of a symmetry along a longitudinal axis is evident and enhanced by the concatenation of the circular slabs of the central motif in the choir: although the drawing is not present throughout the nave, this linear arrangement seems to maintain a marked orientation in the direction of the altar, according to a consolidated Cosmatesque tradition. On closer inspection, however, this is not a bilateral symmetry (i.e. a reflection), but a rotational symmetry with reference to each part that makes up the *guilloche* drawing. The transept pavement, instead, appears as a set of panels side by side, for which there is no common regulatory geometry: the design of the floor in this area suggests a sense of staticity rather than movement, and the inlaid rectangular panels seem to outline a colourful marble carpet.

This particular arrangement, due to the evident contrast with the traditional pre-Cosmatesque and Cosmatesque structures, led to the hypothesis of a reassembly of the pavement between the fourteenth and fifteenth centuries. In this period, the most significant panels of an ancient damaged floor would have been used, arranging them in a way that, although not casual, did not respond to the canons of Cosmatesque art (Severino 2011, pp. 20–21). In consideration of the uncertainties related to the spatial organicity of the floor design as a whole–which would make any geometric, iconological and iconographic interpretation artificial–attention is therefore focused on the individual panels for which the geometric-configurative structure is undoubtedly intact. The variety of shapes used, clearly refers to the tradition of Cosmate masters: triangles, squares, rectangles, circles are the basic elements of compositions that see complex motifs emerge, linked to a singular ability in the aggregation of the various elements. In this way, star or hexagonal drawings and interweaving of straight or curved lines that support circular elements, define decorative patterns which, while keeping to a notoriously widespread syntax, appear in many respects as unique (Figs. 6–7).

Fig. 6. Salerno. Cathedral of San Matteo. Geometric interpretation of some decorative motifs with modular groups in the choir and transept floor. (Drawings by the authors).

Certainly the compositions were the result of constructive practice and laying methods (Williams 2004) which allowed often extraordinary intuitions. It should be noted, however, that, in the pavement of the cathedral, the coexistence of different patterns appears evident: some characterized by textures with larger marble elements of an evidently pre-Cosmatesque style, and others made with a more refined technique (Severino 2011, p. 24). Consistent with the Cosmatesque ratio and the ideological assumptions of Byzantine and Islamic art–evident sources of inspiration–the Salerno floor therefore proposes the organic assembly of several different motifs in which, however, the design is always supported by precise geometric correspondences. Next to the typical *guilloche* and *quincunx* and the curvilinear bands that follow their roundels, there are numerous decorative frames, with linear development drawing, and inlaid panels–typical of

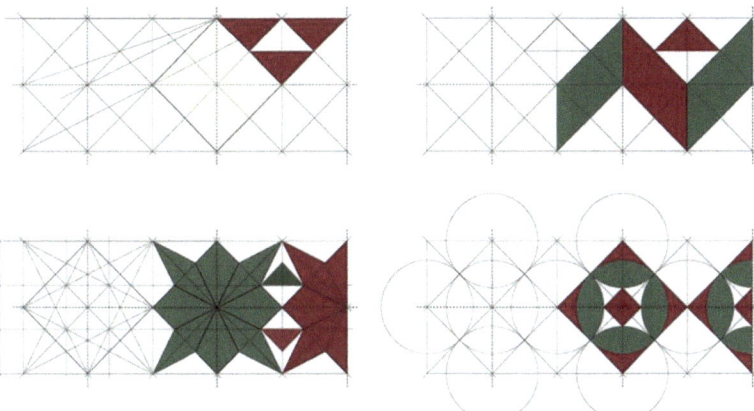

Fig. 7. Salerno. Cathedral of San Matteo. Geometric interpretation of some decorative motifs with linear development in the choir and transept floor. (Graphic elaboration by the authors).

the Cosmatesque floors partition patterns–in which geometric elements with periodic development instead tessellate the whole surface.

The filling areas between the curvilinear motifs and the rectangular frames that delimit them also represent significant opportunities for formal invention.

All the motifs are always based on one-dimensional or two-dimensional symmetries, but–similarly to what has already been highlighted in specific research on Cosmatesque floors–the symmetry of the starting element often limits the 24 possible symmetry groups (17 in the two dimensions and 7 unidirectional) reducing them only to 10 in the plane (Moran and Williams 2007). The basic modules (or primary tiles) are in fact those of the equilateral and isosceles triangle (derived from the division of the square according to the diagonals), the square, the rectangle, the hexagon, the circle (all symmetrical in themselves). From their arrangement derive blocks that define more complex schemes, whose theoretical drawing allows to obtain tesserae sometimes with a rectangular, parallelogram, *vesica piscis* shape. The decorative patterns each show their own internal geometric logic: this is generally structured on central or rotational symmetries, mostly combined with linear transformations of a translatory type by periodic repetition. Single unit modules therefore give rise, aggregated together, to the definition of a tile which is then repeated on the surface of the panel in one or two directions. Triangles and all quadrilateral shapes allow to obtain reiterations on the plane without overlapping or empty spaces; this does not always happen with hexagonal or octagonal blocks whose composition implies, in some cases, the need to close the tessellation of the surface by using complementary tesserae. Translation, rotation, reflection and glide reflection movements regulate all linear and two-dimensional symmetries (Fig. 8).

Compared to other similar examples, the flooring of the Salerno cathedral is characterized by panels which, with few exceptions, always have an internal drawing model that "*determina l'andamento delle fasce marmoree, regola la disposizione dei dischi porfirei e divide la superficie in più campi riempiti con diversi motivi ornamentali*" (Longo 2010, p. 180).

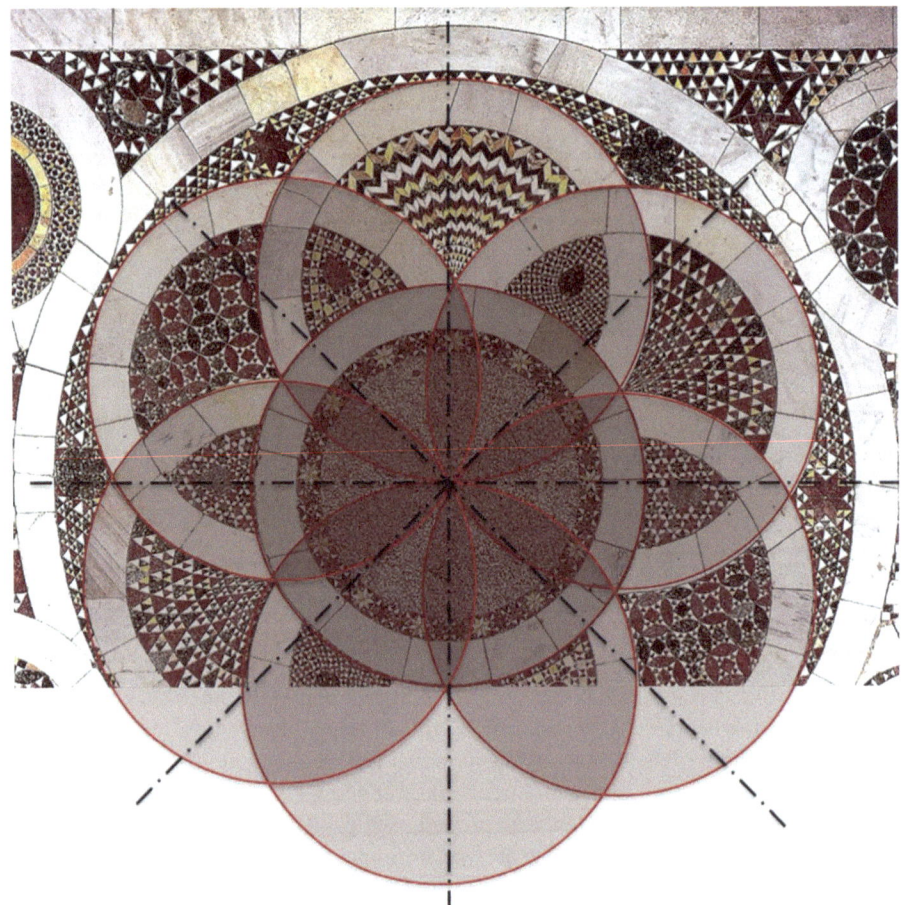

Fig. 8. Salerno. Cathedral of San Matteo. Graphic study of the rotational symmetry of one of the decorative motifs of the cathedral floor. (Graphic elaboration by the authors).

These motifs are therefore related to the space they occupy, often using modules that adapt to the background to be filled. If a great variety of themes is generated starting from basic modules that strongly recall the typical Cosmatesque repertoire (for example the numerous *ad quadratum* and *ad triangulum* schemes) it is possible to find absolutely new patterns in different panels. This is the case of motifs with intertwined strips that evoke the interwoven textures of the Arab style; the "star of David" with a strong Islamic matrix. Or the rhombus motifs, so-called "*ad cubum*", that recall to three-dimensionality according to typical schemes of the first-century Roman opus sectile. This probably proves the Byzantine and Arab-Norman cultural mixes of the workers who had to work in the Salerno cathedral.

In the refined work of tessellation of the floor surface, the creation of filling motifs of the voids deriving from the installation of larger modular tiles or from the aggregative logic of the blocks is often recurring. It follows, in some cases, the determination of complementary forms (as previously noted) or the decomposition of the modules into congruent sub-modules, so as to dilate the pattern *ad infinitum*. This is the case, for example, of triangular-shaped blocks: these generate self-similar sets that are repeated, in the same way, on different scales according to the fractal scheme known as the "Sierpinski triangle". From a mathematical point of view it can therefore be said that, similarly to what happens for Cosmatesque floors "*l'imperativo di riempire lo spazio spiega [...] l'apparire di un terzo tipo di simmetria [...]: la simmetria di similitudine o, come spesso viene oggi chiamata, la simmetria frattale. Gli spazi, di volta in volta rimasti vuoti [...] vengono riempiti sistematicamente con forme simili di scala più piccola e il risultato può essere un motivo localmente auto-simile*" (Moran and Williams 2007).

A fundamental aspect in the geometric interpretation of patterns is represented by colour. Specifically read in the contrast between different shades, it represents perhaps "*l'elemento più importante nella percezione d'insieme*" (Moran and Williams 2007, p. 22). The orange-yellow-cream (*giallo antico*), red and green of the marbles *tesserae*, together with the greys and whites of the marble bands, give the rhythms of the symmetries not only an ornamental value. Like the modular repetition, the chromatic arrangement of the *tesserae* becomes a founding element in the analysis of the floor design. The presence of polychromy, in fact, allows to give a different interpretation to the geometric-mathematical genesis of the plots. And this not only in relation to the isometric transformations in the plane, whereby translations, rotations, reflections and glide reflections are enriched with distinctive elements, but also in projective terms. In fact, in many cases, some modular subdivisions or repetitions present in the geometric motifs can be read, thanks to the colour, as plane homological transformations with a prevalent recurrence of affine and homothetic transformations.

5 Conclusion

The brief examination conducted on the Salerno cathedral inlaid marble floor aims to highlight the configurative richness of its decorative motifs and the geometric logic underlying the drawing of the plots. The intent is, on the one hand, to document a lesser-known artistic episode, on the other, to allow a comparison between the Salerno decorations and the best-known examples of Cosmatesque style, through a graphic analysis of the geometric language. If, in fact, the corpus of geometric motifs created by Cosmati constitutes a significant research field, and has so far offered numerous studies and classifications on the geometries of decorations (Moran and Williams 2007; Pajares-Ayuela 2001; Williams 1997; Piazzesi et al. 1954), it can be enriched thanks to comparisons with hybrid decorative examples. The Salerno decorations in fact present drawings and structures that, although they strongly recall the Cosmatesque logics, convey decisive Byzantine cultural influences and a clear Arab-Norman matrix. The attention paid to the marble floor inlays, although not aimed at a systematic classification work, had the purpose of describing the syntactic elements at the basis of formal aesthetics. At the same time, they are highlighted the ways in which local craftsmen, according to principles of harmony and order, have synthesized formal repertoires, of different nature, in elegant repetitions of forms. Finally, brief considerations have concerned the use of polychromy which, beyond the extraordinary ornamental value, also acts as a significant element of the patterns' geometric-mathematical structure, in the drawing of the Salerno floor panels too.

References

Bassan, E.: Cosmati. Enciclopedia dell'Arte Medievale. Treccani (1994). http://www.treccani.it/enciclopedia/cosmati_%28Enciclopedia-dell%27-Arte-Medievale%29/. Accessed 24 Oct 2019

Braca, A.: Il duomo di Salerno. Laveglia editore, Salerno (2003)

Carucci, A.: I mosaici salernitani nella storia e nell'arte. Di Mauro editore, Cava de' Tirreni (SA) (1983)

Chiavoni, E., Di Cosimo, B., Cigola, M.: Rappresentazione dei mosaici cosmateschi. Forma, Geometria, Colore. In: Palma Crespo, M., Gutiérrez Carrillo, M.L., García Quesada, R. (eds). Actas del V Congreso Internacional ReUSO "Sobre Una Arquitectura Hecha de Tiempo, Conservación y Contemporaneidad", vol. 2, pp. 523–528. Editorial Universidad de Granada, Granada (2017)

Cigola, M.: Mosaici pavimentali cosmateschi: segni, disegni e simboli. Palladio 11, 101–110 (1993)

Hegel, G.W.F.: Estetica. Einaudi, Torino (1972)

Longo, R.: L'opus sectile nei cantieri normanni: una squadra di marmorari tra Salerno e Palermo. In: Quintavalle, C.A. (ed.) Atti del XII Convegno internazionale "Medioevo: le officine", pp. 179–189. Electa, Milano (2010)

Matthiae, G.: Cosmati. In: Enciclopedia Universale dell'Arte, vol. III, pp. 837–843. Istituto Geografico De Agostini, Novara (1980)

Moran, J., Williams, K.: I pavimenti dei Cosmati. MATEpristem (2007). http://matematica.unibocconi.it/articoli/i-pavimenti-dei-cosmati. Accessed 27 June 2019

Pajares-Ayuela, P.: Cosmatesque Ornament: Flat Polychrome Geometric Patterns in Architecture. W.W. Norton, New York (2001)

Piazzesi, A., Mancini, V., Benevolo, L.: Una statistica del repertorio geometrico dei Cosmati. Quaderni dell'Istituto di storia dell'architettura, V. Bonsignori, Roma (1954)

Severino, N.: Il pavimento cosmatesco del duomo di Salerno. Alla luce di nuove ipotesi storiche ed analisi stilistiche (2011). https://www.academia.edu/39871083/Studi_sui_pavimenti_cosmateschi. Accessed 20 Oct 2019

Williams, K.: The pavements of the Cosmati. Math. Intell. 19(1), 41–45 (1997)

Williams, K.: Architettura e matematica: mille anni di motivi geometrici. In: Seminario di cultura Matematica del Politecnico di Milano (2004). http://fds.mate.polimi.it/file/seminari_cm/seminari/file/Williams-motivi_geometrici.pdf. Accessed 24 Oct 2019

Zarnecki, G.: Normanni. In: Enciclopedia dell'Arte Medievale (1997). http://www.treccani.it/enciclopedia/normanni_%28Enciclopedia-dell%27-Arte-Medievale%29/. Accessed 23 Oct 2019

The Tension Between Pictorial Surface and Pictorial Space as a Source of Emotion in Graphic Expression

Aurelio Vallespín Muniesa[(✉)] [ID]

Architectural Graphic Expression Department, University of Zaragoza, Zaragoza, Spain
aureliov@unizar.es

Abstract. This research focuses on the emotion triggered by paintings through the creation of tension between the pictorial space and the pictorial surface. As both an architect and painter, I am fascinated by this relationship. I believe it is not only an inexhaustible source of artistic inspiration, both architectural and pictorial, but also a means for the evolution of both. Firstly, I will explain what is meant by emotion.

Keywords: Wall space · Pictorial space · Spatial expansion · Pictorial surface · Mural painting

1 Emotion

This definition is based on the following premises: painting and architecture are artistic disciplines; the purpose of art is to stir emotions; and, in particular, amongst these emotions, are those that bring us closer to the sublime. This is aligned with the aesthetic experience of Edmund Burke; for him *the sublime* was a result of the strongest emotions that the mind is capable of feeling. This 18th century author wrote about two types of passions: those belonging to self-preservation and those belonging to society, the former being pain and the latter pleasure. On emotion arising from pain, this being the most powerful and all that is fitting for pain, he wrote that it, "operates in a manner analogous to terror, is a source of the sublime; that is, it is productive of the strongest emotion which the mind is capable of feeling" (Burke 2010 [1757], p. 64). In order to transmit emotions with the utmost intensity, one can resort to darkness and reproduce the description of the King of Terrors by the poet J. Milton, describing it as "significant uncertainty" (Burke 2010 [1757], p. 88).

There are very different ways to trigger the strongest possible emotions that the mind is capable of feeling, via the: Uniformity, vastness, infinity, uniformity, darkness… This research, however, aligned with Burke's notion of "significant uncertainty", focuses on the doubt, perplexity and uncertainty that a work of art can provoke in the observer. This doubt, perplexity and uncertainty, in turn, can be provoked in very different ways, via factors such as time of contemplation, synaesthesia and monochromy, and even a

L. Agustín-Hernández et al. (Eds.): EGA 2020, SSDI 6, pp. 701–711, 2020.
https://doi.org/10.1007/978-3-030-47983-1_62

combination thereof. This paper focuses on the way that connects painting and architecture, the tension that arises between the pictorial space and the pictorial surface, with its architectural facet of the wall space.

2   Pictorial Space in Antiquity and the Renaissance

Yet there came a time when that pictorial space was so embraced by the observer that in order to provoke and disconcert, the opposite effect was enhanced: the pictorial surface. Impressionism is based more on the subject than on the object, on the way objects astonish our first fleeting glimpse, creating an atmosphere that deliberately avoids the earlier depth of pictorial space. Precedents can be seen in the work of Corot. In his later works, such as *The Parc des Lions at Port-Marly* (Fig. 1), space is reduced through the use of colour. The intense light colour behind the trees reduces the space between them instead of enlarging it, highlighting how this light intense colour appears in all the planes denoting depth.

Fig. 1. Camille Corot, 1872. *The Parc des Lions at Port-Marly.* Madrid. Thyssen-Bornemisza Museum.

Monet discovered solutions that allowed him to keep the weight of the painting on the surface, while still reflecting nature, a pictorial surface representing a three-dimensional reality. By means of the application of balanced colour, the general tone is enhanced and reflections of the surrounding colours are introduced in each area, creating a monotonous and flat surface. By the end of his life Monet began to focus his paintings entirely in the foreground (Greenberg 1959 [1956], p. 56).

This is evident in the space in the *Musée de l'Orangerie* in the exact conditions that Monet imposed for the cession of his murals: oval spaces with glass ceilings. On 31

October 1921 Monet wrote a letter to Clemenceau containing significant information, including the following: "I accept the *Orangerie* on the condition that the management of Fine Arts undertakes to do with the works what I deem to be essential." Considering this hypothesis, I have reduced several motifs of the Decorations and I believe I have arrived at a combination that will give a good result conserving the oval shape that I have always liked" (Monet 2010, p. 374). The glass ceiling appears referenced by Monet in a letter written to Alexandre dated 6 July 1921 (Monet 2010, p. 273). (Figure 2), they can be interpreted as a search for lost pictorial space. The murals, conceived as pictorial surfaces, are transformed into pictorial space when placed in a concave space (Fig. 3). The use of glass ceilings also contributes to this. Steinberg (1956, p. 235) comments that the observer can invert his or her position or that of the painting, s/he can also transform the vertical planes into horizontal and the vertical into horizontal, due to the confusion and ambiguity that arises due to the reflections of that which is represented. Perhaps this confusion also encompasses the ambiguity between the pictorial surface and the pictorial space.

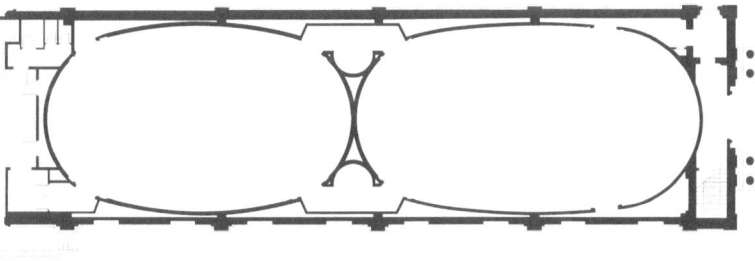

Fig. 2. Musée *de l'Orangerie.* Plan of the layout of the rooms where the panels are installed, by the author.

3   Monet and Impressionism

Yet there came a time when that pictorial space was so embraced by the observer that in order to provoke and disconcert, the opposite effect was enhanced: the pictorial surface. Impressionism is based more on the subject than on the object, on the way objects astonish our first fleeting glimpse, creating an atmosphere that deliberately avoids the earlier depth of pictorial space. Precedents can be seen in the work of Corot. In his later works, such as *The Parc des Lions at Port-Marly* (Fig. 1), space is reduced through the use of colour. The intense light colour behind the trees reduces the space between them instead of enlarging it, highlighting how this light intense colour appears in all the planes denoting depth.

Monet discovered solutions that allowed him to keep the weight of the painting on the surface, while still reflecting nature, a pictorial surface representing a three-dimensional reality. By means of the application of balanced colour, the general tone is enhanced and reflections of the surrounding colours are introduced in each area, creating a monotonous

Fig. 3. Musée *de l'Orangerie*. Installation after the refurbishment in 2006. Photomontage by the author.

and flat surface. By the end of his life Monet began to focus his paintings entirely in the foreground (Greenberg 1959 [1956], p. 56).

This is evident in the space in the Musée de l'Orangerie in the exact conditions that Monet imposed for the cession of his murals: oval spaces with glass ceilings. On 31 October 1921 Monet wrote a letter to Clemenceau containing significant information, including the following: "I accept the Orangerie on the condition that the management of Fine Arts undertakes to do with the works what I deem to be essential." Considering this hypothesis, I have reduced several motifs of the Decorations and I believe I have arrived at a combination that will give a good result conserving the oval shape that I have always liked" (Monet 2010, p. 374). The glass ceiling appears referenced by Monet in a letter written to Alexandre dated 6 July 1921 (Monet 2010, p. 273). (Figure 2), they can be interpreted as a search for lost pictorial space. The murals, conceived as pictorial surfaces, are transformed into pictorial space when placed in a concave space (Fig. 3). The use of glass ceilings also contributes to this. Steinberg (1956, p. 235) comments that the observer can invert his or her position or that of the painting, s/he can also transform the vertical planes into horizontal and the vertical into horizontal, due to the confusion and ambiguity that arises due to the reflections of that which is represented. Perhaps this confusion also encompasses the ambiguity between the pictorial surface and the pictorial space.

4   Cezanne and Cubism

In this period Cézanne, who also aspired to reality without relinquishing sensations as Monet did, enhanced the pictorial surface. However, he did this in a different way, based on visual perception. En Madame Cézanne in a Yellow Chair, (Fig. 4), the horizontal stripe on the wall does not match at its ends, because "if a line passes beneath a wide strip of paper, the two visible segments appear dislocated." (Merleau-Ponty 2012 [1945], p. 41) However, he goes on to say that once these distortions are frozen by repainting them on the canvas, the spontaneous movement in which they pile up in perception stops and a there is a tendency towards the geometric perspective, but with a propensity to reduce space. The Alley at Chantilly (Fig. 5) shows how this tendency to reduce the pictorial space can be further emphasized. Cézanne compresses the first layers of vegetation and brings forward the house above the observer in the farthest plane, thus triggering tension between the space and the pictorial surface.

Fig. 4. Paul Cézanne, 1893–95. *Madame Cézanne in a Yellow Chair.* Private collection.

Picasso and Braque were to continue what Cézanne had begun (Greenberg 1951, p. 70), "at first they were more crucially concerned, in and through their Cubism, with obtaining sculptural results by strictly non sculptural means; that is, with finding for every aspect of three-dimensional vision an explicitly two-dimensional equivalent, regardless of how much verisimilitude might suffer in the process." Greenberg then goes on to say that, "flatness had not only invaded but was threatening to swamp the Cubist picture." (Greenberg 1959, p. 86). "Depicted flatness –that is, the facet-planes–had to be kept separate enough from literal flatness to permit a minimal illusion of three-dimensional space to survive between the two." (Greenberg 1959, p. 87). This illusion of space could be achieved with the use of letters in the paintings, as in Braque's Portuguese, (Fig. 6) where the frontal placement of the letters places them on the literal plane, the pictorial surface, while everything else is set further back in different intermediate planes. It is in this circumstance that collage appears, as a tool similar to lettering, a means to create pictorial space (Greenberg 1959, p. 95), on the pictorial surface, instead of being contained therein. However, at a certain point these mechanisms were used ambiguously, as they could be used to generate a certain amount of space, and could also be used in reverse, some nullifying others.

Fig. 5. Paul Cézanne, 1888. *The Alley at Chantilly.* Toledo. The Toledo Museum of Art.

5 Rothko and American Abstract Expressionism

American abstract expressionism embraced the tension between illusory space and the flat surface as a way of asserting painting in its pure form, rather than camouflaging it, as traditional painting sought to (Greenberg 1960, p. 113). Rothko (1958, p. 183) once said: "My pictures are indeed façades (as they have been called). Sometimes I open one door and one window or two doors and two windows. I do this only through shrewdness. There is more power in telling little than in telling all." This reflection explains the coexistence of surface and illusory space. The façade is understood as pictorial surface and when the window is opened, the pictorial space appears. It even describes surfaces that can be opened to enter space in which one can be inside.

Around this time Rothko began working on the *Seagram* murals, in which the emotions he sought were different. He wanted observers to feel that they are trapped in a room where all the doors and windows are bricked up, so that all they can do is butt their heads forever against the wall, as he felt when he visited Michelangelo's Laurentian Library (Ficher 1970, p. 189–190). If in previous paintings he opened doors and windows, here he closed them. He no longer sought to access the pictorial space, but rather to accentuate its surface.

There is an analogy that can be discerned between Monet and Rothko. If the individual works of the former highlight the surface, the murals in the *Orangerie* point to the space. With Rothko, however, the opposite happens: the individual works refer us to space and the *Seagram* works highlight the surface. The *Seagram* murals were never placed in the *Four Seasons* restaurant on the ground floor of the *Seagram* building. In 1969 Rothko donated nine of the more than 40 canvases to the *Tate Gallery*. The installation was designed by the artist before his death. These murals are currently in the Tate Modern where attempts have been made to recapture that environment (Fig. 7).

Fig. 6. Georges Braque, 1911. *The Portuguese* Basel. Kunstmuseum.

Fig. 7. Georges Braque, 1911. *The Portuguese* Basel. Kunstmuseum.

6   Wall Space

It is also necessary to analyse a special case of the tension between the pictorial surface and the pictorial space: the tension that arises between wall painting and architecture. How the wall surface modifies the architectural space where it is placed, either by merging with the surface or by emphasising it through the painting on it. This is what we will call the wall space.

Riegl (1992, [1901], p. 33), when discussing architecture, states that it is divided into two parts: space and its limits. "The task of architecture is divided into two parts, which necessarily complement and condition each other, but which also somehow oppose each other: the creation of a (closed) space as such and the creation of the limits of this space. Thus the possibility of implementing unilaterally only one part of the task at the expense of the other has been open from the beginning to human artistic practice. It was possible to let the spatial outlines dominate in such a way that the architectural work did become a sculptural work. On the other hand, one could expand the spatial outlines to such a distance that the thought of the infinity and immeasurability of free space was awakened in the spectator." (Riegl 1992, [1901], p. 33). The wall space seeks to study, through the wall painting and the architecture that encompasses it, the relationship between space and its limits, taking into consideration the ambiguity that arises owing to the capacity of the wall surface to be pictorial surface or pictorial space and therefore to be perceived as a surface, more specifically, a limit or space, in the same way that architecture is seen.

7 Galla Placidia

Perhaps the source of my interest in this subject stems from the mosaic surfaces of Galla Placidia in Ravenna. At the age of fourteen years old, they evoked such emotion in me. And this led me to wonder, years later, what could have caused such emotion, a reaction that, indeed, has been reproduced whenever I have returned. I am sure, however, that the subsequent visits triggered a reaction that was more like a memory in the form of an emotional reflex.

Studying this from a perspective of drawing, it is interesting to see how in the supporting plane to the dome in the Mausoleum, the vertical plane in the mosaic is curved so as to serve as a support for the figures and then is lost into infinity. Undoubtedly we are faced with a rather primitive, but also highly effective, way of representing this idea, of generating spatial expansion through drawing. This visual resource has a dual purpose: it ensures that the figures are not left floating on an undefined background, while at the same time it provides the desired spatial expansion (Fig. 8).

In this small mausoleum the apparent space is much larger than the actual space thanks to the designs and, in particular, to the use of background. "From afar, the mosaic sinks into a recess and has those figures that the eye is so consumed by. It is not the form that makes the figure appear. What is it? The form follows the light, the light and the shadows. The eyes of a Byzantine figure are everywhere, or reach everywhere. These eyes from the background are piercing." (Deleuze 2008 [1981], p. 210). The light is projected to the background, which is dark blue, and has a dominating effect over the figures, emphasising the spatial concept. The result is as follows: distance is created from the limits and a spatial expansion takes place, simultaneously producing indefinite limits.

Moreover, given that the drawing is in mosaic form, it has a more luminous quality, one that goes beyond that produced by the colour. This is surface light and is emitted by the reflection of light on the glass. This property of the glass tesserae was fully exploited

Fig. 8. The Mausoleum of Galla Placidia. The curve in the vertical plane creates space. Photograph taken by the author.

Fig. 9. The Mausoleum of Galla Placidia. Red and white tesserae show a combination of polished and matt finishes. Photograph taken by the author.

by the craftsmen, creating a variety of light effects. They combined matt and polished tesserae, creating effects that truly astonish the observer (Fig. 9).

Sometimes the shiny areas are shadows and the luminous areas are matt, lacking shine, as is the case with Theodora's dress or Justinian's robes in 6th-century Basilica of San Vitale. Another effect is achieved by placing the tesserae at different angles so as to emit different sparkles depending on the particular source of the light at any one time. In some murals, tesserae are laid tilted up to 30 degrees from the plane of the wall in which they are set. It is noteworthy that this first glimpse of light requires a non-frontal vision, that is, one that is not perpendicular to the plane of the image, which would be the norm. The use of tesserae with gold or silver leaf is also habitual; in such cases, the light is reflected both on the surface of the glass and on the surface of the inner metallic material.

This light draws the eye to the pictorial surface, and in this instance, being located in an architectural environment, to the surface of the space in which it is installed. The light from the sparkles produced by the mosaics marks the surface and thus helps us define and demarcate the space, which in this case is a mere 15 metres at its highest point. The effect, therefore, is entirely opposite to the previous example illustrated.

8   Conclusion

It is therefore, on the one hand, thanks to the drawing and the arrangement of the figure in the background together with the backdrop, that the space expands and is revealed as indefinite, thus leading us back to the pictorial space. Whereas, on the other hand, it is the reflections of the glass mosaic surfaces, the light, that lead us back to the pictorial surface, allowing us to define to some extent the volume of the building, thus giving us a more realistic perspective of the space and helping us to demarcate it (Fig. 10). That is why it can be said that in Galla Placidia there is a tension between what is drawn by a pictorial space and the light and glassy reflections of the mosaics that lead the eye to the pictorial surface. This tension provokes doubt, perplexity and uncertainty, which is, in turn, what triggers the emotion emanating from the work of art in the observer.

Fig. 10. Mausoleum of Galla Placidia. Floral decoration of the dome is seen in the light of the sparkles and reflections. Photograph taken by the author.

Thus there is a perceptible tension between the pictorial space and the pictorial surface and, as a result, the observer feels perplexity or even fear, emotions necessary to unleash the aforementioned interior processes. And this is a perfect vehicle for its enhancement through the modification that occurs between the wall surface and the architectural space. This is not per se the role of the mosaic, far from it, but without a doubt, this is the one that interests us. It is its capacity to alter space, as another architectural element, one that creates ambiguity through its perception as a surface or as a space.

References

Alberti, L.B.: De la pintura y otros escritos sobre arte. Técnos, Madrid (1999)

Aristoteles: Física. Planeta Agostini, Madrid (1996)

Burke, E.: Indagaciones filosóficas sobre el origen de nuestras ideas acerca de lo sublime y de lo bello. Alianza, Madrid, 2010 (1757)

Deleuze, G.: Pintura. El concepto de diagrama. Cactus, Buenos Aires, 2008 (1981)

Ficher, J.: La butaca: Mark Rothko, retrato del artista enfadado, en Mark Rothko, 2007. Escritos sobre arte (1934–1969), pp. 188–198. Paidós, Barcelona (1970)

Euclides: Óptica. Gredos, Madrid (2000)

Greenberg, C.: Cezanne, en Clement Greenberg, 2002 (1961). Arte y Cultura. Barcelona. Paidós, pp. 63–72 (1951)

Greenberg, C.: El último Monet, en Clement Greenberg, 2002. Arte y Cultura. Barcelona. Paidós, pp. 49–57, 1959 (1956)

Greenberg, C.: Collage, en Clement Greenberg, 2002 (1961). Arte y Cultura. Paidos, Barcelona, pp. 85–99 (1959)

Greenberg, C.: La pintura moderna, en Clement Greenberg, 2006. La pintura moderna y otros ensayos, pp. 111–120. Siruela, Madrid (1960)

Merleau-Ponty, M.: La duda de Cézanne. Casimir, Madrid, 2012 [1945]

Monet, C.: Los años de Giverny Correspondencia. Turner publicaciones, Madrid (2010)

Panofsky, E.: La perspectiva como forma simbólica. Tusquets, Barcelona, 2008 (1927)

Rodman, S.: Notas de una conversación con Mark Rothko, en Mark Rothko, 2007. Escritos sobre arte (1934–1969). Paidós, Barcelona, pp. 176–177 (1956)

Rosenblum, R.: Cubism and Twentieth-century art. Harry N. Abrams, New York, 1976 (1959)

Rothko, M.: Como combinar la arquitectura, la pintura y la escultura, en Mark Rothko, 2007. Escritos sobre arte (1934–1969), p. 120. Paidós, Barcelona (1951)

Rothko, M.: Conferencia en el Pratt Institute, en Mark Rothko, 2007. Escritos sobre arte (1934–1969), pp. 182–186. Paidós, Barcelona (1958)

Steinberg, L.: Monet's Water Lilies en 1976 (1972). Other criteria. Confrontations with Twentieth-Century Art, pp. 235–239. Oxford University Press, New York (1956)

White, J.: Nacimiento y renacimiento del Espacio pictórico. Alianza Forma, Madrid, 1994 (1957)

The Importance of Incorporating the BIM (Building Information Modeling) Methodology into Historic Buildings to Achieve Their Viability

Manuel José Soler Severino[1], Ricardo Santonja Jiménez[1](✉),
and Luis Agustín-Hernández[2](✉)

[1] Polytechnic University of Madrid (UPM), Madrid, Spain
{manueljose.soler,ricardo.santonja}@upm.es
[2] University of Zaragoza (UNIZAR), Zaragoza, Spain
lagustin@unizar.es

Abstract. BIM ("Building Information Modeling"), is a digital information exchange methodology, widely implemented in engineering, architecture and construction worldwide.

Our cultural heritage is a historical good of humanity. When it is deteriorated or in poor condition, it is necessary for its reconstruction or restoration to be in the most faithful and precise way to the original state. Its subsequent maintenance is also imperative for the perfect duration of the good.

The application of BIM methodology in historic buildings, such as the ones that make up the cultural heritage of Aragon, an example would be the gothic churches, is essential to be able to rebuild and maintain them, reinterpreting original information in an accurate way based on digital information models.

Applying this methodology allows us to obtain accurate information by developing a three-dimensional model with parameterized information of its elements with total reliability.

The main objective is to achieve, the application of BIM methodology in the historical buildings of heritage, specifically, in the gothic of Aragon Kingdom.

Another objective is the optimization of both schedule (planning) and costs in these historic buildings.

This is what we call Digital Archeological Architecture (DAA), which could be a new professional field for architects.

Keywords: BIM · Viability · Historic building

1 Introduction and Background

Obtaining digitized information that can be applied to the cultural heritage makes way for the development of a base model (or preliminary model). To which we have been applying information, comparing it to its original state and its successive subsequent modifications.

© The Editor(s) (if applicable) and The Author(s), under exclusive license
to Springer Nature Switzerland AG 2020
L. Agustín-Hernández et al. (Eds.): EGA 2020, SSDI 6, pp. 712–721, 2020.
https://doi.org/10.1007/978-3-030-47983-1_63

This methodological digital system is typically called BIM ("Building Information Modeling"). BIM consists in incorporating specific information in three-dimensional models. This has as one of its main objectives the optimization of costs.

The importance lies in controlling and checking the information that is being introduced into the model.

This will force those responsible for the model to carry out a thorough investigation of the background of the buildings that make up the historical heritage. So that they will be rehabilitated, from the study of the information from existing drawings, pictures, and documentation.

The reconstruction of these historic buildings needs true information in its original state. It is necessary for them to have great maintenance once the restoration is finished to avoid its deterioration since, in addition to the inclement weather, there is also a touristic factor, that affects the use and deterioration.

It is important to incorporate management into these historic buildings. This management will be post-reconstruction or rehabilitation. This agent is a BIM Facility, whose responsibility will be to manage the historic building in the operation phase.

2 State of the Art

There are currently research studies being conducted concerning this methodology. [2] Nieto, J.E., in Ph.D. is an example of these investigations. In it, it establishes BIM as a key to the rehabilitation process of historic buildings of Spanish heritage. The generation of information models must try to incorporate information with the greatest precision to the investigation of the historic building to be rehabilitated or conserved and avoid the loss of them.

It is very important to consider, as described, [8] Tabales, M.A., the archaeological analysis of historic buildings. We must also apply and consider reliable techniques for identifying the information that we should consider when rehabilitating or rebuilding these types of historic buildings.

Regarding the application of BIM, an initial analysis of this methodology will be developed. It should be applied in the management of historic buildings, looking for the particularities that the model should have when applying the BIM methodology.

There is little information on the application of the BIM methodology in historic buildings, especially in achieving viability in terms of schedule, cost, and maintenance.

2.1 "Building Information Modeling"

"BIM", comes from the acronym of the English phrase, 'Building Information Modeling' (Modelado de la Información del Edificio), it could be called MIE in Spanish, although it is known worldwide as BIM.

The purpose of the BIM methodology is the creation and use of digitized, coordinated and related information in an architecture and construction project. Not only in design but also in the structures facilities process and the entire organization of cost deadlines, quality, and sustainability. This way we are able to achieve what we will call "The virtual building".

"The BIM methodology, it is a digital representation of physical, functional, and formal characteristics. To share technology and architectural information, for the management and organization of the building's life cycle, to create a self-managed model called, virtual building."

BIM is an approach to the design, construction, organization and management of building construction.

It is a methodology that focuses on the way of understanding and working in architecture from a different point of view. One that has an integrated model resulting from the work and organization of the participating team.

One might say that BIM methodology is the industrial revolution of the 21st century in terms of the construction industry.

In the construction industry, incompatibility between systems generally prevented project team members from exchanging information accurately and quickly. This is the cause of numerous problems in architectural projects and construction, so are the number of costs and deadlines schedules, and a general lack of coordination in a project.

The adoption of BIM methodology and the use of integrated digital models throughout the life cycle of the building represents a step in the right direction towards the elimination of costs resulting from the interoperability of data. Simply using a digital model is not enough.

New processes must be incorporated and existing ones adapted. Following the analogy of the industrial revolution, it implies a cultural and sociological change in the way of understanding the life cycle of the construction process.

This last statement is vital to understand that we are not talking about new technology or the evolution of the existing one, we are talking about new work processes (Fig. 1).

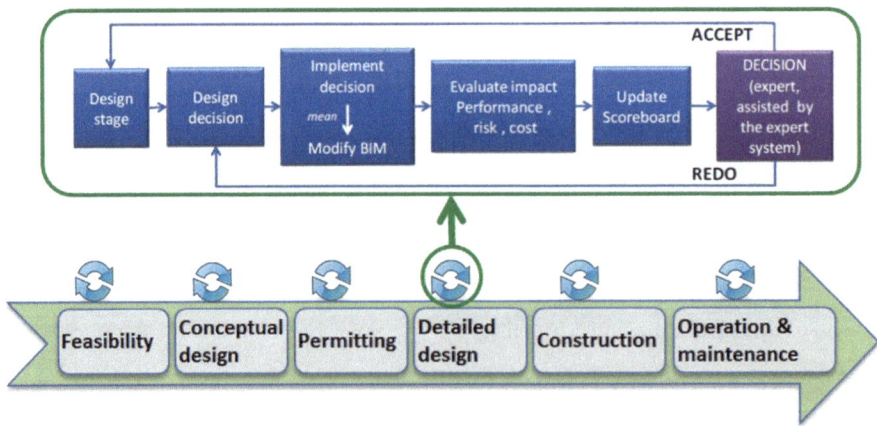

Fig. 1. Loop-based design at a given stage (Mediavilla 2013, life Project cycle)

One of the keys is the identification of the project's life cycle. The objective of the BIM model is to develop a study based on information with sufficient details for its future construction. A goal is to optimize deadlines schedules and costs by applying construction management through, in Spanish (Dirección Integrada de Proyecto, DIP), to improve your maintenance. [8] Mediavilla, A.

The aim of BIM methodology is to implement the information into the project's life cycle with the original methodology without BIM, that forces a greater effort into producing information in the different phases of the project.

In turn, BIM is an information management resource, and as such, it can be used to illustrate the entire building, maintenance, and even demolition process.

BIM methodology is an open project information platform available to all stakeholders in the construction process.

BIM methodology, as a ISO 21500 (International standard for project management), has processes and procedures, as well as areas of knowledge and phases for each deliverable of the process.

BIM is the methodology which allows to share the information effectively and reliably and ISO 21500 (International standard for project management) is responsible for the management processes and resources.

We are talking about new forms and work processes or the need to adapt existing ones.

2.2   Ecclesiastical Gothic Architecture of Aragon Kingdom

The gothic style entered the crown of Aragon from France at the hands of the monks of the Order of Cister. They began to build their monasteries in isolated places corresponding to their religious life of recollection and loneliness.

The first historical building that began to be built in Aragon was the "Royal monastery of Santa Maria de Veruela". Its first stage is Romanesque, moving on to gothic. It is considered the first building of that style in all of the Aragon Kingdom and a future example for the following monasteries.

The Cistercian gothic followed the principles of rigorous asceticism and poverty of order. The buildings were disorganized on the outside and austerely decorated on the inside, where the communal lifestyle took place.

The gothic style rapidly spread throughout the kingdom of Aragon reaching the Mediterranean. Within the denomination of Levantine Gothic or Mediterranean Gothic, we find Catalonian Gothic, Valencian Gothic, and Mallorquin Gothic.

At that time, in addition to having common rulers who were the ones who ordered the construction of many buildings, the construction masters traveled between the three territories to build, copy or b e inspired by what had already been built.

It is important to add the sharing of a common weather and easy access to similar construction materials, hence in the three territories, the style has common characteristics.

Aragon has a big Ecclesiastical cultural heritage. If we analyze the gothic, some buildings at the beginning of the Mudejar style were transformed over the centuries with styles like Renaissance or Baroque, like the Cathedral of San Salvador in Zaragoza or the Cathedral of Palma de Mallorca (formerly belonging to the kingdom of Aragon and considered as a Mediterranean gothic). Both have been declared World Heritage by UNESCO and are considered masterpieces. They need adequate maintenance to facilitate their conservation and must be economically sustainable.

Therefore, a compilation of information about those buildings that are well preserved must be carried out. This way we can apply them to those buildings that due to their difficulties and situations are deteriorated. This is the case of existing buildings throughout the entire Community of Aragon (Fig. 2).

Fig. 2. Ricardo Santonja Jiménez. Cathedral of Palma de Mallorca. 2017. Research Project of the Gothic Churches of Aragon Kingdom. UNIZAR.

By applying BIM methodology, we can optimize the processes, incorporating constructive information from the first stage of its data collection to the management of the historical buildings as a real estate asset. [2] Building Smart.

The archeologist will become a consultant by integrating himself in the process as an interested stakeholder, with the aim of validating the information that can be incorporated into the BIM model to seek its reliability.

3   Methodological Analysis

The methodology applied in this research consisted of developing intervention techniques based on the collection of information from historical archives and through the photography of existing buildings. By this, we can establish general criteria that serve as the basis for subsequent analyses in historic buildings that we intend to rehabilitate.

We have worked with experts in the corresponding subjects, to establish a first model that serves as development.

The data obtained will be parameterized and a type model based on the preliminary information will be created, that model will be used to compare it with the historic buildings under study. [1] Banfi, F.

Once the general model has been implemented and the information obtained developed, the specific BIM model for each particular situation of the historical buildings can be used to be rehabilitated or restored.

So, we will establish a particularized model with the general information of the style, a schedule and construction system, used in that period of time. The main objective is the first BIM model.

3.1   Archeological Architecture Digitized (AAD)

Archeological Architecture Digitized (AAD), it is the term we have implemented in our research group of the UNIZAR, to define a general three-dimensional model with previous information from a historical investigation is incorporated.

The application of this methodology is based on an initial recovery and investigation of the existing documentation (including photographic reports, drawings, and technical data). The main constructive details of the building, researching and reflecting on the information in a three-dimensional model, as if it were the current state of historic buildings.

After obtaining the information, different interventions from the BIM model will be considered, we need these different variants and to be able to make a decision. Expert consultants look at the BIM model that will serve for the restoration or the purpose that is expected to be obtained with the intervention.

Once the final BIM model is obtained, the construction details and facilities and other more specialized aspects are considered by the restorative team. An intervention will be made by the definitive BIM model in a LOD 350 (level of development), and thus obtaining a reliable first model of the historic building (Fig. 3).

Fig. 3. Ricardo Santonja Jiménez. Cathedral of Palma de Mallorca. 2017. Light Study, Research Project of the Gothic Churches of Aragon Kingdom. UNIZAR.

The goal is to find reliable information and incorporate it in a methodical way to the model avoiding loss of information or data being incorrectly introduced into the BIM model, this must be done by the BIM Manager of the process.

The end of the construction phase will reflect a much more developed project, with much more information involved in the BIM model that could be assimilated to an LOD 500 (level of development). This will be a digitized "As built" BIM model, which will be the starting point of the "Book of the Digital Building".

The "Book of the Digital Building" (BDB), will be the final model that incorporates all the information of the final state of the construction and its future interventions. It will be an identification of all the subsequent realization in the building.

These BIM methodologies open a new field of Cultural Heritage analysis and its most reliable restoration and maintenance.

3.2   Cost Management in the Historical Building

The cost management in a BIM model of a historic building should start from the preliminary information obtained in the general model.

There are different types of cost estimates, the difference lies in the information and therefore in its accuracy, the more accurate this is, the more reliable the cost. These estimates will be developed in the different phases of the project's life cycle and will be the instrument to estimate and subsequently manage the cost through the integrated project management based on the BIM model.

The cost management project control is a basic tool for the optimization of them within historic buildings. They help analyze their compliance, so it will be essential to have: a clear definition of the roles and responsibilities in relation to cost planning. Budget development and budget control in each phase of the project should have a responsible of budget maintenance.

The cost control is carried out periodically and linked to the programming of the execution period allows: a dynamic comparison of the committed (real) cost with the expected cost, analyzing the deviations on the budget of each item and the total of the project, including contingencies and non-designed units and redefine the payment policy to obtain the most favorable conditions for the project based on the cash flow forecast.

Therefore, the principles of cost control should be: [7] Soler, M.J.

1. Creating a complete and realistic cost estimate.
2. Identify and try to control the critical variables that can act on cost (qualities, specifications, control requirements, dimensions, etc.)
3. Always pursue the objective of minimizing the total cost as long as it does not affect the quality (historical specifications in the building safety), and keep it in an acceptable range.
4. The entire project team, must we work economically.
5. Evaluate the implications of any changes before making them.
6. Develop a good "Change management").

The cost control will be carried out through the "Change Orders" system and the control of the items of the Project Budget, which will accurately reflect the cost of each change and its influence on the entire project.

The cost control is closely linked to the control of terms, bearing in mind at all times the deviations in costs and terms that are given in the different items.

The instrument for the control of the costs is the accounting of the expenses incurred in the Project, for which the control of all the expenses by items corresponding to the Budget must be kept.

We will have a computerized cost control system to avoid loss of information and its non-optimization.

3.3   The Maintenance of Historical Heritage

The current change in the restoration of buildings leads towards constructing environments with the application of eco-efficient management. The intervention of "Facilities Management" in the subsequent cycle of maintenance of buildings.

In the cultural heritage in general and specifically applied to the Gothic churches of Aragon, we find some constructions that are not prepared for the development of eco-efficient Architecture. They are also not fit to optimize with a reasonable cost with effective parameters of subsequent management of the building and incorporate the fundamental criteria of maintenance.

BIM, will help develop and incorporate the maximum information of the original state and its adaptation to the incorporation of eco-efficient systems to achieve integration with the environment and be able to meet the needs of its users. [5] Rivas-López, J.

This new field appears in the restoration and is not sufficiently investigated at the start of a new line of research. It is very necessary to avoid deterioration over time of the restored asset.

The Gothic churches that integrate the Cultural Heritage of Aragon, require previous digitalization, with a research phase of the existing documentation for the incorporation of the information of all its constructive elements. Adapting the three-dimensional model to these requirements, incorporating in a logical way the new construction systems with clear eco-efficient premises, without distorting its restoration.

The restoration that incorporates the most efficient options must be established, taking into account technological and economic aspects, to avoid extra costs and subsequent restorations in the churches that hinder their conservation from the financial point of view.

4   Conclusions

Restorations are increasingly complex and require a greater number of resources and agents involved.

We are talking about projects that are carried out anywhere in the world and geolocations which are not always possible.

Traditional techniques are no longer suitable for managing these types of projects, due to the deficiencies they present and the complexity of the projects. These techniques show poor collaboration from the design stage.

Using the BIM methodology, errors can be detected in the later stages. In most cases, they are caused by failures in distributed documentation and communication deficiencies. Poor management and organization of information.

The central model offered by BIM shows clear information at all times and a real visualization of what was built. The state participates in the project and plans for collaboration. Integration and coordination among all stakeholders is proposed.

Once the project has been completed, it is possible to link the "as built" model with other types of activities, such as the maintenance of the facilities, thus reducing the costs derived from them.

With the BIM Management methodology and the Guidelines of the International Standard-ISO 21500 (Project management), it is possible to include the "Stakeholders"

- clients, sponsors, architects, engineers, builders, etc. at an earlier stage of the project. [3] Choclan, F.

This way, we can create a "pre-construction" so that risks are minimized (in costs and planning), with the consequent saving in time and costs, quality improvement, Health and Safety, and Sustainability. Managing projects in a standardized way, following the ISO 21500 Guidelines and the tools and techniques of each Organization (Fig. 4).

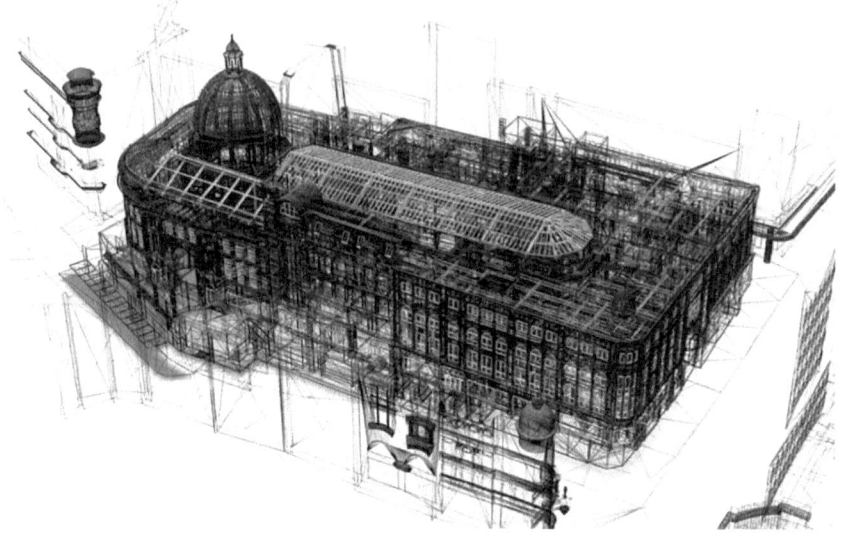

Fig. 4. Esteban Rivas-López. Modelo Cathedral tree-dimensional BIM model. Universidad de Granada, España, 2017.

After this research is done, we can obtain the following conclusions:

- The incorporation of the BIM methodology in the Cultural Heritage of Aragon is essential.
- Digitalization is very important, not exclusively in the reconstruction or rehabilitation phase, but also in the management phase of the finished asset.
- Eco-efficient criteria should be incorporated, without detracting from the historical and artistic point of view to economically adjust subsequent interventions and maintenance.
- A "Book of the digital building" must be made, incorporating information and updating it when it is intervened in them.
- A "new line research" should be created, incorporating these criteria together with the new management models such as the "Project & Construction Management" or the agile methodologies to seek the future effectiveness of these buildings.

References

1. Banfi, F.: Building information modelling - a novel parametric modeling approach based on 3D surveys of historic architecture. In: Ioannides, M., Fink, E., Moropoulou, A. Hagedorn-Saupe, M., Fresa, A., Liestø, G., Rajcic, I., Grussenmeye, P. (eds.) Digital Heritage. Progress in Cultural Heritage: Documentation, Preservation, and Protection, pp. 116–127. Springer, Cham (2016)
2. Building Smart: BIM aplicado al Patrimonio Cultural. Cuadernos uBIM, no. 14, 46 páginas (2018)
3. Choclan, F., Soler M.J., Gonzalez, R.J.: Building Information Management: Gestión con la Norma Internacional ISO 21500. Span. J. BIM Buildingsmart (1401) (2014)
4. Mediavilla, A., Mazza, D., Robert, S., Guigou, C., Martin, J., Pruvost, H., Scherer, R., Ferrando, C., Delponte, E.: The HOLISTEEC platform for building design optimization: an overview. In: Proceedings of the 10th European Conference on Product and Process Modelling (ECPPM), Viena, pp. 893–898 (2014)
5. Nieto, J.E.: Generación de modelos de información para la gestión de una intervención en el patrimonio arquitectónico. Tesis Doctoral, Universidad de Sevilla (2014)
6. Rivas-López, J.: A propósito de BIM y gestión del patrimonio arquitectónico en el contexto actual del sector AEC. Span. J. BIM (1702), 35–43 (2017)
7. Soler, M.J.: Architectural Management, 2nd edición. Bellisco Virtual, Madrid (2018)
8. Tabales, M.A.: Sistemas de análisis arqueológicos de edificios históricos. Universidad de Sevilla, Sevilla (2002)

Author Index

L. Agustín-Hernández et al. (Eds.): EGA 2020, SSDI 6, pp. 723–724, 2020.
https://doi.org/10.1007/978-3-030-47983-1